D0769395

**ss Cataloging in Publication Data**

ractices of commercial construction / Cameron K. Andres, Ronald C.

92-7
Smith, Ronald C. II. Title.
)4

20030429

ephen Helba
Ed Francis
**r:** Linda Cupp
Holly Shufeldt
**ation:** Carlisle Publishers Services
Diane Ernsberger
a Sorrells-Smith

Matt Ottenweller
ark Marsden
es Sans by Carlisle Communications, Ltd. It was printed and bound by Courier Kendallville,
d by Phoenix Color Corp.

---



. Ltd.

Pearson Education Australia Pty. Limited
Pearson Education North Asia Ltd.
Pearson Educación de Mexico, S.A.de C.V.
Pearson Education Malaysia Pte. Ltd.

10 9 8 7 6 5 4 3 2 1
ISBN 0-13-048292-7

# PRINCIPLES A
# PRACTICES
# COMMERC
# CONSTRUC

Seventh Ec

CAMERO
RONA

Inc
ISBI
1. Bu
TH145.A
690—dc2

**Editor in Chief:** St
**Executive Editor:**
**Development Edito**
**Production Editor:**
**Production Coordin**
**Design Coordinator:**
**Cover Designer:** Lind
**Cover Art:** Superstock
**Production Manager:**
**Marketing Manager:** M
This book was set in Sto
Inc. The cover was printe



Pearson Education Ltd.
Pearson Education Singapore Pt
Pearson Education Canada, Ltd.
Pearson Education—Japan

*To the men and women*
*in the construction industry*

# PREFACE

The process of commercial construction brings together designers, engineers, project managers, fabricators, and site personnel, as well as heavy equipment, to build commercial buildings.

Many demands are made on the building process: (1) the project must be completed on time and on budget; (2) it must meet all relevant building codes and environmental concerns; and (3) it must function as intended. Although "form for function" does not fall within the scope of this text, it is important to acknowledge that a building should accommodate, enhance, and enliven its intended use.

A combination of time-honored building practices, space age materials, and computer-based design offers the construction industry an increasingly formula-based approach to construction. The use of standardized building elements such as joists, trusses, beams, concrete floor slabs, curtain walls, and entire building frames has directed the manufacture of preengineered building components, whether they are structural or not. Catalog selection of structural components and non-structural items offer a cost-effective solution for many owners. However, preengineered components increasingly challenge designers and site personnel to pay greater attention to detail.

As the construction process becomes ever more refined, it is essential that manufacturers, suppliers, designers, and contractors speak a common language. That language includes an understanding of common industry terms and definitions, the ability to read and interpret complex drawings and details, and a strong working knowledge of the complete building process. The language of construction also includes measurement of quantities. Measurements on the construction site must be within the allowable tolerances of the prefabricated items to ensure speedy assembly. Drawings, specifications, and details must be clear and accurate. Because the industry relies on both standard and metric units of measure, all construction personnel need to be well versed in both systems. To reinforce this need, the seventh edition of *Principles and Practices of Commercial Construction* provides material dimensions, tables, details, and worked examples in both standard and metric units.

When conditions and need dictate a custom approach, site personnel must be able to respond. The use of traditional materials, such as poured reinforced concrete, require skilled site personnel to plan, monitor, and execute the construction process in a timely manner. It is imperative that students entering the construction field as designers, project managers, and construction supervisors have an understanding of the complete construction process and an ability to meld traditional methods and materials with new concepts. Qualified tradespeople provide the expertise necessary to erect and assemble the structure; they too benefit from an understanding of the entire construction process.

The seventh edition of *Principles and Practices of Commercial Construction* continues in the same vein as the previous edition. It provides the reader with a complete overview of traditional concepts and practices as well as preengineered components that are used in the construction of commercial buildings. From building layout to exterior finishing, this edition describes and illustrates the various stages of the building process. New materials and new concepts are outlined throughout the text, introducing the reader to current practices in commercial construction.

Traditional materials used in the construction of the structural frame—timber, steel, and concrete—are described in detail. The physical properties of each material are discussed, and a description of the structural components made from each material is provided. Tests for quality control and proper methods of application for each material are also outlined.

In response to new ideas, materials, and innovations, diagrams have been revised and photographs have been added or changed throughout the seventh edition. Additional information has been provided on excavation wall bracing in Chapter 3, including guidelines for safe working conditions in trenches. In Chapter 5, additional information on soil stabilization and the use of mini piles has been added. Chapter 6 provides additional calculations of formwork to better illustrate the effects of concrete pressures within wall formwork of different heights. Chapter 7 provides additional information on concrete admixtures for concrete mixes, and Chapter 12 adds additional information on single-ply roofing membranes and insulation requirements for roofs. Chapter 15 revises the discussion on building insulation to reflect current theories and practices in building envelope construction; a discussion on construction sealants and adhesives has also been added. Chapter 16 includes additional information on stucco coatings and the use of wood siding. The discussion on thermal efficiency of windows has also been enhanced. Additional review questions have been added to Chapters 1, 2, 3, and 10.

It is hoped that students engaged in construction-related technologies and trade apprenticeship programs continue to find *Principles and Practices of Commercial Construction* a useful overview of the construction process. An instructor's manual is provided with answers to all review questions and suggestions for additional class exercises and group projects.


## ACKNOWLEDGMENTS

The construction industry is a unique collection of planners, design professionals, manufacturing firms, and tradespeople. The success of the industry is a testimonial to the combined expertise of all these people. A text such as *Principles and Practices of Commercial Construction* would not be possible without the information that is provided by every segment of the industry.

I wish to thank the reviewers: Clark B. Pace, University of Washington; Felix T. Uhlik, Georgia Institute of Technology; W. Ronald Woods, University of North Florida; and David R. Busch, Columbus State Community College. They gave their time and provided many valuable suggestions for the seventh edition of *Principles and Practices of Commercial Construction*. In particular, the author is grateful for the generous contributions of the American Concrete Institute. I thank Cam Colin for his interest, keen eye, and practical suggestions; I thank Marion Andres for writing, editing and proofreading over the years.

To the editors and production staff at Prentice Hall, a sincere "Thank you!" for your ongoing confidence and expertise. Your dedication and energy over the years have been a continuing source of inspiration.



*Cameron K. Andres, P. Eng.*
*Ladysmith, British Columbia, Canada*

# CONTENTS

# 1 SITE INVESTIGATION

The cultural fabric of every nation is interwoven with the structures and monuments that its people have designed and built over the ages. In many instances, these undertakings required long periods of time and resulted in great costs, both in human lives and financial terms.

Each generation of builders has contributed to the construction mosaic by applying the skills passed down from preceding generations to create the many different structures that reflected their needs and expressed their artistic talents. As technological skills advanced, new methods and materials provided designers additional flexibility to express their creative talents. Armed with this legacy, today's builders continue the tradition by striving to build better and more efficient buildings for the benefit of society.

The successful completion of any construction project can be considered as the transformation of an idea, a need, or a vision into something of value and permanence. In today's competitive world, as in the past, to ensure success in such an undertaking, a talented group of designers, tradespeople, and administrators must work together as a cohesive unit with a common cause—to transform the dream into reality.

Every construction project consists of a host of details that must be planned, coordinated, and executed within a predetermined length of time. However, the exercise becomes purely academic if no thought has been given to a location for the final outcome. So, before considering the actual construction methods and materials, let us begin by looking briefly at the concept of land development and the organization of the construction team.

## LAND DEVELOPMENT

In general, all lands available for development come under the authority of a planning board, put in place by government officials, to ensure that development within the area under their jurisdiction meets the requirements of the local planning act. Planning acts can be instituted by federal, state, county, or local officials, and the size of the area controlled by a planning act depends on which level of government is involved. State and provincial planning acts provide general guidelines for land use for large areas of the state or province and deal with long-range planning. A city planning act deals with the area enclosed by the boundaries of the city limits, and its concerns are much more immediate and localized, although long-range plans must also be considered.

The purpose of a planning act is to provide a set of guidelines, in the form of zoning bylaws, to control the development that occurs within the area controlled by the particular planning act. Zoning bylaws are a means for regulating the development of an area by dividing the area into various land uses. In some instances, local and state planning acts are superimposed on one another to coordinate the planning of major highways and utility rights-of-way, which are normally a state or federal concern, with local land-use requirements.

Typical zoning within a city or town consists of areas designated as residential, commercial, industrial, and special use. In turn, each zone can be regulated for population density and the type of activity that is allowed.

An area designated as residential can be restricted to single-family detached housing or can allow high-density multiunit dwellings. Commercial areas can range from small, local strip-mall outlets to a central commercial district. Industrial areas range from warehouse-type operations to the manufacture of heavy equipment and hazardous chemicals.

If an owner has a particular occupancy requirement, the location of the building will be dictated by the local development board. On the other hand, if the owner has a particular site for development, the type of building and occupancy will be regulated by the zoning bylaws within the area.

Before actual construction begins, the owner must obtain a development permit based on preliminary drawings of the proposed structure and its use. The drawings must indicate proposed site entrances and exits, building location, parking provisions, landscaping, and type of occupancy. Once the development board approves the proposed development, the owner must then complete the drawings and apply for a building permit.

To obtain the required permit, complete construction drawings must be submitted to the building inspection department of the local planning board for evaluation and approval. Once this approval has been obtained, a building permit is issued to the site and construction can begin.

## THE CONSTRUCTION TEAM

Each project begins with an owner who has a specific need for a building or considers a building to be a good investment. The owner may be a private citizen, a group of investors, a large corporation, or an elected governing body. The initiative of the owner precipitates the construction process.

To ensure a viable project, the owner must have two important elements in place: (1) appropriate funds for the funding of the project, and (2) a suitable site. Once these items are in place, the owner selects an architectural firm to be responsible for the design and management of the construction project.

Once commissioned, the architect is responsible for the design of the building and the production and coordination of all drawings and specifications—architectural, structural, electrical, and mechanical. In many instances the structural, electrical, and mechanical drawings are sublet to other firms that specialize in these areas. Once all the drawings have been completed, tenders or bids are requested from general contractors. Usually, the general contractors provide a total price for the complete project; however, if the project is large and complex in nature, it may be broken down into several contracts and done in stages.

In any event, the general contractor depends on various subcontractors to supply specialty items such as piling, reinforcing steel, structural steel, precast concrete sections, roofing, cabinets, doors and windows, and electrical and mechanical equipment. The contract is usually awarded to the lowest bidder; however, the owner can select any contractor for any number of reasons.

When a general contractor is awarded a contract, the project then becomes the sole responsibility of that contractor. All work that is done must be in accordance with the drawings and specifications as prepared by the architect. If additional costs arise from omissions or errors on the drawings, these are negotiated with the owner's representative as extras to the contract.

The foregoing process may take months or even years and, while approval from the various regulatory bodies is being sought, a detailed investigation of the site for the design requirements of the proposed structure can begin. To ensure proper planning and design of the proposed structure, as much information must be obtained above and below the surface of the proposed building site as is practical. Geotechnical experts—individuals specializing in soil sampling and testing—are retained by the project managers to establish the parameters that will be used in the design of the building foundations.

The amount of testing done on the site depends on a number of conditions: the size and complexity of the structure, the type of soil encountered, proximity of the proposed structure to existing buildings, and the level of the groundwater table are the more important items that must be considered. The information is then passed on to the structural designers, who must decide on the type and size of foundations that will be used.

Many projects exceed construction designers' budgets and completion dates because of unforeseen problems during the excavation and construction of their foundations. To ensure that these problems are kept to a minimum, a thorough site investigation is a wise investment.

The site evaluation can be considered in two phases: a *primary investigation* and a *secondary investigation*. The primary investigation deals with the evaluation of the physical state of the building site during the design stages of the project; the secondary investigation is done by the general contractor just before the actual construction process begins. Both are of equal importance and must be conducted in detail.

## PRIMARY INVESTIGATION

The primary site investigation is usually done in two stages: (1) a *surface evaluation* of the building site and (2) a *subsurface investigation*. The surface evaluation of the site normally consists of a topographic survey to establish grades for drainage, landscaping requirements, and the placement of services. In addition, the survey information gathered at this time can be used by the contractor to estimate fill and excavation quantities.

Subsurface investigation consists of the evaluation of the soil below the surface to establish criteria for the foundation requirements of the proposed structure. Standard laboratory tests done on subsurface samples provide the necessary data from which the load-bearing properties of the soil can be established. The methods used in obtaining subsurface soil samples vary significantly depending on the complexity and size of the proposed structure.

For relatively shallow foundations, subsurface samples may be obtained from a simple test pit only several feet deep; however, in the case of deep foundations, samples are usually obtained at various depths up to 100 ft. (30 m) or more below the surface. To obtain samples at these depths, a drill rig is used to provide the test borehole, and special methods are used for extracting the required samples. Standard procedures are followed in all stages of the soil investigation to ensure that valid results are obtained from the samples when analyzed in the laboratory.

### Surface Evaluation

In a surface site investigation, the topography of the site is of prime concern to both the designer and the builder. From the designer's point of view, the structure must be designed to complement the site. The builder must be aware of the conditions under which construction must proceed. Is the site relatively level? Will it be necessary to remove large quantities of earth or, alternatively, will extensive fill material be required? Are there any rock outcrops that will require blasting? Is there surface water to be drained and will the site have drainage problems in the future? Has the site been used as a landfill site in the past?

Each condition must be evaluated and dealt with in the surface evaluation, for each will ultimately have a definite bearing on the cost of excavation and the type of foundation required.

### Subsurface Investigation

Large buildings impose substantial loads on their foundations and depend on soil of good bearing capacity—bedrock in some cases—to provide the necessary support. The depth at which this bearing is available will dictate the type and cost of the foundation. If the project is feasible from this standpoint, this investigation will help to estimate the cost and will influence the design of the foundations. Before we consider the methods used in evaluating the subsurface conditions, let us look at the properties of soil in general.

## NATURE OF SOIL

To make use of the samples taken during subsurface exploration, it is necessary to understand something of the nature of soil, types of soil, and how they react under various circumstances. This topic, in all its facets, is a complete study in itself, and we deal here with the subject only as it pertains to construction.

For engineers and architects, *soil* denotes all the fragmented material found in the earth's crust. Included is material ranging from individual rocks of various sizes, through sand and gravel, to fine-grained clays. Whereas particles of sand and gravel are visible to the naked eye, particles of some fine-grained clays cannot be distinguished even when viewed through low-powered microscopes.

All soils are made up of large or small particles derived from one or more of the minerals that make up solid rock. These particles have been transported from their original location by various means. For example, there are notable deposits of *eolian soil* in western North America, which were deposited by wind. There are also numerous deposits of *glacial till*, a mixture of sand, gravel, silt, and clay, moved and deposited by glaciers. Other soils have been deposited by the action of water, whereas others, known as *residual soils*, consist of rock particles that have not been moved from their original location, but are products of the deterioration of solid rock.

Soil types, as determined by particle size, are as follows:

- *Cobbles and boulders:* larger than 3 in. (75 mm) in diameter.
- *Gravel:* smaller than 3 in. (75 mm) and larger than #4 (5 mm) sieve (approximately ¼ in.).
- *Sand:* particles smaller than #4 (5 mm) sieve and larger than #200 (630 μm) sieve (40,000 openings per square inch).
- *Silts:* particles smaller than 0.02 mm and larger than 0.002 mm in diameter.
- *Clays:* particles smaller than 0.002 mm in diameter.

For purposes of establishing the abilities of these soils to safely carry a load, they are classified as *cohesionless soils, cohesive soils, miscellaneous soils,* and *rock.* Table 1-1 lists the allowable bearing values, in pounds per square foot (kiloPascals), for the various types of soil indicated.

Cohesionless soils include sand and gravel—soils in which the particles have little or no tendency to stick together under pressure. Cohesive soils include dense silt, medium dense silt, hard clay, stiff clay, firm clay, and soft clay. The particles of these soils tend to stick together, particularly with the addition of water. Miscellaneous soils include glacial

**Table 1-1**
**RELATIVE BEARING STRENGTHS OF SOILS**

| Type of Soil | Allowable bearing strength kPa | Psf |
|---|---|---|
| Cohesionless soils | | |
| Dense sand, dense sand and gravel | 285 | 6000 |
| Cohesive soils | | |
| Dense silt | 140 | 3000 |
| Medium dense silt | 100 | 2000 |
| Hard clay | 285 | 6000 |
| Stiff clay | 200 | 4000 |
| Firm clay | 100 | 2000 |
| Soft clay | 50 | 1000 |
| Miscellaneous soils | | |
| Dense till | 475 | 10000 |
| Cemented sand and gravel | 960 | 20000 |
| Rock | | |
| Massive | 4800 | 100000 |
| Foliated | 3800 | 80000 |
| Sedimentary | 1900 | 40000 |
| Soft or shattered | 950 | 20000 |

till and conglomerate. The latter is a mixture of sand, gravel, and clay, with the clay acting as a cement to hold the particles together.

Rock is subdivided into *massive, foliated, sedimentary,* and *soft* or *shattered.* Massive rocks are very hard, have no visible bedding planes or laminations, and have widely spaced, nearly vertical or horizontal joints. They are comparable to the best concrete. Foliated rocks are also hard, but have sloping joints, which preclude equal compressive strength in all directions. They are comparable to sound structural concrete. Sedimentary rocks include hard shales, sandstones, limestones, and siltstones, with softer components. Rocks in this category may be likened to good brick masonry. Soft or shattered rocks include those that are soft or broken but not displaced from their natural beds. They do not become plastic when wet and are comparable to poor brick masonry.

One problem caused by soil in building operations relates to backfilling around foundation walls and in service trenches. Once soil has been removed from its original location, it tends to increase in bulk because pressure is no longer keeping the particles closely packed. When this soil is returned to the excavation, it is still in this bulked state, and, unless some special measures are taken during backfilling, time and weather will bring about a return to the original volume, causing shrinkage in the hole. This can be overcome by compacting the soil while it is being replaced.

Tests have shown that mixing a certain amount of water with the soil enables the particles to be packed more closely together so that all of the voids that existed between the particles during the bulked state have been filled. The addition of too much water will tend to separate the particles and so increase the overall volume.

The practical implications are important. If soil can be replaced and packed with this optimum amount of water, no shrinkage will take place later. The optimum moisture condition is usually close to the condition of the soil as it was removed. Thus, soil removed from an excavation should be protected against drying out as much as possible if it is to be used again as backfill. If the moisture content changes, however, the soil should be allowed to dry or water should be added, as the case may be, before it is returned. Soil should be replaced in thin layers, not more than 6 in. (150 mm) thick, to achieve its maximum density. Each layer must be treated by some type of compaction machinery. If this process is followed carefully, it is possible to backfill even very large trenches, so that the soil approximates its original condition, and no appreciable settling of the surface occurs.

## Frost Penetration

Buildings that are located in cold climates or that are subjected to artificially produced temperatures (e.g., ice arenas, cold storage plants, and ice plants) can experience differential movements in their foundations as a result of frost action.

In many instances, when the ground freezes, there are no visible changes in the soil. However, fine-grained soils such as clayey sands and silts are susceptible to volume changes due to freeze-drying. The resulting migration of water through the soil due to capillary action can increase the volume of the soil at the frost line, causing unwanted pressure on floor slabs and building foundations.

When the mean air temperature drops below freezing over a sustained length of time, the depth of frost penetration into the soil increases. In fine-grained soils that have a source of moisture below the frost line, a peculiar phenomenon occurs. At the freezing plane, as the water in the soil turns to ice, the drying action draws water from the unfrozen soil. This water, in turn, freezes and joins the ice crystals above, adding to the growth of a layer of pure ice. This block of ice is known as an ice lens. Because water expands as it freezes, these ice lenses can produce undesirable pressure on foundations. Paved parking areas are susceptible to the effects of ice lenses, for when the ice melts in the spring, the soil under the paving has no bearing strength and a pot hole occurs.

Pressures due to soil volume changes have a wide range and are related to the soil grain size and moisture content. The type of soil is also influential: fine-grained soils develop more pressure than coarser-grained ones. Thus, clay soils develop higher pressure than silts, and silts develop higher pressure than fine sands. In general, coarse sands and gravels do not experience a significant volume change, and saturated soils will experience maximum volume changes, resulting in maximum pressures. As the moisture content drops, the resulting pressure also drops.

In a building site investigation, the possibility of ice lenses forming around foundations or under concrete floor slabs must not be overlooked. The three factors that contribute to frost heaving are: (1) a freezing plane in the soil, (2) a fine-grained soil through which moisture can move, and (3) a supply of water. If any of these conditions can be eliminated, frost heaving can usually be controlled successfully. Remedies to control the amount of frost penetration within the vicinity of the building foundation include the application of insulation around the building foundation, removal and replacement of fine-grained soil with coarse granular material, or controlling the moisture in the soil by paying careful attention to drainage. Implementation of one or a combination of these remedies will usually counter the problem of frost heave on a building site.

## SUBSURFACE SAMPLING

As stated previously, subsurface exploration is of prime importance to designers in gathering information for the design of the building foundations. Neglecting to investigate subsurface conditions adequately before construction begins can result in structural faults that, at a later date, may be very expensive to remedy (Figure 1-1).

The purpose of subsurface exploration is primarily to establish three things: (1) to obtain samples at various depths below the surface for the purpose of laboratory evaluation, (2) to determine the variation of the soil (soil profile) that exists on the site, and (3) to determine the depth at which free water is encountered; that is, the location of the groundwater table.

Usually, a truck-mounted drilling rig (Figure 1-2) is employed to drill the test holes from which samples of relatively undisturbed soil are obtained using

**Figure 1-1** Severe Structural Damage to Building Wall Due to Unstable Foundation Conditions.

**Figure 1-2** Truck-Mounted Drill Rig.

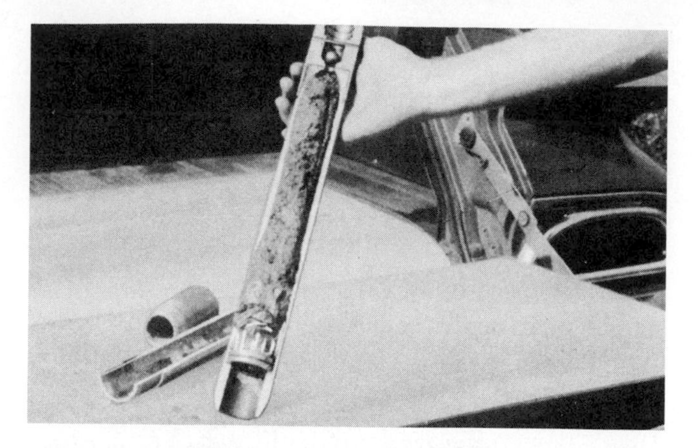

**Figure 1-3** Split Spoon Sampler. (Reprinted by permission of ELE International, Soiltest Products Division.)

**Figure 1-4** Shelby Tubes Used for Obtaining Soil Samples from Test Borehole.

special sampling tools. The two most common are the split spoon sampler (Figure 1-3) and the Shelby tube (Figure 1-4).

The Shelby tube is a thin-walled cylinder that is pressed down into the soil at the bottom of the test hole and then raised with a core of soil inside. The split spoon sampler is used to test the condition of the soil in the bottom of the test hole and to extract a soil sample for testing. By using a standard weight to drive the spoon into the soil, and counting the number of blows required to drive the spoon into the soil, a predetermined distance (usually a total of 18 in. [450 mm] in three 6 in. [150 mm] increments), an immediate indication of the soil-bearing capacity can be established.

Other methods are available for use in the evaluation of subsurface conditions, and the one adopted in any particular case depends on the requirements of the project. In some instances, two or more methods may be employed to ensure a thorough analysis of the soil. These methods include the use of (1) sounding rods, (2) augers, (3) test pits, (4) wash borings, (5) rock drillings, and (6) geophysical instruments.

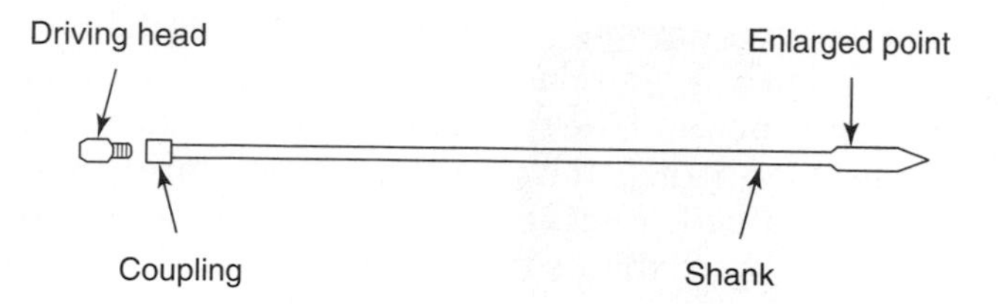


**Figure 1-5** Sounding Rod. (Reprinted by permission of ELE International, Soiltest Products Division.)

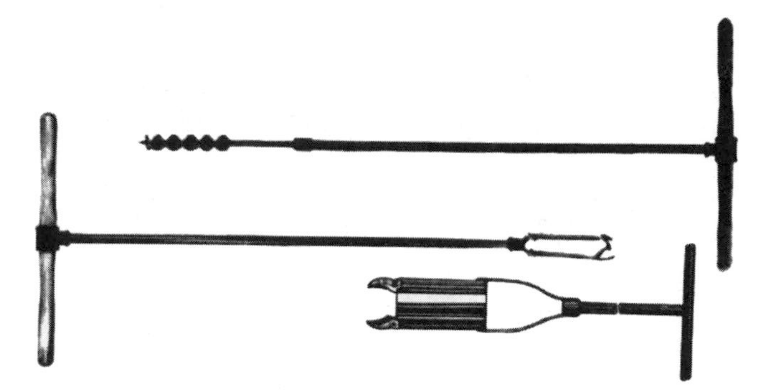

**Figure 1-6** Hand-Operated Augers.

(1) A sounding rod is used to determine the depth below the surface at which rock appears and whether the soil resistance is increasing or decreasing as the test proceeds. This method cannot be used in soil containing rocks of any significant size, because the rod would be deflected from its vertical course.

The rod is of solid steel, from ⅝ to 1½ in. (16 to 38 mm) in diameter, about 5 ft (1500 mm) long. It has a pointed and enlarged head (see Figure 1-5) for easy penetration and reduced friction. The length is increased by adding sections by means of threaded couplings. The top end is fitted with a flat driving cap. The rod may be driven either by hand or by a mechanical driver.

(2) An auger is useful for bringing up samples from relatively shallow depths in soils that are cohesive enough to be retained in the tool while it is being raised to the surface. It consists of a cylinder, usually 2 in. (50 mm) in diameter, with cutting lips on the lower end (see Figure 1-6). It is connected by ordinary couplings to a series of pipe sections. As the auger is turned, layers of earth are peeled off and forced up into the auger cylinder. When the cylinder is full, the auger is brought to the surface, emptied, cleaned, and returned. Power augers bring up what are known as *disturbed auger borings*.

(3) The use of test pits is probably the best method of examining subsurface soil because it is possible, by this method, to examine the layers of

**Figure 1-7** Drill Bits: (a) Diamond Drill; (b) Rotary Bit; (c) Cross-Chopping Bit. (Reprinted by permission of ELE International, Soiltest Products Division.)

earth exactly as they exist. In addition, soil moisture conditions are evident, and load tests can be made at any desired depth. This method is relatively expensive, and the depth to which examination can be carried out is limited. Excavating is usually done by hand, but if a required depth cannot be readily reached by hand, mechanical digging equipment may be used.

(4) The wash boring method requires the use of water; as a result, the borings are in the form of mud. Any given sample may be a mixture of two or more layers of soil. Thus, a particular stratum may not be detected at all.

The equipment for wash boring consists of an outer casing, inside of which is a hollow drill rod with a cross-chopping bit. Boring is done by raising, lowering, and turning the rod while water is forced down the rod and out through ports in the sides of the bit. Loose material is forced up between the drill rod and the outer casing by water pressure and collected through an opening near the top of the casing. This is known as a *wet* sample.

**Figure 1-8** Crawler-Mounted Drill Rig.

(5) A number of systems are employed for drilling rock; among them are *diamond drilling, shot drilling,* and *churn drilling.*

A diamond drill consists of a diamond-studded bit, as shown in Figure 1-7, attached to a core barrel. The barrel is in turn attached to a drill rod mounted in a rig similar to the one shown in Figure 1-8. The drill rod is rotated, and water is forced down through the hollow rod to cool the bit and carry the drill cuttings to the surface. The bit cuts a circular groove, and the core is forced up into the core barrel. When it is of the desired length, the core is broken off by a special device and brought to the surface for inspection. Diamond drilling may be done either vertically or at an angle.

Shot drilling is similar to diamond drilling except for the bit. It consists of a circular, hollow, hard steel bit with a slot around the bottom edge to allow circulation of shot. A flow of chilled steel shot is fed through the drill rod to the bit; as the bit turns, the shot cuts the rock and forms a core, which is forced up into the core barrel. Cores as large as 72 in. (1800 mm) in diameter can be taken with a shot bit.

Churn drilling consists of operating a hard steel chopping bit attached to a drill stem inside a casing. A cable, to which is fastened a set of weights, raises and drops the bit so that it strikes the bottom, chipping the rock. Water is forced down the side of the casing, and the resulting slurry is removed from the hole with a bucket.

**Figure 1-9** Automatic Engineering Seismograph. (Reprinted by permission of ELE International, Soiltest Products Division.)

(6) Two basic types of geophysical instruments are used for shallow 100 ft (30 m or less) subsurface investigation and exploration: refraction seismographs (Figure 1-9) and earth electrical resistivity units. The techniques involved have been known and used successfully for many years for deep exploration in the oil and mineral industries.

Through advances in electronic technology, the heavy, bulky equipment once required for this deep exploration has been reduced to small, compact units that can be easily carried by a two-person crew.

The *seismic refraction theory* is based on the fact that shock waves travel at particular and well-defined velocities through materials of various densities. The denser the material is, the greater the speed. The velocities may range from as low as 600 ft/sec (180 m/sec) in light, dry top soil to 20,000 ft/sec (6000 m/sec) in unseasoned granite. If the speed of the shock wave is known, the type, hardness, and depth of the stratum responsible for the refracted wave may be accurately determined (Figure 1-10).

The *electrical resistivity measurement* method of analysis depends on the ability of earth materials and formations to conduct electrical current, which follows relatively good conductors and avoids poor ones. Conductivity of the material depends on its electrolytic properties. *Conductivity* and *resistivity* vary according to the *presence and quantity of fluids, mineral salt content of the fluids,*

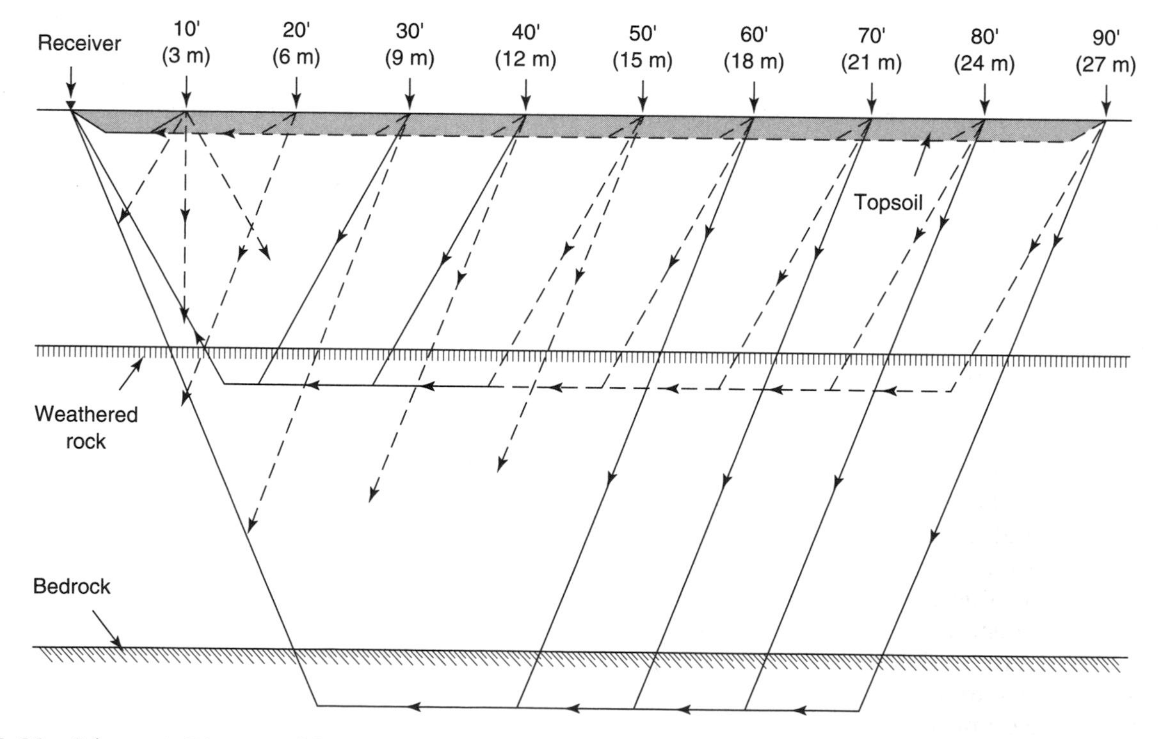


**Figure 1-10** Schematic Diagram of Seismic-Wave Refraction Principle.

**Figure 1-11** Earth Resistivity Measurement Instrument: (a) Instrument in Position for Readings; (b) Diagram of Set in Position for Reading. (Reprinted by permission of ELE International, Soiltest Products Division.)

*volume of pore spaces, pore size and distribution, degree of saturation,* and a number of other factors (Figure 1-11).

The use of this type of equipment does not eliminate the need for test boring. Drilling and sampling are necessary for foundation investigations where accurate information on the bearing capacity of a soil is required and where samples are needed for laboratory analysis. However, the use of electronic equipment may materially reduce the amount of drilling necessary and may help in the intelligent selection of drilling sites.

## Soil Profile

The soil profile is a cross section of the soil based on a visual evaluation of the samples obtained at various depths during the subsurface investigation. The depth to which samples are obtained depends on the size of the building, site conditions, and the type of material encountered during the drilling of the test borehole. If unusual circumstances occur during drilling, a reevaluation of the situation can be made on the site by the drilling team and the geotechnical engineer. In some instances, budgetary constraints determine the depth and number of test holes drilled; however, this approach to subsurface evaluation is shortsighted and in many cases has proven to be false economy.

A field log report for a typical test borehole is shown in Figure 1-12. It is a valuable record of what was discovered below the surface in a particular test hole. It provides a description of the soil types encountered and notes the methods that were used in obtaining the test samples, the depths at which these samples were taken, the condition of the samples, the results of any tests performed on the samples at the time of drilling, and the location where free water was encountered.

## Groundwater Level

Knowing the location of the groundwater level is important for two main reasons. From the contractor's point of view, a high water table means that extra costs will be encountered during the excavation and construction of foundations. For the designer, accurate information is of utmost importance because the location of the water table affects the soil-bearing capacity, which in turn determines the size and type of foundation required for the building.

# SURFACE TESTING

Surface testing is necessary when soil with low bearing values is removed and the resulting excavation must be refilled with new material. The new material must be placed and compacted to a uniform density

Civil Engineering Technology Dep't.
Test Hole Log 91–1

Project: Soils Borehole Demonstration
Test hole no.: 91–1
Driller: Mobile Augers & Research
Depth standpipe slotted: 4.36 m
Depth of water when checked on: 22/02/91
Depth of frozen soil: 1.05 m

Date drilled: 12/02/91
SFC. elev.: 1085.00 m
Rig no.: B61
Logged by: JJH/MJM
Was: 10.27 m
Scale = 1:50

| Elev / Depth | Soil symbols Sampler symbols and field test data | USCS | Soil Description | Remarks | Sample No. | N or % Rec. | PP (Kg/cm²) | WC (%) | DD (kN/m³) | $q_u$ (kPa) |
|---|---|---|---|---|---|---|---|---|---|---|
| 1085 / 0 | | OL | **Topsoil:** Organics, some silt | Rootlets | | | | | | |
| 1084.5 / 0.5 | | ML | **Silt:** Sand & organic mixture Grey brown | | | | | | | |
| 1084 / 1 | 14/6 7/6 6/6 | SM | **Silty Sand:** Uniform, fine-grained Grey brown, some clay | Stiff consistency | SPT1 | 13 | | 7.5 | | 155 |
| 1082.5 / 2.5 | | | | Shelby tube bent on a rock at 2.4 meters | U1 | 50% | | 22.0 | | |
| 1082 / 3 | 5/6 5/6 10/6 | CL | **Clay:** Stratified Silt & sand layers | Stiff consistency; Saturated Silt nodules at 3.0 meters | SPT2 | 15 | | 10.3 | | 187 |
| 1081 / 4 | 1/6 3/6 4/6 | SC | **Clayey sand:** Some silt Light brown | Very stiff consistency | SPT3 | 7 | | 24.7 | | 79 |
| 1079.5 / 5.5 | | VC | **Variable mixed clay:** Silt and sand Stiff consistency | | U2 | 100% | | 24.9 | 16.7 | |
| 1079 / 6 | 3/6 4/6 6/6 | | | | SPT4 | 10 | 3.8 | 19.7 | | 102 |
| 1078 / 7 | 2/6 4/6 6/6 | | | | SPT5 | 10 | | 20.8 | | 105 |

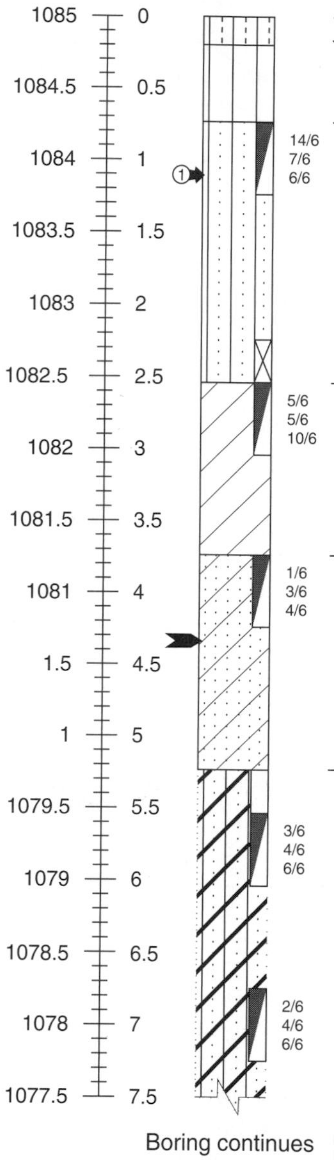

Depth scale: 1085 (0), 1084.5 (0.5), 1084 (1), 1083.5 (1.5), 1083 (2), 1082.5 (2.5), 1082 (3), 1081.5 (3.5), 1081 (4), 1.5 (4.5), 1 (5), 1079.5 (5.5), 1079 (6), 1078.5 (6.5), 1078 (7), 1077.5 (7.5)

Boring continues

End of hole - 10.86 m of standpipe set
Water not checked at time of drilling
(1) symbol shows the depth to which soil was frozen

Page number 1

**Figure 1-12** Soil Log Report.

Civil Engineering Technology Dep't.
Test Hole Log 91–1

Project: Soils Borehole Demonstration
Test hole no.: 91–1
Driller: Mobile Augers & Research
Depth standpipe slotted: 4.36 m
Depth of water when checked on: 22/02/91
Depth of frozen soil: 1.05 m

Date drilled: 12/02/91
SFC. elev.: 1085.00 m
Rig no.: B61
Logged by: JJH/MJM
Was: 10.27 m
Scale = 1:50

| Elev / Depth | Soil symbols Sampler symbols and field test data | USCS | Soil Description | Remarks | Sample No. | N or % Rec. | PP (Kg/cm$^2$) | WC (%) | DD (kN/m$^3$) | $q_u$ (kPa) |
|---|---|---|---|---|---|---|---|---|---|---|
| 1077.5 / 7.5 | | | | | | | | | | |
| 1077 / 8 | | S4 | **SP–SM:** Hard, fine-grained Saturated silt layers | Stratified | U3 | 100% | | 19.4 | 23.4 | |
| 1076.5 / 8.5 | 9/6 15/6 19/6 | | | | SPT6 | 34 | 4.5+ | 20.9 | | 390 |
| 1076 / 9 | | | | | | | | | | |
| 1075.5 / 9.5 | | | | | | | | | | |
| 1075 / 10 | 5/6 12/6 15/6 | | | | SPT7 | 27 | | 27.1 | | 375 |
| 1074.5 / 10.5 | | | | | | | | | | |

Water checked 22/02/91

End of hole - 10.86 m of standpipe set
Water not checked at time of drilling
(1) symbol shows the depth to which soil was frozen

Page number 2

**Figure 1-12** Continued.

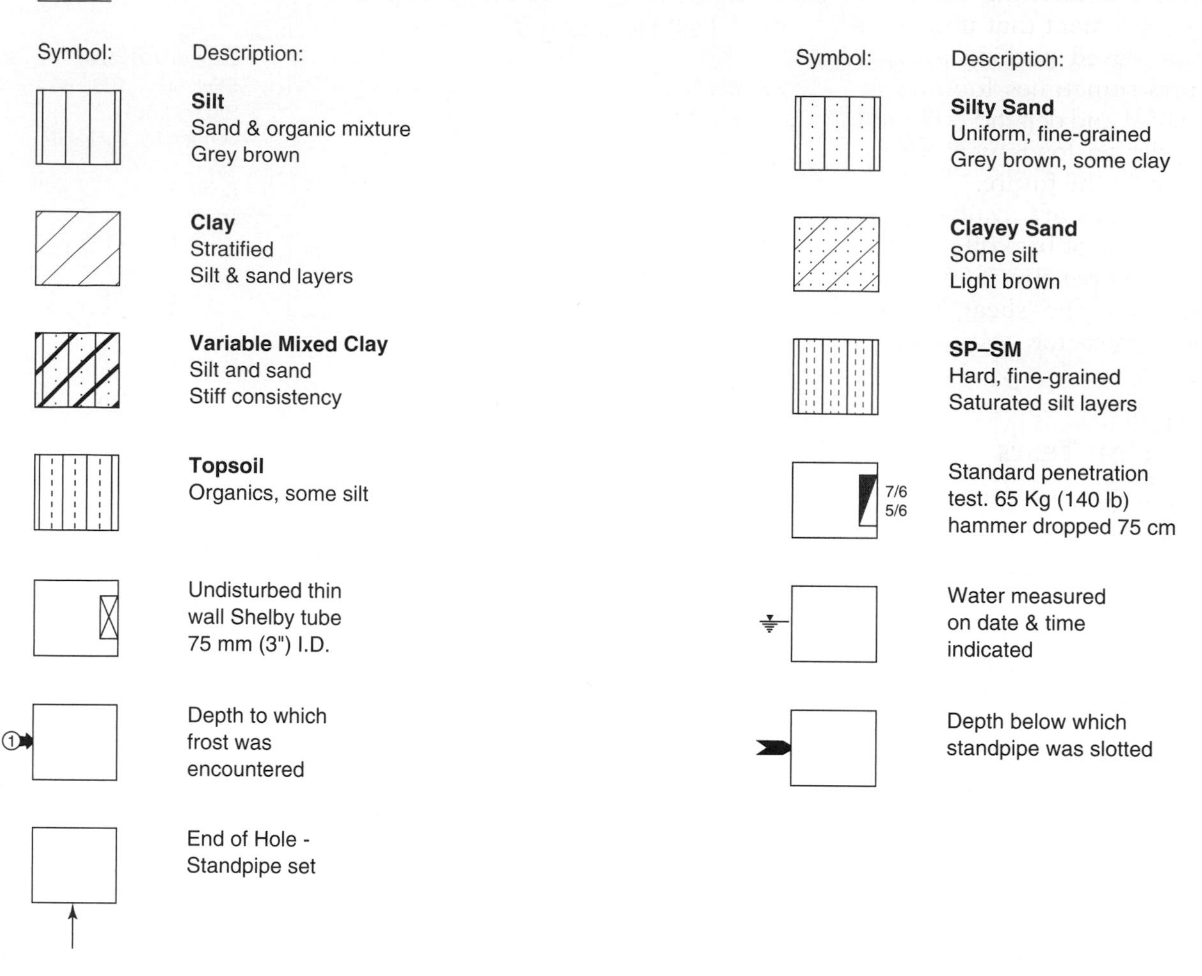


Notes:

1. Exploratory boring was drilled on Feb. 12, 1991, using a hollow-stem 100 mm (4") diameter continuous flight power auger.

2. No free water was encountered at the time of drilling, but when re-checked the following day was at 10.33 m.

3. Results of tests conducted on samples recovered are reported on the logs, using abbreviations some of which are:

DD = Natural dry density (kN/m3)
WC = Natural moisture content (%)
-80 = % passing 80 micron sieve
PP = Pocket penetrometer (Kg/cm2)
$q_u$ = Unconfined compression (kPa)
%Rec. = Percent of sample recovered based on 45 cm (18") long Shelby tube

LL = Liquid limit
PI = Plasticity index
SS = Soluable sulphates
SR = Soil resistivity (ohm-cm)
pH = Soil pH (%)
N = Penetration resistance for last 30 cm (1 ft.) of tube travel

4. This log is subject to the limitations, conclusions, and recommendations in this report.

Page number 3

**Figure 1-12** Continued.

to minimize differential settlements and to ensure that any settlement that may occur is within tolerable limits. Paved parking areas, concrete slabs on grade, and sometimes footings are placed on such new material and depend on its ability to support the required design loads without displaying excessive settlements in the future.

Various tests are available for checking the surface conditions of backfill material: the *penetrometer* for measuring penetration resistance, the *vane tester* for measuring the shear resistance of soils, and *moisture testers* for testing the moisture content of soils.

## Penetration Tests

Penetration tests provide a fast and easy way of determining the bearing capacity of soils. Two approaches have been developed: the *static penetration test* and the *dynamic penetration test*. Both methods measure the resistance of the soil to a device, commonly known as a penetrometer, which is driven into the soil in a predetermined manner. In general, the static test is used in the evaluation of cohesive soils (clays) and the dynamic approach is used in the evaluation of cohesionless soils (gravel deposits).

The instrument developed for the static approach is the *proving ring penetrometer*, shown in Figure 1-13. This instrument has a conelike probe equipped with a proving ring and a dial gauge that indicates soil penetration resistance readings. To determine the penetration resistance at a given location in the soil, the penetrometer is pushed firmly into the soil, at a uniform rate, until the top of the penetration cone is level with the soil surface. The dial is then read and the corresponding penetration load is determined from a calibration chart supplied with the instrument. Correlation between penetration load, bearing capacity, and density of soil is also provided.

The dynamic test, also known as the *standard penetration test*, uses a 140 lb (63.5 kg) weight dropped from a height of 30 inches (762 mm) onto the end of the rod that is attached to the probe. The number of blows, *N*, required to produce a penetration of 1 ft (305 mm) is known as the penetration resistance of the soil. Although the proving ring penetrometer has largely replaced the standard penetration test for surface soil evaluations, the standard penetration test is useful in evaluating soil density in test bore holes.

## Shear Tests

It is often necessary to make tests of soil conditions for foundation work in clay soils; one of these tests is on the shear strength of clay. A small tester used for making such tests is a hand held vane tester, shown in Figure 1-14. It may be used on the surface, in excavations, or with extension rods, down a borehole. It consists of a *four-bladed vane*, connected by a *rod* to a *torsion head* containing a *helical* spring.

To make a test, the vane is pushed into the soil and the torsion head is rotated at the rate of 1 rpm. When the clay shears, the load on the torsion spring is released and a pointer registers the maximum deflection of the spring. The shear strength of the soil is then read from a calibration curve.

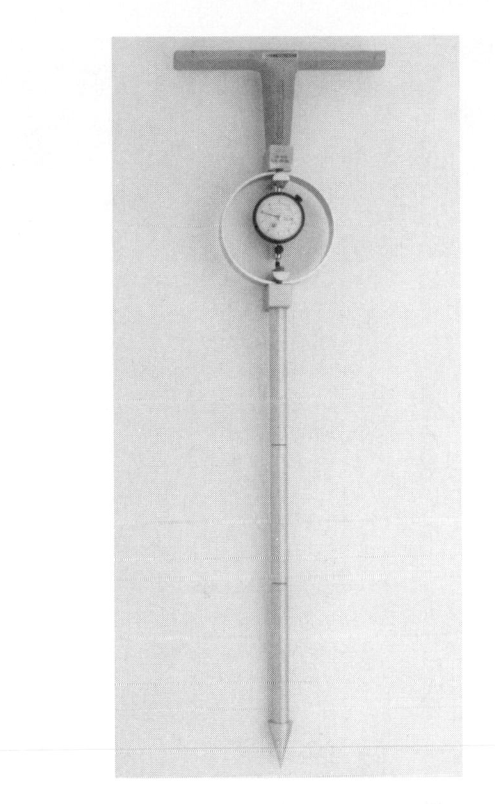

**Figure 1-13** Proving Ring Penetrometer. (Reprinted by permission of ELE International, Soiltest Products Division.)

## Moisture Tests

There are a number of reasons for taking moisture tests on soils during construction and several methods of doing so. One method involves the use of a tester that operates on the principle of a calcium carbide reagent being introduced into the free moisture in the sample. This forms a gas, the amount of which depends on the amount of free moisture in contact with the reagent. The gas is confined in a sealed chamber, and by measuring the gas pressure, the amount of free moisture in the sample is determined.

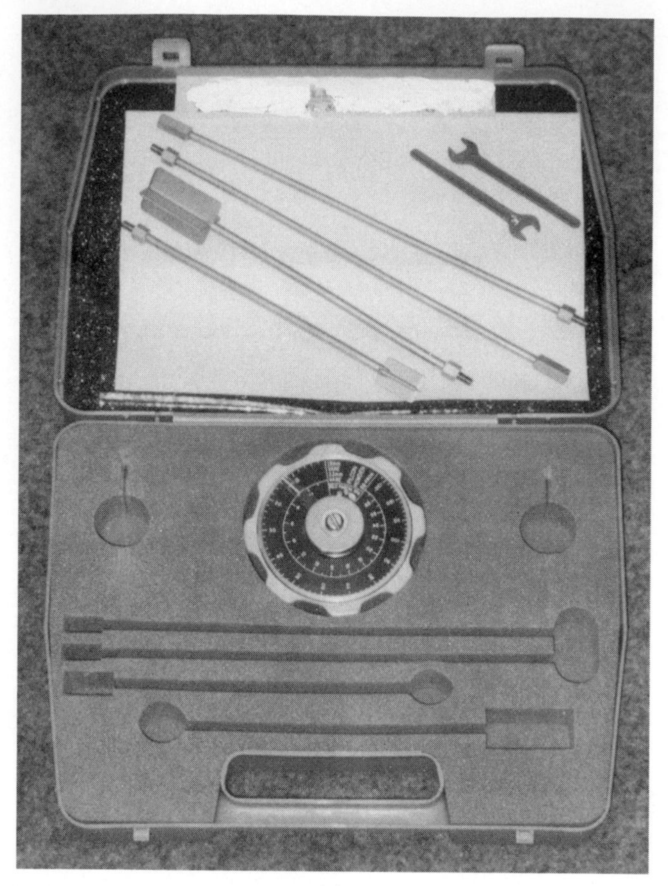

(a)

(b)

**Figure 1-14** (a) Hand-Held Vane Shear Tester Kit. (Reprinted by permission of ELE International, Soiltest Products Division.) (b) Hand-Held Vane Shear Tester. (Reprinted by permission of ELE International, Soiltest Products Division.)

## Nuclear Surface Moisture-Density Tests

To ensure that future differential settlements are kept to a minimum, density and water content must be carefully monitored during the backfilling operation. Although ASTM standards list two methods for determining the optimum density of soils, the tests themselves have little practical value. More importantly, uniform compaction of the base material is much more critical under a slab or pavement than some arbitrary specified value. The contractor must have a means with which to measure densities that is quick, easy, and reliable. One such instrument is the *nuclear density tester.*

Using a gamma source, the nuclear density tester provides a safe, simple, accurate, and quick method for determining the density and moisture content of construction materials such as asphalt, soil, aggregates, and concrete. By employing recent developments in microprocessing, the instrument provides direct digital readouts for wet and dry densities, percent moisture, percent compaction, void ratio, and percent voids in standard or metric units (Figure 1-15). Two methods of testing, the *backscatter method* and the *direct transmission method,* are available to the operator, depending on the type of material and the thickness of the lift to be tested.

The backscatter method is nondestructive and is used primarily for testing material densities near the surface. In this application, both the gamma source and the detectors remain on the surface (Figure 1-16). The backscatter method provides reliable density evaluations to a depth of 3.5 in. (90 mm) and is used primarily for the evaluation of asphaltic concrete. Soil moisture content is also determined in this manner, with both the detectors and the source at the surface.

The direct transmission test requires punching a hole into the material being tested. The advantage of this method is that the operator can select a depth of measurement (Figure 1-17), thus reducing errors due to the surface roughness and chemical composition of the material being tested. This method is usually used to test thick lifts of soil, as well as stone and asphalt.

In most jurisdictions, to ensure the proper use and care of the instrument, a license is required to own and operate this type of tester. Manufacturers and distributors provide recognized courses for operators in the proper use and calibration of this instrument, thus ensuring that the results obtained in the field are accurate and reliable.

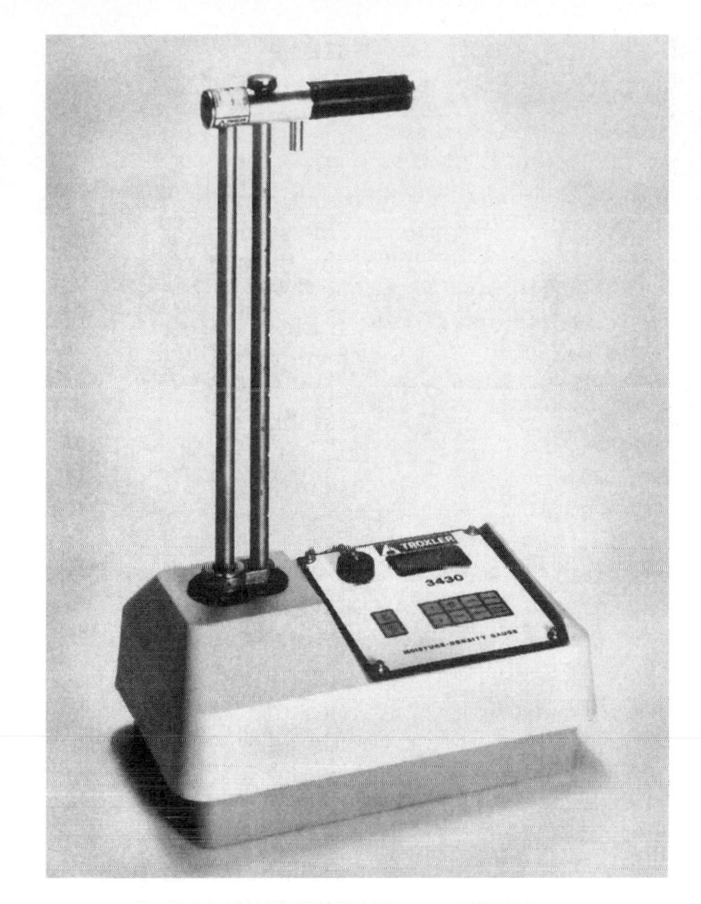


**Figure 1-15** Nuclear Moisture-Density Tester Used in Evaluating the Moisture Content and the Density of Compacted Materials. (Reprinted by permission of Troxler Electronic Laboratories, Inc.)

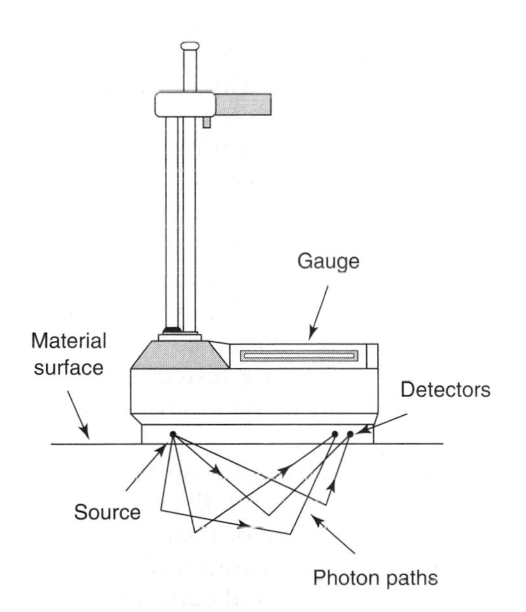


**Figure 1-16** Backscatter Method. (Reprinted by permission of Troxler Electronic Laboratories, Inc.)

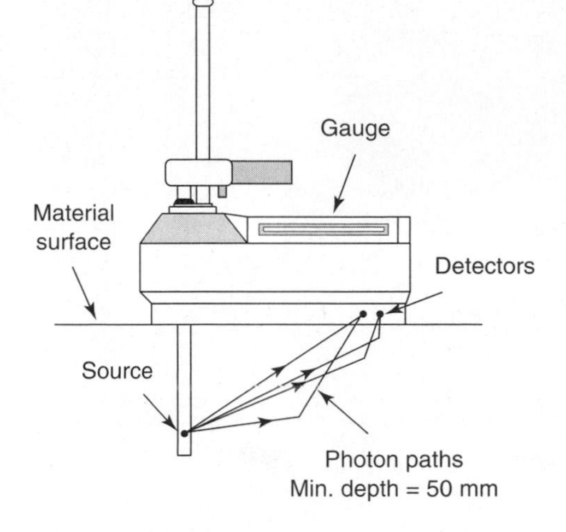


**Figure 1-17** The Direct Transmission Method. (Reprinted by permission of Troxler Electronic Laboratories, Inc.)

## SECONDARY INVESTIGATION

The word "secondary" does not necessarily refer to importance but to chronological sequence. The primary investigations are carried out before plans and specifications are drawn; that is, they are made by the planners. But once the building has been decided on and the tenders let, the prospective contractor will be responsible for additional investigations.

### Access to the Site

It is often advantageous to give thought and study to the most practical and economical routes by which equipment and materials are to be moved to the job site. The contractor will want to know what roads or streets give access and if it is feasible to move the necessary equipment over them. It may be necessary to build a private road.

If a navigable waterway is accessible, is it practical and economical to use for the delivery of equipment and material to the site? Is air transportation available? Fast transportation of urgently needed workers and supplies may be more economical in the long run.

### Availability of Services

One of the most important considerations is the availability of electrical power at the job site. Practically, no modern construction job can be carried out without it. Is power close at hand or must it be brought in over long distances? Will the power line

**Figure 1-18** Proper Protection for the Public around a Building Site is Paramount.

be supplied by the power company or will this be the responsibility of the construction company?

Gas is another important commodity at a construction site, for both heating and cooking. If there is gas in the area, the location and size of the main must be known in order to bring gas to the site.

Water must be available when a construction job begins and the builder must ascertain whether there is a water main in the vicinity. If not, some other method of supplying water must be found. This may be accomplished by means of a well, pipeline, or tank trucks.

## Site Safety and Local Building Bylaws

When a contractor receives a building permit for a construction project, the permit is based on the condition that compliance with all local building bylaws during the course of construction will be met. The major concerns of local officials, while construction is in progress, are public safety and the possible detrimental effects on traffic flow caused by street closures. Although the wording of local bylaws may vary to meet specific needs, typical requirements in the construction plan include:

1. A plan of the construction operation, indicating the extent and length of time of street closures.
2. Enclose the building site with proper fencing (Figure 1-18), and provide appropriate overhead protection for pedestrian traffic.
3. Proper barricades and shoring around all excavations.
4. Scaffolding to be designed, erected, and approved by qualified individuals.

Government agencies such as the Occupational Safety & Health Administration (OSHA) of the U. S. Department of Labor, as well as state and provincial bodies such as the Workmen's Compensation Board, provide guidelines for all aspects of safety on the construction site. OSHA provides standards and regulations for every stage of construction, from basic soil classification to handling hazardous materials. Standards and specifications for the erection and maintenance of scaffolding, ladders, and hoisting equipment are also provided as are guidelines for individual trades such as structural steel erection. Safety guidelines for site personnel involved in dangerous activities such as blasting are also covered. Site visits and enforcement of all regulations is done on a regular basis.

Depending on the size of the construction project, a full-time safety officer may be on site as a requirement of the local safety board or workmen's compensation board. Regular site visits by the local safety board officer ensure that all safety regulations are observed. To ensure that all site personnel are safety conscious, safety meetings are held at regular intervals by the general contractor or project manager. Their frequency is determined by the complexity and type of work being done on the site. Typical safety concerns on the building site include:

1. First aid kits located at strategic locations on the site.
2. Supervisory personnel trained in first aid.
3. Approved shoring for all open trenching.
4. Appropriate barriers around all openings in floors and roofs. Nonessential openings in floors and roofs must be securely covered.
5. All ladders and stairs must be fitted with appropriate handrails and secured to walls.
6. All hoisting equipment, such as tower cranes and construction elevators, must meet local safety standards and must be maintained on a regular basis.
7. Proper storage of all sheathing materials to ensure that they do not become airborne.
8. All individuals working on the site must be equipped with hard hats, safety glasses, safety boots, lifting harnesses, and gloves.
9. Appropriate respirators and a fresh air supply must be provided for crews working in enclosed spaces.

It must be emphasized that safety has to be everyone's concern. All personnel must be aware of the dangers that exist on the work site. Casual laborers and young individuals, such as summer students, are often unprepared for the pace of the work. Many workers are not instructed in proper lifting techniques and lack a general awareness of the dangers that exist when working around heavy equipment. Good employers ensure that all new personnel are assigned to work with someone who has experience and are given proper instruction in the operation of any power equipment.

## Local Labor Supply

An important consideration when planning a construction job is the availability of skilled and unskilled labor. Contractors usually provide their own staff of technical workers (engineers, estimators, and superintendents) but usually rely on local labor for the remainder. In some areas, unskilled labor will be in adequate supply, whereas skilled workers such as carpenters, bricklayers, electricians, and plumbers will have to be brought in. In other instances, the entire labor force may have to be imported, whereas in other situations an adequate supply of labor of all kinds will be locally available. Importing help may mean that the contractor will have to feed and house these workers during construction at additional expense.

## Site Conditions

The prospective builder will need to know the type of soil that will be encountered during excavation and the effects of groundwater on the excavated area. The builder should be able to obtain this information from the planner; if it is not available, subsurface exploration will have to be carried out.

## Local Weather Conditions

Climate plays an important part when planning a construction job. Is the area subject to a great deal of rain, is it unusually dry, or just average? The answers to these questions will have an effect on how equipment and materials are to be stored, what types of equipment will work best under those conditions, and whether special construction schedules will have to be developed.

Is the job in an area where a long winter season is likely to be encountered? If so, this will affect the types of equipment to be used, the heating and lighting requirements, and the proper protection that must be provided to carry on winter work.

Unforeseen inclement weather may disrupt construction schedules such that completion dates cannot be met. This could result in penalties, which may significantly increase the overall cost of the project. In short, the more a builder learns about the conditions to be met during the operation, the more likely the job can be completed with a minimum of trouble. The builder will be able to arrive at a realistic cost figure, plan a reasonably smooth working schedule, and complete the building in a satisfactory manner.

## Review Questions

1. What two items must an owner provide to ensure a reasonable chance of success in the construction of a new building?
2. What are the major responsibilities of the architect?
3. Outline two basic differences between a primary and a secondary site investigation.
4. What two items might be found on the soil profile that could affect the type of foundation used in a structure?

**5.** What four pieces of information may be found in a field log report concerning the soil test samples?

**6.** Explain clearly **(a)** the principle on which a refraction seismograph instrument works, and **(b)** the principle involved in the operation of an earth resistivity meter.

**7.** Give a brief explanation of each of the following terms: **(a)** *complementary* usage of soil testing methods, **(b)** bulking of soil, **(c)** residual soil, and **(d)** frost heaving in soil.

**8.** Give an explanation for the following statement: "As a general rule, the resistivity readings for sand and gravel will be high and uniform, whereas those for bedrock will be high and erratic."

**9.** Explain **(a)** what is meant by the *shear strength* of clay, **(b)** what is meant by *degree of compaction,* **(c)** why it is important to know the degree of compaction in certain backfilling operations, and **(d)** why moisture control is important in compacting soil.

**10.** State the prime purpose for zoning bylaws within a city or town.

**11.** State the difference between cohesive and cohesionless soils.

**12.** State the three conditions that contribute to frost heaving.

**13.** Give an example where the use of a nuclear density tester would be advantageous to the contractor.

**14.** What benefit does a Shelby tube have over an ordinary auger when obtaining soil samples from a drill hole?

**15.** Give three reasons for obtaining subsurface soil samples on a building site.

**16.** In Figure 1-12, at what depth was water encountered for the given test hole? Convert the reading in metres to feet.

# 2 SITE LAYOUT

Once the building site has been established, it is necessary to locate the building precisely within the boundaries of the site and to indicate to what depth excavation must be performed. When the site is not level, excavation depths will vary, and these variations must be shown.

Included among the drawings for a building is the site plan [see Figures 2-1 (a) and (b)], which shows the property lines, available utilities, location of trees, and the like, and the location of the building. Approaches and slopes of finished grades are usually shown. In addition, a *bench mark* or *datum point* is often indicated on the plan. A bench mark is a point of known elevation, established by registered survey and marked by a brass plate on a post at or near ground level or by a brass plug set into a building, bridge, or other permanent structure, near ground level. A datum point is a reference point, usually given an arbitrary elevation, for example, 100 ft or 0.0 ft (100.00 m or 0.0 m) such as a manhole cover or some permanent surface at or near ground level on an adjacent building or other structure.

## SURVEYING INSTRUMENTS

Surveying instruments are used to run building control lines, lay out angles, and determine the various differences in elevation required during the construction of a building and its foundations. These precision instruments incorporate a telescope, leveling devices, horizontal and vertical cross hairs, and finely calibrated scales for measuring horizontal and vertical angles.

Three basic types of surveying instruments are available for use in the construction industry: the *level,* also known as a builders' level; the *level transit;* and the *theodolite* or *transit.* Each type of instrument is available in a wide variety, from basic mechanical models to sophisticated electronic ones. Electronic models with software packages that contain various surveying programs, data storage capabilities, and remote entry infrared keyboards are available. These are known as *total station* surveying instruments. These instruments have the hardware to transmit field data from the building site directly to the office, thus saving time and decreasing the chance of errors occurring when the field data is transcribed to the building plans. Surveying instruments equipped in this fashion are used to their best advantage in topographic surveys and the accurate positioning of structures. In spite of all this sophistication, the use and operation of all surveying instruments require the understanding of basic surveying principles.

### The Level

Two varieties of levels are shown in Figures 2-2 and 2-3. The level operates in a horizontal plane only, and its use is limited to operations that can be done in that plane, such as measuring horizontal angles and differences in elevation. All levels have the same basic parts: a leveling mechanism, a level vial with bubble, a telescope with an objective lens, an eyepiece and focusing knob, a horizontal circle with vernier scale, a mechanism for fine adjustments of horizontal movement, and cross-hair focusing. Figure 2-2 provides a good illustration of the various

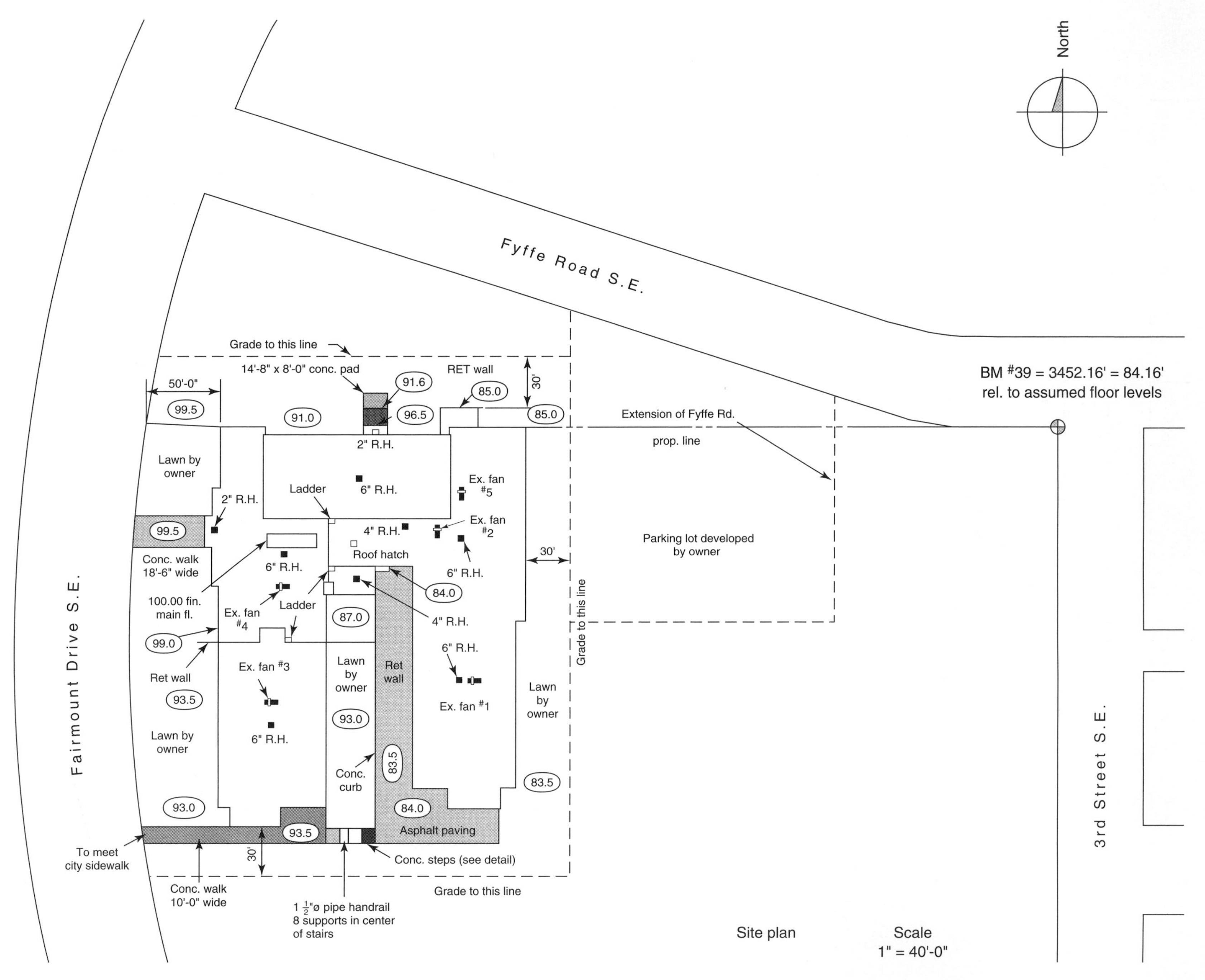


**Figure 2-1** (a) Site Plan Using Standard Units.

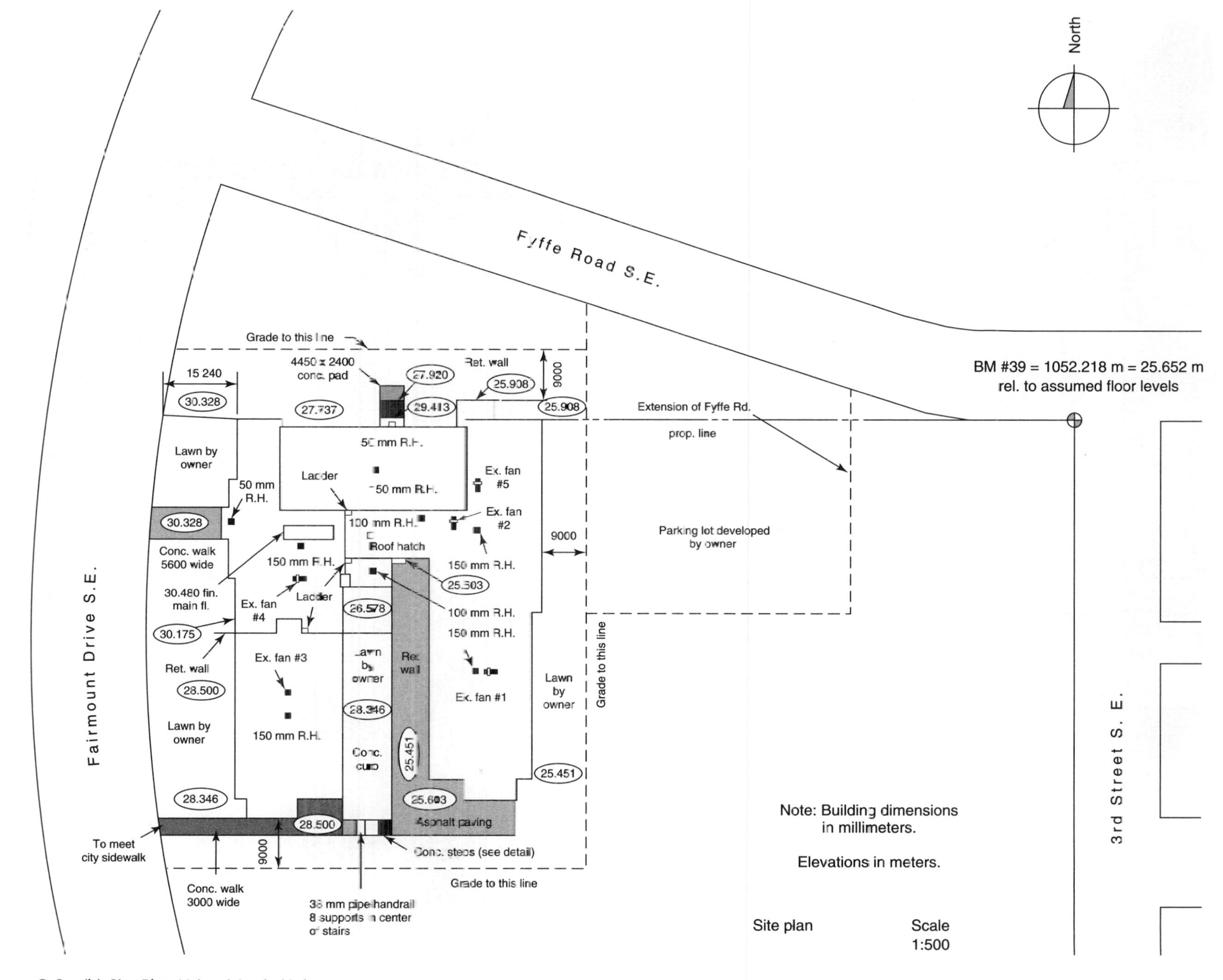


**Figure 2-1** (b) Site Plan Using Metric Units.

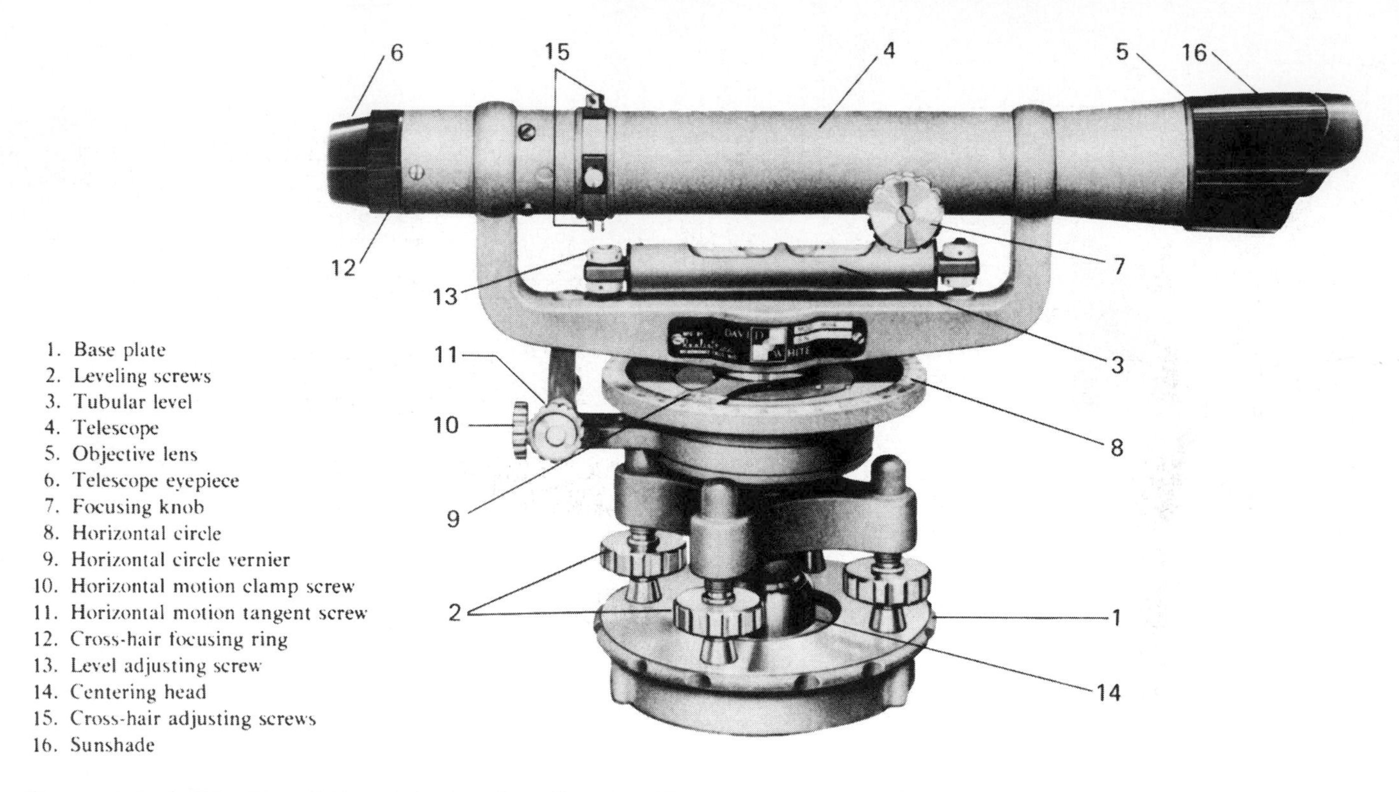


**Figure 2-2** Builders' Level, Four-Point Leveling. (Reprinted by permission of CST/Berger–David White.)

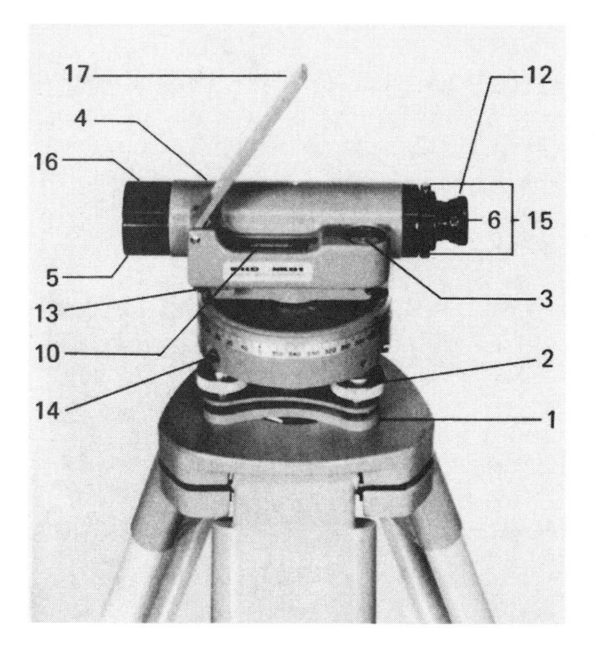


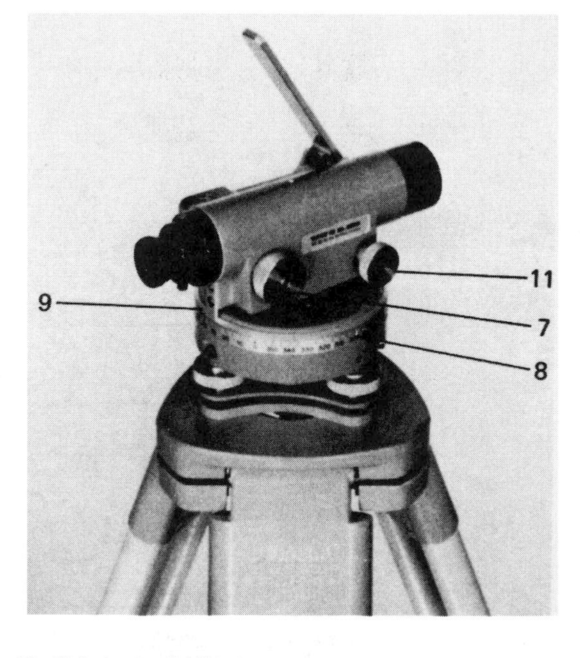


1. Base plate
2. Leveling screw
3. Circular level
4. Telescope
5. Objective lens
6. Telescope eyepiece
7. Focusing knob
8. Horizontal circle
9. Horizontal circle vernier
10. Tubular level
11. Horizontal motion tangent screw
12. Cross-hair focusing ring
13. Tubular level adjusting screw
14. Adjusting screw for leveling screw
15. Cross-hair adjusting screws
16. Sunshade
17. Bubble viewing mirror

**Figure 2-3** Builders' Level, Three-Point Leveling. (Reprinted by permission of CST/Berger–David White.)

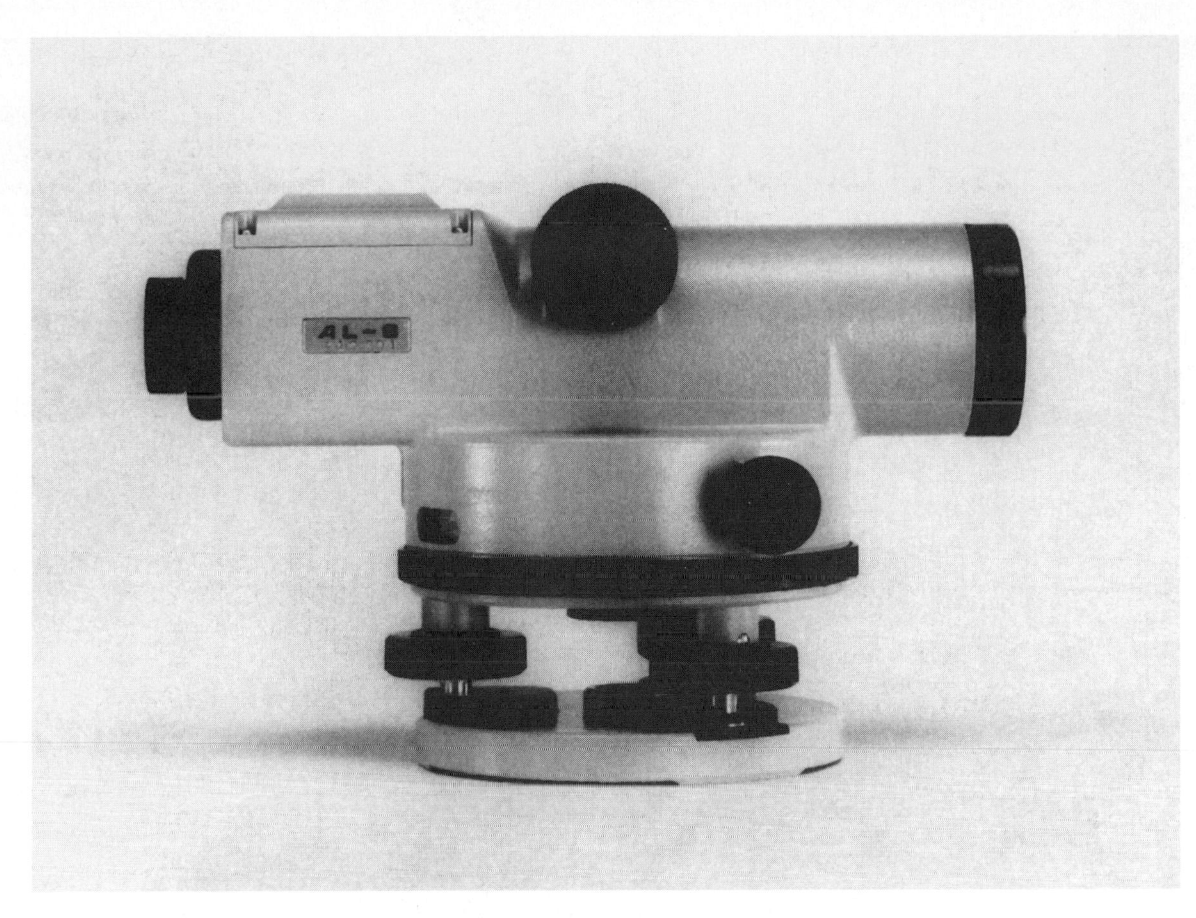


**Figure 2-4** The AL-6 Automatic Level. (Reprinted by permission of Pentax Canada Inc.)

components of a level; newer models have been streamlined and reduced in size and weight.

## Automatic Level

Figure 2-4 shows the latest development in automatic levels. This type of level has a self-leveling feature, which makes its setup much faster and ensures that the instrument remains level during use. Stadia marks on the cross hairs allow the instrument user to establish horizontal distances when sighting on a special rod known as a *stadia rod*. Although relatively durable, this type of level must be used with care to avoid damaging the self-leveling mechanism.

## Electronic Levels

The latest generation of survey level is the electronic variety, which incorporates the latest in microchip technology. This instrument is both self-leveling and self-reading. All the operator must do is sight on the rod and press a button; in approximately 4 sec, the electronic equipment on the level does the rest (Figure 2-5).

The electronic circuitry ensures superb accuracy and ease of use, thereby reducing the chance for error. The rod reading and horizontal distance are displayed on a small screen (Figure 2-5) in either metric or standard units. Elevations are calculated, and readings can be stored automatically in the data storage unit. Power is provided by a nickel–cadmium rechargeable internal battery good for one day of work. Because of its speed, accuracy, and data storage capabilities, the electronic level is excellent for topographical surveys, positioning of pipelines, monitoring of structural deformations, and highway construction applications.

## Laser Level

The laser level (Figure 2-6) is a direct development of space-age technology. The construction laser is a low-energy laser and should not be confused with the high-energy lasers used for other purposes. Although the laser is completely safe to use, looking into the laser beam itself should be avoided as damage to the eye could result—an effect like looking directly into the sun or any other bright light.

(a)

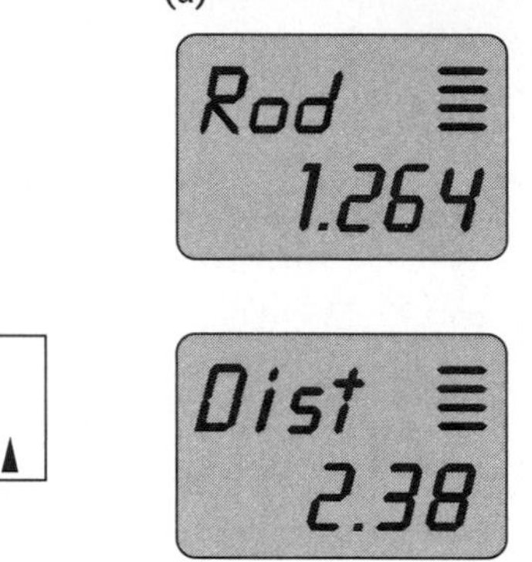


(b)

**Figure 2-5** (a) State-of-the-Art Wild NA2000 Electronic Level; (b) Reading and Distance Measurements are Automatically Calculated Electronically within 4 Seconds. (Reprinted by permission of Leica Geosystems.)

The laser level produces a controlled beam of red light 3/16 in. to 3/8 in. (5 to 10 mm) in diameter by exciting the electrons in a helium–neon gas with electrical energy. The beam can be made to rotate about the instrument, producing a plane of light. If the instrument is level, the plane of light can be used as a reference plane for leveling.

When an object such as a survey rod intersects the plane of light, the laser becomes visible as a red line. The person holding the rod can then take a reading, without the need for another person at the instrument. Because the laser is a beam of light that can be aimed in a particular direction, it provides a perfect reference line for use in construction applications that require strict alignment. Typical applications are horizontal leveling, vertical alignment, and grade control for laying of pipe.

**Horizontal Leveling.** Once the instrument is leveled and the laser beam rotated, the resulting plane of light created by the laser can be used as a reference for aligning screeds for concrete floors (Figure 2-7), pouring concrete slabs, and installing suspended ceilings (Figure 2-8).

**Figure 2-6** Laser Level.

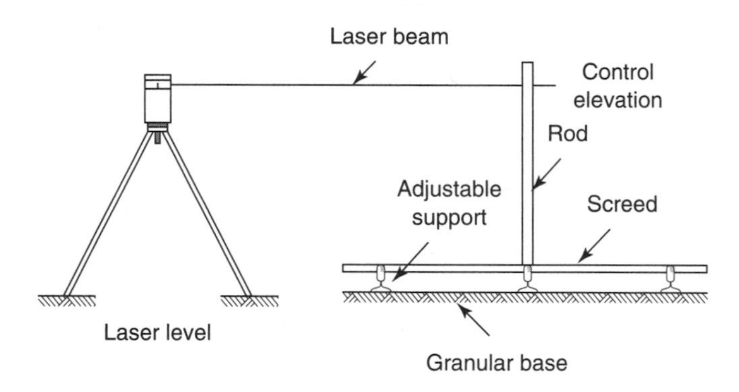


**Figure 2-7** Leveling Slab Screeds for Slab on Grade.

**Vertical Alignment.** The instrument head can be rotated 90° so that the resulting plane is now vertical. The plane of light can then be used for the alignment of formwork, as for high-rise construction. In some applications the laser is set inside the building to monitor the vertical alignment of the structure as each new floor is added (Figure 2-9).

**Grade Alignment.** Another popular use for the laser is in aligning pipes at specific grades to ensure proper drainage. The laser beam is preset at a specific grade and then set in the trench at one end of the pipeline. Pipe sections are then laid by lining up the center of the pipe, with the aid of a target, on the laser beam (Figure 2-10). When the pipe is of a large diameter, an alternative approach is to place the preset laser level on top of the pipe and align the top of the pipe with respect to the laser beam.

Levels can be taken using a normal surveying rod or with the use of a special rod that has a traveling laser-sensitive target (Figure 2-11). The target moves up and down the rod until it intersects the rotating beam of light. Once the plane of light strikes the target, the target locks on the light beam, allowing an accurate rod reading to be taken.

## Level Transit

Besides turning in a horizontal plane, a level transit will also tilt through a vertical arc, up to 45° from the horizontal position (see Figure 2-12). Thus it is more versatile and makes possible such operations as measuring vertical angles and running lines. It has all the basic parts of a builders' level plus those involved with vertical movement.

## Theodolite or Transit

The telescope on a theodolite may be rotated 360° in a vertical plane (Figure 2-13). The instrument is designed primarily for measuring horizontal and vertical angles; consequently, the graduations on the scales are finer, providing for greater accuracy in

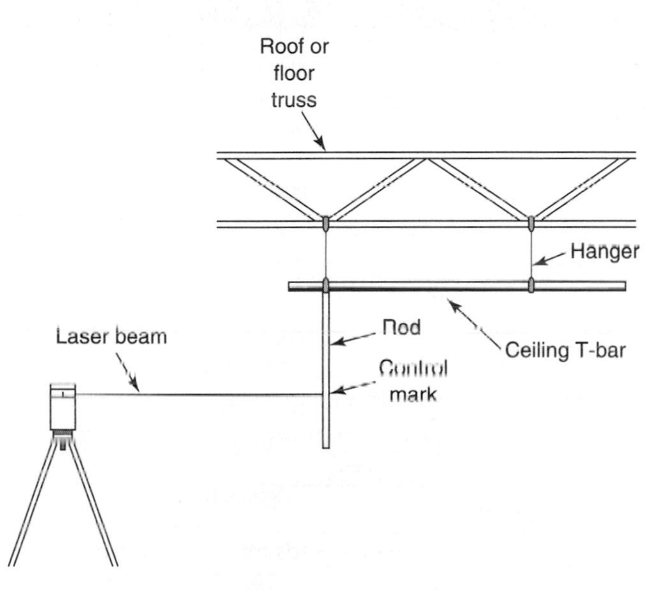


**Figure 2-8** Using Laser for Ceiling Installation.

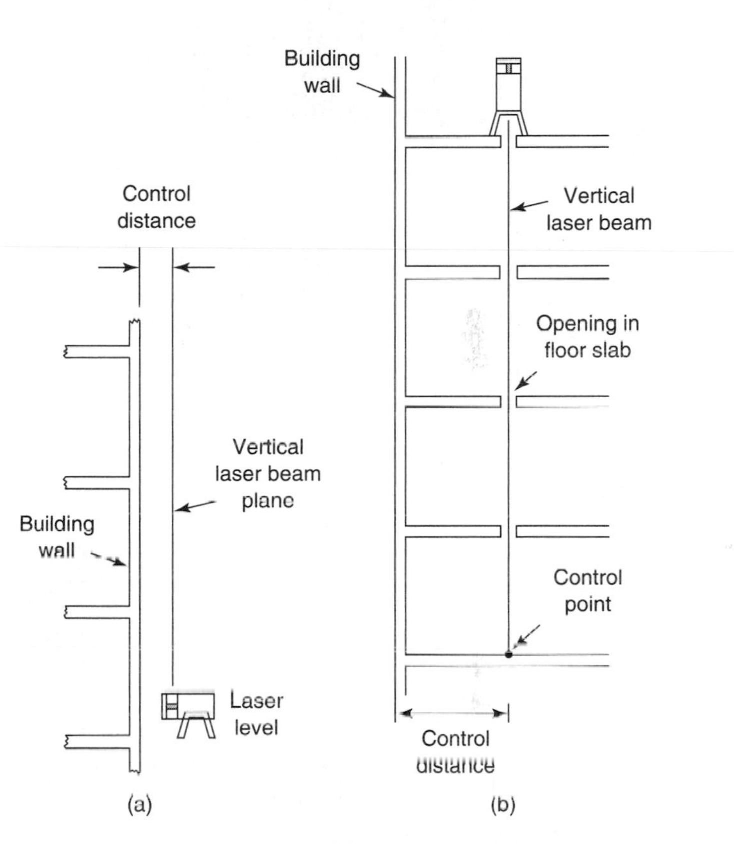


**Figure 2-9** Laser Level Used for Vertical Alignments of Structures: (a) Laser Level Outside the Structure; (b) Laser Level Inside the Structure.

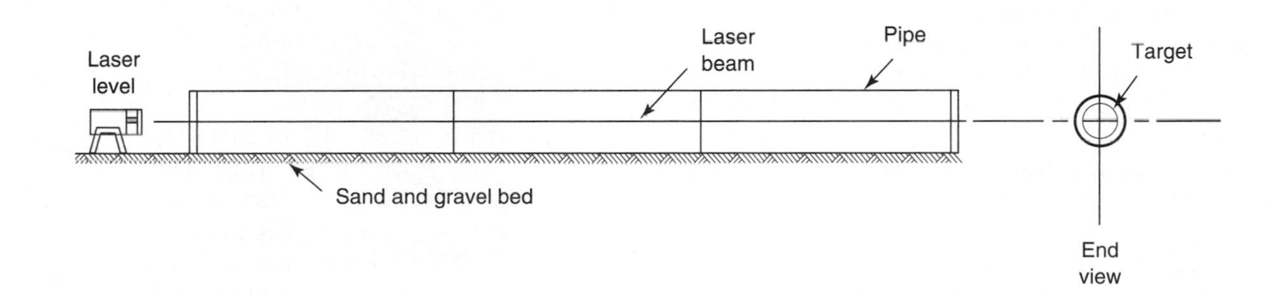

**Figure 2-10** Pipe Alignment Using Laser.

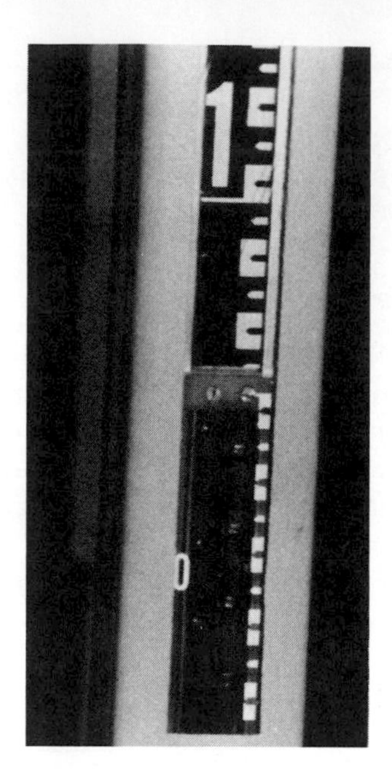

**Figure 2-11** Laser-Sensitive Target on Survey Rod.

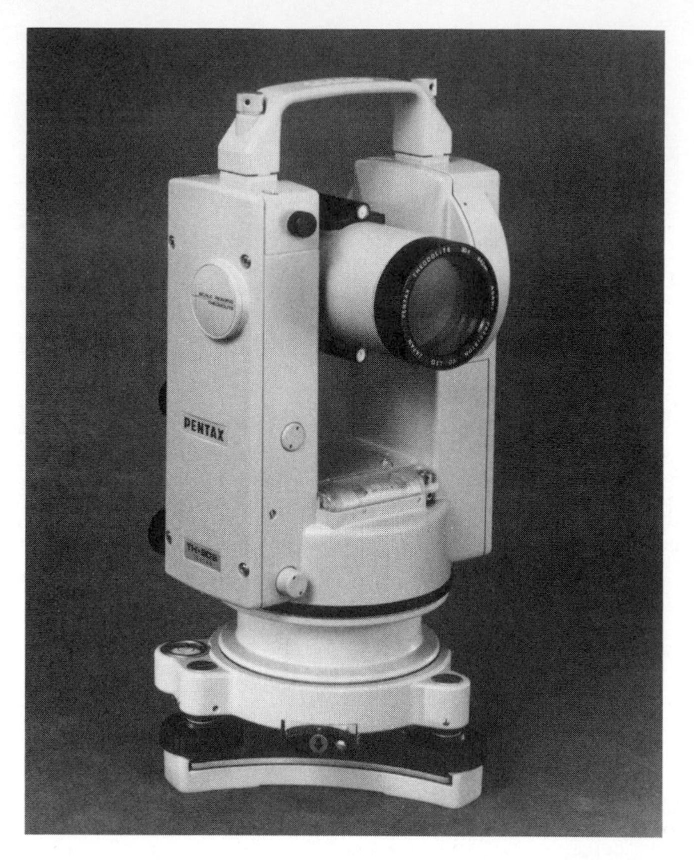


**Figure 2-13** Latest Generation of Theodolite Available with Internal Lighting System and Optical Plummet. (Reprinted by permission of Pentax Canada Inc.)

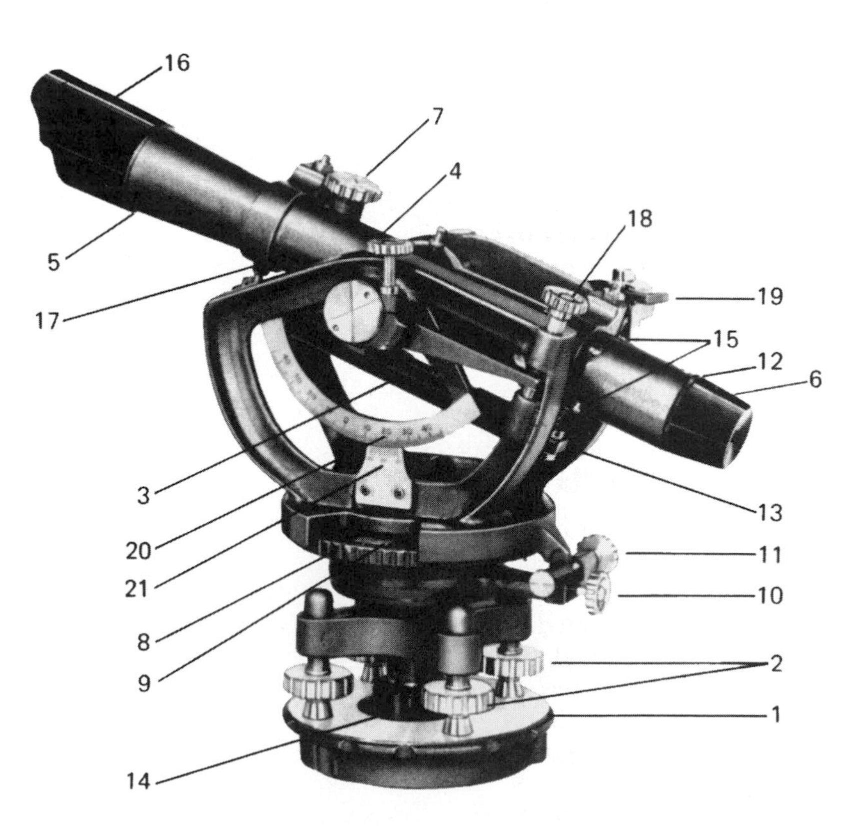


1. Base plate
2. Leveling screws
3. Tubular level
4. Telescope
5. Objective lens
6. Telescope eyepiece
7. Focusing knob
8. Horizontal circle
9. Horizontal circle vernier
10. Horizontal motion clamp screw
11. Horizontal motion tangent screw
12. Cross-hair focusing ring
13. Level adjusting screw
14. Centering head
15. Cross-hair adjusting screws
16. Sunshade
17. Vertical motion clamp screw
18. Vertical motion tangent screw
19. Locking lever
20. Vertical arc
21. Vertical vernier

**Figure 2-12** Level Transit. (Reprinted by permission of CST/Berger–David White.)

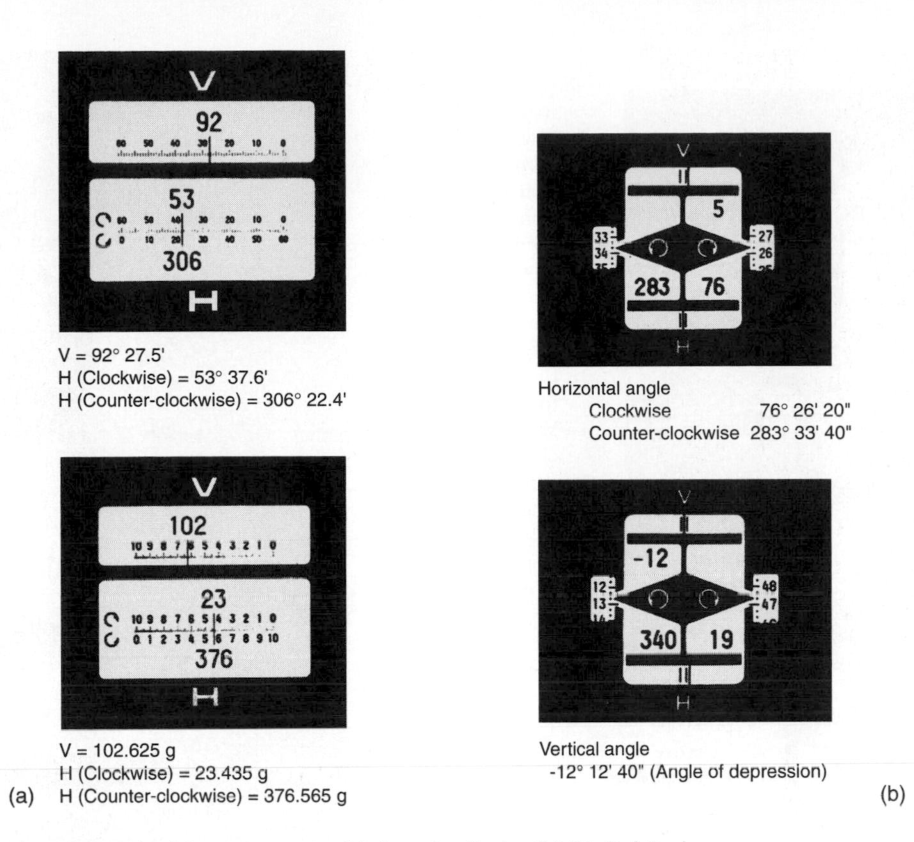


**Figure 2-14** Examples of Angular Measurements: (a) Angular Scale; (b) Digital Scale.

angular readings than is possible with most builders' levels or level transits. The latest generation of theodolites allows readings to be taken from scales within the telescope, rather than from exterior graduated plates. An optical plummet, in place of a plumb bob, is provided for positioning the instrument over a point.

Angular measurements can be made in degrees or gradients (400 grads in a complete circle). Readings may be taken from an angular scale, as shown in Figure 2-14(a), or in digital form, as shown in Figure 2-14(b). Angle reading accuracy up to 20 sec (one-third of a minute) on the digital scale can be easily achieved by the instrument user.

## Instrument Parts

The basic parts of a builders' level are indicated in Figures 2-2 and 2-3. All have a *base plate* (1), by which it is attached to a tripod and upon which the instrument is mounted. It is leveled by means of *leveling screws* (2), and the centering of a bubble in a *circular* or *tubular level vial* (3) indicates when the *telescope* (4) is level. Some instruments have a second level tube (see Figure 2-3) for more accurate leveling of the telescope.

At the front end of the telescope is an *objective lens* (5) and at the rear a *telescope eyepiece* (6), which houses horizontal and vertical cross hairs. A *focusing knob* (7) allows the operator to adjust the object focus to suit his or her eyesight.

A *horizontal circle* (8), graduated in degrees, is mounted below the telescope, and the telescope and frame, with an attached *vernier scale* (9) containing an index pointer or mark, revolve on it, making possible the measurement of horizontal angles, usually to within 5' of a degree. The instrument can be held in any position by a *horizontal motion clamp screw* (10) in Figure 2-2 and brought into fine adjustment by a *horizontal motion tangent screw* (11).

The cross hairs can be brought into sharp, clear focus by a *cross-hair focusing ring* (12) and adjusted with relation to one another by the *cross-hair adjusting screws* (15).

The *centering head* (14) (see Figure 2-2) allows the instruments to be moved laterally within the confines of the cutout circle in the base plate. This facilitates positioning the instrument over a point. A *sunshade* (16) cuts down glare while sighting, and the *bubble-viewing mirror* (17) in Figure 2-3 allows the operator to see the telescope level bubble while sighting through the eyepiece.

The level transit, shown in Figure 2-12, has all the parts of the builders' level. In addition, it has a *locking lever* (19) to hold the telescope in a horizontal

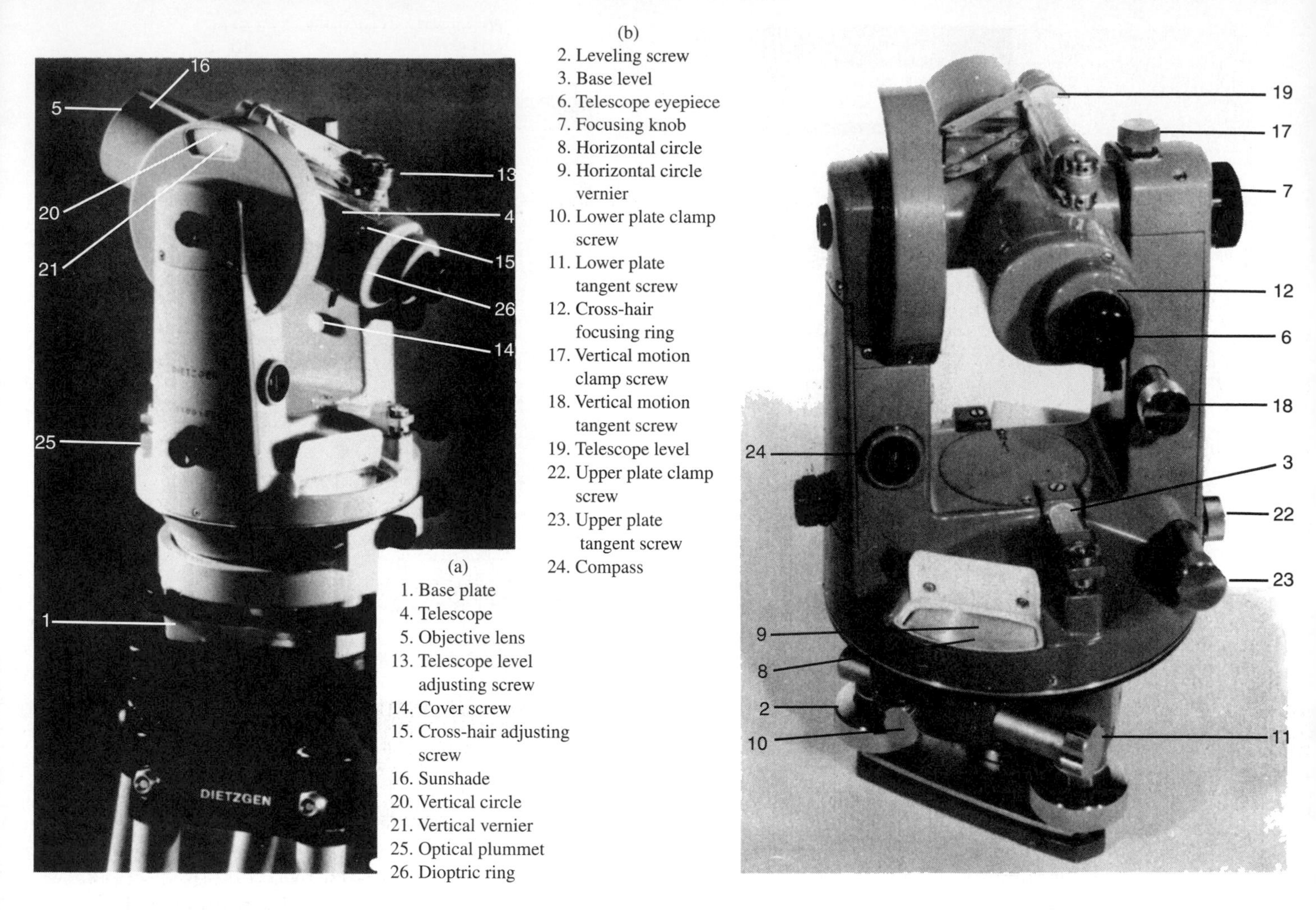


**Figure 2-15** Transit Three-Point Leveling. (Reprinted by permission of Dietzgen.)

position and, when it is disengaged, a *vertical motion clamp screw* (17) to hold the telescope in any desired position. A *vertical motion tangent screw* (18) allows fine vertical motion adjustments. A *vertical arc* (20) marked off in degrees, and a *vertical vernier* (21) containing an index mark and a minute scale, provide the means for reading the angles of depression or elevation (below or above the horizontal), usually to within 5′ of a degree.

A transit (see Figure 2-15) has *base levels* (3) used for leveling the whole instrument and a *telescope level* (19) for fine leveling of the telescope itself.

In place of a vertical arc, a transit has a full *vertical circle* (20), in graduations of ⅓° or ½° (20′ or 30′ divisions) and a *vernier* (21), which allows a reading to within 20″ or 30″ of a degree. The horizontal circle and its vernier will be similarly graduated.

## MEASURING ANGLES WITH BUILDERS' LEVEL OR LEVEL TRANSIT

With many levels and level transits, angles can be measured to within one-twelfth (5′) of a degree. The horizontal circle and the vertical arc are marked off in degrees and numbered every 10° (see Figure 2-16). The vernier is marked into twelve divisions of 5′ each,

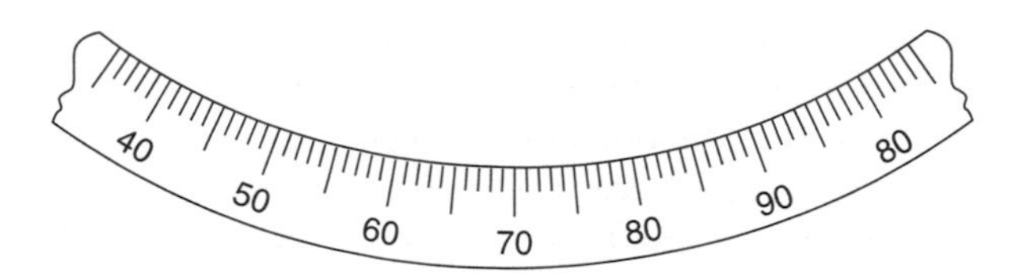


**Figure 2-16** Horizontal Circle Scale.

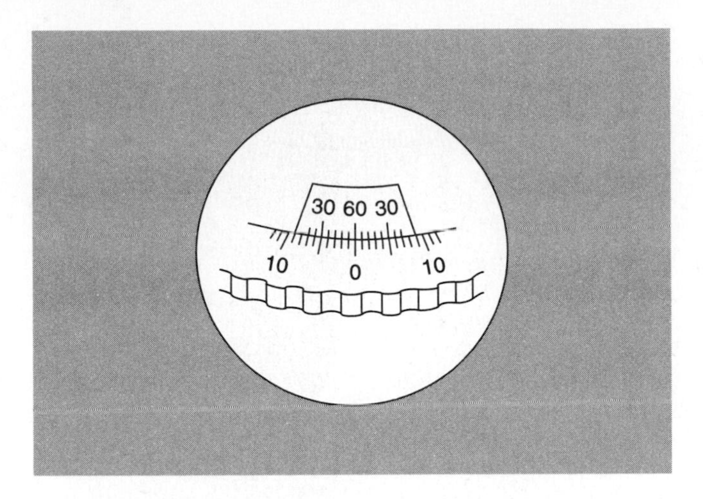


**Figure 2-17** Vernier Scale.

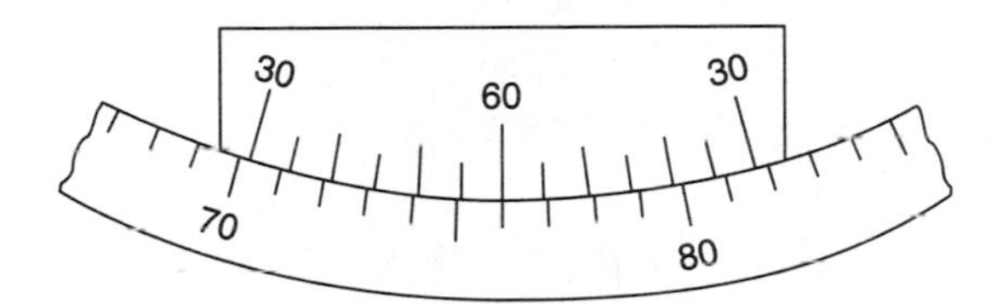


**Figure 2-18** Reading of 76°0′.

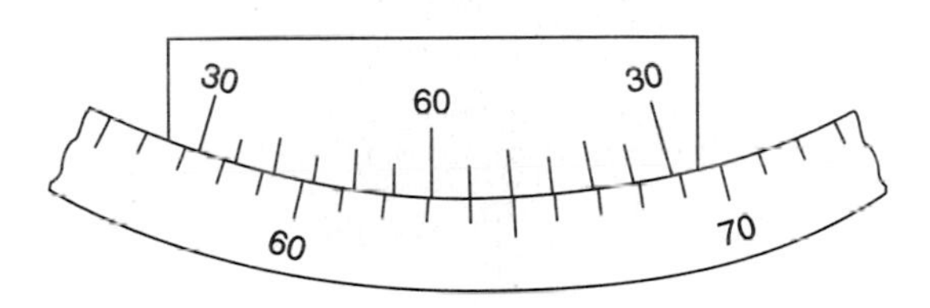


**Figure 2-19** Reading of 63°5′.

the twelve representing 1° (60′). These twelve spaces are equal to eleven spaces (11°) on the circle or arc. The vernier has zero (marked 60) in the center and reads both right and left to 30′ (see Figure 2-17).

The vernier makes it possible to take a reading to within 5′ (one-twelfth of a degree). If the zero (60) on the vernier coincides *exactly* with a degree mark on the circle, then the reading will be $x$°0′. Figure 2-18 shows a reading of 76°0′. But if the zero on the vernier comes *between* two degree marks on the circle, rely on the vernier to read the fraction of a degree involved.

Remember, each space on the vernier is one-twelfth smaller than a space on the circle. When zero on the vernier is just enough past a degree mark to allow the first line on the vernier to the right or left of zero to coincide with a line on the circle, the vernier zero must be one-twelfth of a degree (5′) past the degree mark. Therefore, the reading would be $x$°5′. Figure 2-19 illustrates a reading of 63°5′. If the second line on the vernier coincides with a degree mark, it means that the vernier zero is two-twelfths

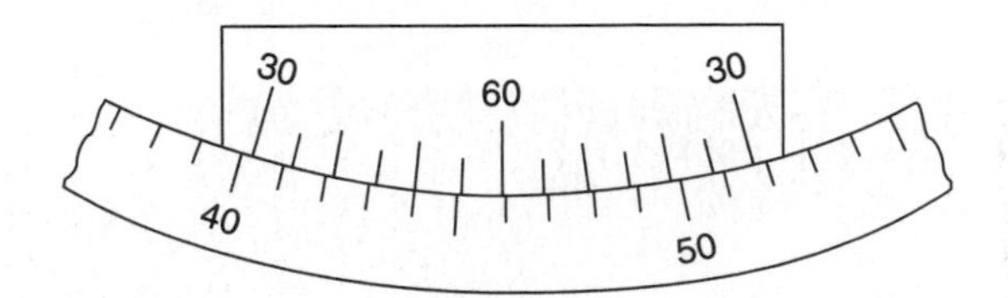


**Figure 2-20** Reading of 45°50′.

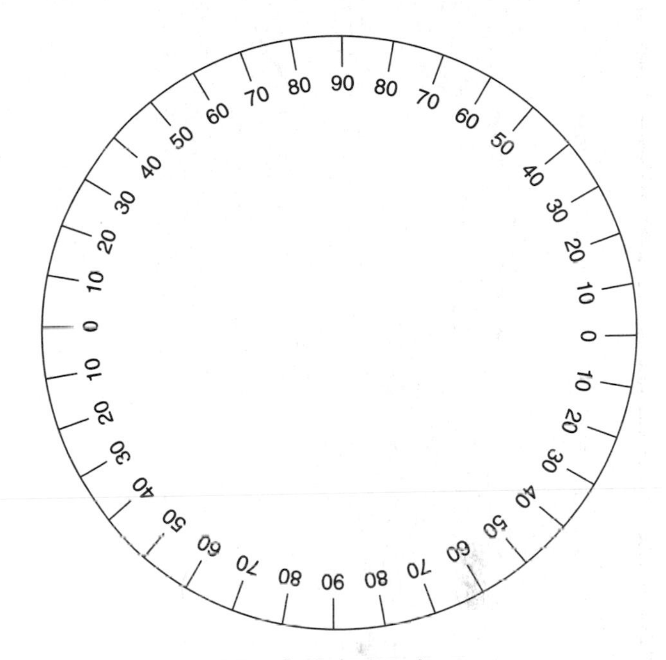


**Figure 2-21** Horizontal Circle Quadrants.

of a degree (10′) past the degree reading. In other words, to read the number of minutes by which the vernier zero is past a degree mark, find a line on the vernier that coincides with a line on the circle. If none is found between 0 and 30, in the direction being read, go back to the 30 at the opposite end of the vernier and read toward 60 again. Remember that each space on the vernier, to the right or left, equals 5′. Figure 2-20 shows a reading of 45°50′.

The horizontal circle is marked off in quadrants, with the degrees reading from 0 to 90, then down to 0 again, repeated in the two other quadrants (see Figure 2-21).

## MEASURING ANGLES WITH A TRANSIT

Because a transit's primary purpose is to measure angles, horizontal or vertical, the circles on a transit are usually more finely graduated than those on a level or level transit. The circle may be numbered by one of the systems shown in Figure 2-22 or by a variation of either. Each degree will be divided into two or three divisions (30′ or 20′ intervals), as indicated

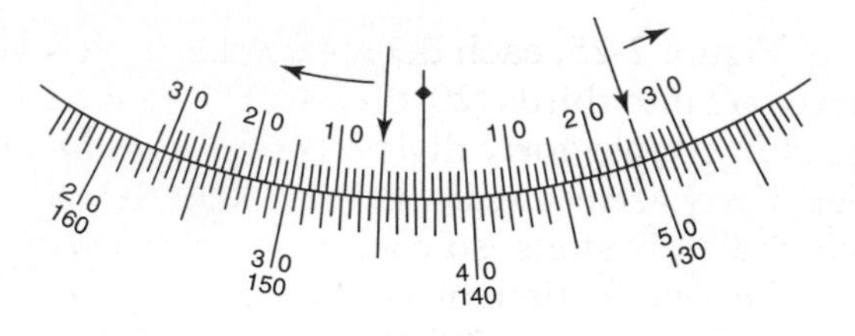

**Figure 2-24** Vernier Reading to 1′.

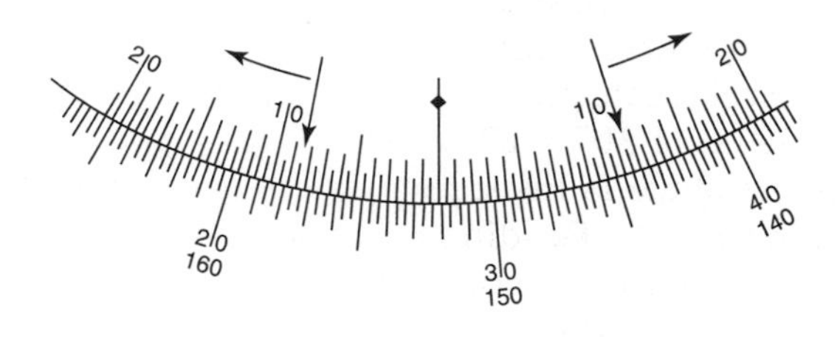

**Figure 2-25** Vernier Reading to $^1/_2$′.

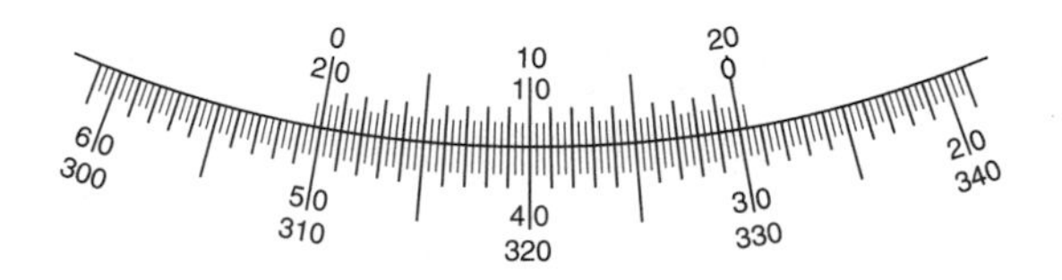

**Figure 2-26** Vernier Reading to 20″.

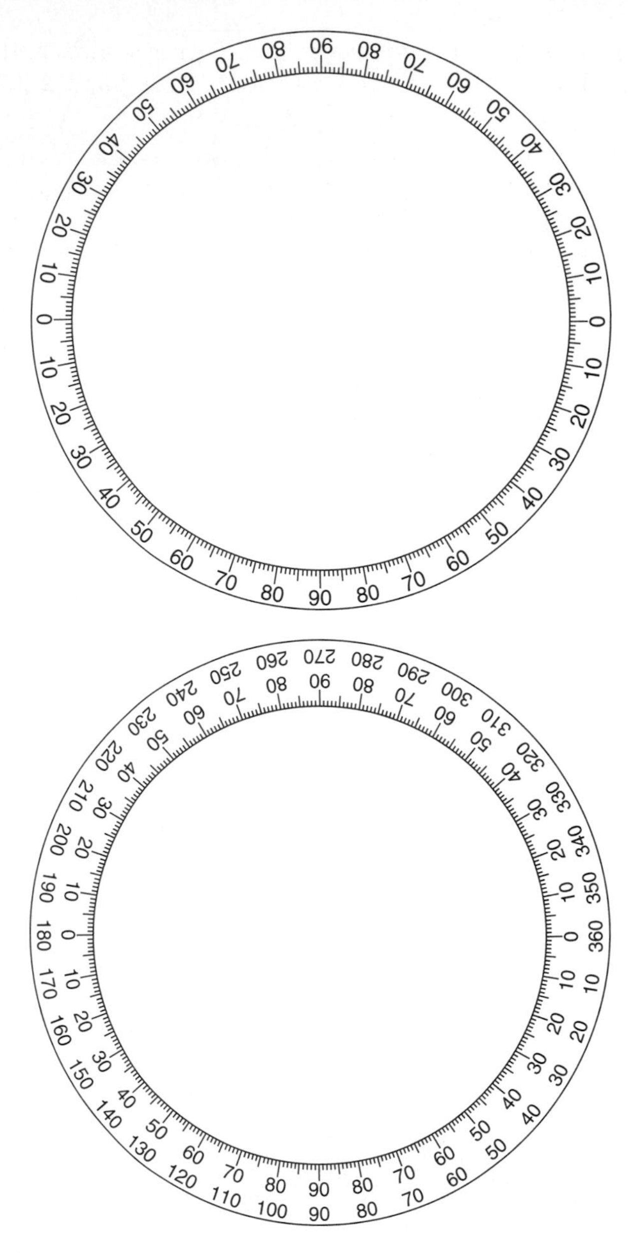

**Figure 2-22** Transit Circles.

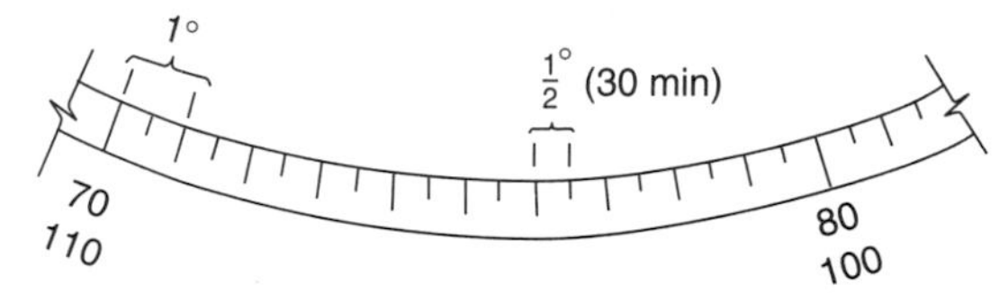


Section of transit circle, graduated into 30 min. intervals.

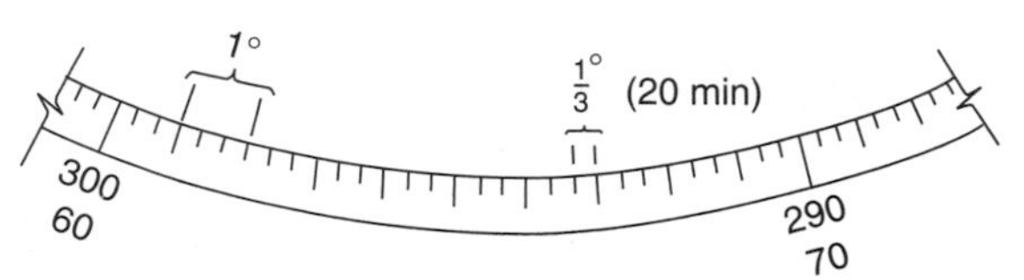

**Figure 2-23** Graduations on Transit Circles.

in Figure 2-23. Similarly, the vernier will be graduated into 1′, 30″, or 20″ intervals (see Figures 2-24, 2-25, and 2-26). It is important to remember that with scales that have a double row of numbers, the circle scales and the vernier must be read in the direction of increasing numbers on the circle.

In Figure 2-24, the circle scale is divided into ½° or 30′ divisions and the vernier into thirty divisions on each side of the index mark. Following the vernier principle described in the foregoing section, thirty vernier scale divisions are equal to twenty-nine circle scale divisions, and each vernier scale division will be equal to one-thirtieth of a circle scale division or 1′.

If the angle is turned in a clockwise direction, the scale of increasing numbers will be the lower one, so the vernier index will be past the 142°30′ graduation on the circle. Then, reading clockwise on the vernier, the fifth vernier mark (5′ mark) coincides with a scale graduation. The correct reading will then be 142°30′ + 5′ = 142°35.′

If the angle is turned in a counterclockwise direction, the reading will be taken in the direction of increasing numbers in the upper row on the circle. In this case, the vernier index is past the 37° mark on the circle scale. Now, reading the vernier in the same direction, the 25′ vernier mark coincides with a circle scale graduation. The correct reading will be 37° + 25′ = 37°25′.

In Figure 2-25, each degree on the circle scale is marked off into thirds (20′ divisions), and the vernier is graduated into forty divisions on each side of the index. Forty vernier scale divisions equal thirty-nine circle scale divisions, so each vernier scale division is equal to one-fortieth of a circle scale division: one-fortieth of 20′ = $^{1}/_{2}$′ or 30″.

If the angle is turned counterclockwise, the vernier index is past 27°40′. Reading counterclockwise on the vernier, the 11$^{1}/_{2}$′ vernier mark coincides with a circle scale graduation. Thus, the reading will be 27°40′ + 11′30″ = 27°51′30″.

If the angle is turned clockwise, the vernier index is past 152°. Reading clockwise on the vernier, the 8$^{1}/_{2}$′ vernier mark coincides with a circle graduation. Therefore, the reading will be 152°+ 8′30″ = 152°8′30″.

## SETTING UP A LEVEL OR LEVEL TRANSIT

To use a leveling instrument, it must be mounted on a *tripod* (see Figures 2-27 and 2-28) and properly adjusted before use. Proceed as follows:

1. Check the tripod to see that all screws, nuts, and bolts are properly tightened (see Figures 2-27 and 2-28).
2. Set the tripod in the desired location and spread the legs so that the eyepiece of the telescope will be in a comfortable position for sighting. A spacing of about one meter for the legs is generally satisfactory. The spacing of the legs should be adjusted so that the tripod head is as level as can be accomplished by eye.
3. Push the legs firmly into the ground. On a paved surface, be sure the points will hold securely.
4. Tighten the wing nuts at the top of the tripod legs into a reasonably firm position.
5. Remove the cap from the tripod head and place it in the instrument box. Lift the instrument from the box by the frame, *not by the telescope*, and set it on the tripod head. Make sure the horizontal clamp screw is loose.
6. Holding the level by the upper structure, screw the leveling head firmly into place.

**Figure 2-27** Instrument Tripod.

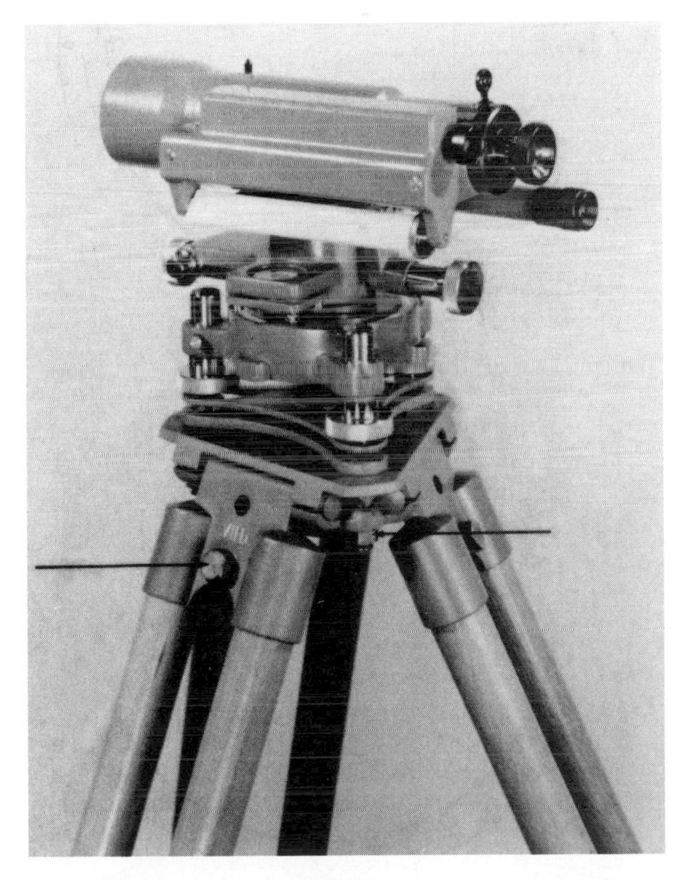

**Figure 2-28** Instrument Mounted on Tripod. (Reprinted by permission of Leica Geosystems.)

## Leveling the Instrument

Leveling is the most important operation in preparing to use the instrument. It is good practice to adopt the following procedure for four-point leveling:

1. Ensure that the locking levers holding the telescope in a horizontal position are locked if a level transit is being used.
2. Loosen two adjacent leveling screws, as shown in Figure 2-29(a).
3. Turn the telescope so that it is parallel to two opposing screws (Figure 2-29(b)).
4. Bring the bubble to the center of the level tube (see Figure 2-30) by loosening one screw and tightening the opposite one. Try to turn them approximately the same amount simultaneously. Figure 2-31 shows how to hold and turn the screws for adjustment. The pressure of the screws against the leveling head should be light and only finger tight.

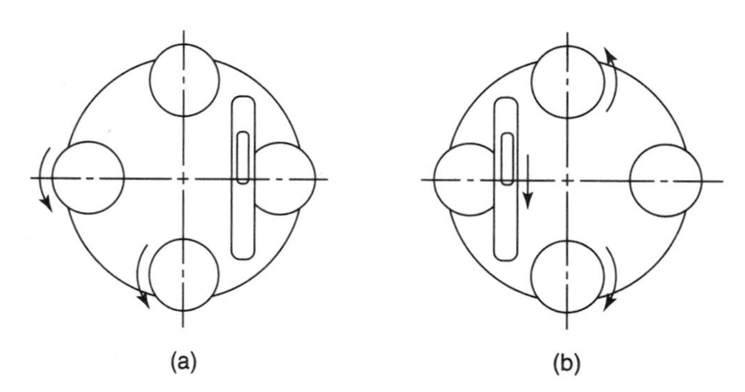


**Figure 2-29** Turning Level Screws.

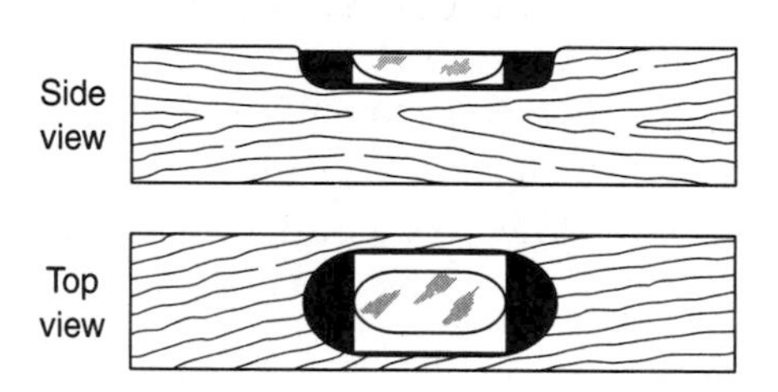


**Figure 2-30** Level Bubble Centered.

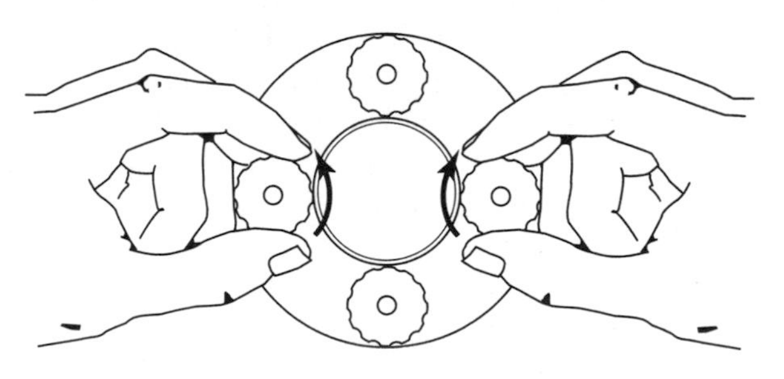

**Figure 2-31** Adjusting Level Screws.

5. Move the telescope 90° over the other pair of screws and repeat step 4.
6. Turn back over the first pair of screws and center the bubble again.
7. Turn back to the second pair of screws and check the levelness.
8. Turn the telescope 180° over the same pair of screws as in step 6. The bubble should now be centered in this or any other position of the telescope.

## Three-Point Leveling

For instruments with only a circular level vial, leveling is a simple operation. The instrument is leveled by adjusting the leveling screws in pairs until the *circular level bubble* is brought to the exact center of the circular vial.

For instruments that have a second, tubular telescope level (see Figure 2-3), the procedure is as follows:

1. Level the instrument by using the leveling screws and the circular level bubble.
2. Turn the instrument until the telescope level is parallel to one pair of leveling screws.
3. Turn that pair of screws by the same amount, in opposite directions, until the telescope level bubble is accurately centered.
4. Rotate the instrument 180°. If the bubble does not remain centered, correct half of the deviation by adjusting the level screws and the other half by adjusting the level-adjusting screw (see 14, Figure 2-3).
5. Turn the instrument 90° so that one end of the telescope lies over the third level screw and the other end between the first two. If the bubble is not centered, center it by adjusting the third level screw.

Some instruments with a second level have a *split bubble system* in that level (see Figure 2-32). The telescope bubble is viewed through a viewing microscope and, if the telescope is not level, the bubble will appear to be split, as in Figure 2-33(a). When the two halves are made to coincide, as in Figure 2-33(b), the bubble is centered. The procedure is as follows:

1. Center the lower split level bubble by adjusting the level screws, as described.
2. Bring the two halves of the split bubble into coincidence by adjustment of the *tilting screw* (see 3, Figure 2-32).

## Sighting Adjustments

The instrument is now ready for sighting adjustments. These may vary slightly for different individuals.

**Figure 2-32** Level with Split Bubble System: 1, Level Cover Plate; 2, Bubble Viewing Eyepiece; 3, Tilting Screw, 4, Level Screw. (Reprinted by permission of Leica Geosystems.)

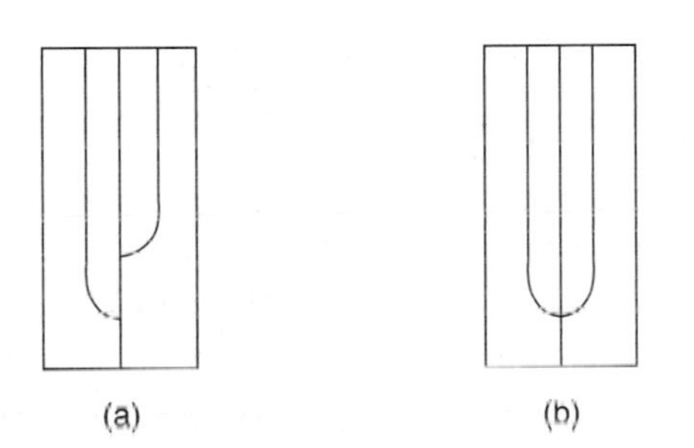


**Figure 2-33** Split Telescope Bubble.

1. Aim the telescope at the object and sight it first along the top of the tube.
2. Then look through the eyepiece and adjust the focus by turning the focusing knob until the object is clear.
3. Now turn the eyepiece cap or focusing ring until the cross hairs appear sharp and black.
4. Adjust the telescope until the object is centered as closely as possible. *Do not put your hands on the tripod.* Tighten the horizontal clamp screw.
5. Center the object exactly by bringing the vertical cross hair into final position with the horizontal motion tangent screw.

## Collimation Adjustments

The horizontal cross hair is on the *line of sight,* which should be on a true horizontal plane for any direction in which the telescope is pointed. The vertical distance from the ground on which the tripod stands or from a stake from which measurements are being taken to that line of sight is called the *height of instrument* (H.I.).

*Collimation* of a leveling instrument means checking and adjusting to ensure that the line of sight through the telescope is parallel to a true horizontal plane. To check and make this adjustment, proceed as follows:

1. Drive two stakes, A and B, into the ground approximately 100 ft (30 m) apart.
2. Set the instrument up close enough to stake A so that the telescope eyepiece will be within 4 in. (100 mm) of a level rod held on the stake. Level the instrument very carefully.
3. Sight through the *objective end* of the telescope and, using a pencil held on the rod, determine the H.I. at stake A.
4. Move the rod to stake B and, using a rod target, determine the distance from the top of stake B to the line of sight.
5. Move the instrument to a point beside stake B, and repeat steps 3 and 4 to find an H.I. and a rod reading on stake A.
6. Calculate the true difference in elevation between stakes A and B. The difference is equal to the average of the differences taken with the instrument at the two locations, A and B.
7. Calculate the amount by which the line of sight is high or low from stake B. Depending on whether stake A is higher or lower than B, add or subtract the true difference in elevation from step 6 to or from the H.I. at stake B to find the correct rod reading at A.
8. To make the correction, set the correct reading on a target rod held at stake A. Make the adjustment to the cross hairs, with the instrument sitting in position at B. Move the cross-hair ring up or down until the horizontal cross hair exactly coincides with the correct rod reading set on the target. The cross-hair ring is moved up or down by the adjustment of the upper and lower cross-hair ring adjustment screws. Turn each by the same amount so that the same tension is maintained on the screws.

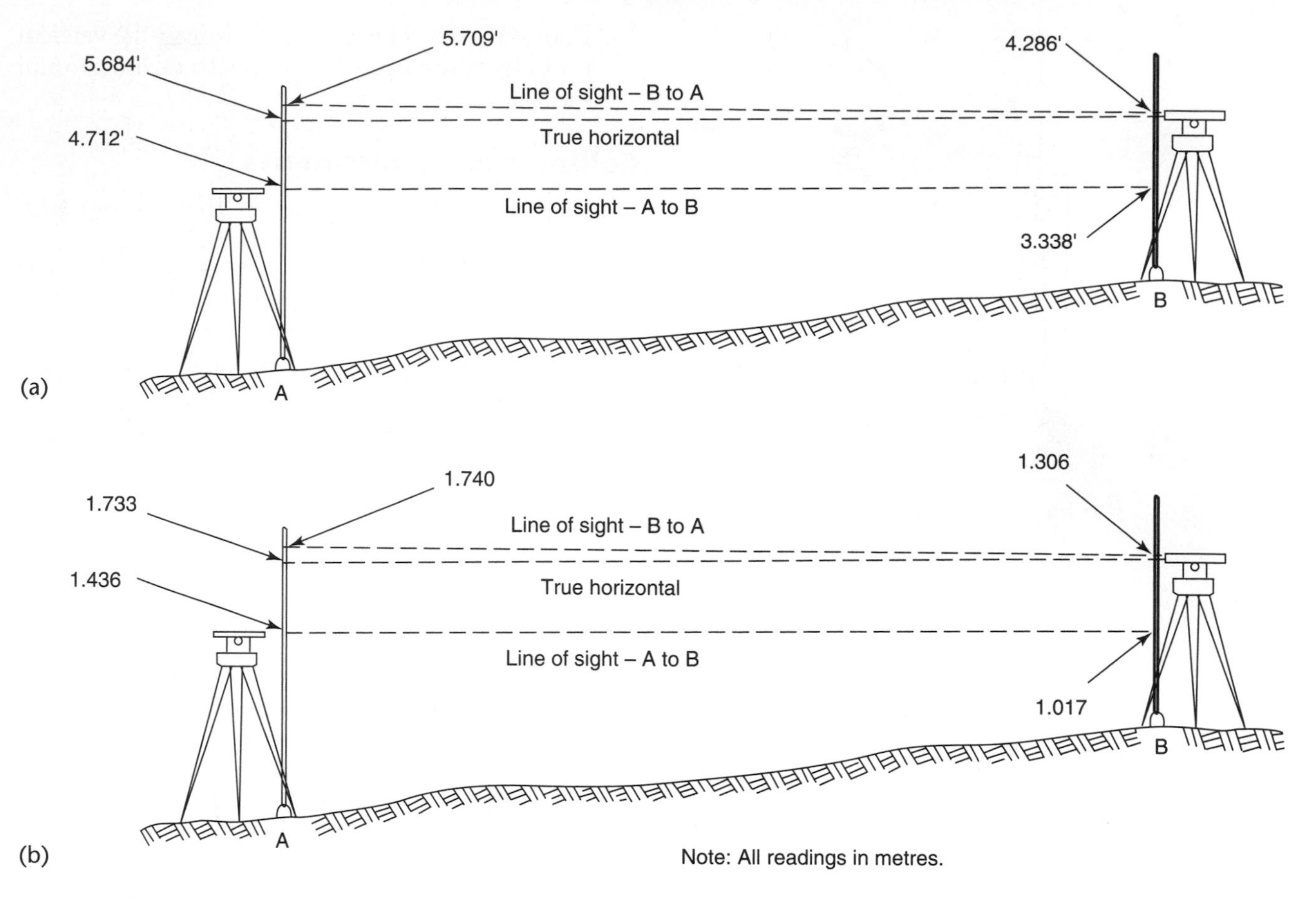


**Figure 2-34** (a) Example of Instrument Collimation Using Standard Units. (b) Example of Instrument Collimation Using Metric Units.

---

*Example:*

Figure 2-34(a) illustrates the steps and readings taken to make the collimation adjustments using standard units when checking the accuracy of a leveling instrument.

1. H.I. at stake A = 4.712 ft
2. Rod reading at B, from A = 3.338 ft
3. Apparent difference in elevation = 1.374 ft
4. H.I. at stake B = 4.286 ft
5. Rod reading at A, from B = 5.709 ft
6. Apparent difference in elevation = 1.423 ft
7. True difference in elevation between A and B

$$= \frac{(1.374 + 1.423)}{2} = 1.398 \text{ ft}$$

8. Correct rod reading at A = 4.286 + 1.398 = 5.684 ft
9. The line of sight was high by 5.709 − 5.684 = 0.025 ft

Figure 2–34(b) illustrates the readings taken in metric units and the calculations appear as follows:

1. H.I. at stake A = 1.436 m
2. Rod reading at B, from A = 1.017 m
3. Apparent difference in elevation = 0.419 m
4. H.I. at stake B = 1.306 m
5. Rod reading at A, from B = 1.740 m
6. Apparent difference in elevation = 0.434 m
7. True difference in elevation between A and B

$$= \frac{(0.419 + 0.434)}{2} = 0.427 \text{ m}$$

8. Correct rod reading at A = 1.306 + 0.427 = 1.733 m
9. The line of sight was high by 1.740 − 1.733 = 0.007 m

---

## LEVELING RODS AND THEIR USES

One important use of a leveling instrument is to find the difference in grade or elevation between two points. To do this, it is necessary to use a *leveling rod* and, to use a leveling rod, you must be able to read it properly.

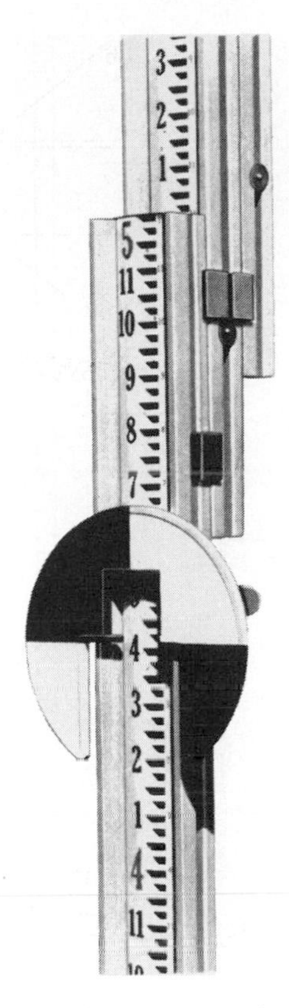

**Figure 2-35** Section of Leveling Rod with Target.

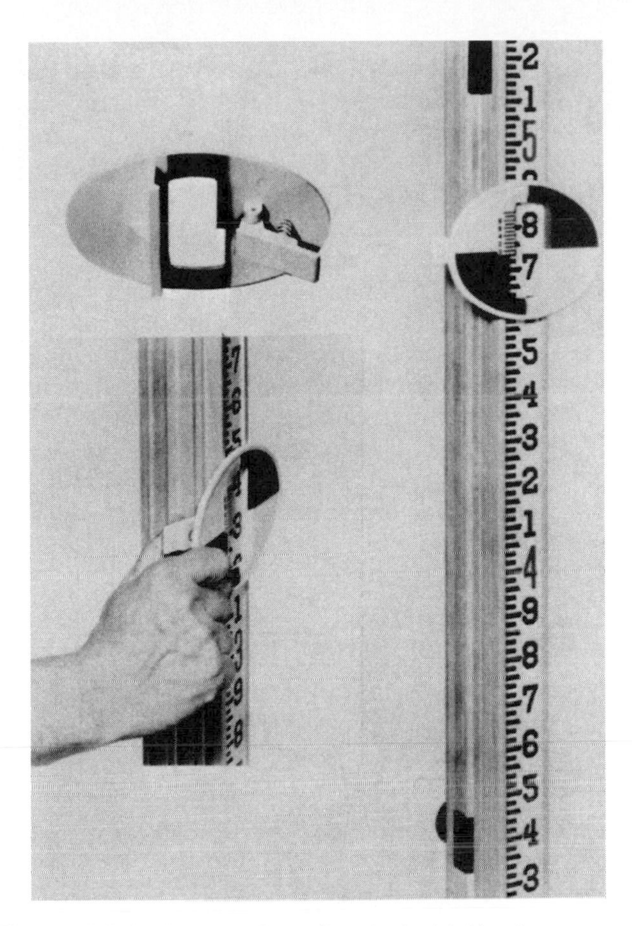

**Figure 2-36** Section of Leveling Rod with Vernier. (Reprinted by permission of CST/Berger–David White.)

A section of a leveling rod is shown in Figure 2-35. Some rods are equipped with a *target* (see Figure 2-35), which can be moved up or down the rod until the horizontal target line corresponds to the instrument's line of sight (horizontal cross hair). The target may be mounted on a bracket held in place by friction or, on some rods, the target is equipped with some type of clamp, one of which is illustrated in Figure 2-36.

Rods may be graduated in feet and inches (see Figure 2-35), or they may be graduated in feet and decimal fractions of a foot (see Figure 2-36). In the latter case, the target on such a rod may have a vernier scale on it that will allow a rod reading to be made to three decimal places (thousandths of a foot). Metric rods, on the other hand, may be graduated in millimeters (1/1000 of a meter), in 5-mm increments, or in 10-mm increments, depending on the accuracy required for the survey being done.

## Rod Graduations

Rods using standard units of measure are graduated in 1-ft intervals, marked in large numbers (see Figure 2-36). Each 1-ft division is graduated into ten equal parts, each one-tenth of a foot, numbered 1 through 9 (see Figure 2-37(a)). Next, each one-tenth of a foot is graduated into ten equal parts, each equal to one-hundredth of a foot (see Figure 2-37(a)). Each one-hundredth of a foot is represented by either a black or a white band; consequently, you read to either the top or bottom of a black band to determine the correct reading in hundredths of a foot (see Figure 2-37(a)).

The metric rod shown in Figure 2-37(b) is graduated in 5-mm increments, the distance between each point being 5 mm. To indicate the number of meters above the ground that the reading is taken, dots are placed beside the numbers representing tenths of a meter. In Figure 2-37(b), the numeral 9 has three dots below it, indicating that the reading represents 3.900 m. The numeral 1 has four dots below it, indicating that it represents a reading of 4.100 m. Some typical readings are also shown, based on the 5-mm increment rod graduations.

## Elevations

The difference between the rod readings at two locations will be the difference in elevation or grade, which is illustrated in Figures 2-38(a) and (b). To read this difference, set the instrument in a convenient location from which both points can be seen, preferably about midway between them. Level the

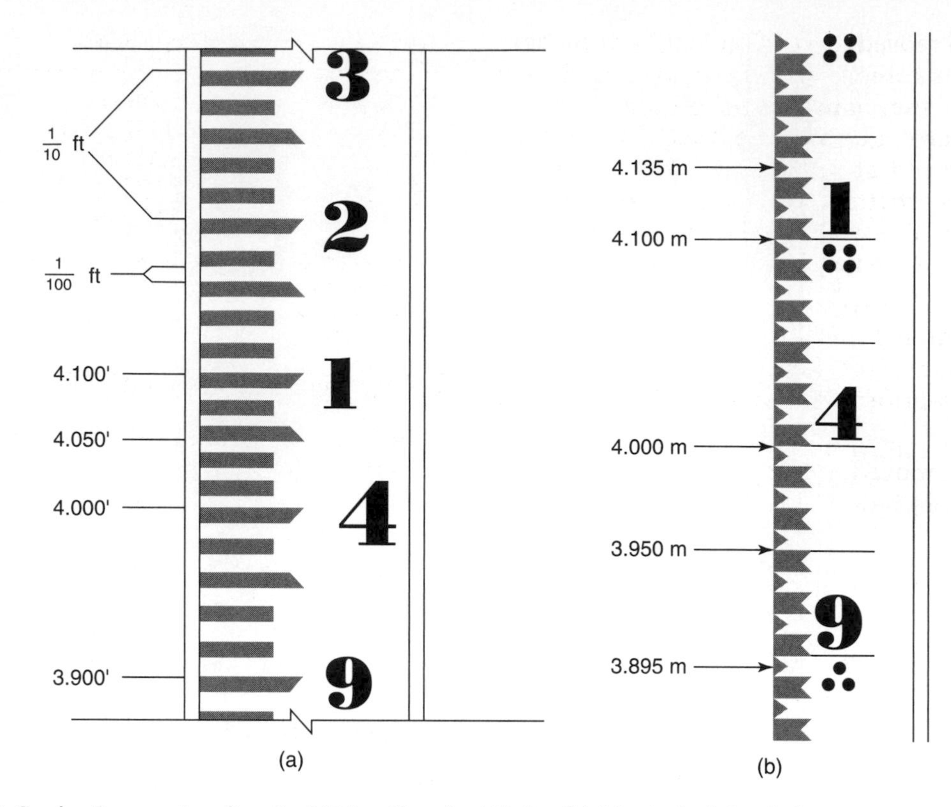


**Figure 2-37** (a) Graduations on Leveling Rod Using Standard Units; (b) Metric Rod Gradations.

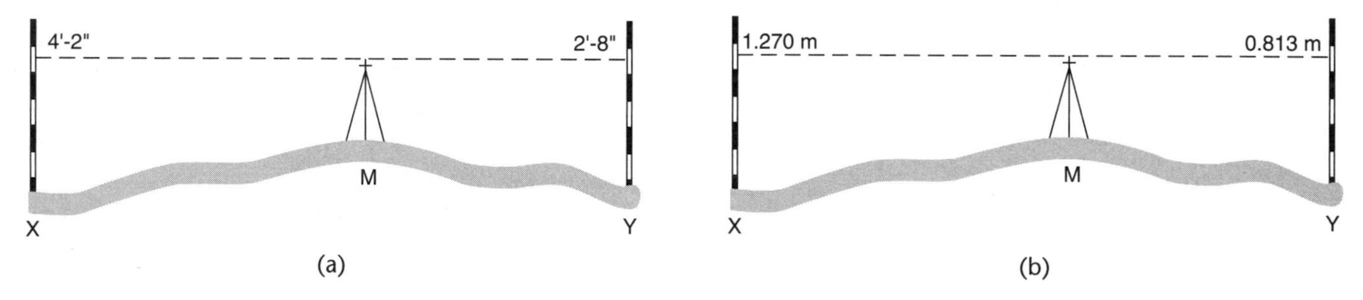


**Figure 2-38** (a) Difference in Grade, Standard Units. (b) Difference in Grade, Metric Units.

instrument as explained previously. Have someone hold a rod vertically at one of the two points, or *stations,* and sight on it. Record the reading at which the horizontal cross hair cuts the graduations on the rod. Have the rod held over the second station and swing the instrument around without otherwise disturbing it and take another reading on the rod. The difference between the two readings is the amount that one station is above or below the other.

For example, Figure 2-38(a), the reading at Station X is 4 ft 2 in. and at Station Y, 2 ft 8 in. The difference between these two readings, 1 ft 6 in., is the amount by which Station Y is higher in elevation than Station X. In other words, there is a difference in grade between the two points of 18 in. In Figure 2-38(b), the reading at Station X is 1.270 m and at Station Y, 0.813 m. The difference between these two readings is 0.457 m and Station Y is higher in elevation than Station X by 457 mm. Applying this approach, the elevations of any number of stations can be compared.

Another use of a leveling instrument is to find the elevation at a particular point or points, knowing the elevation of some designated point, a bench mark. Proceed as follows:

1. Set up the instrument at a convenient point between the bench mark and the unknown elevation, where the rod (held on the bench mark) will be in sight. Level the instrument.
2. Take a sight on the rod and record the reading. This is called a *backsight.* This reading, added to the bench mark elevation, is the H.I.

3. Have the rod moved to a convenient location between the instrument and the unknown elevation. Loosen the clamp screw and swivel the instrument around so that a reading can be taken on the rod at its new location. This is a *foresight.* Record that reading and subtract it from the H.I. The result is the elevation of the point on which the rod rests, Station 1.
4. Now move the instrument to a new position between Station 1 and the unknown elevation and take a backsight. Add that backsight reading to the elevation of Station 1 and you have a new H.I.
5. Have the rod moved to a new station and take a foresight. From it establish the elevation of Station 2.
6. This procedure is repeated until the final station reaches the unknown elevation.

Notice that a backsight was added to a known elevation to obtain the H.I. It is thought of as a *plus* quantity. A foresight is subtracted from the H.I. to obtain a station elevation. It is thought of as a *minus* quantity. In reality, this continuous process of adding and subtracting is not necessary. The backsights are merely all recorded as being plus, and their sum is a plus quantity. The foresights are all recorded as minus, and their sum is a minus quantity. The difference between these two totals is net plus or minus quantity, depending on which is larger. The difference, added to or subtracted from the original bench mark, will give the elevation of the point in question.

For example, assume the elevation of Point *A* is to be referenced to a bench mark having a recorded elevation of 2642.62 ft. Point *A* is far enough away from the bench mark that three stations are required to get from the bench mark to point *A*. The readings in the log book would appear as follows:

| Set Up | Backsight | Foresight |
|---|---|---|
| 1 | 4.26 ft | 3.19 ft |
| 2 | 6.19 ft | 8.27 ft |
| 3 | 5.35 ft | 6.92 ft |
| 4 | 3.68 ft | 7.75 ft |
| | +19.48 ft | −26.13 ft |

Net difference = 26.13 − 19.48 = −6.65 ft.
Elevation of point *A* = 2642.62 − 6.65 = 2635.97 ft.

Using metric units, based on a recorded bench mark elevation of 808.471 m, the log book entries would appear as follows:

| Set Up | Backsight | Foresight |
|---|---|---|
| 1 | 1.298 m | 0.973 m |
| 2 | 1.887 m | 2.521 m |
| 3 | 1.631 m | 2.109 m |
| 4 | 1.122 m | 2.362 m |
| | +5.938 m | −7.965 m |

Net difference = −7.965 + 5.938 = 22.027 m
Elevation of point *A* = 805.471 − 2.027 = 803.444 m

## TRANSIT APPLICATIONS

Level transits and theodolites are more versatile than the builders' level in that they can be used for measuring vertical angles and provide a means for establishing and maintaining vertical lines of reference. The following pages outline some of the more practical uses of level transits and theodolites on the construction site as well as methods used to check the accuracy of the line of sight of the instrument.

### Running Straight Lines

One particular use of a level transit is for running straight lines. A straight line must be run from a point in a given direction or to another point.

1. Set the instrument up over the starting point. To do this, use the plumb bob in the instrument box. Attach its cord to the eyelet in the middle of the centering head. Adjust the position of the tripod legs until the plumb bob is as nearly centered as possible over the starting point.
2. Loosen two adjacent screws and move the centering head within its limits until the plumb bob is aligned exactly over the center of the starting point.
3. Level the instrument and unlock the locking levers.
4. Sight on the second point given or point the telescope in the given direction and lock it in that position with the horizontal clamp screw.
5. By tilting the telescope up or down, a number of positions can be sighted and marked on the ground, all of which will be in a straight line from the starting point (see Figure 2-39).

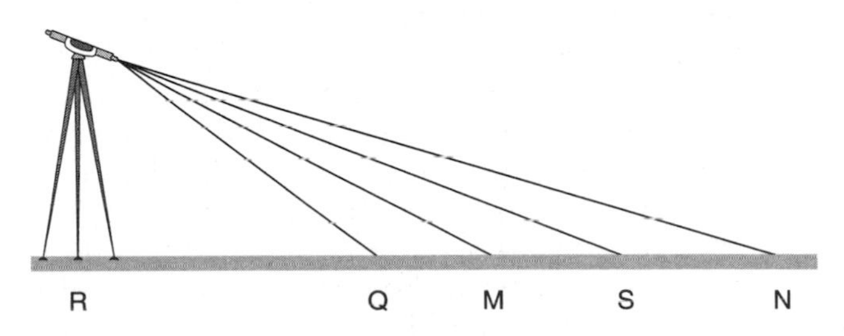


**Figure 2-39** Running a Straight Line.

## Measuring Angles

A level or level transit is readily used for laying out angles or for measuring a given angle. To lay out an angle, the intersection of the two legs must be known, as well as the direction of one leg, and the size of the angle.

1. Set up the instrument over the point of the angle, using the plumb bob as previously described, and level the instrument.
2. Unlock the locking levers and sight on a point (a stake) on the given leg of the angle. Record the reading on the horizontal circle or turn the horizontal circle to a zero reading. The bottom edge of the horizontal circle is knurled, and the circle can be turned by applying light pressure with a finger.
3. Loosen the horizontal clamp screw, and turn the instrument through the required angle. It can be brought to the required angle by using the horizontal tangent screw. Tighten the clamp screw again.
4. Run a line in this direction for the required distance.

For measuring angles already laid out, the procedure is much the same. Figure 2-40 shows a plot outline for which the angles may have to be measured.

1. Set up the instrument at Station 1 and center it over the station pin.
2. Level the instrument and release the locking levers.

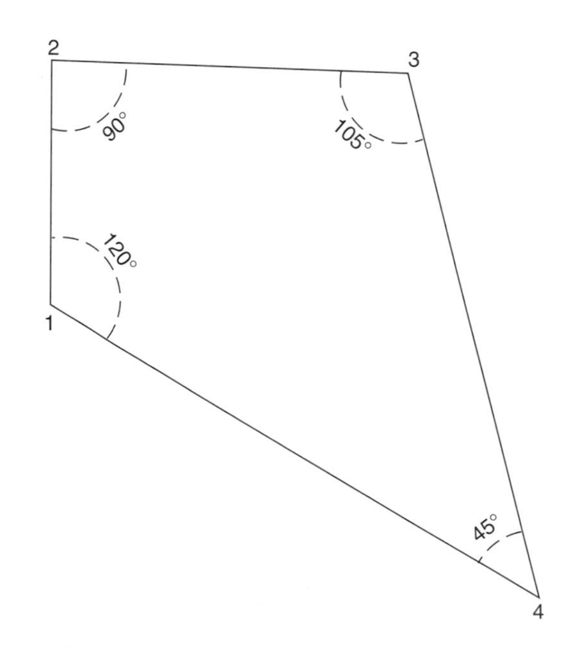


**Figure 2-40** Measuring Angles.

3. Sight on Station 2, centering the vertical cross hair on the station pin. Set the horizontal circle to zero or record the reading.
4. Turn the instrument and sight accurately on Station 4. Record the new reading on the circle. In this case it is 120°.
5. Now set up over Station 2, sight back to Station 1, and around to Station 3. Read the angle, 90°.
6. Repeat the procedure at Stations 3 and 4.

## Sights in Vertical Plane

A level transit is required for taking vertical sights, such as plumbing, building walls, piers, columns, doorways, and windows. First, level the instrument; then release the locking levers that hold the telescope in the level position. Swing the telescope vertically and horizontally until the line to be established is directly on the vertical cross hair. Tighten the horizontal clamp screw. If the telescope is now rotated up or down, each point cut by the vertical cross hair is in a vertical plane with the starting point (see Figure 2-41).

## Leveling the Transit

1. Set up the tripod with the legs extended to the approximate height required and the head in an approximately level position.
2. Place the instrument on the tripod head and turn it down snugly or fasten it in place with the bolt, depending on the type of instrument.
3. Place the telescope tube in the horizontal position and rotate it until it is approximately parallel with one pair of leveling screws.

For transits with four-point leveling, proceed as described for leveling other instruments. For transits with three-point leveling:

4. Use the two leveling screws and the plate vials to make necessary adjustments to bring the bubble to the center of the telescope vial.

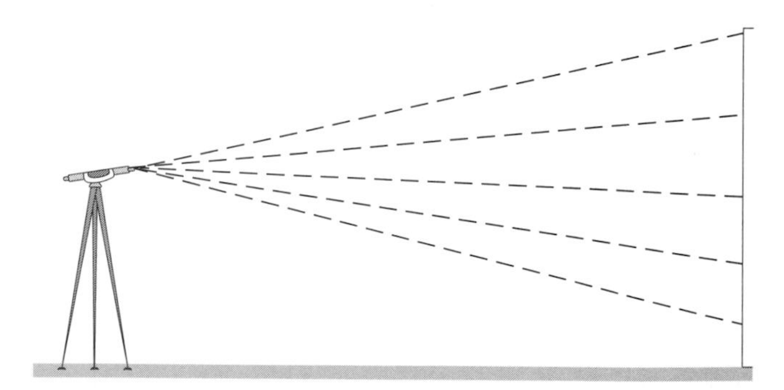

**Figure 2-41** Establishing a Vertical Line.

5. Rotate the instrument 180° and make the necessary adjustments to center the bubble.
6. Rotate to the original position and check.
7. Rotate the telescope tube 90° to place it directly over the third leveling screw. Adjust that screw to bring the bubble to center.
8. Rotate 180° and make the necessary adjustments.
9. Bring the instrument back to its original position for a final check.

### Collimation of Transit

To collimate the instrument is to make the line of sight perpendicular to the horizontal axis *of the instrument.* If the instrument is not collimated, a straight line cannot be extended by direct means. The recommended procedure is as follows:

1. Set the instrument up where clear sights of approximately 200 ft (60 m) are available on both sides and carefully level the instrument.
2. Sight the instrument accurately on a point on stake A (see Figure 2-42). Lock the horizontal motion.
3. *Plunge* the instrument (rotate the telescope on its horizontal axis so that the positions of the telescope and level tube are reversed).
4. Sight on stake B and set a point on it to coincide with the vertical cross hair.
5. Leaving the telescope in the reversed position, loosen the horizontal clamp screw and rotate the instrument until the vertical cross hair coincides with the point on stake A. Use the horizontal tangent screw for the final alignment.
6. Leave the horizontal motion clamped and plunge the telescope again (bring it back to its original position).
7. Sight in the direction of stake B. If the vertical cross hair coincides with the point previously set on stake B, no adjustment is necessary.
8. If they do not coincide, set stake C and a point on it that *does* coincide with the vertical cross hair.
9. Because double-reversing is involved, the apparent error is four times the actual error. Therefore, set stake D and mark a point on it that is one-fourth of the distance from stake C to stake B.
10. Shift the cross-hair ring laterally by adjusting the capstan screws at the ends of the horizontal cross hair until the vertical cross hair coincides with the point on stake D.

For further adjustments to this or any other instrument, consult the manufacturer's manual.

## LAYOUT BEFORE EXCAVATION

The boundaries of the lot on which a building is to be constructed should be established by markers, called *monuments,* set by a registered surveyor. The site plan shows the positions of the boundary lines, relative to the streets, roads, and the like (see Figures 2-43(a) and (b)). The contractor locates the boundary markers and, by running lines between them, can establish the boundaries. Now, from the distances given on the site plan and using a tape, the building can be laid out by driving stakes at the corners and on the column center lines (see Figure 2-44). The accuracy of the staking may be checked by measuring the diagonals of the rectangles concerned. Each pair will be exactly equal if the stakes are set in their correct position.

These corner stakes will be lost during excavating operations, and it is therefore necessary to record their position beforehand. This is done by means of *batter boards* on which you mark points that are on the various building and column lines. Batter boards are horizontal bars fastened to posts, which are set up singly or in pairs around the important points in the building (see Figure 2-45). The basic purpose of such staking is to enable workers to locate batter boards correctly. They must be set far enough back from the stakes to ensure that excavating will not disturb them.

Lines running between designated points on two opposite batter boards represent a building line (outside of wall frame, center line of column footings,

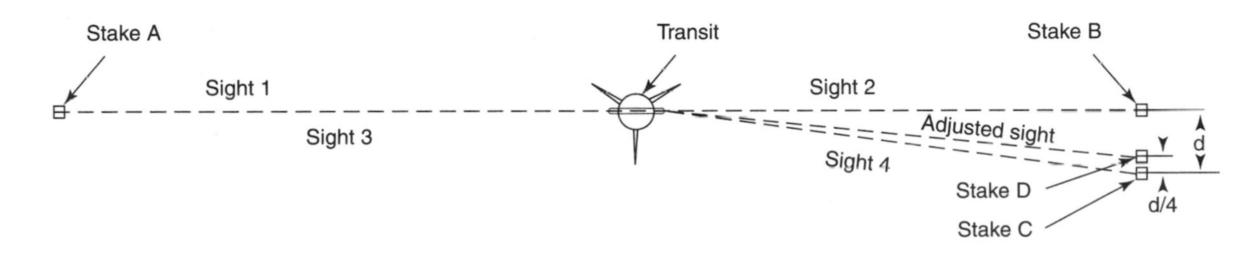


**Figure 2-42** Collimating Transit.

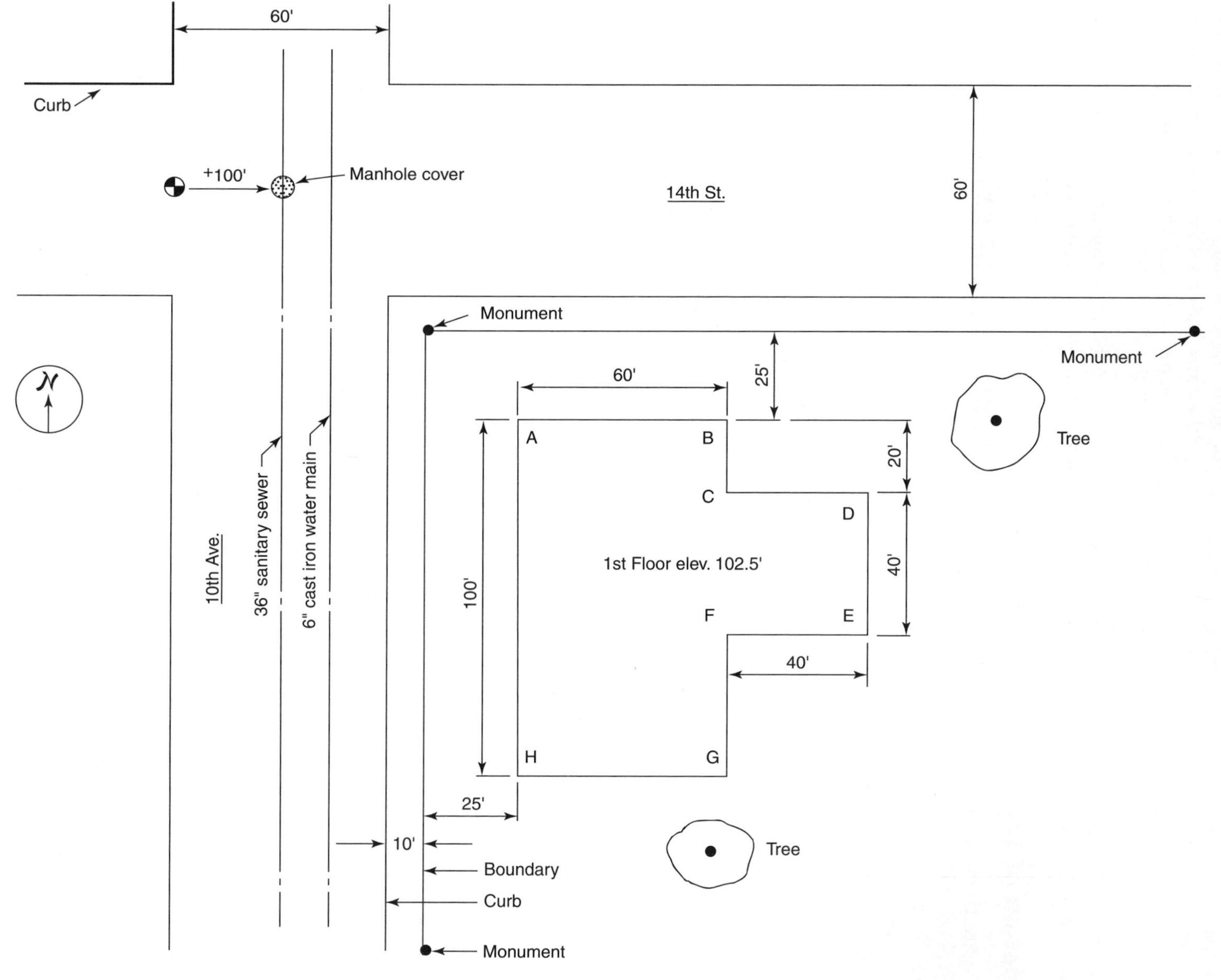


**Figure 2-43** (a) Site Plan, Standard Units.

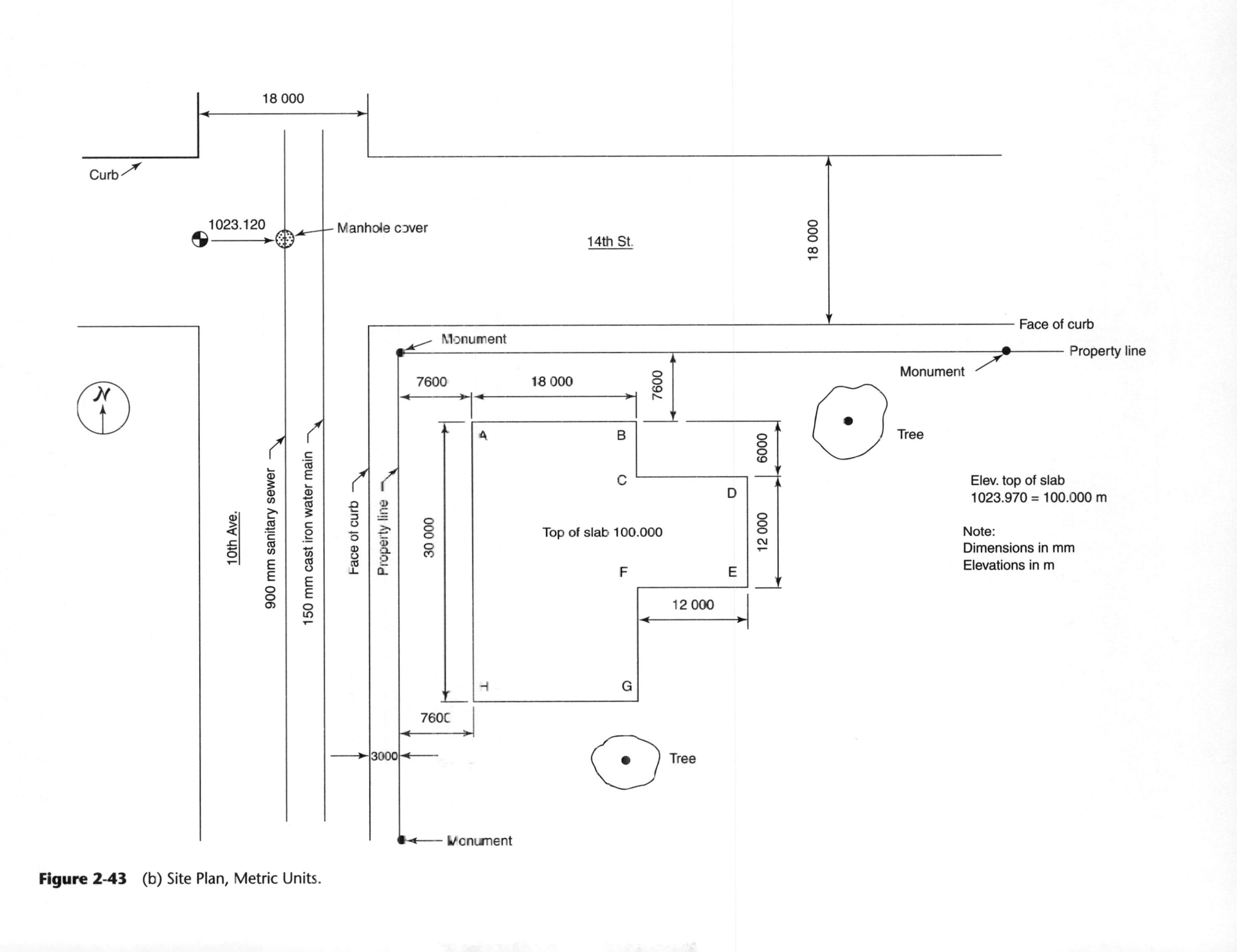


**Figure 2-43** (b) Site Plan, Metric Units.

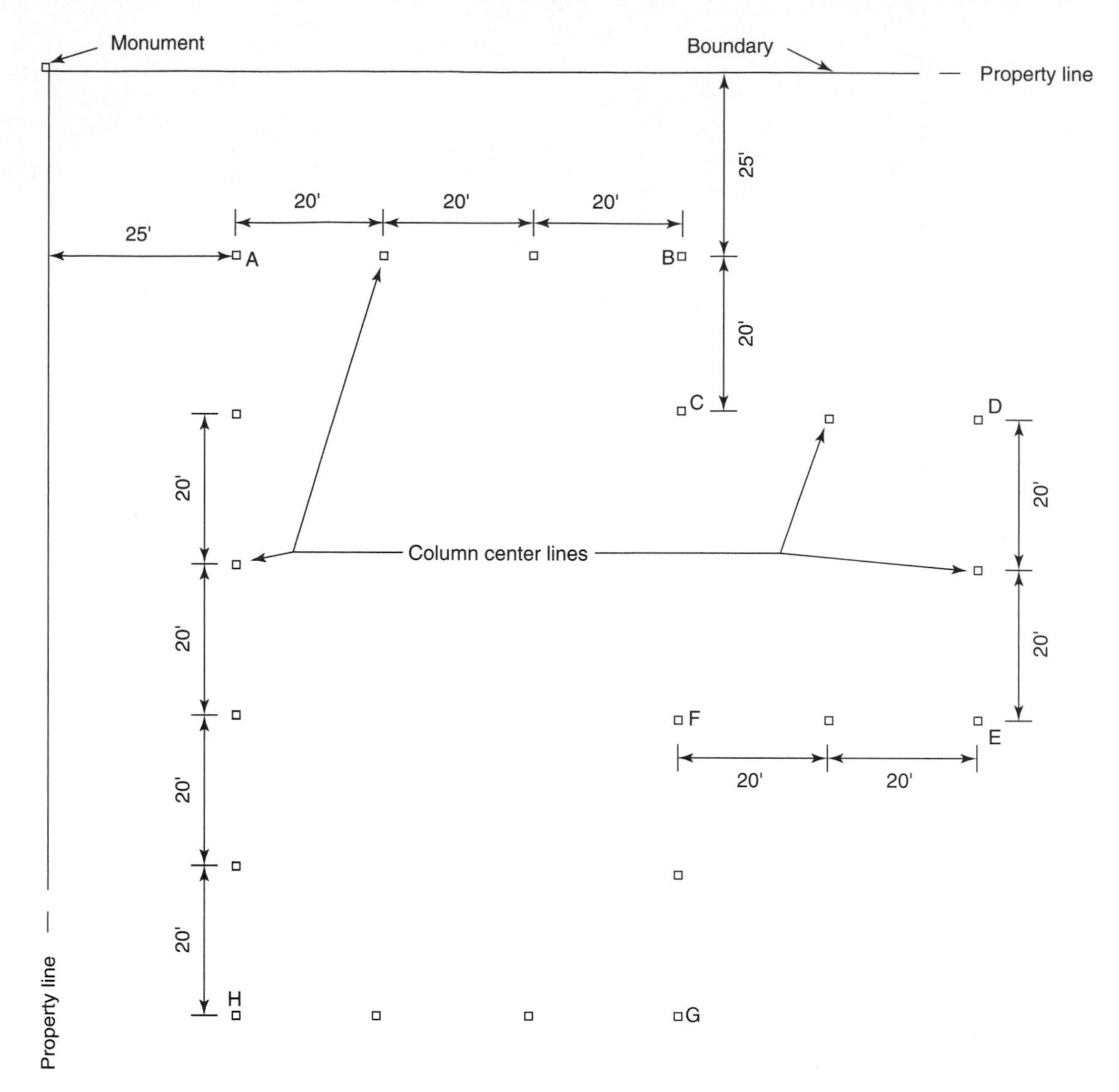


**Figure 2-44** Building Stakes.

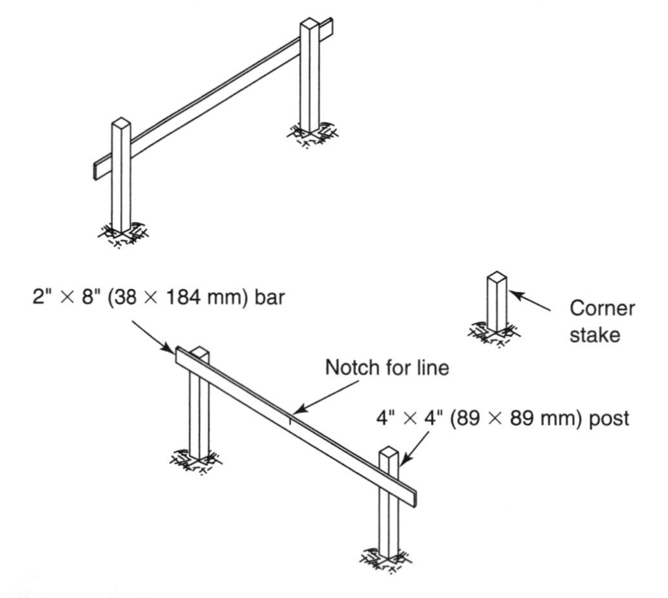


**Figure 2-45** Batter Boards.

and so on). Two intersecting lines represent a building corner, and a plumb bob dropped at that intersection will pinpoint a corner on the excavation floor.

Because wire lines are often used to withstand the strain of pulling them taut, strong batter boards and posts are required. The posts may be timbers 4 in. × 4 in. (100 mm × 100 mm or larger), small steel beam sections, or pipe driven firmly into the ground and braced if necessary. The horizontal bars are then bolted or welded to the posts, making sure that all are at the same level. This level should coincide with a floor level of the building, if possible, or at least bear some definite relation to it. The elevation should be plainly marked on one or more of the batter boards. It is good practice to check these elevations during excavation and foundation work to make sure that there has been no shift or displacement.

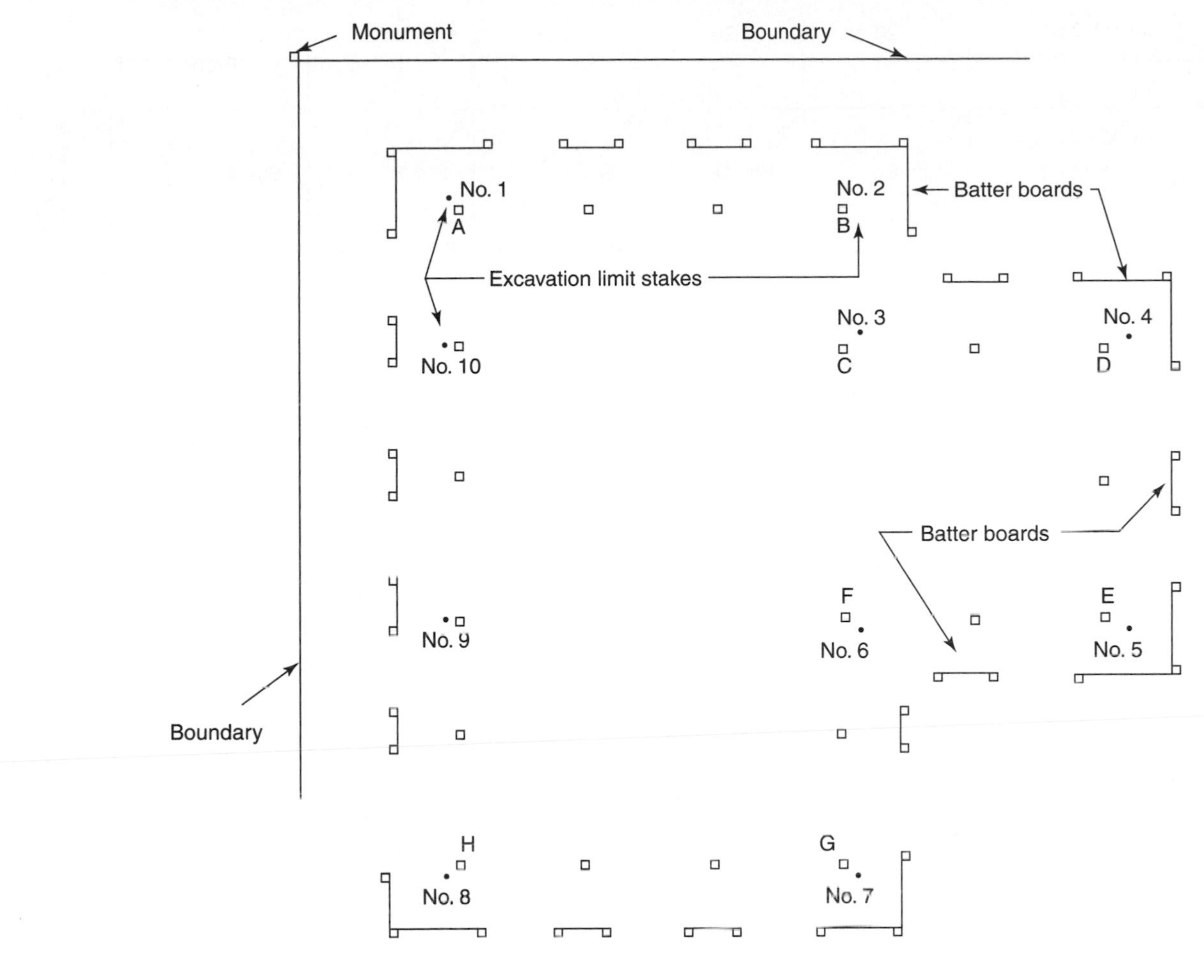


**Figure 2-46** Batter Boards in Place.

The limits of the excavation now have to be staked out. The excavation will usually extend 2 ft (600 mm) or more beyond the boundaries of the building itself—to the payline—to allow room to work outside the forms. Excavation limit stakes are accordingly driven at the required distance from the corner stakes, as shown in Figure 2-46.

## PRELIMINARY LAYOUT PROBLEMS

### Setting Batter Boards to a Definite Level

For the building indicated in Figure 2-43(a), it has been decided to set the batter boards to the top of slab elevation, shown on the plan as 102.5 ft. Proceed as follows:

1. Set up the leveling instrument so that sights can be taken on the bench mark and the batter boards and level it.
2. Take a backsight on the bench mark (given elevation of 100.00 ft). Suppose that the rod reading is 4.86 ft. The H.I. is then 100.00 + 4.86 = 104.86 ft.
3. The difference between the H.I. and the batter board elevation is 104.86 − 105.50 = 2.36 ft. Set the target on the rod at 2.36 ft.
4. Now place the rod alongside a post of the batter board at A, Figure 2-46, and move it up or down until the cross hair coincides with the target. Mark the height of the bottom of the rod on the post.
5. Fasten one end of a batter board to the post with its top edge at that mark. Level the batter board and fasten to the second post.
6. Set the remainder of the batter boards at exactly the same level, using the instrument to check their height and levelness.

The procedure is the same in metric units. The slab elevation for the building shown in Figure 2-43 (b) has been set at 1023.970 m. Assuming that the batter

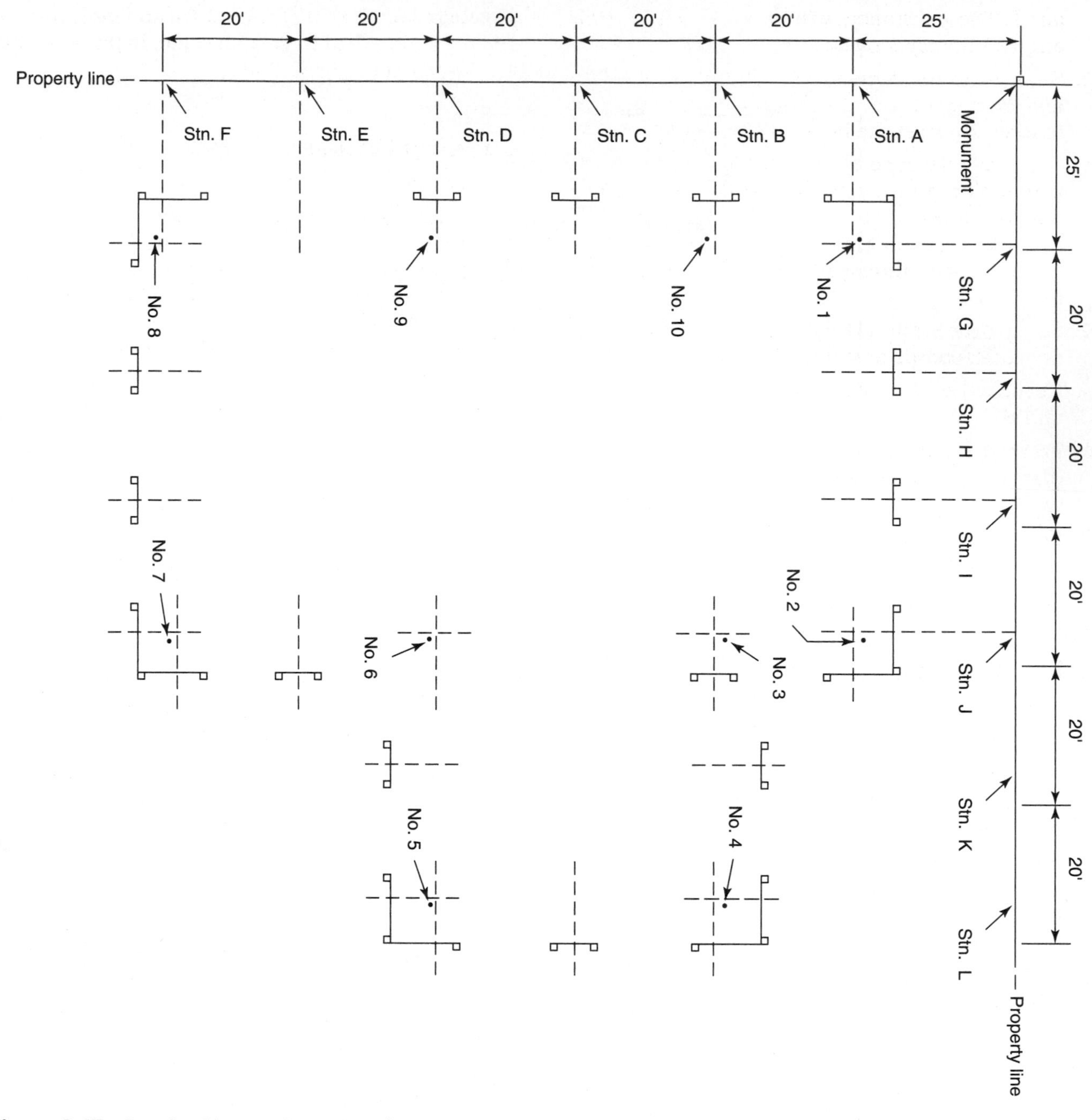

**Figure 2-47** Running Lines on Batter Boards.

boards are to be set at the slab elevation, the first three steps appear as follows:

1. Set up the leveling instrument so that sights can be taken on the bench mark and the batter boards and level it.
2. Take a backsight on the bench mark (elevation given as 1023.120 m). Suppose that the rod reading is 1.481 m. The H.I. is then 1023.120 + 1.481 = 1024.601 m.
3. The difference in the H.I. and the batter board elevation is 1024.601 − 1023.970 = 0.631 m. Set the target on the rod at 0.631 m or 631 mm.

Steps Four, Five, and Six are the same as in the foregoing example.

## Projecting Building Lines onto Batter Boards

Once the batter boards are securely in place, the building line locations can now be established and marked on the batter boards bars. Using the batter board layout shown in Figure 2-47, the approach is as follows:

1. From the boundary intersections, accurately measure the distances to Stations A, B, C, D, E,

and F. These distances will be 25, 45, 65, 85, 105, and 125 ft, respectively.

2. Set up the instrument over Station A, level it, and backsight on the boundary intersection marker. Now set the horizontal circle to zero.
3. Turn the telescope to 90° and tighten the horizontal clamp screw. Sight across the intervening batter boards at stakes 1 and 2 and mark carefully on each batter board where the vertical cross hairs intersect the top edge of the batter board.
4. Set up over Station B and mark the building line on batter boards at stakes 10 and 4.
5. Repeat the procedure at Stations C, D, E, and F.
6. From the boundary intersection, measure the distances to Stations G, H, I, J, K, and L. They will be 25, 45, 65, 85, 105, and 125 ft, respectively.
7. From these stations mark the remainder of the batter boards.

## Marking the Depth of Cut on Excavation Limit Stakes

Figure 2-48 shows the excavation limit stakes for the building in Figure 2-43(b). Assume the foundation plan indicates that the tops of column footings are to have an elevation of +1019.920 m. The excavation is to be taken 250 mm below that level.

1. Set up the instrument at a convenient location from which it is possible to sight on the bench mark and the limit stakes.
2. Find the differences in elevation between the bench mark and each limit stake and record these differences. Suppose that they are as follows:

| Stake No. | Elevation (m) | Difference (m) |
|---|---|---|
| 1 | 1023.575 | +0.455 |
| 2 | 1022.875 | −0.245 |
| 3 | 1022.815 | −0.305 |
| 4 | 1022.520 | −0.600 |
| 5 | 1022.050 | −1.070 |
| 6 | 1022.360 | −0.760 |
| 7 | 1022.595 | −0.525 |
| 8 | 1022.815 | −0.305 |
| 9 | 1023.120 | 0.000 |
| 10 | 1023.365 | +0.245 |

3. Calculate the difference in elevation between the bottom of the excavation and the bench mark. In this instance the difference will be 1023.120 − (1019.920 − 0.250) = 3.450 m.
4. Calculate the cut at each stake and indicate it on the stake. The cut at each stake, in this example, is shown in Figure 2-49.

## Running Grades

Sometimes it is necessary to run *grade lines*—lines that have a uniform rate of rise throughout their length—*on a construction site.* In standard units, grade lines are usually expressed as a given number of feet of rise per 100 ft of length and are usually expressed as a percentage. For example, a 1% grade means that there is a rise of 1 ft for every 100 ft of horizontal distance.

In metric units, slope is usually expressed as a pure number ratio, that is, as a nondimensional ratio. For example, a 1:100 grade represents a uniform vertical change in elevation of 1 unit for every 100 horizontal units. Using metres as the measure of length, a 1:100 grade indicates a 1-m change in elevation for every 100 m of horizontal distance.

One method of obtaining a uniform drop in any given distance is as follows:

1. Drive a stake to the required height at the start of the proposed line, set up the instrument over this stake, and level it.
2. Measure (with the rod or a tape) the height of the telescope above the stake; suppose, in this example, that it is 3 ft 8 in. (1.120 m).
3. Align the telescope in the direction of the line; this may be done by sighting on the rod held at the far end of the line.
4. Measure 100 ft (30 m) along the line and set another stake.
5. Raise the target the distance the drop is to be, for example, 6 in. (150 mm). This will make the target read 1.270 m in this example.
6. Drive the stake until the target is on the cross hair when the rod is held on the stake. The line between the stakes now drops 6 in. in 100 ft (150 mm in 30.0 m).
7. Set the target back to its original reading, 3 ft 8 in. (1.120 m), and incline the telescope downward until the cross hair is again on the target. Tighten the vertical motion clamp screw.

Stakes set anywhere along this line and driven in until the cross hair is on the target when the rod is held on the stake will now be on the same grade.

When building lines must be laid out on a site that is surrounded by other buildings or if the building takes up most of the lot, the use of batter boards may not be practical. In this situation, column lines and any other necessary control lines and elevations

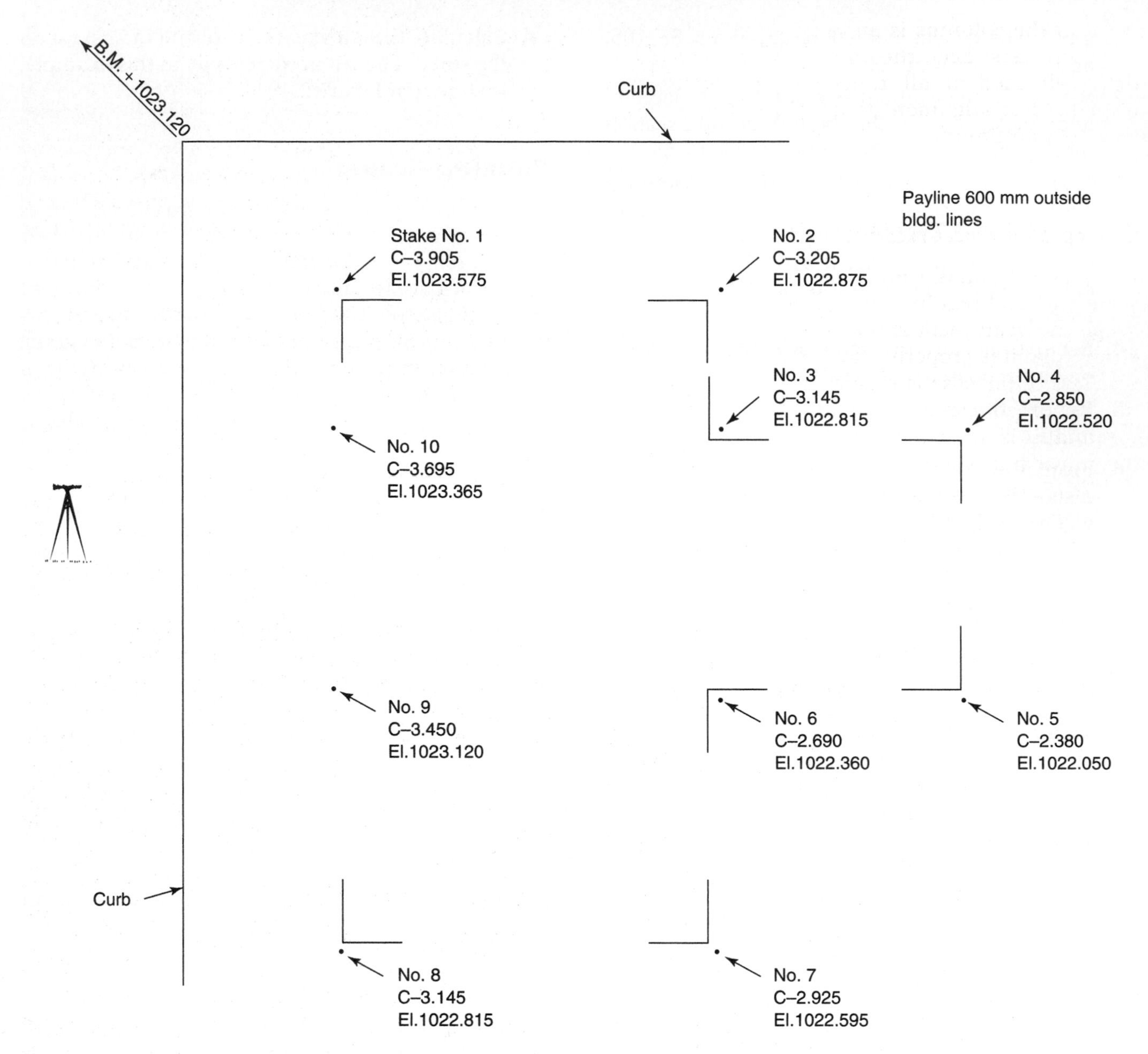


**Figure 2-48** Excavation Limit Stakes.

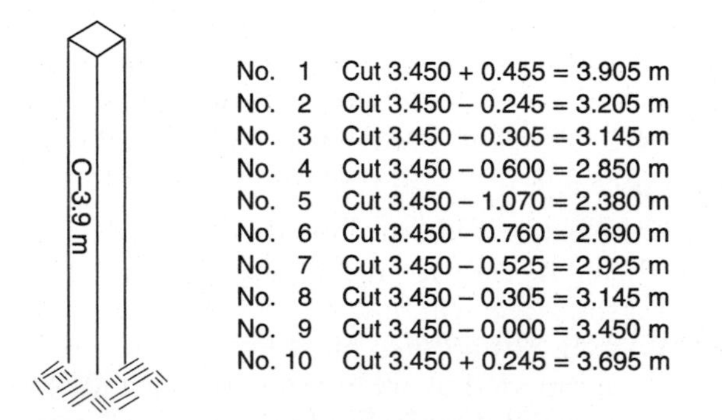

C–3.9 m
No. 1 Cut 3.450 + 0.455 = 3.905 m
No. 2 Cut 3.450 – 0.245 = 3.205 m
No. 3 Cut 3.450 – 0.305 = 3.145 m
No. 4 Cut 3.450 – 0.600 = 2.850 m
No. 5 Cut 3.450 – 1.070 = 2.380 m
No. 6 Cut 3.450 – 0.760 = 2.690 m
No. 7 Cut 3.450 – 0.525 = 2.925 m
No. 8 Cut 3.450 – 0.305 = 3.145 m
No. 9 Cut 3.450 – 0.000 = 3.450 m
No. 10 Cut 3.450 + 0.245 = 3.695 m


**Figure 2-49** Excavation Stake Marked for Cut.

are marked on existing buildings or structures surrounding the site. The bench mark used for the control of the survey should be located at some distance from the site to ensure that it is not disturbed by the excavation.

Large buildings require constant monitoring, and a survey crew should be checking lines and elevations continually. Grades must be checked before and after excavation, and footing elevations must be checked during forming and before concrete is poured.

Once the concrete has been poured and set, column and wall locations are marked on the footings. Slab elevations are marked on columns as the building frame progresses, and vertical alignment of the columns is an ongoing task. The survey crew must keep the construction superintendent informed at all times of any changes in elevations or alignment to ensure that errors are kept to a minimum.

## CARE OF INSTRUMENTS

Builders' levels and transit levels are precision instruments and require particular care in handling. When the instrument is not attached to the tripod, ensure that it is properly placed in its carrying case. One end of the case is marked *eyepiece end* so that the instrument can always be fitted in properly. Be sure that it is not subjected to shocks during packing, unpacking, and transportation.

If the instrument gets wet, let it dry, preferably in a relatively dust-free environment. If it has become dusty, clean the dust from the lens with the brush provided. Never rub the optical surfaces. Have the instrument checked and cleaned periodically by a reliable instrument repair shop.

When the instrument is attached to its tripod, carry both as an incorporated unit. Grasp the tripod about two-thirds of the way toward the top, closing the legs, and tip it slightly so that it rests comfortably against the shoulder. Make sure that the instrument is not jarred or bumped.

Do not stand the instrument on a surface into which the metal leg tips cannot bite. If necessary, use a wooden tripod stand such as that illustrated in Figure 2-50.

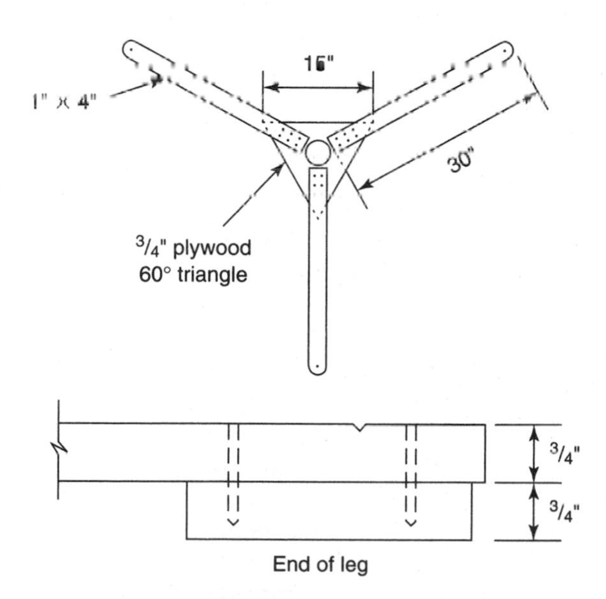


**Figure 2-50** Tripod Stand.

## REVIEW QUESTIONS

**1.** What is the purpose of a site plan?

**2.** Describe the leveling procedures for both three- and four-point leveling.

**3.** Briefly describe the purpose of each of the following parts of a level: **(a)** horizontal motion tangent screw, **(b)** centering head, **(c)** vernier scale, **(d)** vertical motion clamp screw, and **(e)** adjustment screw at the end of the bubble tube.

**4.** What is meant by the *elevation* of an object?

**5.** Define the following: **(a)** backsight, **(b)** H.I., **(c)** turning point, and **(d)** station.

**6.** A bench mark is recorded as 1656.121′. What is the elevation of Point A, measured from this bench mark, if the recorded readings were as follows:

| | T.P.1 | T.P.2 | T.P.3 | T.P.4 | T.P.5 | T.P.6 |
|---|---|---|---|---|---|---|
| Backsight | 2.56′ | 6.63′ | 5.16′ | 4.20′ | 3.92′ | 6.08′ |
| Foresight | 3.12′ | 2.18′ | 6.31′ | 7.65′ | 1.58′ | 2.35′ |

**7.** Determine the height of a communications tower, assuming that the transit is set up 100 ft from the base and the level line of sight is 4.6 ft above grade. When sighting on the top of the tower, the angle of inclination was found to be 34°27′.

**8.** A bench mark is recorded as 504.786 m. What is the elevation of Point A, measured from this bench mark, if the recorded readings in meters are as follows:

| | T.P.1 | T.P.2 | T.P.3 | T.P.4 | T.P.5 | T.P.6 |
|---|---|---|---|---|---|---|
| Backsight | 0.780 | 2.021 | 1.280 | 1.195 | 1.853 | |
| Foresight | 0.951 | 0.664 | 1.923 | 0.482 | 0.716 | |

**9.** We wish to determine the height of a chimney. The instrument is set up 30 m from the base and a level sight is to be 1.400 m above ground level. When sighting on the top of the chimney from the same point, the angle of inclination is found to be 31°23′. What is the height of the chimney?

**10.** The elevations of the corner stakes at the site of an excavation are found to be **(a)** 102.6′, **(b)** 99.7′, **(c)** 94.8′, and **(d)** 95.5′ relative to a datum point of +100.0′. Calculate the amount of *cut* to be shown on each stake if the bottom of the excavation is to be 11 ft below datum.

**11.** The elevations of the corner stakes at the site of an excavation are found to be **(a)** 31.272 m, **(b)** 30.389 m, **(c)** 28.295 m, and **(d)** 29.108 m relative to a datum point of +30.000 m; calculate the amount of cut to be shown on each stake if the bottom of the excavation is to be 3.353 m below datum.

**12.** Give the reading indicated on the section of scale shown (see Figure 2-51) if the angle was turned clockwise.

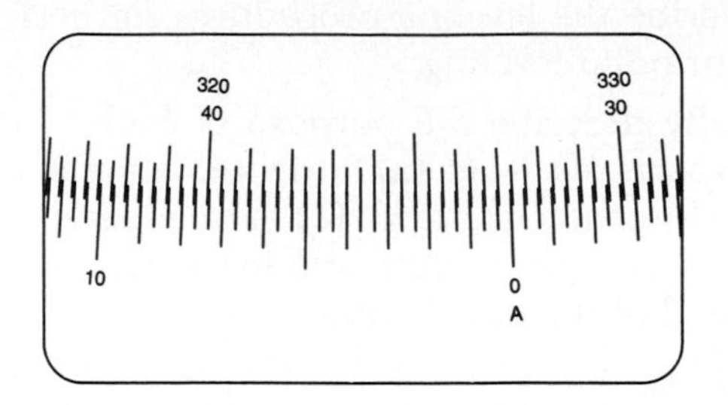


**Figure 2-51**

**13.** What is the basic difference between a builder's level and a transit?

**14.** What is the reason for setting up batter boards when laying out a building?

**15.** Explain the term *payline.*

**16.** Give one advantage that a laser level has over a survey level when used for establishing a level plane. An example of a level plane is a suspended ceiling in a building.

# 3 Excavations and Excavating Equipment

A soils report provides the necessary geotechnical information required by foundation engineers to establish the type of foundation that will be appropriate for the proposed building. The type of foundation depends on the conditions that are found at the site, and the magnitude and distribution of design loads imposed by the building superstructure on the supporting soil. Many soils are naturally consolidated. If left undisturbed, they provide a good base for building foundations. When subsurface water is encountered, special measures must be taken in the design and construction of the building foundations to ensure that the building will not sustain damage due to water infiltration and excessive differential settlements.

Site excavation prepares the building site for all future construction activity. Initial site preparation can begin with the removal of all topsoil and stockpiling it for later use as landscaping material. The building site may need enhancing to ensure proper drainage around the building and in the parking areas, and trenches for services must be dug. In areas where cold temperatures affect the soil, foundations must be placed below the frost line. In many locations, sufficient soil bearing values are found only at some distance below the surface of the ground, and the overlying material must be removed before the placement of the building footings can proceed. On restricted sites, parking levels placed below ground require, in some instances, an excavation several stories deep. Areas below grade that contain shops, services, transportation access, and pedestrian links to other buildings require extensive excavation.

The rate of excavation is contingent upon the weather. Lengthy delays can be costly. Rain may make digging difficult and a change in excavation equipment may be necessary. Time lost at this early stage may delay the completion of the overall project. Today's contractor has a wide choice of excavation equipment and, with good planning, can effectively deal with almost any situation.

## EFFECTS OF SOIL TYPES ON EXCAVATING EQUIPMENT

The type of soil at the building site will be the major factor in determining the size and type of excavating equipment needed for the job. If the soil is loose or noncohesive (dry sand or gravel, for instance), a hydraulic excavator can be used effectively to excavate and load the excess material onto trucks. In many instances, the initial removal of the topsoil is done with a bulldozer. If the soil must be removed from the site, a loader is used to load the material onto a truck. On large sites, an access road or ramp is normally provided (Figure 3-1), to remove the equipment once the excavation is complete. Track-mounted hydraulic excavators are most effective in areas with limited accessibility, and have the added ability to dig below their own level (Figure 3-2), making them more versatile than other excavators. More cohesive soils require more power and, in some cases, a bulldozer equipped with a ripper will be used to break up the soil so that an excavator or loader can be used for the removal of the loosened material. For large sites, where large quantities of soil need to be moved from one area to another, scrapers may be employed.

Soils that are too soft or wet to support machinery traveling over them may be excavated by

**Figure 3-1** Access Ramp into Building Site Excavation.

**Figure 3-2** Excavator Clearing Out Excavation for Footing.

means of a crawler crane equipped with a dragline bucket or clamshell. These attachments take advantage of the long reach provided by the crane boom, allowing the crane to operate from a stable location. Because of its limited mobility, this method of excavation is used primarily in dredging canals, riverbeds, and harbors.

When narrow trenches or individual holes are required for footings or piers, a tractor equipped with a backhoe or a hydraulic excavator can be used. Trenchers are used for narrow excavations, such as service lines, where soil conditions permit.

The removal and loading of boulders and broken rock is executed by large excavators and loaders. Typical examples of these machines used on the construction site will be described in some detail in subsequent paragraphs.

## REMOVING GROUNDWATER

One serious problem encountered regularly during digging operations is the presence of water in the excavation. This may be caused by rain, melting snow, an underground stream, or the fact that the water table in the area is high. The *water table* is the normal level of groundwater, and if that level is close to the surface, excavating will allow the groundwater to seep and collect in the excavation.

There are two principal methods of getting rid of water. One is to use a pump (or pumps) to empty the excavation or to keep it dry. This method is particularly useful in getting rid of collected rain water and runoff or in removing the water from an uncovered underground stream. Figure 3-3 illustrates

**Figure 3-3** Pump Draining Excavation. (Reprinted by permission of Homelite.)

**Figure 3-4** Excavation Surrounded by Well-Point System. (Reprinted by permission of Moretrench American Corporation.)

the removal of collected rain water from a pier excavation by pump. Pumps are also used to get rid of seepage water. A sump, into which the water collects and from which the pump can then remove it, is required.

Another method is to lower the water table in the area in which the excavating will take place. This is done by a *dewatering system,* which involves sinking a series of *well-points* around the area and extracting the water by suction pump. Figure 3-4 shows an excavation surrounded by a dewatering system to keep the working area dry. Figure 3-5 illustrates how a typical well-point system would be laid out, and Figure 3-6 shows a section through such a system.

A system consists of a series of well-points (see Figure 3-7), each with a riser pipe and a swing joint connecting to a common header or manifold. In

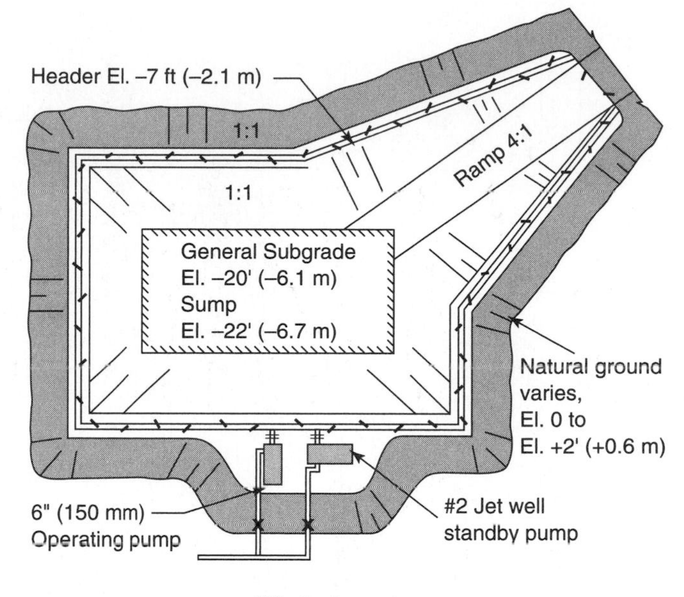


**Figure 3-5** Typical Well-Point Layout Plan. (Reprinted by permission of Moretrench American Corporation.)

Figure 3-8, a profile of a well-point system shows the setup. Notice the normal water level and the predrained water level brought about by the water extraction.

The header pipe is exhausted by a combination pumping unit, which is a centrifugal pump continuously primed by a positive displacement vacuum pump. Atmospheric pressure forces water through the ground to the well-point screens, into the well-points, up the riser and swings, and through the header to the pump. Any air entering the system is separated in the float chamber and passed to the vacuum pump. The water passes to the centrifugal pump, which discharges it to a drainage system (see Figure 3-9).

In easily drained, coarse soils, well-points are driven directly into the soil. In fine, dense soils, they are generally set in a column of sand, as shown in Figure 3-10, which helps to prevent the well-point screen from becoming clogged with silt.

The well-point is a hollow, perforated tube, covered with a stainless steel filter screen that allows water to enter the well-point but keeps out sand and silt (see Figure 3-7). The bottom end is serrated to enable it to penetrate the soil more easily. In the bottom end is a chamber containing a ball valve with a wooden center so that it will float. When water is being drawn from the well-point, the ball floats up against a ring valve at the top of the chamber and prevents sand from being drawn into the system.

When placing the well-point in the ground, a jet of water can be forced down through the pipe,

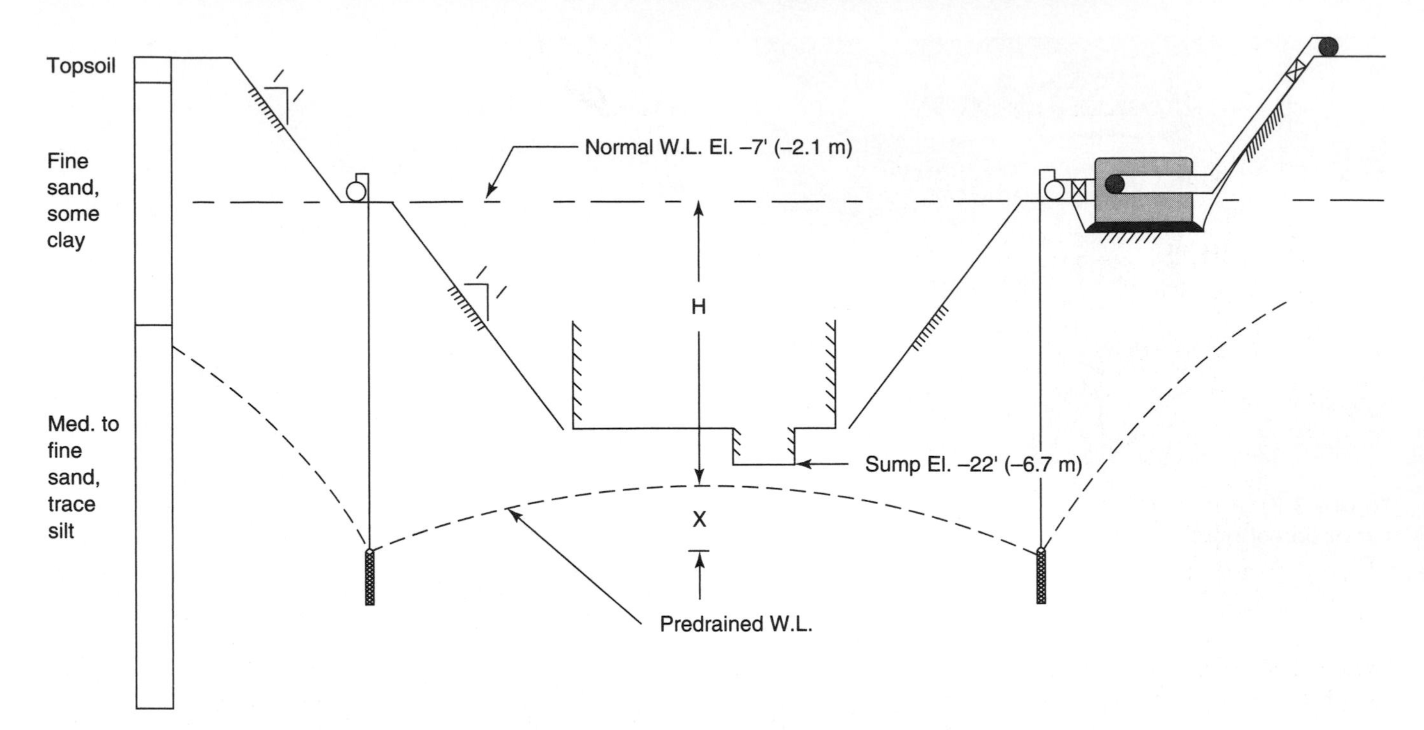

**Figure 3-6** Typical Well-Point Layout Section. (Reprinted by permission of Moretrench American Corporation.)

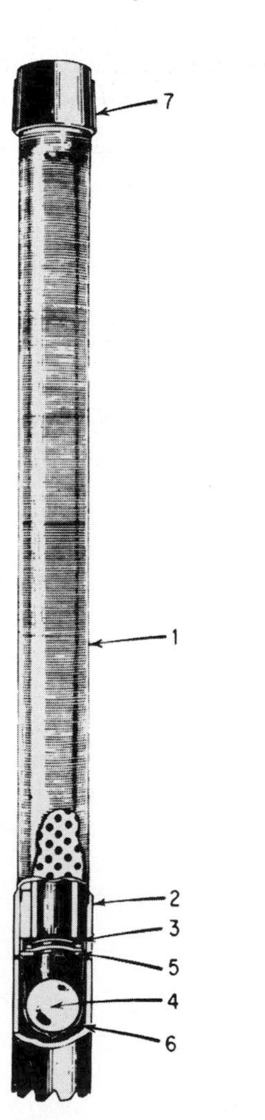


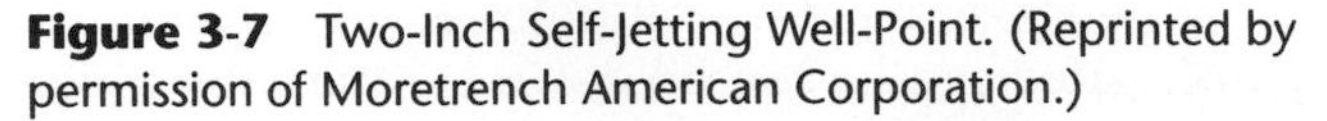
**Figure 3-7** Two-Inch Self-Jetting Well-Point. (Reprinted by permission of Moretrench American Corporation.)

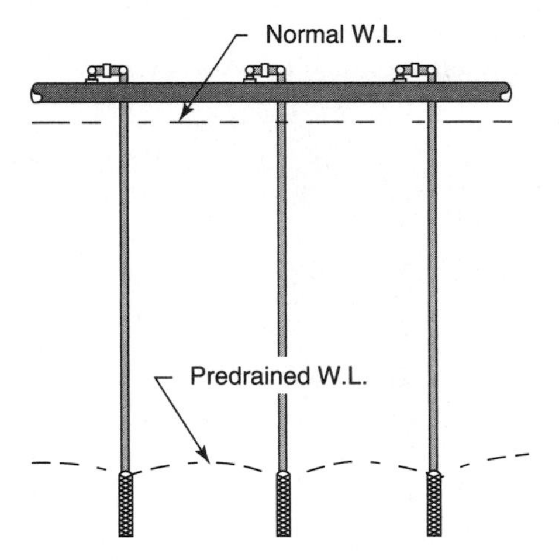


**Figure 3-8** Profile of Well-Point System. (Reprinted by permission of Moretrench American Corporation.)

loosening the soil around the tip and making penetration easier. During this operation, the ball valve is forced down into the retainer basket at the bottom of the chamber. Figure 3-11 shows a well-point being jetted into place.

Another method for controlling water levels and seepage is the use of horizontal perforated drains. Installation of these lines can be done after the building foundation has been completed (Figure 3-12), just before backfilling is to begin. The drainage lines may be tied to a sump or fed into a storm sewer.

**Figure 3-9** Dewatering Pump. (Reprinted by permission of Moretrench American Corporation.)

**Figure 3-11** Jetting a Well-Point. (Reprinted by permission of Moretrench American Corporation.)

**Figure 3-10** Properly Sanded Well-Point. (Reprinted by permission of Moretrench American Corporation.)

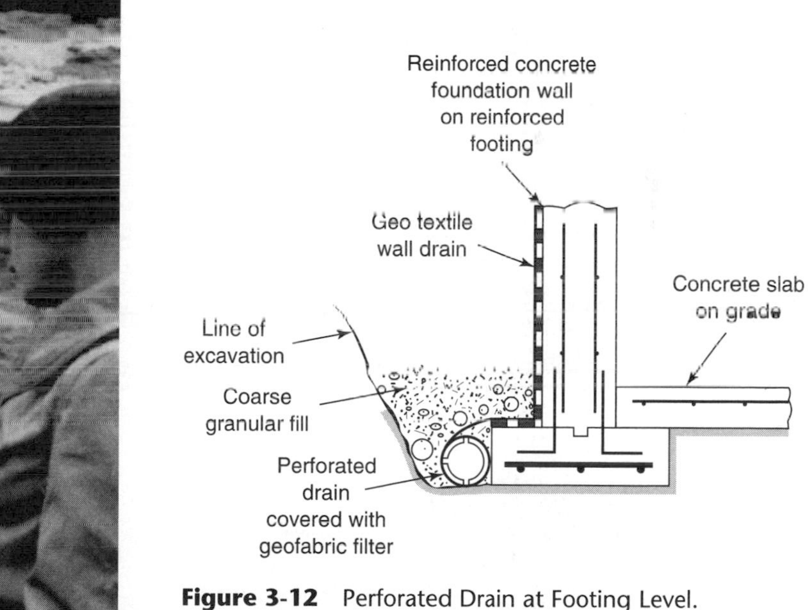


**Figure 3-12** Perforated Drain at Footing Level.

Vertical wall drains can also be used in conjunction with these drainage lines to remove moisture that may accumulate on the wall exterior. The use of geotextile filters over perforated drains is recommended to prevent premature clogging of the drain perforations due to sand fines.

## GENERAL AND SPECIAL EXCAVATIONS

Excavating operations are of two types: *general* and *special.* General excavations include all of the work, other than rock excavation, that can be carried out by mechanical equipment. For example, excavating done with draglines, shovels, or scrapers, loading with payloader or clamshell, and hauling by truck all come under the heading of general excavation. Special excavation includes the work that must be done by blasting, by special machines, by hand, or by a combination of hand and machine. The contractor should be fully aware of the types of excavation that must be dealt with to estimate the cost properly and to ensure that the right kind of equipment is on hand to do the best job.

## ESTIMATING AMOUNT OF MATERIAL TO BE REMOVED

Before an excavation is begun, its cost must be estimated. This estimate should be based on the cubic yards of material to be dug and hauled away. Plans show the area occupied by the building and the depth to which the excavation must be carried. Because additional space is required to work around the outside of forms in an excavation, a line is drawn at a reasonable distance, usually 2 ft (600 mm) outside the building line, and the excavating contractor is paid for the material excavated to this line, the *payline.*

When soil is disturbed from its original condition, broken up, and moved, it experiences a change in volume (usually as an increase). This increase in volume is termed *soil swell* or *bulking.* During excavation, this increase in volume must be taken into account when calculating the volume of material that has to be moved. Most soils will swell or bulk between 20% and 50%. Table 3-1 provides typical percentages of bulking encountered with common soil types.

**Table 3-1**
SOIL SWELL PERCENTAGE

| Soil Type | Percentage of Swell from Compact State |
|---|---|
| Silt | 20 |
| Clay | 25 |
| Sand and gravel | 50 |
| Loam | 25 |
| Stone | 50 |

When dealing with earthwork calculations, the volume of a material is related to its condition or state. Soil in its undisturbed state, or in its *in bank* condition, has a volume termed its *bank cubic yard* (*bank cubic metre*). Because excavated or disturbed soil generally experiences an increase in volume, the excavated volume will be greater than the in-bank condition. The resulting volume of disturbed or excavated soil is measured as a *loose cubic yard* (*loose cubic metre*). On the other hand, when a soil is compacted, whether in its natural condition or whether it is newly placed, the compaction process will increase the density of the material and decrease the volume that the soil occupied before the compaction was done. A soil that has been compacted is measured as a *compacted cubic yard* (*compacted cubic metre*). Knowing this, it becomes clear that when a contractor is in the process of calculating costs for the excavation, loading, hauling, and placement of materials, allowances must be made for these differences in volume. Examples illustrating the effects of these differences are provided in the section dealing with the estimating and hauling of excavated materials later in this chapter.

If the soil is cohesive enough or if some means of supporting the sides of the excavation are to be used, the compact volume to be removed is simply a matter of length × width × depth. But for loose, sliding soils, the excavation must have sloping sides, usually with a 1:1 slope. This fact must be taken into account when estimating the volume of material to be removed. A formula based on the frustum of a pyramid provides relatively accurate results in calculating the volume of the excavation in such cases, and can be expressed as

$$\frac{H}{3}(A + B + \sqrt{A \times B})$$

where $A$ = top area
$B$ = bottom area
$H$ = vertical height

***Example:***

A rectangular excavation measures 20 ft by 30 ft at surface grade. If the slope of the sides is 0.5:1 and the excavation is 12 ft deep, calculate the amount of material that has been removed in cubic yards.

***Solution:***

Apply the formula $\frac{H}{3}(A + B + \sqrt{A \times B})$.

where $A$ = top area
= 20 × 30
= 600 sq ft

Depth of excavation: $H$ = 12 ft

$B$ = bottom area
= [20 − (12/2) × 2] × [30 − (12/2) × 2]
= 8 × 18
= 144 sq ft

Volume of material removed:

$$V = \frac{12}{3} \times (600 + 144 + \sqrt{600 \times 144})$$

= 4151.76 cu ft

Volume in cubic yards = 4151.76/27
= 153.77 cubic yd

---

***Example:***

A rectangular excavation measures 6 m by 9 m at surface grade. If the slope of the sides is 0.5:1 and the excavation is 3.6 m deep, calculate the amount of material that has been removed in cubic meters.

***Solution:***

Apply the formula $\frac{H}{3}(A + B + \sqrt{A \times B})$

where $A$ = top area
= 6 × 9
= 54 m²

Depth of excavation: $H$ = 3.6 m

$B$ = bottom area
= [6 − (3.6/2) × 2] × [9 − (3.6/2) × 2]
= 2.4 × 5.4
= 12.96 $m^2$, say 13 $m^2$

Volume of material removed:

$$V = \frac{3.6}{3} \times (54 + 13 + \sqrt{54 \times 13})$$

= 112.2 $m^3$

---

Another common method that is used to calculate the volume of an excavation is known as the *average end area* method. This approach, because of its simplicity, is a popular method for calculating volumes. However, in most instances, the calculated volume is on the high side. The volume calculation can be expressed as

$$V = \frac{(A_1 + A_2)}{f\,2} \times L$$

where $A_1$ and $A_2$ are the end areas (or $A_1$ can be the top area and $A_2$ can be the bottom area) of the excavation, and $L$ is the perpendicular distance between the two areas.

---

***Example:***

Using the dimensions of the excavation in the preceding example, calculate the volume of the excavation using the average end area method.

***Solution:***

Top area $A_1$ − 600 sq ft
Bottom area $A_2$ = 144 sq ft
Distance between the two areas, $L$ = 12 ft

$$\text{Volume } V = \frac{(600 + 144) \times 12}{2}$$

= 4464 cu ft

Note that the result is slightly higher than in the previous calculation.

---

When the surface of a building site is level, the volume of an excavation can be calculated exactly by dividing the excavation into a series of prisms and pyramids and applying the appropriate mathematical formula to calculate the volume of each shape.

---

***Example;***

Using standard mensuration formulae, calculate the resulting volume of the 20-ft by 30-ft by 12-ft deep excavation in the previous example,

***Solution:***

Applying standard mathematical formulae, the volume calculation may be done as follows:

Volume based on the of floor of excavation (rectangle)
= 8 × 18 × 12
= 1,728 cu ft

Volume of side slopes (right triangles)
= (6/2 × 12) × (8 + 18) × 2
= 1,872 cu ft

Volume of corners (pyramids) = 6 × 6 × 12/3 × 4
= 576 cu ft

Total volume in cubic yards
= (1728 + 1872 + 576)/27 = 155 cu yds

Comparing the results of the various methods, the results obtained by the first two approximate methods correspond favorably with the mathematical approach.

---

**Figure 3-13** Tower Crane in a Downtown Construction Site.

## CRANES

Before discussing machines used specifically for excavating, it is appropriate to deal with some aspects of the crane, a machine that has become indispensable on the building site. The crane was developed to provide efficient lifting capability on the construction site and, subsequently, has been adapted to other uses, such as excavating and pile driving.

The development of the crane is based on its predecessor, the derrick. The derrick was originally the only piece of equipment available for loading and unloading material on the construction site. Although the derrick is still used for lifting heavy loads, its setup time is lengthy and it cannot be moved without disassembly. The need for additional mobility on and off the site and for shorter setup times promoted the development of the modern-day crane.

Two basic types of cranes are commonly available for use on the construction site: the *stationary crane* and the *mobile crane*. There are several variations of each type, and each is available in a wide range of lifting capacities and boom lengths, thus providing the contractor with a generous selection of options.

The stationary crane, more commonly known as the *tower crane* (Figure 3-13), comes in two basic variations. In the one type, the tower of the crane is supported at ground level on a concrete foundation as illustrated in Figure 3-14(a). Heavy anchor bolts as shown in Figure 3-14(b) are used to secure the tower legs to the foundation. Extensions are added to the tower to increase its height as construction proceeds. For the second type, the tower is supported by the building frame and, as the building rises, the tower

(a)

(b)

**Figure 3-14** (a) Tower Crane Tower Supported on Concrete Pad. (b) Anchor Bolts Secure Crane Tower to Concrete Pad.

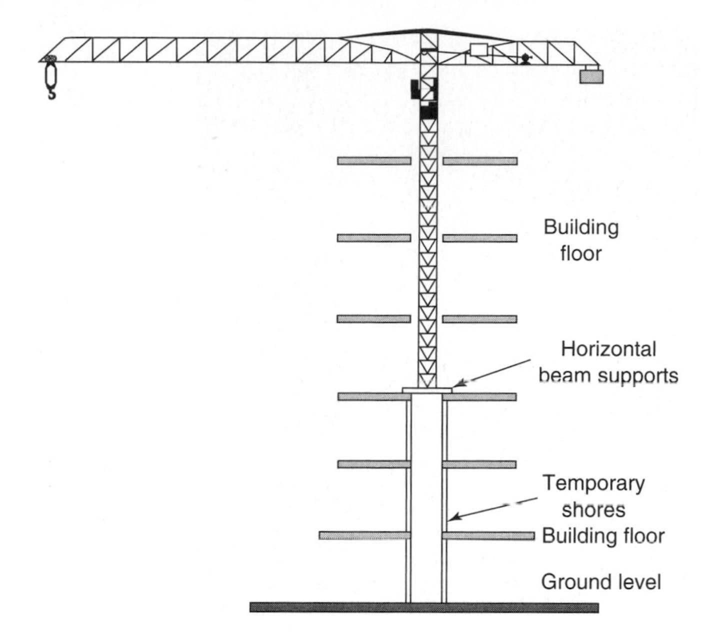

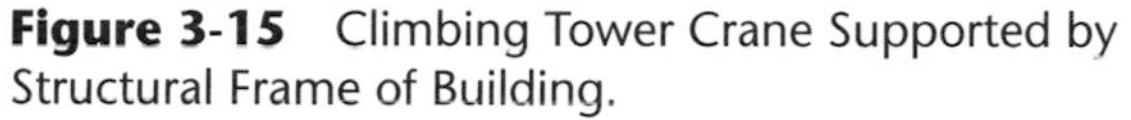

**Figure 3-15** Climbing Tower Crane Supported by Structural Frame of Building.

hoists itself up the building frame using the lower floors of the building for its support (Figure 3-15). Tower cranes have one advantage over mobile cranes in that the operator's cab is placed at the top of the tower, providing the operator an unobstructed view of the construction site. Tower cranes are usually set up at a location on the construction site from where all corners of the building can be reached, often in the elevator shaft of the building being constructed.

The main drawback of tower cranes is that once positioned they are not moved. At the completion of the building, they must be disassembled and lowered to the ground in relatively small sections by a mobile crane or, in extreme situations, by a small derrick.

The development of the mobile crane has also taken two directions with reference to mobility: the *crawler crane* and the *truck crane*. The crawler crane has a carrier equipped with steel tracks, much like a military tank or crawler tractor (Figure 3-16). Once assembled, it has mobility on the job site; however, because of its slow speeds and large bulk, the crane cannot move from one site to another without some disassembly and the use of low-bed trucks to transport the pieces to another site.

To improve the mobility of the crane from one site to another, trucklike carriers with rubber tires and an additional cab were developed to allow the operator to drive the crane from site to site (Figure 3-17). This makes the crane truly mobile in that the carrier can now travel on public roads and highways. However, the boom still has to be shortened to a manageable length and the excess pieces moved on a separate truck. To improve mobility and to reduce the boom rigging time, lattice booms have now been

**Figure 3-16** Manitowoc Model 21000 Lattice Crawler Crane Has a Base Capacity of 834 US Tons and Can Reach to 640 feet (195 metres) with the Luffing Jib Extension Installed. Crawlers of This Size Are Designed Modularly to Allow the Machines to Break Down for Shipping. (Reprinted by permission of Grove Worldwide.)

replaced with hydraulic booms, allowing the crane to be completely self contained (Figure 3-18).

The one advantage the crawler crane has over the rubber-tired crane is that it can lift relatively heavy loads without the use of stabilizers or outriggers. The rubber-mounted crane must be set in a level position and stabilized with hydraulic outriggers (Figure 3-19) before it can lift any significant load with appreciable reach.

In some instances, site conditions require that equipment or materials be moved quickly from one location to another. "Pick-and-carry" type cranes that can be operated and driven from a single cab (Figure 3-20) are very useful in these situations.

## Crane Principles

A crane consists primarily of a power unit mounted on a carrier, with a hoist, a boom, and control cables for raising and lowering the load and boom (Figure 3-21 on Page 59). A *gantry* is used (Figure 3-22 on Page 59) to provide better boom support and reduce forces in

**Figure 3-17** Lattice Boom Crane on Carrier with Rubber Tires.

**Figure 3-18** The TMS 900E Truck Mounted Crane Has a 90 US Ton Capacity and with All of Its Inserts and Extensions Installed Can Reach a Maximum Tip Height of 237 feet (72 metres). (Reprinted by permission of Grove Worldwide.)

**Figure 3-19** Grove RT913E Rough Terrain Crane with a 130 US Ton Capacity. Rough Terrain Cranes Are Used on Unimproved Surfaces. Typically They Have Two Axles and Have to Be Transported to a Job Site. (Reprinted by permission of Grove Worldwide.)

**Figure 3-20** A Pick-and-Carry Rubber-Mounted Crane.

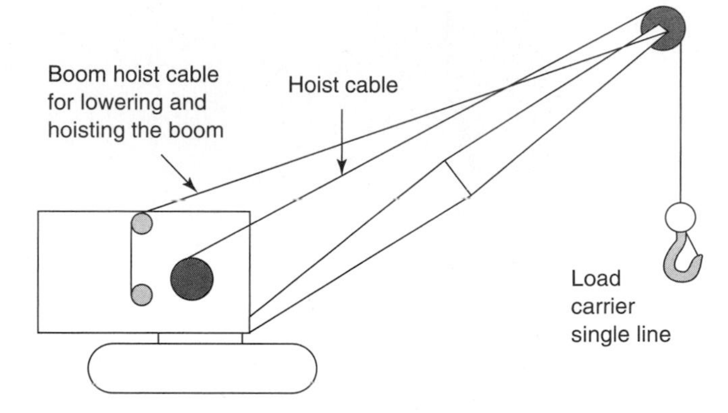


**Figure 3-21** Diagram of Crane. (Reprinted by permission of Power Crane & Shovel Association.)

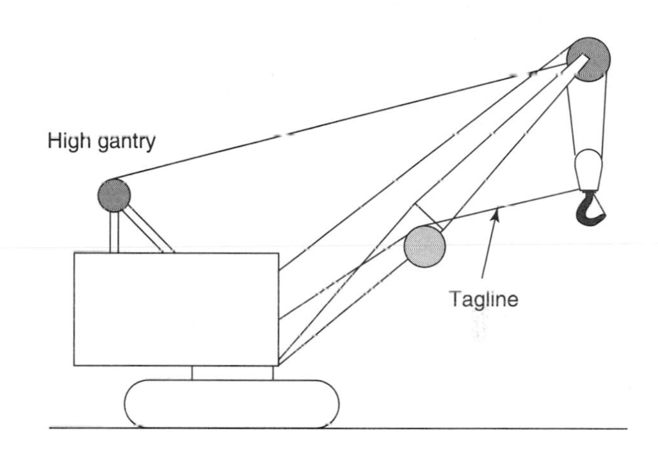


**Figure 3-22** Crane with Gantry. (Reprinted by permission of Power Crane & Shovel Association.)

the boom by increasing the angle between the boom and the boom hoist cable or pendant. For greater reach, in the case of lattice booms, an extension can be added to the boom by inserting an additional section of boom known as a *boom insert* between the upper and lower ends. On hydraulic cranes (Figure 3-23), the crane boom is extended by hydraulics. A *jib,* an extension to the end of the boom (see Figure 3-24), is used for extending the height to which loads can be lifted; it can be added to a lattice boom or a hydraulic boom. A jib decreases the lifting capacity of the crane and should be used with caution.

The load capacity of a crane is based on two considerations: (1) the stability of the crane, and (2) the strength of its components. Depending on the circumstances under which a load is lifted, either of these can govern the safe lifting load of the crane. Loading charts are provided with each crane and must be adhered to religiously. In general, the load capacity of a crane depends on the following:

1. *Stability of the footing:* Crawler-mounted cranes are much more stable than rubber-mounted cranes because of their weight and lower center

**Figure 3-23** Hydraulic Crane with Extended Boom. Note the Jib Being Carried on the Main Section of the Boom.

of gravity. Rubber-mounted cranes must be leveled and supported completely by their outriggers, which must be fully extended and positioned firmly on stable ground.

2. *Strength of the boom and its supporting hardware:* This is one of the major governing factors in establishing *load ratings,* which are based on the strength of the boom under ideal conditions. Should the load be lifted off center or the angle of the boom be decreased, the load capacity of the crane is greatly decreased.
3. *Hydraulic pressure limits* (for hydraulic cranes): These limits are usually controlled by the strength of the hydraulics. Load charts should never be exceeded.
4. *Counterweight:* This is added to the after-end of the machine; manufacturers' specifications provide standard and maximum counterweights and give crane ratings for both. Counterweights may be increased to a specified maximum, but the operating radius *must not* exceed that given by the manufacturer.
5. *Quadrant in which it is working:* Crane capacity is also affected by the quadrant over which it is lifting (Figure 3-25), and it is important not to

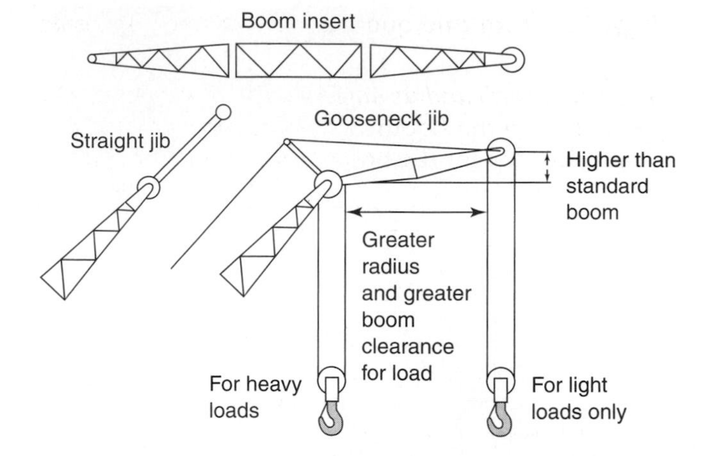


**Figure 3-24** Boom Extensions. (Reprinted by permission of Power Crane & Shovel Association.)

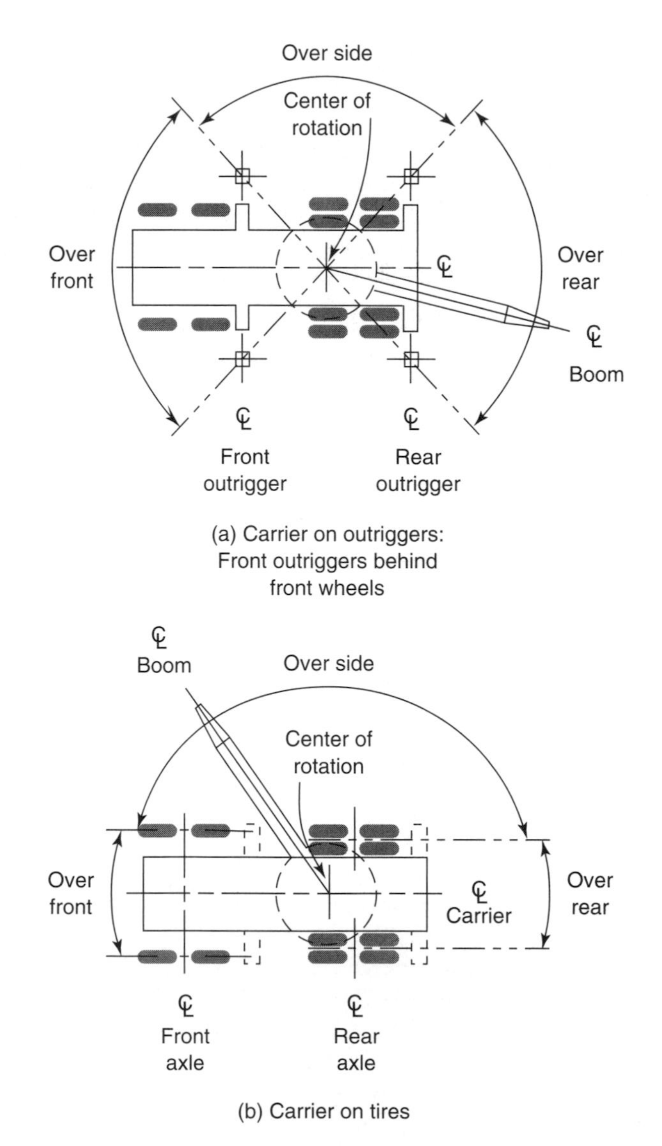


**Figure 3-25** Operating Quadrants of a Crane. Note the Decrease of Front and Rear Lifting Ranges in (b) When the Outriggers Are Not Employed Compared with (a).

swing from one quadrant to another of lesser rating.

6. *Boom length and its angle with the horizontal:* The shallower the boom's angle, the less load it can lift. The longer the boom, the less load it can lift.

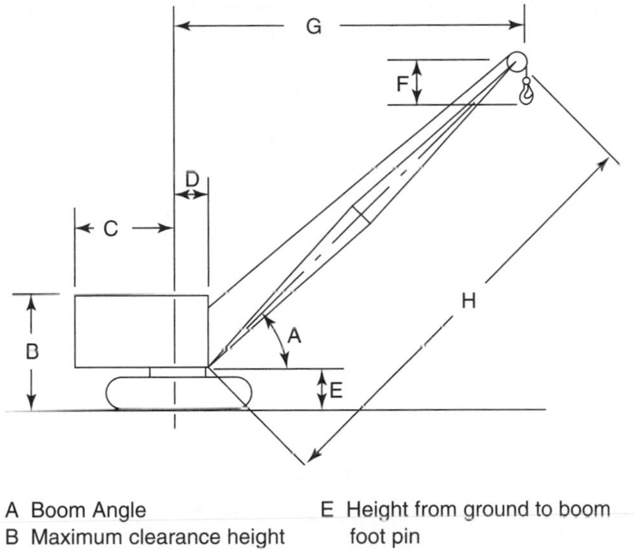


A Boom Angle
B Maximum clearance height of cab
C Maximum radius of tail swing
D Center of rotation to boom foot pin
E Height from ground to boom foot pin
F Distance from center of boom point sheave to bottom of hook
G Clearance radius of boom
H Length of boom

Working radius, not shown in above diagram, is of course of primary importance in crane selection.

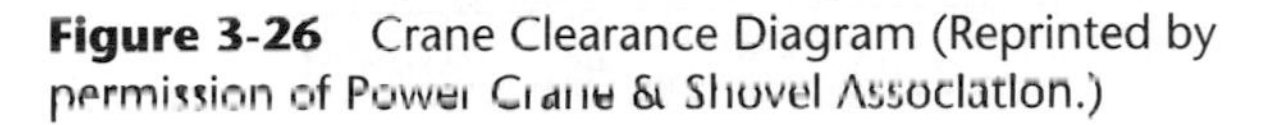

**Figure 3-26** Crane Clearance Diagram (Reprinted by permission of Power Crane & Shovel Association.)

Various crane clearances that are of importance when selecting a crane are given in Figure 3-26.

Cranes have been adapted for many uses, and a wide variety of attachments have been developed for various applications, as illustrated in Figure 3-27. Those used for excavation will be discussed later.

## EXCAVATING MACHINES

The past two decades have brought great advances in the design and quality of excavating equipment available to excavating contractors. Using the latest in materials and technology, larger, faster, and more effective machines have been built to meet the demand for improved efficiency on the construction site.

A rather unique innovation that has given excavating contractors more control over the grading and excavation process is the use of a laser level. As we discussed in Chapter 2, a laser level creates a level plane over an entire building site. This plane can be used as a reference elevation to control the depth of an excavation or the height of a finished grade. By mounting a laser-sensitive rod to the hydraulic controls for the blade of a bulldozer or grader, the depth of the cut or the height of the finished grade can be maintained to very close tolerances. By tilting the plane of the laser off the horizontal, exact slopes can be maintained over large areas, such as parking lots, to ensure that proper drainage patterns are achieved. The following pages provide an overview of some of the excavating and earth moving equipment currently used in the construction industry.

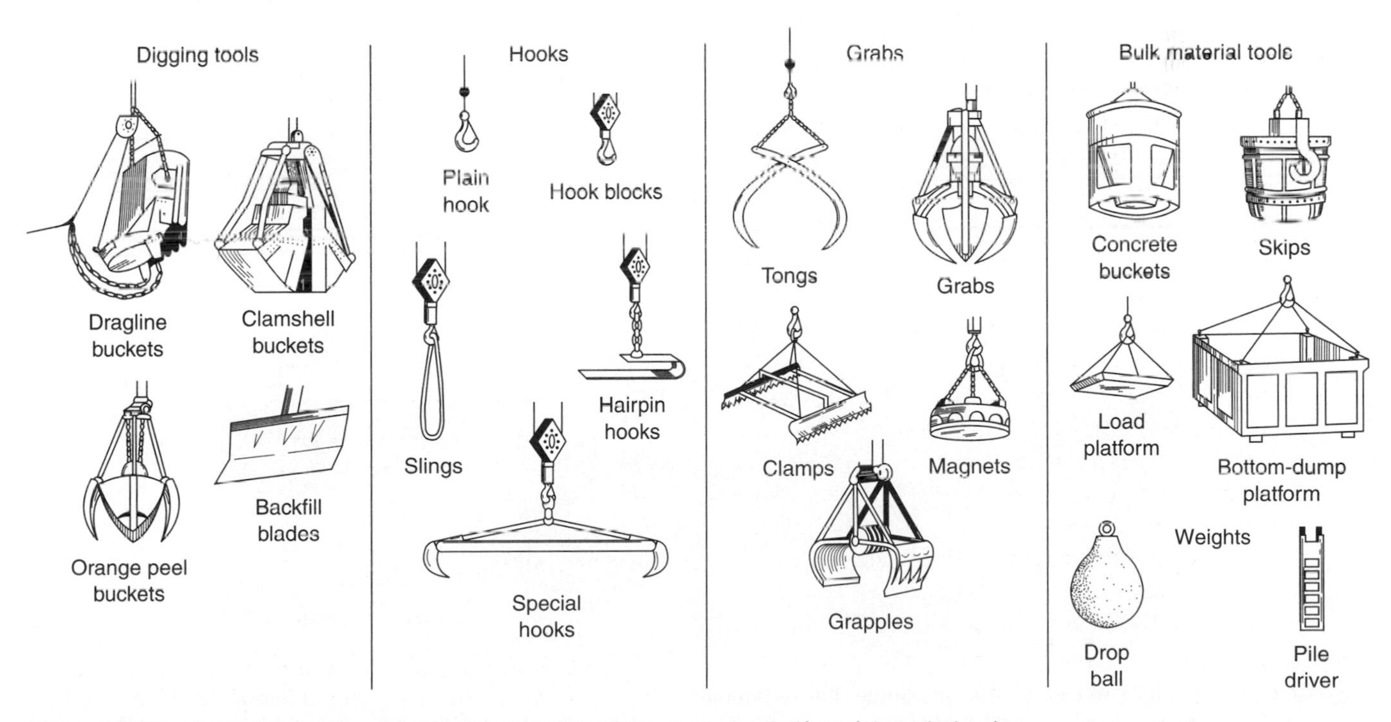


**Figure 3-27** Crane Tools. (Reprinted by permission of Power Crane & Shovel Association.)

## Bulldozers and Graders

A *bulldozer* can be defined as a track-mounted tractor, equipped with a blade, used for clearing and leveling uneven terrain in preparation for the construction of a highway or building. Today's bulldozers (Figure 3-28) certainly fit the description. If the blade is mounted perpendicular to the line of travel, the machine is a true bulldozer; if the blade is set at an angle to the line of travel, it is known as an angle dozer. A *grader* (Figure 3-29) can be considered a light-duty angle dozer.

Bulldozers and angle dozers have limited use for excavating by themselves; however, they are able to loosen and remove soil from its original position and push it to an area where it can be stored or loaded

**Figure 3-28** D11N Caterpillar Track-Type Tractor with Ripper. Blade Width Up to 20 ft 10 in. (6358 mm), with a Capacity of 42.4 cu yd (32.4 m$^3$). (Reprinted Courtesy of Caterpillar Inc.)

**Figure 3-29** Grader Used for Leveling Granular Base Material.

onto trucks and removed. Because of the power and weight of a bulldozer, it is most valuable for beginning an excavation, for stripping valuable topsoil from the site, or for site leveling. In deep excavations, a bulldozer's usefulness is reduced to loosening material and readying it for removal.

Graders are used in leveling hauled materials to some specified elevation. Their use is limited in that they cannot move deep layers of material in a single pass. In a construction setting, they are used primarily in leveling parking areas and grading the building site to provide proper drainage.

## Loaders

The development of rubber-tired loaders, with their responsive maneuverability and relatively large buckets (Figure 3-30), has contributed immensely to the efficiency of loading material onto trucks. A loader can excavate loose soils, but its chief use is loading excavated material into trucks for removal. Crawler-type tractors fitted with buckets (Figure 3-31) can also be used as loaders; however, their loading cycle times are much slower.

**Figure 3-30** Volvo Wheel Loader. Maximum Bucket Size, 18.3 cu yds (14.0 m³). (Reprinted by permission of Volvo Construction Equipment North America, Inc.)

**Figure 3-31** Crawler Loader.

## Scrapers

Elevating scrapers have established an important position in the earthmoving field, particularly when large volumes of materials must be excavated and deposited some distance away. A scraper is a combination machine in that it can load, haul, and discharge material all under its own power. Mounted on large tires, it is fast and highly maneuverable. A scraper can be powered by one or two tractors, front and back, or, in the case of large machines, each wheel can have its own power unit. In some instances, when the ground has a high clay content, a crawler tractor may be used as a pusher during the loading operation (Figure 3-32).

## Hydraulic Excavators

Although much heavy digging and loading in the past was done by the power shovel (Figure 3-33), these mechanical monsters have been replaced on the construction site by new efficient machines commonly known as hydraulic excavators (Figure 3-34). Using a pulling action rather than a pushing action on the bucket, powerful hydraulics, instead of cables, operate the boom and bucket. Because the operation of a hydraulic system is much more responsive than the old cable and pulley system, the working ranges have increased significantly (Figure 3-35). With this improved maneuverability and, thus, greater efficiency in the excavating and loading cycle, this new generation of

**Figure 3-32** Wheeled Scraper.

**Figure 3-33** Power Shovel. (Reprinted by permission of Badger Equipment Co.)

**Figure 3-34** 320 L Excavator.

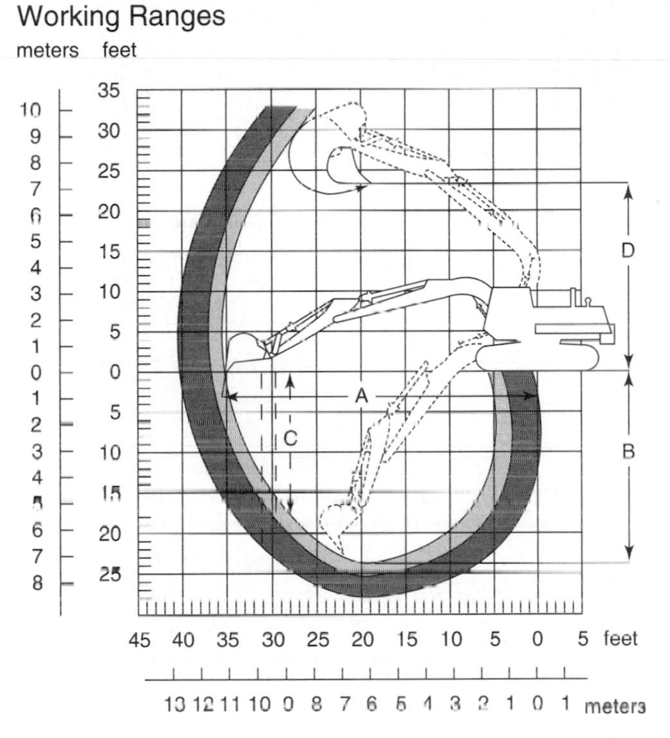


| | Stick Length | | | | | | | |
|---|---|---|---|---|---|---|---|---|
| | *2440 mm/8' | | *2900 mm/9'6" | | *3660 mm/12' | | **2900 mm/9'6" | |
| A Maximum reach at ground level | 11.0 m | 36'0" | 11.4 m | 37'5" | 12.1 m | 39'8" | 11.4 m | 37'5" |
| B Maximum digging depth | 7.1 m | 23'2" | 7.5 m | 24'8" | 8.3 m | 27'2" | 7.5 m | 24'8" |
| C Maximum vertical wall | 4.8 m | 15'11" | 5.2 m | 17'2" | 5.9 m | 19'5" | 5.3 m | 17'3" |
| D Maximum dump height | 6.2 m | 20'4" | 6.35 m | 20'10" | 6.6 m | 21'6" | 6.35 m | 20'10" |

* Equipped with 1980 mm/78" tip radius bucket.

** 2900 mm/9'6" stick on special application boom and equipped with 1980 mm/78" tip radius bucket.

**Figure 3-35** Working Ranges of the 235C Excavator. (Reprinted Courtesy of Caterpillar Inc.)

**Figure 3-36** Tractor-Mounted Backhoe Equipped with a Front-End Loader. Note the Tamping Attachment on the Ground.

excavating machines has become the contractor's choice for most construction site excavations.

The large range of sizes provided by manufacturers ensures that an appropriate machine size is available for the job at hand. The size of an excavator selected for any particular job is based on a number of factors, including the following:

1. Capacity of the bucket
2. Type of soil being extracted
3. Radius required to reach the digging area
4. Radius required to reach the hauling unit
5. Distance required to reach the stock pile
6. Clearance height required to reach over a hauling unit
7. Physical clearances of the machine when working in confined areas

## Backhoes

The backhoe is an example of a power shovel that evolved to meet certain excavation requirements. Mounted on the rear of a tractor (Figure 3-36), the unit can be moved from site to site over fairly long distances without the use of a transporter. Most units come equipped with a loading bucket on the front of the tractor to provide additional versatility.

Because of its size, the backhoe is used for excavations where space is limited and the excavation is relatively narrow. Typical uses include trenching for utility lines, digging drainage ditches and basements, sloping and grading embankments, and reshaping existing roadside ditches.

A truck-mounted variation of the backhoe is shown in Figure 3-37. The bucket is mounted on a telescopic boom, allowing for an extended reach during excavation. This unit is most useful in emergency situations in that it is self-contained and can be on site in a minimum of time.

## Dragline

One of the attachments used on a crane boom is a *dragline*. As illustrated in Figure 3-38, the dragline attachment consists of a dragline bucket, hoist cable, drag cable, and fairlead. The machine is operated by pulling the bucket toward the power unit and regulating the digging depth by adjusting the tension on the hoist cable.

The dragline has a wide range of operations. It remains on firm or undisturbed ground and digs below its own level, backing away from the excavation as the material is removed. Although it does not dig to as accurate a grade as a power shovel or a pull shovel, it has a larger working range, an advantage in

**Figure 3-37** Truck-Mounted Excavator with Hydraulic Telescopic Boom.

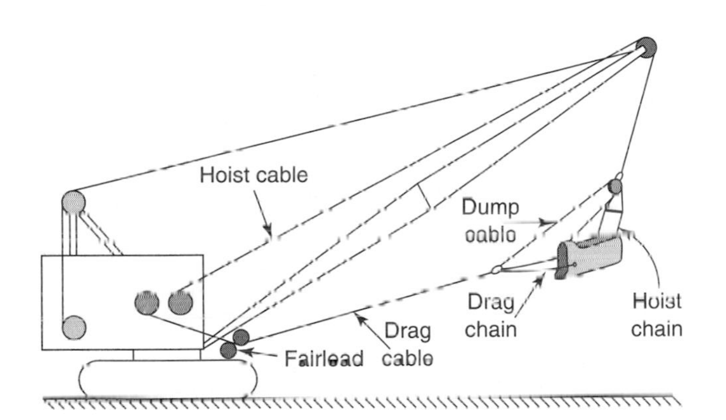


**Figure 3-38** Basic Parts of Dragline. (Reprinted by permission of Power Crane & Shovel Association.)

many situations. Figure 3-39 outlines the dragline working range.

In addition to its large working range, the dragline has the further advantage of being suited to digging in excavations below water level and in mud or quicksand. Figure 3-40 illustrates a few typical uses.

## Clamshell

The crane boom is often used with a hinged bucket, called a *clamshell,* for vertical excavating below ground level and for handling bulk materials such as sand and gravel.

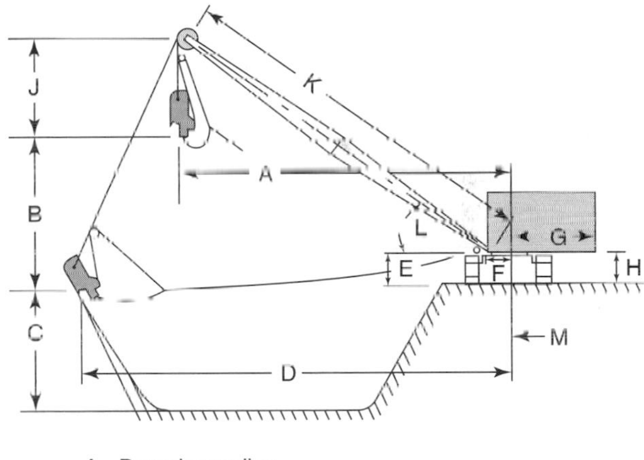


A Dumping radius
B Dumping height
C Maximum digging depth
D Digging reach (depends on conditions and operator's skill)
E Distance from ground to boom foot pin
F Distance from center of rotation to boom foot pin
G Rear-end radius of counterweight
H Ground clearance
J Length of bucket (depends on size and make)
K Boom length
L Boom angle
M Centerline of rotation

**Figure 3-39** Dragline Working Range. (Reprinted by permission of Power Crane & Shovel Association.)

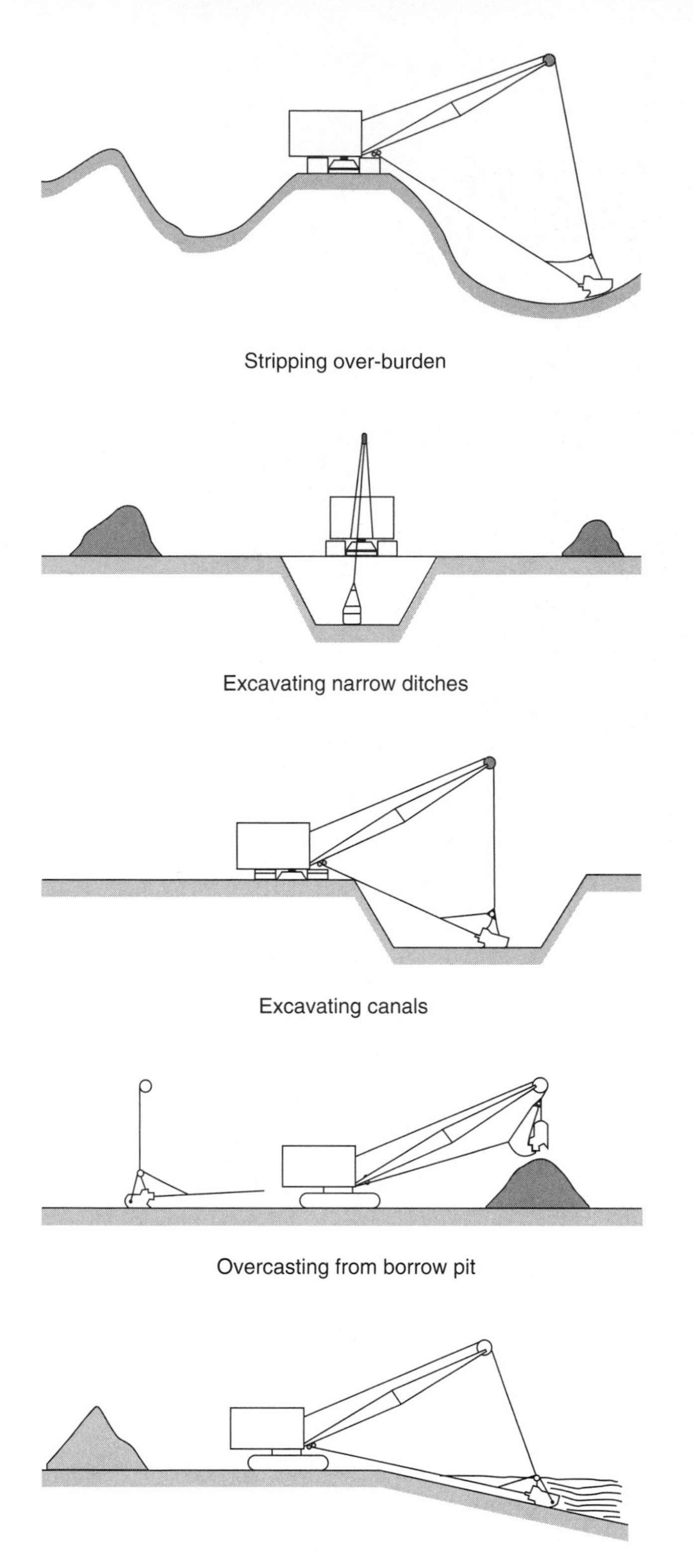


**Figure 3-40** Typical Dragline Operation. (Reprinted by permission of Power Crane & Shovel Association.)

The clamshell bucket consists of two scoops hinged in the center, with holding arms connecting the headblock to the outer ends of the scoops. A closing line is reeved through blocks on the hinge and the head so that the bucket closes when it is taken in. The bucket

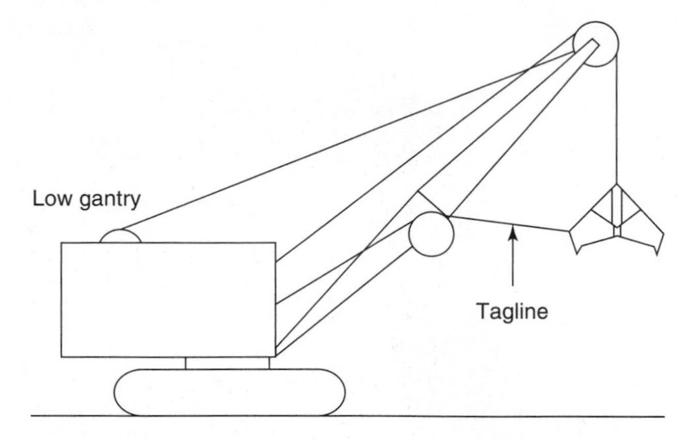


**Figure 3-41** Clamshell Bucket with Tagline. (Reprinted by permission of Power Crane & Shovel Association.)

is opened by releasing the closing and hoisting line while holding the bucket with the holding and lowering line. Clamshells are usually equipped with a tagline to control the swing of the bucket (see Figure 3-41).

Buckets are available in a wide variety of sizes and types. A heavy-duty type is available for digging, and lighter-weight ones for general-purpose work or handling light materials. Some buckets have removable teeth, used when digging in hard material, whereas others have special cutting lips.

## TRUCKS

No discussion of excavating machinery would be complete without mentioning the hauling units, usually trucks, with bodies made for that specific purpose. Figure 3-42 illustrates an off-highway heavy-duty hauler, with a protective canopy over the cab and a tapered back for easy dumping. Haulers of this type, with hauling capacities of 120 cu yd (92 $m^3$) and more, make short work of moving excavated material. Figure 3-43 illustrates a wheeled loader loading an articulated hauler that can provide power to all six axles when the terrain is rough or wet.

Trucks of this size may be classified in a number of ways, including the following:

1. Size and type of engine
2. Number of gears
3. Two-, four-, or six-wheel drive
4. Method of dumping
5. Capacity
6. Net weight and gross weight

Truck hauling capacities can be estimated for a particular job once site conditions are established. The type of material to be hauled, hauling distances, the type of terrain over which the truck must travel and, in some instances, the height above sea level all

**Figure 3-42** Volvo A35D Articulated Hauler. (Reprinted by permission of Volvo Construction Equipment of North America, Inc.)

**Figure 3-43** Articulated Hauler Being Loaded by a Wheeled Loader. (Reprinted by permission of Volvo Construction Equipment of North America, Inc.)

play a part in calculating the efficiency of the hauling unit. Turbocharged units are not affected by elevations up to 10,000 ft (3,000 meters); however, trucks with normal carburation systems will lose power at about 3% for every 1,000 ft (300 meters) of elevation gain.

Once a particular size of a hauling unit has been selected, the determination of the production capacity begins with the calculation of travel times for loaded and unloaded conditions. Manufacturers' specification sheets provide rated values for engine power, body-hoist speed, payload, service capacities, net mass of the unit, and the gross mass of the unit with payload. Manufacturers' computer programs relating *vehicle mass, coefficient of traction,* and *gradeability* to *rimpull values* and *speed,* provide specific load hauling data for each piece of equipment. These values provide the basis for the calculation of production capacities of a truck unit as dictated by site conditions.

Cycle travel time is based on the attainable speeds of the truck for loaded and unloaded conditions. Attainable speeds are affected by four factors: distance traveled, road resistance, grade resistance or assistance, and rimpull. Maximum traction (usable rimpull) is defined as the product of the total force on the drive axles and the coefficient of traction. In equation form it can be written as

$$RP_u = F \times CT$$

where $RP_u$ = maximum usable rimpull in pounds
$F$ = load on drive axles in pounds
$CT$ = coefficient of traction.

In metric, the units for usable rimpull and load on the drive axles can be expressed in kilograms or kilonewtons.

The coefficient of traction is based on the type of material over which the truck must travel in the loaded condition. In the case of rubber-tired vehicles, it represents the coefficient of friction for rubber tires on a particular road surface. Maximum *usable* rimpull represents the maximum amount of pull that can be developed before the drive wheels begin to slip.

The maximum *available* rimpull that a truck can develop is dictated by the size of the engine and the rimpull it can develop at stall. The *required* rimpull is based on the anticipated road conditions and is dependent on the weight of the truck, the rolling resistance, and the grade of the road. In equation form, it can be expressed as

$$RP_r = GW \times TR$$

where $RP_r$ = required rimpull in lb
$GW$ = gross loaded weight in lb
$TR$ = total travel resistance
= percent road grade + percent rolling resistance

When the truck is going downhill, the road grade aids the truck and the total travel resistance is reduced, resulting in a lower required rimpull. The total travel resistance in this instance can become travel assistance; that is, the truck will tend to roll down the grade and braking action will be required to maintain a speed that does not exceed the maximum operating speed of the truck. This travel assistance can be expressed as

$$TA = \text{percent road grade} - \text{percent rolling resistance}$$

Total cycle time is based on the attainable speed of the truck in each direction of travel. In the loaded condition, the attainable speed of the truck is usually slower than it is on the return trip when the truck is empty. Attainable speeds are based on the maximum speed of the truck in each direction modified by the attainable speed factor. The shorter the distance, the lower the attained speed because of lost time due to shifting, acceleration, deceleration, and stopping. Typical speed factors for converting maximum speeds to attainable speeds range from 0.4 to 0.85 depending on hauling conditions.

---

***Example:***

Determine the maximum usable rimpull for a truck with a gross loaded weight of 40 tons traveling on dry hard clay loam having a coefficient of traction of 0.60. The load on the drive axle is 60,000 lb. Determine the required rimpull assuming that the loaded truck must move the load up a 6% grade and the rolling resistance is found to be 2.5%. Based on a computer program estimate of 15 mph maximum speed in the loaded condition and an attainable speed factor of 0.66, calculate the attainable speed of the truck.

***Solution:***

Maximum usable rimpull in pounds is

$$60{,}000 \times 0.60 = 36{,}000 \text{ lb}$$

The required rimpull based on the given road conditions is

$$RP_r = GW \times TR$$

where $RP_r$ = required rimpull in lb
$GW$ = 80,000 lb
$TR$ = total travel resistance
= percent road grade + percent rolling resistance
$TR = 6 + 2.5 = 8.5\%$

Required rimpull

$$RP = 80{,}000 \times 8.5/100 = 6{,}800 \text{ lb}$$

The attainable speed is the product of the maximum speed and the attainable speed factor, or

$$0.66 \times 15 = 9.9 \text{ mph}$$

---

***Example:***

Based on the conditions in the preceding example, calculate the total cycle time for the truck, assuming a maximum unloaded speed of 30 mph down the slope, a resulting speed factor of 0.75, and a hauling distance of 2 miles one way. The net weight of the truck is 30,000 lb.

***Solution:***

The maximum attainable speed downhill is

$$0.75 \times 30 = 22.5 \text{ mph}.$$

The total cycle time is

$$\frac{2 \times 60}{9.9} + \frac{2 \times 60}{22.5} = 17.5 \text{ min}$$

---

***Example:***

Determine the maximum usable rimpull that a truck with a gross loaded mass of 60,000 kg can develop when traveling on soft clay having a coefficient of traction of 0.60. The load on the drive axle is 20,400 kg. Calculate the required rimpull assuming the truck must move up a 10% grade and the rolling resistance is found to be 5%. Based on the performance chart maximum speed of 10.5 km/hr and a speed factor of 0.5, calculate the attainable speed of the truck up the grade.

***Solution:***

Maximum usable rimpull expressed in kilograms is based on the load on the drive axle; in this case, it was given as 20,400 kg.

$$\begin{aligned}\text{Usable rimpull} &= 20{,}400 \times 0.60 \\ &= 12{,}240 \text{ kg}\end{aligned}$$

Expressed in kilonewtons, usable rimpull becomes

$$\frac{12{,}240 \times 9.81}{1{,}000} = 120 \text{ kN}$$

The required rimpull becomes

$$\text{RP} = \frac{60{,}000 \times 9.81}{1{,}000} \times \frac{(10 + 5)}{100} = 88 \text{ kN}$$

Using a factor of 0.5 for the foregoing example, the attainable speed becomes

$$0.5 \times 10.5 = 5.25 \text{ km/hr}.$$

---

***Example:***

Assuming a net mass of the truck as 30,000 kg, a rolling resistance of 3%, and a maximum speed of 40 km/hr on the return run, calculate the rimpull value in kilograms for the unloaded condition. Using the attainable speed calculated for the 10% road grade in the previous example, calculate the total cycle time based on a one-way distance of 3 km.

***Solution:***

Because the truck is now traveling down the slope, the road grade works in favor of the truck and the rimpull calculation actually determines the amount of braking action required to limit the speed of the truck. The travel assistance, in percent, is calculated as

$$TA = \text{road grade} - \text{rolling resistance}$$

The resulting rimpull is calculated as

$$RP = NW \times TA$$

where $RP$ = rimpull in kilograms
$NW$ = net or empty weight of truck, 30,000 kg
$TA$ = travel assistance

For the given values, travel assistance becomes

$$TA = 10 - 3 = 7\%$$

and the resulting required rimpull becomes

$$\begin{aligned}RP &= 30{,}000 \times \frac{7.0}{100} \\ &= 2{,}100 \text{ kg}\end{aligned}$$

Total cycle time is calculated by using the attainable speed for each leg of the hauling distance and determining the time it takes to cover the total distance. In the loaded condition, the maximum attainable speed was calculated as 5.25 km/hr. In the unloaded condition, the maximum attainable speed becomes

$$0.3 \times 40 = 12 \text{ km/hr}$$

$$\begin{aligned}\text{Total cycle time} &= \frac{3 \times 60}{5.25} + \frac{3 \times 60}{12} \\ &= 49.3 \text{ min}\end{aligned}$$

The preceding calculations are very simple examples of the types of calculations that are done when estimating production capacities of hauling units. With the aid of computer programs and manufacturers' specifications, these types of calculations can be made for any size of truck or hauling unit under any set of circumstances.

---

## CHOOSING EXCAVATION EQUIPMENT

A number of factors must be considered when deciding what type or types of excavating machinery will do the job most efficiently. The first point to consider is the volume of material to be removed. This will influence the size of the machines to be used. The depth of the excavation will determine the height to which material must be lifted to get it into trucks.

The disposal of the excavated material may influence the type of machinery used. If it can be deposited on a spoil bank at the site, one machine may do both digging and depositing. The distance from excavation to spoil bank is also a factor. On the other hand, if the material must be removed from the site, hauling units and special loading equipment may be required.

The type of soil to be excavated will influence the kind of equipment selected. Some machines work well in loose, dry soils but not in wet or highly compacted earth. The time allowed for excavation will also affect the type, size, and amount of equipment to be used. With such a wide range of equipment available for earth excavating and moving, the contractor must also have some method of evaluating each type and size of machine.

Many factors have a bearing on the type and size of machine that will be considered. The major considerations are:

1. Type of material to be excavated and hauled;
2. Site conditions;
3. Distance of haul;
4. Time allowed for job completion; and
5. Contract price.

With these requirements in mind, the contractor now must select equipment to complete the work within the required time span and yet realize the projected profit.

To estimate the hourly production of a piece of equipment, use the following formula:

$$P = \frac{E \times I \times H}{C}$$

where $P$ = production, cu yd/hr ($m^3$/hr) (in-bank)
$E$ = machine efficiency, min/hr
$I$ = shrinkage factor for loose material
$H$ = heaped capacity of machine, cu yd($m^3$)
$C$ = cycle time of the machine, min

The production or volume of material that a piece of equipment can move is based on the volume occupied by the material in its natural state or, as it is commonly referred, the *in-bank* condition of the material. Most materials, when disturbed, increase in volume, some by as much as 50% (see Table 3-1). To allow for this increase in volume, the shrinkage factor $I$ is applied to the heaped capacity $H$ of the earthmover to reduce the load to the in-bank condition. In other words, the in-bank volume is equal to the machine heaped capacity multiplied by the shrinkage factor. In equation form:

$$\text{in-bank machine capacity} = H \times I$$

The shrinkage factor $I$ can be calculated by the formula

$$I = \frac{1}{1 + \frac{\%\ \text{swell}}{100}}$$

---

***Example:***

Calculate the shrinkage factor for clay based on the values in Table 3-1.

***Solution:***

From Table 3-1, the percent swell for clay is 25%.

$$I = \frac{1}{1 + \frac{25}{100}}$$
$$= 0.8$$

---

As no machine is 100% efficient, the efficiency of the machine must also be included in the calculations. Average efficiencies for various types of equipment are as follows:

- *Crawler tractor equipment:* 50 min/hr
- *Rubber-tired hauling units:* 45 min/hr
- *Large rubber-tired loaders and dozers:* 45 min/hr
- *Small rubber-tired loaders:* 50 min/hr

That is, for every hour on the job, a piece of equipment is between 75% and 85% efficient. Some equipment can be as much as 90% efficient, depending on the material being handled.

The *cycle time* of a piece of equipment is based on the time required to obtain its load, move it to its dumping point, and return to the loading point. The cycle time is based on the sum of two factors: *travel time factors* and *fixed time factors.*

Travel time factors relate to the movement of the machine (usable speeds and pulls, load resistance, travel resistance, and machine weight). Fixed time factors are based on the characteristics built into the machine (spotting, loading, turning, dumping, and reversing). Therefore, the total cycle time is equal to cycle travel time plus cycle fixed time, or

$$C = CT + CF$$

The travel time ($CT$) in minutes can be calculated by

$$CT = \frac{D}{S \times 88}$$

where $D$ = distance traveled, feet
$S$ = speed, miles per hour
88 = distance moved, feet per minute when traveling 1 mph,

In metric units the formula is very similar and is expressed as

$$CT = \frac{D}{S \times 16.67}$$

where $D$ = distance traveled, m
$S$ = speed, km/hr
16.67 = distance moved, m/min, when traveling 1 km/hr

The fixed time ($CF$) depends on the piece of equipment being considered. All major suppliers have charts and data for the various pieces of equipment, which enable the contractor to establish these factors. Fixed cycle times can vary from 0.25 min to over 5 min, depending on the circumstances.

---

***Example:***

A crawler-mounted power shovel having a 1½-cu yd bucket is to load well-blasted rock into trucks. If the machine efficiency is 50 min/hr, establish the total cycle time. Use Table 3-2 for production data.

***Solution:***

From Table 3-1, the percent increase in volume for rock is 50%. The resulting shrinkage factor is

$$I = \frac{1}{1 + \frac{50}{100}} = 0.67$$

Using the production formula,

$$P = \frac{E \times I \times H}{C}$$

Rewrite in terms of $C$:

$$C = \frac{E \times I \times H}{P}$$

From Table 3-2(a), for a shovel with a 1½-cu yd bucket, $P$ = 180 cu yd/hr. With $I$ = 0.67, $H$ = 2 cu yd, and $E$ = 50 min/hr.

$$C = \frac{50 \times 0.67 \times 2}{180}$$

$$= 0.37 \text{ min}$$

Such a short cycle time indicates that ideal conditions exist for the power shovel (see conditions for Table 3-2(a)), and the total cycle time calculated is based solely on the cycle fixed time. If the shovel was required to travel some distance between loading and dumping, the cycle time would increase correspondingly.

---

***Example:***

A crawler-mounted power shovel having a 1.5 $m^3$ bucket is to load poorly-blasted rock into trucks. If the machine efficiency is 50 min/hr, establish the total cycle time. Use Table 3-2(b) for production data.

***Solution:***

From Table 3-1, the percent increase in volume for rock is 50%. The resulting shrinkage factor is

$$I = \frac{1}{1 + \frac{50}{100}} = 0.67$$

Using the production formula,

$$P = \frac{E \times I \times H}{C}$$

Rewrite in terms of $C$:

$$C = \frac{E \times I \times H}{P}$$

From Table 3-2(b), for a shovel with a 1.5 $m^3$ bucket, $P$ = 122 $m^3$/hr with $I$ = 0.67, $H$ = 1.5 $m^3$, and $E$ = 50 min/hr:

$$C = \frac{50 \times 0.67 \times 1.5}{122}$$

$$= 0.49 \text{ min}$$

Again, the cycle time is relatively short. However, the impact of the poorly blasted rock is reflected in the lower production value for the shovel, probably due to a longer loading time.

---

***Example:***

A rubber-tired scraper with a heaped capacity of 30 cu yd has a cycle fixed time of 1.0 min (loading, dumping, and turning). If the average speed of the scraper is 3.5 mph and it must travel a 0.75-mile round trip, calculate the anticipated production of the machine if it is hauling clay with a swell factor of 25%. Assume a machine efficiency of 50 min/hr.

***Solution:***

$$\text{Production } P = \frac{E \times I \times H}{C}$$

**Table 3-2(a)**
**HOURLY SHOVEL HANDLING CAPACITY FOR SHOVEL DIPPER SIZES (CUBIC YARDS)**

| Class of Material | ⅜ | ½ | ¾ | 1 | 1¼ | 1½ | 1¾ | 2 | 2½ |
|---|---|---|---|---|---|---|---|---|---|
| Moist loam or sandy clay | 85 | 115 | 165 | 205 | 250 | 285 | 320 | 355 | 405 |
| Sand and gravel | 80 | 110 | 155 | 200 | 230 | 270 | 300 | 330 | 390 |
| Good common earth | 70 | 95 | 135 | 175 | 210 | 240 | 270 | 300 | 350 |
| Clay, hard, tough | 50 | 75 | 110 | 145 | 180 | 210 | 235 | 265 | 310 |
| Rock, well-blasted | 40 | 60 | 95 | 125 | 155 | 180 | 205 | 230 | 275 |
| Common earth, with rock and roots | 30 | 50 | 80 | 105 | 130 | 155 | 180 | 200 | 245 |
| Clay, wet and sticky | 25 | 40 | 70 | 95 | 120 | 145 | 165 | 185 | 230 |
| Rock, poorly blasted | 15 | 25 | 50 | 75 | 95 | 115 | 140 | 160 | 195 |

[a]Conditions:
1. Cu yd bank measurement per hour.
2. Suitable depth of cut for maximum effect.
3. Continuous loading with full dipper.
4. 90° swing, grade-level loading.
5. All materials loaded into hauling units.

*Source:* Reproduced by permission of Power Crane and Shovel Association.

**Table 3-2(b)**
**HOURLY SHOVEL HANDLING CAPACITY FOR SHOVEL DIPPER SIZES (CUBIC METRES)**

| Class of Material | 0.30 | 0.40 | 0.60 | 0.80 | 1.00 | 1.15 | 1.33 | 1.50 | 1.90 |
|---|---|---|---|---|---|---|---|---|---|
| Moist loam or sandy clay | 65 | 88 | 126 | 157 | 191 | 218 | 245 | 271 | 310 |
| Sand and gravel | 61 | 84 | 119 | 153 | 176 | 206 | 229 | 252 | 298 |
| Good common earth | 54 | 73 | 103 | 134 | 161 | 184 | 206 | 229 | 268 |
| Clay, hard, tough | 38 | 57 | 84 | 111 | 138 | 161 | 180 | 203 | 237 |
| Rock, well-blasted | 31 | 46 | 73 | 96 | 119 | 138 | 157 | 176 | 210 |
| Common earth, with rock and roots | 23 | 38 | 61 | 80 | 99 | 119 | 138 | 153 | 187 |
| Clay, wet and sticky | 19 | 31 | 54 | 73 | 92 | 111 | 126 | 141 | 176 |
| Rock, poorly blasted | 11 | 19 | 38 | 57 | 73 | 88 | 107 | 122 | 149 |

[a]Conditions:
1. Cubic metre bank measurement per hour.
2. Suitable depth of cut for maximum effect.
3. Continuous loading with full dipper.
4. 90° swing, grade-level loading.
5. All materials loaded into hauling units.

*Source:* Original data provided by the Power Crane and Shovel Association; metric conversion by author.

Total cycle time = cycle travel time + cycle fixed time or

$$C = CT + CF$$

$$\text{Cycle travel time } CT = \frac{D}{S \times 88} = \frac{0.75 \times 5{,}280}{3.5 \times 88} = 12.9 \text{ min}$$

Cycle fixed time $CF = 1.0$ min

Total cycle time $C = 12.9 + 1.0 = 13.9$ min

$$\text{Shrinkage factor } I = \frac{1}{1 + \frac{\%\text{ swell}}{100}} = \frac{1}{1 + \frac{25}{100}} = 0.80$$

Given $E = 50$ min/hr and $H = 30$ cu yd:

$$P = \frac{50 \times 0.80 \times 30}{13.9} = 86.3 \text{ cu yd/hr (in bank)}$$

---

***Example:***

A rubber-tired scraper with a heaped capacity of 22 m$^3$ has a cycle fixed time of 1.0 min (loading, dumping, and turning). If the average speed of the scraper is 5.5 km/hr and it must travel a 1.2 km round trip, calculate the anticipated production of the machine if it is hauling clay with a swell factor of 25%. Assume a machine efficiency of 50 min/hr.

***Solution:***

$$\text{Production} \quad P = \frac{E \times I \times H}{C}$$

Total cycle time = cycle travel time + cycle fixed time or

$$C = CT + CF$$

$$\text{Cycle time travel } CT = \frac{D}{S \times 16.67}$$

$$= \frac{1.2 \times 1{,}000}{5.5 \times 16.67}$$

$$= 13.0 \text{ min}$$

Cycle fixed time $CF = 1.0$ min

Total cycle time $C = 13 + 1.0 = 14.0$ min

$$\text{Shrinkage factor } I = \frac{1}{1 + \frac{\%\text{ swell}}{100}}$$

$$= \frac{1}{1 + \frac{25}{100}}$$

$$= 0.8$$

Given $E = 50$ min/hr and $H = 22\ \text{m}^3$:

$$P = \frac{50 \times 0.80 \times 22}{14}$$

$$= 62.9\ \text{m}^3/\text{hr (in bank)}$$

## PROTECTION OF EXCAVATIONS

The protection of an excavation means taking steps to ensure that sidewall cave-ins do not occur, necessitating costly work in removing fallen earth or causing damage to forms or injury to workers. The amount of protection required depends on the *type of soil being excavated,* the *depth of the excavation,* the *level of the water table,* the *type of foundation* to be built, and the *available space* around the excavation.

The sides of an excavation dug in dry, compact, stable soil may need no protection at all, if it is not too deep. In the excavation shown in Figure 3-44, for example, the sides are standing unprotected. Notice that near the top, where the soil is looser, the sides have been sloped to prevent cave-in. It is also evident that there is sufficient area for the excavation sides to be kept back from the main working area. One reason for the stability of the soil in this case is that the area has been kept dry by a dewatering system.

**Figure 3-44** Freestanding Excavation Walls. (Reprinted by permission of Moretrench American Corporation.)

**Figure 3-45** Excavation to Property Lines. (Reprinted by permission of Spencer, White & Prentis Foundation Corporation.)

In less stable soils the excavation must be protected against cave-in, and this can be done in one of two ways. One is to slope the sides until the angle of repose for that particular soil is reached. In some cases this may mean a slope of 1:1, or 45°. This method may not be possible where space is limited or when the extra cost of excavation is prohibitive. The alternative is to provide temporary support for the earth walls.

Many large buildings being built in the urban areas of the country today have large, heavy foundations, which are often constructed in a deep excavation. In addition, many are built in congested or restricted areas (see Figure 3-45), making very little more space available for the excavation than will be occupied by the building. The foundations of many large buildings are located in difficult earth, along waterways, in low-lying areas, or on filled land. These factors make it necessary that large, deep excavations (see Figure 3-46), often extending to the property limits, such as the excavation shown in Figure 3-47, be carefully planned.

Most methods used for supporting the earth walls around the perimeter of an excavation involve a system of sheeting supported by vertical supports. The basic systems are *interlocking steel sheet piling, steel soldier piles with horizontal timber sheeting* (see Figure 3-48), and *concrete slurry walls*. Only in very shallow excavations would ordinary plank sheeting be used.

**Figure 3-46** Deep Excavation.

**Figure 3-47** Steel WF Sections and Timber Planks Used in the Protection of an Existing Building Foundation at the Edge of a New Excavation.

**Figure 3-48** I-shaped Steel Sections and Horizontal Timbers Restrain the Soil at the Edge of the Excavation.

## Interlocking Sheet Piling

Interlocking steel sheet piling (Figure 3-49) is used in situations when seepage into the excavation must be controlled (see Figure 3-50), or when structural elements must be constructed below the water table. It can be used as a permanent installation or removed once the work has been completed. The driving of sheet piling is difficult in dense, rocky material and extraction may be a major problem once construction is completed. To ensure that maximum watertightness is achieved between sections, proper alignment of the sections during placement is critical.

In applications where the groundwater level is high or the excavation is near the water's edge, sufficient bracing must be provided to ensure that the sheet piling is not forced into the excavation by the resulting horizontal pressure. However, in most instances, when the piling is properly driven, substantial pressures can be tolerated without additional bracing.

## Steel Soldier Piles

The use of soldier piles and horizontal timber sheeting is an economical method for retaining excavation walls and is adaptable to many situations. The method consists of driving steel W shapes, 6 to 10 ft (1.8 to 3 m) o.c., around the perimeter of the excavation and placing heavy wood planks, 3 to 4 in. (75 to 100 mm) in thickness, horizontally between the flanges (see Figure 3-51). If the planks are spaced 1 to 2 in. (25 to 50 mm) apart, the buildup of water pressure behind the sheeting will be eliminated, and the earth, remaining drier, will have increased shear strength, thus reducing the tendency to slide. Such

**Figure 3-49** Interlocking Steel Sheet Piling.

**Figure 3-50** Steel Sheet Piling Providing Protection for Excavation Near Water's Edge.

soldier piles can penetrate denser material than sheet piling and can be extended by welding, if necessary.

In some cases, the soldier piles are installed in drilled holes, which are backfilled with lean concrete after the timber has been installed. Concrete planks may replace timber ones, if necessary.

If the earth will stand unsupported for a span of about 3 ft (900 mm), an alternative to using sheeting is to place piles on those centers and spray the wall surface with gunite to hold the earth in place. Wire mesh may be used in place of gunite, in which case the surface may be sprayed with an asphalt emul-

**Figure 3-51** Soldier Piles and Horizontal Timbers Protect Foundation Formwork.

sion. In such situations, concrete piles may be substituted for steel ones.

## Concrete Slurry Walls

The technique of building slurry walls involves the casting of concrete walls, from 18 in. to 5 ft (450 mm to 1.5 m) in thickness and up to 400 ft (120 m) in depth below grade around one or more sides of an excavation. The cast-in-place, reinforced concrete wall is placed in sections or *panels,* usually not exceeding 25 ft (7.5 m) in length and the full depth required. The completed wall may serve a dual purpose by acting as a retaining wall during the excavating phase and while the interior foundations are being built and later being used as the permanent exterior foundation walls for the building.

The wall may be built in consecutive or alternating panels, and the procedure will vary somewhat, depending on which system is used. Under the consecutive method (see Figure 3-52), after the first panel is complete, others are constructed next to it, on either side and, except for the first one, all the panels will be the same with regard to reinforcing and end forming.

When the alternating panel method is used, *primary panels* are completed with sections of equal size left unexcavated between them (see Figure 3-53). When the primary panels are finished, the *secondary*

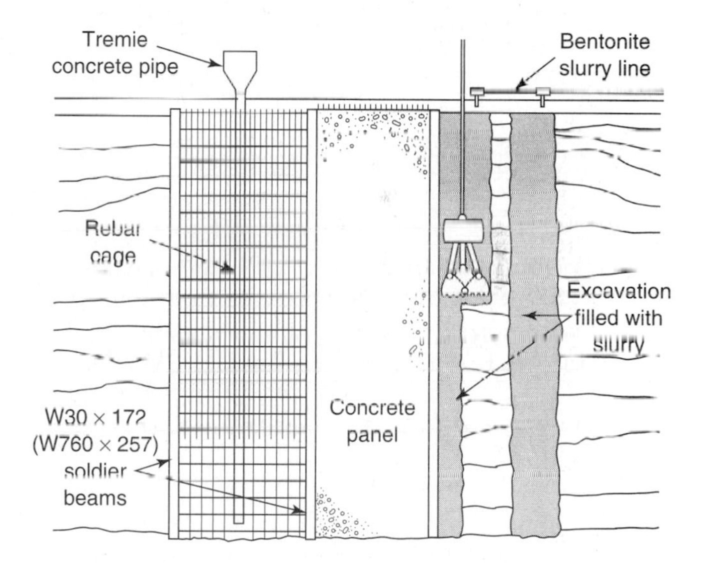


**Figure 3-52** Consecutive Slurry Wall Panels under Construction. (Reprinted by permission of I.C.O.S. Corp. of America.)

*panels* are excavated and concreted to form a continuous wall.

The excavation consists of a trench, with the choice of digging equipment depending on the soil involved. In soft to medium hard, loose, or cohesive soils, the popular digging tool is a clamshell bucket, similar to those illustrated in Figure 3-54. The alignment of the trench is controlled by two concrete guide walls, about 1 ft (300 mm) wide and 3 to 5 ft

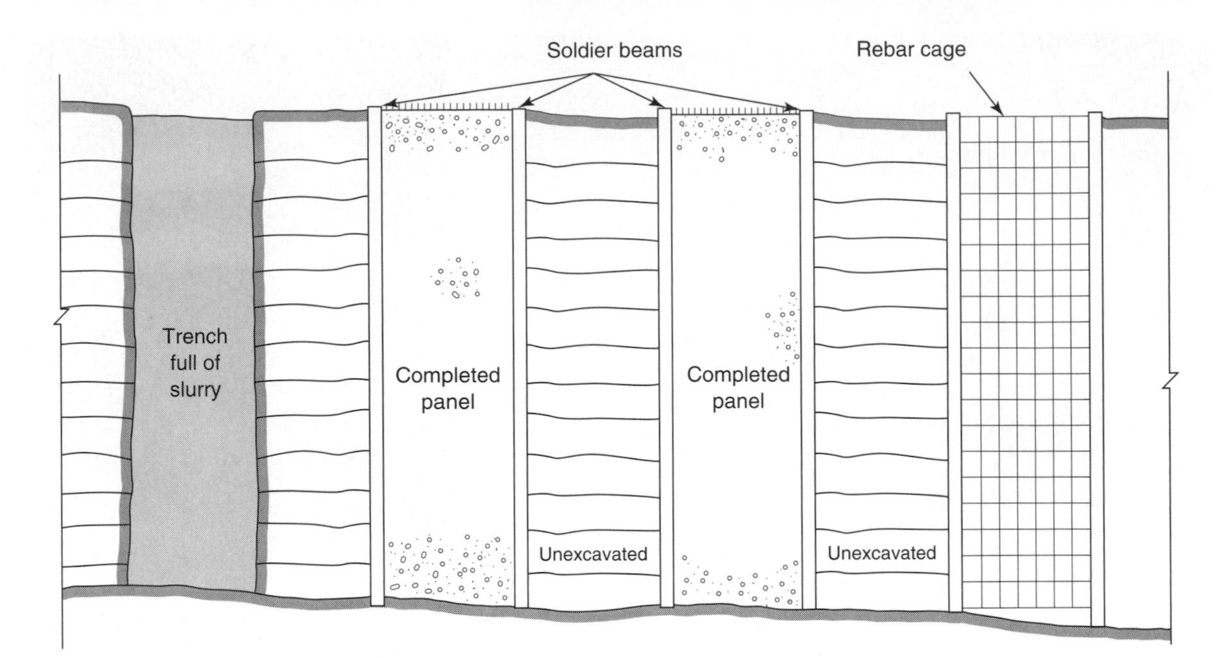


**Figure 3-53** Alternating Slurry Wall Panels Under Construction.

(900 to 1,500 mm) deep, spaced apart the width of the trench and placed at grade level.

As material is excavated from the trench, bentonite slurry is immediately piped in, with the level of the slurry being maintained close to the surface and at least 2 ft (600 mm) higher than the highest level of groundwater to produce a positive static pressure on the walls of the excavation.

Under one method of construction, when one primary panel is completely excavated, two large-diameter pipes, called *end pipes,* are placed, one at each end of the trench. A rigid, steel reinforcing cage, fabricated on the site, is lifted and placed into the slurry-filled trench. Concrete is then placed in the panel by the tremie method, displacing the slurry, which is pumped into tanks to be reused. When the concrete has partially set, the end pipes are slowly withdrawn, leaving a semicircular concrete key at each end of the panel. In some cases, predriven steel HP shapes may be used to form a mechanical connection between panels.

Another method of providing panel end connections and reinforcing at the same time is illustrated in Figure 3-55. Two long W shapes, with a depth equal to the width of the trench and a length equal to the depth of the trench, are welded together with a reinforcing steel cage between them (see Figure 3-56). This unit is lowered into the slurry in the trench, and concrete is placed by tremie. If panels are being placed by the alternating method, only the steel reinforcing cage is required in the secondary panels.

In the consecutive panel method, only one end beam and a reinforcing cage are required for each successive panel after the first one is finished. In Figure 3-57, such an end beam is being installed. Notice that on its outside face the space between the flanges of the end beam has been filled with pieces of thick-foamed insulation. This prevents any concrete that escapes around the edges of the end beam from adhering to the face and thus reducing the concrete bond for the next panel. Before the next panel is placed, the packing is removed by ripping claws attached to one end of a clamshell bucket.

In soils containing boulders or similar obstructions, a series of holes is predrilled at short intervals along a panel, and then the material can be excavated with one of the heavy clamshell buckets illustrated.

The slurry mixture is made of bentonite and water, with a relative density of 1.05 to 1.10. It penetrates around soil particles and gels when left undisturbed. The gelled slurry maintains the soil particles in position by adhesion and creates a zone of stable soil on both sides of the trench. The lateral depth of soil so stabilized depends on the permeability of the soil and will range from several feet in loose material to very little in dense clay. The fact that the reinforcing is installed in slurry does not appreciably affect its performance, and good bond stresses with the concrete are obtained.

Slurry walls are unlikely to be competitive in price with steel sheeting or soldier piles in soils in which driving is relatively easy or in areas that may be easily dewatered and excavated. They do, however, provide a viable alternative when land is wet, the digging is difficult, or the depth is beyond an economical limit for sheeting or piles.

A new technique has recently been developed in which reinforced precast concrete walls are installed

(a)

(b)

(c)

**Figure 3-54** (a) Special Clamshell Bucket; (b) 9-Ton Clamshell Bucket; (c) Bucket between Concrete Guide Walls. (Reprinted by permission of I.C.O.S. Corp. of America.)

in the slurry-filled trenches. The slurry, in this case, is made up of a mixture of bentonite, cement, and additives with water. It will remain fluid until after the precast walls have been positioned, but sets up later to hold the panels in place. The panels have a tongue-and-groove joint to make a continuous wall, and anchors may be cast into them, similar to standard slurry walls.

## Top-Down Construction

A recent innovation used in conjunction with the slurry diaphragm wall method is the concept of top-down construction. Several variations of this method have been used depending on site conditions; however, its execution is based on a combination of foundation engineering and mining construction techniques that allows for the simultaneous construction of the superstructure and the below-grade excavation of the building.

First, the building perimeter is enclosed with slurry walls, which will also serve as the exterior foundation walls of the structure. To provide support for the interior columns of the superstructure, caissons with temporary steel liners are drilled to a load-bearing stratum and filled with concrete to the underside of the

**Figure 3-55** End Beams and Reinforcing Cage Welded Together. (Reprinted by permission of I.C.O.S. Corp. of America.)

**Figure 3-56** Unit Being Lowered into Trench. (Reprinted by permission of I.C.O.S. Corp. of America.)

lowest floor slab in the building (see Figure 3-58). Load-bearing column sections are then inserted into the caissons and the hollow shafts are backfilled with sand. The caisson liners are then removed.

**Figure 3-57** End Beam Being Lowered into Trench. (Reprinted by permission of I.C.O.S. Corp. of America.)

Before excavation begins, a mud slab is poured at grade level on which the ground-floor structural slab is poured. The mud slab provides a base on which to pour the structural floor slab and ensures a smooth finish on the underside. Openings are provided in the structural slab to allow equipment to excavate and remove the material below the slab. Excavation continues until the next floor level is reached, where the cycle is then repeated. Structural steel beams with metal decking and a concrete slab may be used instead of the reinforced concrete structural slab with equally good results.

Several advantages can be realized using this approach to the excavation of deep basements. Where the site is surrounded by other buildings, the procedure minimizes the possibility of settlement during open excavation; the slurry wall can be built in one continuous operation from existing grade, reducing construction time; and the slurry wall provides its own waterproofing and allows for the removal of water within the building perimeter without disturbing the level of the water table outside.

In addition, as the excavation proceeds downward, the floor slabs provide lateral support for the slurry walls, thus eliminating the need for tieback anchors. Because the load-bearing columns are in place, the

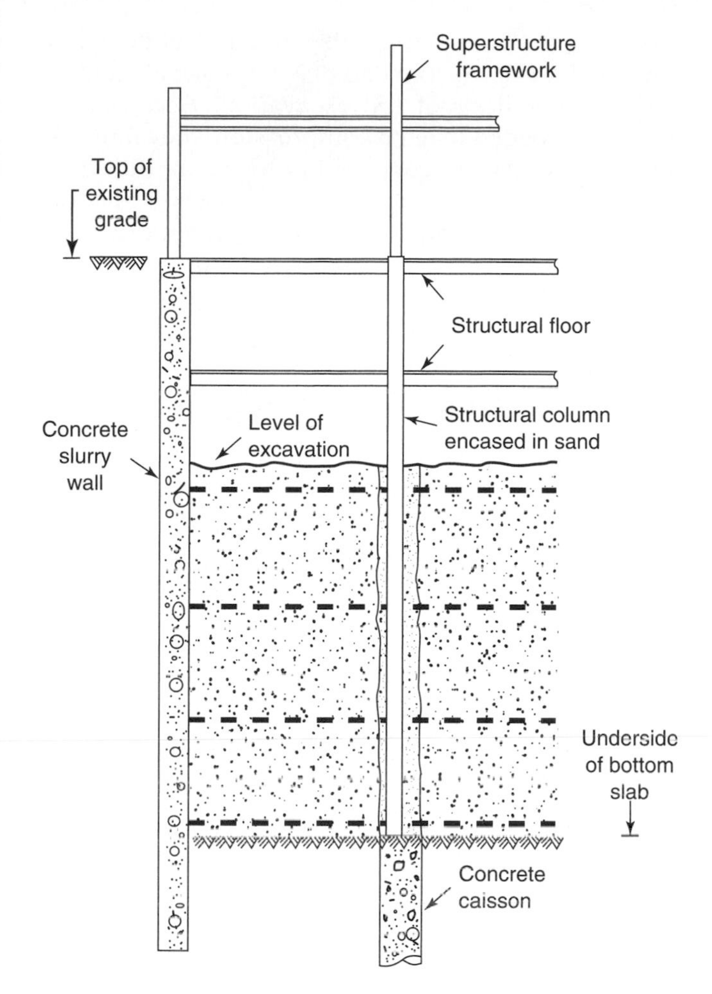


**Figure 3-58** Top-Down Construction.

erection of the structural frame above ground can proceed at the same time that the mining operation is in progress below grade, thus allowing two construction crews to work at the same time in opposite directions.

# EXCAVATION WALL BRACING

No matter what restraining system is used to stabilize the sides of an excavation, the walls will have to be braced against the lateral pressures exerted by the surrounding soil and water, as well as any superimposed loads due to construction equipment. The importance of proper bracing cannot be overemphasized. As with all components of the shoring system, the bracing must be carefully designed and installed to do its job properly. Failure of the braces will, at best, cause a good deal of reconstruction and expense. At the worst, it will cause injury and death to workers.

Government agencies such as OSHA, the National Bureau of Standards, the Bureau of Reclamation, and The Army Corps of Engineers provide guidelines for the proper selection, design, and construction of all shoring components.

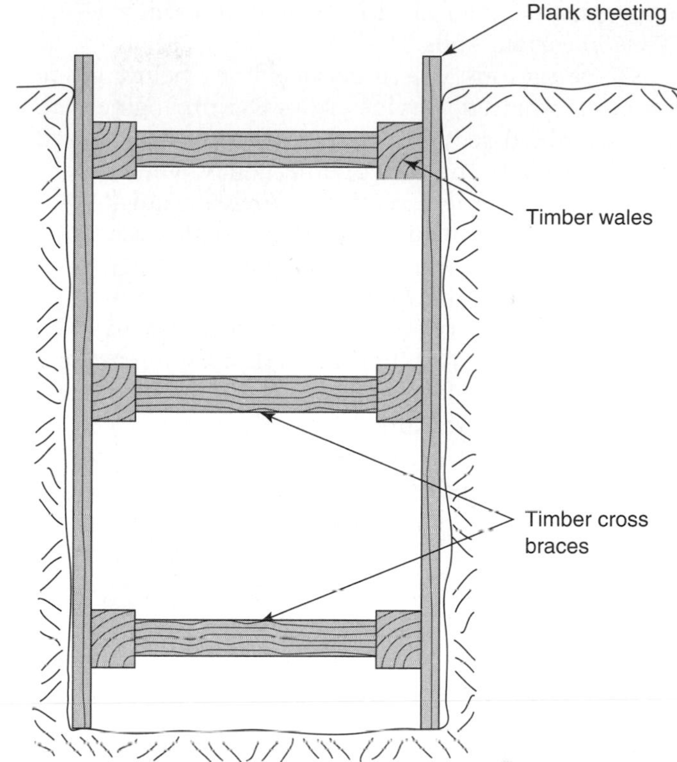


**Figure 3-59** Timber Shoring for Trenches.

## Shoring for Trenches

Trenches are used for the placement of building services such as water, sewer, and electricity. In some instances, reinforced concrete service tunnels, used to enclose the service lines, are constructed in a trench. A trench is a special type of excavation in that it is relatively narrow in relation to its depth. Trenches can vary in depth from 4 ft (1.2 m) to 20 ft (6 m) and can be as narrow as 2 ft (600 mm).

Because trenches are relatively narrow and deep, they pose a significant hazard to site personnel who must work in the confined space. In the event of a collapse, injury and death to individuals working in the trench are common occurrences. It is an understatement to say that proper design and construction of the shoring and bracing are of the utmost importance.

Materials for shoring vary depending on the soil conditions and the depth and width of the trench. For relatively narrow trenches, timber shoring and bracing can be used effectively. A common approach to timber shoring is to line each side of the trench with vertical plank sheeting secured to timber wales (see Figure 3-59). Timber cross braces of appropriate length and size are then wedged between the wales to restrain the horizontal pressure of the soil. For ease of installation, trench jacks, rather than timbers, may also be used as bracing members. OSHA provides excellent guidelines and data for the selection of appropriate materials for typical trenching situations in various soils.

In situations where service lines below grade must be repaired, service crews use preengineered, preassembled sections of shoring that are lowered into the trench. In times of emergency, when time is of the essence, these sections provide a quick solution and ensure maximum safety for the crew during the course of the repairs. The need for proper design and installation of all shoring cannot be overstated. Supervisory personnel must be well versed in proper installation of shoring to ensure a safe working environment for the crew who must work in the excavation.

Where soil is extremely unstable due to water seepage, other methods of shoring can be used. Steel sheet piling can be used effectively to control water seepage, however, it is much more costly to install and remove. Stabilizing the soil by freezing is also a viable alternative in extreme cases.

## Internal Sloping Braces

Where lateral restraint is required quickly or for a short period of time, a practicable method of supporting the walls of an excavation is to use *internal sloping braces.* Steel or timber columns spaced at intervals around the perimeter of the excavation are held vertical by one or more such brace. The top end of the brace is secured to the top of the vertical column and, if necessary, another brace is attached at mid height. The bottom end of each brace is then secured to the floor of the excavation. Because the braces are loaded in axial compression, they must be relatively robust in cross section to ensure that they do not buckle under the lateral thrust of the restrained soil. This method of bracing is not practical for deep excavations because the bracing system becomes cluttered and hinders the movement of equipment and materials.

## Horizontal Diagonal Bracing

One result of using internal bracing is that it interferes with the excavating process. One way of partially overcoming this problem is to replace the top sloping braces with two or more rows of *horizontal, diagonal corner braces,* as illustrated in Figure 3-60. In this case, the bracing is all steel pipe, the bottom ones sloping and the top ones (two rows of pipe) set diagonally into the corners. Such diagonal bracing should not be used where the angles are greater than 90°.

## Tiebacks

With the development of rock and earth tiebacks, internal bracing systems have been replaced, to a great extent, with *tieback bracing systems.* Although they usually are more expensive to install,

**Figure 3-60** Excavation with Horizontal Bracing. (Reprinted by permission of Spencer, White & Prentis Foundation Corporation.)

the main advantage of a tieback bracing system over internal bracing is that it provides a clear working space within the excavation (see Figure 3-45). Basically, a tieback system involves the use of steel rod anchors grouted into the earth or rock outside the excavation wall. The other end is secured to the restraining structural sections of the excavation wall (see Figure 3-61). In the excavation illustrated in Figure 3-46, the rod tiebacks extend through horizontal double-channel wales to which they are anchored using a large plate washer and nut. Tieback systems are most effective in firm ground such as dense clay, hardpan, glacial till, or rock. Difficulties may be experienced in soft clays or unconsolidated granular materials where anchoring the rods may be difficult.

One disadvantage with tieback systems is that they normally extend beyond the property lines of the building site. Care must be taken when installing the anchors to ensure that existing subterranean service lines around the building site perimeter are not disrupted. Permission must be obtained from all affected individuals before any such work begins.

Once again it must be remembered that tieback systems extend beyond the property lines, and their installation may be hindered by existing structures and services. A thorough site investigation must be conducted prior to any construction.

## Rock Tiebacks

When rock tiebacks are designated, 4- to 6-in. (100- to 150-mm) pipe is driven through the earth overburden at a 45° angle until rock is reached using conventional drilling equipment. Then a 3½-inch (90 mm) hole, up

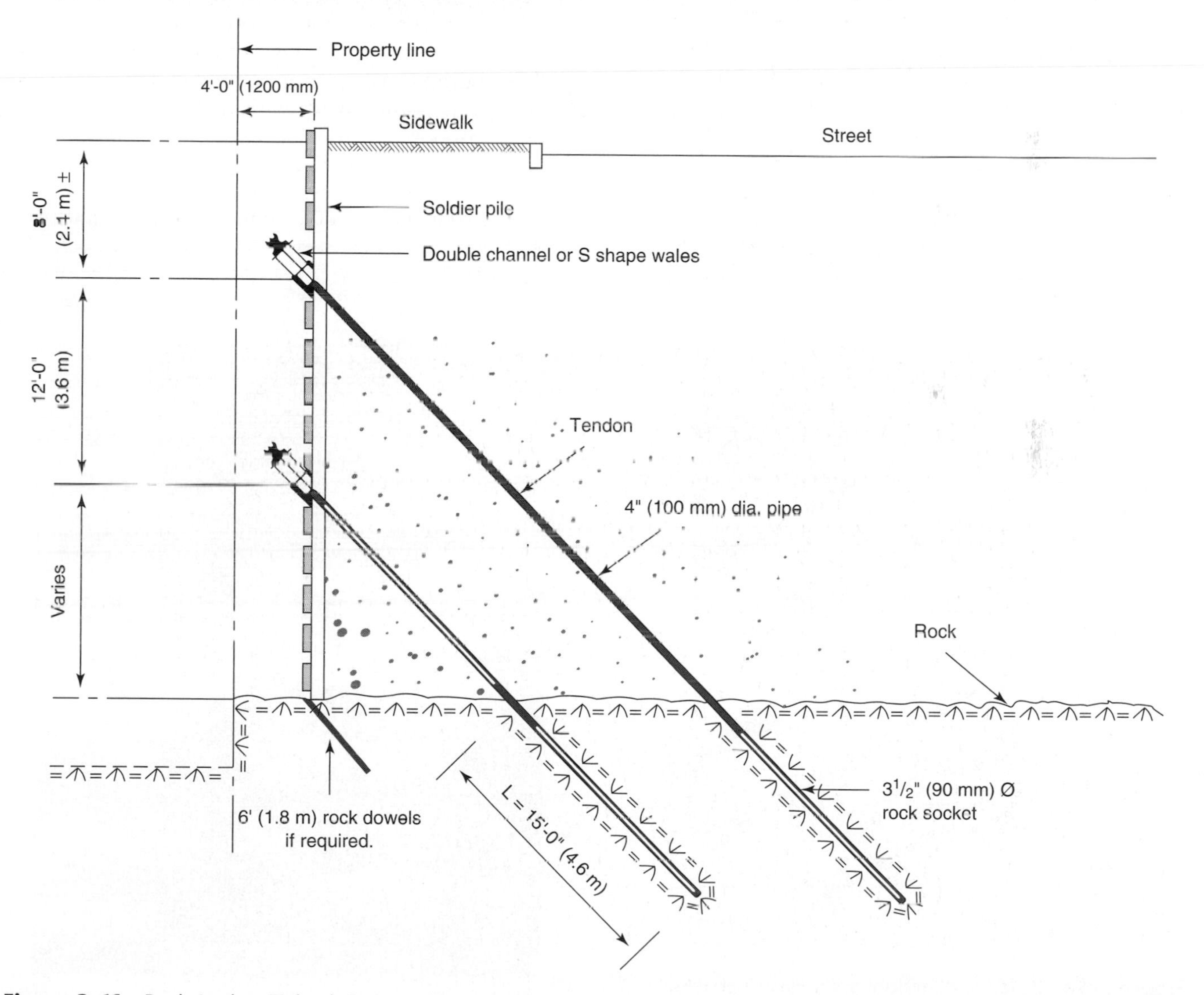


**Figure 3-61** Rock Anchor Tieback System. (Reprinted by permission of Spencer, White & Prentis Foundation Corporation.)

to 15 ft (5 m) long, is drilled into the rock, and high tensile rod or wire strands are grouted into the hole and tested to a predetermined working stress (see Figure 3-61). The top end of the rod or wire may be anchored to a concrete wall, passed through a hole in a soldier pile, and anchored on the inside or passed through sheet piling and anchored to a waler.

### Earth Tiebacks

Earth tiebacks are used where no rock is present for anchorage. They are more difficult to use because of the lower anchorage values, but they use a flatter angle of penetration, 15° to 30° from the horizontal being common. There are two types of earth tiebacks: *large diameter* and *small diameter.* Large-diameter tiebacks require a tie hole about 1 ft (300 mm) in diameter, to 65 ft (20 m) deep, belled at the bottom (see Figure 3-62). When the required depth is reached, rod or cable is inserted into the hole, which is then grouted to the *slip plane* (see Figure 3-62). In some cases the bell at the end of the hole is eliminated, the hole is made longer, and then the ties depend on the friction between concrete and earth for their anchor.

Small-diameter tiebacks are used in earth that tends to cave during drilling. This technique consists of installing 4 in. (100 mm)-diameter steel casing using rotary drilling equipment. When the casing has reached the required depth, rod or cable is inserted and grouted to the slip plane, the casing being extracted during the grouting operation. This leaves the grout in contact with the earth to provide a friction anchor. In some cases, the casing may be left in the ground, and anchorage is provided by the friction between earth and steel.

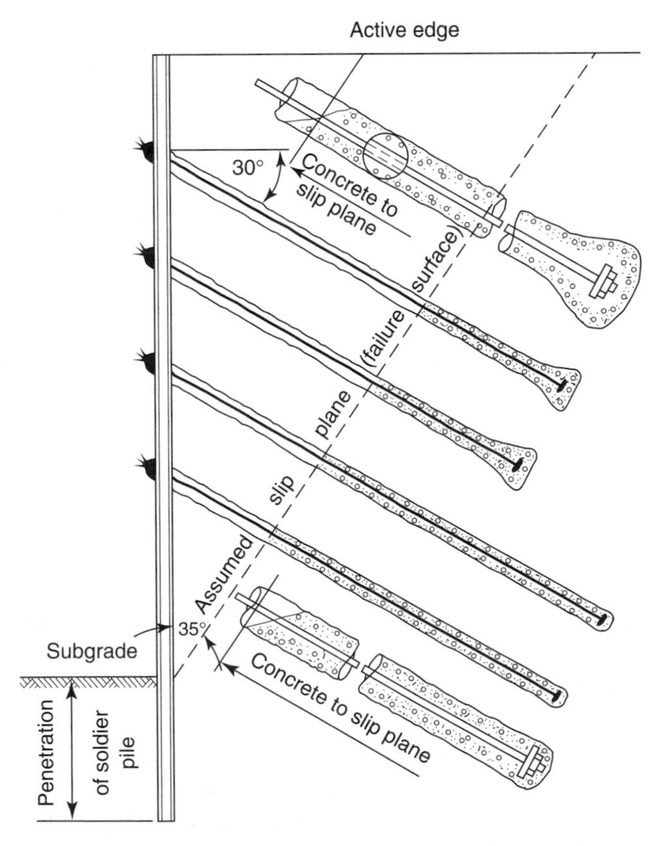


**Figure 3-62** Belled and Straight Shaft Earth Tiebacks. (Reprinted by permission of Spencer, White & Prentis Foundation Corporation.)

## PROTECTION OF ADJACENT BUILDINGS

In many instances, it is necessary to carry the excavation for a new building right up to the foundations of one or more existing buildings (see Figure 3-63). This presents a problem, particularly if the new excavation is to be deeper than the foundations of the existing building. Part of the support for that foundation will be removed, and it is the responsibility of the builder to protect the building against movement caused by settlement during and after construction of the new building. Temporary support may be provided by *shoring* or *needling,* if the building is relatively light, whereas permanent support is provided by *underpinning,* or ex-

**Figure 3-63** Support for Existing Building Foundations During New Excavation.

tending the old foundation down to a new bearing level. In the case of heavy buildings, underpinning will usually be installed during an early stage of excavating, without shoring or needling.

## Shoring

Shoring is the simplest method of providing temporary support for an existing building, but it is only useful for light structures. It involves only the use of shoring members and jacks (see Figure 3-64). If they are short, the shores may be timber; otherwise, they will be HP shapes or steel pipe. Screw jacks are most suitable if the pressure is to remain for an extended period of time.

There must be solid bearing for the end of the shoring member near the top of the wall being supported. One simple method is to use a waler bolted to the face of the wall, as illustrated in Figure 3-64.

The lower end of each shore rests on a jack head. Each jack rests on a *crib* large enough to bear the load safely. Be sure to set the crib at the proper angle to ensure that jacking pressure is in direct line with the shore.

The disadvantages of the shoring method of support should be kept in mind. First, it is only practical for light buildings. Second, the height to which shores will reach effectively is limited, and therefore the method is not practical for tall buildings. Third, the lower ends of shores extend into the new working area and are an inconvenience to workers and machine operators.

## Needling

Needling involves the use of a *needle beam* that is thrust through the wall of the building being supported. The inner end stands on solid blocking, while the outer end is supported by a post resting on a jack. Figure 3-65 illustrates how a needle beam is used. The post jacks must be set in pits dug in the working area to the depth of the new excavation. Work is then carried out around them until permanent support can be provided for the old building.

## Underpinning

Underpinning is the provision of permanent support for existing buildings by extending their foundations to a new, lower level containing the desired bearing stratum. This may be necessary for a number of reasons. The removal of part of the supporting soil by a new, adjacent excavation may make underpinning necessary. Also, the addition of new loads to an existing structure, either in the form of new machinery or by adding extra floors to the structure, may make additional support necessary. Underpinning may also be indicated if the existing building shows signs of sinking or settling due to poor bearing soil.

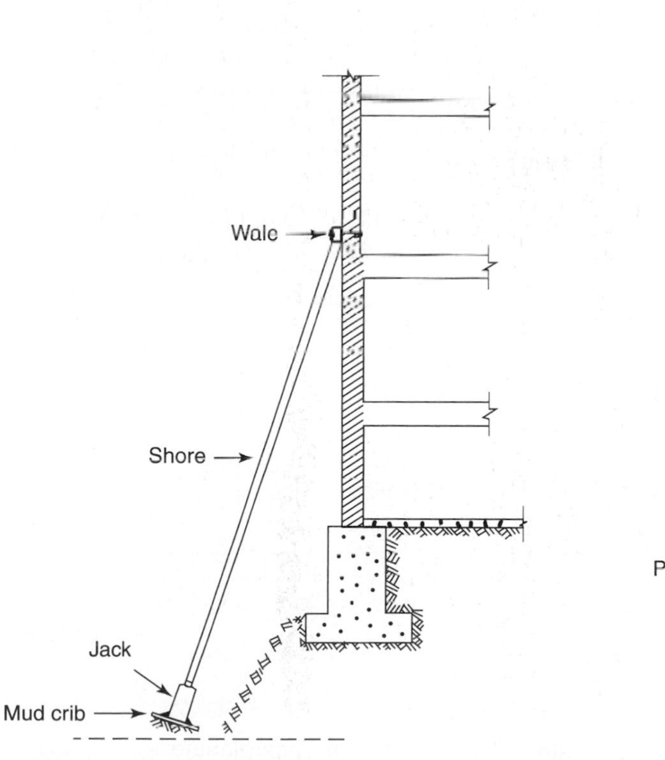


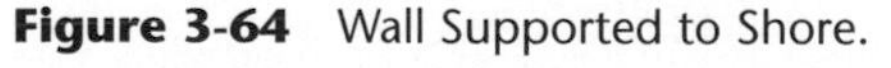
**Figure 3-64** Wall Supported to Shore.

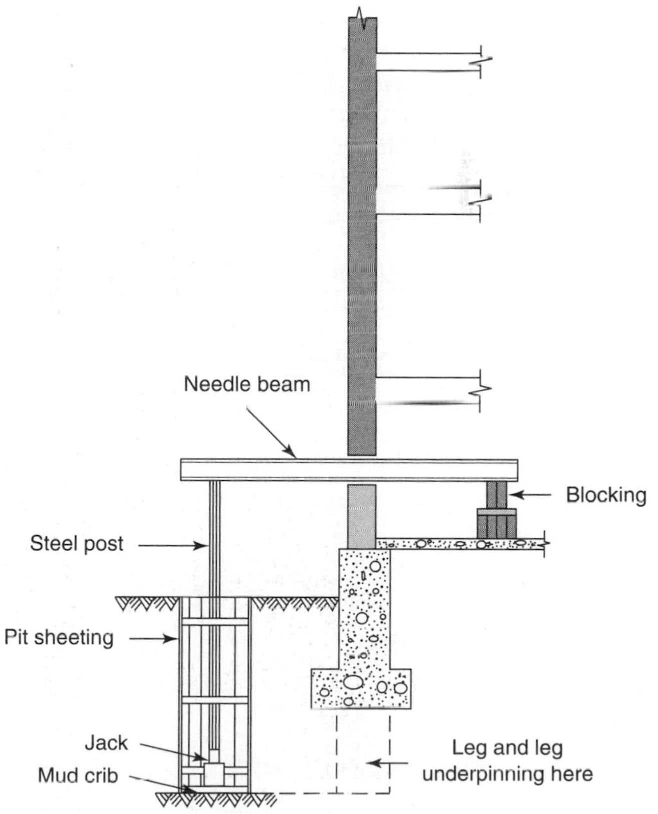


**Figure 3-65** Wall Supported by Needle Beam.

Underpinning may be provided in a number of ways. The method used depends on such factors as freedom to work, building loads involved, depth to new bearing, and type of earth encountered.

## Leg-and-Leg Underpinning

Under this method, the underpinning is placed in spaced sections, or *legs,* with relatively small sections of earth being removed from beneath the old foundation at spaced intervals, leaving the remainder intact to temporarily support the building. Forms are set, and concrete legs are placed in each such section. When the concrete has hardened, the space between the top of the leg and the underside of the old foundation is dry packed and wedged, or sometimes just wedged, to transfer the load to the new bearing member.

A new section is then excavated and formed alongside each completed leg and new legs are placed. Adjoining legs should be tied together with *keys* or *steel dowels.* This process is repeated until the entire length of the wall has been underpinned. Forms for such underpinning are simple, as illustrated in Figure 3-66.

## Pit Underpinning

This type of underpinning is similar to leg-and-leg, the difference being in the depth to which the work is carried. When the new bearing level is a considerable distance below the old foundation, in excess of 5 or 6 ft (1.5 or 1.8 m), *spaced pits* may be dug under it (similar to the technique employed in leg-and-leg), cribbed, and filled with concrete. When the concrete has hardened, the space above is dry packed and wedged to transfer loads to the new columns. The second series of pits is then dug alongside the completed ones, and so on, until the underpinning is completed. Depths of 45 to 50 ft (13.5 to 15 m) may be reached by this method.

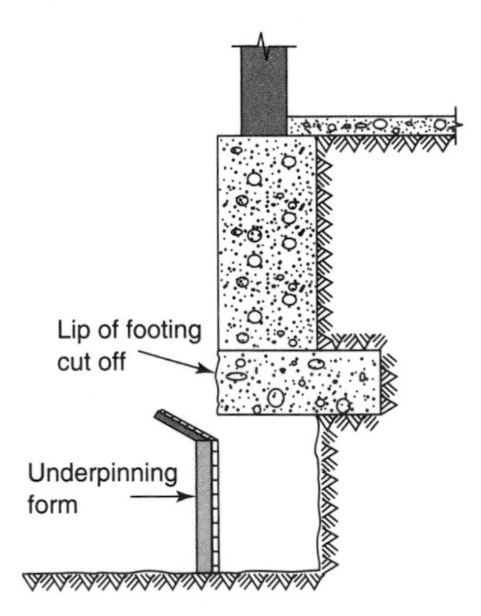


**Figure 3-66** Short Underpinning Form.

## Jacked Cylinder Underpinning

Sectional steel cylinders are jacked down beneath the foundation to be underpinned until they reach a suitable bearing. Then they are cleaned out and filled with concrete. A concrete cap, similar to a pile cap, is then formed and cast between the tops of the cylinders and the underside of the foundation. When the cap has hardened, the space between it and the foundation is dry packed so that the loads are transferred to the underpinnings.

## Pretest Underpinning

When forced into the ground under a load, any type of footing builds up a resistance beneath it, which finally prevents any further penetration. Tests indicate that when this happens under a test load and then that load is released, the footing will *rebound,* and if the load is reapplied, an equivalent resistance will finally be created, but at a lower level than before. Tests further indicate that rebound follows each release of load and that equal resistance under subsequent reloading is found only at greater depth.

It follows that, if any type of underpinning is test-loaded for adequacy and the load released before the cap or other connection is placed between it and the foundation, the underpinning will sink again when the foundation load comes on and further settlement will occur before it develops the resistance created under the test load. To overcome this, the *pretest* technique has been developed.

Sectional steel cylinders are jacked down beneath the foundation, cleaned out, and filled with concrete, as previously described. When the concrete is ready, the cylinder is test-loaded to an overload capacity, usually 50% in excess of the permanent load, using two jacks as illustrated in Figure 3-67. While the full test pressure is maintained on the jacks, a short steel column is placed between the top of the

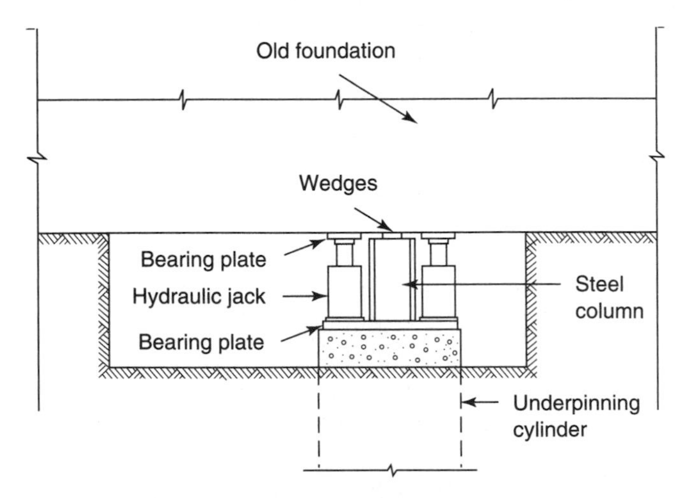


**Figure 3-67** Pretest of Underpinning.

cylinder and the underside of the foundation, and the load of the foundation is permanently transferred to the underpinning cylinder by steel wedges.

## ROCK EXCAVATION

Excavating sometimes includes the removal of solid rock, and this involves blasting, a job that is normally carried out by specialists. It is important, however, that the basic principles of rock removal by blasting be understood by those involved with the construction job.

In the building construction industry, blasting is used for such purposes as rock excavation, demolition work on buildings and foundations, stump clearing, and the breaking up of boulders too large to handle. The techniques of blasting have been made possible through the development of gunpowder and the refinement of explosives such as dynamite.

## BLASTING PRINCIPLES

Explosives commonly used in commercial blasting are practically all solid–solid or solid–liquid mixtures capable of rapid and violent decomposition, with resultant conversion into large volumes of gas. Decomposition of a *high* explosive such as dynamite takes place with extreme rapidity, whereas a *low* explosive such as black blasting powder decomposes more slowly, simulating rapid burning. As a result, high explosives are called *detonating* explosives and low explosives are referred to as *deflagrating* explosives.

The most widely used high explosive in the construction field is dynamite. It is basically composed of liquid nitroglycerin as a *sensitizer,* sawdust or wood pulp as a *liquid absorber,* and *oxygen supplier* such as sodium nitrate, and a small amount of *antacid* material such as zinc oxide or calcium carbonate. Several types of dynamite exist, the common ones being (1) *straight,* (2) *extra* or *ammonia,* (3) *gelatin,* and (4) *permissive* dynamite.

Dynamite is produced in various grades, depending on the percentage (by weight) of nitroglycerin the material contains. Thus, a 30% grade contains 30% nitroglycerin, and a 60% grade contains 60% nitroglycerin. This does not mean that 60% dynamite is twice as strong as 30% dynamite; in fact, it is only about 1⅓ times as strong.

For practical use, dynamite is rolled into cartridges covered with wax-impregnated, 70 lb (32-kg) manila paper. Cartridges range from ⅞ to 8 in. (22 to 200 mm) in diameter and from 8 to 24 in. (200 to 600 mm) in length.

To produce an explosion, dynamite may be ignited in any one of three ways, depending on the type of dynamite used and the general conditions at the time of blasting. One method is to use a safety fuse connected directly to the cartridge. Another is to use a blasting cap at the end of a fuse, and the third is the use of electric blasting caps (see Figure 3-68).

A safety fuse consists of black powder protected by a flexible fabric tube. Safety fuses are available in two burning rates: 1 yd (900 mm) in 120 sec or 1 yd (900 mm) in 90 sec. Fuses are generally lit with matches, and their burning time provides an opportunity to clear the blast area.

Many dynamites cannot be ignited by the burning action of a fuse and must be set off by a primary blast wave. This is done by placing a blasting cap on the end of a safety fuse and inserting the cap into a cartridge. The burning fuse ignites the cap, which detonates and in turn sets off the dynamite.

In many cases, safety fuses and detonators are replaced by electric blasting caps. These are small copper cylinders about 1⅛ in. (28 mm) long and ⅜ in. (9.5 mm) in diameter containing a small explosive charge. Two conducting wires lead into one end and are connected by a resistance bridge. When an electric current passes across this bridge, the heat generated ignites the charge, which in turn detonates the main charge. Two kinds of blasting caps known as *delay caps* are also made—there is the ordinary delay cap, which allows approximately 0.5 sec between groups or firing periods, and the millisecond delay cap. In the latter, various delays are available, ranging from 0.025 to 0.1 sec.

Electric blasting circuits may be wired in series, in parallel, or in series–parallel (see Figure 3-69) (employed when a large number of caps are to be set off). Circuits are energized either by batteries or by a blasting machine. Batteries are used only when a blasting machine is not available. The latter produces the desired current and produces it every time the machine is used. Three types of blasting machines are the twist handle, rack bar, and condenser discharge.

Excavating with explosives is carried out primarily by drilling holes in the rock, loading the holes with

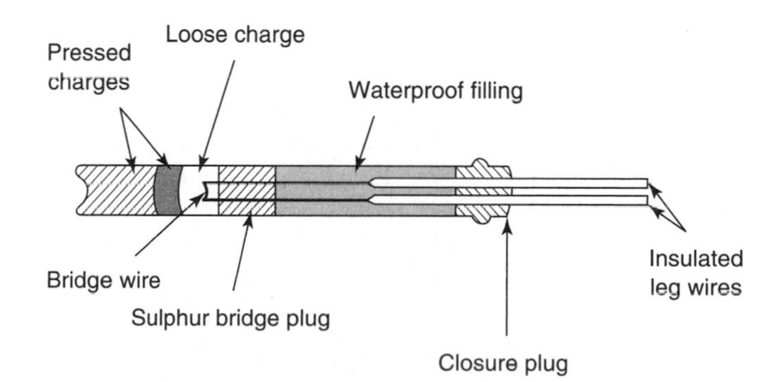


**Figure 3-68** Diagram of Electric Blasting Cap.

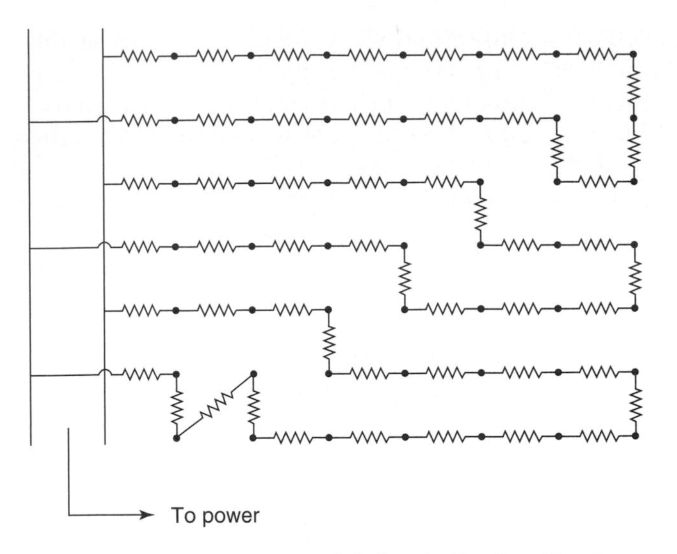


**Figure 3-69** Series-in-Parallel Electric Blasting Circuit.

**Figure 3-70** Truck-Mounted Drill Drilling at an Angle.

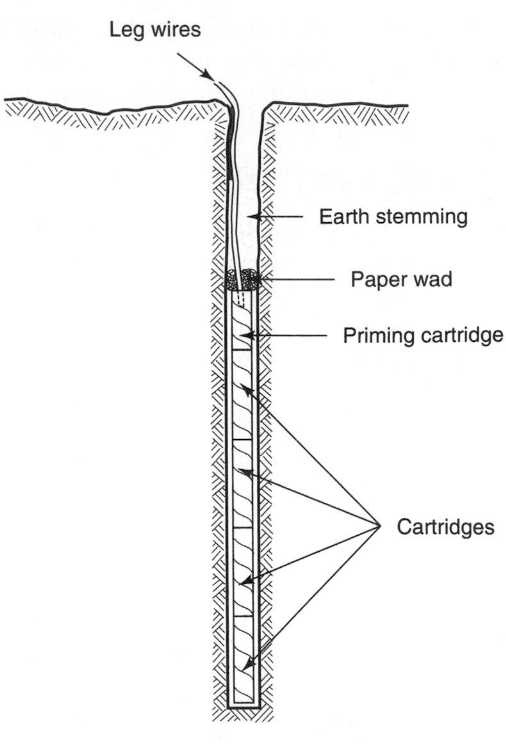


**Figure 3-71** Section through Loaded and Stemmed Hole.

explosives, and firing the charge. Holes are drilled horizontally, vertically, or at an angle (Figure 3-70) depending on circumstances. Holes of all diameters and for different explosives are loaded basically in the same manner.

When drilling is completed, holes should be cleaned with a jet of compressed air and tamped with a wooden pole. If they are to remain unused for some time after drilling, they should be sealed with paper plugs to keep them clean. Loading a hole consists of placing a primer charge and a number of cartridges in the hole and tamping them into place. A cap is placed in the primer charge by piercing a cartridge with a wooden poker and embedding the cap in the hole, allowing the leg wires to project. The primer charge is lowered into the hole, and additional charges are then placed and tamped. The cartridges are connected by slitting the end of each as it is placed. The pressure caused by tamping causes the slits to open and dynamite to bleed from one cartridge to another. The primer cartridge is not slit.

After the hole has been loaded to the desired depth, it is *stemmed* with an inert material such as sand or clay. The stemming confines the blast and prevents blowouts through the hole. During the placing, tamping, and stemming operations, the leg wires from the blasting cap must be held against one side of the hole to prevent their being damaged. Figure 3-71 shows a vertical section through a loaded and stemmed hole.

To achieve the best results, an organized and calculated pattern of holes must be set up and loaded with either instantaneous or delay caps. Instantaneous loading and shooting do not permit control of the direction of rock throw and tend to produce muck, which is often too large to handle. With the use of delay time caps, the direction of rock throw is controlled, and fragmentation is generally improved. Figure 3-72 shows two steps in a blasting operation, one at the height of the explosion and the other at its conclusion.

In Figure 3-73, a typical delay pattern is illustrated. Notice that the longer delays are used in the back rows of the holes and along the sides, and that the short-delay caps are used in the bore holes close to the working face. The short-delay charges move the rock near the face and thus provide space for the material moved by the charges primed with long-delay caps.

(a)

(b)

**Figure 3-72** Controlled Blasting: (a) Explosion at Maximum; (b) Blast Completed.

| | | | | | | | | | | | |
|---|---|---|---|---|---|---|---|---|---|---|---|
| 8 | 7 | 6 | 5 | 4 | 4 | 4 | 4 | 5 | 6 | 7 | 8 |
| 7 | 6 | 5 | 4 | 3 | 3 | 3 | 3 | 4 | 5 | 6 | 7 |
| 6 | 5 | 4 | 3 | 2 | 2 | 2 | 2 | 3 | 4 | 5 | 6 |
| 5 | 4 | 3 | 2 | 1 | 1 | 1 | 1 | 2 | 3 | 4 | 5 |
| 4 | 3 | 2 | 1 | 0 | 0 | 0 | 0 | 1 | 2 | 3 | 4 |

Note: numbers refer to milliseconds delay

Free face

**Figure 3-73** Typical Delay Pattern.

In some cases, large boulders or rock masses left after the primary blast may require secondary blasting for complete breakage. Secondary blasting may be done in three ways: by (1) *mudcapping,* (2) *blockholing,* or (3) *snakeholing.*

In mudcapping, the required number of cartridges are tied together in a bundle and placed on top of the boulder or rock pile. They are covered with a piece of waterproof paper and a layer of mud approximately 6 in. (150 mm) thick. The mud contains the blast and directs the shock wave into the rock.

If extremely hard rock such as granite or taprock does not respond to mudcapping, blockholing must be used. This consists of drilling one or more holes, depending on the size, and loading, stemming, and firing them as in primary blasting.

Snakeholing consists of tunneling under a boulder and placing a charge there to lift it and break it up. The excavated earth is replaced in the hole as *stemming* (see Figure 3-74).

In primary blasting, the *burden* and the *spacing* of the holes are extremely important. The spacing is the distance between holes in a line, and the burden is the distance of a hole from the free face.

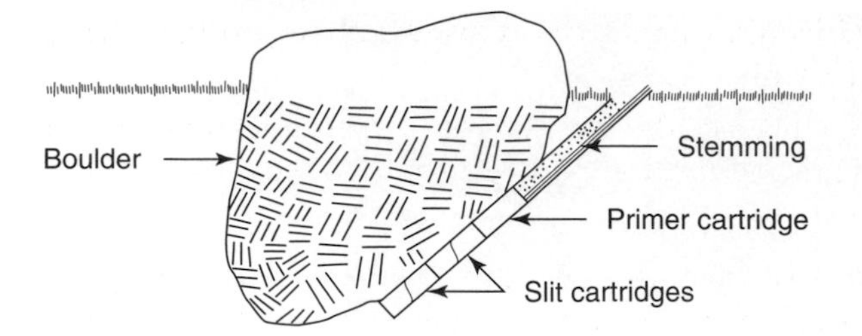


**Figure 3-74** Snakeholing.

Hole sizes for general blasting range from 1 to 1½-in. (25 mm to 40 mm) in diameter and are drilled to a maximum of 20 ft (6 m). To illustrate how to plan a drilling pattern and how to determine the amount of explosive required for a given solution, 1⅛-in. (28-mm) diameter drill holes will be considered.

The maximum amount of explosive that can be placed in a hole 1⅛-in. in diameter and 1 foot deep is 0.69 lb. Under free-breaking conditions, rock normally requires a load of 1 lb of explosive per cubic yard. At that rate, the volume of rock that can be broken up with 0.69 lb of explosive is 0.69 cu yd, or 18.6 cu ft. A slab of rock 1-ft thick that produces 18.6 cu ft will have a square surface dimension of $4.3 \times 4.3$ ft. So if a number of holes 1-ft deep were drilled in a square pattern, the maximum spacing would be 4.3 ft in any direction.

Applying similar assumptions using metric units, based on a 28-mm diameter hole, the maximum amount of explosive that can be placed in a hole 28-mm in diameter, 300 mm deep is 0.31 kg. Under free breaking conditions, 0.6 kg of explosive is required to break up 1 cubic metre of rock. The volume of rock that can be broken up with 0.31 kg of explosive is 0.5 m$^3$. A slab of rock 300-mm thick that produces 0.5 m$^3$ will have a square surface dimension of 1.3 m by 1.3 m. The resulting maximum spacing becomes 1.3 m in any direction. The general spacing and burden limit in instantaneous blasting is 4.3 ft (1.3 m), although this space may vary to some degree with the use of delay caps.

A variety of patterns may be developed by the use of delay caps. One basic pattern is illustrated in Figure 3-75, where a shot is to be fired into a bank with both ends closed, leaving a vertical face.

The three top center holes, loaded with #0 delay caps, open the shot and release the burden on the #1 delay holes below and on each side. The burden of each successive hole is free-faced by the firing of the charge in front of it. This free-facing of the burden of every charge is the basic principle behind delay pattern blasting. The general load factor will be 1.0 lb of explosive per cubic yard (0.6 kg of explosive per cubic meter) of rock, but the #0 delay charges probably have a load factor of 1.25 lb/cu yd (0.75 kg/m$^3$), to produce a good, clean start for the pattern. Figure 3-72 shows a bank shot, such as that previously described, being fired.

When there is danger of damage occurring because of rock throw, blasting mats should be used. The two most-often-used types are woven rope and wire mats. Wire mats must be weighted down to contain rock during a blast, whereas rope mats are heavy enough by themselves.

Safety is a prime factor in blasting operations, and the following are fifteen general safety rules to practice when handling explosives.

1. Smoking must not be tolerated in the blasting area.
2. Only hardwood or nonspark pokers must be used to pierce cartridges.

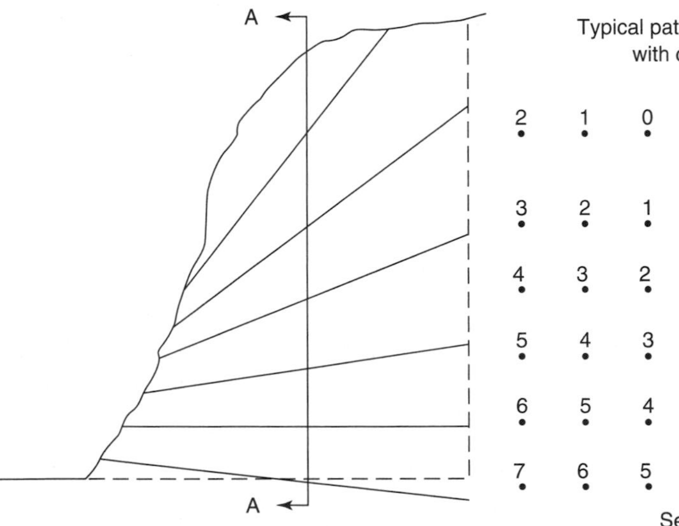


**Figure 3-75** Typical Pattern for Bank Shot with Closed Ends.

3. Circuits should always be checked before firing.
4. It must be made certain that the area has been cleared before firing a shot.
5. Shunts should always be used when setting up an electric circuit.
6. When electrical storms are in the area, all electric blasting should be stopped.
7. Only nonsparking tape should be used for measuring hole depths during loading.
8. The *shucking* of dynamite (removal from the cartridge) should be avoided unless absolutely necessary. Shucked dynamite should not be tamped.
9. Inventory should be taken before and after firing of a shot to guard against mislaid explosives and caps.
10. Only blasting galvanometers should be used for testing the circuit.
11. Loaded hole areas should be clearly marked with appropriate signs.
12. Piles of cartridges should never be left beside a large-diameter hole.
13. Holes must be cool prior to loading. This should be checked.
14. The primer cartridge should never be forced into a hole and should never be tamped.
15. Storage or dry-cell batteries should never be used to activate electric blasting circuits.

## REVIEW QUESTIONS

**1.** A service trench is to be 30 ft long, 10 ft deep, and 8 ft wide. Assuming that the ends are vertical and the sides have a slope of 1:1, use the average end method to calculate the volume of the excavation in cu ft.

**2.** An excavation for a square footing is 3.5 m deep. The bottom of the excavation is four sq m. If the sides have a 0.5:1 slope, use the frustum formula to calculate the volume of excavation in cu m.

**3.** An excavation measuring 160 ft by 120 ft at the top has sides that slope in at a rate of 1.5:1. If the excavation is 15 ft deep, use the frustum formula to calculate the volume of the excavation in cu yd.

**4.** A trench 115 m in length has vertical ends and is 5 m deep. If the bottom of the trench is 4 m in width and the sides slope out at a rate of 0.75:1, use the average end area method to calculate the resulting volume in cu m.

**5.** What are the two basic types of cranes that are used on construction sites?

**6.** What advantage does a tower crane have over a mobile crane on a construction site? What disadvantage?

**7.** List six factors that have an effect on the lifting ability of a crane.

**8.** What are the operational differences between a dragline bucket and a clamshell bucket?

**9.** What six items can be used to classify trucks?

**10.** A rubber-tired scraper with an efficiency of 50 min/hr has a heaped capacity of 20 cu yd. If it is to move material with a swell factor of 25%, what amount of in-bank material can it move in 1 hr, assuming it has a total cycle time of 5 min?

**11.** A front-end loader is to load trucks with granular material that has a swell factor of 15%. If the loader has a heaped capacity of 2.5 cu yd, calculate the total cycle time for the loader if it can move 40 cu yd of in-bank material in 1 hr with an efficiency of 51 min/hr.

**12.** A rubber-tired scraper with an efficiency of 50 min/hr has a heaped capacity of 20 $m^3$. If it is moving material with a swell factor of 20%, what amount of in-bank material can it move in 1 hr if it has a total cycle time of 6 min?

**13.** A front-end loader is loading trucks with granular material that has a swell factor of 15%. If the loader has a heaped capacity of 2.25 $m^3$, calculate the total cycle time for the loader if it can move 38 $m^3$ of in-bank material in 1 hr with an efficiency of 50 min/hr.

**14.** Answer briefly **(a)** When is dewatering equipment necessary? **(b)** Name the five basic parts of a dewatering system. **(c)** What is a possible alternative to the use of dewatering equipment?

**15.** Explain what is meant by **(a)** *special* excavating work, **(b)** soil swell, and **(c)** payline.

**16.** **(a)** Outline the limitations to the use of shoring as a means of protecting adjacent buildings. **(b)** Outline the procedure for carrying out pretest underpinning.

**17.** Explain the principle involved in **(a)** the use of bentonite slurry in excavating trenches, and **(b)** the use of tremie for placing concrete in a slurry-filled trench.

**18.** By means of neat diagrams, illustrate **(a)** the use of internal sloping braces to support a wall of sheet piling, **(b)** belled earth tiebacks to support soldier piles, and **(c)** the use of horizontal diagonal braces as support for excavation protection walls.

**19.** What are the advantages of top-down construction when excavating on a site with limited access?

**20.** The working face of an open pit consists of soft shale overlaid by hard rock. If the entire face is to be moved in a single blasting operation, describe briefly how the holes should be loaded.

# 4 FOUNDATION LAYOUT

Upon completion of the site investigation and design of the foundations, actual site work can begin. In many instances, preliminary excavation must be done on the site before the actual layout of the building foundations begins. Site access may be required, existing trees may require removal or relocation, and site leveling may be necessary to provide drainage.

The layout of the foundation structure must be carried out with great accuracy to ensure that its various elements are placed exactly as called for in the building plans. A set of building plans includes a foundation plan, on which is shown the position of piles, footings, column bases, slab stiffeners, with all elevations and necessary details.

Figures 4-2(a) and (b) illustrate the foundation plans in standard units and metric units for the buildings shown in Figures 4-1(a) and (b). These plans illustrate a typical shallow foundation consisting of wall, pilaster, and footing foundations with slabs on grade in the basement area. Footings consist of isolated square footings under the columns and strip footings under the walls that are enlarged at pilaster locations. Details indicating elevations, wall thickness, footing sizes, and steel reinforcing bars are shown.

Figures 4-4(a) and (b) illustrate a deep foundation consisting of grade beams, pile caps, and piles for the office building shown in Figures 4-3(a) and (b). In this instance, the building superstructure rests on the grade beams and pile caps and all building loads are transferred to solid bearing through the perimeter and interior piles.

## LOCATING POINTS FROM BATTER BOARD LINES

The first step in laying out the foundation, after completing the excavation, is to rig the batter board lines. They may be either cord or wire, depending on their length, but they must always be strung very tightly.

A plumb bob dropped from the intersection of two corner batter board lines (see Figure 4-5) to the excavation floor establishes a corner position. A stake is driven at that point and a nail placed in the top of the stake so that it coincides exactly with the point of the plumb bob. All of the outside corners are established in the same way. The perimeter of the building is thus outlined, as shown in Figures 4-6(a) and (b). The accuracy of the staking should now be checked by measuring the diagonals of the rectangles involved. If there is no error, the diagonals will be exactly equal.

A further check may be made by measuring the angles with the leveling instrument. For example, for the layout shown in Figure 4-6(a), proceed as follows:

1. Set up the instrument over Stake A, center over the pin, and level it. Backsight on the pin in Stake B. Set the horizontal circle to 0.
2. Turn the telescope and sight on Stake D. Calculate the size of the angle *DAX* by trigonometry, using a right triangle and solving for tan angle *DAX*, from natural trigonometric functions. In this case,

$$\tan DAX = \frac{20}{100} = 0.200; \text{ angle } DAX = 11°18'$$

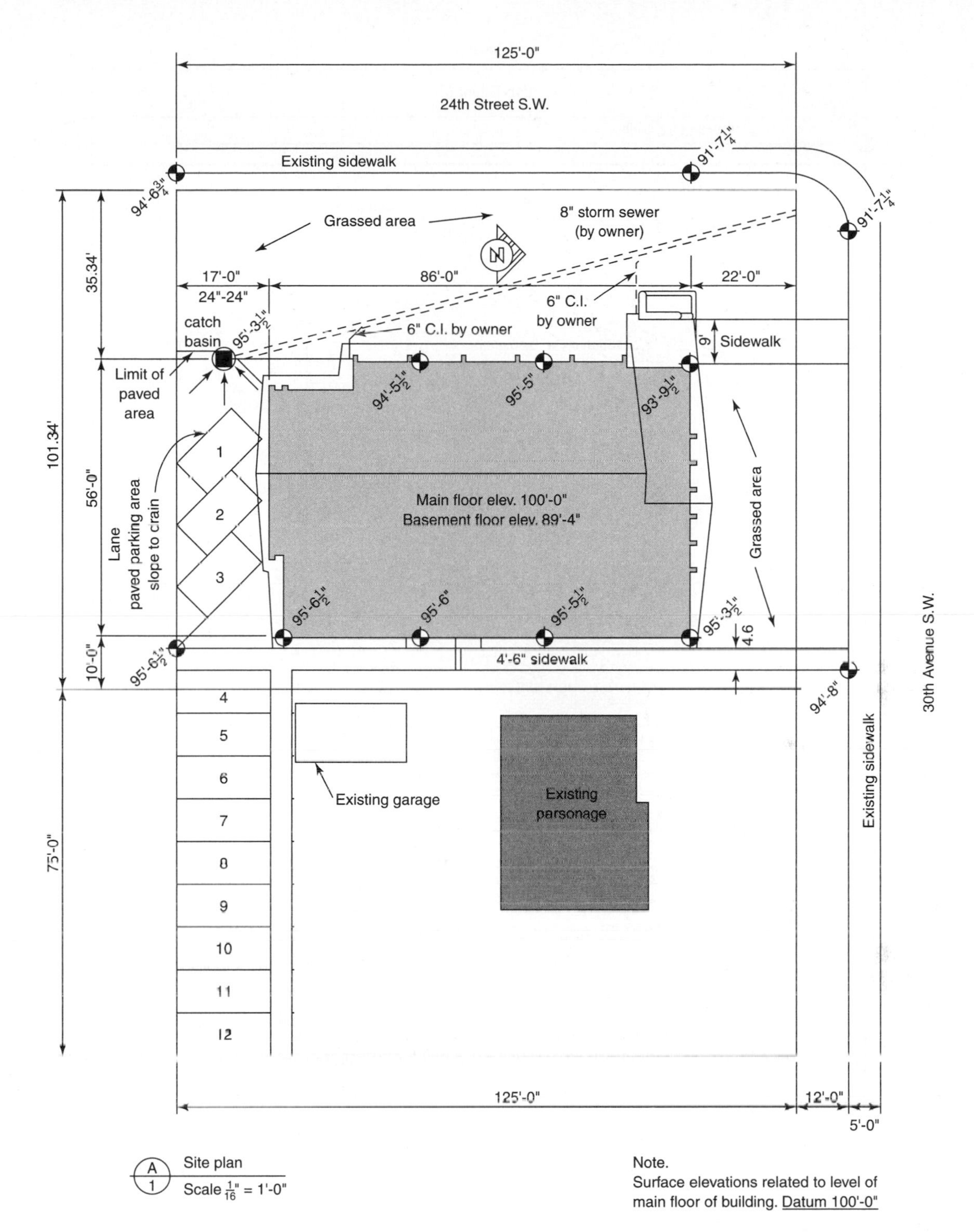


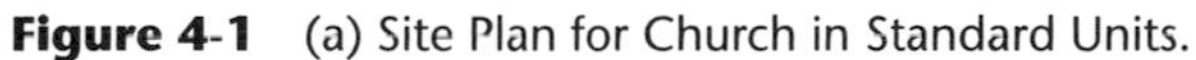

**Figure 4-1** (a) Site Plan for Church in Standard Units.

**3.** Turn again and sight on Stake C. By the same process,

$$\tan CAB = \frac{20}{60} = 0.333; \text{ angle } CAB = 18°26'$$

**4.** Sight on Stakes E, F, G, and H in order.

(a) $\tan EAX = \dfrac{60}{100} = 0.60; \text{ angle } EAX = 30°58'$

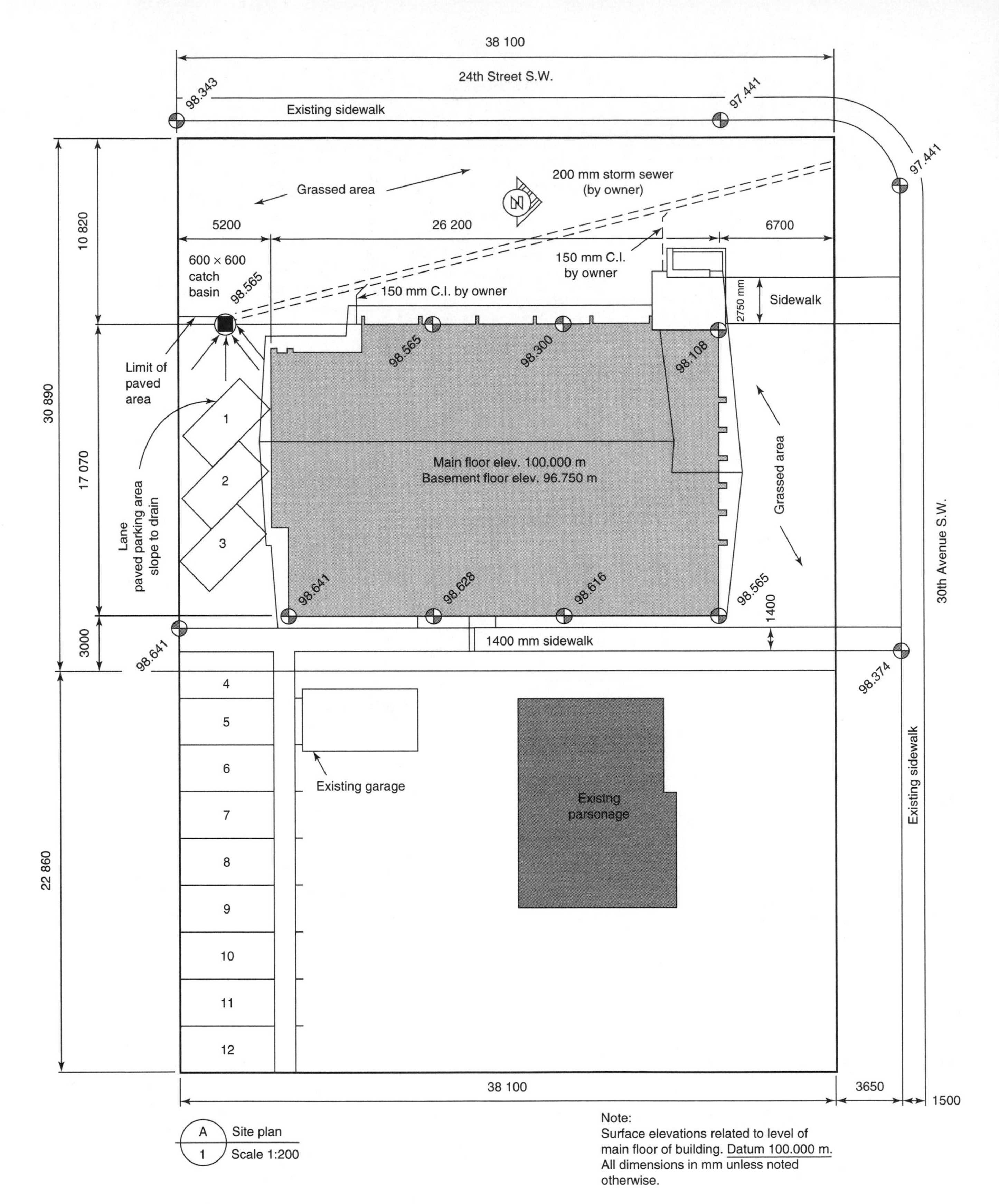


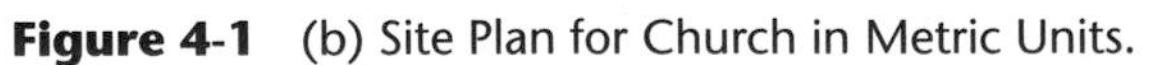

**Figure 4-1** (b) Site Plan for Church in Metric Units.

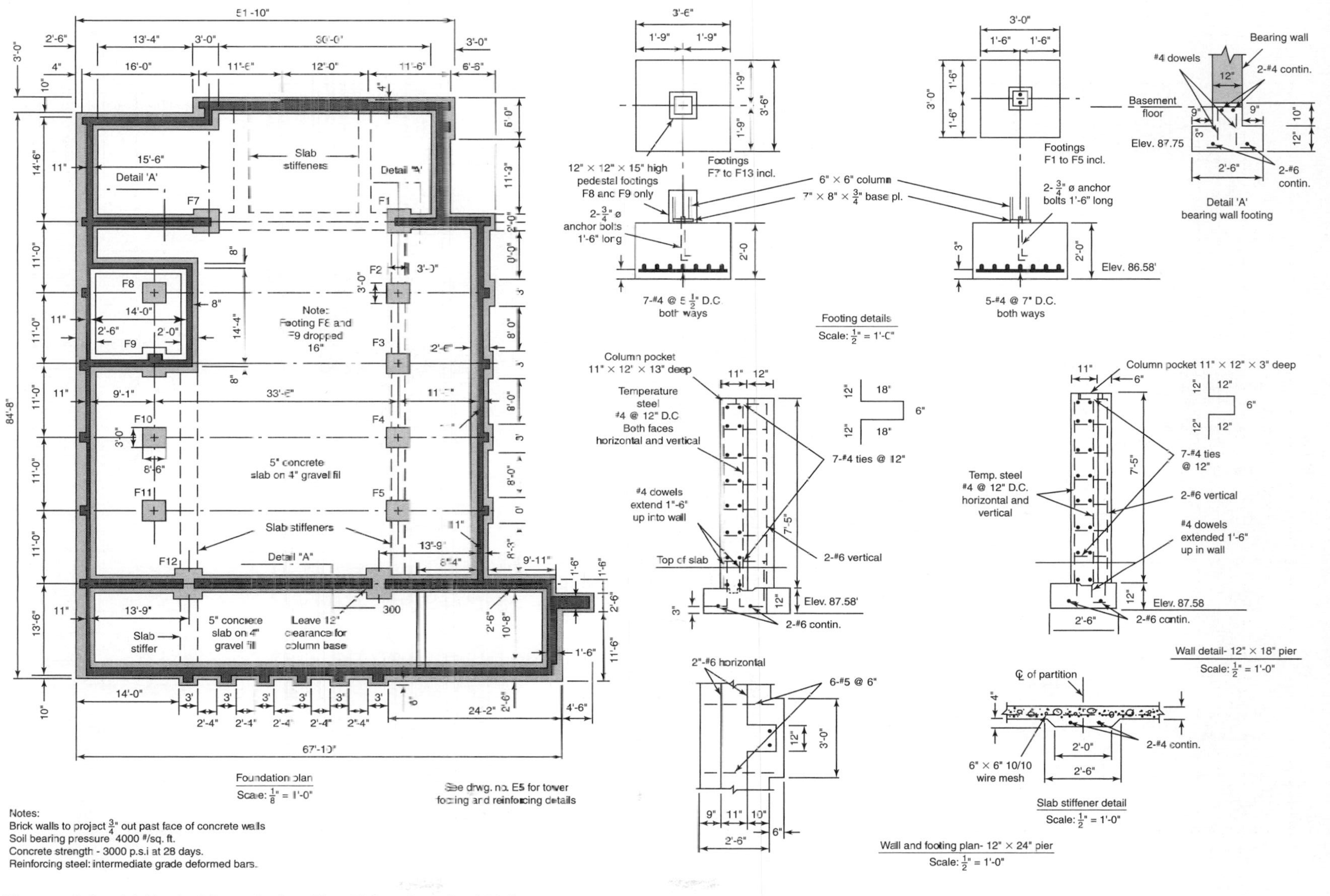

**Figure 4-2** (a) Typical Foundation Plan Using Standard Units.

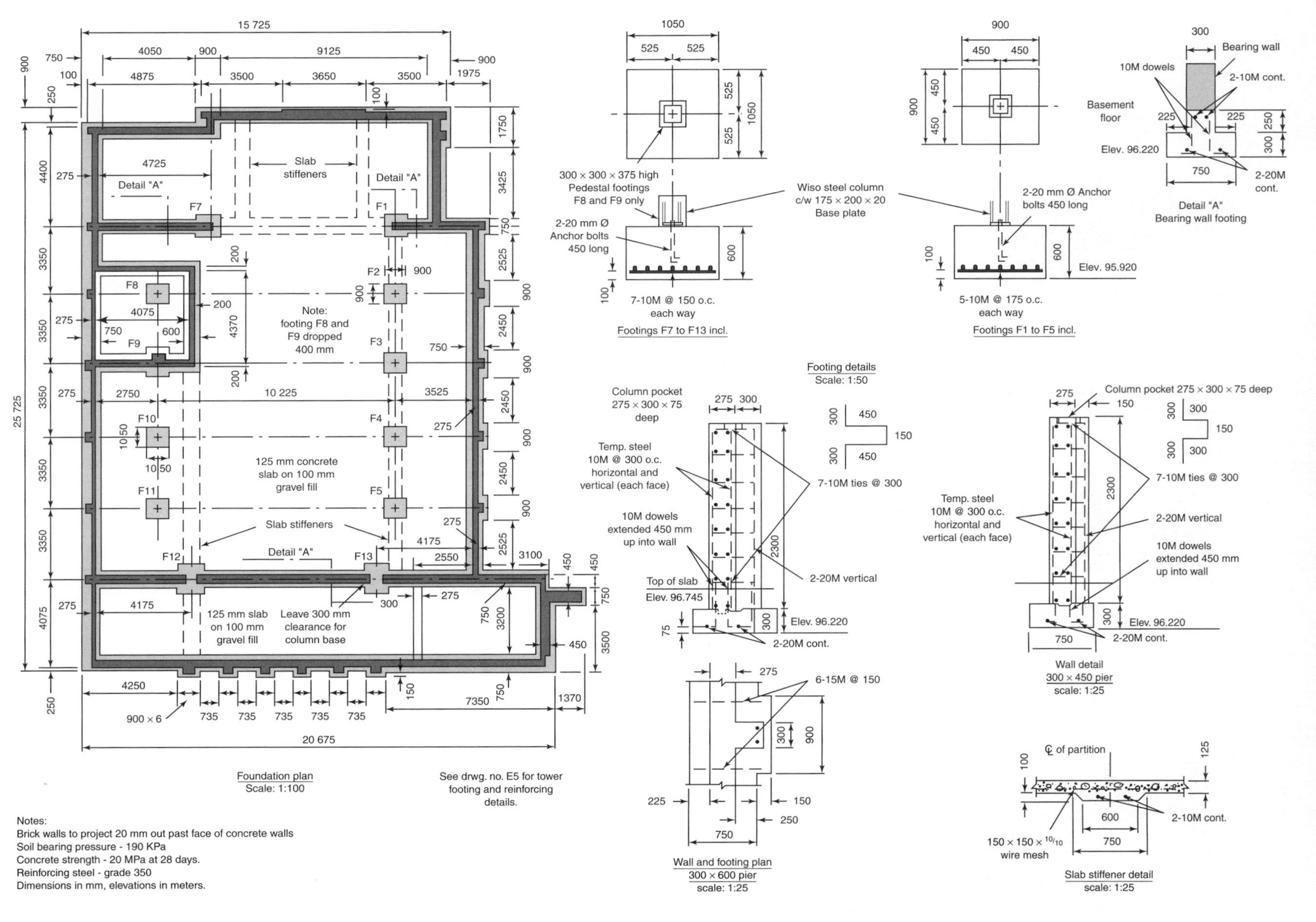

**Figure 4-2** (b) Typical Foundation Plan Using Metric Units.

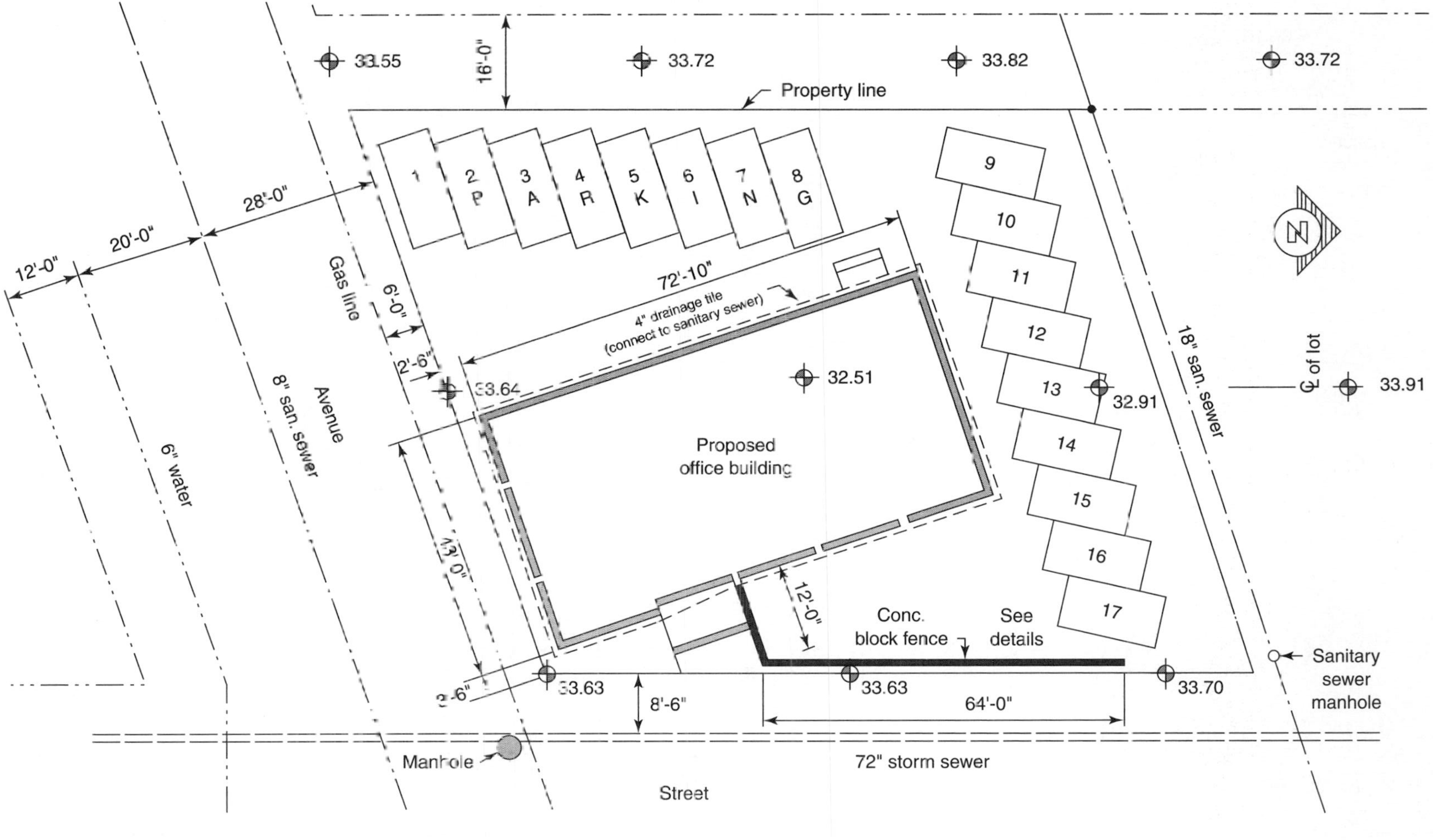


**Figure 4-3** (a) Site Plan for Office Building in Standard Units.

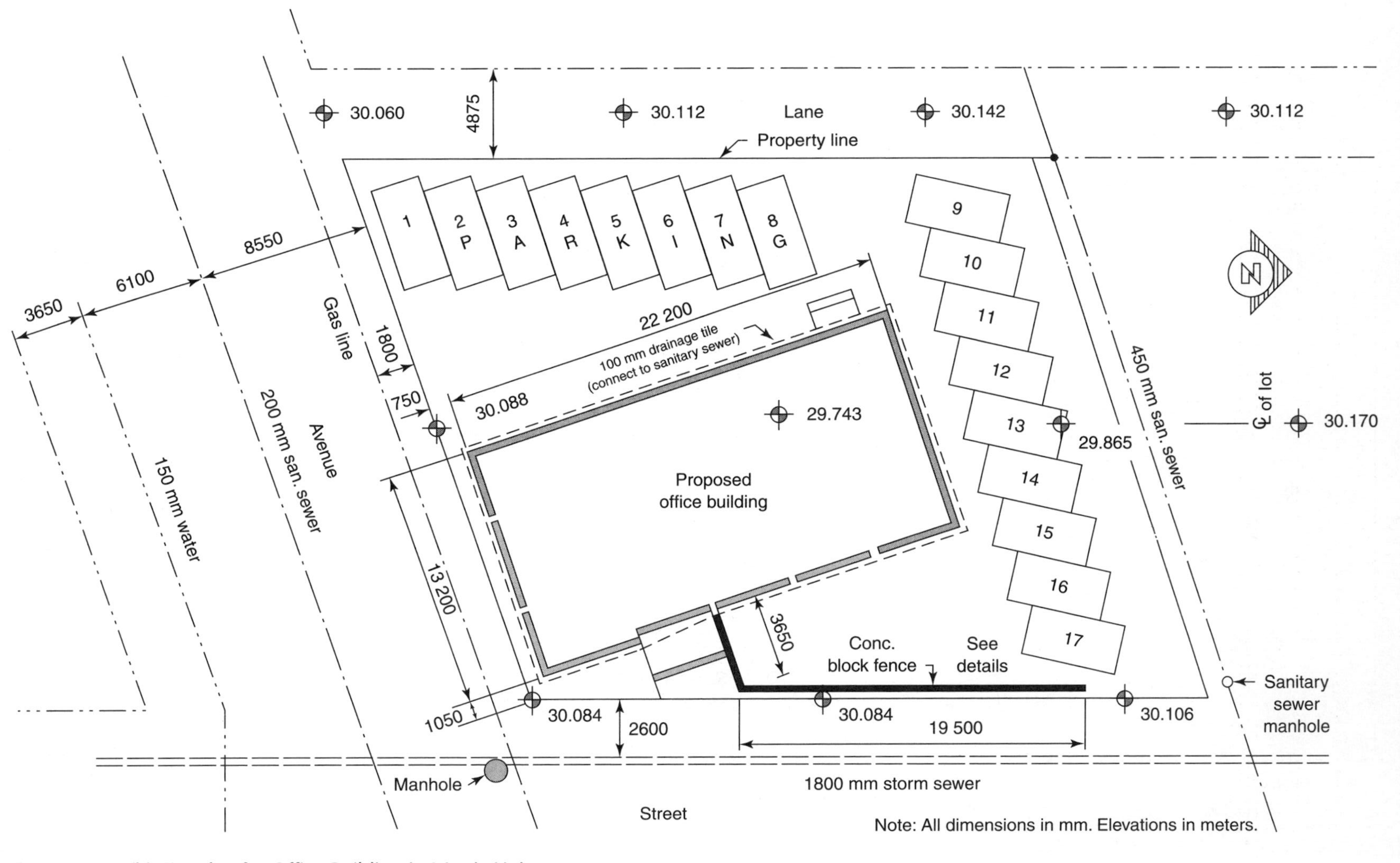

**Figure 4-3** (b) Site Plan for Office Building in Metric Units.

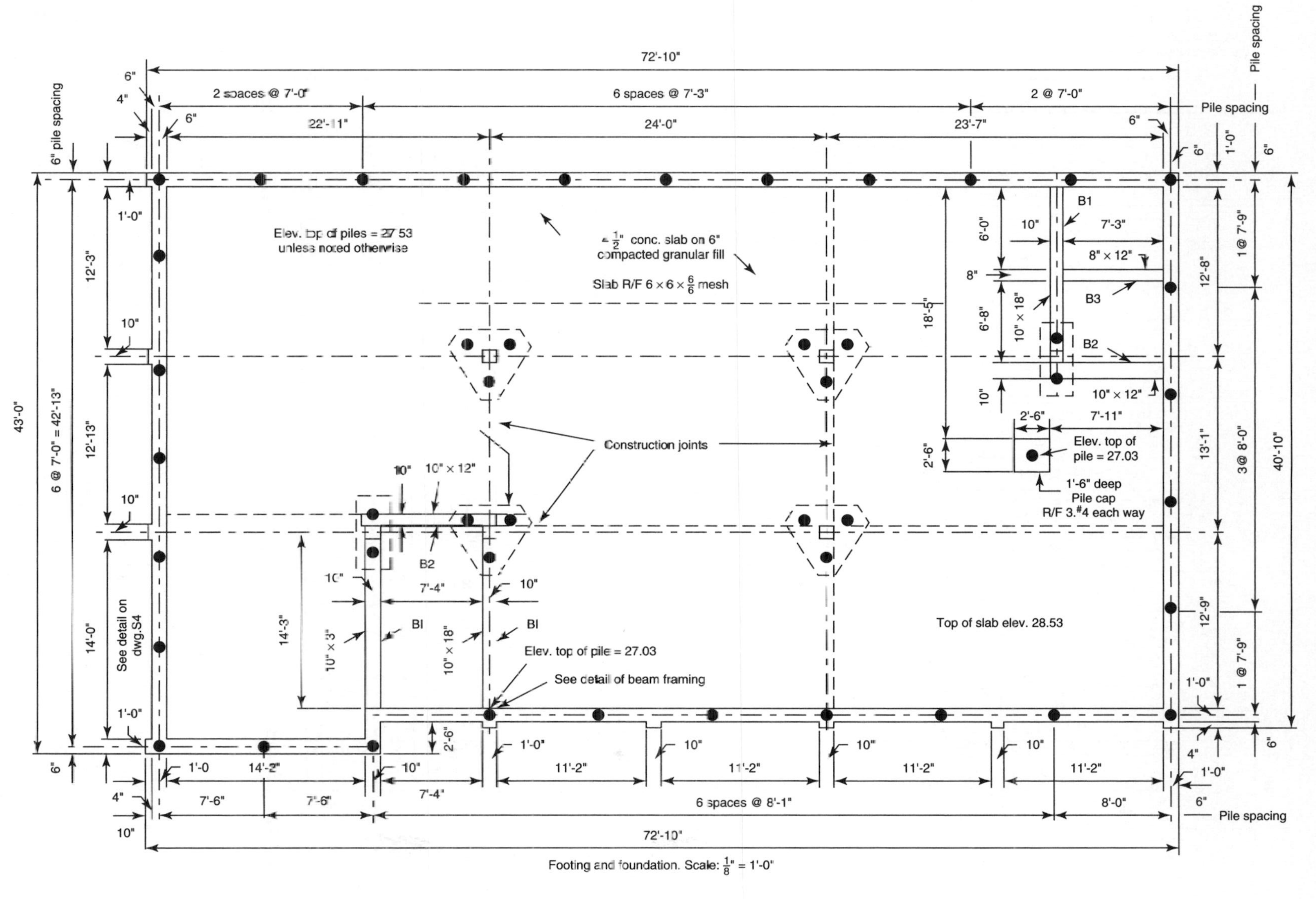


**Figure 4-4** (a) Pile Foundation Plan, Standard Units.

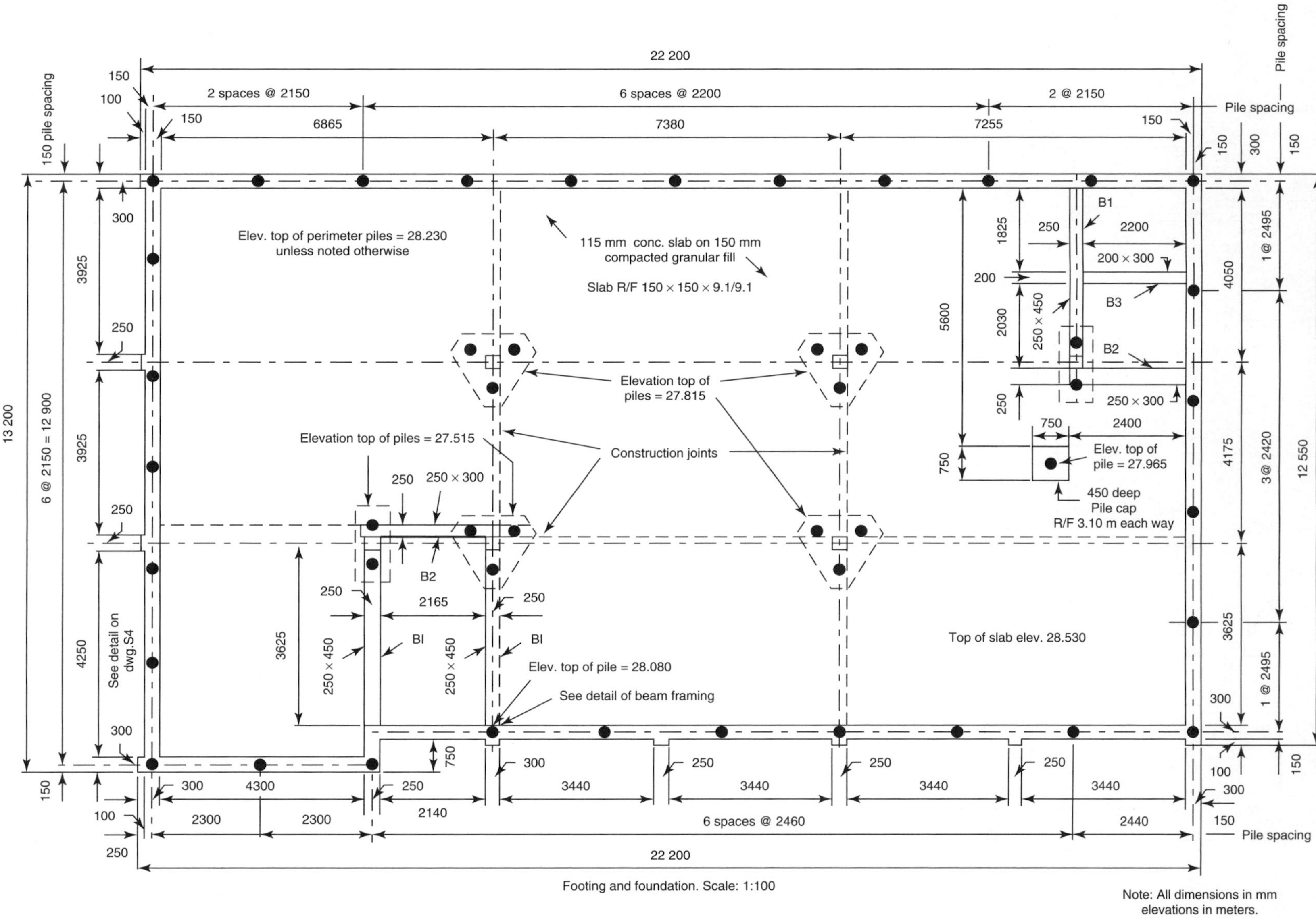

**Figure 4-4** (b) Pile Foundation Plan, Metric Units.

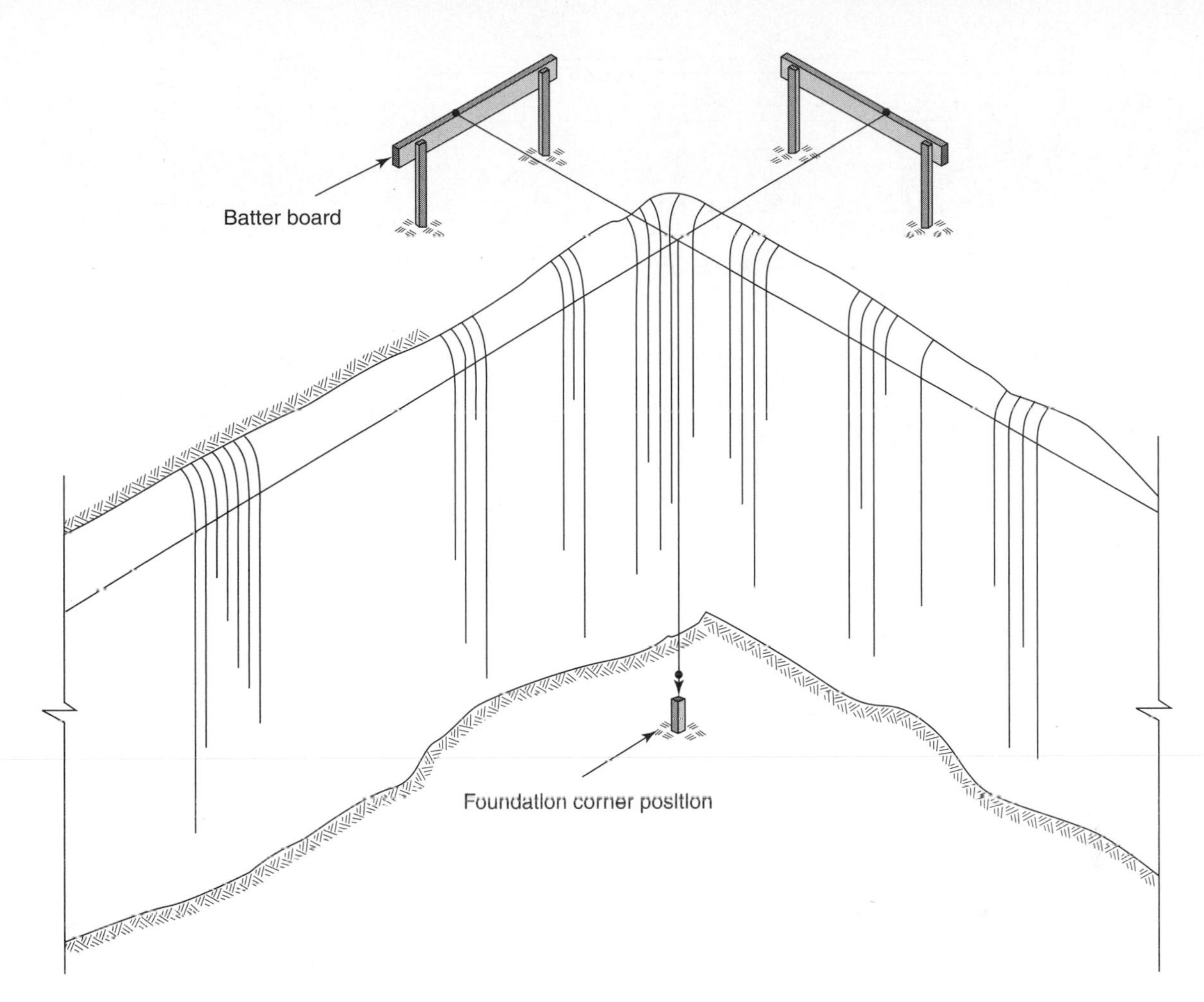


**Figure 4-5** Establishing a Foundation Corner Position from Batter Board Lines.

(b) $\tan FAB = \dfrac{60}{60} = 1.0$; angle $FAB = 45°0'$

(c) $\tan GAB = \dfrac{100}{60} = 1.667$; angle $GAB = 59°2'$

(d) angle $HAB = 90°\ 0'$

Applying the same procedure to the layout in Figure 4-6(b), the calculations are as follows:

1. $\tan DAX = \dfrac{6\,000}{30\,000} = 0.200$; angle $DAX = 11°18'$
2. $\tan CAB = \dfrac{6\,000}{18\,000} = 0.333$; angle $CAB = 18°26'$
3. $\tan EAX = \dfrac{18\,000}{30\,000} = 0.60$; angle $EAX = 30°58'$
4. $\tan FAB = \dfrac{18\,000}{18\,000} = 1.0$; angle $FAB = 45°0'$
5. $\tan GAB = \dfrac{30\,000}{18\,000} = 1.667$; angle $GAB = 59°2'$
6. angle $HAB = 90°\ 0'$

Column center lines can now be established. Drop a plumb bob from line intersections (see Figure 4-7) on opposite sides of the building and drive stakes at those points. A line strung between these stakes represents the column center line, and the positions of column footing centers can be measured along this line and indicated by driving stakes at each point.

## LAYOUT FOR FOOTING FORMS

There are a number of methods for establishing the outside line of exterior footing forms. If the footings are not large, the projection of the footing beyond the outside wall line can be measured easily. For example, in Figure 4-2(a), the bearing wall footing extends 9 in. beyond the wall. From the corner stake already established, measure off a square with 9-in. sides (see Figure 4-8), and drive a stake with one face on the line and one the thickness of the footing form material beyond the intersecting line (see Figure 4-9). This stake should be driven or cut off exactly at the elevation shown for the footings. For example, in Figure 4-2(a), the elevation of the bottom of footings F1 to F5 is shown as 86.58 ft and the footings are 2 ft thick. Therefore, the elevation of the top of the stakes will be

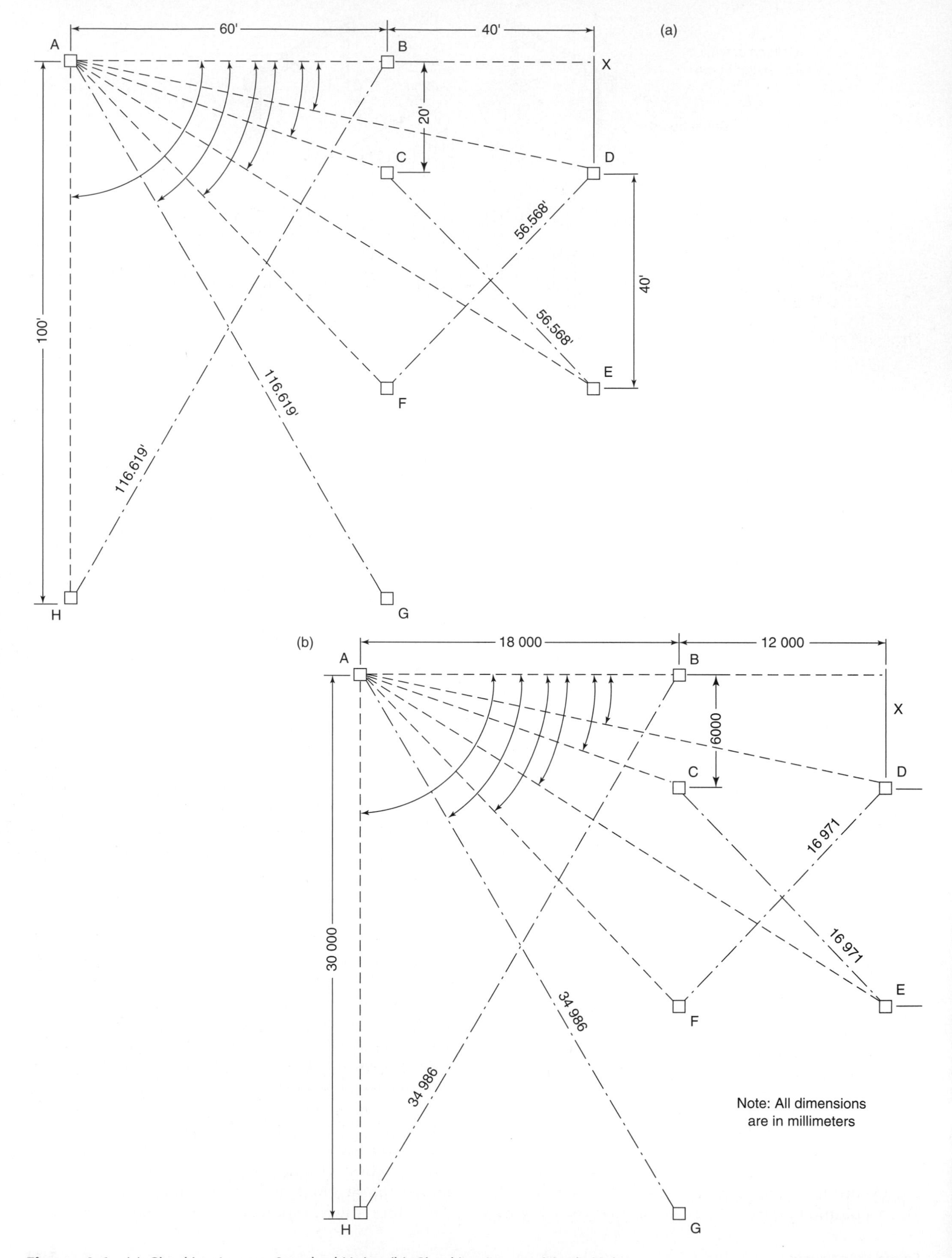


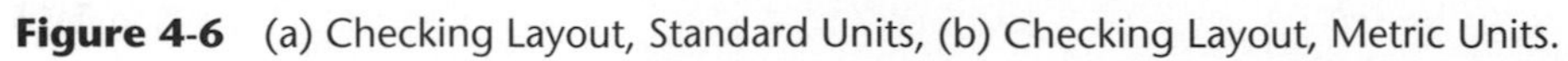

**Figure 4-6** (a) Checking Layout, Standard Units, (b) Checking Layout, Metric Units.

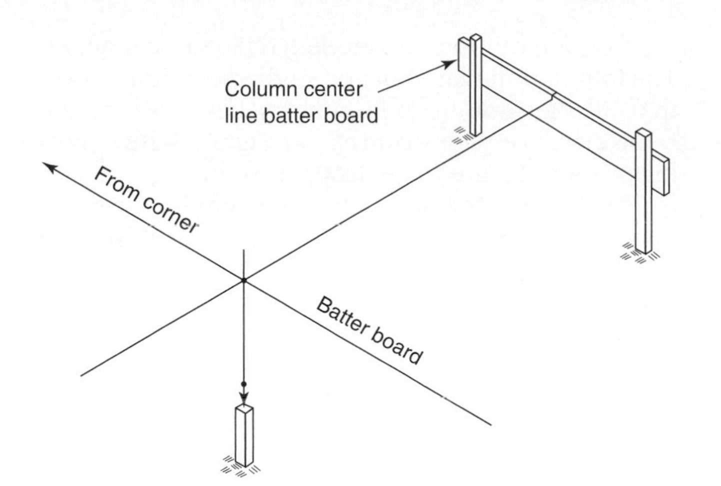


**Figure 4-7** Locating Column Center Line Stake.

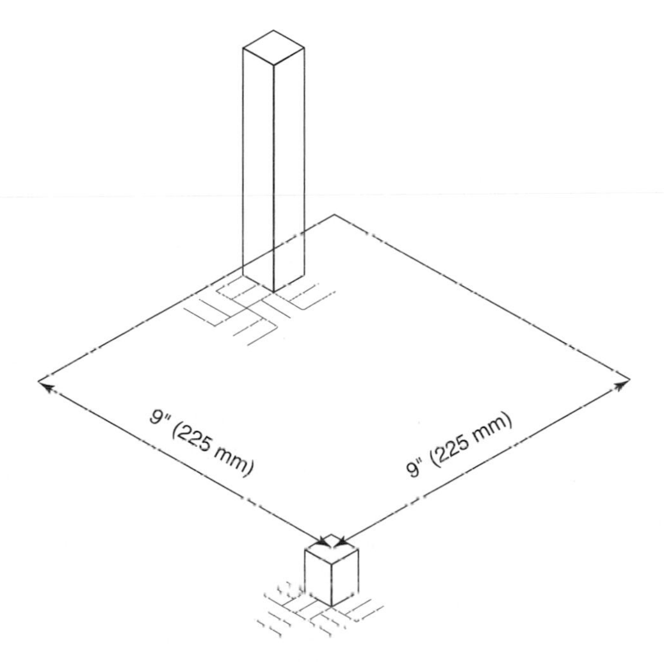


**Figure 4-8** Locating Outside Footing Stake.

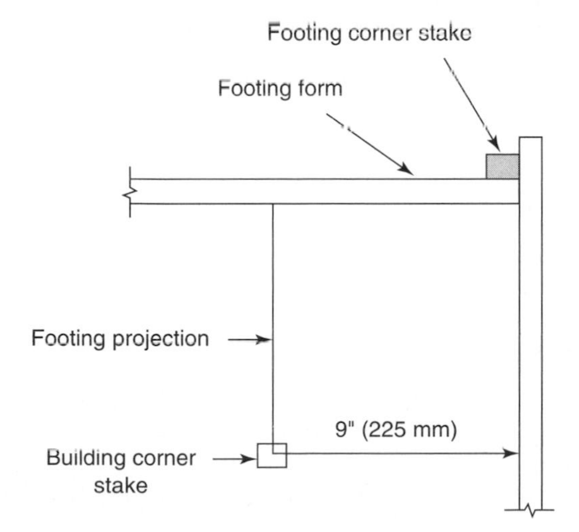

**Figure 4-9** Footing Corner Stake Position.

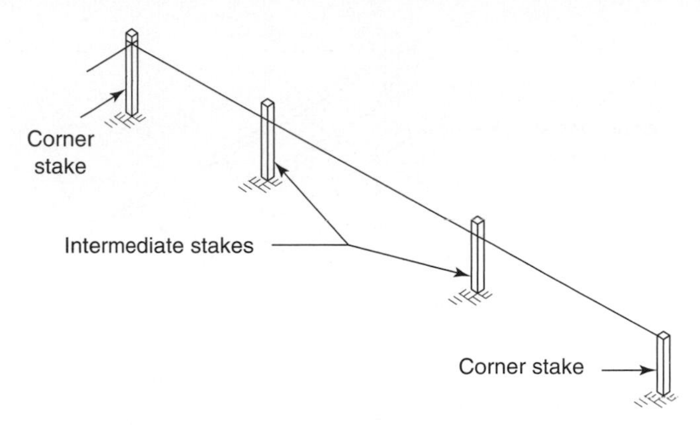


**Figure 4-10** Line for Outside of Footing.

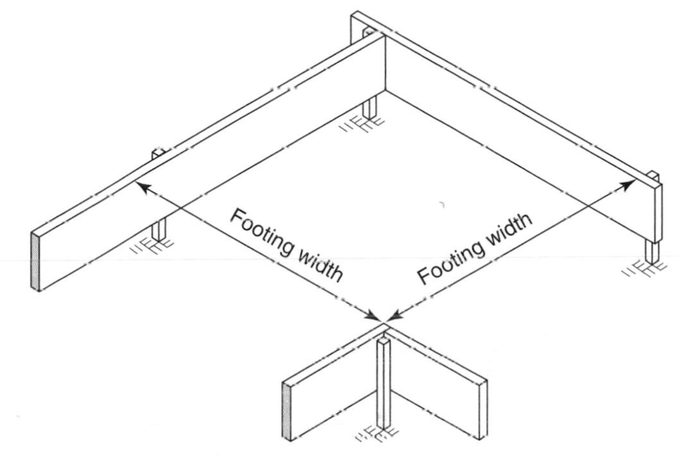


**Figure 4-11** Inside Footing Stake Position.

88.58 ft. In Figure 4-2(b), the elevation to the underside of footings F1 to F5 is indicated 95.920 m and the footings are 600 mm thick.

Stakes are similarly placed at each outside corner, and lines are strung between them to guide the driving of intermediate stakes (Figure 4-10). Stakes for the inside form are located by measuring from this line. The inside corner stake must be set away from the line in both directions, as shown in Figure 4-11.

If footings are wide, their inside and outside extremities may be located on batter boards. The position of inside and outside forms can then be determined from batter board lines (see Figure 4-12).

If column footings are not large, they can be positioned as in Figure 4-13. Here the footing form has been made up and set over the center line stake. It is positioned correctly by measuring from the stake and the line and finally leveled. Large column footings may be located by double batter board lines, in a manner similar to that previously described. Double intersecting lines will establish all four corners of a column footing.

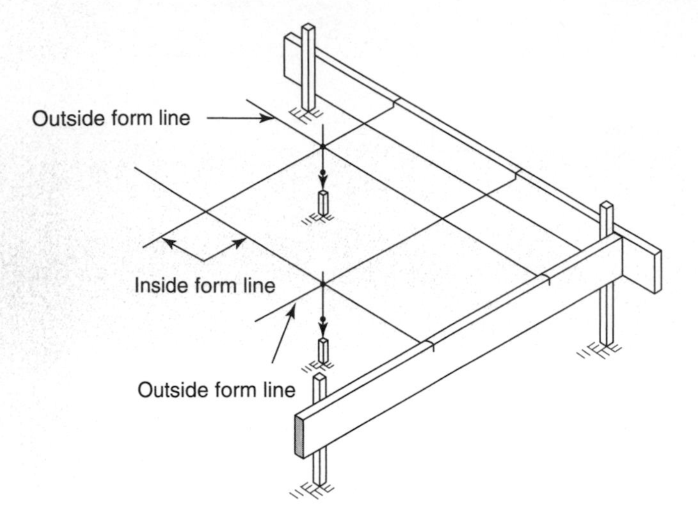


**Figure 4-12** Double Batter Board Lines.

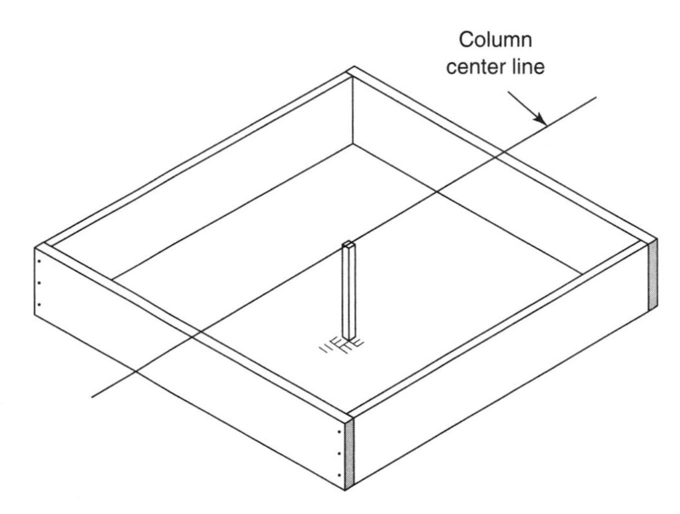


**Figure 4-13** Positioning Column Footing Form.

## LEVELING OF FOOTING FORMS

The first stake driven to a known elevation is now used as a reference point to level the remainder of the footing stakes or the top of footing forms. Set up the instrument in a convenient location within the perimeter of the building. Take a backsight on a rod held on the first stake and tighten the target in position. Now have the rod held alongside the next stake to be leveled and have it moved up or down until the horizontal line on the target coincides with the cross hair. Mark the stake at the bottom of the rod. The form can then be fastened to the stake so that its top edge coincides with the mark. If the forms themselves are being leveled, hold the rod on the top edge and raise or lower the form until it shows level. It is fastened in place when level.

Column footings are leveled in the same manner. The foundation plan must be studied carefully to see that all the footings and column bases are at the same elevation. For example, in Figure 4-2(a), footings F8 and F9 are to be dropped 16 in.

The layout for perimeter pile centers (see Figure 4-4) may be carried out by first establishing and staking the centers of the corner piles from batter board lines. Lines may now be strung between these points and the location of the remainder of the perimeter piles found by measuring along the lines. Pile spacings are shown on the foundation plan. In the particular case shown in Figure 4-4(b), 14 in. (350 mm) cast-in-place concrete piles are specified.

The column piles may now be located by measurement. The elevation of the tops of all piles is indicated on the foundation plan and must be checked as the piles are placed.

## BENCH MARKS AND DEEP EXCAVATIONS

It is easy to establish an elevation at the bottom of a shallow excavation (no deeper than the rod being used) from a bench mark at ground level. For deeper excavations, however, other methods must be used. One method is outlined in the following example. Remember that the *angles of depression* must be read accurately. Consider the excavation shown in Figure 4-14(a). Let us suppose that for this particular building the elevation of the top of wall footings is to be +76.5 ft with relation to a bench mark of +100.0 ft established at ground level. The excavation is 106 ft wide and 25 ft deep (for a 100-ft wide building). The elevation of the excavation floor must be established. To establish a reference elevation on the excavation floor, proceed as follows:

1. Set up and level the instrument at a convenient location near the edge of the excavation (see Figure 4-14(a).
2. Take a backsight on the bench mark. Suppose that the reading is 4.76 ft. Then the resulting H.I. is 104.76 ft.
3. Pick a convenient spot on the floor of the excavation and drive a stake at that point (Point C in Figure 4-14(a)). Lower the telescope and sight on the top of the stake. The angle of depression is found to be 17°30′.
4. Measure a definite distance (30 ft in this case) in the same line of sight and drive a second stake at D, level with the first. Let the angle of depression to the top of Stake D be 25°20′. Both of these readings must be very accurate.

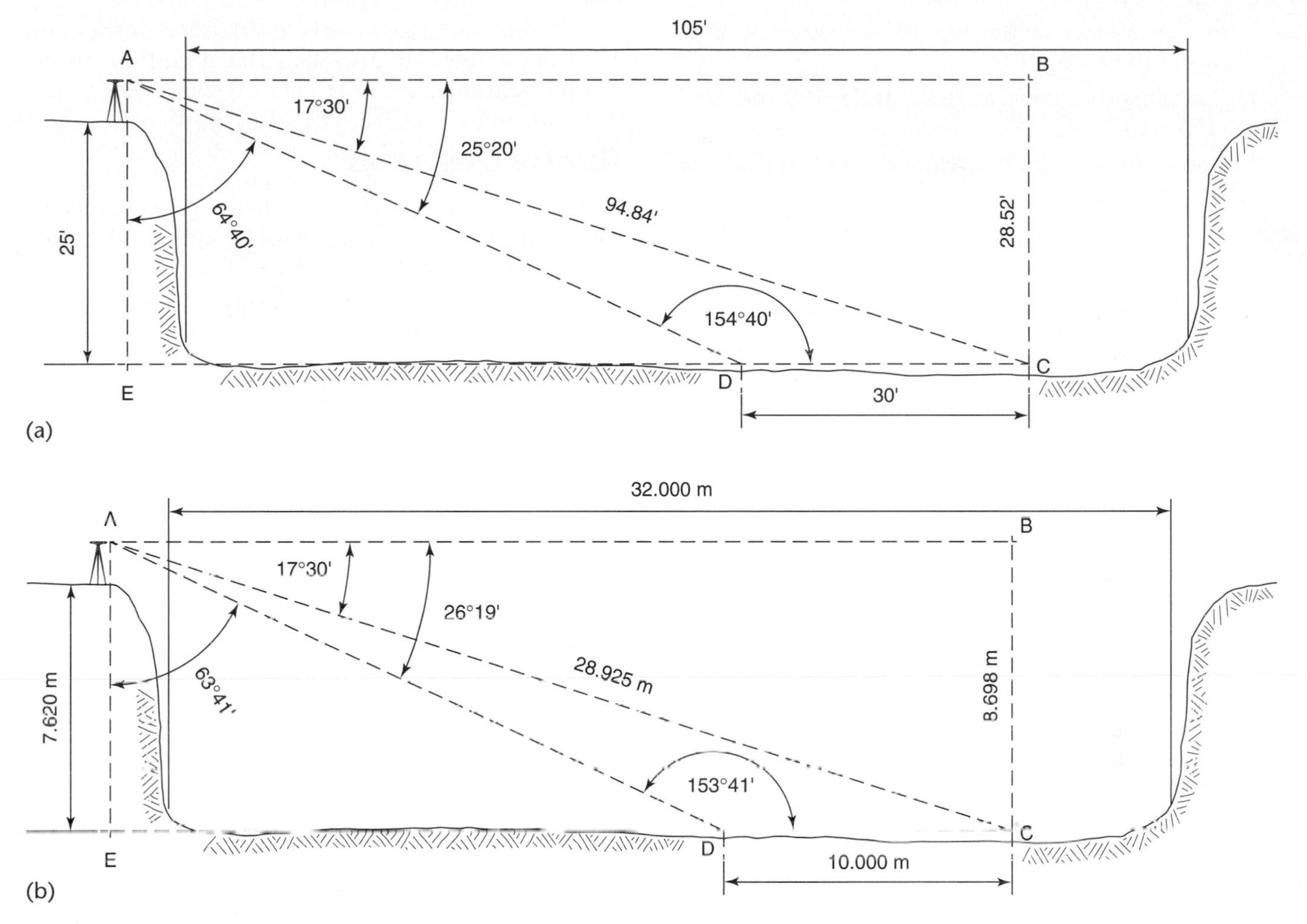


**Figure 4-14** (a) Elevation in Deep Excavation, Feet. (b) Elevation in Deep Excavation, Metres.

5. Calculate the size of angles *DAC* (7°50′) and *ADC* (154°40′).
6. Apply the sine rule to triangle *ADC* to find the length of side *AC*.

$$\frac{30}{\text{sine } 7°50'} = \frac{AC}{\text{sine } 154°40'}$$

$$30 \times 0.42788 = AC \times 0.13629$$

$$AC = \frac{30 \times 0.42788}{0.13629} = 94.84 \text{ ft}$$

7. With respect to triangle *ABC*, *BC/AC* = sine 17°30′; therefore,

$$\frac{BC}{94.84} = 0.30071$$

$$BC = 0.30071 \times 94.84 = 28.52 \text{ ft}$$

8. The elevation of the top of Stake C, relative to the bench mark, will be 104.76 − 28.52 = 76.24 ft.

Stake C may now be used as a control mark from which to establish the elevation of footing forms, and the like, for the foundation.

Using metric units, the excavation dimensions are shown in Figure 4-14(b). Let us assume that for this building the elevation of the top of wall footings is to be +92.800 m with reference to a bench mark of +100.000 m established at ground level. The excavation is 32 m wide and 7.62 m deep (for a 30-m wide building). To establish the reference elevation, step one is the same as in the preceding example. The calculations resulting from the given information are as follows:

1. The backsight reading is 1.451 m. The resulting H.I. is 101.451 m.
2. The angle of depression is found to be 17°30′ on top of the stake at Point C in Figure 4-14(b).
3. Measure a definite distance (10 m in this case) in the same line of sight and drive a second stake at D, level with the first. Assume the angle

of depression to the top of the stake at D is found to be 26°19′.

4. Calculate the size of angles *DAC* (8°49′) and *ADC* (153°41′).
5. Apply the sine law to triangle *ADC* to find the length of side *AC*.

$$\frac{10.000}{\text{sine } 8°49'} = \frac{AC}{\text{sine } 153°41'}$$

$$10.000 \times 0.44333 = AC \times 0.15327$$

$$AC = \frac{10.000 \times 0.44333}{0.15327} = 28.925 \text{ m}$$

6. With respect to triangle *ABC*, *BC/AC* = sine 17°30′; therefore,

$$\frac{BC}{28.925} = 0.30071$$

$$BC = 0.30071 \times 28.924 = 8.698 \text{ m.}$$

7. The elevation of the top of Stake C, relative to the bench mark, will be

$$101.451 - 8.698 = 92.753 \text{ m}$$

Stake C may now be used as a control point from which to establish the elevation of the foundations.

## Review Questions

**1.** Give three reasons for establishing column center lines rather than outside lines of footings when laying out column footing locations.

**2.** Explain why it may be preferable to have both inside and outside form lines of wide footings established on batter boards.

**3.** Why should footing form stakes be cut off to elevation?

**4.** Suggest an alternative method to the use of batter boards and lines that might be used to transfer points to the footings in an excavation.

**5.** Suggest an alternative method of establishing the elevation at the bottom of an excavation from a datum point at ground level.

# 5 FOUNDATIONS

## NEED FOR DEEP FOUNDATIONS

The foundation of a building is generally regarded as that part of the structure that transmits the superimposed load of the building to the supporting soil. The foundation must be proportioned to ensure that the superimposed building loads do not exceed the load-bearing capacity of the soil and that differential settlements are kept to a minimum. If the load-bearing characteristics of the soil are good, the depth of the foundation will depend primarily on the building design; that is, the depth of the basement and the arrangement of the structural framing. If the soil is unstable, then the foundation must penetrate the poor soil to a more stable stratum.

Normally, soils near the surface are less stable and less densely compacted than the material some distance below the surface. By removing the upper layer of unstable soil, the building weight, if distributed evenly, can be supported by the more densely compacted material that has been exposed. When good soil-bearing properties are found near the surface, the foundation can be placed at some minimum depth below grade, usually dictated by the building design or by the depth of frost penetration.

A foundation that is a relatively short distance below finished grade is termed a *shallow foundation*. Shallow foundations are usually made of cast-in-place reinforced concrete and consist of isolated pad footings to support column loads, continuous strip footings under load-bearing walls, or a variation of the two to ensure the proper distribution of loads. However, the term is relative as there is no hard rule for determining when a shallow foundation becomes a deep foundation.

Foundations that must resist large gravity loads and load reversals due to the overturning effects of wind loads require greater foundation depths to maintain the stability of the superstructure. For large buildings, it is not uncommon for portions of their foundations to extend to bedrock to develop adequate support for the imposed loads. This type of foundation can be termed a *deep foundation* and is accomplished through the use of piles, caissons, deep wall foundations, mats and, in some cases, a combination of mats and piles.

The type of foundation used in any particular case depends on the size and distribution of building loads, the depth from the ground surface to the stable layer of bearing material, the type of material through which the foundation must pass, and the location of the structure with respect to other buildings (see Figure 5–1).

## TYPES OF PILES

Piles can be classified by pile material, method of placement, and method of load transfer. Pile sections are fashioned from the traditional building materials: wood, steel, and concrete. They may be composite in nature, using different materials for different portions of the pile. A pile with a wooden lower section and a concrete upper section is the most common composite type.

Two methods are used for placing piles: driving and pouring. Wood, steel, and precast concrete

**Figure 5-1** Building Site Surrounded by Heavy Buildings.

sections are driven; concrete piles and caissons are poured into predrilled shafts that may or may not be lined with steel casings. The type of pile placement is usually determined by the soil conditions on the site and the magnitude of the loads that must be supported.

When considering load transfer characteristics, there are four variations: *bearing piles, friction piles, friction plus bearing,* and *sheet piles*. The bearing pile transfers loads through the unstable surface soils to the denser, more stable soils below. Loads are carried vertically through the pile, and the load capacity of the pile depends on the end cross-sectional area and the bearing strength of the material on which it bears. Bedrock is usually considered the ideal material on which to bear piles; however, this is often not practical because of depth, and less dense material must be used as the bearing medium. To increase the load-bearing capacity of bearing piles, their ends are belled to increase the end-bearing area.

The friction pile, on the other hand, does not necessarily reach high-bearing materials, but depends on the frictional resistance developed between the soil through which it passes and the surface of the pile. Thus, the load capacity of the friction pile depends on the surface area of the pile in contact with the soil and the shear strength of the soil. Piles of this type are used in areas that have deep deposits of clayey soils, and they may be poured in place or driven. Driven piles usually develop all of their load capacity in this manner; however, some end bearing may be realized because of the consolidation of the soil around the end of the pile due to the driving process. Cast-in-place concrete friction piles can also take advantage of end bearing if the pile tip terminates in a relatively dense layer of soil.

Sheet piles are normally not intended for vertical loads but are designed to resist horizontal pressure. The amount of pressure that they can resist is related directly to the depth to which they are driven. The principal uses for sheet piling are to hold back earth embankments, to stabilize the sides of excavations, or to serve as temporary cofferdams.

## Parts of a Typical Pile

Regardless of the type of material from which they are made, all pile sections have certain basic parts. Figure 5–2 illustrates a typical pile and its parts. The following definitions apply to any such pile:

- *Head:* the upper part in final position.
- *Foot:* the lower part in final position.
- *Tip:* the small end before or after it is placed in position.
- *Butt:* the large end before or after it is placed in position.
- *Pile ring:* a wrought iron or steel hoop that is placed on the head of the pile to prevent cracking, brooming, or splitting.

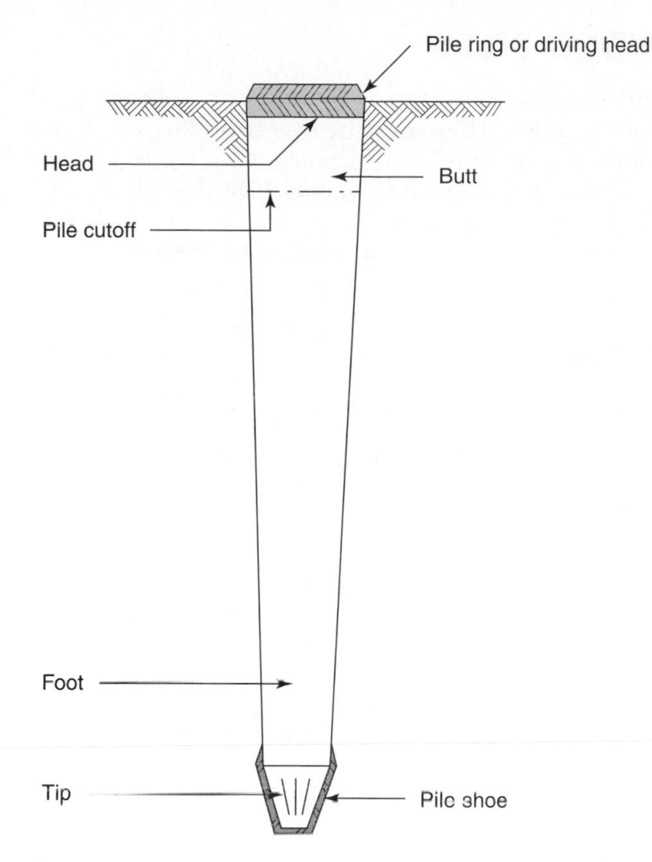


**Figure 5-2** Parts of a Typical Pile.

- *Driving head:* a device placed on the head of a pile to receive hammer blows and to protect it from injury while it is being driven. A driving head may be used instead of a pile ring.
- *Pile cutoff:* the portion of the pile that is removed after completion of driving.
- *Pile shoe:* a metal cone placed on the tip of a pile to protect it from cracking or splitting. The pile shoe also helps the pile to penetrate such materials as riprap, coarse gravel, shale, or hardpan.

## BEARING CAPACITY OF PILE FOUNDATIONS

The design of pile foundations consists of two general steps: (1) the selection and design of the piles and the driving equipment to be used, and (2) the study of the soils to which the loads are transmitted. The allowable bearing resistances for soils range from 1 ton/sq ft (95 kN/m$^2$) for clay up to approximately 30 tons/sq ft (2,900 kN/m$^2$) for granite bedrock. The average compressive bearing values for gravel soils range between 5 to 6 tons/sq ft (450 to 575 kN/m$^2$). The supporting strength of a pile is therefore proportional to its size and to the strength of the soil into which it is driven.

The following examples demonstrate the approach that is taken to compute the load resistance developed by a driven wooden pile, in both standard and metric units.

---

***Example:***

A 35-ft wooden pile having an average diameter of 12 in. and a tip diameter of 8 in. is driven into soil that has an allowable bearing capacity of 4 tons/sq ft and a shear strength of 700 lb/sq ft. Determine the safe carrying load of the pile assuming the top 6 ft does not resist any load due to unstable soil conditions.

***Solution:***

End-bearing capacity = allowable soil bearing capacity × tip area of the pile

$$\text{End area} = \frac{22}{7} \times \left(\frac{4}{12}\right)^2$$
$$= 0.35 \text{ sq ft}$$

$$\text{End-bearing capacity} = 8{,}000 \times 0.35$$
$$= 2{,}800 \text{ lb}$$

$$\text{Effective pile surface area} = \frac{22}{7} \times \frac{12}{12} \times (35 - 6)$$
$$= 91.14 \text{ sq ft.}$$

$$\text{Shear capacity of soil} = 91.14 \times 700$$
$$= 63{,}800 \text{ lb}$$

$$\text{Total pile capacity} = 2800 + 63{,}800$$
$$= 66{,}600 \text{ lb}$$

---

***Example:***

A 10.7 m wooden pile having an average diameter of 300 mm and a tip diameter of 200 mm is driven into soil that has an allowable bearing capacity of 380 kN/m$^2$ and a shear strength of 33.5 kN/m$^2$. Determine the safe carrying load of the pile assuming the top 2 m do not resist any load due to unstable soil conditions.

***Solution:***

End-bearing capacity = allowable soil-bearing capacity × tip area of the pile

$$\text{End area} = \frac{22}{7} \times \frac{(100)^2}{(1{,}000)^2}$$
$$= 0.0314 \text{ m}^2$$

$$\text{End-bearing capacity} = 380 \times 0.0314$$
$$= 11.94 \text{ kN}$$

$$\text{Effective pile surface area} = \frac{22}{7} \times \frac{300}{1{,}000} \times (10.7 - 2)$$
$$= 8.20 \text{ m}^2$$

$$\text{Shear resistance of soil} = 8.20 \times 33.5$$
$$= 274.7 \text{ kN}$$
$$\text{Total resistance of pile} = 11.94 + 274.7$$
$$= 286.64 \text{ kN}$$

## Pile Caps

Piles supporting gravity loads can be used individually or in groups. If two or more piles are used to support loads, the piles must be tied together at the surface with a pile cap. A *pile cap* is a cast-in-place reinforced concrete pad (see Figure 5–3 (b)), of sufficient thickness, poured over the top of the pile group so that the supported load is shared by each supporting pile.

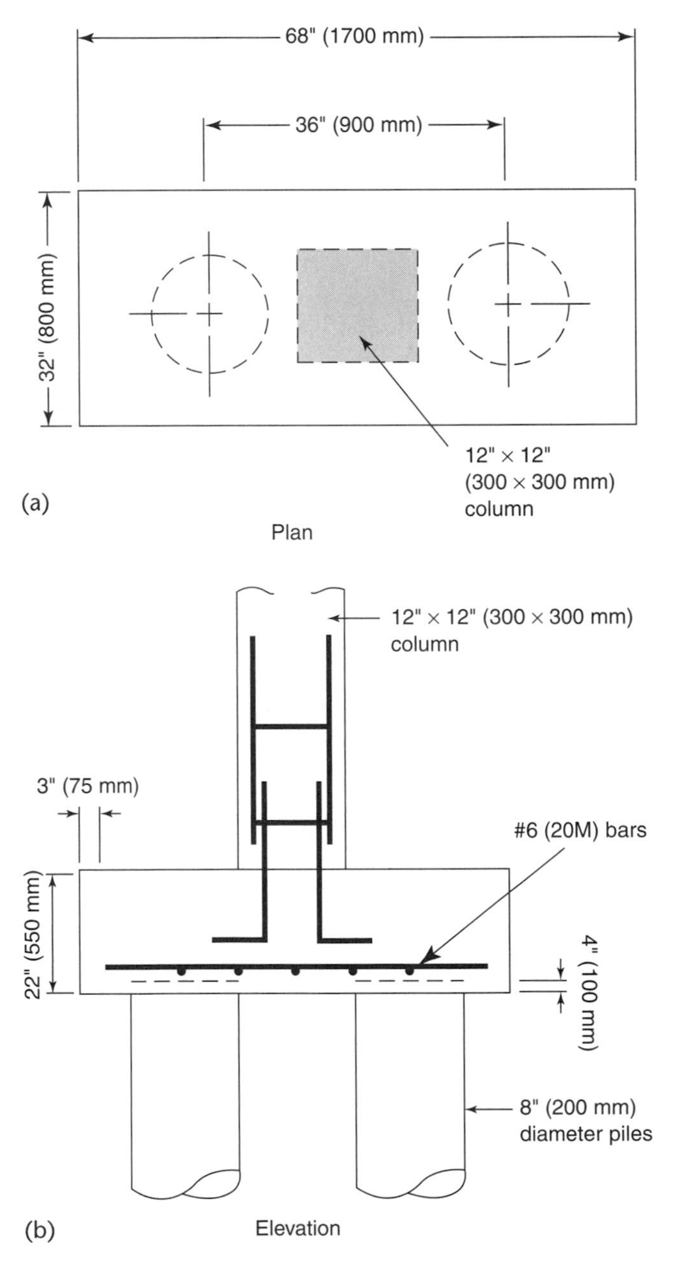

**Figure 5-3** (a) Typical Dimensions of a Pile Cap Supporting a 12 × 12 in. (300 × 300 mm) Concrete Column on Two 8-in. (200-mm)-Diameter Piles. (b) Poured Pile Cap Ready for Column Reinforcing.

# WOOD PILES

Considerable selection is necessary to obtain good wooden piles. They must be free from large or loose knots, decay, splits, and shakes. Crooks and bends should be not more than one-half of the pile diameter at the middle of the bend. Pile sweep should be limited so that (1) for piles less than 70 ft (21 m) in length, a straight line joining the midpoint of the butt and the midpoint of the tip does not pass through the surface of the pile; (2) for piles 70 to 80 ft (21 to 24 m) in length, a similar straight line does not lie more than 1 in. (25 mm) outside the surface of the pile; and (3) for piles more than 80 ft (24 m) in length, a similar straight line does not lie more than 2 in. (50 mm) outside the surface of the pile. The taper should be uniform from tip to butt.

Pile lengths are available in increments of 12 in. (300 mm). The minimum tip diameter is normally 6 in. (150 mm), and the maximum butt diameter for any length is 20 in. (500 mm).

Various species of trees are used as piles, but because of availability, long straight lengths, and their ready acceptance of preservatives, the most commonly used are southern pine, red pine, lodgepole pine, Douglas fir, western hemlock, and larch. In the past, timber piles were pressure treated primarily with creosote; however, new waterborne preservatives such as ammoniacal copper arsenate and chromated copper arsenate are now being used (see Figure 5–4).

Some advantages of using wood piles are as follows:

1. Wood piles have an indefinite life expectancy when placed under water or driven below groundwater level.
2. Wood piles are light.

**Figure 5-4** Driven Pressure-Treated Timber Piles Used in Sandy Soil.

3. In many areas, wood piles are readily available, relatively inexpensive, and easy to transport.
4. Wood piles produce greater skin friction than piles of most other materials.

However, wood piles are subject to attack by insects, marine borers, and fungi unless treated. They have a lower resistance to driving forces than other types of piles and have a tendency to split or splinter while being driven. Wood piles support a smaller load than other types of comparable size, which means using more piles and larger footings.

## CONCRETE PILES

There are two principal types of concrete piles: *cast in place* and *precast*. The cast-in-place pile is formed in the ground, in the position in which it is to be used. The precast pile is usually cast in a factory, where prestressing techniques can be employed, and after curing it is driven or jetted like a wood pile.

Cast-in-place piles are divided into two general groups, the *shell* type and the *shell-less* type. Shell type piles are made by first driving a steel shell or casing into the ground, filling it with concrete, and leaving the shell in place. The shell acts as a form and prevents mud and water from mixing with the concrete. Such piles may or may not be reinforced, depending on circumstances. Shell type piles are useful where the soil is too soft to form a hole for an uncased pile or where the soil is hard to compress and would deform an uncased pile.

Shells may be cylindrical or tapered, with smooth or corrugated outside surfaces. One type of tapered shell produced in sections is known as a *step-taper* pile (see Figure 5–5).

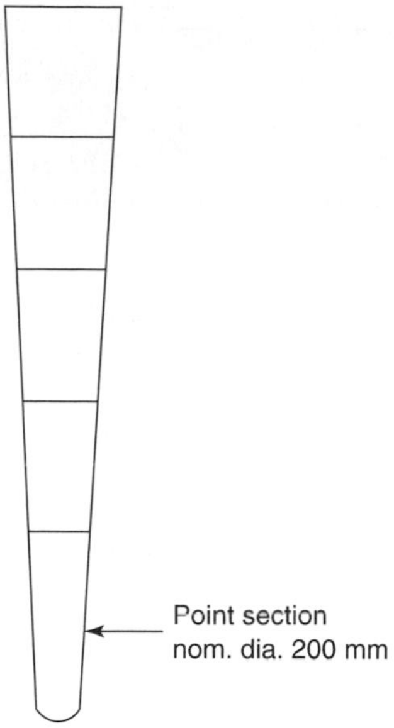

Point section
nom. dia. 200 mm

**Figure 5-5** Typical Step-Taper Pile.

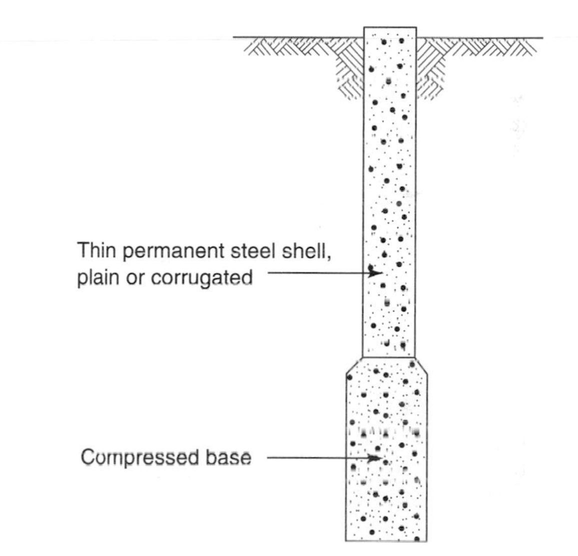


**Figure 5-6** Cased Pile with Compressed Base.

A number of variations of shell type piles are made. One is a cased pile with a compressed base section, as illustrated in Figure 5–6. It is used where the side support of the soil is so light that the pile must be used as a column or when the pile is designed to meet lateral forces arising from eccentric loading.

Another variation is the *button-bottom* pile shown in Figure 5–7. It is used where an increase in end-bearing area is required. The enlarged bottom eliminates side friction support unless the soil is highly compacted around the pile (see Figure 5–8).

*Swaged* piles, shown in Figure 5–9, are used where driving is difficult or where it is necessary to have watertight shells in place before the concrete is

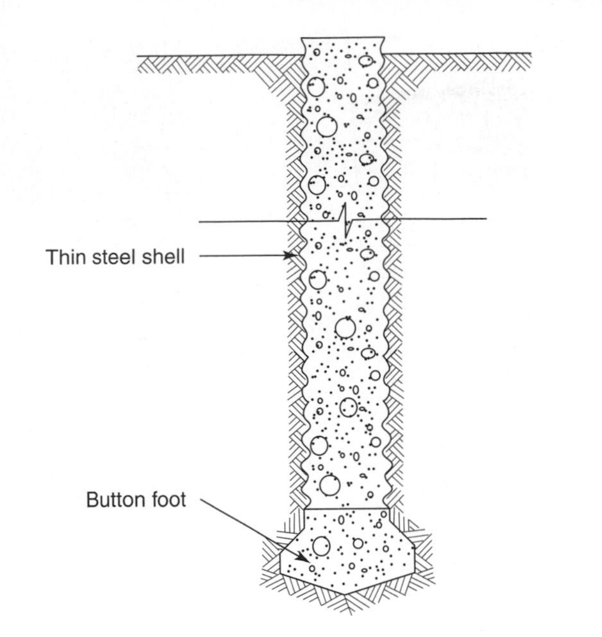


**Figure 5-7** Button-Bottom Cased Concrete Pile.

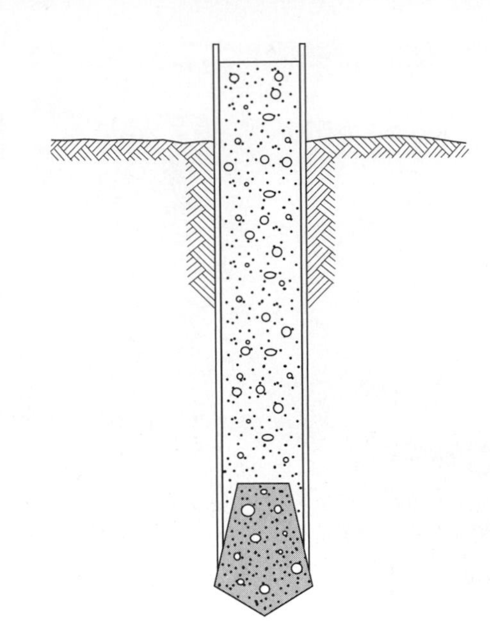

**Figure 5-9** Swaged Pile.

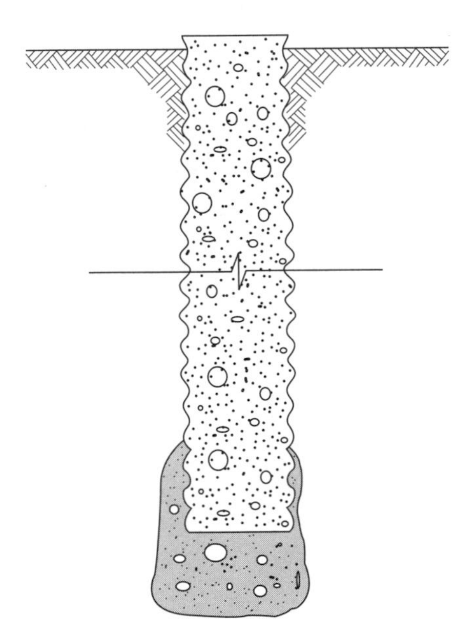

**Figure 5-8** Cased, Pedestaled Concrete Pile.

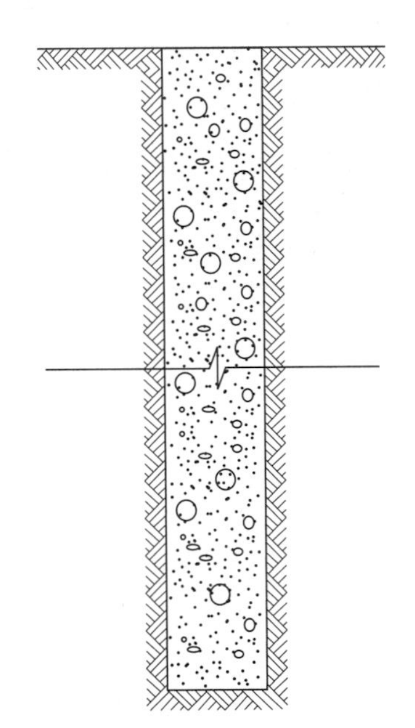

**Figure 5-10** Simple Shell-Less Pile.

placed. This type of pile consists of a steel shell driven over a conical precast concrete end plug.

Shell-less piles are made by driving a steel pipe, fitted with a special end or tapered shoe, into the ground for the full depth of the pile. The pipe is then pulled up, leaving the shoe at the bottom, and the hole is filled with concrete. This type of pile is satisfactory where soil is cohesive enough so that a reasonably smooth inside surface is maintained when the shell is removed. Figure 5–10 illustrates the simplest type of shell-less pile. Sometimes concrete is poured as the shell is being lifted. This eliminates some of the possibility of earth becoming mixed with the concrete.

Variations of the shell-less pile are also made. One involves the use of a bored hole rather than a punched hole. Figure 5–11(a) shows a drilling rig in place on a job site ready to begin drilling. Figure 5–11(b) shows a casing in place ready for concrete. In this instance, the casing will be withdrawn after the concrete has been placed. Figure 5–11(c) illustrates a typical reinforcing cage for a poured pile. Piles that must provide extra bearing resistance are belled at the bottom. Figure 5–12 illustrates the belling attachment used to increase the bearing area of the end. Figure 5–13 shows an auger loaded with earth. Figure 5–14 illustrates a vertical section through a completed belled pile.

(a)

(b)

(c)

**Figure 5-11** (a) Drilling for Cased Pile. (Reprinted by permission of Calweld, Inc.) (b) Casing for Pile. Note the Auger in the Foreground Used for Drilling the Hole. (c) Reinforcing Cage for Pile.

**Figure 5-12** Belling Equipment for Pile Bottoms.

**Figure 5-13** Drill Loaded with Earth.

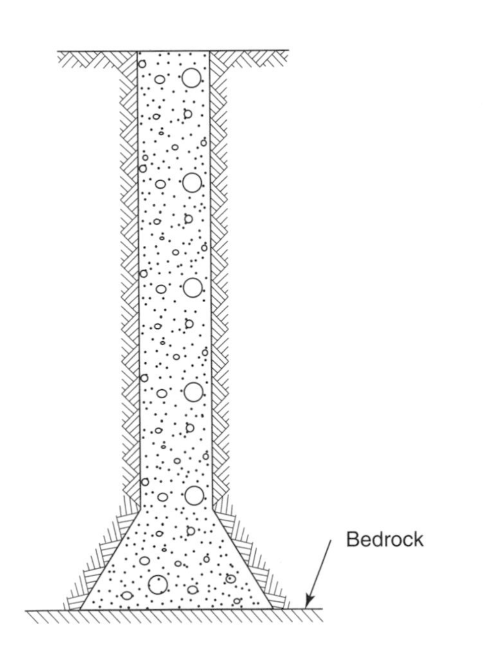


**Figure 5-14** Section Through Completed Belled Pile.

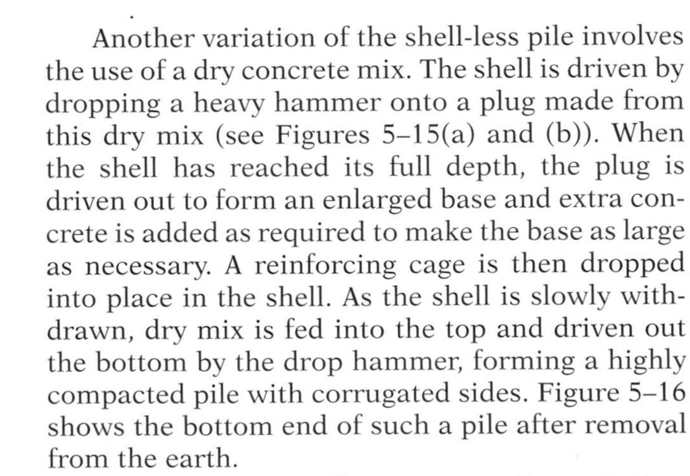

Another variation of the shell-less pile involves the use of a dry concrete mix. The shell is driven by dropping a heavy hammer onto a plug made from this dry mix (see Figures 5–15(a) and (b)). When the shell has reached its full depth, the plug is driven out to form an enlarged base and extra concrete is added as required to make the base as large as necessary. A reinforcing cage is then dropped into place in the shell. As the shell is slowly withdrawn, dry mix is fed into the top and driven out the bottom by the drop hammer, forming a highly compacted pile with corrugated sides. Figure 5–16 shows the bottom end of such a pile after removal from the earth.

Precast concrete piles are normally made in a casting yard under controlled conditions. This allows not only the development of high-strength concrete but also flexibility in design, reinforcing, length, and so on.

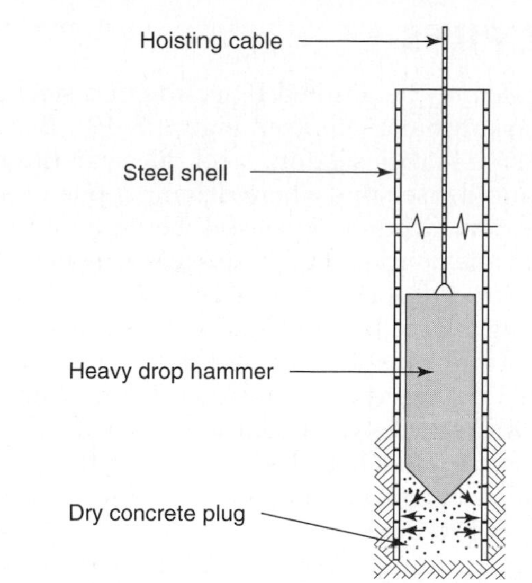


(b)

**Figure 5-15** (a) Driving a Shell with Drop Hammer and Concrete Plug. (b) Typical Steel Shell with Drop Hammer.

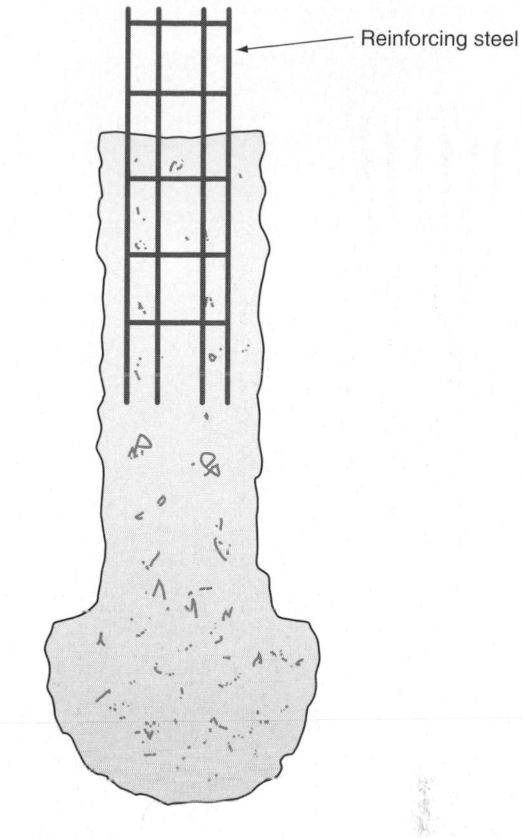


**Figure 5-16** Bottom End of Compacted Shell-less Pile.

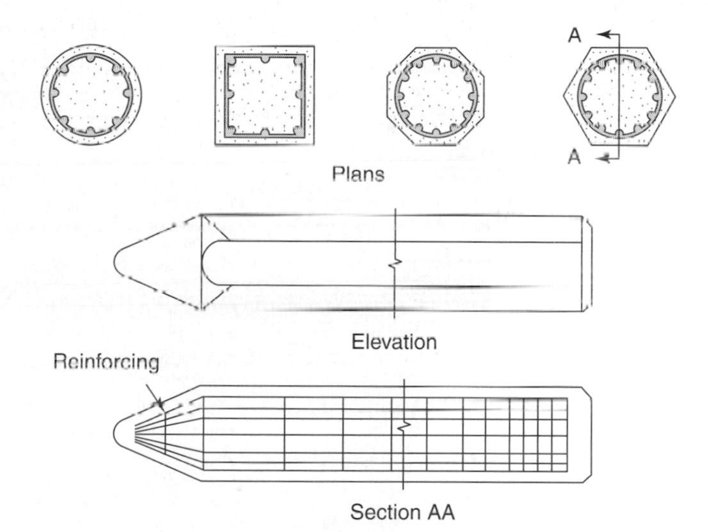


**Figure 5-17** Precast Pile Shapes.

Precast reinforced piles are made in round, square, hexagonal, and octagonal shapes, as illustrated in Figure 5–17. Very long piles are made by casting hollow reinforced sections, usually 16 ft (4.8 m) in length, and joining the sections together by stressed steel cables.

Precast concrete piles usually need assistance when being driven, especially in sand, and one method of providing assistance is to use a water jet. One type of precast pile has been developed especially for installation by jetting. It is tapered, octagonal in cross section, and has one or two vertical grooves in each face. The center is hollow, allowing water to be forced down to loosen the soil at the tip. The grooves allow the jetted water to return to the surface (see Figure 5–18).

Precast piles may be impregnated with asphalt to help prevent damage by seawater, spray, and the like. Piles are pressure treated with hot asphalt at 450°F to 500°F (230°C to 260°C), and penetrations of up to 1½ in. (40 mm) are obtained.

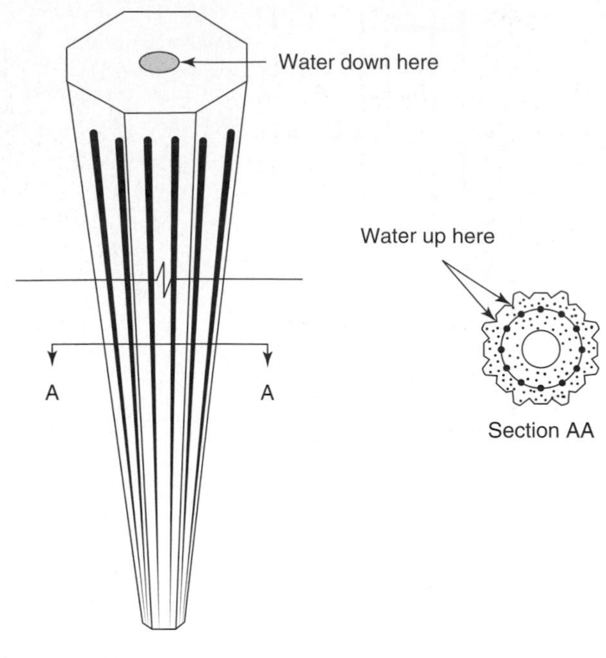


**Figure 5-18** Precast Jetting Pile.

## STEEL PILES

A steel pile may be a rolled H section or a steel pipe known as a tubular pile (see Figure 5–19). Because of their small cross section, steel piles can often be driven into dense soils where driving a pile of solid cross section through dense soil would be difficult. However, one point should be considered when driving long structural shapes: the possibility of the pile tip striking a large boulder below the surface. There have been many instances in which long steel pilings have been deflected by massive boulders below the surface and, as a result of continued driving, the pile tip has worked itself to the surface of the ground again. This, of course, makes it necessary to extract the pile and to determine the limits of the boulder before proceeding. This operation can become rather expensive and points out the advisability of having a clause in the contract to protect the builder from such contingencies.

Because steel piles develop their load resistance primarily in friction, they are usually driven in clusters. Once each pile develops a certain re-

**Figure 5-19** Air Hammer Driving Steel HP-Shape Piles. (Reprinted by permission of Bethlehem Steel Co.)

**Figure 5-20** Hollow Pipe Sections with H Sections Welded to the Ends for Use as Piles.

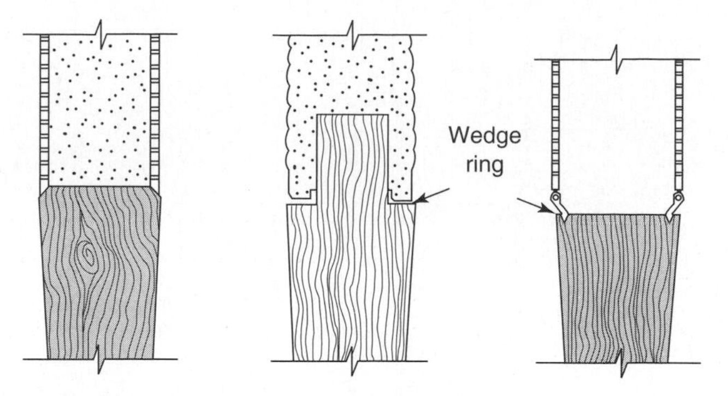


**Figure 5-21** Composite Pile Connection.

quired load resistance, the top is cut off to some predetermined elevation. The cluster is then covered with a pile cap, which helps to distribute the supported load uniformly to all the piles in the cluster.

To aid the driving of a hollow pipe through rocky terrain and to prevent the pipe filling with earth, a protective tip may be welded onto the leading end (see Figure 5–20).

Tubular piles vary in diameter from 8 to 72 in. (200 mm to 1,800 mm). They may be driven from the top, using a drop or mechanical hammer, or from the bottom, using a drop hammer falling on a concrete plug. The pipe is usually filled with concrete, which means that if the pipe is driven open-ended, the earth must be removed from the inside by a water jet.

## COMPOSITE PILES

Two materials may sometimes be combined to make up a single pile. The most common combination is wood and concrete, and the result is known as a composite pile.

This type of pile may be used to advantage under several conditions. One is if the permanent water head is not more than 70 ft (21 m) below ground level. This is about the length limit for the upper concrete section of the pile. Another is where the use of wood piles alone would necessitate 10 ft (3 m) or more of dry excavation or as little as 4 ft (1,200 mm) of very wet excavation. This can be eliminated by the use of a composite pile. A third situation is when the overall length of the pile is so great that it would be economically impractical to obtain or handle either straight wood or concrete piles. Cast-in-place concrete piles of up to approximately 70 ft (21 m) in length can rest on top of wood piles of any obtainable length. Wood piles are difficult to handle and expensive when longer than 80 ft (24 m). Precast concrete piles are very expensive in lengths beyond 55 ft (17 m).

The wooden pile is driven to ground level, and the head of the pile is fitted with a steel casing. Driving continues by means of a mandrel or core until the required depth is reached. The mandrel is then withdrawn and the shell filled with concrete.

The shell is fitted to the wooden pile head in several different ways. One method is to set the end of the shell over the slightly tapered end of the pile head. Another method is to make a tenon with a square shoulder at the end of the pile, set a sealing ring over the tenon, and rest the shell on the ring. Another method involves the use of a wedge ring, which is forced into the flat top of the pile head. Figure 5–21 illustrates these three methods of connection.

A number of piles are sometimes driven close together in a group, to be later capped with a common concrete cap. Several points pertaining to this particular operation warrant special consideration.

First, it is generally recognized that when a number of piles are driven close together the load-bearing value of each is reduced. As a result, a number of building codes have established a minimum center-to-center spacing for grouped piles. These vary from 2 ft 6 in. to 5 ft (750 mm to 1,500 mm), with the latter considered less desirable.

There are precautions to be taken in the use of piles and limitations to their use. For example, when piles are driven close together, those driven last may cause those driven first to heave. This may necessitate some redriving but, in any case, a check should be made on the elevation of the tops of the piles that are driven first.

Driving piles in the proper order will help to avoid the heaving problem. The piles at the center of the group should be driven first, followed in succession by those closer to the outside. Also, when piles are driven in a group, those driven last are generally more difficult to drive and pull than the earlier ones.

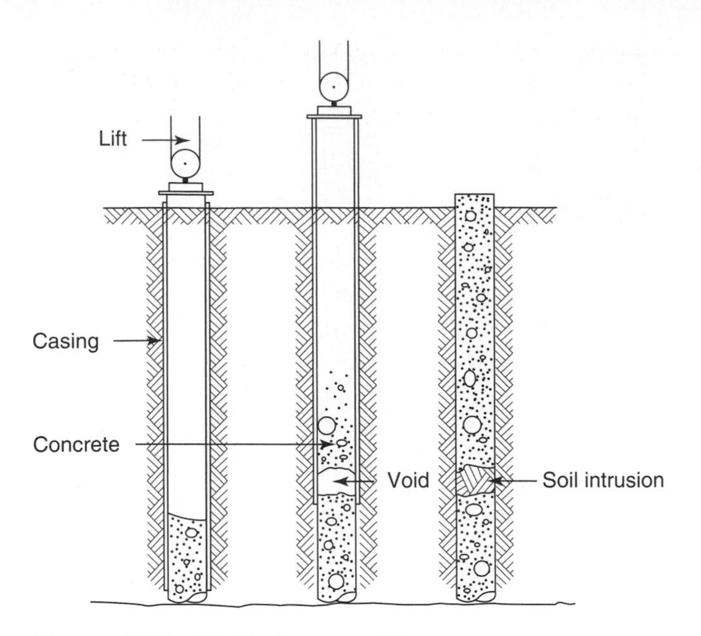


**Figure 5-22** Void in Concrete Pile.

This is due to the increased compaction of the soil as it is displaced by the piles.

Driving a large number of piles in a group, in an attempt to accommodate a very heavy load, can cause disturbance to the surrounding area. If they are displacement piles, the heaving caused by this disturbance can be quite destructive to the foundations of buildings in the near vicinity.

When shell-type piles are being used, the increased compaction due to group driving may result in damage (distortion or crushing) to shells already in place. As a result, it may be difficult to place concrete or the passage may even be completely blocked.

One source of trouble with shell-less piles is related to the practice of pulling the shell as the concrete is being placed. Concrete may adhere to the inside of the shell if it is not clean or if the weather is hot and some initial set takes place before the shell is withdrawn. This may result in the concrete pile being pulled apart (see Figure 5–22) and the space filling with soil. Thus, the bearing capabilities of the pile are destroyed.

## PILE DRIVERS

Wood, precast concrete, steel H piles, and large tubular piles are driven into the ground by a pile driver striking the pile head. Small tubular piles and shells for cast-in-place concrete piles are more often pulled into the ground by a core. The core is forced into the earth by a tapered drop hammer and pulls the shell or tube with it.

**Figure 5-23** Diesel Hammer Driving Steel Bearing Pile.

Pile drivers may consist of a *drop, mechanical,* or *vibratory hammer.* A drop hammer is the simplest type of machine, consisting of a heavy weight, lifted by a cable and guided by leads, which is allowed to drop freely on the pile head.

The hammer of a mechanical driver operates like a piston actuated by steam, compressed air (see Figure 5–19), or the internal combustion of diesel fuel (see Figure 5–23).

A vibratory hammer is secured to the head of a pile and operates by delivering vibrations to the pile head in up-and-down cycles at the rate of 100 cps (see Figure 5–24). These vibrations set up waves of compression in the pile, which in turn produce minute amounts of expansion and contraction. As the pile expands longitudinally, it displaces soil at the pile tip, and the weight of the pile, hammer, and equipment forces the pile into the tiny void. At the same time as the pile expands in length, it contracts slightly in diameter and thus relieves the friction between the earth and the pile surface. This relief of friction helps the pile to move downward.

**Figure 5-24** Vibro Driver Used in Driving Pile Shell.

These movements are very small but occur 100 times every second; as a result, penetration may actually be quite rapid. For example, a pile that might require an hour to be set with a steam hammer can be driven in 2 or 3 minutes by a vibratory hammer. This type of equipment performs best in sandy soil but may be used in silt, clay, or soil containing layers of gravel.

Assistance may have to be provided when piles are being driven. This is particularly true when driving precast piles in granular soils. The high friction rate makes driving very difficult. The assistance is usually in the form of a water jet that softens the soil at the pile tip. Water is jetted down the hollow center and returns to the surface around the outside of the pile.

Figure 5–15 (a) illustrates the principle involved in *pulling* a tubular pile or pile shell into the ground. A tapered drop hammer is dropped on a core made of dry concrete mix, driving it into the ground. At the same time, the hammer's tapered point forces the concrete against the inside wall of the shell, creating tremendous friction. As a result, the friction between the core and the tube pulls the tube down as the core is driven into the ground.

## CAISSONS

The increase in column loads in many modern buildings, due to the wider spacing of columns in order to get more uninterrupted floor space and, in some cases, the depth required to reach good bearing strata, has made pile groups impractical in these situations. In their place, many designers are using *caissons*. Strictly speaking, a caisson is a shell or box or casing that, when filled with concrete, will form a structure similar to a cast-in-place pile, but is larger in diameter. Over the years, however, the term has come to mean the complete bearing unit.

### Bore Caissons

The development of boring tools for large-diameter holes and the equipment to handle them has made the use of *bored caissons* much more common. A bored caisson is one in which a hole of the proper size is bored to depth. A cylindrical casing or caisson is usually set into the hole. The use of heavy equipment has made it possible to bore 10-ft (3-m) diameter holes to a depth of over 150 ft (45 m). Figure 5–25 illustrates such equipment, in this case a 6 ft (1,800-mm) drilling bucket equipped with reamers to enlarge the diameter of the hole.

These boring machines are not suitable in areas where there is wet, granular material, silt, or boulders. The bored caisson method works best in cohesive soil in which the drilled hole will remain open until a caisson can be installed in it or until concrete can be placed into the drilled excavation (see Figure 5–26).

In some cases, drilled caissons may be belled (see Figure 5–14) to provide greater bearing area at the bottom. Belling equipment (see Figure 5–12) is used to a limit of approximately a 20–ft (6–m) diameter bell. Hand digging is necessary for larger bells and to clean up machine-dug bells for proper bearing of the concrete on the rock or hardpan.

If there is a problem with water at the top of the rock surface, belling becomes very difficult, and large, straight-sided caissons are used. The bottom edge of such a caisson is fitted with tungsten carbide teeth and becomes a core barrel that can be rotated into the rock to effect a seal and allow work at the bottom of the caisson. It is possible to bite into the rock as much as 10 ft (3 m) to try to seal off water or the inflow of wet silt.

**Figure 5-25** Six-foot (1.8-Metre) Drilling Bucket with Reamers. (Reprinted by permission of Calweld, Inc.)

**Figure 5-26** Steel Casing Being Used to Line Drilled Caisson Shaft.

## Jacked-In Caissons

In materials such as silt, wet sand, or gravel, which present difficulties in installing bored caissons, a caisson may be forced into the ground by jacks, while at the same time it is rotated back and forth to reduce the side friction. The material inside the caisson is then excavated. In some cases, excavating will be carried out while the caisson is being forced down, thus making the sinking operation easier.

## Gow Caisson

The gow caisson is made in tapered sections (see Figure 5–27), each of which fits inside the one above it. A pit is dug and the first section of steel casing placed in it. It is forced into the ground by driving and excavating inside the cylinder at the same time. A cylinder 2 in. (50 mm) smaller in diameter is then placed inside the first and driven to its full depth. This continues until the full depth of the caisson is reached, at which time the excavation is belled to provide a greater bearing area.

## Drilled-In Caissons

Where loads to be carried are very large, the depth to bearing rock is excessive, or the type of material en-

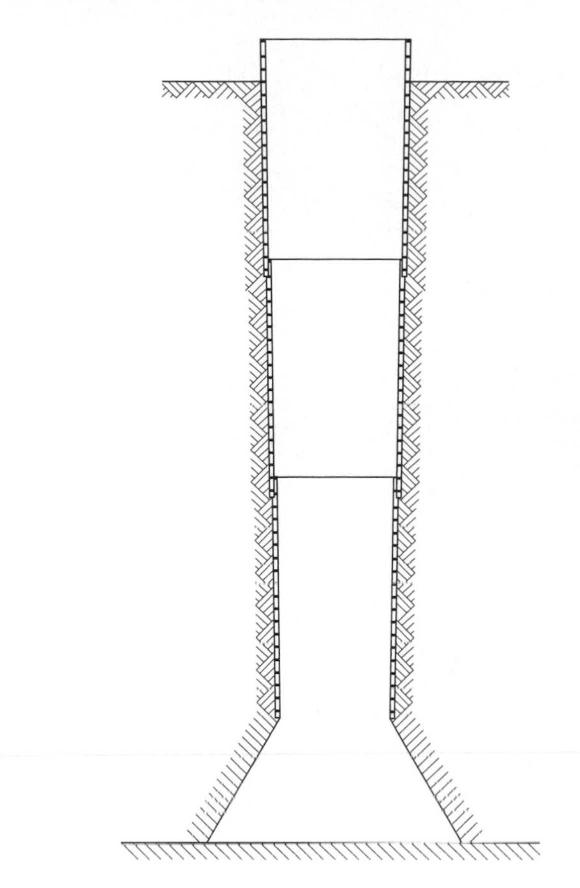

**Figure 5-27** Gow Caisson.

countered is unsuitable for bored caissons, drilled-in caissons may be used. Such a caisson is installed by driving a steel pipe, 24 to 42 in. (600 to 1,200 mm) in diameter and ½ in. (12.5 mm) in thickness, to rock, using a pile hammer. The pipe has a 1¼ in. (32-mm) steel shoe welded to the bottom to reinforce the tip and is excavated by auger, by blowing with compressed air, or by a cable drill rig fitted with suction buckets. Such caissons have been installed to depths of over 200 ft (60 m).

When bedrock is reached, a socket that is slightly smaller in diameter than the caisson shell is churn-drilled into the rock, and heavy steel HP shape or the equivalent in reinforcing rods is installed to extend from the top of the caisson to the bottom of the socket. The entire assembly is then concreted in place, producing a fixed-end caisson.

### Slurry Caisson

The technique of using slurry walls (described in detail in Chapter 3) has now been applied to the construction of deep piers or caissons. A rectangle pit is dug to rock, using bentonite slurry to keep the sides of the excavation from collapsing. The pit is then filled with concrete by the tremie method to form a caisson.

## COFFERDAMS

Although caissons and cofferdams fulfill the same general function, a *caisson* is a permanent structure used for protection while excavating and as a form for concrete, whereas a *cofferdam* is a temporary boxlike structure used to hold back water or earth while work is being done inside it; it is later removed.

A cofferdam consists of sections of sheeting driven side by side to form a watertight unit. The sheeting forms a box large enough to work inside. The sections are usually driven in water or mud, which is removed from the inside once the cofferdam is complete. If the bottom is too soft, a layer of concrete may be placed over it to provide a firm base. Whatever permanent structure is required can then be built inside the cofferdam, which is removed when the work has been completed.

## SPREAD FOUNDATIONS

When stable soil with good bearing values is found within the building site and building loads are well distributed, spread foundations can be used to support the proposed structure. The purpose of a spread foundation is to spread the concentrated loads of the structure over areas large enough to ensure that the bearing capacity of the soil is not exceeded. Spread foundations may be isolated to support concentrated loads from columns or strip footings under walls (see Figure 5–28), or they can be designed as one large monolithic pad under the entire building area.

When unusual framing requirements in the building must be supported or if nonuniform soil bearing characteristics are encountered at the site, piles can be used in conjunction with the pad. There are a number of variations for spread footings, and the one that will be used in any particular case depends on the soil conditions and load concentrations.

Foundations poured as monolithic pads are commonly known as *mat foundations* (Figure 5–29). Mat foundations range in thickness from 3 to 8 ft (900 to 2,400 mm) depending on the loads that must be supported. Reinforcing bars are provided in layers throughout the mat to control cracking of the concrete and to minimize differential settlements within the structure. These types of foundations are used where deep basements (usually for parking) are incorporated into the building design.

In some cases, monolithically poured beams and slabs are used to reduce the overall thickness of the mat. The beams usually run at right angles to one another creating a gridlike pattern. The slab may be poured at the bottom or at the top of the beams, depending on the structural framing of the building (Figure 5–30).

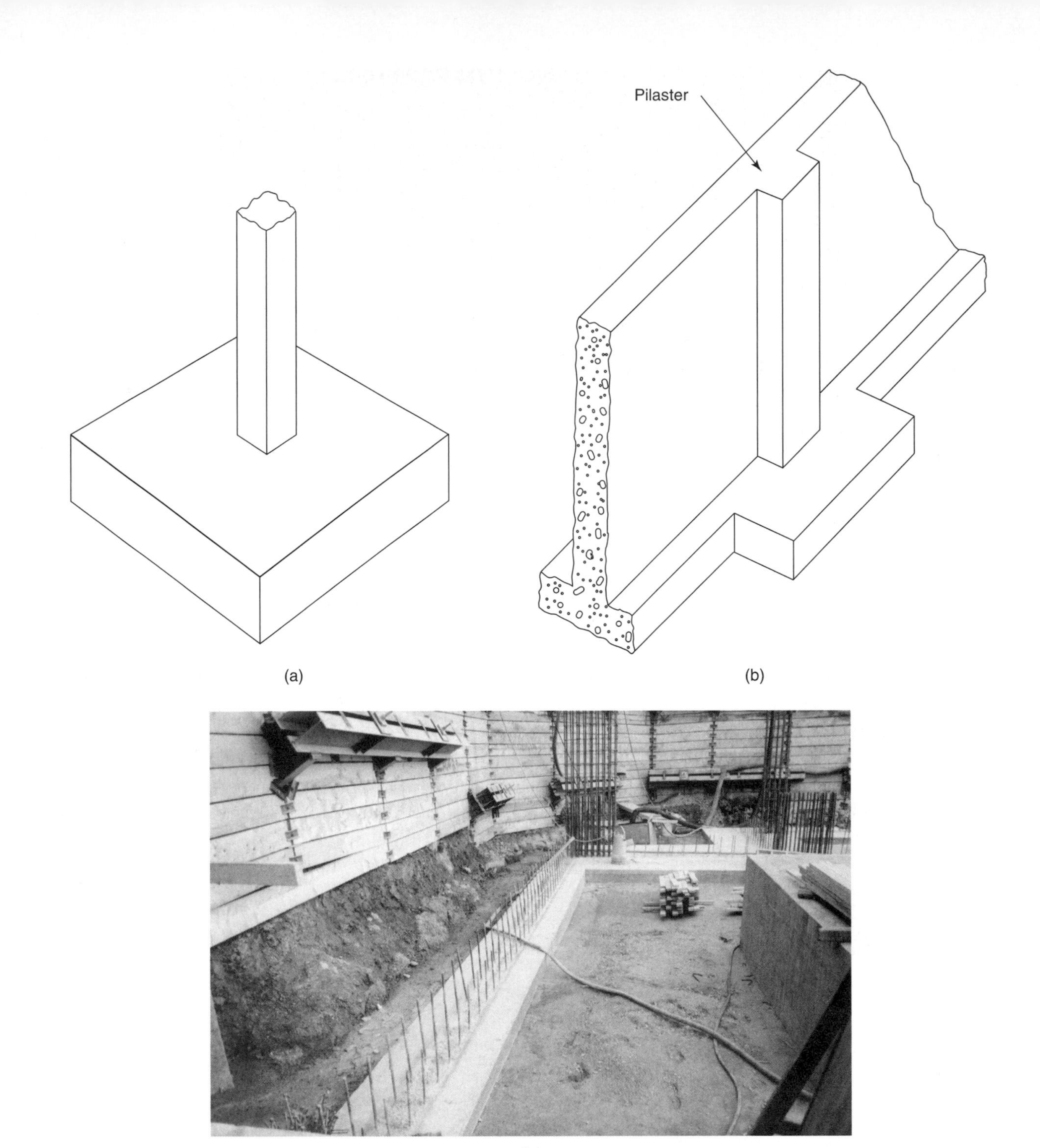


**Figure 5-28** Spread Footings: (a) Isolated Spread Footing Supporting Column; (b) Continuous Spread Footing Under Wall, Enlarged at Pilaster; (c) Typical Continuous Wall Footing.

Another type of mat foundation uses a relatively thin slab with an increase in thickness at column locations (Figure 5–31). The thickened portion spreads the concentrated load over a greater area, much like an isolated spread footing.

When existing footings must be underpinned or increased in size, *grillage footings* can be used effectively for the distribution of building loads by providing additional bearing area. If the overall thickness of the footing is a concern, the use of grillage footings may be appropriate as their overall depth is relatively small. Grillage footings consist of tiers of steel sections placed at right angles to one another, fastened together with bolts and spacers, and capped with a steel plate, which serves as the column base (see Figure 5–32). The

**Figure 5-29** Preparing a Mat Foundation. (Reprinted by permission of Bethlehem Steel Co.)

entire assembly is then encased with concrete if the footing is to be permanent. The concrete stabilizes the entire assembly, while the steel sections provide the necessary flexural strength to resist the pressure of the soil.

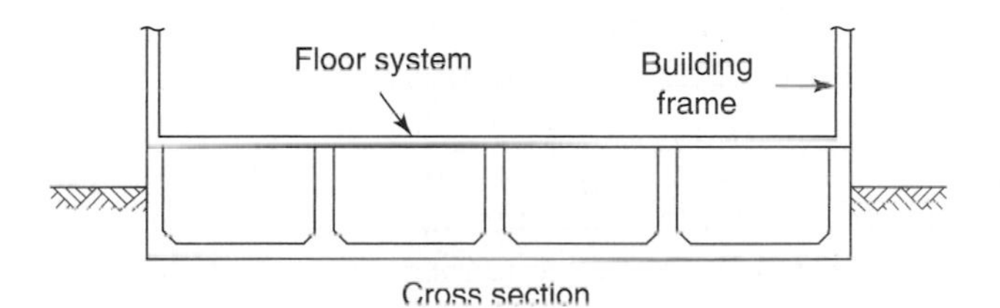


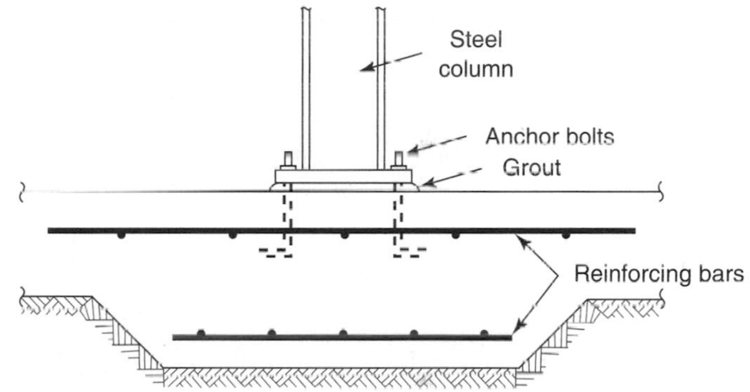


**Figure 5-31** Mat Foundation Thickened under Column.

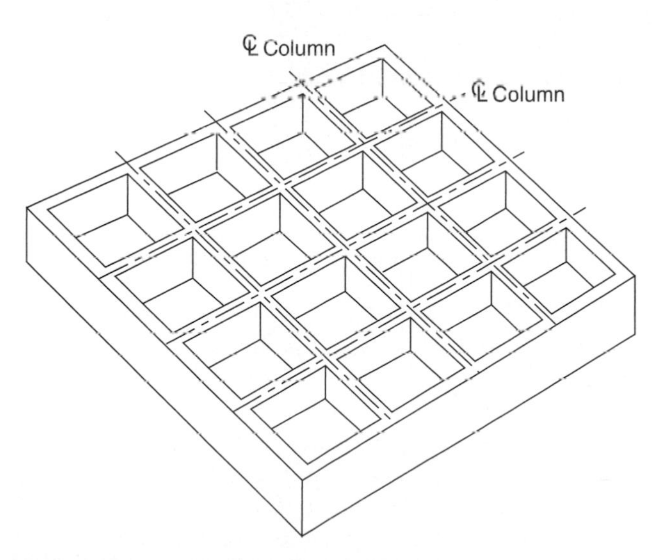


**Figure 5-30** Ribbed Mat Foundation.

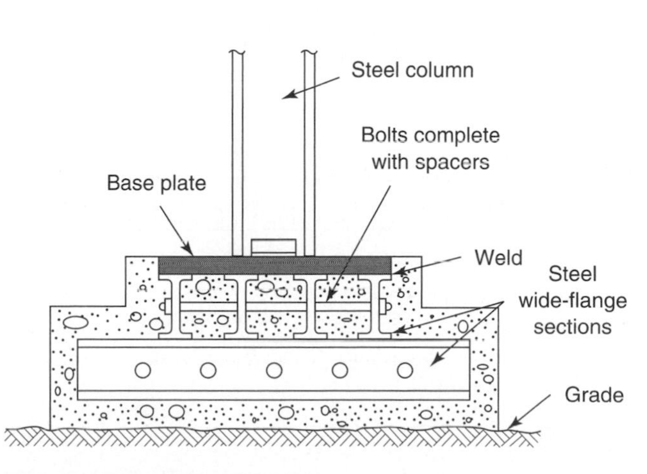


**Figure 5-32** Grillage Footing.

## SLURRY WALL FOUNDATIONS

The perimeter slurry walls, described in Chapter 3, may be used as foundation walls as well as protection of an excavation during the excavating stage. If the wall is to be used as part of the foundation, its thickness will have to be designed accordingly, but otherwise the procedure is the same as for an excavation slurry wall.

## EARTHQUAKE RESISTING FOUNDATIONS

The primary aim of a structural engineer, when designing a building in an earthquake zone, is to minimize the loss of life. The secondary aim is to minimize the structural damage to the building. Damage from an earthquake usually occurs in two stages—the initial damage to the building frame due to the shock waves of the quake and secondary effects such as fire and flooding due to the damage incurred by the service lines to the building.

Major loss of life occurs when collapse of a building is sudden and progressive. Buildings constructed to less stringent standards are susceptible to this type of failure. In areas where earthquakes occur, such as the western coast of North America, as well as Mexico and Japan, buildings must be designed to withstand the effects of lateral loads generated by vibrations within the soil during an earthquake. Non-load-bearing and load-bearing walls of unreinforced masonry are susceptible to failure due to their low resistance to lateral shear loads. Structural steel frames with connections that are designed primarily for gravity loads will fail prematurely in an earthquake. Precast concrete frames with pin-type connections are prone to sudden collapse during earthquake activity. Reinforced concrete building frames with columns that have insufficient lateral reinforcing are prone to shear failures from the effects of earthquake loads, as was evident in recent earthquakes in California. From experience, designers now agree that a degree of flexibility within the superstructure, combined with connections that resist rotation, provide the best protection from sudden building collapse. Studies done on newer buildings that have experienced earthquake loading, when designed in accordance with the latest building codes, have performed well no matter what material was used for the structural frame.

The shock waves produced during an earthquake can have a detrimental effect on soil bearing properties to the extent that a fine-grained soil will liquefy and lose all its bearing strength. Although building foundations can be designed to accommodate changes in soil properties, the effects of the resulting horizontal forces transmitted to the superstructure must also be considered (Figure 5–33). A feasible approach to control such forces is to isolate the building superstructure from the foundation.

Various approaches have been devised to separate the building superstructure from the foundation. Mechanical anti-sway devices that break up the harmonic motion within the building frame have proven satisfactory. Large bearing pads of tough plastic placed between column bases and the foundation have also been used. Thick rubber pads used as isolators (Figure 5–34) have proven to be most successful in absorbing the horizontal forces that impact the building frame. These isolator pads have also been used in bracing members to add additional flexibility to the superstructure (Figure 5–35).

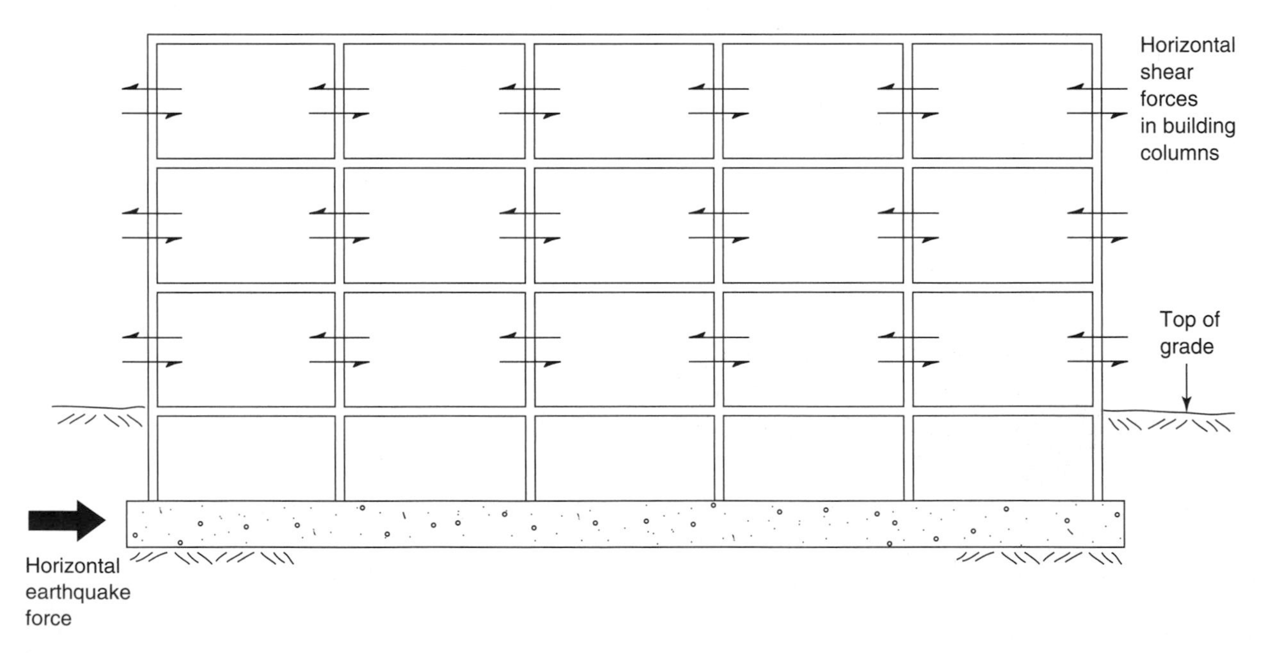


**Figure 5-33** Horizontal Force in Building Columns Due to Earthquake Action.

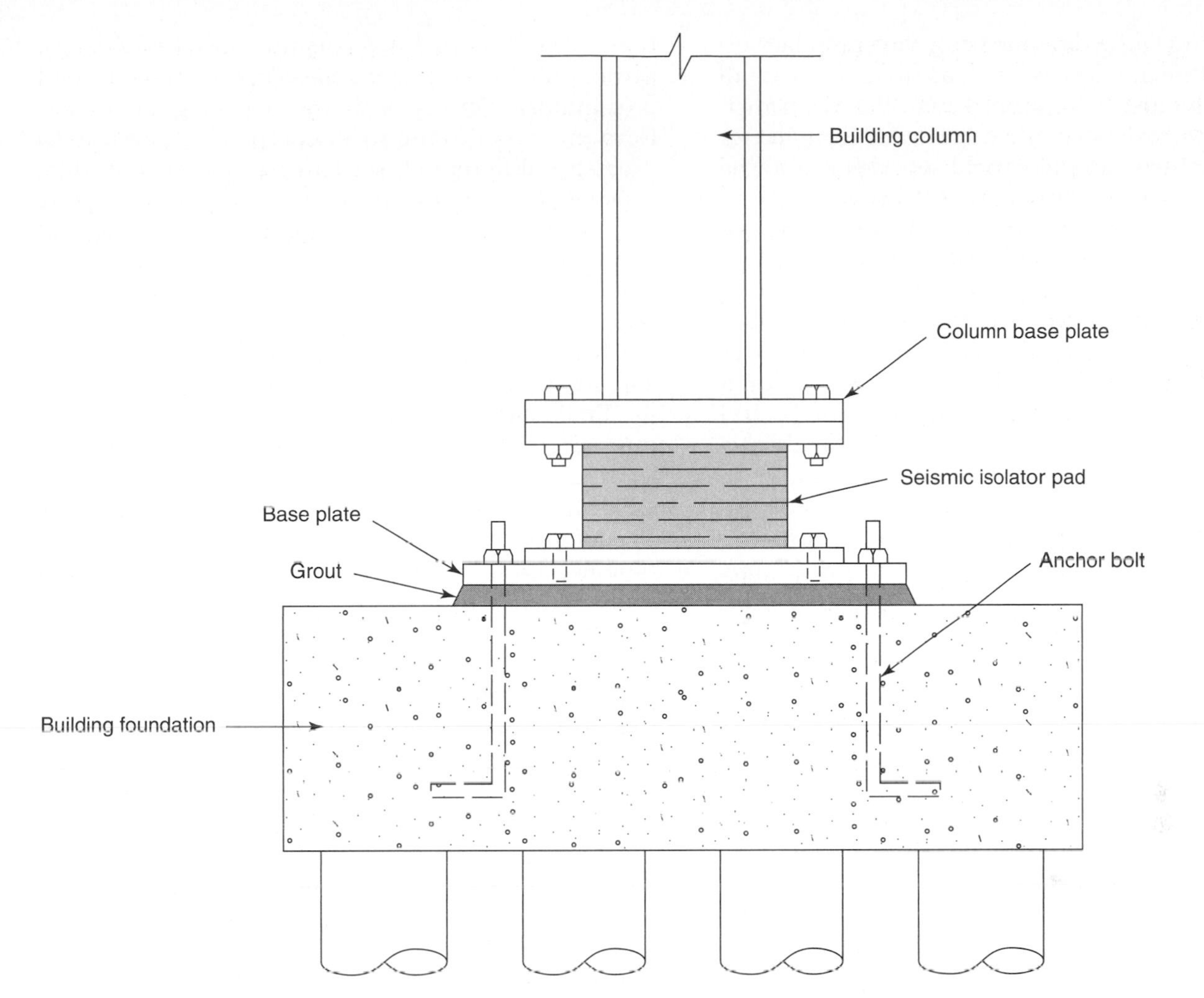


**Figure 5-34** Seismic Isolator Pad for Building Column.

**Figure 5-35** Seismic Isolator Pad in Building Cross Bracing.

To minimize the damage from the secondary effects of earthquake forces, such as fire and flood, all service lines must be equipped with flexible transitions or automatic shut off valves. Gas lines, water lines, sewer lines, and electrical services can all be effectively protected using these features.

Computer programs provide the structural engineer the means to model and analyze entire structures with various loading scenarios, allowing for better designs and use of structural materials. Ongoing research in this area has contributed to better building codes, providing designers with building guidelines that result in less property damage and minimize the loss of life. In many cities, buildings of historical value are now being upgraded to present-day codes. This type of retrofitting becomes much more complex and expensive in that existing materials, building elements, and all connections must be analyzed and reinforced accordingly, without significant changes to the appearance of the building.

## SOIL ENHANCEMENT FOR FOUNDATIONS

The deep foundation systems described in the foregoing pages are based on techniques that rely on soils that are relatively stable and have the strength characteristics necessary to support superimposed loads. This traditional approach to foundation construction requires drilling or excavating to a depth where bedrock can be reached as the bearing stratum or where naturally consolidated soils have the required bearing strength and stability. In areas where good soil characteristics may be lacking, these traditional methods may be too expensive or time consuming. An alternative to the traditional approach for the construction of building foundations in areas with poor soil is that of soil enhancement.

Soil enhancement, also known as soil modification, is used to improve the bearing capacity and the settlement characteristics of the soil, in situ, rather than removing it or passing through it with the foundation. Approaches that have been developed include compaction, preloading or precompression, drainage, vibratory densification, grouting, and chemical stabilization. All of these methods increase the density of the soil and improve bearing values.

In building construction, compaction of soil in situ or new fill material is done by mechanical rollers or vibratory compactors. However, this approach is only effective for relatively shallow depths. If new material must be applied over existing soil, it must be placed in relatively thin layers (6 in. (150 mm)) and each layer is compacted to some specified density. Structural elements, such as slabs on grade, depend on a uniform density of the supporting grade to perform properly. Nonuniform compaction can lead to excessive differential settlements under the slab, causing the slab to crack. Standard bearing tests such as the standard Proctor test (ASTM D698) and the modified Proctor test (ASTM D1557) have been developed to determine soil densities, however, instruments such as the *nuclear density tester* discussed in Chapter 1 provide fast and reliable density values and are preferred by most contractors.

Preloading and drainage are used to control or minimize settlements caused by future loads. Preloading consists of stockpiling material over the area that requires consolidation, and is time related. Draining of the area using sand drains reduces the time needed to achieve the consolidation process in some instances. Grouting is used where additional soil shear strength is required and is often used in conjunction with friction piles. Chemical stabilization reduces the expansion of surface soils due to freeze-thaw action. It is used to stabilize soil under interior and exterior concrete slabs on grade and under asphalt pavements in parking lots. Vibratory densification methods are used to advantage where the entire building site requires soil enhancement to depths that make methods such as compaction and preloading impractical.

### Minipiles

A relatively new innovation in foundation construction, used in conjunction with soil enhancement, is the *minipile*. The minipile, also known as a micropile, is a small diameter pile that has been developed for foundation upgrades under existing structures where access and headroom is limited. Because they are relatively small, their installation reduces the amount of waste material that must be removed, and minimizes the rerouting of existing installations such as underground utilities. Minipiles can be driven, drilled, or grouted. Grouting methods such as compaction grout, jet grout, post grouting, and pressure grouting enhance the load resistance of the piles by modifying the soil immediately around the pile, allowing them to be used in almost any type of soil and loading condition. They can be placed as single units or in clusters, depending on load requirements. Design loads ranging from 3 tons (25 kN) to 200 tons (1800 kN) or more can be supported.

### Vibroflotation Foundation

Vibroflotation works best in sandy soils; it employs mechanical vibration of soil beneath the surface and

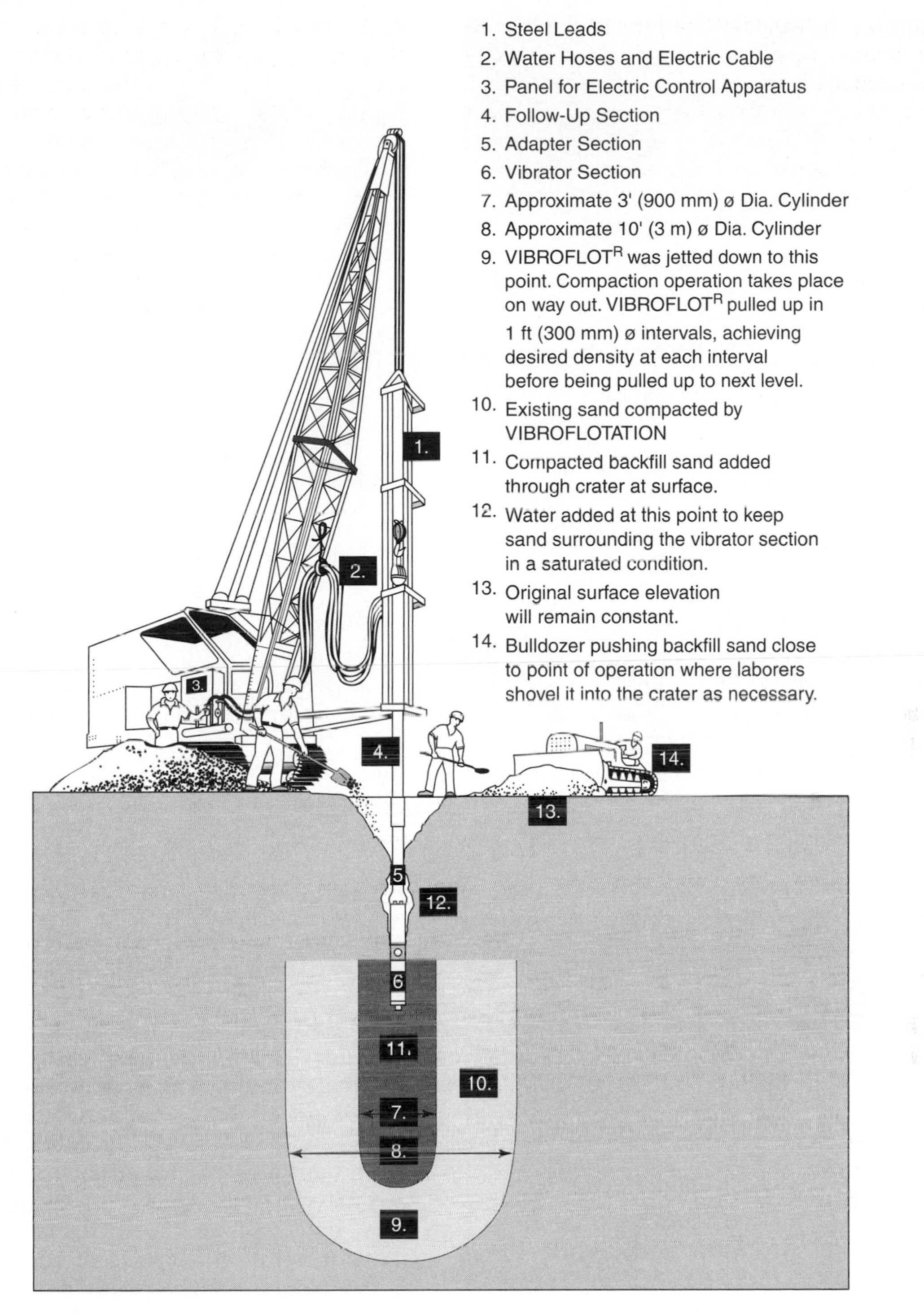
1. Steel Leads
2. Water Hoses and Electric Cable
3. Panel for Electric Control Apparatus
4. Follow-Up Section
5. Adapter Section
6. Vibrator Section
7. Approximate 3' (900 mm) ø Dia. Cylinder
8. Approximate 10' (3 m) ø Dia. Cylinder
9. VIBROFLOT^R was jetted down to this point. Compaction operation takes place on way out. VIBROFLOT^R pulled up in 1 ft (300 mm) ø intervals, achieving desired density at each interval before being pulled up to next level.
10. Existing sand compacted by VIBROFLOTATION
11. Compacted backfill sand added through crater at surface.
12. Water added at this point to keep sand surrounding the vibrator section in a saturated condition.
13. Original surface elevation will remain constant.
14. Bulldozer pushing backfill sand close to point of operation where laborers shovel it into the crater as necessary.
1.
2.
3.
4.
5
12.
6
11.
10.
7.
8.
9.
13.
14.

**Figure 5-36** Typical Vibrating Machine in Action.

simultaneous saturation with water to *move, shake,* and *float* the soil particles into a dense state.

The machine used consists of a long shaft with a vibrating head on the end, mounted on a crane. The head is jetted into the ground to the required depth and slowly withdrawn, compacting the existing soil while additional material is backfilled through the crater created at the surface.

Figure 5–36 illustrates a typical machine in action. It is positioned over the area to be compacted, and the lower water jet is turned on (see Figure 5–37). Water is pumped in faster than it can drain away into the subsoil, creating a *quick* condition beneath the vibrator, which allows it to penetrate the soil by its own weight and vibration. The initial action creates a crater at the surface, about 3 ft

**Figure 5-37** Vibrator Ready for Action with Water Jet On. (Courtesy Hayward Baker Inc.)

**Figure 5-38** Vibrator Inserted in Soil. (Courtesy Hayward Baker Inc.)

(900 mm) in diameter, as shown in Figure 5–38. On typical sites, the machine can penetrate from 15 to 25 ft (4.5 to 7.5 m) in approximately 2 minutes. The average depth of compaction is approximately 15 ft (4.5 m), but depths to 70 ft (20 m) can be compacted successfully.

**Figure 5-39** Fine Material Being Brought to the Surface. (Courtesy Hayward Baker Inc.)

Having reached the required depth, water is switched from the lower to the upper jet and withdrawal begins, with the compaction taking place during this time. The machine compacts by simultaneous vibrations and saturation. The compactor vibrates granular soil with approximately 10 tons (90 kilonewtons) of centrifugal force into a dense mass; at the same time, the excess water floats the finest particles to the surface and washes them away (see Figure 5–39). Granular material is added from the surface to replace the material washed away and to compensate for the increased density of the in situ soil. The vibrator is raised slowly at a rate of approximately 1 ft/min (300 mm/min) and the hole backfilled at the same time (see Figure 5–40), so the entire depth of soil is compacted into a hard core or column.

A single compaction produces a cylinder of highly densified material with great bearing capacity. Density is uniform above and below the water table and may be varied to fit the job requirement. The compacted cylinder is between 6 and 10 ft (1.8 and 3 m) in diameter, about 10% of which is sand

**Figure 5-40** Soil Stabilization As a Result of the Vibration Process. (Courtesy Hayward Baker Inc.)

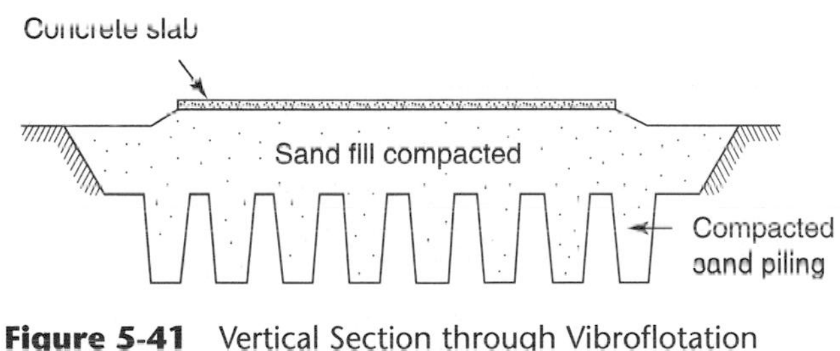


**Figure 5-41** Vertical Section through Vibroflotation Foundation.

added from the surface. Figure 5–41 illustrates a typical compacted site.

## Vibroreplacement Process

In the vibroreplacement process, the soil is first displaced by the vibrator; if there is a high moisture content, holes are excavated by a drilling bucket or other suitable means. The holes are backfilled in stages with coarse material (gravel or crushed stone), which is simultaneously compacted by the vibrator. As a result, the soil strength is increased by (1) replacing a portion of the existing soil with a high-strength, compacted column of stone; (2) compacting and adding stone to an outer layer of the existing soil; and (3) consolidating a further outer volume of soil, due to the horizontal component of the force exerted on the column of stone.

## Vibratory Probe Soil Compaction

In vibratory tube compaction, a tubular probe is driven into the soil by means of a vibratory driver, similar to that shown in Figure 5–24, and extracted again by the same means. The vibratory action of the probe causes the soil particles to readjust themselves into a denser state. Soil inside the probe becomes densified as the probe is driven and extracted and, upon extraction, a highly densified sand pile remains. A secondary effect is compaction, which also takes place outside the probe. As a result, properly determined spacing of the probes will allow densification of the soil over the entire area involved (see Figure 5–42).

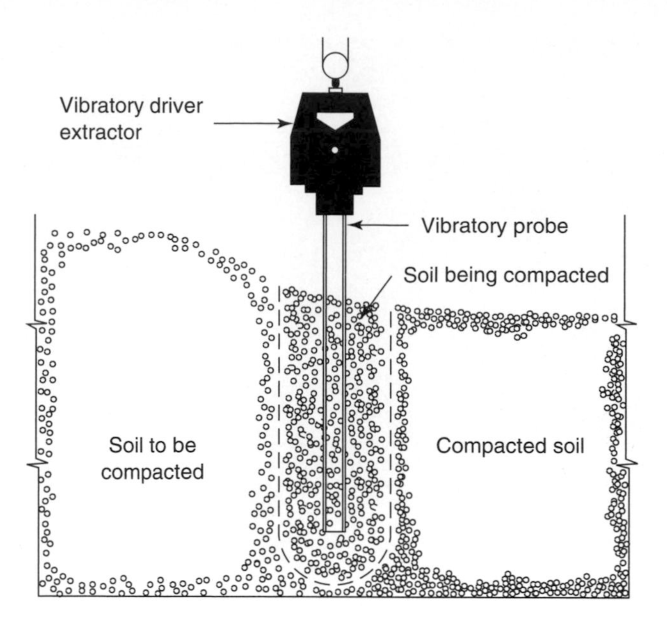


**Figure 5-42** Vibratory Probe Soil Compaction. (Reprinted by permission of L. B. Foster Co.)

## REVIEW QUESTIONS

**1.** What two conditions usually determine the depth of a foundation under a building?

**2.** What two conditions can influence the depth of a foundation?

**3.** List three types of deep foundations.

**4.** What three characteristics can be used to classify piles?

**5.** List four ways that piles can support load.

**6.** What is the purpose of **(a)** a pile ring, **(b)** a pile shoe, and **(c)** pile driver leads?

**7.** Calculate the load capacity of a precast concrete pile that is 14 in. in diameter and is driven into soil having a shear strength of 500 lb/sq ft, and a bearing resistance of 2.5 tons per square foot. Assume the top 5 feet of the pile cannot develop any shear resistance. Total pile length is 35 feet.

**8.** Calculate the load resistance of a precast concrete pile that is 350 mm in diameter and is driven into soil having a shear strength of 20 kPa and a bearing resistance of 240 kPa. Assume the top 1.5 m of the pile cannot develop any shear resistance. Total pile length is 12 m.

**9.** Describe briefly how a steel pile shell is driven with a drop hammer and concrete plug.

**10.** List four different types of caissons and explain how each differs from the rest.

**11.** Explain the purpose of **(a)** a cofferdam, **(b)** jetting a pile, and **(c)** a mat foundation.

**12.** Explain clearly the difference between the vibroflotation process and the vibroreplacement process.

**13.** Describe briefly how a sand pile is produced. Illustrate by means of a neat diagram.

**14.** Why is it good practice to check the elevation of the first piles in a group while the subsequent piles are being driven?

**15.** Outline the advantages of a slurry wall foundation.

# 6 FORMWORK

All concrete sections made with poured-in-place concrete require some temporary means of support for the freshly mixed, plastic concrete. Whether the concrete is placed in a factory setting, as in the case of precast sections, or poured in place on the building site, some means of support is necessary to hold the concrete in place during its curing period. This temporary framing is known as *formwork*.

In addition to providing the necessary support for the freshly mixed concrete while it is curing, formwork must be capable of supporting the construction crew and any other applied loads during the construction period. Formwork also provides support for the reinforcing bars, which are placed according to the structural drawings and must be maintained in the correct location during the pouring and consolidation of the plastic concrete. Correct location of reinforcing bars is imperative to ensure that the structural section will develop its design strength after curing.

Because of their temporary nature, forms must be designed and built so that they are lightweight, strong, and easily removed from the concrete sections that they surround. In a precast concrete plant setting, forms must be able to withstand a good number of reuses without losing their shape. On the job site, formwork is built according to the requirements of the structural drawings and must be easily dismantled and moved as construction of the building advances.

## FORM MATERIALS

Because the plastic concrete must be supported until it develops sufficient strength to support its own weight, materials used as formwork must have sufficient strength to provide the necessary support during this critical period. Materials currently being used in the building of forming sections have been selected primarily for strength, durability, weight, and ease of assembly.

### Wood

Wood in the form of dimensioned lumber and plywood sheathing is the most widely used material for building forms because of its good strength-to-weight ratio, workability, relative low cost, and reusability (see Figure 6-1). Dimensioned lumber used in form building is usually of the softwood variety because of its availability and good strength. Spruce, pine, fir, and western hemlock are the species most commonly used in this application. Although dimensioned lumber has been used as sheathing material in the past, plywood sheets have replaced it in this application, and dimensioned lumber is now used primarily for framing, bracing, and shoring.

The development of exterior-grade plywoods has provided a strong and durable product that is ideal for sheathing in formwork. With its large, smooth surface, strength, resistance to change in shape when wet, and ability to withstand rough usage without splitting, plywood is indispensable as a form sheathing material.

Overlaid plywood is frequently used when smooth, grainless surfaces are required and when the number of reuses is to be increased. The use of phenolic impregnated paper or fiberglass-reinforced, high-density polyethylene sheet bonded to the plywood surface has improved the resistance of the sheathing to water absorption and improved stripping and cleaning for reuse.

An additional advantage of using plywood as sheathing in formwork is its ability to bend, thus making it possible to produce smooth, curved surfaces

**Figure 6-1** Dimensioned Lumber Used as Structural Supports for Plywood Sheathing in the Construction of a Wall Form.

(see Figure 6-2). The degree of bending depends on the thickness of the plywood and on whether it is curved across the face grain or parallel to it. Table 6-1 gives typical minimum bending radii for plywood panels. Panels with clear, straight grain will permit greater bending than those shown in the table.

Thicker plywood may be bent by cutting saw kerfs across the inner face at right angles to the curve.

**Figure 6-2** Intricate Formwork Can Be Achieved Using Plywood Sheathing.

The kerf should cut three plies of a five-ply sheet or five plies of a seven-ply sheet. They should be spaced so as to just close up when the plywood is bent to the correct radius (see Figure 6-3). However, in applications when structural integrity of the material is required, manufacturers do not recommend the kerfing of plywood.

Another method of curving thick face material is to use two or more thicknesses of ¼-in. (6-mm) plywood, bent one at a time.

## Steel

Steel is widely used as a material for making forms because of its strength and durability. Steel angles and bars are used extensively as supporting members for form panels (see Figure 6-4) faced with plywood sheathing. Steel corner pieces used in connecting panels are popular because of their strength and reusability. Sheet steel sections are used for standard form shapes such as pan forms used in one- and two-way ribbed slabs (see Figure 6-5). Where large spans

**Table 6-1**
**MINIMUM BENDING RADII FOR PLYWOOD PANELS**

| Plywood Thickness | | Minimum Bend Radius (mm) | | Minimum Bend Radius (in.) | |
|---|---|---|---|---|---|
| **(mm)** | **(in.)** | **Perpendicular to Face Grain** | **Parallel to Face Grain** | **Perpendicular to Face Grain** | **Parallel to Face Grain** |
| 6, 7.5 | 1/4, 5/16 | 1500 | 700 | 60 | 28 |
| 8, 9.5 | 5/16, 3/8 | 2400 | 1000 | 96 | 40 |
| 11, 12.5 | 7/16, 1/2 | 3600 | 2400 | 144 | 96 |
| 14, 15.5 | 9/16, 5/8 | 4800 | 3600 | 192 | 144 |
| 18.5, 19 | 3/4 | 6000 | 4800 | 240 | 192 |
| 20.5 | 13/16 | 7000 | 5800 | 280 | 232 |

*Source:* Reprinted with permission of the Council of Forest Industries of British Columbia.

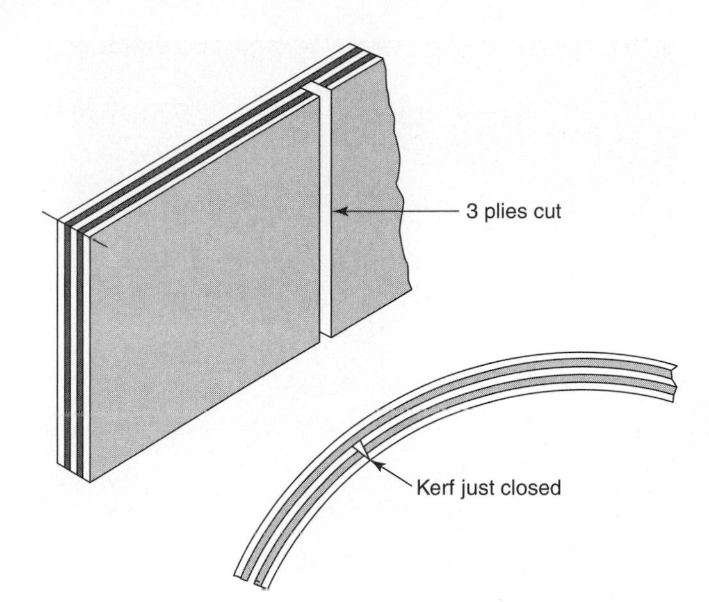


**Figure 6-3** Saw Kerfing Plywood for Bending.

**Figure 6-4** Steel Column Clamps on Column Form.

are encountered, structural steel sections can be used as framing members to support formwork sections.

## Plastic

Fiberglass-reinforced plastic is another material that has been used successfully as a forming material. Because of its moldability, light weight, strength, and toughness, it is used in components where reusability is a factor. Standard form shapes such as *dome pans* for two-way ribbed slabs (see Figure 6-6) are its most common application.

## Aluminum

Although expensive compared with the other materials, aluminum is used as a structural framing material because of its light weight. Extruded I-shaped sections are used as beams for supporting slab formwork and as wales on wall formwork. The I-shaped sections have been specifically developed so that plywood sheathing can be fastened to them without using special connectors (see Figure 6-7). Greater spans and loads can be supported by these sections compared with those of wood. Aluminum shores have also been developed as an alternative to those of wood and steel.

**Figure 6-5** Metal Pan Forms.

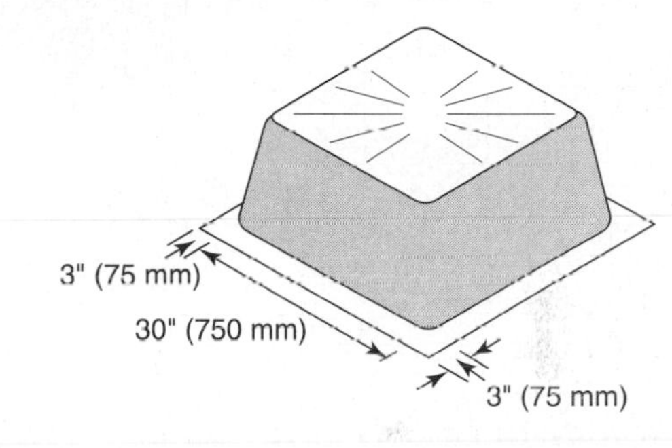


**Figure 6-6** Dome Pan.

## Form Liners

Form liners serve two purposes: (1) to improve stripping of the form from the concrete surface without damaging the form material or the concrete, and (2) to produce a desired texture on exposed concrete surfaces. Plywood, steel, and fiberglass are the usual materials used to provide a smooth finish on the concrete. To produce a certain texture, plastic and rubber liners are commonly used because of their versatility and toughness. PVC plastic liners can be shaped to create an infinite number of designs and, with the use of form-releasing agents to improve stripping, can be reused in many cases. Polystyrene form liners can also be used to provide various intricate designs; however, they are limited to a single application (Figure 6-8).

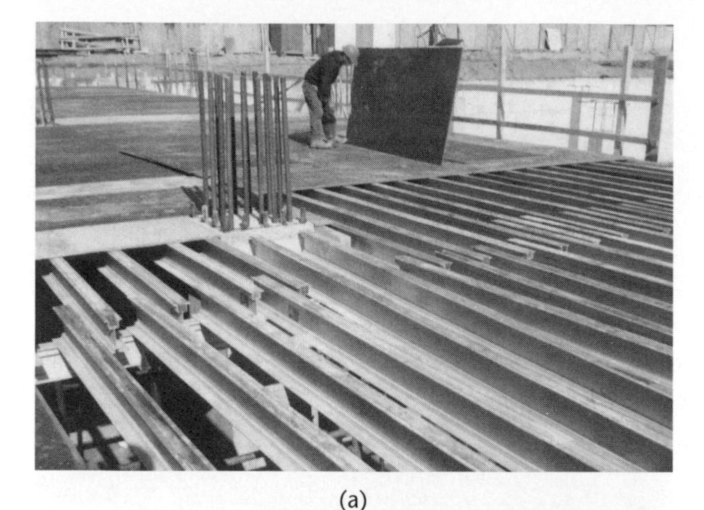

(a)

(b)

**Figure 6-7** (a) Aluminum I-Shaped Sections Used as Beams under Slab Formwork. (b) Aluminum I-Shaped Sections. Note Wood Nailer in Chord.

## Insulating Board

In some applications, insulating board of dense polystyrene can be placed into the forms and integrated into the concrete wall section to provide additional insulation value to the wall. The material is fastened to one or both form faces; when the form is removed, the insulation remains, either bonded to the concrete or held in place by clips.

If placed on an interior wall face, the insulation must be covered by a protective layer of wallboard as a finish. If used on the exterior of a foundation wall, any portion of the insulation extending above grade must be covered with a cement parging to protect it from the sun's ultraviolet rays.

**Figure 6-8** Pattern on Concrete Surface Resulting from Use of Form Liner.

# FORMWORK PRINCIPLES

The principles behind good formwork are based on the same basic frame theories used in the design and construction of permanent structural frames. Formwork must be able to withstand construction forces that, in many respects, can be more severe than those experienced by the completed structure. It is imperative that each component of the formwork be erected according to the formwork drawings to ensure that all construction loads are safely supported.

Although formwork is temporary in nature, the methods used in building formwork must adhere to all of the code specifications that apply to the particular material being used. Each component of the form must be able to support its load from two points of view: (1) strength, based on the physical properties of the material used, and (2) serviceability, the ability of the selected sections to resist the anticipated loads without exceeding deflection limits.

Typical deflection limits for the various components are usually a maximum of span/360, but not to exceed $\frac{1}{16}$-in. (1.5 mm) for sheathing and $\frac{1}{4}$-in. (6 mm) for joists and beams. These limits ensure that the resulting concrete sections will be straight once the forms are removed.

From a construction point of view, formwork must be economical and practical to build. A number of important factors should be kept in mind when designing and constructing formwork.

1. Forms must be strong enough to withstand the pressure of plastic concrete and to maintain their shape during the concrete placing operation.
2. Forms must be tight enough to prevent wet concrete from leaking through joints and causing unsightly fins and ridges.
3. Forms must be as simple to build as circumstances will allow.
4. Forms must be easy to handle on the job.
5. Form sections must be of a size that can be lifted into place without too much difficulty and transported from one job to another if necessary.
6. Forms must be made to fit and fasten together with reasonable ease.
7. The design must be such that the forms, or sections of them, may be removed without damage to the concrete or to themselves.
8. Forms must be made so that workers can handle them safely.

## FORMWORK LOADS AND PRESSURES

The basic consideration in form design is strength the forms' ability to support, without excessive deflections, all loads and forces imposed during construction. Two types of problems arise in formwork design: (1) horizontal forms must support gravity loads based on the mass of the concrete, the construction crew and equipment, and the weight of the formwork itself, and (2) vertical forms must primarily resist lateral pressures due to a particular height of plastic concrete. Wall and column forms are examples where lateral concrete pressures are a prime concern, while formwork supporting a structural slab must be designed to sustain gravity loads.

In addition to concrete pressures and gravity loads, all formwork must be able to resist the lateral forces due to wind and forces caused by power equipment such as power buggies during concrete placement.

### Formwork Loads

Two types of gravity loads are considered in the design calculations: *dead loads* and *live loads*. The weight of the formwork and the plastic concrete are considered dead loads, while workers and equipment are considered as live loads. The plastic concrete can vary in mass from 40 to 375 lb/cu ft (645 to 6,025 kg/m$^3$), depending on the type of aggregates used in the mix. Normal-density concrete made from natural sand and gravel aggregates has a unit mass ranging between 145 to 150 lb/cu ft (2,330 to 2,400 kg/m$^3$), including the reinforcement. The formwork itself can vary from 3 to 15 pounds per square foot (0.15 to 0.75 kN/m$^2$).

The American Concrete Institute Committee 347 recommends a minimum live load of 50 psf (2.4 kPa) for work crews and equipment and 75 psf (3.6 kPa) if power buggies are used to transport the concrete, to allow for the extra weight of runways and impact loads. A minimum total load of 100 psf (4.8 kPa) for concrete and equipment is recommended and 125 psf (6.0 kPa) if power buggies are used.

Adequate bracing must be provided for all formwork to ensure lateral stability of the formwork in all directions during and after the pour. Wind loads must be considered on vertical formwork in particular; however, formwork supporting slabs must also be stabilized against the forces due to wind and equipment. Wind forces on wall forms can be based on local codes; however, a minimum of 15 psf (0.72 kN/m$^2$) applied to the vertical projection of the formwork is recommended by ACI Committee 347. Bracing for wall forms should be based on a minimum of 100 pounds per lineal foot (1.5 kN/m) applied at the top of the wall. Slab forms should be braced for 100 plf (1.5 kN/m) or 2% of the dead load, whichever is greater, applied to that portion of the formwork under a single pour. If unusual circumstances are encountered, a more thorough analysis of the situation must be made.

### Concrete Pressure on Formwork

The lateral pressure exerted by plastic concrete on vertical formwork is rather complex in nature and is affected by several factors. The freshly placed concrete initially acts as a liquid, exerting fluid or hydrostatic pressure against the vertical form. Because hydrostatic pressure at any point in a liquid is the result of the weight of the fluid above, the density of the concrete mix influences the magnitude of the force acting on the form. But, because fresh concrete is a composite material rather than a true liquid, the laws of hydrostatic pressure apply only approximately and only before the concrete begins to set.

The rate of placement also affects lateral pressure. The greater the height to which concrete is placed while the whole mass remains in the liquid stage, the greater the lateral pressure at the bottom of the form.

The temperatures of concrete and atmosphere affect the pressure because they affect the setting time. When these temperatures are low, greater heights can be placed before the concrete at the bottom begins to stiffen, and greater lateral pressures are therefore built up.

Vibration increases lateral pressures because the concrete is consolidated and acts as a fluid for the full depth of vibration. This may cause increases of up to 20% in pressures over those incurred by spading. Other factors that influence lateral pressure include the consistency or fluidity of the mix, the maximum aggregate size, and the amount and location of reinforcement.

The ACI Committee 347 has developed formulas for calculating the maximum lateral pressure at any elevation in the form. They are based on prescribed conditions of concrete temperature, rate of placement, slump of concrete, mass of concrete, and vibration. They may be used for internally vibrated concrete of normal density, placed at not more than 10 ft/hr (3 m/hr), with no more than 4 in. (100 mm) of slump. Vibration depth is limited to 4 ft (1,200 mm) below the concrete surface. The two formulas for wall form design based on standard units are as follows:

1. For walls, with a rate of pour not exceeding 7 ft per hour,

$$p = 150 + \frac{9{,}000R}{T}$$

Maximum pressure = 2,000 psf or $150h$, whichever is less.

2. For walls, with rate of pour greater than 7 ft per hour,

$$p = 150 + \frac{43{,}400}{T} + \frac{2{,}800R}{T}$$

Maximum pressure = 2,000 psf or $150h$, whichever is less.

The variables in the formulas represent the following quantities:

$p$ = maximum lateral pressure, psf

$R$ = rate of placement, ft/hr

$T$ = temperature of concrete as placed in the forms, °F

$h$ = maximum height of fresh concrete in the form, ft

Table 6-2(a) provides the maximum lateral pressure used in designing wall forms, using standard units, based on the two foregoing formulas.

The two formulas for wall form design based on metric units are as follows:

1. For walls, with a rate of pour not exceeding 2 m per hour,

$$p = 7.2 + \frac{785R}{T + 17.8}$$

Maximum pressure = 96 MPa or $23.5h$, whichever is less.

2. For walls, with rate of pour greater than 2 metres per hour,

$$p = 7.2 + \frac{1{,}156}{T + 17.8} + \frac{244R}{T + 17.8}$$

Maximum pressure = 96 kPa or $23.5h$ kPa, whichever is less.

The variables in the formulas represent the following quantities:

$p$ = maximum lateral pressure, kPa

$R$ = rate of placement, m/hr

$T$ = temperature of concrete as placed in the forms, °C

$h$ = maximum height of fresh concrete in the form, metres

Table 6-2(b) illustrates the resulting lateral pressures used in designing wall forms, based on metric units.

Column forms are frequently filled to their full height within the time that it takes the concrete to begin to set. In addition, vibration may extend the full depth of the form, resulting in higher lateral pressures than those incurred in wall forms.

For column forms, the formula used for concrete pressure calculations, as recommended by A.C.I. Committee 347, is the same as that for walls having a pour rate of not more than 7 ft per hour, taking the aforementioned factors into consideration:

$$p = 150 + \frac{9{,}000R}{T}$$

With maximum pressure = 3000 psf or $150h$, whichever is less (see Table 6-3a).

In metric, the formula for walls having a rate of pour not greater than 2 metres per hour is used for column pressure calculations.

$$p = 7.2 + \frac{785R}{T + 17.8}$$

With maximum pressure being 144 kPa or $23.5h$, whichever is less. Table 6-3(b) provides values of lateral pressure to be used in the design of column forms based on the above formula.

The maximum height of a single lift of concrete recommended for a column pour is 18 ft (5.5 m) within a 2-hour period. The recommended minimum pressure to be used in design calculations for both walls and columns is 600 psf (29 kPa).

## Rate of Placement

The rate of pour depends on the type of equipment that is being used to distribute the concrete. A concrete pump is the most efficient way to place con-

**Table 6-2(a)**
**MAXIMUM PRESSURE FOR DESIGN OF WALL FORMS (PSF)[a]**

| Rate of Placement, R (ft/hr) | Maximum Lateral Pressure, p (psf) for Indicated Temperature (°F) | | | | | |
|---|---|---|---|---|---|---|
| | 90 | 80 | 70 | 60 | 50 | 40 |
| 1 | 250 | 262 | 278 | 300 | 330 | 375 |
| 2 | 350 | 375 | 407 | 450 | 510 | 600 |
| 3 | 450 | 488 | 536 | 600 | 690 | 825 |
| 4 | 550 | 600 | 664 | 750 | 870 | 1050 |
| 5 | 650 | 712 | 793 | 900 | 1050 | 1275 |
| 6 | 750 | 825 | 921 | 1050 | 1230 | 1500 |
| 7 | 850 | 938 | 1050 | 1200 | 1410 | 1725 |
| 8 | 881 | 973 | 1090 | 1246 | 1466 | 1795 |
| 9 | 912 | 1008 | 1130 | 1293 | 1522 | 1865 |
| 10 | 943 | 1043 | 1170 | 1340 | 1578 | 1935 |

[a]Do not use design pressures in excess of 2000 psf or 150 × height (in ft) of fresh concrete in forms, whichever is less.
*Source:* Reprinted by permission of American Concrete Institute.

**Table 6-2(b)**
**MAXIMUM PRESSURE FOR DESIGN OF WALL FORMS (kPA)[a]**

| Rate of Placement R (m/hr) | Maximum Lateral Pressure, p (kPa), for Indicated Temperature (°C) | | | | | |
|---|---|---|---|---|---|---|
| | 30 | 25 | 20 | 15 | 10 | 5 |
| 0.25 | 11.3 | 11.8 | 12.4 | 13.2 | 14.3 | 15.8 |
| 0.50 | 15.4 | 16.4 | 17.6 | 19.2 | 21.3 | 24.4 |
| 0.75 | 19.5 | 21.0 | 22.8 | 25.1 | 28.4 | 33.0 |
| 1.00 | 23.6 | 25.5 | 28.0 | 31.1 | 35.4 | 41.6 |
| 1.25 | 27.7 | 30.1 | 33.2 | 37.1 | 42.5 | 50.2 |
| 1.50 | 31.8 | 34.7 | 38.4 | 43.1 | 49.6 | 58.8 |
| 1.75 | 35.9 | 39.3 | 43.5 | 49.1 | 56.6 | 67.5 |
| 2.00 | 40.0 | 43.9 | 48.7 | 55.1 | 63.7 | 76.1 |
| 2.25 | 42.9 | 47.0 | 52.3 | 59.2 | 68.5 | 82.0 |
| 2.50 | 44.1 | 48.5 | 53.9 | 61.0 | 70.7 | 84.7 |
| 2.75 | 45.4 | 49.9 | 55.5 | 62.9 | 72.9 | 87.3 |
| 3.00 | 46.7 | 51.3 | 57.1 | 64.8 | 75.1 | 90.0 |

[a]Do not use design pressure in excess of 95.8 kPa or 23.5 × height (in metres) of fresh concrete in forms, whichever is less.
*Source:* Reprinted by permission of the American Concrete Institute; metric conversion by author.

crete in that the concrete can be poured continuously. In the case of walls and columns, this can result in relatively high concrete pressures on the formwork. Rate of pour is normally expressed in feet (meters) of concrete poured per hour. Typical examples for calculating pour rates are now given.

### *Example:*

Calculate the rate of pour for a section of a shear wall on the twelfth floor of a high-rise building. Assume the distance from the ground to the top of the wall form is 168 ft. A tower crane, using two buckets, each with a capacity of 1.5 cu yd, has a rate of travel of 90 ft/min up and 120 ft/min down. Assume pickup time is 20 sec and dump time is 5 min. Pour conditions are:

wall height = 14 ft
wall thickness = 10 in.
wall length = 60 ft

***Solution:***

Cycle time:

| | | |
|---|---|---|
| time to travel up | $\frac{168}{90}$ | = 1.87 min |
| time to travel down | $\frac{168}{120}$ | = 1.40 min |
| pickup time | $\frac{20}{60}$ | = 0.33 min |
| dump time | | = 5.00 min |
| Total | | 8.60 min |

**Table 6-3(a)**
**MAXIMUM LATERAL PRESSURE FOR DESIGN OF COLUMN FORMS (psf)[a]**

| Rate of Placement R (ft/hr) | Maximum Lateral Pressure, p (psf), for Indicated Temperature (°F) | | | | | |
|---|---|---|---|---|---|---|
| | 90 | 80 | 70 | 60 | 50 | 40 |
| 1 | 250 | 262 | 278 | 300 | 330 | 375 |
| 2 | 350 | 375 | 407 | 450 | 510 | 600 |
| 3 | 450 | 488 | 536 | 600 | 690 | 825 |
| 4 | 550 | 600 | 664 | 750 | 870 | 1050 |
| 5 | 650 | 712 | 793 | 900 | 1050 | 1275 |
| 6 | 750 | 825 | 921 | 1050 | 1230 | 1500 |
| 7 | 850 | 938 | 1050 | 1200 | 1410 | 1725 |
| 8 | 950 | 1050 | 1178 | 1350 | 1590 | 1950 |
| 9 | 1050 | 1163 | 1307 | 1500 | 1770 | 2175 |
| 10 | 1150 | 1275 | 1435 | 1650 | 1950 | 2400 |
| 11 | 1250 | 1388 | 1564 | 1800 | 2130 | 2625 |
| 12 | 1350 | 1500 | 1693 | 1950 | 2310 | 2850 |
| 13 | 1450 | 1613 | 1822 | 2100 | 2490 | 3000 |
| 14 | 1550 | 1725 | 1950 | 2250 | 2670 | |
| 16 | 1750 | 1950 | 2207 | 2550 | 3000 | |
| 18 | 1950 | 2175 | 2464 | 2850 | | |
| 20 | 2150 | 2400 | 2721 | 3000 | | |
| 22 | 2350 | 2625 | 2979 | | | |
| 24 | 2550 | 2850 | 3000 | | | |
| 26 | 2750 | 3000 | | | | |
| 28 | 2950 | | (3000 psf maximum) | | | |
| 30 | 3000 | | | | | |

[a]Do not use design pressures in excess of 3000 psf or 150 × height of fresh concrete in forms, whichever is less.
*Source:* Reprinted by permission of American Concrete Institute.

Rate of delivery in cu yd/hr:

$$\frac{60}{8.60} \times 1.5 = 10.46 \text{ cu yd/hr}$$

Rate of pour calculations:

Volume of concrete to be poured:

$$14 \times \frac{10}{12} \times 60 \times \frac{1}{27} = 25.93 \text{ cu yd}$$

Time required to pour 25.93 cu yd of concrete:

$$\frac{25.93}{10.46} = 2.48 \text{ hr}$$

Rate of pour in feet per hour:

$$\frac{14}{2.48} = 5.65 \text{ ft/hr}$$

---

***Example:***

A 10-in.-thick parking structure wall has a total length of 150 ft and a height of 16 ft. Calculate the rate of pour based on a concrete pump capacity of 1,200 cu ft/hr.

***Solution:***

Since a concrete pump normally delivers concrete continuously, only the capacity of the pump need be considered in the calculation of the rate of pour.

Volume of pour:

$$\frac{10}{12} \times 150 \times 16 = 2{,}000 \text{ cu ft}$$

Time required to pour wall:

$$\frac{2{,}000}{1{,}200} = 1.67 \text{ hr}$$

Rate of pour in feet per hour:

$$\frac{16}{1.67} = 9.6 \text{ ft/hr}$$

---

***Example:***

Calculate the rate of pour for a section of an elevator shaft in the central core on the tenth floor of a high-rise building. Assume the distance from the ground to the top of the wall form is 36.6 m. A tower crane, using two buckets, each with a capacity of 1 m$^3$, has a rate of travel of

**Table 6-3(b)**
**MAXIMUM LATERAL PRESSURE FOR DESIGN OF COLUMN FORMS (kPa)[a]**

| Rate of Placement R (m/hr) | Maximum Lateral Pressure, p (kPa), for Indicated Temperature (°C) | | | | | | |
|---|---|---|---|---|---|---|---|
| | 5 | 10 | 15 | 20 | 25 | 30 | 35 |
| 0.25 | 15.81 | 14.26 | 13.18 | 12.39 | 11.79 | 11.31 | 10.92 |
| 0.50 | 24.41 | 21.32 | 19.17 | 17.58 | 16.37 | 15.41 | 14.63 |
| 0.75 | 33.02 | 28.38 | 25.15 | 22.78 | 20.96 | 19.52 | 18.35 |
| 1.00 | 41.63 | 35.44 | 31.13 | 27.97 | 25.54 | 23.62 | 22.07 |
| 1.25 | 50.24 | 42.50 | 37.12 | 33.16 | 30.13 | 27.73 | 25.78 |
| 1.50 | 58.84 | 49.56 | 43.10 | 38.35 | 34.71 | 31.83 | 29.50 |
| 1.75 | 67.45 | 56.62 | 49.08 | 43.54 | 39.30 | 35.94 | 33.22 |
| 2.00 | 76.06 | 63.67 | 55.07 | 48.73 | 43.88 | 40.05 | 36.93 |
| 2.25 | 84.67 | 70.73 | 61.05 | 53.93 | 48.47 | 44.15 | 40.65 |
| 2.50 | 93.27 | 77.79 | 67.03 | 59.12 | 53.05 | 48.26 | 44.37 |
| 2.75 | 101.88 | 84.85 | 73.02 | 64.31 | 57.64 | 52.36 | 48.09 |
| 3.00 | 110.49 | 91.91 | 79.00 | 69.50 | 62.22 | 56.47 | 51.80 |
| 3.25 | 119.10 | 98.97 | 84.98 | 74.69 | 66.81 | 60.57 | 55.52 |
| 3.50 | 127.70 | 106.03 | 90.97 | 79.89 | 71.39 | 64.68 | 59.24 |
| 3.75 | 136.31 | 113.09 | 96.95 | 85.08 | 75.98 | 68.78 | 62.95 |
| 4.00 | | 120.15 | 102.93 | 90.27 | 80.56 | 72.89 | 66.67 |
| 4.50 | | 134.27 | 114.90 | 100.65 | 89.74 | 81.10 | 74.10 |
| 5.00 | | | 126.86 | 111.04 | 98.91 | 89.31 | 81.54 |
| 5.50 | | | 138.83 | 121.42 | 108.08 | 97.52 | 88.97 |
| 6.00 | | | | 131.80 | 117.25 | 105.74 | 96.40 |
| 6.50 | | | | 142.19 | 126.42 | 113.95 | 103.84 |
| 7.00 | | | | | 135.59 | 122.16 | 111.27 |
| 7.50 | | | | | | 130.37 | 118.71 |
| 8.00 | (143.6 kPa maximum) | | | | | 138.58 | 126.14 |
| 8.50 | | | | | | | 133.57 |
| 9.00 | | | | | | | 141.01 |
| 9.50 | | | | | | | |

[a]Do not use design pressures in excess of 144 kN/m² or 23.5 × height (in metres) of fresh concrete in forms, whichever is less.
*Source:* Reprinted by permission of American Concrete Institute; metric conversion by author.

30 m/min up and 45 m/min down. Assume pickup time is 20 sec and dump time is 3 min. Pour conditions are

wall height = 3.6 m
wall thickness = 250 mm
wall length 30.5 m

***Solution:***

The rate of pour depends on the total time that it takes for one cycle of the crane to pick up a bucket, unload it, and return it to the ground. Using two buckets, the filling time can be discounted as it occurs while the other bucket is being emptied.

Cycle time:

$$\text{time to travel up } \frac{36.6}{30} = 1.22 \text{ min}$$

$$\text{time to travel down } \frac{36.6}{45} = 0.81 \text{ min}$$

$$\text{pickup time } \frac{20}{60} = 0.33 \text{ min}$$

$$\text{dump time } = \underline{3.00 \text{ min}}$$

$$\text{Total } 5.36 \text{ min}$$

Rate of delivery in m³/hr:

$$\frac{60}{5.36} \times 1.0 = 11.20 \text{ m}^3/\text{hr}$$

Rate of pour calculations:

Volume of concrete to be poured:

$$3.6 \times \frac{250}{1{,}000} \times 30.5 = 27.45 \text{ m}^3$$

Time required to pour 27.45 m³ of concrete:

$$\frac{27.45}{11.2} = 2.45 \text{ hr}$$

Rate of pour in meters per hour:

$$\frac{3.6}{2.45} = 1.47 \text{ m/hr}$$

***Example:***

A 300-mm-thick basement wall has a total length of 48.8 m and a height of 3 m. Calculate the rate of pour based on a concrete pump capacity of 38 m³/hr.

***Solution:***

Because a concrete pump normally delivers concrete continuously, only the capacity of the pump need be considered in the calculation of the rate of pour.

Volume of pour:

$$\frac{300}{1{,}000} \times 3.0 \times 48.8 = 43.92 \text{ m}^3$$

Time required to pour wall:

$$\frac{43.92}{38} = 1.2 \text{ hr}$$

Rate of pour in meters per hour:

$$\frac{3}{1.2} = 2.5 \text{ m/hr}$$

---

## Concrete Pressure Calculations

Once the rate of pour has been calculated and the temperature of the concrete determined, the pressure calculations proceed. The following examples illustrate the pressures that may result based on typical pour rates and pour temperatures. It is important to remember that the formulas developed by the ACI are based on normal-density concrete of normal slump.

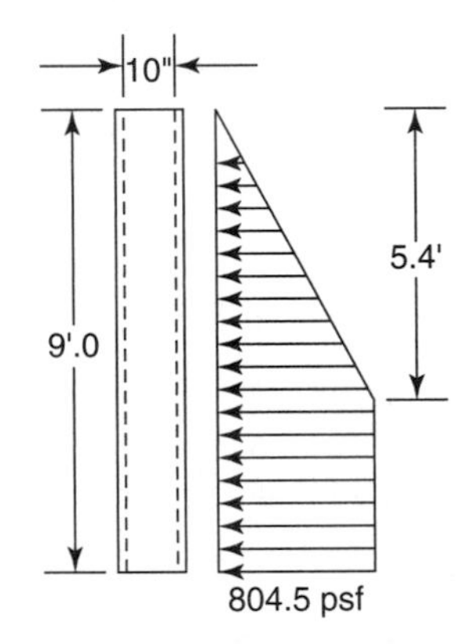


**Figure 6-9a** Wall Pressure Distribution Due to Plastic Concrete in psf.

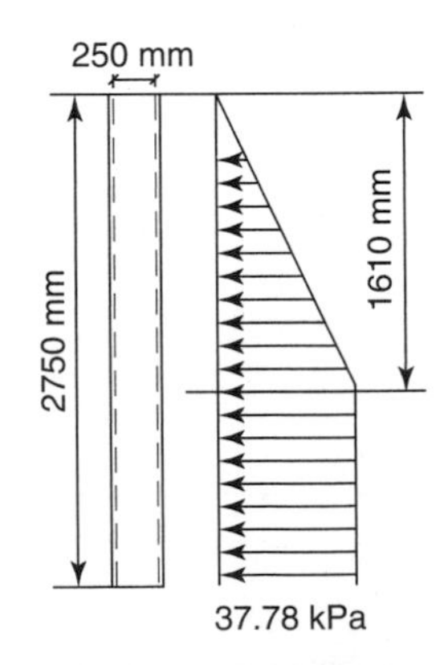


**Figure 6-9b** Wall Pressure Distribution Due to Plastic Concrete in kPa.

---

***Example:***

A 10-in.-thick wall, 50 ft in length and 9 ft high, is poured at the rate of 4 ft/hr. Calculate the maximum pressure developed in the concrete form assuming a concrete temperature of 55°F. Determine the height to which this maximum pressure will extend.

***Solution:***

Applying ACI formula for concrete pressure in wall forms (rate of pour less than 7 ft/hr),

$$\begin{aligned} p &= 150 + \frac{9{,}000 \times R}{T} \\ &= 150 + \frac{9{,}000 \times 4}{55} \\ &= 804.5 \text{ psf} \end{aligned}$$

Check: Maximum pressure is 2,000 psf or $150h$, whichever is less:

$$\begin{aligned} 150h &= 150 \times 9 \\ &= 1{,}350 \text{ psf} > 804.5 \text{ psf} \end{aligned}$$

Maximum pressure is as calculated from formula.

To determine the height of the maximum pressure, the term $150h$ represents the pressure of concrete in a liquid state and can be equated to the maximum pressure calculated.

$$p = 150h$$

which can be written in terms of $h$ as

$$h = \frac{p}{150}$$

where $h$ represents the distance from the top of the concrete pour to where the calculated maximum pressure begins (see Figure 6-9(a)). In this example, the distance from the top of the concrete pour to point of maximum pressure is

$$\begin{aligned} h &= \frac{804.5}{150} \\ &= 5.4 \text{ ft} \end{aligned}$$

---

***Example:***

A concrete column 18 in. by 18 in. square, 24 ft high, is poured in 15 min. Calculate the resulting pressure distribution on the formwork, assuming that the concrete temperature is 50°F.

***Solution:***

In this case, the rate of pour in feet per hour must be calculated before the form pressure can be considered.

Volume of concrete poured:

$$\frac{18}{12} \times \frac{18}{12} \times 24 \times \frac{1}{27} = 2 \text{ cu yd}$$

Rate of pour in cubic yards per hour:

$$\frac{2 \times 60}{15} = 8 \text{ cu yd/hr}$$

Rate of pour in feet per hour:

$$\frac{24 \times 60}{15} = 96 \text{ ft/hr}$$

Using the ACI formula for concrete pressure,

$$p = 150 + \frac{9{,}000R}{T}$$

but not to exceed 3,000 psf or $150h$.

Inserting values

$$p = 150 + \frac{9{,}000 \times 96}{50}$$
$$= 17{,}430 \text{ psf}$$

Check maximum pressure:

$$150h = 150 \times 24$$
$$= 3{,}600 \text{ psf} > 3{,}000 \text{ psf}$$

Maximum pressure is 3,000 psf.

Height from top to where maximum pressure extends is

$$h = \frac{p}{150}$$
$$= \frac{3{,}000}{150}$$
$$= 20 \text{ ft}$$

See Figure 6-10(a).

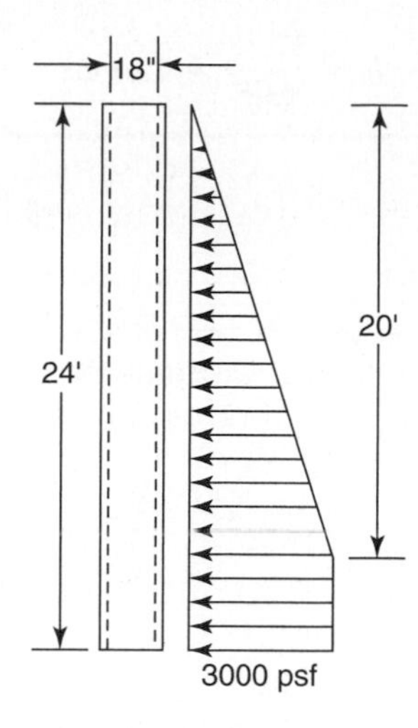


**Figure 6-10a** Column Pressure Distribution Due to Plastic Concrete in psf.

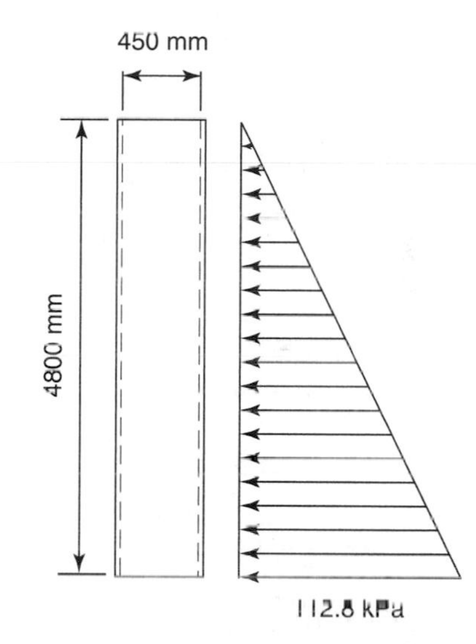


**Figure 6-10b** Column Pressure Distribution Due to Plastic Concrete in kPa.

***Example:***

A 250-mm-thick wall, 15 m in length, and 2.75 m high is poured at the rate of 1.2 m/hr. Calculate the maximum pressure developed in the concrete form assuming a concrete temperature of 13°C. Determine the height to which this maximum pressure will extend.

***Solution:***

Applying ACI formula for concrete pressure in wall forms for a rate of pour not exceeding 2 m/hr,

$$p = 7.2 + \frac{785R}{T + 17.8}$$
$$= 7.2 + \frac{785 \times 1.2}{(13 + 17.8)}$$
$$= 37.78 \text{ kN/m}^2 \text{ (kPa)}$$

Check: Maximum pressure is 144 kN/m$^2$ or $23.5h$, whichever is less

$$23.5h = 23.5 \times 2.75$$
$$= 64.63 \text{ kPa} > 37.78 \text{ kPa}$$

To determine the height of the maximum pressure, the term $23.5h$ represents the pressure of concrete in a liquid state and can be equated to the maximum pressure calculated,

$$p = 23.5h$$

which can be written in terms of $h$ as

$$h = \frac{p}{23.5}$$

where $h$ represents the distance from the top of the concrete pour to where the calculated maximum pressure begins (see Figure 6-9(b)). In this example, the distance from the top of the concrete pour to the point of maximum pressure is

$$h = \frac{37.78}{23.5} = 1.61 \text{ m}$$

---

***Example:***

A concrete column 450 mm × 450 mm square, 4.8 m high, is poured in 10 min. Calculate the resulting pressure distribution on the formwork, assuming the concrete temperature is 10°C.

***Solution:***

In this case, the rate of pour in meters per hour must be calculated before the form pressure can be considered.

Volume of concrete poured:

$$\frac{450}{1{,}000} \times \frac{450}{1{,}000} \times 4.8 = 0.972 \text{ m}^3$$

Rate of pour in cubic meters per hour:

$$\frac{0.972 \times 60}{10} = 5.83 \text{ m}^3\text{/hr}$$

Rate of pour in meters per hour:

$$\frac{4.8 \times 60}{10} = 28.8 \text{ m/hr}$$

Using the ACI formula for concrete pressure in columns

$$p = 7.2 + \frac{785R}{T + 17.8}$$

but not to exceed 144 kPa or $23.5h$.

Inserting values

$$p = 7.2 + \frac{785 \times 28.8}{10 + 17.8}$$

$$= 820.4 \text{ kPa}$$

Check maximum pressure

$$p = 23.5 \times 4.8$$

$$= 112.8 \text{ kPa} < 144 \text{ kPa}$$

Maximum pressure is 112.8 kPa.

Height from top to where maximum pressure extends is

$$h = \frac{p}{23.5}$$

$$h = \frac{112.8}{23.5} = 4.8 \text{ m}$$

Due to the rate of pour achieved, the maximum pressure in this calculation occurs at the bottom of the column (see Figure 6-10(b)).

---

## FORMWORK DESIGN

In order to take full advantage of sawn lumber sections and plywood sheeting as formwork materials in the construction of wall and column formwork, two methods of framing are common: (1) forms without studs and (2) forms with studs. For design purposes, formwork without studs (see Figure 6-11) can be divided into the following components.

1. Form sheathing
2. Wales
3. Ties
4. Lateral bracing

To ensure the integrity of the frame, each component within the frame must be capable of resisting the forces due to the plastic concrete. The amount of load on each component of the form is related to the arrangement and spacing of the various parts.

As an example, consider the wall form in Figure 6-11. The first component that is considered is the sheathing. The concrete exerts a pressure on the sheathing based on the pour conditions. Because the sheathing has a limited capacity to resist this pressure, it must be reinforced by the wales. These wales, in turn, have only so much strength (based on their section properties), so they in turn must be restrained. In this case it is the form tie that provides the final means of restraint for the structure. To ensure that the entire structure is properly aligned vertically, lateral bracing anchored to the ground is provided. If all of the components are properly proportioned, the form will perform as required.

When the concrete pressure within the formwork increases to a point where the spacing of the wales be-

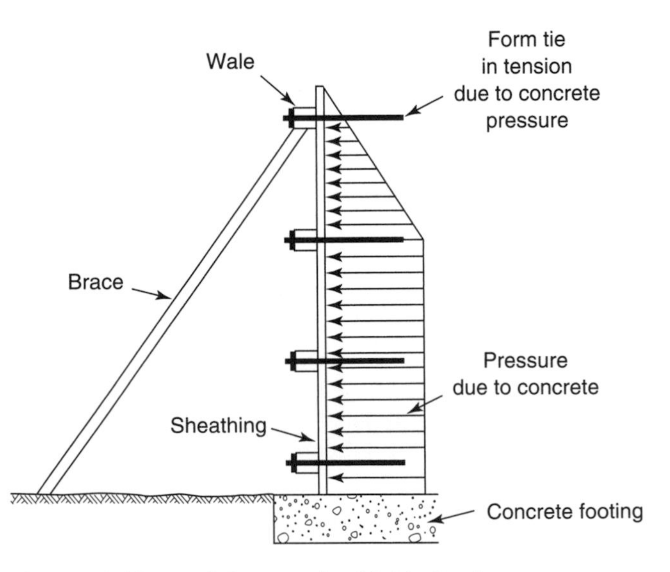


**Figure 6-11** Wall Formwork with No Studs.

comes impractical, studs are added to the framework of the form to support the sheathing (see Figure 6-12). The breakdown of the forming system then becomes

1. Form sheathing
2. Studs
3. Wales
4. Ties
5. Lateral bracing

In this situation, the studs can be placed at some reasonable spacing, based on the strength of the sheathing. Using the strong axis of the studs to advantage, the spacing of the wales can now be increased. However, this will result in greater forces on the wales and they must be designed accordingly. Typical wales are composed of double 2″ × 4″s (38 × 89s) as illustrated in Figure 6-13. When lumber sections become ineffective, double-channel steel sections or I-shaped aluminum sections are used as wales.

When a flexural section (beam) supporting a uniformly distributed load is continuous over a number of supports, the resulting reactions imposed on the supports vary with the number of spans over which the beam is continuous. Figure 6-14 illustrates reactions for continuous flexural members supporting a uniformly distributed load over varying numbers of supports. Note that the maximum reaction occurs at the first support in from the discontinuous end of the beam section.

Forms supporting gravity loads are also composed of components similar to those for walls and columns. The basic components are as follows:

1. Form sheathing
2. Joists (similar to wales)
3. Beams
4. Shores (similar to ties)
5. Mud sills
6. Lateral bracing

Consider the formwork supporting the slab in Figure 6-15. The gravity loads imposed on the form must be transferred from one component to another until they are safely transmitted to the ground. Again, the sheathing is the first component that must be considered. Because it has a limited ability to support load, it must be supported by joists much like a floor system. Joist spacing is based on the strength of the sheathing. Because of their larger cross section, joists can support loads over longer distances. However,

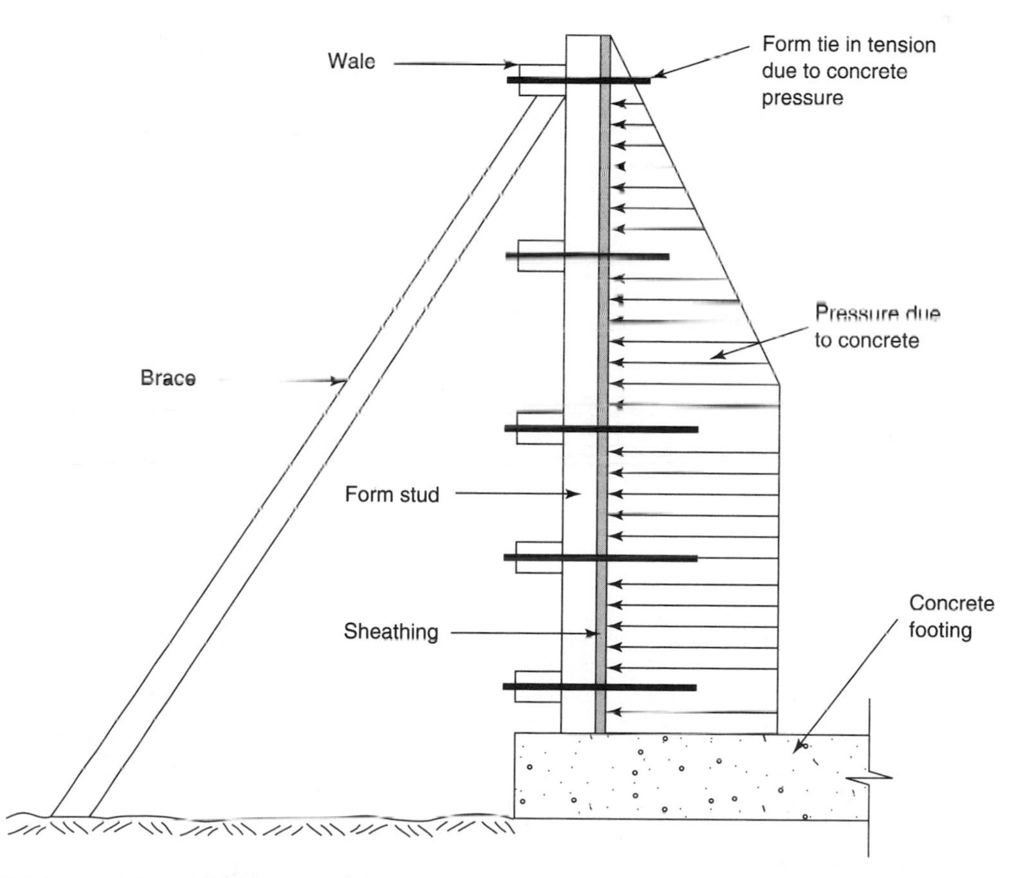


**Figure 6-12** Wall Formwork with Stud.

**Figure 6-13** Double 2″ × 4″s (38 × 89s) Used as Wales.

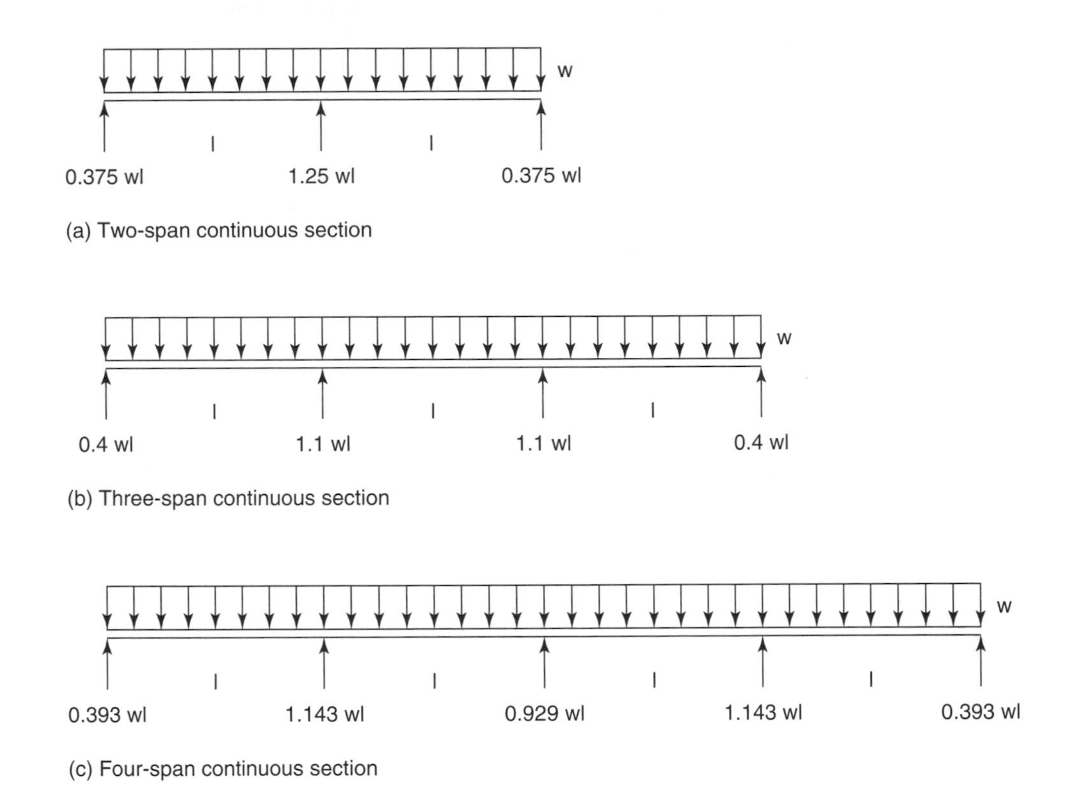


**Figure 6-14** Reactions of Continuous Sections Supporting Uniformly Distributed Loads. (a) Two-span Continuous Section; (b) Three-span Continuous Section; (c) Four-span Continuous Section.

they in turn must be supported. Beams, which are larger than the joists, are used as supports for the joists. The beams' ability to support load determines the spacing of the shores. The shores then must be of sufficient size to support the loads imposed by the beams. Finally, the mud sill must be of sufficient size to transfer the load from the shore to the supporting soil without undue settlement. Each component of the form must do its part to support the applied gravity loads. To ensure the lateral stability of the entire system, bracing must be provided in each direction.

## Formwork Tables

To facilitate the design of formwork components, the American Concrete Institute has developed a series

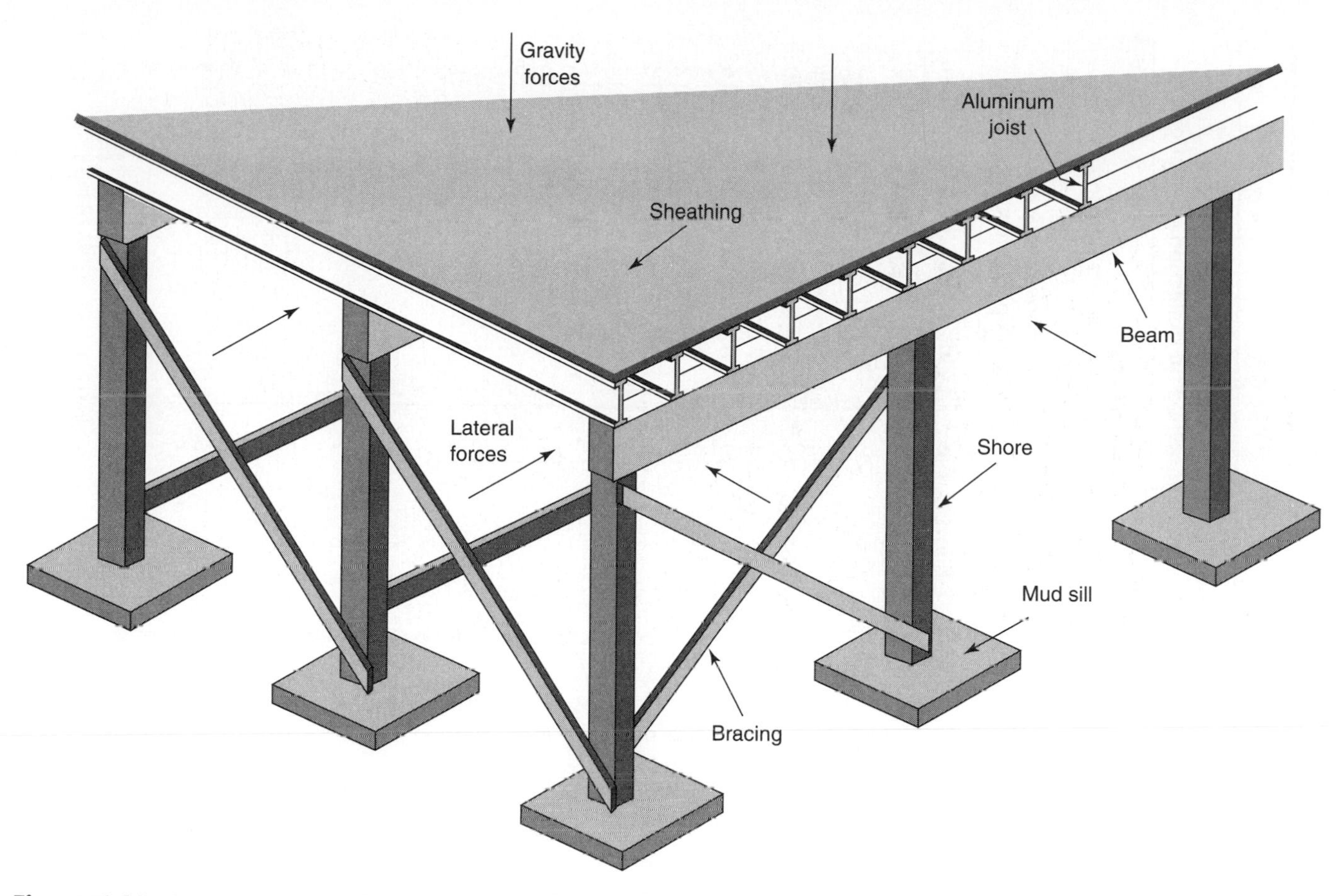


**Figure 6-15** Forces Acting on Floor Slab Formwork.

of tables that provide a designer with a quick method for selecting the appropriate size of section for a particular loading situation. Tables 6-4 to 6-13 are examples, reprinted with the permission of the ACI.

These tables are provided as a reference for the student to illustrate the selection of various sections for concrete formwork and represent a small sample of material strengths that are available for formwork components. These tables should not be used in actual formwork calculations without a thorough understanding of their limitations. The type of material, nature of the load, number of reuses, moisture conditions, and deflection limitations all have a bearing on the size of section and the spacing of supports.

If lumber is used, its grade and species must be known. Different species of lumber have different strengths. Plywood, for example, has a higher strength when the face grain is parallel to the span than when it is perpendicular to the span. The strength of wood is also decreased by absorption of moisture, and this fact must be taken into account in the design of formwork.

Form loading is the prime concern when designing formwork. It must be established as accurately as possible to ensure that the formwork will support the required loads and yet not be overdesigned to such an extent that it will add unnecessary costs to the job. The duration of loading and the number of reuses anticipated for the forms have a direct bearing on the final allowable loads that can be safely supported.

Deflections in the formwork must also be carefully considered. The final shape of the structure depends on the shape of the forms and, if excessive deflections occur during the placing of the concrete, the final outcome may be embarrassing, if not totally unacceptable. In extreme cases, collapse of the formwork during the placing of the concrete may occur because of excessive deflections.

Symbols used in the formwork tables are defined as follows:

- $F_c$, compression parallel to grain
- $E'$, modulus of elasticity
- $F'_b$, bending stress
- $F'_v$, horizontal shear stress

The following examples illustrate the use of the formwork tables in determining the required spacings of supports for formwork components based on particular material selections.

**Table 6-4(a)**

**SAFE SPACING, IN INCHES, OF SUPPORTS FOR PLYWOOD SHEATHING, CONTINUOUS OVER FOUR OR MORE SUPPORTS[a]**

Maximum deflection 1/360 of span, but not more than 1/16 in.

| Pressure or load of concrete, pounds per square foot | Stresses and spans for short duration loads, for all sanded grades of Group 1 plywood. *E* modified for deflection calculations. | | | | | | | | Stresses and spans for long-term loading, for all sanded grades of Group 1 plywood. *E* modified for deflection calculations. | | | | | | | |
|---|---|---|---|---|---|---|---|---|---|---|---|---|---|---|---|---|
| | $F'_b$ = 1930 psi; rolling shear = 72 psi; $E'$ = 1,500,000 psi | | | | | | | | $F'_b$ = 1545 psi; rolling shear = 57 psi; $E'$ = 1,500,000 psi | | | | | | | |
| | sanded thickness, face grain parallel to span | | | | sanded thickness, face grain perpendicular to span | | | | sanded thickness, face grain parallel to span | | | | sanded thickness, face grain perpendicular to span | | | |
| | ½ in. | ⅝ in. | ¾ in. | 1 in. | ½ in. | ⅝ in. | ¾ in. | 1 in. | ½ in. | ⅝ in. | ¾ in. | 1 in. | ½ in. | ⅝ in. | ¾ in. | 1 in. |
| 75 | 20 | 23 | 26 | 31 | 10 | 14 | 18 | 25 | 20 | 23 | 26 | 31 | 10 | 14 | 18 | 25 |
| 100 | 18 | 21 | 24 | 29 | 9 | 13 | 17 | 23 | 18 | 21 | 24 | 29 | 9 | 13 | 17 | 23 |
| 125 | 16 | 20 | 23 | 27 | 8 | 12 | 15 | 22 | 16 | 20 | 23 | 27 | 8 | 12 | 15 | 22 |
| 150 | 15 | 18 | 21 | 26 | 8 | 11 | 14 | 21 | 15 | 18 | 21 | 26 | 8 | 11 | 14 | 21 |
| 175 | 15 | 17 | 20 | 25 | 7 | 10 | 14 | 20 | 15 | 17 | 20 | 25 | 7 | 10 | 14 | 20 |
| 200 | 14 | 17 | 19 | 24 | 7 | 10 | 13 | 19 | 14 | 17 | 19 | 24 | 7 | 10 | 13 | 19 |
| 300 | 12 | 15 | 17 | 22 | 6 | 9 | 11 | 16 | 12 | 14 | 16 | 20 | 6 | 9 | 11 | 16 |
| 400 | 11 | 13 | 15 | 20 | 5 | 8 | 10 | 15 | 10 | 12 | 14 | 18 | 5 | 8 | 10 | 15 |
| 500 | 10 | 12 | 14 | 18 | 5 | 7 | 10 | 14 | 9 | 11 | 12 | 16 | 5 | 7 | 10 | 14 |
| 600 | 10 | 11 | 13 | 16 | 5 | 7 | 9 | 13 | 9 | 10 | 11 | 14 | 5 | 7 | 9 | 13 |
| 700 | 9 | 11 | 12 | 15 | 5 | 7 | 9 | 12 | 8 | 9 | 10 | 13 | 5 | 7 | 8 | 12 |
| 800 | 8 | 10 | 11 | 14 | 4 | 6 | 8 | 12 | 7 | 9 | 10 | 12 | 4 | 6 | 7 | 12 |
| 900 | 8 | 9 | 10 | 13 | 4 | 6 | 8 | 11 | 7 | 8 | 9 | 12 | 4 | 5 | 7 | 10 |
| 1000 | 7 | 9 | 10 | 12 | 4 | 6 | 7 | 11 | 7 | 8 | 9 | 11 | 4 | 5 | 6 | 10 |
| 1100 | 7 | 8 | 9 | 12 | 4 | 6 | 7 | 11 | 6 | 8 | 8 | 11 | 4 | 5 | 6 | 9 |
| 1200 | 7 | 8 | 9 | 11 | 4 | 5 | 6 | 10 | 6 | 7 | 8 | 10 | 4 | 4 | 5 | 8 |
| 1300 | 6 | 8 | 9 | 11 | 4 | 5 | 6 | 9 | 5 | 7 | 8 | 9 | 4 | 4 | 5 | 8 |
| 1400 | 6 | 7 | 8 | 10 | 4 | 5 | 6 | 9 | 5 | 6 | 7 | 9 | — | 4 | 5 | 7 |
| 1500 | 6 | 7 | 8 | 10 | 4 | 4 | 5 | 8 | 5 | 6 | 7 | 8 | — | 4 | 5 | 7 |
| 1600 | 6 | 7 | 8 | 9 | — | 4 | 5 | 8 | 5 | 6 | 7 | 8 | — | 4 | 4 | 7 |
| 1700 | 5 | 6 | 7 | 9 | — | 4 | 5 | 7 | 4 | 5 | 6 | 7 | — | 4 | 4 | 6 |
| 1800 | 5 | 6 | 7 | 9 | — | 4 | 5 | 7 | 4 | 5 | 6 | 7 | — | — | 4 | 6 |
| 1900 | 5 | 6 | 7 | 8 | — | 4 | 5 | 7 | 4 | 5 | 6 | 7 | — | — | 4 | 6 |
| 2000 | 5 | 6 | 7 | 8 | — | 4 | 4 | 7 | 4 | 5 | 6 | 7 | — | — | 4 | 6 |
| 2200 | 4 | 5 | 6 | 7 | — | 4 | 4 | 6 | 4 | 5 | 5 | 6 | — | — | 4 | 5 |
| 2400 | 4 | 5 | 6 | 7 | — | — | 4 | 6 | 4 | 4 | 5 | 6 | — | — | — | 5 |
| 2600 | 4 | 5 | 5 | 6 | — | — | 4 | 5 | — | 4 | 5 | 5 | — | — | — | 5 |
| 2800 | 4 | 4 | 5 | 6 | — | — | 4 | 5 | — | 4 | 4 | 5 | — | — | — | 4 |
| 3000 | 4 | 4 | 5 | 6 | — | — | — | 5 | — | 4 | 4 | 5 | — | — | — | 4 |

[a]Above solid line, deflection controls span. Below dashed line, rolling shear governs. Between the lines, bending controls. Spans are given center to center of supports, assuming 1½-in. support width for shear spans. If supports of a different width are used, detailed calculations should be made to check spans in the range now shown as controlled by shear.

*Source:* Reproduced by permission of the American Concrete Institute.

**Table 6-4(b)**

**SAFE SPACING, IN MILLIMETRES, OF SUPPORTS FOR PLYWOOD SHEATHING, CONTINUOUS OVER FOUR OR MORE SUPPORTS[a]**

Maximum deflection 1/360 of span, but not more than 1.5 mm.

| Pressure or load of concrete, kPa | Stresses and spans for short-duration loads, for all sanded grades of Group 1 plywood. *E* modified for deflection calculations. | | | | | | | | Stresses and spans for long-term loading, for all sanded grades of Group 1 plywood. *E* modified for deflection calculations. | | | | | | | |
|---|---|---|---|---|---|---|---|---|---|---|---|---|---|---|---|---|
| | $F'_c$ = 13.3 MPa; rolling shear = 0.50 MPa; *E* = 10.34 GPa | | | | | | | | $F'_b$ = 10.65 MPa; rolling shear = 0.40 MPa; *E'* = 10.34 GPa | | | | | | | |
| | sanded thickness, face grain parallel to span (mm) | | | | sanded thickness, face grain perpendicular to span (mm) | | | | sanded thickness, face grain parallel to span (mm) | | | | sanded thickness, face grain perpendicular to span (mm) | | | |
| | 12.5 | 15.5 | 18.5 | 25 | 12.5 | 15.5 | 18.5 | 25 | 12.5 | 15.5 | 18.5 | 25 | 12.5 | 15.5 | 18.5 | 25 |
| 3.59 | 500 | 575 | 650 | 775 | 250 | 350 | 450 | 625 | 500 | 575 | 650 | 775 | 250 | 350 | 450 | 625 |
| 4.79 | 450 | 525 | 600 | 725 | 225 | 325 | 425 | 575 | 450 | 525 | 600 | 725 | 225 | 325 | 425 | 575 |
| 5.99 | 400 | 500 | 575 | 675 | 200 | 300 | 375 | 550 | 400 | 500 | 575 | 675 | 200 | 300 | 375 | 550 |
| 7.18 | 375 | 450 | 525 | 650 | 200 | 275 | 350 | 525 | 375 | 450 | 525 | 650 | 200 | 275 | 350 | 525 |
| 8.38 | 375 | 425 | 500 | 625 | 175 | 250 | 350 | 500 | 375 | 425 | 500 | 625 | 175 | 250 | 350 | 500 |
| 9.58 | 350 | 425 | 475 | 600 | 175 | 250 | 325 | 475 | 350 | 425 | 475 | 600 | 175 | 250 | 325 | 475 |
| 14.36 | 300 | 375 | 425 | 550 | 150 | 225 | 275 | 400 | 300 | 350 | 400 | 500 | 150 | 225 | 275 | 400 |
| 19.15 | 275 | 325 | 375 | 500 | 125 | 200 | 250 | 375 | 250 | 300 | 350 | 450 | 125 | 200 | 250 | 375 |
| 23.94 | 250 | 300 | 350 | 450 | 125 | 175 | 250 | 350 | 225 | 275 | 300 | 400 | 125 | 175 | 250 | 350 |
| 28.73 | 250 | 275 | 325 | 400 | 125 | 175 | 225 | 325 | 225 | 250 | 275 | 350 | 125 | 175 | 225 | 325 |
| 33.52 | 225 | 275 | 300 | 375 | 125 | 175 | 225 | 300 | 200 | 225 | 250 | 325 | 125 | 175 | 200 | 300 |
| 38.30 | 200 | 250 | 275 | 350 | 100 | 150 | 200 | 300 | 175 | 225 | 250 | 300 | 100 | 150 | 175 | 300 |
| 43.09 | 200 | 225 | 250 | 325 | 100 | 150 | 200 | 275 | 175 | 200 | 225 | 300 | 100 | 125 | 175 | 250 |
| 47.88 | 175 | 225 | 250 | 300 | 100 | 150 | 175 | 275 | 175 | 200 | 225 | 275 | 100 | 125 | 150 | 250 |
| 52.67 | 175 | 200 | 225 | 300 | 100 | 150 | 175 | 275 | 150 | 200 | 200 | 275 | 100 | 125 | 150 | 225 |
| 57.46 | 175 | 200 | 225 | 275 | 100 | 125 | 150 | 250 | 150 | 175 | 200 | 250 | 100 | 100 | 125 | 200 |
| 62.24 | 150 | 200 | 225 | 275 | 100 | 125 | 150 | 225 | 125 | 175 | 200 | 225 | 100 | 100 | 125 | 200 |
| 67.03 | 150 | 175 | 200 | 250 | 100 | 125 | 150 | 225 | 125 | 150 | 175 | 225 | | 100 | 125 | 175 |
| 71.82 | 150 | 175 | 200 | 250 | 100 | 100 | 125 | 200 | 125 | 150 | 175 | 200 | | 100 | 125 | 175 |
| 76.61 | 150 | 175 | 200 | 225 | | 100 | 125 | 200 | 125 | 150 | 175 | 200 | | 100 | 100 | 175 |

[a]Above the solid line spans are governed by deflection; below the dashed line spans are governed by rolling shear. Between the lines, bending governs. Spans are given center to center of supports. Spans controlled by shear are based on 38 mm wide supports. If supports are a different width, detailed calculations should be made to check spans controlled by shear.

*Source:* Original data provided by the American Concrete Institute; metric conversion by author.

**Table 6-5(a)**

**SAFE SPACING, IN INCHES, OF SUPPORTS FOR PLYWOOD SHEATHING WITH ONLY TWO POINTS OF SUPPORT[a]**

Maximum deflection 1/360 of span, but not more than 1/16 in.

| Pressure or load of concrete, pounds per square foot | Stresses and spans for short duration loads, for all sanded grades of Group 1 plywood. *E* modified for deflection calculations. | | | | | | | | Stresses and spans for long-term loading, for all sanded grades of Group 1 plywood. *E* modified for deflection calculations. | | | | | | | |
|---|---|---|---|---|---|---|---|---|---|---|---|---|---|---|---|---|
| | $F'_b$ = 1930 psi; rolling shear = 72 psi; $E'$ = 1,500,000 psi | | | | | | | | $F'_b$ = 1545 psi; rolling shear = 57 psi; $E'$ = 1,500,000 psi | | | | | | | |
| | sanded thickness, face grain parallel to span | | | | sanded thickness, face grain perpendicular to span | | | | sanded thickness, face grain parallel to span | | | | sanded thickness, face grain perpendicular to span | | | |
| | ½ in. | ⅝ in. | ¾ in. | 1 in. | ½ in. | ⅝ in. | ¾ in. | 1 in. | ½ in. | ⅝ in. | ¾ in. | 1 in. | ½ in. | ⅝ in. | ¾ in. | 1 in. |
| 75 | 16 | 19 | 22 | 26 | 8 | 11 | 15 | 21 | 16 | 19 | 22 | 26 | 8 | 11 | 15 | 21 |
| 100 | 14 | 17 | 20 | 25 | 7 | 10 | 13 | 19 | 14 | 17 | 20 | 24 | 7 | 10 | 13 | 19 |
| 125 | 13 | 16 | 18 | 23 | 7 | 9 | 12 | 18 | 13 | 16 | 18 | 23 | 7 | 9 | 12 | 18 |
| 150 | 13 | 15 | 17 | 22 | 6 | 9 | 12 | 17 | 13 | 15 | 17 | 22 | 6 | 9 | 12 | 17 |
| 175 | 12 | 14 | 16 | 21 | 6 | 8 | 11 | 16 | 12 | 14 | 16 | 21 | 6 | 8 | 11 | 16 |
| 200 | 11 | 14 | 16 | 20 | 6 | 8 | 11 | 15 | 11 | 14 | 16 | 20 | 6 | 8 | 11 | 15 |
| 300 | 10 | 12 | 14 | 18 | 5 | 7 | 9 | 13 | 10 | 12 | 14 | 18 | 5 | 7 | 9 | 13 |
| 400 | 9 | 11 | 12 | 16 | 4 | 6 | 8 | 12 | 9 | 11 | 12 | 16 | 4 | 6 | 8 | 12 |
| 500 | 8 | 10 | 11 | 15 | 4 | 6 | 8 | 11 | 8 | 10 | 11 | 14 | 4 | 6 | 8 | 11 |
| 600 | 8 | 9 | 11 | 14 | 4 | 6 | 7 | 11 | 8 | 9 | 10 | 13 | 4 | 6 | 7 | 11 |
| 700 | 8 | 9 | 10 | 13 | 4 | 5 | 7 | 10 | 7 | 8 | 9 | 12 | 4 | 5 | 7 | 10 |
| 800 | 7 | 9 | 10 | 12 | 4 | 5 | 7 | 10 | 7 | 8 | 9 | 11 | 4 | 5 | 7 | 10 |
| 900 | 7 | 8 | 9 | 12 | — | 5 | 6 | 9 | 6 | 7 | 8 | 10 | — | 5 | 6 | 9 |
| 1000 | 7 | 8 | 9 | 11 | — | 5 | 6 | 9 | 6 | 7 | 8 | 10 | — | 5 | 6 | 9 |
| 1100 | 6 | 8 | 8 | 11 | — | 5 | 6 | 9 | 6 | 7 | 7 | 9 | — | 5 | 6 | 9 |
| 1200 | 6 | 7 | 8 | 10 | — | 4 | 6 | 8 | 5 | 6 | 7 | 9 | — | 4 | 6 | 8 |
| 1300 | 6 | 7 | 8 | 10 | — | 4 | 6 | 8 | 5 | 6 | 7 | 9 | — | 4 | 6 | 8 |
| 1400 | 6 | 7 | 7 | 9 | — | 4 | 6 | 8 | 5 | 6 | 7 | 8 | — | 4 | 5 | 8 |
| 1500 | 5 | 6 | 7 | 9 | — | 4 | 5 | 8 | 5 | 6 | 6 | 8 | — | 4 | 5 | 8 |
| 1600 | 5 | 6 | 7 | 9 | — | 4 | 5 | 8 | 5 | 6 | 6 | 8 | — | 4 | 5 | 7 |
| 1700 | 5 | 6 | 7 | 9 | — | 4 | 5 | 7 | 5 | 5 | 6 | 8 | — | 4 | 5 | 7 |
| 1800 | 5 | 6 | 7 | 8 | — | 4 | 5 | 7 | 4 | 5 | 6 | 7 | — | 4 | 5 | 7 |
| 1900 | 5 | 6 | 6 | 8 | — | 4 | 5 | 7 | 4 | 5 | 6 | 7 | — | 4 | 4 | 7 |
| 2000 | 5 | 6 | 6 | 8 | — | 4 | 5 | 7 | 4 | 5 | 6 | 7 | — | — | 4 | 6 |
| 2200 | 4 | 5 | 6 | 7 | — | 4 | 5 | 7 | 4 | 5 | 5 | 7 | — | — | 4 | 6 |
| 2400 | 4 | 5 | 6 | 7 | — | 4 | 4 | 7 | 4 | 5 | 5 | 6 | — | — | 4 | 6 |
| 2600 | 4 | 5 | 5 | 7 | — | — | 4 | 6 | 4 | 4 | 5 | 6 | — | — | 4 | 5 |
| 2800 | 4 | 5 | 5 | 7 | — | — | 4 | 6 | 4 | 4 | 5 | 6 | — | — | — | 5 |
| 3000 | 4 | 5 | 5 | 6 | — | — | 4 | 6 | — | 4 | 5 | 6 | — | — | — | 5 |

[a]Above solid line, deflection controls span. Below dashed line, rolling shear governs. Where there is no dashed line, bending controls below the solid line. Bending also controls spans shown between the dashed and solid lines. Spans are given center to center of supports, assuming 1½ in. support width for shear spans. If supports of a different width are used, detailed calculations should be made to check spans in the range now shown as controlled by shear.

*Source:* Reproduced by permission of the American Concrete Institute.

**Table 6-5(b)**

**SAFE SPACING, IN MILLIMETRES, OF SUPPORTS FOR PLYWOOD SHEATHING WITH ONLY TWO POINTS OF SUPPORT[a]**

Maximum deflection 1/360 of span, but not to exceed more than 1.5 mm.

| Pressure or load of concrete, kPa | Stresses and spans for short duration loads, for all sanded grades of Group 1 plywood. *E* modified for deflection calculations. | | | | | | | | Stresses and spans for long-term loading, for all sanded grades of Group 1 plywood. *E* modified for deflection calculations. | | | | | | | |
|---|---|---|---|---|---|---|---|---|---|---|---|---|---|---|---|---|
| | $F_b$ = 13.3 MPa; rolling shear = 0.50 MPa; $E'$ = 10.34 GPa | | | | | | | | $F'_b$ = 10.65 MPa; rolling shear = 0.40 MPa; $E'$ = 10.34 GPa | | | | | | | |
| | sanded thickness, face grain parallel to span (mm) | | | | sanded thickness, face grain perpendicular to span (mm) | | | | sanded thickness, face grain parallel to span (mm) | | | | sanded thickness, face grain perpendicular to span (mm) | | | |
| | 12.5 | 15.5 | 18.5 | 25 | 12.5 | 15.5 | 18.5 | 25 | 12.5 | 15.5 | 18.5 | 25 | 12.5 | 15.5 | 18.5 | 25 |
| 3.59 | 400 | 475 | 550 | 650 | 200 | 275 | 375 | 525 | 400 | 475 | 550 | 650 | 200 | 275 | 375 | 525 |
| 4.79 | 350 | 425 | 500 | 625 | 175 | 250 | 325 | 475 | 350 | 425 | 500 | 600 | 175 | 250 | 325 | 475 |
| 5.99 | 325 | 400 | 450 | 575 | 175 | 225 | 300 | 450 | 325 | 400 | 450 | 575 | 175 | 225 | 300 | 450 |
| 7.18 | 325 | 375 | 425 | 550 | 150 | 225 | 300 | 425 | 325 | 375 | 425 | 550 | 150 | 225 | 300 | 425 |
| 8.38 | 300 | 350 | 400 | 525 | 150 | 200 | 275 | 400 | 300 | 350 | 400 | 525 | 150 | 200 | 275 | 400 |
| 9.58 | 275 | 350 | 400 | 500 | 150 | 200 | 275 | 375 | 275 | 350 | 400 | 500 | 150 | 200 | 275 | 375 |
| 14.36 | 250 | 300 | 350 | 450 | 125 | 175 | 225 | 325 | 250 | 300 | 350 | 450 | 125 | 175 | 225 | 325 |
| 19.15 | 225 | 275 | 300 | 400 | 100 | 150 | 200 | 300 | 225 | 275 | 300 | 400 | 100 | 150 | 200 | 300 |
| 23.94 | 200 | 250 | 275 | 375 | 100 | 150 | 200 | 275 | 200 | 250 | 275 | 350 | 100 | 150 | 200 | 275 |
| 28.73 | 200 | 225 | 275 | 350 | 100 | 150 | 175 | 275 | 200 | 225 | 250 | 325 | 100 | 150 | 175 | 275 |
| 33.52 | 200 | 225 | 250 | 325 | 100 | 125 | 175 | 250 | 175 | 200 | 225 | 300 | 100 | 125 | 175 | 250 |
| 38.30 | 175 | 225 | 250 | 300 | 100 | 125 | 175 | 250 | 175 | 200 | 225 | 275 | 100 | 125 | 175 | 250 |
| 43.09 | 175 | 200 | 225 | 300 | | 125 | 150 | 225 | 150 | 175 | 200 | 250 | | 125 | 150 | 225 |
| 47.88 | 175 | 200 | 225 | 275 | | 125 | 150 | 225 | 150 | 175 | 200 | 250 | | 125 | 150 | 225 |
| 52.67 | 150 | 200 | 200 | 275 | | 125 | 150 | 225 | 150 | 175 | 175 | 225 | | 125 | 150 | 225 |
| 57.46 | 150 | 175 | 200 | 250 | | 100 | 150 | 200 | 125 | 150 | 175 | 225 | | 100 | 150 | 200 |
| 62.24 | 150 | 175 | 200 | 250 | | 100 | 150 | 200 | 125 | 150 | 175 | 225 | | 100 | 150 | 200 |
| 67.03 | 150 | 175 | 175 | 225 | | 100 | 150 | 200 | 125 | 150 | 175 | 200 | | 100 | 125 | 200 |
| 71.82 | 125 | 150 | 175 | 225 | | 100 | 125 | 200 | 125 | 150 | 150 | 200 | | 100 | 125 | 200 |
| 76.61 | 125 | 150 | 175 | 225 | | 100 | 125 | 200 | 125 | 150 | 150 | 200 | | 100 | 125 | 175 |

[a]Above the solid line spans are governed by deflection; below the dashed line spans are governed by rolling shear. Between the lines, bending governs. Spans are given center to center of supports. Spans controlled by shear are based on 38-mm wide supports. If supports are a different width, detailed calculations should be made to check spans controlled by shear.

*Source:* Original data provided by the American Concrete Institute; metric conversion by author.

**Table 6-6(a)**

**SAFE SPACING (INCHES) OF SUPPORTS FOR JOISTS, STUDS, OR OTHER BEAM COMPONENTS OF FORMWORK, CONTINUOUS OVER THREE OR MORE SPANS[a]**

Spacing | Spacing | Spacing

Maximum deflection is 1/360 of the span or ¼ in., whichever is smaller.

| Uniform load, lb per lineal ft (equals uniform load on forms times spacing between joists or studs, ft) | $F'_b$ varies with member | | | | | | | $E' = 1{,}300{,}000$ psi | | | | | | | $F'_v = 140$ psi | |
|---|---|---|---|---|---|---|---|---|---|---|---|---|---|---|---|---|
| | Nominal size of S × S lumber | | | | | | | | | | | | | | | |
| | 2 × 4 | 2 × 6 | 2 × 8 | 2 × 10 | 3 × 4 | 3 × 6 | 3 × 8 | 3 × 10 | 4 × 2 | 4 × 4 | 4 × 6 | 4 × 8 | 6 × 2 | 6 × 4 | 8 × 2 | 10 × 2 |
| | $F'_b$ psi | | | | | | | | | | | | | | | |
| | 1280 | 1100 | 1020 | 940 | 1280 | 1100 | 1020 | 940 | 1400 | 1280 | 1110 | 1110 | 1270 | 1160 | 1170 | 1120 |
| 100 | 69 | 100 | 127 | 154 | 82 | 118 | 146 | 175 | 40 | 92 | 129 | 158 | 46 | 103 | 50 | 55 |
| 200 | 48 | 71 | 90 | 110 | 63 | 92 | 116 | 142 | 31 | 73 | 108 | 133 | 36 | 85 | 40 | 43 |
| 300 | 40 | 58 | 73 | 90 | 51 | 75 | 94 | 116 | 27 | 60 | 88 | 117 | 32 | 72 | 35 | 38 |
| 400 | 32 | 50 | 63 | 78 | 44 | 65 | 82 | 100 | 23 | 52 | 77 | 101 | 28 | 62 | 31 | 34 |
| 500 | 27 | 42 | 55 | 69 | 40 | 58 | 73 | 90 | 21 | 47 | 69 | 90 | 25 | 56 | 28 | 31 |
| 600 | 23 | 37 | 48 | 62 | 34 | 53 | 67 | 82 | 19 | 43 | 63 | 82 | 23 | 51 | 25 | 28 |
| 700 | 21 | 33 | 44 | 56 | 30 | 48 | 62 | 76 | 17 | 40 | 58 | 76 | 21 | 47 | 23 | 26 |
| 800 | 19 | 30 | 40 | 51 | 27 | 43 | 57 | 71 | 15 | 36 | 54 | 71 | 20 | 44 | 22 | 24 |
| 900 | 18 | 28 | 37 | 47 | 25 | 40 | 52 | 66 | 14 | 32 | 51 | 67 | 19 | 42 | 21 | 23 |
| 1000 | 17 | 26 | 35 | 44 | 23 | 37 | 48 | 62 | 13 | 30 | 47 | 62 | 18 | 40 | 20 | 22 |
| 1100 | 16 | 25 | 33 | 42 | 22 | 34 | 45 | 58 | 12 | 28 | 44 | 58 | 17 | 38 | 19 | 21 |
| 1200 | 15 | 24 | 31 | 40 | 21 | 32 | 43 | 54 | 11 | 26 | 41 | 54 | 16 | 36 | 18 | 20 |
| 1300 | 15 | 23 | 30 | 38 | 20 | 31 | 41 | 52 | 11 | 25 | 39 | 51 | 15 | 35 | 17 | 19 |
| 1400 | 14 | 22 | 29 | 37 | 19 | 29 | 39 | 49 | 10 | 23 | 37 | 48 | 14 | 33 | 17 | 18 |
| 1500 | 14 | 21 | 28 | 36 | 18 | 28 | 37 | 47 | 10 | 22 | 35 | 46 | 13 | 31 | 16 | 18 |
| 1600 | 13 | 21 | 27 | 35 | 17 | 27 | 36 | 45 | 9 | 21 | 33 | 44 | 13 | 29 | 15 | 17 |
| 1700 | 13 | 20 | 26 | 34 | 17 | 26 | 34 | 44 | 9 | 20 | 32 | 42 | 12 | 28 | 15 | 17 |
| 1800 | 12 | 20 | 26 | 33 | 16 | 25 | 33 | 42 | 8 | 20 | 31 | 41 | 12 | 27 | 14 | 16 |
| 1900 | 12 | 19 | 25 | 32 | 16 | 25 | 32 | 41 | 8 | 19 | 30 | 39 | 11 | 26 | 14 | 16 |
| 2000 | 12 | 19 | 25 | 31 | 15 | 24 | 31 | 40 | 8 | 18 | 29 | 38 | 11 | 25 | 13 | 15 |
| 2100 | 12 | 18 | 24 | 31 | 15 | 23 | 31 | 39 | 8 | 18 | 28 | 37 | 10 | 24 | 13 | 15 |
| 2200 | 11 | 18 | 24 | 30 | 14 | 23 | 30 | 38 | 7 | 17 | 27 | 36 | 10 | 23 | 12 | 15 |
| 2300 | 11 | 18 | 23 | 30 | 14 | 22 | 29 | 37 | 7 | 17 | 27 | 35 | 10 | 23 | 12 | 14 |
| 2400 | 11 | 17 | 23 | 29 | 14 | 22 | 29 | 36 | 7 | 17 | 26 | 34 | 9 | 22 | 11 | 14 |
| 2500 | 11 | 17 | 23 | 29 | 14 | 21 | 28 | 36 | 7 | 16 | 25 | 33 | 9 | 21 | 11 | 13 |
| 2600 | 11 | 17 | 22 | 28 | 13 | 21 | 28 | 35 | 7 | 16 | 25 | 33 | 9 | 21 | 11 | 13 |
| 2700 | 11 | 17 | 22 | 28 | 13 | 21 | 27 | 34 | 7 | 15 | 24 | 32 | 9 | 20 | 11 | 13 |
| 2800 | 11 | 17 | 22 | 28 | 13 | 20 | 27 | 34 | 7 | 15 | 24 | 31 | 9 | 20 | 10 | 12 |
| 2900 | 10 | 16 | 22 | 27 | 13 | 20 | 26 | 33 | 6 | 15 | 23 | 31 | 8 | 19 | 10 | 12 |
| 3000 | 10 | 16 | 21 | 27 | 12 | 20 | 26 | 33 | 6 | 15 | 23 | 30 | 8 | 19 | 10 | 12 |
| 3200 | 10 | 16 | 21 | 27 | 12 | 19 | 25 | 32 | 6 | 14 | 22 | 29 | 8 | 18 | 9 | 11 |
| 3400 | 10 | 16 | 20 | 26 | 12 | 19 | 24 | 31 | 6 | 14 | 22 | 28 | 8 | 18 | 9 | 11 |
| 3600 | 10 | 15 | 20 | 26 | 12 | 18 | 24 | 30 | 6 | 13 | 21 | 28 | 7 | 17 | 9 | 10 |
| 3800 | 10 | 15 | 20 | 25 | 11 | 18 | 23 | 30 | 6 | 13 | 20 | 27 | 7 | 16 | 8 | 10 |
| 4000 | 9 | 15 | 20 | 25 | 11 | 17 | 23 | 29 | 5 | 13 | 20 | 26 | 7 | 16 | 8 | 9 |
| 4500 | 9 | 14 | 19 | 23 | 11 | 17 | 22 | 28 | 5 | 12 | 19 | 25 | 6 | 15 | 8 | 9 |
| 5000 | 9 | 14 | 18 | 22 | 10 | 16 | 21 | 27 | 5 | 12 | 18 | 24 | 6 | 14 | 7 | 8 |

[a]Span values above solid line are controlled by deflection. Within the dashed box horizontal shear governs span. Elsewhere bending controls span.

*Source:* Reproduced by permission of the American Concrete Institute.

**Table 6-6(b)**

**SAFE SPACING (MM) OF SUPPORTS FOR JOISTS, STUDS, OR OTHER BEAM COMPONENTS OF FORMWORK, CONTINUOUS OVER THREE OR MORE SPANS[a]**

Maximum deflection = span/360, but not to exceed 6 mm.

| Uniform load, kN/m (equals uniform load on forms times spacing between joists or studs, metres) | $F'_b$ varies with member | | | | | $E' = 8.97$ GPa | | | | | $F'_v = 0.97$ MPa | |
|---|---|---|---|---|---|---|---|---|---|---|---|---|
| | Nominal size of S4S lumber | | | | | | | | | | | |
| | 38 × 89 | 38 × 140 | 38 × 184 | 38 × 235 | 64 × 89 | 64 × 140 | 64 × 184 | 64 × 235 | 89 × 89 | 89 × 140 | 89 × 184 | 235 × 38 |
| | $F'_b$ (MPa) | | | | | | | | | | | |
| | 8.8 | 7.7 | 7.0 | 6.5 | 8.8 | 7.7 | 7.0 | 6.5 | 8.8 | 7.7 | 7.7 | 7.7 |
| 1.5 | 1725 | 2500 | 3175 | 3850 | 2050 | 2350 | 3650 | 4375 | 2300 | 3225 | 3950 | 1375 |
| 2.9 | 1200 | 1775 | 2250 | 2750 | 1575 | 2300 | 2900 | 3550 | 1825 | 2700 | 3325 | 1075 |
| 4.4 | 1000 | 1450 | 1825 | 2250 | 1275 | 1375 | 2350 | 2900 | 1500 | 2200 | 2925 | 950 |
| 5.8 | 800 | 1250 | 1575 | 1950 | 1100 | 1525 | 2050 | 2500 | 1300 | 1925 | 2525 | 850 |
| 7.3 | 675 | 1050 | 1375 | 1725 | 1000 | 1450 | 1825 | 2250 | 1175 | 1725 | 2250 | 775 |
| 8.8 | 575 | 925 | 1200 | 1550 | 850 | 1325 | 1675 | 2050 | 1075 | 1575 | 2050 | 700 |
| 10.2 | 525 | 825 | 1100 | 1400 | 750 | 1200 | 1550 | 1900 | 1000 | 1450 | 1900 | 650 |
| 11.7 | 475 | 750 | 1000 | 1275 | 675 | 1075 | 1425 | 1775 | 900 | 1350 | 1775 | 600 |
| 13.1 | 450 | 700 | 925 | 1175 | 625 | 1000 | 1300 | 1650 | 800 | 1275 | 1675 | 575 |
| 14.6 | 425 | 650 | 875 | 1100 | 575 | 925 | 1200 | 1550 | 750 | 1175 | 1550 | 550 |
| 16.1 | 400 | 625 | 825 | 1050 | 550 | 850 | 1125 | 1450 | 700 | 1100 | 1450 | 525 |
| 17.5 | 375 | 600 | 775 | 1000 | 525 | 800 | 1075 | 1350 | 650 | 1025 | 1350 | 500 |
| 19.0 | 375 | 575 | 750 | 950 | 500 | 775 | 1025 | 1300 | 625 | 975 | 1275 | 475 |
| 20.4 | 350 | 550 | 725 | 925 | 475 | 725 | 975 | 1225 | 575 | 925 | 1200 | 450 |
| 21.9 | 350 | 525 | 700 | 900 | 450 | 700 | 925 | 1175 | 550 | 875 | 1150 | 450 |
| 23.3 | 325 | 525 | 675 | 875 | 425 | 675 | 900 | 1125 | 525 | 825 | 1100 | 425 |
| 24.8 | 325 | 500 | 650 | 850 | 425 | 650 | 850 | 1100 | 500 | 800 | 1050 | 425 |
| 26.3 | 300 | 500 | 650 | 825 | 400 | 625 | 825 | 1050 | 500 | 775 | 1025 | 400 |
| 27.7 | 300 | 475 | 625 | 800 | 400 | 625 | 800 | 1025 | 475 | 750 | 975 | 400 |
| 29.2 | 300 | 475 | 625 | 775 | 375 | 600 | 775 | 1000 | 450 | 725 | 950 | 375 |
| 30.6 | 300 | 450 | 600 | 775 | 375 | 575 | 775 | 975 | 450 | 700 | 925 | 375 |
| 32.1 | 275 | 450 | 600 | 750 | 350 | 575 | 750 | 950 | 425 | 675 | 900 | 375 |
| 33.6 | 275 | 450 | 575 | 750 | 350 | 550 | 725 | 925 | 425 | 675 | 875 | 350 |
| 35.0 | 275 | 425 | 575 | 725 | 350 | 550 | 725 | 900 | 425 | 650 | 850 | 350 |
| 36.5 | 275 | 425 | 575 | 725 | 350 | 525 | 700 | 900 | 400 | 625 | 825 | 325 |
| 37.9 | 275 | 425 | 550 | 700 | 325 | 525 | 700 | 875 | 400 | 625 | 825 | 325 |
| 39.4 | 275 | 425 | 550 | 700 | 325 | 525 | 675 | 850 | 375 | 600 | 800 | 325 |
| 40.9 | 275 | 425 | 550 | 700 | 325 | 500 | 675 | 850 | 375 | 600 | 775 | 300 |
| 42.3 | 250 | 400 | 550 | 675 | 325 | 500 | 650 | 825 | 375 | 575 | 775 | 300 |
| 43.8 | 250 | 400 | 525 | 675 | 300 | 500 | 650 | 825 | 375 | 575 | 750 | 300 |
| 46.7 | 250 | 400 | 525 | 675 | 300 | 475 | 625 | 800 | 350 | 550 | 725 | 275 |
| 49.6 | 250 | 400 | 500 | 650 | 300 | 475 | 600 | 775 | 350 | 550 | 700 | 275 |
| 52.5 | 250 | 375 | 500 | 650 | 300 | 450 | 600 | 750 | 325 | 525 | 700 | 250 |
| 55.5 | 250 | 375 | 500 | 625 | 275 | 450 | 575 | 750 | 325 | 500 | 675 | 250 |
| 58.4 | 225 | 375 | 500 | 625 | 275 | 425 | 575 | 725 | 325 | 500 | 650 | 225 |
| 65.7 | 225 | 350 | 475 | 575 | 275 | 425 | 550 | 700 | 300 | 475 | 625 | 225 |
| 73.0 | 225 | 350 | 450 | 550 | 250 | 400 | 525 | 675 | 300 | 450 | 600 | 200 |

[a]Span values above the solid line are governed by deflection. Values within dashed line box are spans governed by shear. Elsewhere bending governs span.

*Source:* Original values provided by the American Concrete Institute; metric conversion by author.

## Table 6-7(a)

### SAFE SPACING (INCHES) OF SUPPORTS FOR JOISTS, STUDS, OR OTHER BEAM COMPONENTS OF FORMWORK, SINGLE SPAN[a]

Spacing

Maximum deflection is 1/360 of the span or ¼ in., whichever is smaller.

| Uniform load, lb per lineal ft (equals uniform load on forms times spacing between joists or studs, ft) | $F'_b$ varies with member | | | | | | | $E' = 1{,}300{,}000$ psi | | | | | | $F'_v = 140$ psi | | |
|---|---|---|---|---|---|---|---|---|---|---|---|---|---|---|---|---|
| | Nominal size of S × S lumber | | | | | | | | | | | | | | | |
| | 2 × 4 | 2 × 6 | 2 × 8 | 2 × 10 | 3 × 4 | 3 × 6 | 3 × 8 | 3 × 10 | 4 × 2 | 4 × 4 | 4 × 6 | 4 × 8 | 6 × 2 | 6 × 4 | 8 × 2 | 10 × 2 |
| | $F'_b$ psi | | | | | | | | | | | | | | | |
| | 1280 | 1100 | 1020 | 940 | 1280 | 1100 | 1020 | 940 | 1400 | 1280 | 1110 | 1110 | 1270 | 1160 | 1170 | 1120 |
| 100 | 56 | 89 | 109 | 131 | 67 | 101 | 124 | 149 | 32 | 75 | 110 | 135 | 37 | 87 | 41 | 44 |
| 200 | 43 | 63 | 80 | 98 | 53 | 82 | 104 | 125 | 25 | 59 | 92 | 114 | 30 | 69 | 32 | 35 |
| 300 | 35 | 52 | 66 | 80 | 46 | 67 | 85 | 104 | 22 | 52 | 79 | 103 | 26 | 60 | 28 | 31 |
| 400 | 31 | 45 | 57 | 69 | 40 | 58 | 73 | 90 | 20 | 47 | 69 | 90 | 23 | 55 | 26 | 28 |
| 500 | 27 | 40 | 51 | 62 | 35 | 52 | 66 | 80 | 19 | 42 | 61 | 81 | 22 | 50 | 24 | 26 |
| 600 | 25 | 37 | 46 | 57 | 32 | 47 | 60 | 73 | 17 | 38 | 56 | 74 | 20 | 46 | 22 | 24 |
| 700 | 23 | 34 | 43 | 53 | 30 | 44 | 55 | 68 | 16 | 35 | 52 | 68 | 19 | 42 | 21 | 23 |
| 800 | 22 | 32 | 40 | 49 | 28 | 41 | 52 | 63 | 15 | 33 | 48 | 64 | 18 | 40 | 20 | 22 |
| 900 | 20 | 30 | 38 | 46 | 26 | 39 | 49 | 60 | 14 | 31 | 46 | 60 | 17 | 37 | 18 | 20 |
| 1000 | 19 | 28 | 36 | 44 | 25 | 37 | 46 | 57 | 13 | 30 | 43 | 57 | 16 | 35 | 17 | 19 |
| 1100 | 18 | 27 | 34 | 42 | 24 | 35 | 44 | 54 | 13 | 28 | 41 | 55 | 15 | 34 | 17 | 18 |
| 1200 | 17 | 26 | 33 | 40 | 23 | 33 | 42 | 52 | 12 | 27 | 40 | 52 | 14 | 32 | 16 | 18 |
| 1300 | 16 | 25 | 31 | 39 | 22 | 32 | 41 | 50 | 12 | 26 | 38 | 50 | 14 | 31 | 15 | 17 |
| 1400 | 15 | 24 | 30 | 37 | 21 | 31 | 39 | 48 | 11 | 25 | 37 | 48 | 13 | 30 | 15 | 16 |
| 1500 | 15 | 23 | 29 | 36 | 20 | 30 | 38 | 46 | 11 | 24 | 35 | 47 | 13 | 29 | 14 | 16 |
| 1600 | 14 | 22 | 28 | 35 | 19 | 29 | 37 | 45 | 10 | 23 | 34 | 45 | 13 | 28 | 14 | 15 |
| 1700 | 14 | 22 | 28 | 34 | 19 | 28 | 36 | 44 | 10 | 23 | 33 | 44 | 12 | 27 | 13 | 15 |
| 1800 | 14 | 21 | 27 | 33 | 18 | 27 | 35 | 42 | 10 | 22 | 32 | 43 | 12 | 26 | 13 | 14 |
| 1900 | 13 | 21 | 26 | 32 | 17 | 27 | 34 | 41 | 9 | 21 | 31 | 41 | 12 | 26 | 13 | 14 |
| 2000 | 13 | 20 | 25 | 31 | 17 | 26 | 33 | 40 | 9 | 21 | 31 | 40 | 11 | 25 | 12 | 14 |
| 2100 | 13 | 20 | 25 | 30 | 16 | 25 | 32 | 39 | 9 | 20 | 30 | 39 | 11 | 24 | 12 | 13 |
| 2200 | 12 | 19 | 24 | 30 | 16 | 25 | 31 | 38 | 8 | 19 | 29 | 39 | 11 | 24 | 12 | 13 |
| 2300 | 12 | 19 | 24 | 29 | 16 | 24 | 31 | 37 | 8 | 19 | 29 | 38 | 10 | 23 | 12 | 13 |
| 2400 | 12 | 18 | 23 | 28 | 15 | 24 | 30 | 37 | 8 | 18 | 28 | 37 | 10 | 23 | 11 | 12 |
| 2500 | 12 | 18 | 23 | 28 | 15 | 23 | 29 | 36 | 8 | 18 | 27 | 36 | 10 | 22 | 11 | 12 |
| 2600 | 12 | 18 | 22 | 27 | 15 | 23 | 29 | 35 | 8 | 18 | 27 | 35 | 10 | 22 | 11 | 12 |
| 2700 | 11 | 17 | 22 | 27 | 14 | 22 | 28 | 35 | 7 | 17 | 26 | 35 | 10 | 22 | 11 | 12 |
| 2800 | 11 | 17 | 21 | 26 | 14 | 22 | 28 | 34 | 7 | 17 | 26 | 34 | 9 | 21 | 10 | 12 |
| 2900 | 11 | 17 | 21 | 26 | 14 | 22 | 27 | 33 | 7 | 16 | 25 | 34 | 9 | 21 | 10 | 11 |
| 3000 | 11 | 16 | 21 | 25 | 14 | 21 | 27 | 33 | 7 | 16 | 25 | 33 | 9 | 20 | 10 | 11 |
| 3200 | 11 | 16 | 20 | 25 | 13 | 20 | 26 | 32 | 7 | 16 | 24 | 32 | 9 | 20 | 10 | 11 |
| 3400 | 10 | 15 | 19 | 24 | 13 | 20 | 25 | 31 | 6 | 15 | 24 | 31 | 8 | 19 | 9 | 10 |
| 3600 | 10 | 15 | 19 | 23 | 12 | 19 | 24 | 30 | 6 | 15 | 23 | 30 | 8 | 19 | 9 | 10 |
| 3800 | 10 | 15 | 18 | 23 | 12 | 19 | 24 | 29 | 6 | 14 | 22 | 29 | 8 | 18 | 9 | 10 |
| 4000 | 10 | 14 | 18 | 22 | 12 | 18 | 23 | 28 | 6 | 14 | 22 | 29 | 8 | 18 | 9 | 10 |
| 4500 | 9 | 13 | 17 | 21 | 11 | 17 | 22 | 27 | 6 | 13 | 20 | 27 | 7 | 17 | 8 | 9 |
| 5000 | 9 | 13 | 16 | 20 | 11 | 16 | 21 | 25 | 5 | 12 | 19 | 26 | 7 | 16 | 8 | 9 |

[a]Span values above solid line are controlled by deflection. Within the dashed box horizontal shear governs span. Elsewhere bending controls span.

*Source*: Reproduced by permission of the American Concrete Institute.

## Table 6-7(b)

## SAFE SPACING (MM) OF SUPPORTS FOR JOISTS, STUDS, OR OTHER BEAM COMPONENTS OF FORMWORK, SINGLE SPAN[a]

Maximum deflection = span/360, but not to exceed 6 mm.

| Uniform load, kN/m (equals uniform load on forms times spacing between joists or studs, metres) | $F'_b$ varies with member size | | | | | $E' = 8.97$ GPa | | | | | $F'_v = 0.97$ MPa | |
|---|---|---|---|---|---|---|---|---|---|---|---|---|
| | Nominal size of S4S lumber | | | | | | | | | | | |
| | 38 × 89 | 38 × 140 | 38 × 184 | 38 × 235 | 64 × 89 | 64 × 140 | 64 × 184 | 64 × 235 | 89 × 89 | 89 × 140 | 89 × 184 | 235 × 38 |
| | $F'_b$ (MPa) | | | | | | | | | | | |
| | 8.8 | 7.7 | 7.0 | 6.5 | 8.8 | 7.7 | 7.0 | 6.5 | 8.8 | 7.7 | 7.7 | 7.7 |
| 1.5 | 1400 | 2225 | 2725 | 3275 | 1675 | 2525 | 3100 | 3725 | 1875 | 2750 | 3375 | 1100 |
| 2.9 | 1075 | 1575 | 2000 | 2450 | 1325 | 2050 | 2600 | 3125 | 1475 | 2300 | 2850 | 875 |
| 4.4 | 875 | 1300 | 1650 | 2000 | 1150 | 1675 | 2125 | 2600 | 1300 | 1975 | 2575 | 775 |
| 5.8 | 775 | 1125 | 1425 | 1725 | 1000 | 1450 | 1825 | 2250 | 1175 | 1725 | 2250 | 700 |
| 7.3 | 675 | 1000 | 1275 | 1550 | 875 | 1300 | 1650 | 2000 | 1050 | 1525 | 2025 | 650 |
| 8.8 | 625 | 925 | 1150 | 1425 | 800 | 1175 | 1500 | 1825 | 950 | 1400 | 1850 | 600 |
| 10.2 | 575 | 850 | 1075 | 1325 | 750 | 1100 | 1375 | 1700 | 875 | 1300 | 1700 | 575 |
| 11.7 | 550 | 800 | 1000 | 1225 | 700 | 1025 | 1300 | 1575 | 825 | 1200 | 1600 | 550 |
| 13.1 | 500 | 750 | 950 | 1150 | 650 | 975 | 1225 | 1500 | 775 | 1150 | 1500 | 500 |
| 14.6 | 475 | 700 | 900 | 1100 | 625 | 925 | 1150 | 1425 | 750 | 1075 | 142 | 475 |
| 16.1 | 450 | 675 | 850 | 1050 | 600 | 875 | 1100 | 1350 | 700 | 1025 | 1375 | 450 |
| 17.5 | 425 | 650 | 825 | 1000 | 575 | 825 | 1050 | 1300 | 675 | 1000 | 1300 | 450 |
| 19.0 | 400 | 625 | 775 | 975 | 550 | 800 | 1025 | 1250 | 650 | 950 | 1250 | 425 |
| 20.4 | 375 | 600 | 750 | 925 | 525 | 775 | 975 | 1200 | 625 | 925 | 1200 | 400 |
| 21.9 | 375 | 575 | 725 | 900 | 500 | 750 | 950 | 1150 | 600 | 875 | 1175 | 400 |
| 23.3 | 350 | 550 | 700 | 875 | 475 | 725 | 925 | 1125 | 575 | 600 | 1125 | 375 |
| 24.8 | 350 | 550 | 700 | 850 | 475 | 700 | 900 | 1100 | 575 | 825 | 1100 | 375 |
| 26.3 | 350 | 525 | 675 | 825 | 450 | 675 | 875 | 1050 | 550 | 800 | 1075 | 350 |
| 27.7 | 325 | 525 | 650 | 800 | 425 | 675 | 850 | 1025 | 525 | 775 | 1025 | 350 |
| 29.2 | 325 | 500 | 625 | 775 | 425 | 650 | 825 | 1000 | 525 | 775 | 1000 | 350 |
| 30.6 | 325 | 500 | 625 | 750 | 400 | 625 | 800 | 975 | 500 | 750 | 975 | 325 |
| 32.1 | 300 | 475 | 600 | 750 | 400 | 625 | 775 | 950 | 475 | 725 | 975 | 325 |
| 33.6 | 300 | 475 | 600 | 725 | 400 | 600 | 775 | 925 | 475 | 725 | 950 | 325 |
| 35.0 | 300 | 450 | 575 | 700 | 375 | 600 | 750 | 925 | 450 | 700 | 925 | 300 |
| 36.5 | 300 | 450 | 575 | 700 | 375 | 575 | 725 | 900 | 450 | 675 | 900 | 300 |
| 37.9 | 300 | 450 | 550 | 675 | 375 | 575 | 725 | 875 | 450 | 675 | 875 | 300 |
| 39.4 | 275 | 425 | 550 | 675 | 350 | 550 | 700 | 875 | 425 | 650 | 875 | 300 |
| 40.9 | 275 | 425 | 525 | 650 | 350 | 550 | 700 | 850 | 425 | 650 | 850 | 300 |
| 42.3 | 275 | 425 | 525 | 650 | 350 | 550 | 675 | 825 | 400 | 625 | 850 | 275 |
| 43.8 | 275 | 400 | 525 | 625 | 350 | 525 | 675 | 825 | 400 | 625 | 825 | 275 |
| 46.7 | 275 | 400 | 500 | 625 | 325 | 500 | 650 | 800 | 400 | 600 | 800 | 275 |
| 49.6 | 250 | 375 | 475 | 600 | 325 | 500 | 625 | 775 | 375 | 600 | 775 | 250 |
| 52.5 | 250 | 375 | 475 | 575 | 300 | 475 | 600 | 750 | 375 | 575 | 750 | 250 |
| 55.5 | 250 | 375 | 450 | 575 | 300 | 475 | 600 | 725 | 350 | 550 | 725 | 250 |
| 58.4 | 250 | 350 | 450 | 550 | 300 | 450 | 575 | 700 | 350 | 550 | 725 | 250 |
| 65.7 | 225 | 325 | 425 | 525 | 275 | 425 | 550 | 675 | 325 | 500 | 675 | 225 |
| 73.0 | 225 | 325 | 400 | 500 | 275 | 400 | 525 | 625 | 300 | 475 | 650 | 225 |

[a]Values above the solid line are governed by deflection. Within the dashed box, horizontal shear governs. Elsewhere bending controls span.

*Source:* Original values provided by the American Concrete Institute; metric conversion by author.

# Table 6-8(a)

## SAFE SPACING (INCHES) OF SUPPORTS FOR JOISTS, STUDS, OR OTHER BEAM COMPONENTS OF FORMWORK, CONTINUOUS OVER TWO SPANS[a]

Maximum deflection is 1/360 of the span or ¼ in., whichever is smaller.

| Uniform load, lb per lineal ft (equals uniform load on forms times spacing between joists or studs, ft) | $F'_b$ varies with member | | | | | | $E' = 1,300,000$ psi | | | | | | | $F'_v = 140$ psi | | |
|---|---|---|---|---|---|---|---|---|---|---|---|---|---|---|---|---|
| | Nominal size of S × S lumber | | | | | | | | | | | | | | | |
| | 2 × 4 | 2 × 6 | 2 × 8 | 2 × 10 | 3 × 4 | 3 × 6 | 3 × 8 | 3 × 10 | 4 × 2 | 4 × 4 | 4 × 6 | 4 × 8 | 6 × 2 | 6 × 4 | 8 × 2 | 10 × 2 |
| | $F'_b$ psi | | | | | | | | | | | | | | | |
| | 1280 | 1100 | 1020 | 940 | 1280 | 1100 | 1020 | 940 | 1400 | 1280 | 1110 | 1110 | 1270 | 1160 | 1170 | 1120 |
| 100 | 61 | 90 | 113 | 139 | 79 | 116 | 146 | 179 | 42 | 94 | 137 | 168 | 50 | 109 | 55 | 59 |
| 200 | 43 | 63 | 80 | 98 | 56 | 82 | 104 | 127 | 30 | 66 | 97 | 128 | 35 | 79 | 39 | 43 |
| 300 | 35 | 52 | 66 | 80 | 46 | 67 | 85 | 104 | 24 | 54 | 79 | 104 | 29 | 65 | 32 | 35 |
| 400 | 31 | 45 | 57 | 69 | 40 | 58 | 73 | 90 | 21 | 47 | 69 | 90 | 25 | 56 | 28 | 31 |
| 500 | 26 | 40 | 51 | 62 | 35 | 52 | 66 | 80 | 19 | 42 | 61 | 81 | 22 | 50 | 25 | 27 |
| 600 | 23 | 36 | 46 | 57 | 32 | 47 | 60 | 73 | 17 | 38 | 56 | 74 | 20 | 46 | 23 | 25 |
| 700 | 20 | 32 | 42 | 53 | 29 | 44 | 55 | 68 | 16 | 35 | 52 | 68 | 19 | 42 | 21 | 23 |
| 800 | 19 | 29 | 39 | 49 | 27 | 41 | 52 | 63 | 15 | 33 | 48 | 64 | 18 | 40 | 20 | 22 |
| 900 | 17 | 27 | 36 | 46 | 24 | 38 | 49 | 60 | 13 | 31 | 46 | 60 | 17 | 37 | 18 | 20 |
| 1000 | 16 | 26 | 34 | 43 | 23 | 36 | 46 | 57 | 12 | 29 | 43 | 57 | 16 | 35 | 17 | 19 |
| 1100 | 16 | 24 | 32 | 41 | 21 | 33 | 44 | 54 | 12 | 27 | 41 | 55 | 15 | 34 | 17 | 18 |
| 1200 | 15 | 23 | 31 | 39 | 20 | 32 | 42 | 52 | 11 | 25 | 40 | 52 | 14 | 32 | 16 | 18 |
| 1300 | 14 | 22 | 29 | 38 | 19 | 30 | 39 | 50 | 10 | 24 | 38 | 49 | 14 | 31 | 15 | 17 |
| 1400 | 14 | 22 | 28 | 36 | 18 | 29 | 38 | 48 | 10 | 23 | 36 | 47 | 13 | 30 | 15 | 16 |
| 1500 | 13 | 21 | 27 | 36 | 17 | 27 | 36 | 46 | 9 | 22 | 34 | 45 | 13 | 29 | 14 | 16 |
| 1600 | 13 | 20 | 27 | 34 | 17 | 26 | 35 | 44 | 9 | 21 | 33 | 43 | 12 | 28 | 14 | 15 |
| 1700 | 13 | 20 | 26 | 33 | 16 | 25 | 34 | 43 | 9 | 20 | 31 | 41 | 12 | 27 | 13 | 15 |
| 1800 | 12 | 19 | 25 | 32 | 16 | 25 | 33 | 42 | 8 | 19 | 30 | 40 | 11 | 26 | 13 | 14 |
| 1900 | 12 | 19 | 25 | 32 | 15 | 24 | 32 | 40 | 8 | 19 | 29 | 38 | 11 | 25 | 13 | 14 |
| 2000 | 12 | 18 | 24 | 31 | 15 | 23 | 31 | 39 | | 18 | 28 | 37 | 10 | 24 | 12 | 14 |
| 2100 | 11 | 18 | 24 | 30 | 14 | 23 | 30 | 38 | 7 | 17 | 27 | 36 | 10 | 23 | 12 | 13 |
| 2200 | 11 | 18 | 23 | 30 | 14 | 22 | 29 | 37 | 7 | 17 | 27 | 35 | 10 | 23 | 12 | 13 |
| 2300 | 11 | 17 | 23 | 29 | 14 | 22 | 29 | 37 | 7 | 17 | 26 | 34 | 9 | 22 | 11 | 13 |
| 2400 | 11 | 17 | 23 | 28 | 14 | 21 | 28 | 36 | 7 | 16 | 25 | 33 | 9 | 21 | 11 | 12 |
| 2500 | 11 | 17 | 22 | 28 | 13 | 21 | 27 | 35 | 7 | 16 | 25 | 33 | 9 | 21 | 11 | 12 |
| 2600 | 11 | 17 | 22 | 27 | 13 | 20 | 27 | 34 | 7 | 15 | 24 | 32 | 9 | 20 | 10 | 12 |
| 2700 | 10 | 16 | 22 | 27 | 13 | 20 | 27 | 34 | 6 | 15 | 24 | 31 | 8 | 20 | 10 | 12 |
| 2800 | 10 | 16 | 21 | 26 | 13 | 20 | 26 | 33 | 6 | 15 | 23 | 31 | 8 | 19 | 10 | 12 |
| 2900 | 10 | 16 | 21 | 26 | 12 | 19 | 26 | 33 | 6 | 15 | 23 | 30 | 8 | 19 | 10 | 11 |
| 3000 | 10 | 16 | 21 | 25 | 12 | 19 | 25 | 32 | 6 | 14 | 22 | 30 | 8 | 18 | 9 | 11 |
| 3200 | 10 | 16 | 20 | 25 | 12 | 19 | 25 | 31 | 6 | 14 | 22 | 29 | 8 | 18 | 9 | 11 |
| 3400 | 10 | 15 | 19 | 24 | 12 | 18 | 24 | 31 | 6 | 13 | 21 | 28 | 7 | 17 | 9 | 10 |
| 3600 | 10 | 15 | 19 | 23 | 11 | 18 | 24 | 30 | 6 | 13 | 21 | 27 | 7 | 17 | 8 | 10 |
| 3800 | 9 | 15 | 18 | 23 | 11 | 17 | 23 | 29 | 5 | 13 | 20 | 26 | 7 | 16 | 8 | 10 |
| 4000 | 9 | 14 | 18 | 22 | 11 | 17 | 23 | 28 | 5 | 12 | 20 | 26 | 7 | 16 | 8 | 9 |
| 4500 | 9 | 13 | 17 | 21 | 10 | 16 | 22 | 27 | 5 | 12 | 19 | 25 | 6 | 15 | 7 | 9 |
| 5000 | 9 | 13 | 16 | 20 | 10 | 16 | 21 | 25 | 5 | 11 | 18 | 24 | 6 | 14 | 7 | 8 |

[a]Span values above solid lines are controlled by deflection. Spans within the dashed box are controlled by shear. Elsewhere bending controls span.

*Source:* Reproduced by permission of the American Concrete Institute.

Table 6-8(b)

SAFE SPACING (MM) OF SUPPORTS FOR JOISTS, STUDS, OR OTHER BEAM COMPONENTS OF FORMWORK, CONTINUOUS OVER TWO SPANS[a]

Maximum deflection = span/360, but not to exceed 6 mm.

| Uniform load, kN/m (equals uniform load on forms times spacing between joists or studs, metres) | $F'_b$ varies with member size | | | | $E' = 8.97$ GPa | | | | | | $F'_v = 0.97$ MPa | |
|---|---|---|---|---|---|---|---|---|---|---|---|---|
| | Nominal size of S4S lumber | | | | | | | | | | | |
| | 38 × 89 | 38 × 140 | 38 × 184 | 38 × 235 | 64 × 89 | 64 × 140 | 64 × 184 | 64 × 235 | 89 × 89 | 89 × 140 | 89 × 184 | 235 × 38 |
| | $F'_b$ (MPa) | | | | | | | | | | | |
| | 8.8 | 7.7 | 7.0 | 6.5 | 8.8 | 7.7 | 7.0 | 6.5 | 8.8 | 7.7 | 7.7 | 7.7 |
| 1.5 | 1525 | 2250 | 2825 | 3475 | 1975 | 2900 | 3650 | 4475 | 2350 | 3425 | 4200 | 1475 |
| 2.9 | 1075 | 1575 | 2000 | 2450 | 1400 | 2050 | 2600 | 3175 | 1650 | 2425 | 3200 | 1075 |
| 4.4 | 875 | 1300 | 1650 | 2000 | 1150 | 1675 | 2125 | 2600 | 1350 | 1975 | 2600 | 875 |
| 5.8 | 775 | 1125 | 1425 | 1725 | 1000 | 1450 | 1825 | 2250 | 1175 | 1725 | 2250 | 775 |
| 7.3 | 650 | 1000 | 1275 | 1550 | 875 | 1300 | 1650 | 2000 | 1050 | 1525 | 2025 | 675 |
| 8.8 | 575 | 900 | 1150 | 1425 | 800 | 1175 | 1500 | 1825 | 950 | 1400 | 1850 | 625 |
| 10.2 | 500 | 800 | 1050 | 1325 | 725 | 1100 | 1375 | 1700 | 875 | 1300 | 1700 | 575 |
| 11.7 | 475 | 725 | 975 | 1225 | 675 | 1025 | 1300 | 1575 | 825 | 1200 | 1600 | 550 |
| 13.1 | 425 | 675 | 900 | 1150 | 600 | 950 | 1225 | 1500 | 775 | 1150 | 1500 | 500 |
| 14.6 | 400 | 650 | 850 | 1075 | 575 | 900 | 1150 | 1425 | 725 | 1075 | 1425 | 475 |
| 16.1 | 400 | 600 | 800 | 1025 | 525 | 825 | 1100 | 1350 | 675 | 1025 | 1375 | 450 |
| 17.5 | 375 | 575 | 775 | 975 | 500 | 800 | 1050 | 1300 | 625 | 1000 | 1300 | 450 |
| 19.0 | 350 | 550 | 725 | 950 | 475 | 750 | 975 | 1250 | 600 | 950 | 1225 | 425 |
| 20.4 | 350 | 550 | 700 | 900 | 450 | 725 | 950 | 1200 | 575 | 900 | 1175 | 400 |
| 21.9 | 325 | 525 | 675 | 900 | 425 | 675 | 900 | 1150 | 550 | 850 | 1125 | 400 |
| | | | | | | | | | | 0 | 0 | 0 |
| 23.3 | 325 | 500 | 675 | 850 | 425 | 650 | 875 | 1100 | 525 | 825 | 1075 | 375 |
| 24.8 | 325 | 500 | 650 | 825 | 400 | 625 | 850 | 1075 | 500 | 775 | 1025 | 375 |
| 26.3 | 300 | 475 | 625 | 800 | 400 | 625 | 825 | 1050 | 475 | 750 | 1000 | 350 |
| 27.7 | 300 | 475 | 625 | 800 | 375 | 600 | 800 | 1000 | 475 | 725 | 950 | 350 |
| 29.2 | 300 | 450 | 600 | 775 | 375 | 575 | 775 | 975 | 450 | 700 | 925 | 350 |
| 30.6 | 275 | 450 | 600 | 750 | 350 | 575 | 750 | 950 | 425 | 675 | 900 | 325 |
| 32.1 | 275 | 450 | 575 | 750 | 350 | 550 | 725 | 925 | 425 | 675 | 875 | 325 |
| 33.6 | 275 | 425 | 575 | 725 | 350 | 550 | 725 | 925 | 425 | 650 | 850 | 325 |
| 35.0 | 275 | 425 | 575 | 700 | 350 | 525 | 700 | 900 | 400 | 625 | 825 | 300 |
| 36.5 | 275 | 425 | 550 | 700 | 325 | 525 | 675 | 875 | 400 | 625 | 825 | 300 |
| 37.9 | 275 | 425 | 550 | 675 | 325 | 500 | 675 | 850 | 375 | 600 | 800 | 300 |
| 39.4 | 250 | 400 | 550 | 675 | 325 | 500 | 675 | 850 | 375 | 600 | 775 | 300 |
| 40.9 | 250 | 400 | 525 | 650 | 325 | 500 | 650 | 825 | 375 | 575 | 775 | 300 |
| 42.3 | 250 | 400 | 525 | 650 | 300 | 475 | 650 | 825 | 375 | 575 | 750 | 275 |
| 43.8 | 250 | 400 | 525 | 625 | 300 | 475 | 625 | 800 | 350 | 550 | 750 | 275 |
| 46.7 | 250 | 400 | 500 | 625 | 300 | 475 | 625 | 775 | 350 | 550 | 725 | 275 |
| 49.6 | 250 | 375 | 475 | 600 | 300 | 450 | 600 | 775 | 325 | 525 | 700 | 250 |
| 52.5 | 250 | 375 | 475 | 575 | 275 | 450 | 600 | 750 | 325 | 525 | 675 | 250 |
| 55.5 | 225 | 375 | 450 | 575 | 275 | 425 | 575 | 725 | 325 | 500 | 650 | 250 |
| 58.4 | 225 | 350 | 450 | 550 | 275 | 425 | 575 | 700 | 300 | 500 | 650 | 225 |
| 65.7 | 225 | 325 | 425 | 525 | 250 | 400 | 550 | 675 | 300 | 475 | 625 | 225 |
| 73.0 | 225 | 325 | 400 | 500 | 250 | 400 | 525 | 625 | 275 | 450 | 600 | 200 |

[a]Values above solid line are spans governed by deflection. Values within dashed-line box are spans governed by shear. All other values are governed by bending.

*Source:* Original values were provided by the American Concrete Institute; metric conversion by author.

# Table 6-9(a)

## SAFE SPACING (INCHES) OF SUPPORTS FOR DOUBLE WALES CONTINUOUS OVER THREE OR MORE SPANS[a]

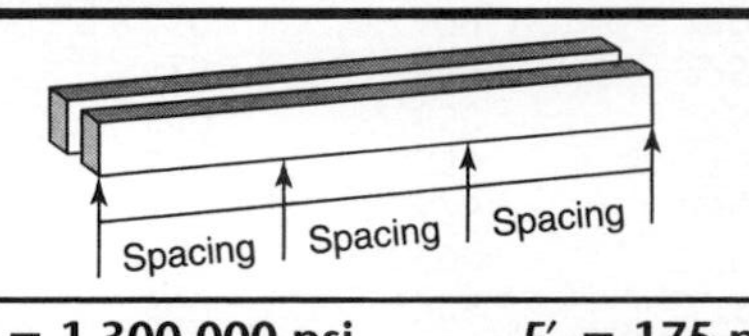


Maximum deflection is 1/360 of the span or ¼ in., whichever is smaller.

| Uniform load, lb per lineal ft (equals uniform load, psi, on forms times spacing of wales in ft) | $F'_b$ varies with member | | | $E' = 1{,}300{,}000$ psi | | | | $F'_v = 140$ psi | | $F'_b$ varies with member | | | $E' = 1{,}300{,}000$ psi | | | | $F'_v = 175$ psi | |
|---|---|---|---|---|---|---|---|---|---|---|---|---|---|---|---|---|---|---|
| | Nominal size of S4S lumber | | | | | | | | | Nominal size of S4S lumber | | | | | | | | |
| | 2 × 4 | 2 × 6 | 2 × 8 | 3 × 4 | 3 × 6 | 3 × 8 | 4 × 4 | 4 × 6 | 4 × 8 | 2 × 4 | 2 × 6 | 2 × 8 | 3 × 4 | 3 × 6 | 3 × 8 | 4 × 4 | 4 × 6 | 4 × 8 |
| | $F'_b$ psi | | | | | | | | | $F'_b$ psi | | | | | | | | |
| | 1280 | 1110 | 1020 | 1280 | 1110 | 1020 | 1280 | 1110 | 1110 | 1590 | 1380 | 1280 | 1590 | 1380 | 1280 | 1590 | 1380 | 1380 |
| 100 | 88 | 124 | 152 | 100 | 141 | 173 | 109 | 153 | 188 | 88 | 124 | 152 | 100 | 141 | 173 | 109 | 153 | 188 |
| 200 | 69 | 100 | 127 | 82 | 118 | 146 | 92 | 129 | 158 | 70 | 104 | 128 | 82 | 118 | 146 | 92 | 129 | 158 |
| 300 | 56 | 82 | 104 | 72 | 106 | 132 | 81 | 116 | 143 | 61 | 91 | 116 | 72 | 107 | 132 | 81 | 116 | 143 |
| 400 | 48 | 71 | 90 | 63 | 92 | 116 | 73 | 108 | 133 | 54 | 79 | 100 | 65 | 100 | 122 | 73 | 108 | 133 |
| 500 | 43 | 63 | 80 | 56 | 82 | 104 | 66 | 97 | 126 | 48 | 71 | 90 | 61 | 91 | 116 | 68 | 102 | 126 |
| 600 | 40 | 58 | 73 | 51 | 75 | 94 | 60 | 88 | 117 | 44 | 65 | 82 | 57 | 83 | 106 | 64 | 98 | 120 |
| 700 | 35 | 54 | 68 | 47 | 69 | 87 | 56 | 82 | 108 | 41 | 60 | 76 | 53 | 77 | 98 | 61 | 91 | 116 |
| 800 | 32 | 50 | 63 | 44 | 65 | 82 | 52 | 77 | 101 | 38 | 56 | 71 | 49 | 72 | 92 | 58 | 85 | 112 |
| 900 | 29 | 45 | 60 | 42 | 61 | 77 | 49 | 72 | 95 | 34 | 53 | 67 | 47 | 68 | 86 | 55 | 81 | 106 |
| 1000 | 27 | 42 | 55 | 40 | 58 | 73 | 47 | 69 | 90 | 32 | 50 | 64 | 44 | 65 | 82 | 52 | 76 | 101 |
| 1100 | 25 | 39 | 51 | 37 | 55 | 70 | 45 | 65 | 86 | 29 | 46 | 61 | 42 | 62 | 78 | 50 | 73 | 96 |
| 1200 | 23 | 37 | 48 | 34 | 53 | 67 | 43 | 63 | 82 | 27 | 43 | 57 | 40 | 59 | 75 | 48 | 70 | 92 |
| 1300 | 22 | 35 | 46 | 32 | 50 | 64 | 41 | 60 | 79 | 26 | 41 | 54 | 38 | 57 | 72 | 46 | 67 | 88 |
| 1400 | 21 | 33 | 44 | 30 | 48 | 62 | 40 | 58 | 76 | 25 | 39 | 51 | 36 | 55 | 69 | 44 | 65 | 85 |
| 1500 | 20 | 32 | 42 | 29 | 45 | 60 | 37 | 56 | 74 | 23 | 37 | 48 | 34 | 53 | 67 | 43 | 62 | 82 |
| 1600 | 19 | 30 | 40 | 27 | 43 | 57 | 36 | 54 | 71 | 22 | 35 | 46 | 33 | 51 | 65 | 41 | 60 | 80 |
| 1700 | 19 | 29 | 38 | 26 | 41 | 54 | 34 | 53 | 69 | 21 | 34 | 44 | 31 | 49 | 63 | 40 | 59 | 77 |
| 1800 | 18 | 28 | 37 | 25 | 40 | 52 | 32 | 51 | 67 | 21 | 32 | 43 | 30 | 47 | 61 | 39 | 57 | 75 |
| 1900 | 17 | 27 | 36 | 24 | 38 | 50 | 31 | 49 | 64 | 20 | 31 | 41 | 28 | 45 | 59 | 37 | 55 | 73 |
| 2000 | 17 | 26 | 35 | 23 | 37 | 48 | 30 | 47 | 62 | 19 | 30 | 40 | 27 | 43 | 57 | 36 | 54 | 71 |
| 2200 | 16 | 25 | 33 | 22 | 34 | 45 | 28 | 44 | 58 | 18 | 29 | 38 | 26 | 40 | 53 | 33 | 52 | 68 |
| 2400 | 15 | 24 | 31 | 21 | 32 | 43 | 26 | 41 | 54 | 17 | 27 | 36 | 24 | 38 | 50 | 31 | 48 | 64 |
| 2600 | 15 | 23 | 30 | 20 | 31 | 41 | 25 | 39 | 51 | 16 | 26 | 34 | 23 | 36 | 47 | 29 | 46 | 60 |
| 2800 | 14 | 22 | 29 | 19 | 29 | 39 | 23 | 37 | 48 | 16 | 25 | 33 | 22 | 34 | 45 | 27 | 43 | 57 |
| 3000 | 14 | 21 | 28 | 18 | 28 | 37 | 22 | 35 | 46 | 15 | 24 | 31 | 21 | 32 | 43 | 26 | 41 | 54 |
| 3200 | 13 | 21 | 27 | 17 | 27 | 36 | 21 | 33 | 44 | 15 | 23 | 30 | 20 | 31 | 41 | 25 | 39 | 52 |
| 3400 | 13 | 20 | 26 | 17 | 26 | 34 | 20 | 32 | 42 | 14 | 22 | 29 | 19 | 30 | 39 | 24 | 37 | 49 |
| 3600 | 12 | 20 | 26 | 16 | 25 | 33 | 20 | 31 | 41 | 14 | 22 | 29 | 18 | 29 | 38 | 23 | 36 | 47 |
| 3800 | 12 | 19 | 25 | 16 | 25 | 32 | 19 | 30 | 39 | 13 | 21 | 28 | 18 | 28 | 37 | 22 | 35 | 46 |
| 4000 | 12 | 19 | 25 | 15 | 24 | 31 | 18 | 29 | 38 | 13 | 21 | 27 | 17 | 27 | 36 | 21 | 33 | 44 |
| 4200 | 12 | 18 | 24 | 15 | 23 | 31 | 18 | 28 | 37 | 13 | 20 | 27 | 17 | 26 | 35 | 21 | 32 | 43 |
| 4400 | 11 | 18 | 24 | 14 | 23 | 30 | 17 | 27 | 36 | 13 | 20 | 26 | 16 | 26 | 34 | 20 | 31 | 41 |
| 4600 | 11 | 18 | 23 | 14 | 22 | 29 | 17 | 27 | 35 | 12 | 19 | 26 | 16 | 25 | 33 | 19 | 31 | 40 |
| 4800 | 11 | 17 | 23 | 14 | 22 | 29 | 17 | 26 | 34 | 12 | 19 | 25 | 16 | 24 | 32 | 19 | 30 | 39 |
| 5000 | 11 | 17 | 23 | 14 | 21 | 28 | 16 | 25 | 33 | 12 | 19 | 25 | 15 | 24 | 31 | 18 | 29 | 38 |

[a]Span values above solid line are governed by deflection. Span values within dashed box are governed by horizontal shear. Elsewhere bending governs span.

*Source:* Reproduced by permission of the American Concrete Institute.

## Table 6-9(b)

## SAFE SPACING (MM) OF SUPPORTS FOR DOUBLE WALES CONTINUOUS OVER THREE OR MORE SPANS[a]

Maximum deflection = span/360, but not to exceed 6 mm.

| Uniform load, kN/m (equals uniform pressure, kN/m2 on forms times spacing of wales in metres) | $F'_b$ varies with member size | | $E' = 8.97$ GPa | | $F'_v = 0.97$ MPa | | $F'_b$ varies with member size | | $E' = 8.97$ GPa | | $F'_v = 1.21$ MPa | |
|---|---|---|---|---|---|---|---|---|---|---|---|---|
| | Actual size of S4S lumber used double | | | | | | Actual size of S4S lumber used double | | | | | |
| | 38 × 89 | 38 × 140 | 38 × 184 | 89 × 89 | 89 × 140 | 89 × 184 | 38 × 89 | 38 × 140 | 38 × 184 | 89 × 89 | 89 × 140 | 89 × 184 |
| | $F'_b$ (MPa) | | | | | | $F'_b$ (MPa) | | | | | |
| | 8.8 | 7.7 | 7.0 | 8.8 | 7.7 | 7.7 | 8.8 | 7.7 | 7.0 | 8.8 | 7.7 | 7.7 |
| 1.5 | 2200 | 3100 | 3800 | 2725 | 3825 | 4700 | 2200 | 3100 | 3800 | 2725 | 3825 | 4700 |
| 2.9 | 1725 | 2500 | 3175 | 2300 | 3225 | 3950 | 1750 | 2600 | 3200 | 2300 | 3225 | 3950 |
| 4.4 | 1400 | 2050 | 2600 | 2025 | 2900 | 3575 | 1525 | 2275 | 2900 | 2025 | 2900 | 3575 |
| 5.8 | 1200 | 1775 | 2250 | 1825 | 2700 | 3325 | 1350 | 1975 | 2500 | 1825 | 2700 | 3325 |
| 7.3 | 1075 | 1575 | 2000 | 1650 | 2425 | 3150 | 1200 | 1775 | 2250 | 1700 | 2550 | 3150 |
| 8.8 | 1000 | 1450 | 1825 | 1500 | 2200 | 2925 | 1100 | 1625 | 2050 | 1600 | 2450 | 3000 |
| 10.2 | 875 | 1350 | 1700 | 1400 | 2050 | 2700 | 1025 | 1500 | 1900 | 1525 | 2275 | 2900 |
| 11.7 | 800 | 1250 | 1575 | 1300 | 1925 | 2525 | 950 | 1400 | 1775 | 1450 | 2125 | 2800 |
| 13.1 | 725 | 1125 | 1500 | 1225 | 1800 | 2375 | 850 | 1325 | 1675 | 1375 | 2025 | 2650 |
| 14.6 | 675 | 1050 | 1375 | 1175 | 1725 | 2250 | 800 | 1250 | 1600 | 1300 | 1900 | 2525 |
| 16.0 | 625 | 975 | 1275 | 1125 | 1625 | 2150 | 725 | 1150 | 1525 | 1250 | 1825 | 2400 |
| 17.5 | 575 | 925 | 1200 | 1075 | 1575 | 2050 | 675 | 1075 | 1425 | 1200 | 1750 | 2300 |
| 19.0 | 550 | 875 | 1150 | 1025 | 1500 | 1975 | 650 | 1025 | 1350 | 1150 | 1675 | 2200 |
| 20.4 | 525 | 825 | 1100 | 1000 | 1450 | 1900 | 625 | 975 | 1275 | 1100 | 1625 | 2125 |
| 21.9 | 500 | 800 | 1050 | 925 | 1400 | 1850 | 575 | 925 | 1200 | 1075 | 1550 | 2050 |
| 23.3 | 475 | 750 | 1000 | 900 | 1350 | 1775 | 550 | 875 | 1150 | 1025 | 1500 | 2000 |
| 24.8 | 475 | 725 | 950 | 850 | 1325 | 1725 | 525 | 850 | 1100 | 1000 | 1475 | 1925 |
| 26.3 | 450 | 700 | 925 | 800 | 1275 | 1675 | 525 | 800 | 1075 | 975 | 1425 | 1875 |
| 27.7 | 425 | 675 | 900 | 775 | 1225 | 1600 | 500 | 775 | 1025 | 925 | 1375 | 1825 |
| 29.2 | 425 | 650 | 875 | 750 | 1175 | 1550 | 475 | 750 | 1000 | 900 | 1350 | 1775 |
| 32.1 | 400 | 625 | 825 | 700 | 1100 | 1450 | 450 | 725 | 950 | 825 | 1300 | 1700 |
| 35.0 | 375 | 600 | 775 | 650 | 1025 | 1350 | 425 | 675 | 900 | 775 | 1200 | 1600 |
| 37.9 | 375 | 575 | 750 | 625 | 975 | 1275 | 400 | 650 | 850 | 725 | 1150 | 1500 |
| 40.9 | 350 | 550 | 725 | 575 | 925 | 1200 | 400 | 625 | 825 | 675 | 1075 | 1425 |
| 43.8 | 350 | 525 | 700 | 550 | 875 | 1150 | 375 | 600 | 775 | 650 | 1025 | 1350 |
| 46.7 | 325 | 525 | 675 | 525 | 825 | 1100 | 375 | 575 | 750 | 625 | 975 | 1300 |
| 49.6 | 325 | 500 | 650 | 500 | 800 | 1050 | 350 | 550 | 725 | 600 | 925 | 1225 |
| 52.5 | 300 | 500 | 650 | 500 | 775 | 1025 | 350 | 550 | 725 | 575 | 900 | 1175 |
| 55.4 | 300 | 475 | 625 | 475 | 750 | 975 | 325 | 525 | 700 | 550 | 875 | 1150 |
| 58.4 | 300 | 475 | 625 | 450 | 725 | 950 | 325 | 525 | 675 | 525 | 825 | 1100 |
| 61.3 | 300 | 450 | 600 | 450 | 700 | 925 | 325 | 500 | 675 | 525 | 800 | 1075 |
| 64.2 | 275 | 450 | 600 | 425 | 675 | 900 | 325 | 500 | 650 | 500 | 775 | 1025 |
| 67.1 | 275 | 450 | 575 | 425 | 675 | 875 | 300 | 475 | 650 | 475 | 775 | 1000 |
| 70.0 | 275 | 425 | 575 | 425 | 650 | 850 | 300 | 475 | 625 | 475 | 750 | 975 |
| 73.0 | 275 | 425 | 575 | 400 | 625 | 825 | 300 | 475 | 625 | 450 | 725 | 950 |

[a]Above the solid lines span values are governed by deflection. Span values within the dashed box are governed by horizontal shear. Elsewhere bending governs the span.

*Source:* Original data provided by the American Concrete Institute; metric conversion by author.

## Table 6-10(a)
## SAFE SPACING (INCHES) OF SUPPORTS FOR DOUBLE WALES, SINGLE SPAN[a]



Maximum deflection is 1/360 of the span or ¼ in., whichever is smaller.

| Uniform load, lb per lineal ft (equals uniform load, psi, on forms times spacing of wales in ft) | $F'_b$ varies with member | $E' = 1{,}300{,}000$ psi | | $F'_v = 140$ psi | | | | | | $F'_b$ varies with member | $E' = 1{,}300{,}000$ psi | | $F'_v = 175$ psi | | | | | |
|---|---|---|---|---|---|---|---|---|---|---|---|---|---|---|---|---|---|---|
| | Nominal size of S4S lumber | | | | | | | | | Nominal size of S4S lumber | | | | | | | | |
| | 2 × 4 | 2 × 6 | 2 × 8 | 3 × 4 | 3 × 6 | 3 × 8 | 4 × 4 | 4 × 6 | 4 × 8 | 2 × 4 | 2 × 6 | 2 × 8 | 3 × 4 | 3 × 6 | 3 × 8 | 4 × 4 | 4 × 6 | 4 × 8 |
| | $F'_b$ psi | | | | | | | | | $F'_b$ psi | | | | | | | | |
| | 1280 | 1110 | 1020 | 1280 | 1110 | 1020 | 1280 | 1110 | 1110 | 1590 | 1380 | 1280 | 1590 | 1380 | 1280 | 1590 | 1380 | 1380 |
| 100 | 71 | 106 | 130 | 84 | 120 | 148 | 93 | 131 | 161 | 71 | 106 | 130 | 84 | 120 | 148 | 93 | 131 | 161 |
| 200 | 56 | 89 | 109 | 67 | 101 | 124 | 75 | 110 | 135 | 56 | 89 | 109 | 67 | 101 | 124 | 75 | 110 | 135 |
| 300 | 49 | 73 | 93 | 58 | 91 | 112 | 65 | 99 | 122 | 49 | 77 | 99 | 58 | 91 | 112 | 65 | 99 | 122 |
| 400 | 43 | 63 | 80 | 53 | 82 | 104 | 59 | 92 | 114 | 45 | 70 | 90 | 53 | 83 | 105 | 59 | 92 | 114 |
| 500 | 39 | 57 | 72 | 49 | 73 | 93 | 55 | 87 | 108 | 42 | 63 | 80 | 49 | 77 | 99 | 55 | 87 | 108 |
| 600 | 35 | 52 | 66 | 46 | 67 | 85 | 52 | 79 | 103 | 39 | 58 | 73 | 46 | 73 | 94 | 52 | 81 | 103 |
| 700 | 33 | 48 | 61 | 42 | 62 | 78 | 49 | 73 | 97 | 37 | 54 | 68 | 44 | 69 | 88 | 49 | 77 | 99 |
| 800 | 31 | 45 | 57 | 40 | 58 | 73 | 47 | 69 | 90 | 34 | 50 | 64 | 42 | 65 | 82 | 47 | 74 | 96 |
| 900 | 29 | 42 | 53 | 37 | 55 | 69 | 44 | 65 | 85 | 32 | 47 | 60 | 40 | 61 | 77 | 45 | 71 | 93 |
| 1000 | 27 | 40 | 51 | 35 | 52 | 66 | 42 | 61 | 81 | 31 | 45 | 57 | 39 | 58 | 73 | 44 | 68 | 90 |
| 1100 | 26 | 38 | 48 | 34 | 49 | 62 | 40 | 58 | 77 | 29 | 43 | 54 | 38 | 55 | 70 | 42 | 65 | 86 |
| 1200 | 25 | 37 | 46 | 32 | 47 | 60 | 38 | 56 | 74 | 28 | 41 | 52 | 36 | 53 | 67 | 41 | 62 | 82 |
| 1300 | 24 | 35 | 45 | 31 | 45 | 57 | 37 | 54 | 71 | 27 | 39 | 50 | 35 | 51 | 64 | 40 | 60 | 79 |
| 1400 | 23 | 34 | 43 | 30 | 44 | 55 | 35 | 52 | 68 | 26 | 38 | 48 | 33 | 49 | 62 | 39 | 58 | 76 |
| 1500 | 22 | 33 | 41 | 29 | 42 | 53 | 34 | 50 | 66 | 25 | 37 | 46 | 32 | 47 | 60 | 38 | 56 | 74 |
| 1600 | 22 | 32 | 40 | 28 | 41 | 52 | 33 | 48 | 64 | 24 | 35 | 45 | 31 | 46 | 58 | 37 | 54 | 71 |
| 1700 | 21 | 31 | 39 | 27 | 40 | 50 | 32 | 47 | 62 | 23 | 34 | 44 | 30 | 44 | 56 | 36 | 52 | 69 |
| 1800 | 20 | 30 | 38 | 26 | 39 | 49 | 31 | 46 | 60 | 23 | 33 | 42 | 29 | 43 | 55 | 35 | 51 | 67 |
| 1900 | 19 | 29 | 37 | 26 | 38 | 48 | 30 | 45 | 59 | 22 | 32 | 41 | 29 | 42 | 53 | 34 | 50 | 65 |
| 2000 | 19 | 28 | 36 | 25 | 37 | 46 | 30 | 43 | 57 | 22 | 32 | 40 | 28 | 41 | 52 | 33 | 48 | 64 |
| 2200 | 18 | 27 | 34 | 24 | 35 | 44 | 28 | 41 | 55 | 20 | 30 | 38 | 27 | 39 | 49 | 31 | 46 | 61 |
| 2400 | 17 | 26 | 33 | 23 | 33 | 42 | 27 | 40 | 52 | 19 | 29 | 37 | 25 | 37 | 47 | 30 | 44 | 58 |
| 2600 | 16 | 25 | 31 | 22 | 32 | 41 | 26 | 38 | 50 | 18 | 28 | 35 | 24 | 36 | 46 | 29 | 42 | 56 |
| 2800 | 15 | 24 | 30 | 21 | 31 | 39 | 25 | 37 | 48 | 18 | 27 | 34 | 24 | 35 | 44 | 28 | 41 | 54 |
| 3000 | 15 | 23 | 29 | 20 | 30 | 38 | 24 | 35 | 47 | 17 | 26 | 33 | 23 | 33 | 42 | 27 | 39 | 52 |
| 3200 | 14 | 22 | 28 | 19 | 29 | 37 | 23 | 34 | 45 | 16 | 25 | 32 | 22 | 32 | 41 | 26 | 38 | 50 |
| 3400 | 14 | 22 | 28 | 19 | 28 | 36 | 23 | 33 | 44 | 16 | 24 | 31 | 21 | 31 | 40 | 25 | 37 | 49 |
| 3600 | 14 | 21 | 27 | 18 | 27 | 35 | 22 | 32 | 43 | 15 | 24 | 30 | 21 | 30 | 39 | 25 | 36 | 48 |
| 3800 | 13 | 21 | 26 | 17 | 27 | 34 | 21 | 31 | 41 | 15 | 23 | 29 | 20 | 30 | 38 | 24 | 35 | 46 |
| 4000 | 13 | 20 | 25 | 17 | 26 | 33 | 21 | 31 | 40 | 14 | 22 | 28 | 19 | 29 | 37 | 23 | 34 | 45 |
| 4200 | 13 | 20 | 25 | 16 | 25 | 32 | 20 | 30 | 39 | 14 | 22 | 28 | 19 | 28 | 36 | 23 | 33 | 44 |
| 4400 | 12 | 19 | 24 | 16 | 25 | 31 | 19 | 29 | 39 | 14 | 21 | 27 | 18 | 28 | 35 | 22 | 33 | 43 |
| 4600 | 12 | 19 | 24 | 16 | 24 | 31 | 19 | 29 | 38 | 13 | 21 | 27 | 18 | 27 | 34 | 22 | 32 | 42 |
| 4800 | 12 | 18 | 23 | 15 | 24 | 30 | 18 | 28 | 37 | 13 | 20 | 26 | 17 | 26 | 33 | 21 | 31 | 41 |
| 5000 | 12 | 18 | 23 | 15 | 23 | 29 | 18 | 27 | 36 | 13 | 20 | 25 | 17 | 26 | 33 | 21 | 31 | 40 |

[a]Span values above solid line are governed by deflection. Span values within dashed box are governed by horizontal shear. Elsewhere bending governs span.

*Source:* Reproduced by permission of the American Concrete Institute.

Table 6-10(b)

SAFE SPACING (MM) OF SUPPORTS FOR DOUBLE WALES, SINGLE SPAN[a]

Spacing

Maximum deflection = span/360, but not to exceed 6 mm.

| Uniform load, kN/m (equals uniform pressure, kN/m2 on forms times spacing of wales in metres) | $F'_b$ varies with member size | | $E' = 8.97$ GPa | | $F'_v = 0.97$ MPa | | $F'_b$ varies with member size | | $E' = 8.97$ GPa | | $F'_v = 1.21$ MPa | |
|---|---|---|---|---|---|---|---|---|---|---|---|---|
| | Actual size of S4S lumber used double | | | | | | Actual size of S4S lumber used double | | | | | |
| | 38 × 89 | 38 × 140 | 38 × 184 | 89 × 89 | 89 × 140 | 89 × 184 | 38 × 89 | 38 × 140 | 38 × 184 | 89 × 89 | 89 × 140 | 89 × 184 |
| | $F'_b$ (MPa) | | | | | | $F'_b$ (MPa) | | | | | |
| | 8.8 | 7.7 | 7.0 | 8.8 | 7.7 | 7.7 | 8.8 | 7.7 | 7.0 | 8.8 | 7.7 | 7.7 |
| 1.5 | 1775 | 2650 | 3250 | 2325 | 3275 | 4025 | 1775 | 2650 | 3250 | 2325 | 3275 | 4025 |
| 2.9 | 1400 | 2225 | 2725 | 1875 | 2750 | 3375 | 1400 | 2225 | 2725 | 1875 | 2750 | 3375 |
| 4.4 | 1225 | 1825 | 2325 | 1625 | 2475 | 3050 | 1225 | 1925 | 2475 | 1625 | 2475 | 3050 |
| 5.8 | 1075 | 1575 | 2000 | 1475 | 2300 | 2850 | 1125 | 1750 | 2250 | 1475 | 2300 | 2850 |
| 7.3 | 975 | 1425 | 1800 | 1375 | 2175 | 2700 | 1050 | 1575 | 2000 | 1375 | 2175 | 2700 |
| 8.8 | 875 | 1300 | 1650 | 1300 | 1975 | 2575 | 975 | 1450 | 1825 | 1300 | 2025 | 2575 |
| 10.2 | 825 | 1200 | 1525 | 1225 | 1825 | 2425 | 925 | 1350 | 1700 | 1225 | 1925 | 2475 |
| 11.7 | 775 | 1125 | 1425 | 1175 | 1725 | 2250 | 850 | 1250 | 1600 | 1175 | 1850 | 2400 |
| 13.1 | 725 | 1050 | 1325 | 1100 | 1625 | 2125 | 800 | 1175 | 1500 | 1125 | 1775 | 2325 |
| 14.6 | 675 | 1000 | 1275 | 1050 | 1525 | 2025 | 775 | 1125 | 1425 | 1100 | 1700 | 2250 |
| 16.0 | 650 | 950 | 1200 | 1000 | 1450 | 1925 | 725 | 1075 | 1350 | 1050 | 1625 | 2150 |
| 17.5 | 625 | 925 | 1150 | 950 | 1400 | 1850 | 700 | 1025 | 1300 | 1025 | 1550 | 2050 |
| 19.0 | 600 | 875 | 1125 | 925 | 1350 | 1775 | 675 | 975 | 1250 | 1000 | 1500 | 1975 |
| 20.4 | 575 | 850 | 1075 | 875 | 1300 | 1700 | 650 | 950 | 1200 | 975 | 1450 | 1900 |
| 21.9 | 550 | 825 | 1025 | 850 | 1250 | 1650 | 625 | 925 | 1150 | 950 | 1400 | 1850 |
| 23.3 | 550 | 800 | 1000 | 825 | 1200 | 1600 | 600 | 875 | 1125 | 925 | 1350 | 1775 |
| 24.8 | 525 | 775 | 975 | 800 | 1175 | 1550 | 575 | 850 | 1100 | 900 | 1300 | 1725 |
| 26.3 | 500 | 750 | 950 | 775 | 1150 | 1500 | 575 | 825 | 1050 | 875 | 1275 | 1675 |
| 27.7 | 475 | 725 | 925 | 750 | 1125 | 1475 | 550 | 800 | 1025 | 850 | 1250 | 1625 |
| 29.2 | 475 | 700 | 900 | 750 | 1075 | 1425 | 550 | 800 | 1000 | 825 | 1200 | 1600 |
| 32.1 | 450 | 675 | 850 | 700 | 1025 | 1375 | 500 | 750 | 950 | 775 | 1150 | 1525 |
| 35.0 | 425 | 650 | 825 | 675 | 1000 | 1300 | 475 | 725 | 925 | 750 | 1100 | 1450 |
| 37.9 | 400 | 625 | 775 | 650 | 950 | 1250 | 450 | 700 | 875 | 725 | 1050 | 1400 |
| 40.9 | 375 | 600 | 750 | 625 | 925 | 1200 | 450 | 675 | 850 | 700 | 1025 | 1350 |
| 43.8 | 375 | 575 | 725 | 600 | 875 | 1175 | 425 | 650 | 825 | 675 | 975 | 1300 |
| 46.7 | 350 | 550 | 700 | 575 | 850 | 1125 | 400 | 625 | 800 | 650 | 950 | 1250 |
| 49.6 | 350 | 550 | 700 | 575 | 825 | 1100 | 400 | 600 | 775 | 625 | 925 | 1225 |
| 52.5 | 350 | 525 | 675 | 550 | 800 | 1075 | 375 | 600 | 750 | 625 | 900 | 1200 |
| 55.4 | 325 | 525 | 650 | 525 | 775 | 1025 | 375 | 575 | 725 | 600 | 875 | 1150 |
| 58.4 | 325 | 500 | 625 | 525 | 775 | 1000 | 350 | 550 | 700 | 575 | 850 | 1125 |
| 61.3 | 325 | 500 | 625 | 500 | 750 | 975 | 350 | 550 | 700 | 575 | 825 | 1100 |
| 64.2 | 300 | 475 | 600 | 475 | 725 | 975 | 350 | 525 | 675 | 550 | 825 | 1075 |
| 67.1 | 300 | 475 | 600 | 475 | 725 | 950 | 325 | 525 | 675 | 550 | 800 | 1050 |
| 70.0 | 300 | 450 | 575 | 450 | 700 | 925 | 325 | 500 | 650 | 525 | 775 | 1025 |
| 73.0 | 300 | 450 | 575 | 450 | 675 | 900 | 325 | 500 | 625 | 525 | 775 | 1000 |

[a]Above the solid lines span values are governed by deflection. Span values within the dashed box are governed by horizontal shear. Elsewhere bending governs the span.

*Source:* Original data provided by the American Concrete Institute; metric conversion by author.

# Table 6-11(a)
## SAFE SPACING (INCHES) OF SUPPORTS FOR DOUBLE WALES, CONTINUOUS OVER TWO SPANS[a]

Spacing Spacing

Maximum deflection is 1/360 of the span or ¼ in., whichever is smaller.

| Uniform load, lb per lineal ft (equals uniform load, psf, on forms times spacing of wales in ft) | $F'_b$ varies with member | | | $E' = 1{,}300{,}000$ psi | | | $F'_v = 140$ psi | | | $F'_b$ varies with member | | | $E' = 1{,}300{,}000$ psi | | | $F'_v = 175$ psi | | |
|---|---|---|---|---|---|---|---|---|---|---|---|---|---|---|---|---|---|---|
| | Nominal size of S4S lumber | | | | | | | | | Nominal size of S4S lumber | | | | | | | | |
| | 2 × 4 | 2 × 6 | 2 × 8 | 3 × 4 | 3 × 6 | 3 × 8 | 4 × 4 | 4 × 6 | 4 × 8 | 2 × 4 | 2 × 6 | 2 × 8 | 3 × 4 | 3 × 6 | 3 × 8 | 4 × 4 | 4 × 6 | 4 × 8 |
| | $F'_b$ psi | | | | | | | | | $F'_b$ psi | | | | | | | | |
| | 1280 | 1110 | 1020 | 1280 | 1110 | 1020 | 1280 | 1110 | 1110 | 1590 | 1380 | 1280 | 1590 | 1380 | 1280 | 1590 | 1380 | 1380 |
| 100 | 87 | 127 | 160 | 106 | 149 | 184 | 116 | 163 | 200 | 94 | 132 | 162 | 106 | 149 | 184 | 116 | 163 | 200 |
| 200 | 61 | 90 | 113 | 79 | 116 | 146 | 94 | 137 | 168 | 68 | 100 | 127 | 88 | 126 | 155 | 97 | 137 | 168 |
| 300 | 50 | 73 | 93 | 65 | 95 | 120 | 77 | 112 | 148 | 56 | 82 | 104 | 72 | 106 | 134 | 85 | 124 | 152 |
| 400 | 43 | 63 | 80 | 56 | 82 | 104 | 66 | 97 | 128 | 48 | 71 | 90 | 62 | 91 | 116 | 74 | 108 | 141 |
| 500 | 39 | 57 | 72 | 50 | 73 | 93 | 59 | 87 | 114 | 43 | 63 | 80 | 56 | 82 | 104 | 66 | 97 | 127 |
| 600 | 35 | 52 | 66 | 46 | 67 | 85 | 54 | 79 | 104 | 39 | 58 | 73 | 51 | 75 | 95 | 60 | 88 | 116 |
| 700 | 33 | 48 | 61 | 42 | 62 | 78 | 50 | 73 | 97 | 37 | 54 | 68 | 47 | 69 | 88 | 56 | 82 | 108 |
| 800 | 31 | 45 | 57 | 40 | 58 | 73 | 47 | 69 | 90 | 34 | 50 | 64 | 44 | 65 | 82 | 52 | 76 | 101 |
| 900 | 28 | 42 | 53 | 37 | 55 | 69 | 44 | 65 | 85 | 32 | 47 | 60 | 42 | 61 | 77 | 49 | 72 | 95 |
| 1000 | 26 | 40 | 51 | 35 | 52 | 66 | 42 | 61 | 81 | 31 | 45 | 57 | 39 | 58 | 73 | 47 | 68 | 90 |
| 1100 | 24 | 38 | 48 | 34 | 49 | 62 | 40 | 58 | 77 | 28 | 43 | 54 | 38 | 55 | 70 | 45 | 65 | 86 |
| 1200 | 23 | 36 | 46 | 32 | 47 | 60 | 38 | 56 | 74 | 27 | 41 | 52 | 36 | 53 | 67 | 43 | 62 | 82 |
| 1300 | 21 | 34 | 44 | 31 | 45 | 57 | 37 | 54 | 71 | 25 | 39 | 50 | 35 | 51 | 64 | 41 | 60 | 79 |
| 1400 | 20 | 32 | 42 | 29 | 44 | 55 | 35 | 52 | 68 | 24 | 37 | 48 | 33 | 49 | 62 | 39 | 58 | 76 |
| 1500 | 20 | 31 | 40 | 28 | 42 | 53 | 34 | 50 | 66 | 23 | 36 | 46 | 32 | 47 | 60 | 38 | 56 | 74 |
| 1600 | 19 | 29 | 39 | 27 | 41 | 52 | 33 | 48 | 64 | 22 | 34 | 45 | 31 | 46 | 58 | 37 | 54 | 71 |
| 1700 | 18 | 28 | 37 | 25 | 40 | 50 | 32 | 47 | 62 | 21 | 33 | 43 | 30 | 44 | 56 | 36 | 52 | 69 |
| 1800 | 17 | 27 | 36 | 24 | 38 | 49 | 31 | 46 | 60 | 20 | 32 | 42 | 29 | 43 | 55 | 35 | 51 | 67 |
| 1900 | 17 | 27 | 35 | 24 | 37 | 48 | 30 | 45 | 59 | 19 | 30 | 40 | 28 | 42 | 53 | 34 | 50 | 65 |
| 2000 | 16 | 26 | 34 | 23 | 36 | 46 | 29 | 43 | 57 | 19 | 29 | 39 | 27 | 41 | 52 | 33 | 48 | 64 |
| 2200 | 16 | 24 | 32 | 21 | 33 | 44 | 27 | 41 | 55 | 18 | 28 | 37 | 25 | 39 | 49 | 31 | 46 | 61 |
| 2400 | 15 | 23 | 31 | 20 | 32 | 42 | 25 | 40 | 52 | 17 | 26 | 35 | 23 | 37 | 47 | 30 | 44 | 58 |
| 2600 | 14 | 22 | 29 | 19 | 30 | 39 | 24 | 38 | 49 | 16 | 25 | 33 | 22 | 35 | 46 | 28 | 42 | 56 |
| 2800 | 14 | 22 | 28 | 18 | 29 | 38 | 23 | 36 | 47 | 15 | 24 | 32 | 21 | 33 | 44 | 27 | 41 | 54 |
| 3000 | 13 | 21 | 27 | 17 | 27 | 36 | 22 | 34 | 45 | 15 | 23 | 31 | 20 | 32 | 42 | 25 | 39 | 52 |
| 3200 | 13 | 20 | 27 | 17 | 26 | 35 | 21 | 33 | 43 | 14 | 23 | 30 | 19 | 30 | 40 | 24 | 38 | 50 |
| 3400 | 13 | 20 | 26 | 16 | 25 | 34 | 20 | 31 | 41 | 14 | 22 | 29 | 19 | 29 | 38 | 23 | 36 | 48 |
| 3600 | 12 | 19 | 25 | 16 | 25 | 33 | 19 | 30 | 40 | 14 | 21 | 28 | 18 | 28 | 37 | 22 | 35 | 46 |
| 3800 | 12 | 19 | 25 | 15 | 24 | 32 | 19 | 29 | 38 | 13 | 21 | 27 | 17 | 27 | 36 | 21 | 34 | 44 |
| 4000 | 12 | 18 | 24 | 15 | 23 | 31 | 18 | 28 | 37 | 13 | 20 | 27 | 17 | 26 | 35 | 21 | 33 | 43 |
| 4200 | 11 | 18 | 24 | 14 | 23 | 30 | 17 | 27 | 36 | 13 | 20 | 26 | 16 | 26 | 34 | 20 | 32 | 42 |
| 4400 | 11 | 18 | 23 | 14 | 22 | 29 | 17 | 27 | 35 | 12 | 19 | 26 | 16 | 25 | 33 | 19 | 31 | 40 |
| 4600 | 11 | 17 | 23 | 14 | 22 | 29 | 17 | 26 | 34 | 12 | 19 | 25 | 16 | 24 | 32 | 19 | 30 | 39 |
| 4800 | 11 | 17 | 23 | 14 | 21 | 28 | 16 | 25 | 33 | 12 | 19 | 25 | 15 | 24 | 31 | 18 | 29 | 38 |
| 5000 | 11 | 17 | 22 | 13 | 21 | 27 | 16 | 25 | 33 | 12 | 18 | 24 | 15 | 23 | 31 | 18 | 28 | 37 |

[a]Span values above solid line are governed by deflection. Span values within dashed box are governed by horizontal shear. Elsewhere bending governs span.

*Source:* Reproduced by permission of the American Concrete Institute.

## Table 6-11(b)
## SAFE SPACING (MM) OF SUPPORTS FOR DOUBLE WALES CONTINUOUS OVER TWO SPANS[a]

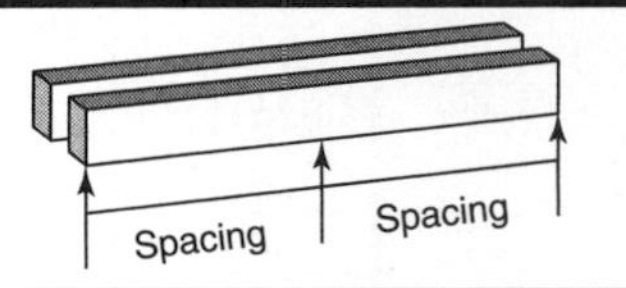


Maximum deflection = span/360, but not to exceed 6 mm.

| Uniform load, kN/m (equals uniform pressure, kN/m2 on forms times spacing of wales in metres) | $F'_b$ varies with member size | | $E' = 8.97$ GPa | | $F'_v = 0.97$ MPa | | $F'_b$ varies with member size | | $E' = 8.97$ GPa | | $F'_v = 1.21$ MPa | |
|---|---|---|---|---|---|---|---|---|---|---|---|---|
| | Actual size of S4S lumber used double | | | | | | Actual size of S4S lumber used double | | | | | |
| | 38 × 89 | 38 × 140 | 38 × 184 | 89 × 89 | 89 × 140 | 89 × 184 | 38 × 89 | 38 × 140 | 38 × 184 | 89 × 89 | 89 × 140 | 89 × 184 |
| | $F'_b$ (MPa) | | | | | | $F'_b$ (MPa) | | | | | |
| | 8.8 | 7.7 | 7.0 | 8.8 | 7.7 | 7.7 | 8.8 | 7.7 | 7.0 | 8.8 | 7.7 | 7.7 |
| 1.5 | 2175 | 3175 | 4000 | 2900 | 4075 | 5000 | 2350 | 3300 | 4050 | 2900 | 4075 | 5000 |
| 2.9 | 1525 | 2250 | 2825 | 2350 | 3425 | 4200 | 1700 | 2500 | 3175 | 2425 | 3425 | 4200 |
| 4.4 | 1250 | 1825 | 2325 | 1925 | 2800 | 3700 | 1400 | 2050 | 2600 | 2125 | 3100 | 3800 |
| 5.8 | 1075 | 1575 | 2000 | 1650 | 2425 | 3200 | 1200 | 1775 | 2250 | 1850 | 2700 | 3525 |
| 7.3 | 975 | 142 | 1800 | 1475 | 2175 | 2850 | 1075 | 1575 | 2000 | 1650 | 2425 | 3175 |
| 8.8 | 875 | 1300 | 1650 | 1350 | 1975 | 2600 | 975 | 1450 | 1825 | 1500 | 2200 | 2900 |
| 10.2 | 825 | 1200 | 1525 | 1250 | 1825 | 2425 | 925 | 1350 | 1700 | 1400 | 2050 | 2700 |
| 11.7 | 775 | 1125 | 1425 | 1175 | 1725 | 2250 | 850 | 1250 | 1600 | 1300 | 1900 | 2525 |
| 13.1 | 700 | 1050 | 1325 | 1100 | 1625 | 2125 | 800 | 1175 | 1500 | 1225 | 1800 | 2375 |
| 14.6 | 650 | 1000 | 1275 | 1050 | 1525 | 2025 | 775 | 1125 | 1425 | 1175 | 1700 | 2250 |
| 16.0 | 600 | 950 | 1200 | 1000 | 1450 | 1925 | 700 | 1075 | 1350 | 1125 | 1625 | 2150 |
| 17.5 | 575 | 900 | 1150 | 950 | 1400 | 1850 | 675 | 1025 | 1300 | 1075 | 1550 | 2050 |
| 19.0 | 525 | 850 | 1100 | 925 | 1350 | 1775 | 625 | 975 | 1250 | 1025 | 1500 | 1975 |
| 20.4 | 500 | 800 | 1050 | 875 | 1300 | 1700 | 600 | 925 | 1200 | 975 | 1450 | 1900 |
| 21.9 | 500 | 775 | 1000 | 850 | 1250 | 1650 | 575 | 900 | 1150 | 950 | 1400 | 1850 |
| 23.3 | 475 | 725 | 975 | 825 | 1200 | 1600 | 550 | 850 | 1125 | 925 | 1350 | 1775 |
| 24.8 | 450 | 700 | 925 | 800 | 1175 | 1550 | 525 | 825 | 1075 | 900 | 1300 | 1725 |
| 26.3 | 425 | 675 | 900 | 775 | 1150 | 1500 | 500 | 800 | 1050 | 875 | 1275 | 1675 |
| 27.7 | 425 | 675 | 875 | 750 | 1125 | 1475 | 475 | 750 | 1000 | 850 | 1250 | 1625 |
| 29.2 | 400 | 650 | 850 | 725 | 1075 | 1425 | 475 | 725 | 975 | 825 | 1200 | 1600 |
| 32.1 | 400 | 600 | 800 | 675 | 1025 | 1375 | 450 | 700 | 925 | 775 | 1150 | 1525 |
| 35.0 | 375 | 575 | 775 | 625 | 1000 | 1300 | 425 | 650 | 875 | 750 | 1100 | 1450 |
| 37.9 | 350 | 550 | 725 | 600 | 950 | 1225 | 400 | 625 | 825 | 700 | 1050 | 1400 |
| 40.9 | 350 | 550 | 700 | 575 | 900 | 1175 | 375 | 600 | 800 | 675 | 1025 | 1350 |
| 43.8 | 325 | 525 | 675 | 550 | 850 | 1125 | 375 | 575 | 775 | 625 | 975 | 1300 |
| 46.7 | 325 | 500 | 675 | 525 | 825 | 1075 | 350 | 575 | 750 | 600 | 950 | 1250 |
| 49.6 | 325 | 500 | 650 | 500 | 775 | 1025 | 350 | 550 | 725 | 575 | 900 | 1200 |
| 52.5 | 300 | 475 | 625 | 475 | 750 | 1000 | 350 | 525 | 700 | 550 | 875 | 1150 |
| 55.4 | 300 | 450 | 625 | 475 | 725 | 950 | 325 | 525 | 675 | 525 | 850 | 1100 |
| 58.4 | 300 | 400 | 600 | 450 | 700 | 925 | 325 | 500 | 675 | 525 | 825 | 1075 |
| 61.3 | 275 | 450 | 600 | 425 | 675 | 900 | 325 | 500 | 650 | 500 | 800 | 1050 |
| 64.2 | 275 | 450 | 575 | 425 | 675 | 875 | 300 | 475 | 650 | 475 | 775 | 1000 |
| 67.1 | 275 | 425 | 575 | 425 | 650 | 850 | 300 | 475 | 625 | 475 | 750 | 975 |
| 70.0 | 275 | 425 | 575 | 400 | 625 | 825 | 300 | 475 | 625 | 450 | 725 | 950 |
| 73.0 | 275 | 425 | 550 | 400 | 625 | 825 | 300 | 450 | 600 | 450 | 700 | 925 |

[a]Above the solid lines span values are governed by deflection. Span values within the dashed box are governed by horizontal shear. Elsewhere bending governs the span.

*Source:* Original data provided by the American Concrete Institute; metric conversion by author.

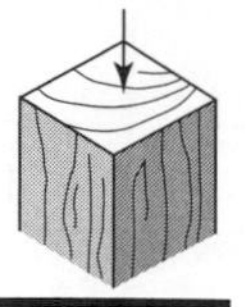

Table 6-12(a)
**ALLOWABLE AXIAL LOAD (POUNDS) ON SIMPLE WOOD SHORES OF THE INDICATED STRENGTH AND EFFECTIVE LENGTH[a]**

| Nominal lumber size, in. | 2 × 4 rough | 2 × 4 S4S | 3 × 4 rough | 3 × 4 S4S | 4 × 4 rough | 4 × 4 S4S | 4 × 2 rough | 4 × 2 S4S | 4 × 3 rough | 4 × 3 S4S | 4 × 6 rough | 4 × 6 S4S |
|---|---|---|---|---|---|---|---|---|---|---|---|---|
| **No. 2 Grade: $F_c$ = 1100 psi $E$ = 1,400,000 $l/d$ = 50 or less** | | | | | | | | | | | | |
| **Effective Length, ft** | | | | | | | **Lacing and/or Bracing Needed*** | | | | | |
| 5 | 1700 | 1300 | 6300 | 5400 | 12500 | 11300 | 5600 | 4900 | 9100 | 8100 | 18900 | 17400 |
| 6 | 1200 | 900 | 4700 | 4000 | 10500 | 9300 | 4600 | 4000 | 7600 | 6700 | 15900 | 14400 |
| 7 | | | 3600 | 3000 | 8500 | 7500 | 3700 | 3200 | 6200 | 5400 | 13000 | 11700 |
| 8 | | | 2800 | 2400 | 6900 | 6100 | 3000 | 2600 | 5000 | 4300 | 10600 | 9500 |
| 9 | | | 2300 | 1900 | 5600 | 4900 | 2500 | 2100 | 4100 | 3500 | 8700 | 7700 |
| 10 | | | 1800 | 1500 | 4700 | 4100 | 2000 | 1800 | 3400 | 2900 | 7200 | 6400 |
| 11 | | | | | 3900 | 3400 | 1700 | 1500 | 2800 | 2400 | 6100 | 5400 |
| 12 | | | | | 3300 | 2900 | 1400 | 1200 | 2400 | 2100 | 5200 | 4600 |
| 13 | | | | | 2900 | 2500 | 1200 | 1100 | 2100 | 1800 | 4400 | 3900 |
| 14 | | | | | 2500 | 2200 | 1100 | 900 | 1800 | 1500 | 3800 | 3400 |
| 15 | | | | | 2200 | | 900 | | 1600 | | 3400 | |
| **No. 2 Grade: $F_c$ = 1250 $E$ = 1,300,000 $l/d$ = 50 or less** | | | | | | | | | | | | |
| **Effective Length, ft** | | | | | | | **Lacing and/or Bracing Needed*** | | | | | |
| 5 | 1600 | 1200 | 6100 | 5200 | 13000 | 11700 | 5800 | 5000 | 9400 | 8300 | 19700 | 18000 |
| 6 | 1100 | 900 | 4500 | 3800 | 10400 | 9200 | 4700 | 4000 | 7600 | 6600 | 15900 | 14300 |
| 7 | | | 3400 | 2900 | 8300 | 7300 | 3700 | 3100 | 6000 | 5200 | 12700 | 11300 |
| 8 | | | 2600 | 2200 | 6600 | 5800 | 3000 | 2500 | 4800 | 4100 | 10200 | 9000 |
| 9 | | | 2100 | 1800 | 5400 | 4700 | 2400 | 2000 | 3900 | 3300 | 8300 | 7300 |
| 10 | | | 1700 | 1400 | 4400 | 3800 | 2000 | 1700 | 3200 | 2700 | 6800 | 6000 |
| 11 | | | | | 3700 | 3200 | 1700 | 1400 | 2700 | 2300 | 5700 | 5000 |
| 12 | | | | | 3100 | 2700 | 1400 | 1200 | 2300 | 1900 | 4800 | 4300 |
| 13 | | | | | 2700 | 2300 | 1200 | 1000 | 1900 | 1700 | 4200 | 3700 |
| 14 | | | | | 2300 | 2000 | 1000 | 900 | 1700 | 1400 | 3600 | 3200 |
| 15 | | | | | 2000 | | 900 | | 1500 | | 3100 | |
| **No. 2 Grade: $F_c$ = 1300 $E$ = 1,600,000 $l/d$ = 50 or less** | | | | | | | | | | | | |
| **Effective Length, ft** | | | | | | | **Lacing and/or Bracing Needed*** | | | | | |
| 5 | 2000 | 1500 | 7200 | 6200 | 14600 | 13200 | 6500 | 5700 | 10600 | 9400 | 22100 | 20200 |
| 6 | 1400 | 1100 | 5400 | 4600 | 12100 | 10800 | 5400 | 4600 | 8800 | 7700 | 18500 | 16700 |
| 7 | | | 4100 | 3500 | 9800 | 8700 | 4400 | 3700 | 7100 | 6200 | 15000 | 13500 |
| 8 | | | 3200 | 2700 | 7900 | 7000 | 3600 | 3000 | 5700 | 5000 | 12200 | 10900 |
| 9 | | | 2600 | 2200 | 6500 | 5700 | 2900 | 2400 | 4700 | 4000 | 10000 | 8900 |
| 10 | | | 2100 | 1800 | 5400 | 4700 | 2400 | 2000 | 3900 | 3300 | 8300 | 7300 |
| 11 | | | | | 4500 | 3900 | 2000 | 1700 | 3300 | 2800 | 6900 | 6100 |
| 12 | | | | | 3800 | 3300 | 1700 | 1400 | 2800 | 2400 | 5900 | 5200 |
| 13 | | | | | 3300 | 2900 | 1500 | 1200 | 2400 | 2000 | 5100 | 4500 |
| 14 | | | | | 2800 | 2500 | 1300 | 1100 | 2100 | 1800 | 4400 | 3900 |
| 15 | | | | | 2500 | | 1100 | | 1800 | | 3900 | |

[a]Values rounded to the nearest 100 lb. R indicates rough lumber; S4S indicates lumber finished on all four sides. Size adjustment applied for No. 2 grades.

*The dimension used in determining $l/d$ is that shown first in the size column. Where this is the larger dimension, the column must be laced (or braced) in the other direction so that $l/d$ in that direction is equal to or less than that used when calculating the loads shown. For the 4 × 2s lacing in the plane of the 2-in. dimension must be at intervals not greater than 0.4 times the unsupported length. For 4 × 3s lacing in the plane of the 3-in. dimension must be at intervals not more than 0.7 times the unsupported length.

*Source:* Reproduced by permission of the American Concrete Institute.

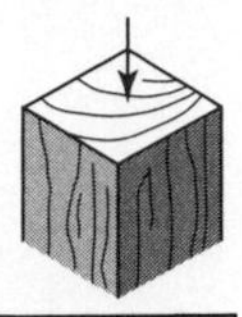

## Table 6-12(b)
## ALLOWABLE AXIAL LOAD (kN) ON SIMPLE WOOD SHORES OF THE INDICATED STRENGTH, BASED ON EFFECTIVE LENGTH*

No. 2 Grade: $F_c$ = 7.59 MPa — $E$ = 9.66 GPa — $l/d$ = 50 or less

| Actual Lumber Size, mm | 38 × 89 | | 64 × 89 | | 89 × 89 | | 89 × 38 | | 89 × 64 | | 89 × 140 | |
|---|---|---|---|---|---|---|---|---|---|---|---|---|
| | R** | S4S** | R | S4S | R | S4S | R | S4S | R | S4S | R | S4S |
| Effective Length, m | | | | | | | Bracing Needed# | | | | | |
| 1.5 | 7.6 | 5.8 | 28.0 | 24.0 | 55.6 | 50.3 | 24.9 | 21.8 | 40.5 | 36.0 | 84.1 | 77.4 |
| 1.8 | 5.3 | 4.0 | 20.9 | 17.8 | 46.7 | 41.4 | 20.5 | 17.8 | 33.8 | 29.8 | 70.7 | 64.1 |
| 2.1 | | | 16.0 | 13.3 | 37.8 | 33.4 | 16.5 | 14.2 | 27.6 | 24.0 | 57.8 | 52.0 |
| 2.4 | | | 12.5 | 10.7 | 30.7 | 27.1 | 13.3 | 11.6 | 22.2 | 19.1 | 47.1 | 42.3 |
| 2.7 | | | 10.2 | 8.5 | 24.9 | 21.8 | 11.1 | 9.3 | 18.2 | 15.6 | 38.7 | 34.2 |
| 3.0 | | | 8.0 | 6.7 | 20.9 | 18.2 | 8.9 | 8.0 | 15.1 | 12.9 | 32.0 | 28.5 |
| 3.4 | | | | | 17.3 | 15.1 | 7.6 | 6.7 | 12.5 | 10.7 | 27.1 | 24.0 |
| 3.7 | | | | | 14.7 | 12.9 | 6.2 | 5.3 | 10.7 | 9.3 | 23.1 | 20.5 |
| 4.0 | | | | | 12.9 | 11.1 | 5.3 | 4.9 | 9.3 | 8.0 | 19.6 | 17.3 |
| 4.3 | | | | | 11.1 | 9.8 | 4.9 | 4.0 | 8.0 | 6.7 | 16.9 | 15.1 |
| 4.6 | | | | | 9.8 | | 4.0 | | 7.1 | | 15.1 | |

No. 2 Grade: $F_c$ = 8.62 MPa — $E$ = 8.97 GPa — $l/d$ = 50 or less

| Actual Lumber Size, mm | 38 × 89 | | 64 × 89 | | 89 × 89 | | 89 × 38 | | 89 × 64 | | 89 × 140 | |
|---|---|---|---|---|---|---|---|---|---|---|---|---|
| | R** | S4S** | R | S4S | R | S4S | R | S4S | R | S4S | R | S4S |
| Effective Length, m | | | | | | | Bracing Needed# | | | | | |
| 1.5 | 7.1 | 5.3 | 27.1 | 23.1 | 57.8 | 52.0 | 25.8 | 22.2 | 41.8 | 36.9 | 87.6 | 80.1 |
| 1.8 | 4.9 | 4.0 | 15.1 | 16.9 | 46.3 | 40.9 | 20.9 | 17.8 | 33.8 | 29.4 | 70.7 | 63.6 |
| 2.1 | | | 11.6 | 12.9 | 36.9 | 32.5 | 16.5 | 13.8 | 26.7 | 23.1 | 56.5 | 50.3 |
| 2.4 | | | 9.3 | 9.8 | 29.4 | 25.8 | 13.3 | 11.1 | 21.4 | 18.2 | 45.4 | 40.0 |
| 2.7 | | | 7.6 | 8.0 | 24.0 | 20.9 | 10.7 | 8.9 | 17.3 | 14.7 | 36.9 | 32.5 |
| 3.0 | | | 7.6 | 6.2 | 19.6 | 16.9 | 8.9 | 7.6 | 14.2 | 12.0 | 30.2 | 26.7 |
| 3.4 | | | | | 16.5 | 14.2 | 7.6 | 6.2 | 12.0 | 10.2 | 25.4 | 22.2 |
| 3.7 | | | | | 13.8 | 12.0 | 6.2 | 5.3 | 10.2 | 8.5 | 21.4 | 19.1 |
| 4.0 | | | | | 12.0 | 10.2 | 5.3 | 4.4 | 8.5 | 7.6 | 18.7 | 16.5 |
| 4.3 | | | | | 10.2 | 8.9 | 4.4 | 4.0 | 7.6 | 6.2 | 16.0 | 14.2 |
| 4.6 | | | | | 8.9 | | 4.0 | | 6.7 | | 13.8 | |

No. 2 Grade: $F_c$ = 8.97 MPa — $E$ = 11.03 GPa — $l/d$ = 50 or less

| Actual Lumber Size, mm | 38 × 89 | | 64 × 89 | | 89 × 89 | | 89 × 38 | | 89 × 64 | | 89 × 140 | |
|---|---|---|---|---|---|---|---|---|---|---|---|---|
| | R** | S4S** | R | S4S | R | S4S | R | S4S | R | S4S | R | S4S |
| Effective Length, m | | | | | | | Bracing Needed# | | | | | |
| 1.5 | 8.9 | 6.7 | 32.0 | 27.6 | 64.9 | 58.7 | 28.9 | 25.4 | 47.1 | 41.8 | 98.3 | 89.8 |
| 1.8 | 6.2 | 4.9 | 24.0 | 20.5 | 53.8 | 48.0 | 24.0 | 20.5 | 39.1 | 34.2 | 82.3 | 74.3 |
| 2.1 | | | 18.2 | 15.6 | 43.6 | 38.7 | 19.6 | 16.5 | 31.6 | 27.6 | 66.7 | 60.0 |
| 2.4 | | | 14.2 | 12.0 | 35.1 | 31.1 | 16.0 | 13.3 | 25.4 | 22.2 | 54.3 | 48.5 |
| 2.7 | | | 11.6 | 9.8 | 28.9 | 25.4 | 12.9 | 10.7 | 20.9 | 17.8 | 44.5 | 39.6 |
| 3.0 | | | 9.3 | 8.0 | 24.0 | 20.9 | 10.7 | 8.9 | 17.3 | 14.7 | 36.9 | 32.5 |
| 3.4 | | | | | 20.0 | 17.3 | 8.9 | 7.6 | 14.7 | 12.5 | 30.7 | 27.1 |
| 3.7 | | | | | 16.9 | 14.7 | 7.6 | 6.2 | 12.5 | 10.7 | 26.2 | 23.1 |
| 4.0 | | | | | 14.7 | 12.9 | 6.7 | 5.3 | 10.7 | 8.9 | 22.7 | 20.0 |
| 4.3 | | | | | 12.5 | 11.1 | 5.8 | 4.9 | 9.3 | 8.0 | 19.6 | 17.3 |
| 4.6 | | | | | 11.1 | | 4.9 | | 8.0 | | 17.3 | |

*Converted values are based on original values that were calculated to the nearest 100 lb.

**R indicates rough lumber, S4S indicates lumber finished on all four sides.

#The dimension used in determining $l/d$ is that shown first in the size column. Where this is the larger dimension, the column must be laced or braced in the other direction so that $l/d$ in that direction is equal to or less than that used when calculating the loads shown. For 89 × 38s, bracing in the plane of the 38-mm dimension must be at intervals not greater than 0.4 times the unsupported length. For 89 × 64s, bracing in the plane of the 64-mm dimension must be at intervals not more than 0.7 times the unsupported length. Size adjustment has been applied for No. 2 grade.

*Source:* Original data provided by the American Concrete Institute; metric conversion by author.

Table 6-13(a)
ALLOWABLE AXIAL LOAD (POUNDS) ON MAXIMUM SHORE AREA IN DIRECT CONTACT WITH WOOD MEMBER BEING SUPPORTED[a]

| Nominal Lumber Size | 2 × 4<br>4 × 2 | | 3 × 4<br>4 × 3 | | 4 × 4 | | 4 × 6 | | 6 × 6 | |
|---|---|---|---|---|---|---|---|---|---|---|
| Area of Cross Sec., sq in. | R<br>5.89 | S4S<br>5.25 | R<br>9.52 | S4S<br>8.75 | R<br>13.14 | S4S<br>12.25 | R<br>20.39 | S4S<br>19.25 | R<br>31.64 | S4S<br>30.25 |
| Allowable Bearing Stress in Supported Member | | | | | | | | | | |
| 375 | 2200 | 2000 | 3600 | 3300 | 4900 | 4600 | 7600 | 7200 | 11900 | 11300 |
| 400 | 2400 | 2100 | 3800 | 3500 | 5300 | 4900 | 8200 | 7700 | 12700 | 12100 |
| 425 | 2500 | 2200 | 4000 | 3700 | 5600 | 5200 | 8700 | 8200 | 13400 | 12900 |
| 450 | 2600 | 2400 | 4300 | 3900 | 5900 | 5500 | 9200 | 8700 | 14200 | 13600 |
| 475 | 2800 | 2500 | 4500 | 4200 | 6200 | 5800 | 9700 | 9100 | 15000 | 14400 |
| 500 | 2900 | 2600 | 4800 | 4400 | 6600 | 6100 | 10200 | 9600 | 15800 | 15100 |
| 525 | 3100 | 2800 | 5000 | 4600 | 6900 | 6400 | 10700 | 10100 | 16600 | 15900 |
| 550 | 3200 | 2900 | 5200 | 4800 | 7200 | 6700 | 11200 | 10600 | 17400 | 16600 |
| 575 | 3400 | 3000 | 5500 | 5000 | 7600 | 7000 | 11700 | 11100 | 18200 | 17400 |
| 600 | 3500 | 3100 | 5700 | 5200 | 7900 | 7300 | 12200 | 11500 | 19000 | 18100 |
| 625 | 3700 | 3300 | 5900 | 5500 | 8200 | 7700 | 12700 | 12000 | 19800 | 18900 |
| 650 | 3800 | 3400 | 6200 | 5700 | 8500 | 8000 | 13300 | 12500 | 20600 | 19700 |
| 700 | 4100 | 3700 | 6700 | 6100 | 9200 | 8600 | 14300 | 13500 | 22100 | 21200 |
| 750 | 4400 | 3900 | 7100 | 6600 | 9900 | 9200 | 15300 | 14400 | 23700 | 22700 |
| 800 | 4700 | 4200 | 7600 | 7000 | 10500 | 9800 | 16300 | 15400 | 25300 | 24200 |
| 850 | 5000 | 4500 | 8100 | 7400 | 11200 | 10400 | 17300 | 16400 | 26900 | 25700 |
| 900 | 5300 | 4700 | 8600 | 7900 | 11800 | 11000 | 18400 | 17300 | 28500 | 27200 |

[a]Calculated to nearest 100 lb. R indicates rough lumber; S4S indicates lumber finished on all four sides.
*Source:* Reproduced by permission of the American Concrete Institute.

### *Example:*

A 12-ft-high wall, 10 in. thick, is to be poured at the rate of 10 ft/hr at a temperature of 75°F. (a) Calculate the maximum concrete pressure on the formwork. (b) Assuming that ¾ in. plywood is to be used, with the face grain parallel to the span, as sheathing with no studs, determine the maximum allowable spacing of the wales based on the maximum concrete pressure. (c) If the ties are to be spaced at 24 in. o.c., calculate the resulting maximum force on the ties if the wales are continuous over three spans.

### *Solution:*

**(a)** For walls poured at 7 to 10 ft/hr, the maximum concrete pressure is

$$p = 150 + \frac{43{,}400}{T} + \frac{2{,}800R}{T}$$

$$= 150 + \frac{43{,}400}{75} + \frac{2{,}800 \times 10}{75}$$

$$= 1{,}102 \text{ psf}$$

Check for maximum pressure: the lesser of 2,000 psf and 150 × 12 = 1,800 psf. Use calculated value of 1,102 psf as maximum pressure.

Table 6-13(b)
ALLOWABLE AXIAL LOAD (kN) ON MAXIMUM SHORE AREA
IN DIRECT CONTACT WITH WOOD MEMBER BEING SUPPORTED[a]

| Actual Lumber Size | 38 × 89<br>89 × 38 | | 64 × 89<br>89 × 64 | | 89 × 89 | | 89 × 140 | | 140 × 140 | |
|---|---|---|---|---|---|---|---|---|---|---|
| Area of Cross Sec., mm² | R<br>3800 | S4S<br>3387 | R<br>6142 | S4S<br>5645 | R<br>8478 | S4S<br>7904 | R<br>13156 | S4S<br>12420 | R<br>20414 | S4S<br>19517 |
| Allowable Bearing Stress in Supported Member, (MPa) | | | | | | | | | | |
| 2.59 | 9.8 | 8.9 | 16.0 | 14.7 | 21.8 | 20.5 | 33.8 | 32.0 | 52.9 | 50.3 |
| 2.76 | 10.7 | 9.3 | 16.9 | 15.6 | 23.6 | 21.8 | 36.5 | 34.2 | 56.5 | 53.8 |
| 2.93 | 11.1 | 9.8 | 17.8 | 16.5 | 24.9 | 23.1 | 38.7 | 36.5 | 59.6 | 57.4 |
| 3.10 | 11.6 | 10.7 | 19.1 | 17.3 | 26.2 | 24.5 | 40.9 | 38.7 | 63.2 | 60.5 |
| 3.28 | 12.5 | 11.1 | 20.0 | 18.7 | 27.6 | 25.8 | 43.1 | 40.5 | 66.7 | 64.1 |
| 3.45 | 12.9 | 11.6 | 21.4 | 19.6 | 29.4 | 27.1 | 45.4 | 42.7 | 70.3 | 67.2 |
| 3.62 | 13.8 | 12.5 | 22.2 | 20.5 | 30.7 | 28.5 | 47.6 | 44.9 | 73.8 | 70.7 |
| 3.79 | 14.2 | 12.9 | 23.1 | 21.4 | 32.0 | 29.8 | 49.8 | 47.1 | 77.4 | 73.8 |
| 3.96 | 15.1 | 13.3 | 24.5 | 22.2 | 33.8 | 31.1 | 52.0 | 49.4 | 81.0 | 77.4 |
| 4.14 | 15.6 | 13.8 | 25.4 | 23.1 | 35.1 | 32.5 | 54.3 | 51.2 | 84.5 | 80.5 |
| 4.31 | 16.5 | 14.7 | 26.2 | 24.5 | 36.5 | 34.2 | 56.5 | 53.4 | 88.1 | 84.1 |
| 4.48 | 16.9 | 15.1 | 27.6 | 25.4 | 37.8 | 35.6 | 59.2 | 55.6 | 91.6 | 87.6 |
| 4.83 | 18.2 | 16.5 | 29.8 | 27.1 | 40.9 | 38.3 | 63.6 | 60.0 | 98.3 | 94.3 |
| 5.17 | 19.6 | 17.3 | 31.6 | 29.4 | 44.0 | 40.9 | 68.1 | 64.1 | 105.4 | 101.0 |
| 5.52 | 20.9 | 18.7 | 33.8 | 31.1 | 46.7 | 43.6 | 72.5 | 68.5 | 112.5 | 107.6 |
| 5.86 | 22.2 | 20.0 | 36.0 | 32.9 | 49.8 | 46.3 | 77.0 | 72.9 | 119.7 | 114.3 |
| 6.21 | 23.6 | 20.9 | 38.3 | 35.1 | 52.5 | 48.9 | 81.8 | 77.0 | 126.8 | 121.0 |

[a]Converted values are based on original values that were calculated to the nearest 100 lb. R indicates rough lumber. S4S indicates lumber finished on all four sides.

*Source:* Original values provided by the American Concrete Institute; metric conversion by author.

**(b)** Assuming that the plywood sheathing will be continuous over four or more supports, using Table 6-4(a), and bending stress $F'_b = 1930$ psi, the maximum spacing of wales is 9 in. for ¾-in. plywood with face grain parallel to the span.

This would be the distance between lines of wales at the bottom of the wall form where the maximum concrete pressure occurs. From previous calculations, we know that the concrete pressure decreases from the bottom to the top of the formwork and the spacing of the wales can be increased accordingly.

**(c)** The resulting maximum load per linear foot of wale is calculated as

$$1{,}102 \times \frac{9}{12} = 827 \text{ pounds per foot}$$

The ties represent the supports for the wales and the tie spacing is the span over which the wales must support the calculated load (see Figure 6-16). In this case, the wales must be sized to support 827 pounds per foot over a 2 ft span.

If the wales are continuous over three spans, based on 24 in. spacing of the ties and applying

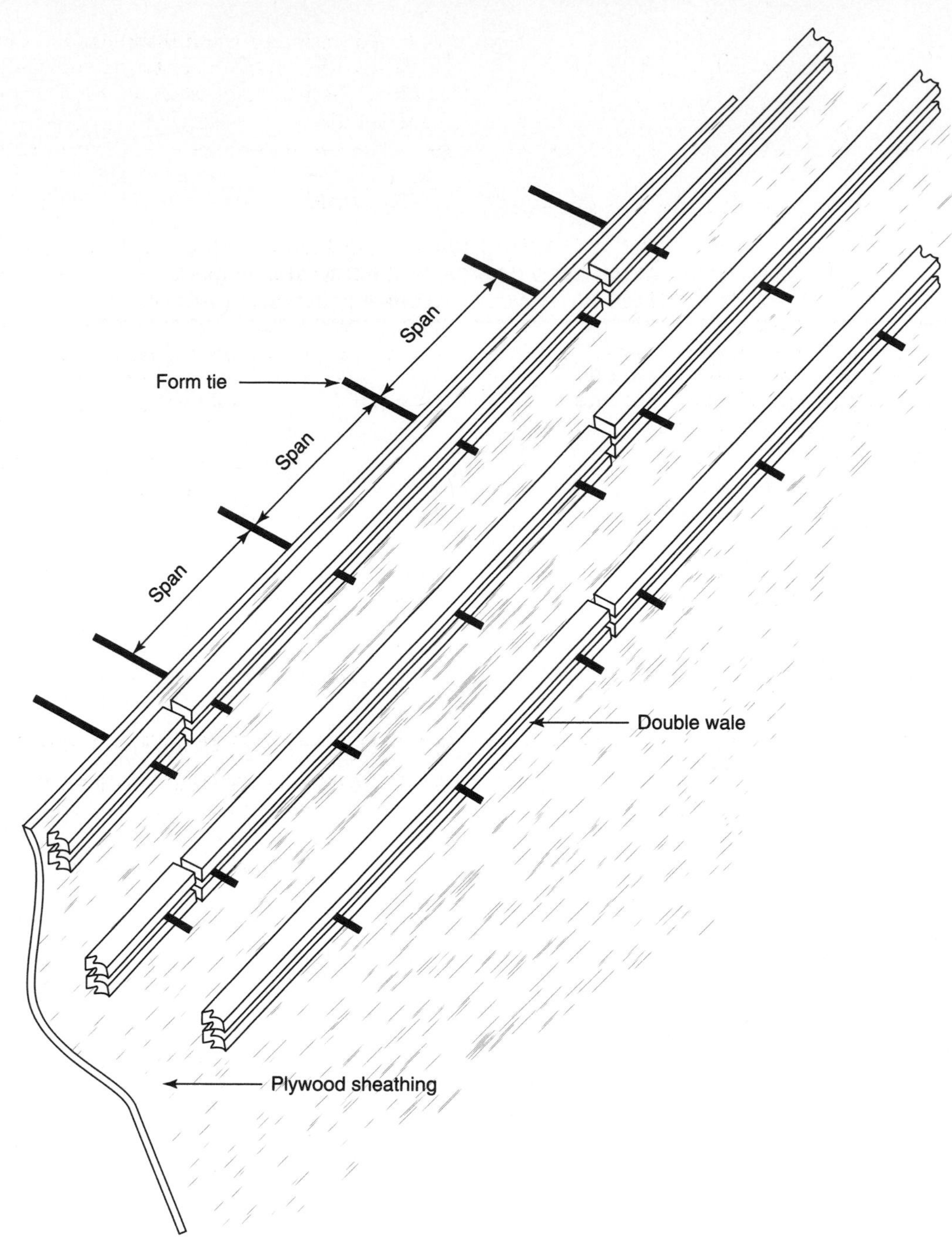


**Figure 6-16** Wales Continuous over Three Spans.

Figure 6-14(b), the maximum load on the ties becomes

$$1.1 \times 827 \times \frac{24}{12} = 1{,}819 \text{ lbs}$$

### *Example:*

A 3-m-high wall 250 mm thick is to be poured at the rate of 2.5 m/hr at a temperature of 20°C. (a) Calculate the maximum concrete pressure on the formwork. (b) Assuming that 18.5-mm plywood is to be used, with the face grain parallel to the span, as sheathing with no studs, determine the maximum allowable spacing of the wales based on the maximum concrete pressure. (c) If the ties are to be spaced at 600 mm o.c., calculate the resulting maximum force on the ties assuming that the wales are to be continuous over three spans.

### *Solution:*

**(a)** For walls poured at a rate greater than 2 m/hr, the maximum pressure is calculated as

$$p = 7.2 + \frac{1{,}156}{T + 17.8} + \frac{244R}{T + 17.8}$$

$$= 7.2 + \frac{1{,}156}{15 + 17.8} + \frac{244 \times 2.5}{15 + 17.8}$$

$$= 61 \text{ kPa}$$

Check maximum pressure: the lesser of 95.8 kPa and 23.5 × 3 = 70.5 kPa. Therefore 61 kPa is maximum pressure.

**(b)** Based on 18.5-mm thick plywood continuous over four or more supports (Table 6-4(b)), with the face grain parallel to the span and $F'_b = 13.3$ MPa, the maximum spacing of wales can be 200 mm.

**(c)** The resulting maximum load per lineal metre on the wales is equal to the maximum concrete pressure × distance between the wales.

$$61 \times \frac{200}{1{,}000} = 12 \text{ kN/m}$$

The resulting maximum load on the ties, assuming that the wales are continuous over three spans (see Figure 6-14(b)), is calculated as 1.10 × the load/metre on the wale × the tie spacing.

$$1.10 \times 12 \times \frac{600}{1{,}000} = 8 \text{ kN}$$

The foregoing calculations illustrated the approach to wall formwork design, assuming no studs. The spacing of the wales was governed by the strength of the plywood sheathing.

The following examples illustrate the practice of using studs in conjunction with the wales. As illustrated in Figure 6-12, the studs are placed between the sheathing and the wales. The spacing of the studs is based on the plywood thickness and direction of the face grain with respect to the studs. The spacing of the wales will now depend on the size and strength of the studs.

---

### *Example:*

A 20-ft-high wall, 12 in. thick, is to be poured at the rate of 8 ft/hr at a temperature of 60°F. (a) Calculate the maximum pressure on the formwork. (b) Using ¾-in.-thick plywood with the face grain parallel to the span, determine the maximum allowable spacing of the studs based on the maximum concrete pressure. (c) Assuming that the studs are to be 2″ × 4″s, with the strong axis resisting the load, determine the spacing of the wales. (d) What spacing of wales could be achieved if the studs were increased to 2″ × 6″s? (e) Based on the calculations in part (d), determine the size of wale that would be required using Table 6-9(a), assuming a wale spacing of 24 inches and tie and stud arrangement as shown in Figure 6-17. (f) Based on the calculations in part (e), determine the resulting load on the ties.

### *Solution:*

**(a)** For walls poured at 7 to 10 ft/hr, maximum pressure is

$$p = 150 + \frac{43{,}400}{T} + \frac{2{,}800R}{T}$$

$$= 150 + \frac{43{,}400}{60} + \frac{2{,}800 \times 8}{60}$$

$$= 1247 \text{ psf}$$

Check maximum pressure: the lesser of 2,000 psf and 150 × 20 = 3,000 psf. Therefore, use 1,247 psf as maximum pressure.

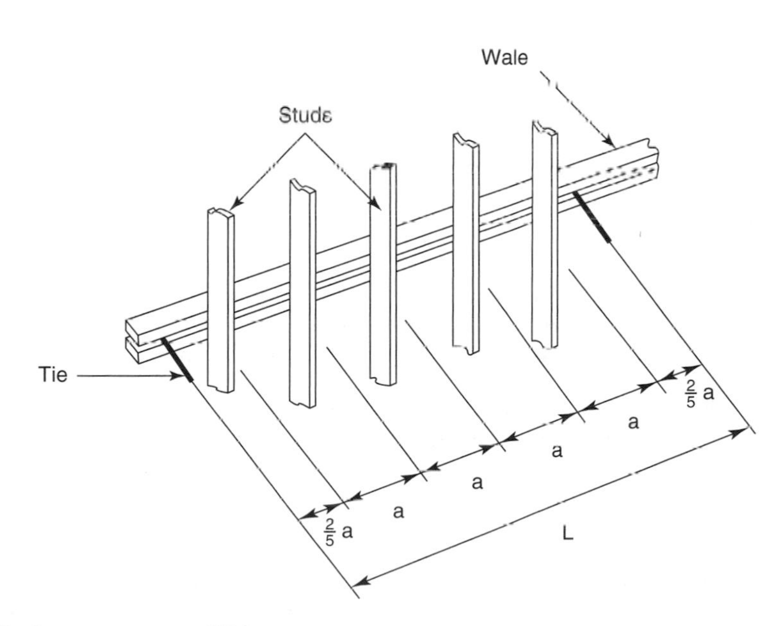


**Figure 6-17** Stud and Tie Arrangement on Wale.

**(b)** Based on plywood sheathing continuous over four or more supports [Table 6-4(a)], face grain parallel to the span and an $F'_b$ = 1,930 psi, the maximum spacing of supports or studs is 9 in. o.c. for $\frac{3}{4}''$ plywood.

**(c)** Load per lineal foot on the studs is

$$1{,}247 \times \frac{9}{12} = 935 \text{ plf}$$

Using Table 6-6(a), 2″ × 4″ studs supported at 16 in. o.c. can support 1,100 lb per foot, which is greater than the 935-plf applied load. The resulting wale spacing that can be used is 16 in. o.c. (stud material $F'_b$ = 1,280 psi or greater).

**(d)** If 2″ × 6″s were to be used as studs, using Table 6-6(a), the spacing of the wales could be increased to 26 in. The capacity of a 2″ × 6″ over a 26-in. span is 1,000 plf, so a good spacing to use for the wales is 24 in. (stud material $F'_b$ = 1,100 psi or greater).

**(e)** Because the load tables are based on uniformly distributed loads, the first requirement is to convert the point loads of the studs to an *equivalent uniformly distributed load* on the wale. Assuming that the studs will be continuous over three or more spans, applying Figure 6-14(c), the resulting reaction or point load $P$ from each stud on the wale can be calculated conservatively as

$$P = 1.15\, w\, l$$

where $w$ = applied load per foot on the stud (plf)
$l$ = spacing of the wales (ft)

The resulting load on the wale from each stud becomes

$$P = 1.15 \times 935 \times \frac{24}{12} = 2{,}151 \text{ lb}$$

If the studs and ties are arranged on the wale as shown in Figure 6-17, with the stud spacing $a$ = 9/12 ft, the equivalent uniformly distributed load, $w_e$, on the wale can be calculated as

$$w_e = \frac{5P}{L}$$

where $w_e$ = equivalent uniformly distributed load on wale (plf)
$P$ = load on the wale from each stud (lb)
$L$ = total distance between ties (ft)

Based on five studs per length of wale and the ties positioned as shown in Figure 6-17, the resulting distance between ties can be calculated as

$$L = (4 \times a) + \frac{4}{5} \times a$$

where $a$ is the distance between studs in feet.

$$L = (4 \times 0.75) + \frac{4}{5} \times 0.75$$
$$= 3.6 \text{ ft}$$

The equivalent uniformly distributed load becomes

$$w_e = \frac{5 \times 2{,}151}{3.6}$$
$$= 2{,}988 \text{ plf}$$

From Table 6-9(a), two 3″ × 8″s ($F'_v$ = 175 psi) would be required for the wale based on the 3.6 ft (43 in.) between ties. As the resulting sections are relatively large, a closer spacing of the ties may be considered.

**(f)** Using an assumption that the wales will be continuous over three spans, the maximum load on the ties based on the 43-in. spacing can be calculated by

$$F_T = 1.15\, w_e \times L$$

where $F_T$ = tension load on tie (lbs)
$w_e$ = equivalent uniformly distributed load on wale (plf)
$L$ = distance between ties (ft)

The maximum load on the ties is calculated as

$$F_T = 1.15 \times 2{,}988 \times 3.6 = 12{,}370 \text{ lb}$$

---

***Example:***

A 6-m-high wall, 300 mm thick, is to be poured at the rate of 1.8 m per hour at a temperature of 15°C. (a) Calculate the maximum pressure on the formwork. (b) Using 18.5-mm thick plywood with the face grain parallel to the span, determine the maximum allowable spacing of the studs based on the maximum concrete pressure. (c) Assuming that 38 × 89 mm studs with the strong axis resisting the load are to be used, determine the allowable spacing of the wales. (d) If 38 × 140 mm studs are to be used, recalculate the spacing of the wales. (e) Based on your calculations in part (d), determine the size of wale that would be required using Table 6-4(b) assuming a wale spacing of 600 mm and a tie and stud arrangement as shown in Fig. 6-17. (f) Using the calculations in part (e), calculate the resulting load on the ties.

***Solution:***

**(a)** For walls poured at a rate of less than 2 m/hr, the maximum pressure is calculated as

$$p = 7.2 + \frac{785R}{T + 17.8}$$
$$p = 7.2 + \frac{785 \times 1.8}{15 + 17.8}$$
$$p = 50.3 \text{ kPa}$$

Check maximum pressure: the lesser of 96 kPa or 23.5 × 6 = 141 kPa. Use 50.3 kPa as the maximum pressure.

**(b)** Based on plywood sheathing continuous over four or more supports [Table 6-4(b)], with the face grain parallel to the span, and an $F'_b$ = 13.3 MPa, 18.5 mm thick plywood can support a maximum pressure of

62.24 kPa at a span of 225 mm. Use 200 mm as a convenient spacing for the studs.

**(c)** The maximum load per lineal meter on the studs based on plywood is calculated as

$$50.3 \times \frac{200}{1,000} = 10.1 \text{ kN/m}$$

Using Table 6-6(b), 38 × 89 studs supported at 525 mm o.c. can resist a load of 10.2 kN/m. It appears that 400 mm spacing of the wales would be appropriate for the studs (stud material $F'_b$ = 8.8 MPa or greater).

**(d)** If 38 × 140 studs were to be used, the distance between wales could be increased to 825 mm (Table 6-6 (b)). A practical spacing of 800 mm could be used (stud material $F'_b$ = 7.7 MPa or greater).

**(e)** Using the convenient spacing of 600 mm for the wales, the maximum load from the studs on the wales can be calculated as

$$P = 1.15\, w\, l$$

based on four-span continuous studs [Figure 6-14(c)]

where $w$ = applied load per meter on the stud (kN/m)
$l$ = spacing of the wales (m)

The maximum concentrated load on the wale from each stud becomes

$$P = 1.15 \times 10.1 \times \frac{600}{1,000}$$

$$= 7 \text{ kN}$$

Using the arrangement for studs and ties as shown in Figure 6-17, the equivalent uniformly distributed load, $w_e$ on the wale from the studs can be calculated by

$$W_e = \frac{5P}{L}$$

where $P$ = load due to the stud on the wale (kN)
$L$ = distance between ties (m)

Using the arrangement of ties and studs as shown in Figure 6-17, the distance between ties can be calculated as

$$L = (4 \times a) + \frac{4}{5} \times a$$

where $a$ is the distance between studs. If the studs are spaced 200 mm o.c., the distance between ties can be calculated as

$$L = (4 \times 200) + \frac{4}{5} \times 200$$

$$= 960 \text{ mm}$$

For a spacing of 960 mm between ties, the resulting equivalent distributed load becomes

$$w_e = \frac{5 \times 7}{960/1,000} = 36.5 \text{ kN/m}$$

Based on a tie spacing of 960 mm, using Table 6-9(b), the size of wale will be two 89 × 140s, using material having an $F'_v$ = 0.97 MPa. These are relatively large sections and a smaller wale spacing and tie spacing could be considered.

**(f)** Assuming that the wales will be continuous over four or more supports, the resulting load on the tie can be calculated as

$$F_T = 1.15\, w_e \times L$$

where $F_T$ = tension load on tie (kN)
$w_e$ = equivalent uniformly distributed load on wale (kN/m)
$L$ = distance between ties (m)

$$F_T = 1.15 \times 36.5 \times \frac{960}{1,000}$$

$$= 40.3 \text{ kN}$$

---

### *Example:*

Plywood ⅝-in. thick is to be used as a form for a concrete floor pour. If the total construction load is anticipated to be 200 psf, determine (a) the required spacing of the beam supports under the plywood. Assume that the plywood face grain is parallel to the span and the plywood is continuous over four or more supports. (b) If 2″ × 4″ sections are to act as supporting beams for the plywood, what maximum spacing of supports will be allowed if the 2″ × 4″s are simply supported?

### *Solution:*

**(a)** From Table 6-4(a), assuming an allowable bending stress of 1930 psi, maximum spacing for ⅝-in. plywood supporting 200 psf is 17 in. Use 16-in. spacing for supporting beams.

**(b)** Assume that the material to be used has a maximum allowable horizontal shear stress $F'_v$ = 140 psi.

Load per foot on joists = load in psf × joist spacing

$$= 200 \times \frac{16}{12} = 267 \text{ plf}$$

From Table 6-7(a), for simply supported sections (2″ × 4″ in size), maximum spacing of supports by interpolation is 30 in.

---

### *Example:*

A 6.1-m-high wall 300 mm thick is to be poured at the rate of 2.4 m/hr at a temp of 16°C. (a) Calculate the maximum pressure on the formwork. (b) Using 18.5-mm plywood with the face grain parallel to the span, determine the maximum allowable spacing of the wales based on the maximum concrete pressure. (c) If the wales are to be double 38 × 89s, what maximum spacing of ties will be allowed if the wales are considered to be continuous over four or more supports? (d) What load will be placed on each tie as a result?

***Solution:***

**(a)** For walls poured at a rate greater than 2 m/hr, the maximum pressure is calculated as

$$p = 7.2 + \frac{1{,}156}{T + 17.8} + \frac{244R}{T + 17.8}$$

$$= 7.2 + \frac{1{,}156}{16 + 17.8} + \frac{244 \times 2.4}{16 + 17.8}$$

$$= 7.2 + 34.20 + 17.33$$

$$= 58.73 \text{ kPa}$$

Check maximum pressure: the lesser of 95.8 kPa and 23.5 × 6.1 = 143.35 kPa. Therefore use 58.73 kPa as maximum pressure.

**(b)** Based on 18.5-mm-thick plywood continuous over four or more supports [Table 6-4(b)], with face grain parallel to the span and an $F'_b$ = 13.3 MPa, the maximum spacing of supports or wales is 225 mm on center.

**(c)** Load per lineal meter on the wales is

$$58.73 \times \frac{225}{1{,}000} = 13.21 \text{ kN/m}$$

From Table 6-9(b) for double wales continuous over three or more spans with an allowable shear stress of 0.97 MPa, the maximum spacing of ties is about 700 mm; use 600 mm.

**(d)** The resulting load on each tie is 13.21 × 600/1,000 = 7.93 kN.

---

The size and spacing of ties should be based on the manufacturer's recommendations or on calculations based on the load per lineal foot on the wales. For example, if the load per lineal foot is found to be 3,000 lb and the ties are spaced 2 ft o.c., the load on each tie will be 3,000 × 2 = 6,000 lb. Based on an allowable tensile stress of 20,000 lb/sq in. for the tie material, the required cross-sectional area of this tie would have to be at least

$$6000/20{,}000 = 0.30 \text{ sq in.}$$

This would require a tie with a diameter of 5/8 in. having an area of 0.31 sq in.

The calculations are very similar in metric units. If the load per lineal metre is found to be 13.5 kN and the ties are spaced 600 mm o.c., the load on each tie will be 13.5 × 0.6 = 8.1 kN. Based on an allowable tensile stress of 135 MPa for the tie material, the required cross-sectional area of this tie would have to be at least

$$8.1 \times 1{,}000/135 = 60 \text{ mm}^2$$

This would require a tie with a diameter of 9 mm with an area of 64 mm$^2$.

---

***Example:***

Plywood, 15.5-mm thick, is to be used as form sheathing for a concrete floor pour. If the total construction load is anticipated to be 9.6 kPa, determine (a) the required spacing of joist supports under the plywood. Assume that the plywood face grain is parallel to the span and the plywood is continuous over four or more supports. (b) If 38 × 140 sections are to act as the supporting joists, what maximum distance between supports can be used for the joists if they are assumed to be single span?

***Solution:***

**(a)** From Table 6-4(b), assuming an allowable bending stress of 13.3 MPa, maximum spacing for 15.5-mm-thick plywood supporting 9.6 kPa is 425 mm. Use 400-mm spacing for the supporting joists.

**(b)** Assume that the material to be used has a maximum allowable bending stress $F'_b$ = 7.7 MPa and an allowable horizontal shear stress of 0.97 MPa.

$$\text{Load per meter on joists} = \text{load in kPa (kN/m}^2\text{)} \times \text{joist spacing (m)}$$
$$= 9.6 \times 400/1{,}000 = 3.84 \text{ kN/m}$$

From Table 6-7(b), for single span sections (38 × 140 in size), maximum span between supports by interpolation is 1400 mm.

---

***Example:***

A concrete slab 8 in. thick is to be supported by timber formwork. If an additional load of 60 psf is allowed for men and equipment, determine a reasonable selection of materials that could be used as formwork to support the given loads.

Determine the resulting loads on the shores, and select the required size of shore assuming a 9-ft height. If the allowable soil bearing capacity is 1,200 psf, calculate the area of the sill required under the shores.

***Solution:***

In a situation such as this, plywood sheathing supported by joists would be used to support the plastic concrete. The spacing of the joists will depend on the thickness of the plywood and the load that must be supported. The first step in the selection of the various components of the formwork is to calculate the anticipated loads.

Load due to plastic concrete (assuming normal density concrete)

$$150 \times 8/12 = 100 \text{ psf}$$

Weight of formwork and construction equipment assumed as 60 psf. Total load to be supported by formwork is

$$100 + 60 = 160 \text{ psf}$$

Using ¾-in. plywood with face grain parallel to the span, and assuming short term loading, from Table 6-4(a), for a distance of 20 in. between supports, the plywood can support 175 psf. Based on this distance between joists, the load in pounds per linear foot on each joist is

$$160 \times 20/12 = 267 \text{ plf}$$

If the joists come in 8 ft lengths, and assuming supporting beams are to be placed at 7 ft on center, using Table 6-7(a), for a span of 84 in. and a load per foot of 267.5 lb, 3″ × 8″ joists would be required. Because the 20-in. spacing makes it awkward to place the 4′ × 8′ plywood sheets, 16-in. spacing would be more appropriate. Based on this spacing, the resulting load per foot on the joists will be

$$160 \times 16/12 = 213 \text{ plf}$$

This closer spacing allows the use of 2″ × 10″ joists [see Figure 6-18(a)].

With the supporting beams spaced 7 ft o.c., the resulting load per lineal foot on a typical interior beam can be calculated as

$$160 \times 7 = 1{,}120 \text{ plf}$$

Assuming shores were placed every 4 ft under the beams, from Table 6-8(a), 3″ × 10″ or 4″ × 8″ sections would be required. Let us use the 4″ × 8″ section for this example (see Figure 6-19(a)). The resulting load on each shore would be

$$1{,}120 \times 4 = 4{,}480 \text{ lb}$$

From Table 6-12(a), using an allowable compression stress parallel to the grain of 1,100 psi, and an unsupported length of 9 ft for the shore, a 4 × 4 S4S section can support 4,900 lb.

The required area of the sill under each shore is based on the bearing capacity of the soil.

$$\text{Required sill area} = \frac{\text{Shore load}}{\text{Soil-bearing capacity}} = \frac{4{,}480}{1{,}200} = 3.73 \text{ sq ft}$$

Use 24″ × 24″ sills.

The bearing capacity of the 4″ × 8″ beam depends on its compressive strength perpendicular to the grain. From Table 6-13(a), based on an assumed bearing stress of 400 psi for the beam, a maximum load of 4,900 lb can be supported on a 4″ × 4″ S4S shore post. This is greater than the 4,480 lb required.

### *Example:*

A concrete slab 200 mm thick is to be supported by timber formwork. If an additional load of 2.9 kPa is allowed for the weight of the formwork, the crew, and equipment, determine a reasonable selection of materials that could be used as formwork to support the given loads.

Determine the resulting loads on the shores, and select the required size of shore assuming a 2.7-m height. If the allowable soil-bearing capacity is 55 kPa, calculate the required area of the sill under the shores.

### *Solution:*

As in the previous example, plywood sheathing supported by sawn lumber joists will be used to support the plastic concrete. The spacing of the joists will depend on the thickness of the plywood and the load that must be supported. Again, the first step in the selection of the various components of the formwork is to calculate the anticipated loads.

Load due to plastic concrete (assuming normal density concrete at 2,400 kg/m$^3$):

$$2{,}400 \times 9.81/1{,}000 = 23.5 \text{ kN/m}^3$$

(converting the mass of concrete to a force exerted by a cubic meter of concrete).

For a 200-mm thick slab, resulting force per square meter is

$$23.5 \times 200/1{,}000 = 4.7 \text{ kN/m}^2$$

(which is equivalent to 4.7 kPa).

Weight of formwork and construction equipment is assumed as 2.9 kPa. Total load to be supported by formwork is

$$4.7 + 2.9 = 7.6 \text{ kPa}$$

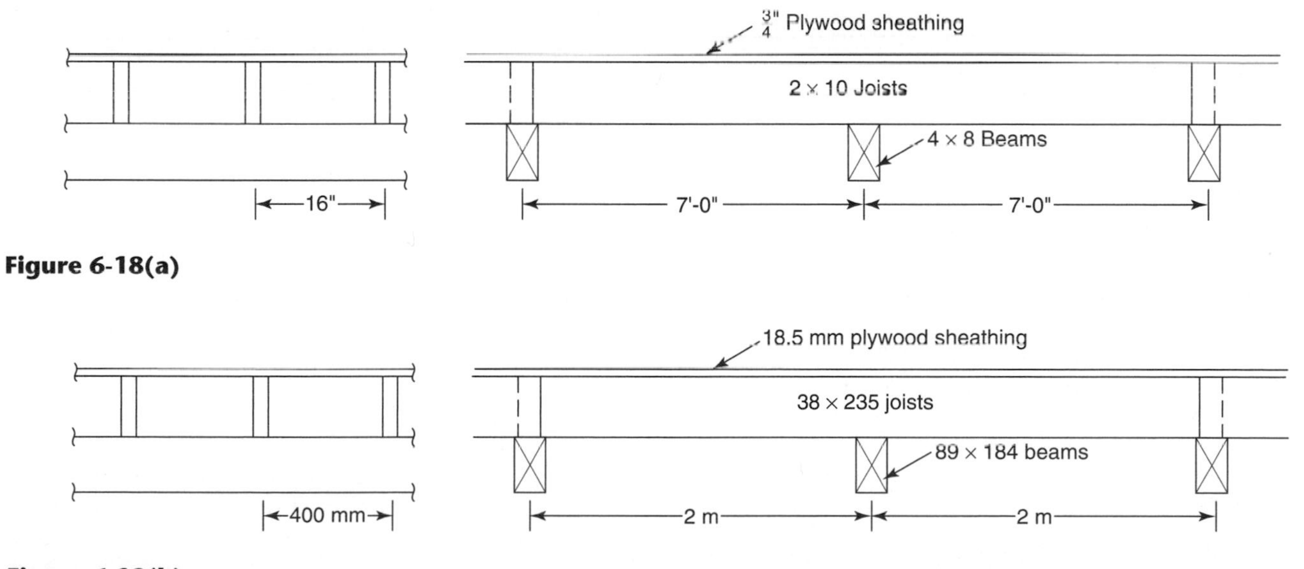


**Figure 6-18(a)**

**Figure 6-18(b)**

Using 18.5-mm plywood with face grain parallel to the span, and assuming short-term loading, from Table 6-4(b), for a distance of 500 mm between supports, the plywood can support 8.38 kPa. For joists spaced 500 mm on center, the load in kilonewtons per linear meter on each joist is

$$7.6 \times 500/1{,}000 = 3.8 \text{ kN/m}$$

If the joists come in 2,400-mm lengths, and assuming supporting beams are to be placed at 2,000 mm on center, using Table 6-7(b), for a span of 2,000 mm and a load per metre of 3.8 kN, 89 × 140 joists or 38 × 235 joists would be required. Because the 500 mm spacing makes it awkward to place the 1,200 mm × 2,400 mm plywood sheets, 400 mm spacing between joists would be more appropriate. Based on this spacing, the resulting load per metre on the joists will be

$$7.6 \times 400/1{,}000 = 3.04 \text{ kN/m}$$

Although the load is smaller at this closer spacing, 38 × 235 joists are still required, as illustrated in Figure 6-18(b).

With the supporting beams spaced 2 m o.c., the resulting load per lineal metre on a typical interior beam can be calculated as

$$7.6 \times 2 = 15.2 \text{ kN/m}$$

Assuming shores were placed every 1,200 mm under the beams (see Figure 6-19(b)), from Table 6-8(b), 64 × 235 or 89 × 184 sections would be required. In this case, select the 89 × 184 section.

The resulting load on each shore would be

$$15.2 \times 1{,}200/1{,}000 = 18.24 \text{ kN}$$

From Table 6-12(b), using an allowable compression stress parallel to the grain of 7.59 MPa, and an unsupported length of 2.7 m for the shore, an 89 × 89 S4S section can support 21.8 kN.

The required area of the sill under each shore is based on the bearing capacity of the soil.

$$\text{Required sill area} = \frac{\text{Shore load}}{\text{Soil-bearing capacity}} = \frac{18.24}{55} = 0.331 \text{ m}^2$$

Use 600 × 600 mm sills.

The bearing capacity of the 89 × 184 mm beam depends on its compressive strength perpendicular to the grain.

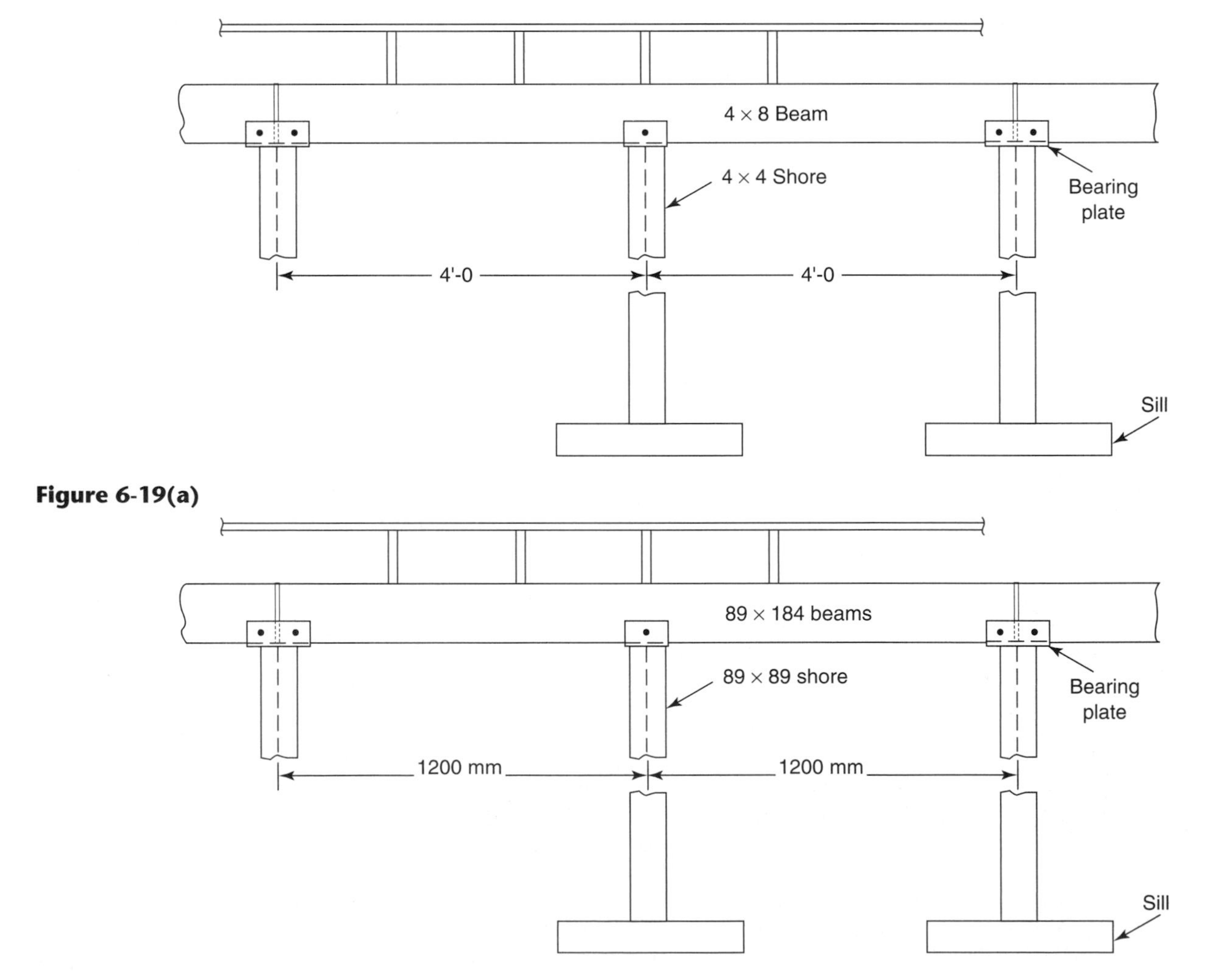


**Figure 6-19(a)**

**Figure 6-19(b)**

From Table 6-13(b), based on a bearing stress of 2.76 MPa for the beam, a maximum load of 21.8 kN can be tolerated when placed on an 89 × 89 S4S post. This is greater than the 18.24 kN required.

If the bearing capacity was less than the applied load, a steel plate would be provided to increase the bearing area so that the applied load could be supported without crushing the beam.

## FOOTING FORMS

Usually, two factors are of prime importance in considering the construction of footings. One is that the concrete must be up to specified strength and the other is that the footings be positioned according to plan. The appearance is seldom of importance as footings are usually below grade.

Old or used material may be employed for footing forms, provided it is sound. A certain amount of tolerance is allowed in footing size and thickness, but reinforcing bars and dowels must be placed as specified. Concrete is sometimes cast against the excavation, but care must be taken that this does not give inferior results, caused by the earth absorbing water from the concrete or by pieces of earth falling into it. Trenches may be lined with wax paper or polyethylene film to prevent this. It is sometimes desirable to form the top 4 in. (100 mm) of a footing cast in earth, as shown in Figure 6-20.

In cases where wall footings are shallow, lateral pressure is small and the forms are simple structures, as illustrated in Figure 6-21. When the soil is firm, the form can be held in place by stakes and braces. If the soil will not hold stakes, the forms may be secured by bracing them against the excavation sides, as shown in Figure 6-22.

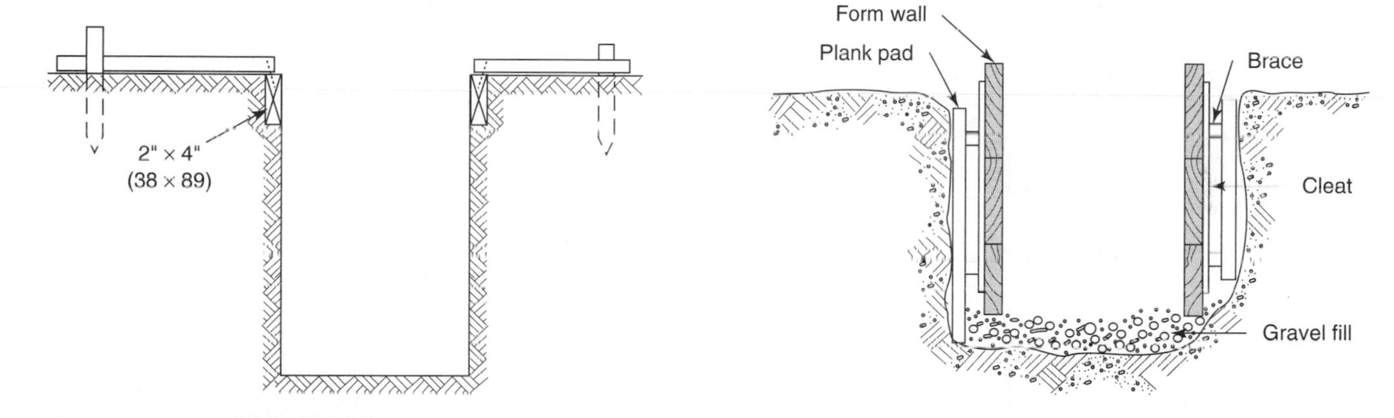


**Figure 6-20** Forming Top of Trench.

**Figure 6-22** Form Braced Against Excavation.

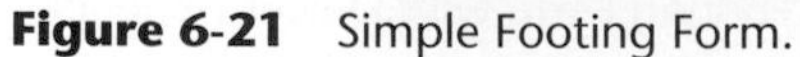

**Figure 6-21** Simple Footing Form.

For deeper wall footings and grade beams (see Figure 6-23), the formwork must be more elaborate and is constructed in much the same manner as wall forms (see next section on wall forms).

When a wall footing is supported by sloping ground, the footing may be stepped longitudinally at intervals; the frequency of the steps depends on the slope. Figure 6-24 illustrates a method of forming such steps.

Simple, rectangular column footing forms may be made as bottomless boxes, constructed in four pieces: two sides and two end sections. Two sections are made the exact dimensions of the footing and the two opposite ones long enough to take care of the thickness of the material and the vertical cleats used for holding the sections together (see Figure 6-25). Ties are used to prevent the sides from bulging under pressure. A template for positioning dowels in the footing can be attached to the form box, as shown in Figure 6-26, which also illustrates an alternative method of tying the form.

Stepped column footings are shown in Figure 6-27. The forms are made as a series of boxes stacked one on top of the other. The upper ones can be supported as shown in Figure 6-27, and if the difference in size between the upper and lower forms is considerable, the bottom one may require a cover. The whole assembly must be weighed or tied down to prevent uplift by freshly placed concrete.

Tapered footing forms are made like a *hopper*. Cleats hold the side and end sections together, and some convenient method must be used to hold the form in place. Two opposite panels are made to the exact dimensions of the footing face, whereas the other two must be enough longer to accommodate the thickness of the plywood and the width of the cleats used (see Figure 6-28). The angles required to cut the panels properly may be obtained as follows, using a framing square to lay them out (see the basic hopper layout in Figure 6-28):

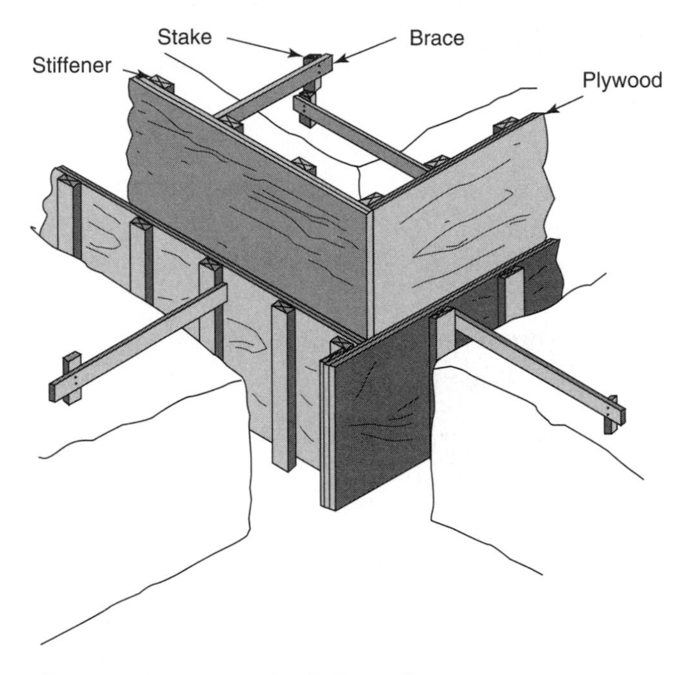


**Figure 6-23** Heavy Grade Beam Form.

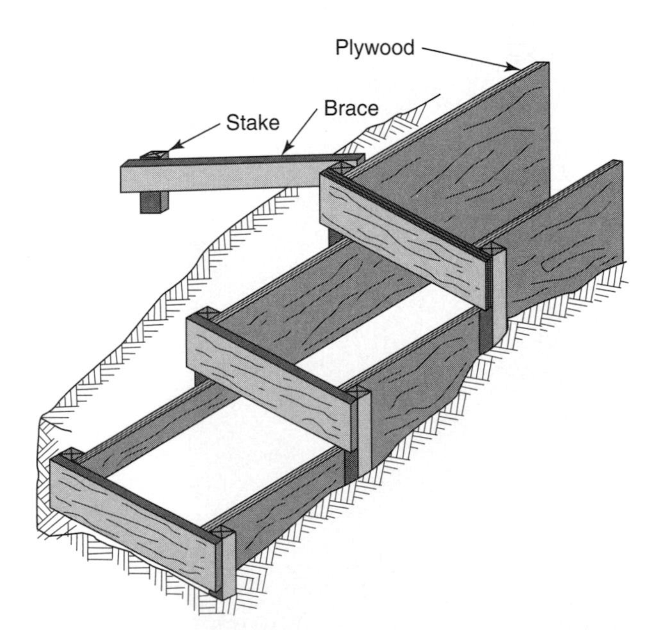


**Figure 6-24** Stepped Wall Footing.

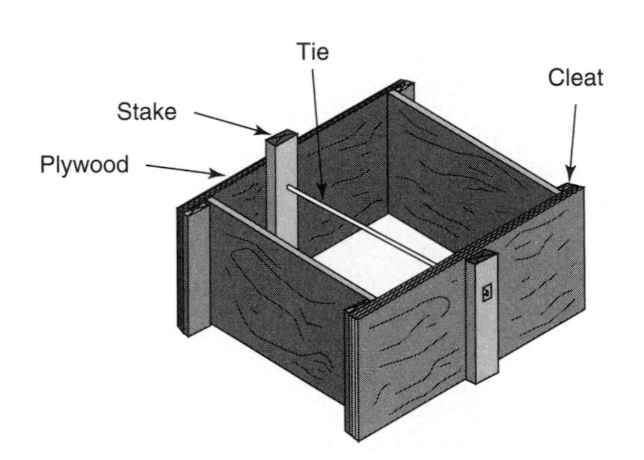


**Figure 6-25** Simple Column Footing Form.

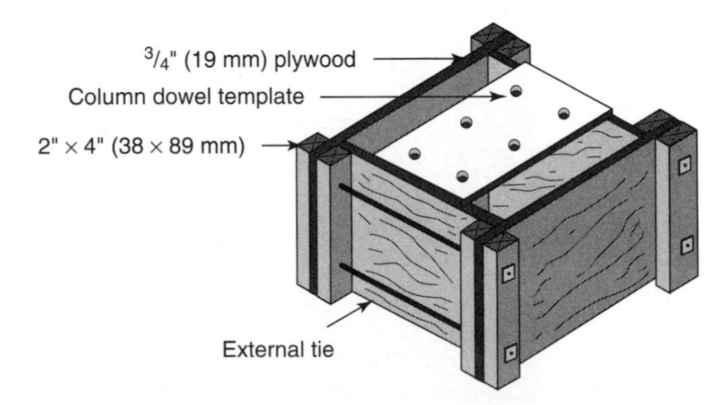


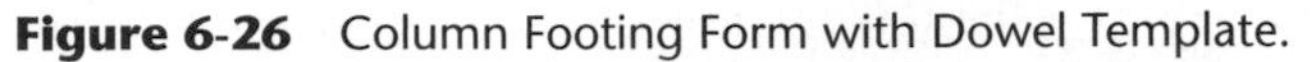

**Figure 6-26** Column Footing Form with Dowel Template.

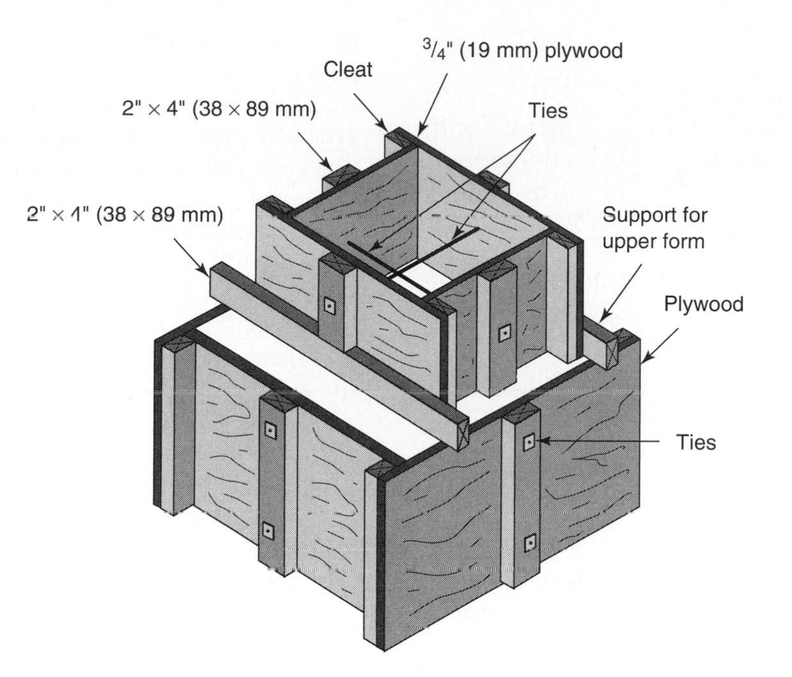

**Figure 6-27** Stepped Column Footing Form.

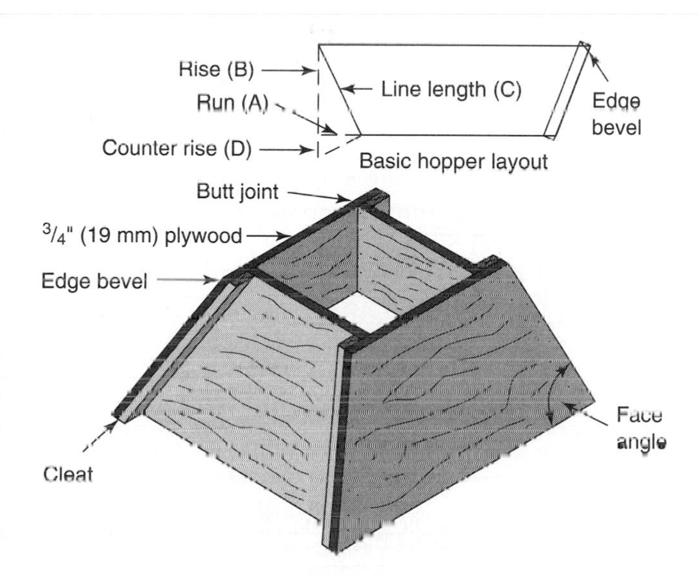

**Figure 6-28** Tapered Footing Form.

- *Butt joint:* Use C and D. Mark on D.
- *Face angle:* Use A and C. Mark on C.
- *Edge bevel:* Use B and A. Mark on A.

It is sometimes necessary to transfer part of the load of one footing to another or to support two columns on a single footing. The footing is then known as a *combined,* or *strap,* footing.

When two columns are supported on a single footing, the form is built the same way as a flat or stepped footing, except that it is longer. When part of the footing load is to be transferred to another pad, a strap or beam must be constructed between the two. When the footing and strap are cast monolithically, the form may be built as shown in Figure 6-29. Depending on the available height, it may not be necessary to set the strap or beam into the footing, as in Figure 6-29.

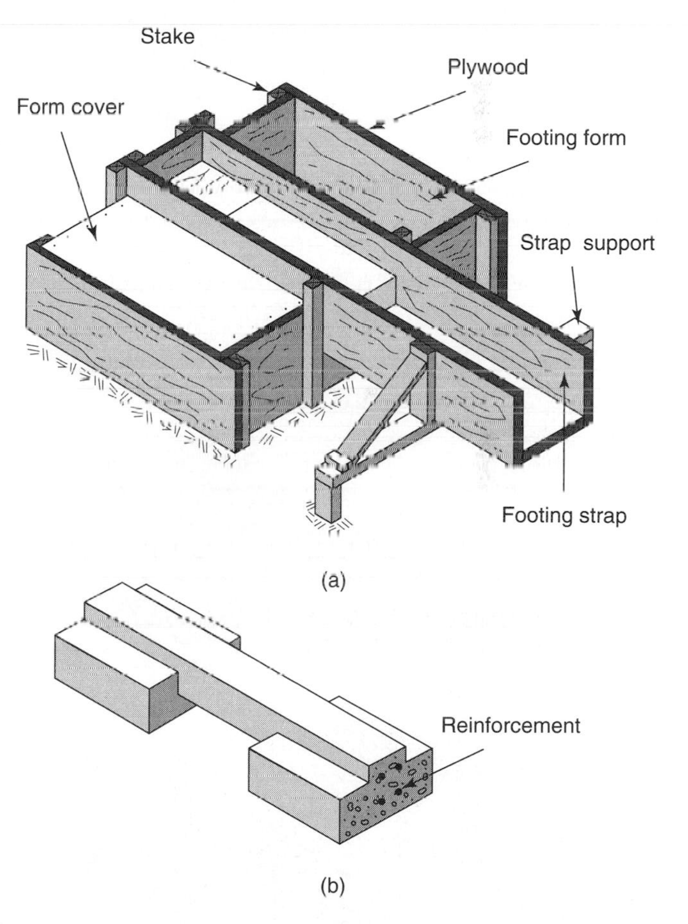


**Figure 6-29** Form for Strap Footing.

## WALL FORMS

Wall forms are made up of five basic parts: (1) *sheathing,* to shape and retain the concrete until it sets; (2) *studs,* to form a framework and support the sheathing; (3) *wales,* to keep the form aligned and support the studs; (4) *braces,* to hold the forms erect under lateral pressure; and (5) *ties* and *spreaders* or *tie–spreader units,* to hold the sides of the forms at the correct spacing (see Figure 6-30).

Wall forms may be built in place or made up of prefabricated panels, depending on the shape of the structure being formed and on whether the form being built can be reused.

### Built-in-Place Forms

Forms are built in place when the design of the structure is such that prefabricated panels cannot be adapted to the shape or when the form is for one use only and the use of prefab panels cannot be economically justified.

The first step in the erection of wall forms of any kind is to locate them properly on the foundation from which the wall will rise. This may be done by anchoring a *sole plate* for either the inside or outside form to the foundation with preset bolts, concrete nails, or power-driven studs. It must be set out from the proposed wall line by the thickness of the sheathing to be used and carefully aligned.

Another method of positioning the form is to anchor a *guide plate* to the foundation in such a position that the forms will be properly located when they are set behind it.

If the form is being built on a sole plate and studs are to be used, they are set on the plate, toe-nailed in place on the required centers, and held vertical by temporary bracing. Plywood sheathing panels are then nailed to the studs, with nails on 12- to 16-in. (300- to 400-mm) centers. The first panel should be set and leveled at the highest point of the foundation to establish alignment for the remainder.

Ties are inserted as sheathing progresses, with the tie holes being located so that the tie ends will meet the wales when they have been installed. Figure 6-31 illustrates some sample types of ties and Figure 6-32 some typical form tie installations.

When one side of the form has been completed, the other may be built in sections and set in place, with the tie ends being threaded through predrilled holes.

If the guide plate method of locating forms is used, sections of inside and outside forms can be built, tied together, and set in position, ready to be bolted or clamped to adjoining sections.

Studs are not always used in form building. Plywood sheets are nailed to a sole plate and lined up with single 2″ × 4″s (38 × 89s). The 2″ 4″s (38 × 89s) rest on brackets that are carried on the tie ends.

Wales are usually doubled for greater strength and spaced to allow the passage of tie ends between them. They may be made up of 2″ × 4″s (38 × 89s) or 2″ × 6″s (38 × 140s) with plywood spacers or, for greater stiffness, a double wale may be made from two standard channels, placed back to back and spaced with bars welded to their flanges (see Figure 6-33). When studs are used in form construction, wales are placed outside of them and held in place by nails, clips, or wale

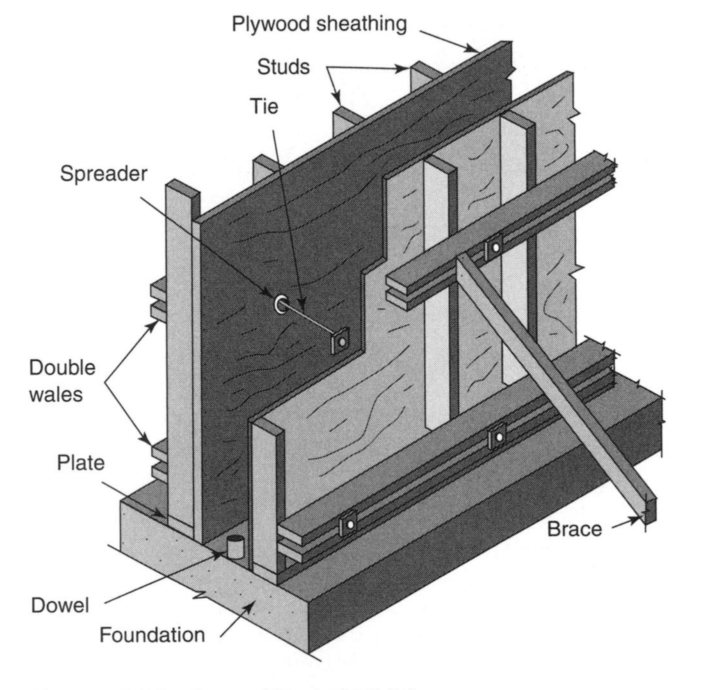


**Figure 6-30** Parts of Typical Wall Form.

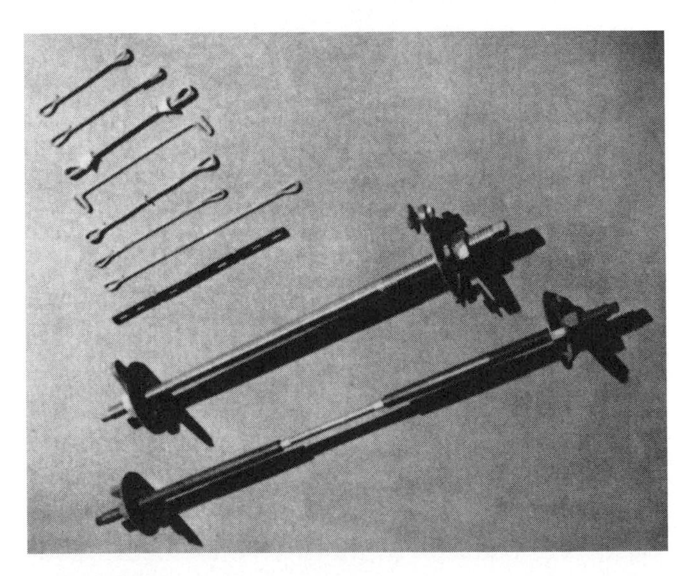

**Figure 6-31** Sample Form Ties.

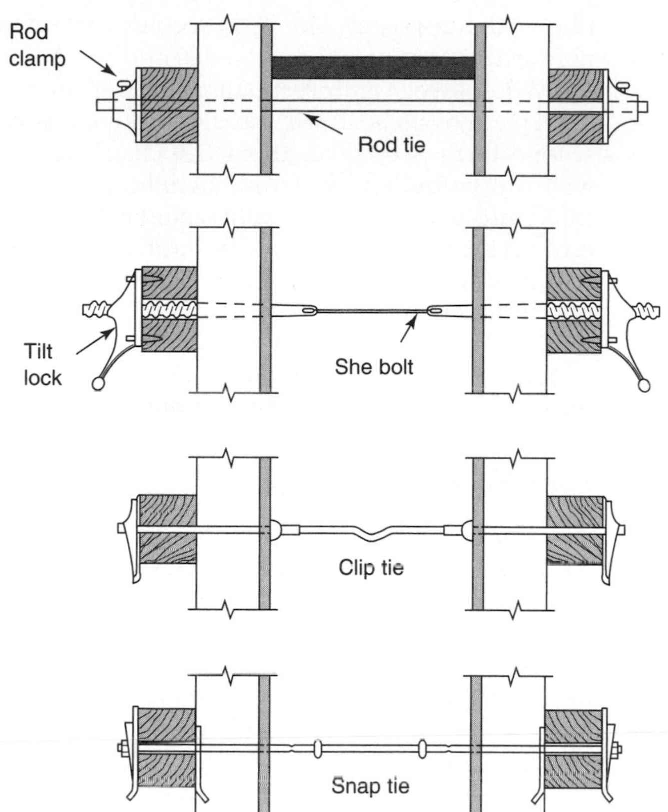


**Figure 6-32** Some Typical Form Tie Installations.

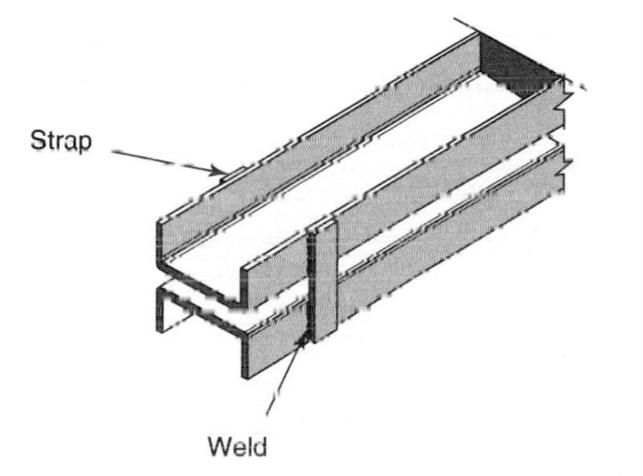


**Figure 6-33** Double-Channel Wale.

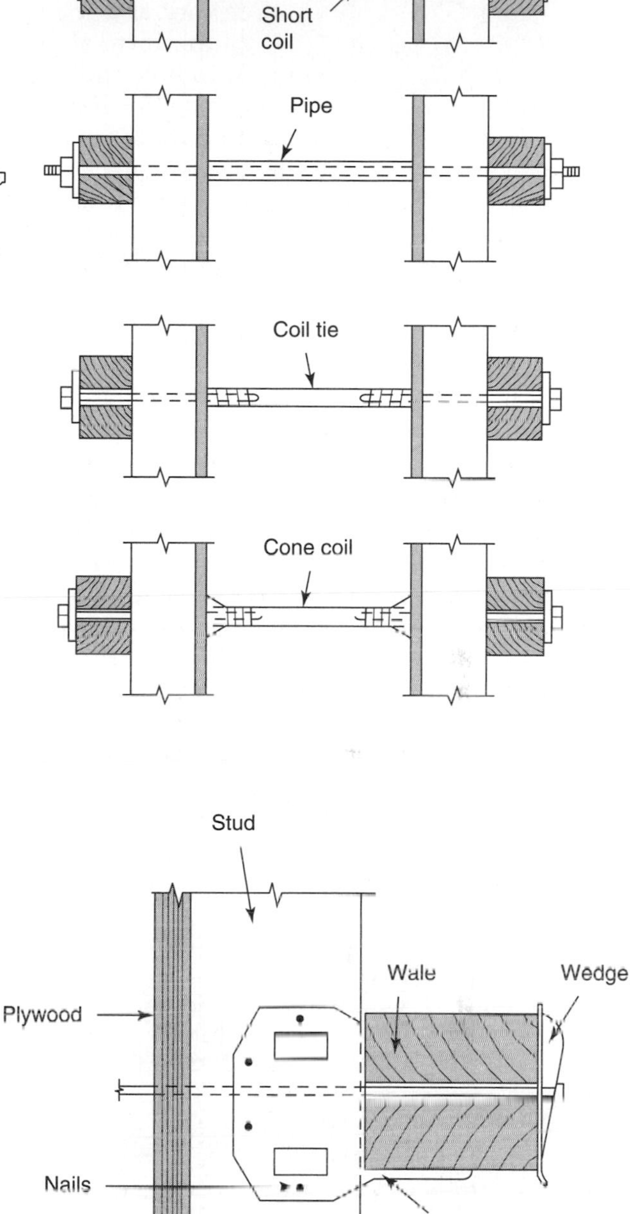


**Figure 6-34** Wale Bracket.

brackets nailed to the studs (see Figure 6-34). When there are no studs, wales are placed against the plywood sheathing. In such a case, *strongbacks*—vertical members tied together in pairs with long ties through the form—are set and braced to provide vertical rigidity.

Depending on the length, braces will consist of single or double 2″ × 4″s (38 × 89s) or 2″ × 6″s (38 × 140s), lengths of pipe, or small S or W shapes. They should extend from the wales or strongbacks to a solid support. That support may be in the form of a well-driven stake (see Figure 6-35), an adjustable brace end anchored to the floor (see Figure 6-36), or a rock anchor grouted into a rock face (see Figures 6-37 and 6-38).

Braces may act in compression only, in tension only, or in both, as when forms are braced on one side only. Heavy wire or cable is suitable for bracing that will be in tension only.

Plumbing is the final operation in building a wall form. Many bracing systems have some means

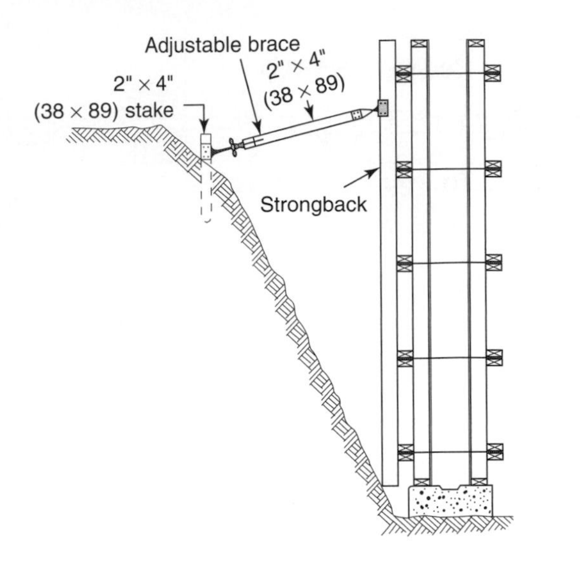


**Figure 6-35** Outside Adjustable Brace.

of adjusting the braces while the form is being plumbed (see Figures 6-35, 6-36, and 6-37). If braces are not adjustable, the wall must be plumbed as the braces are installed and anchored. If one form is plumbed as soon as it is built, there is no need to plumb the opposing one. The second form will be plumbed automatically by the ties and spreaders.

When steel reinforcement is placed in a wall form piece by piece, it will have to be done before the second form is erected. In such a case, the second wall can be built in a flat or inclined position and tilted into place after the reinforcing is complete. If reinforcing mats or cages are prefabricated and then set into a completed form, ties will have to be left out until this is done.

Finally, the ties are tightened. Where through-the-wall tying is impossible or not allowed, external bracing must be provided to securely support both forms.

## Prefabricated Forms

A majority of the formwork being erected is done with the use of prefabricated form panels. These can be reused many times and reduce the time and labor required for erecting forms on the site.

Many types of prefabricated form panels are in use. Contractors sometimes build their own panels from wood framing covered with plywood sheathing. A standard size is 2′ × 8′ (600 × 2,400 mm), but

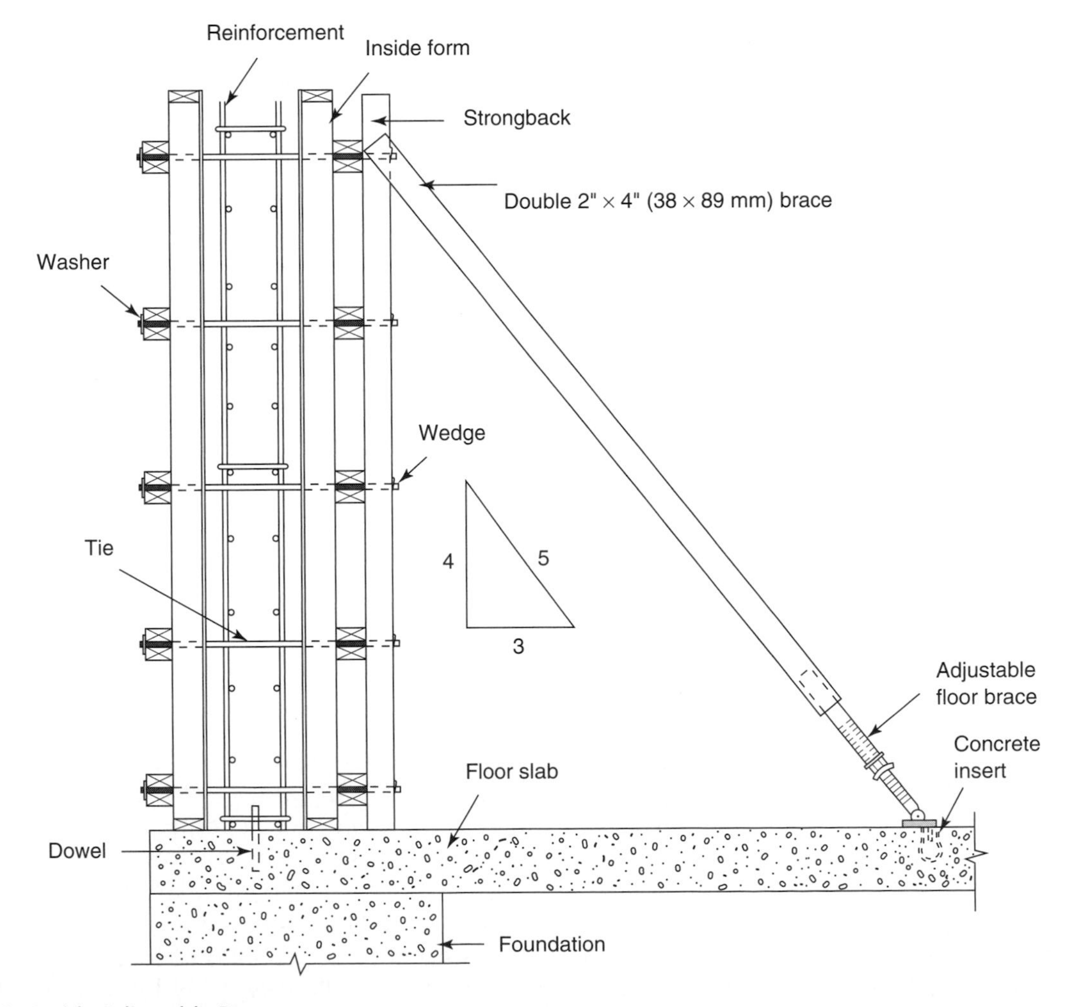


**Figure 6-36** Inside Adjustable Brace.

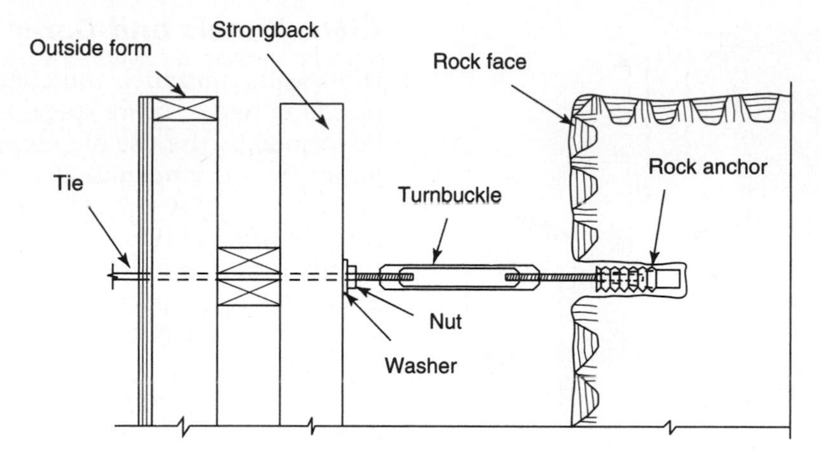


**Figure 6-37** Adjustable Brace to Rock Face.

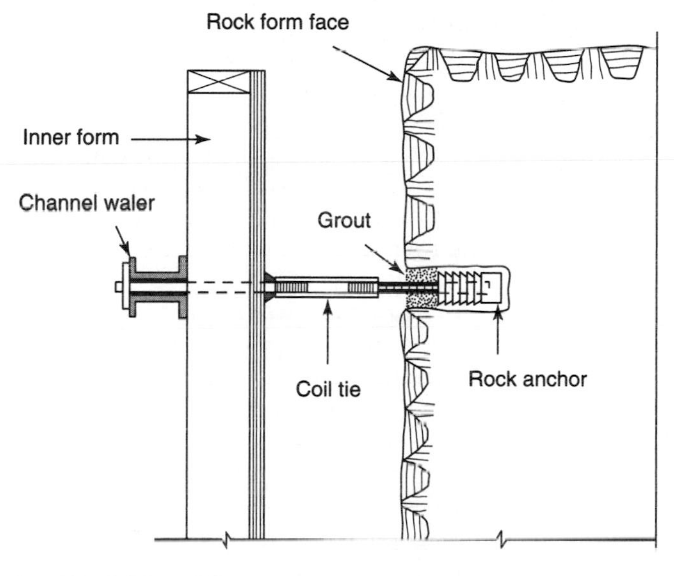


**Figure 6-38** Wall Formed with Single Form.

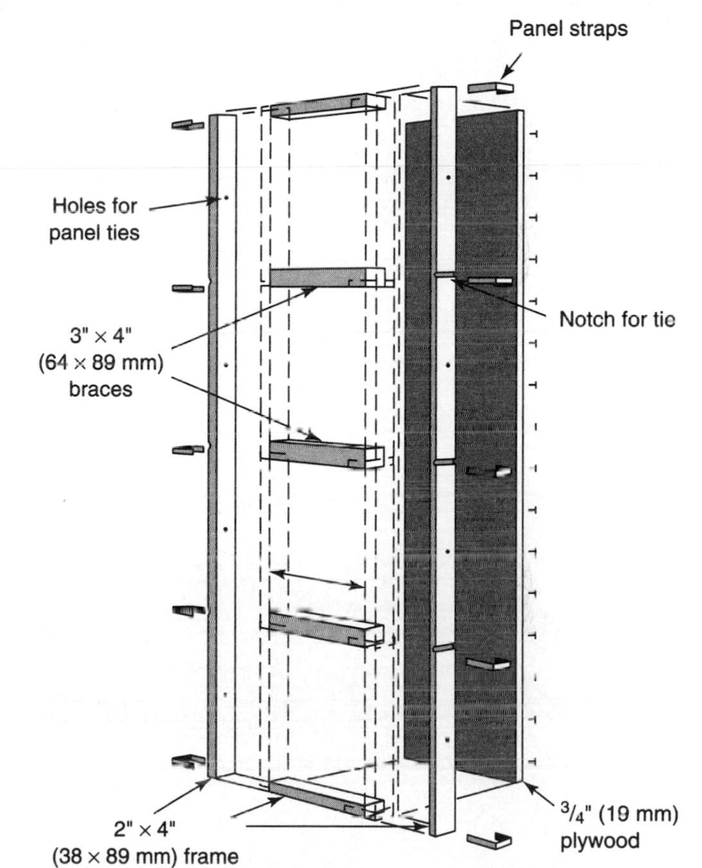


**Figure 6-39** Typical Prefabricated Wooden Form Panel.

panels can be sized to suit any particular situation. Figure 6-39 illustrates a typical 2′ × 8′ (600 × 2,400 mm) panel with metal straps at the corners. Note the grooves in the outside edges through which the ties pass. A simple bolt and wedge device for holding adjacent panels tightly together is shown in Figure 6-40.

Panels made with a metal frame and plywood sheathing are also in common use and are available in a variety of sizes. Special sections are produced to form inside corners, pilasters, and the like (see Figure 6-41). Panels are held together by patented panel clamps. Flat bar ties (which lock into place between panels) eliminate the need for spreaders. Forms are aligned by using one or more doubled rows of 2″ × 4″s (38 × 89s), secured to the forms by a special device attached to the bar ties (see Figure 6-42).

Form panels made completely of steel are also available. A standard size is 24″ × 48″ (600 × 1,200 mm), but various other sizes are also manufactured. Inside and outside corner sections are standard, and insert angles allow odd-sized panels to be made up as desired.

## Giant Panels and Gang Forms

High walls, in which the concrete will have to be placed in two or more stages or *lifts,* will normally be formed by the use of *giant panels* (panels much larger than the normal 2′ × 8′ (600 × 2,400 mm) or 4′ × 8′ (1,200 × 2,400 mm) wood panels or by *gang forming,* making up large panels by fastening together a number of steel-framed panels (see Figure 6-43). A special type of steel-framed panel, shown in Figure 6-44, is made specifically for gang forming.

These large forms are built or assembled on the ground and raised into place by crane. The basic differences between them and regular form panels are their *size,* the *extra bracing* that is required to withstand handling, and the fact that they require *lifting hardware.* Manufacturers of steel-framed panels also supply lifting brackets, which are attached to the top edge of the gang form, but wooden panels have to be fitted with lifting hardware, as illustrated in Figure 6-45.

These forms are built with removable ties so that when the first lift of concrete has been placed and has gained sufficient strength, the bolt ties are removed and the form is lifted to a new position at the top of the first lift. It is held in place by clamping with bolts through the top one or two rows of bolt holes left in the first lift (see Figure 6-46).

Special attention must be given to corners when forms are being erected. These are weak points because the continuity of sheathing and wales is broken. Forms must be pulled tightly

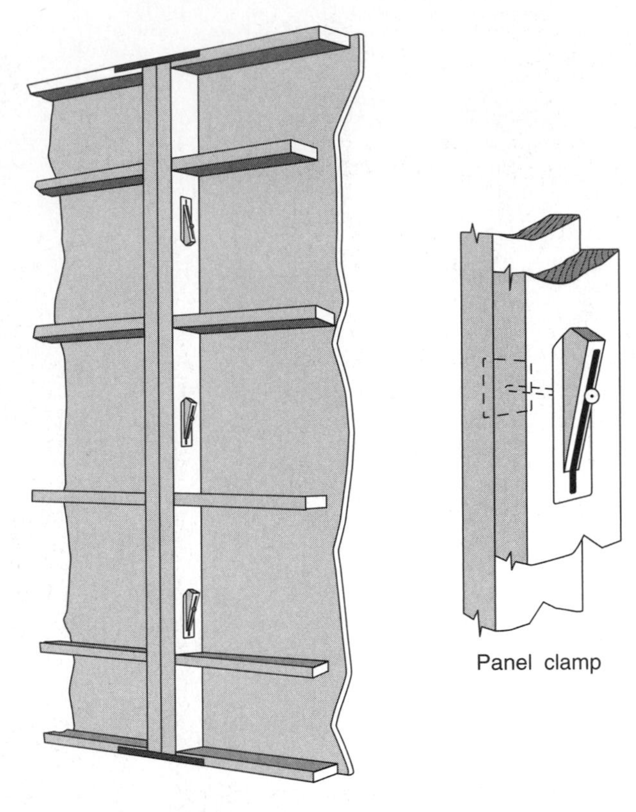


**Figure 6-40** Bolt-and-Wedge Panel Clamp.

**Figure 6-41** Typical Steel Frame Form Panels: (a) Straight Sections; (b) Corner Sections.

**Figure 6-42** Formwork Panels Supported by Double 2″ × 4″ (38 × 89 mm) Wales.

**Figure 6-43** Wall Form Section Made from Ganged Panels.

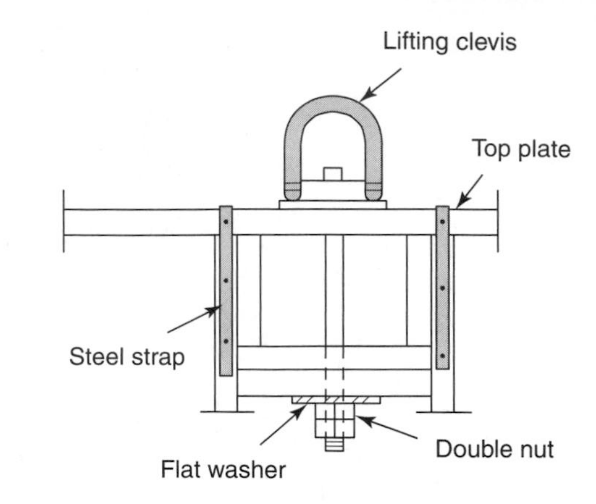


**Figure 6-45** Panel Lifting Hardware.

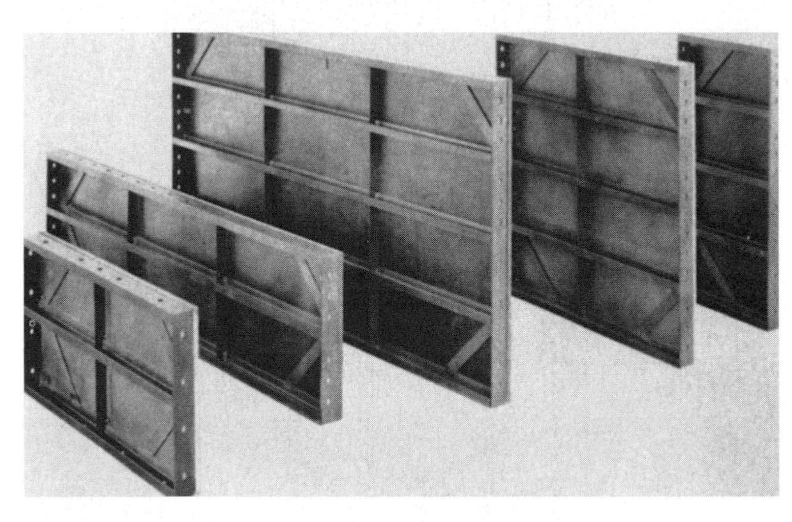

**Figure 6-44** Panel Sections Made for Gang Forming.

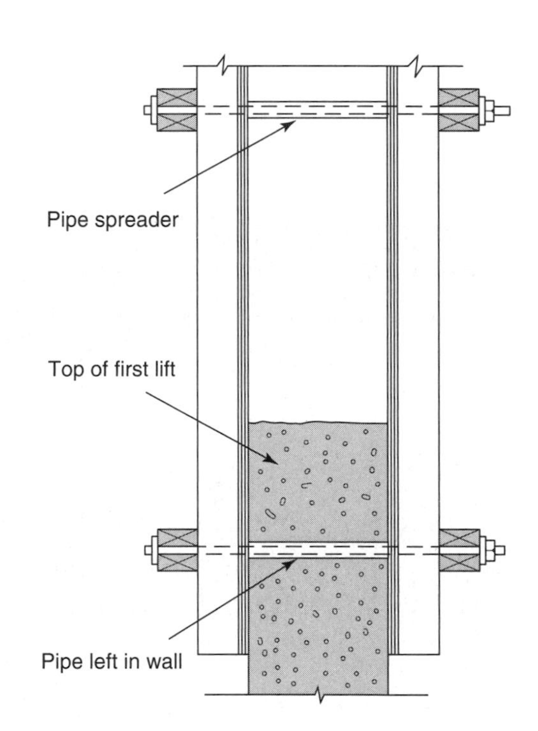


**Figure 6-46** Form Clamped to Top of First Lift.

together at these points to prevent leakage of concrete. One method of doing this is illustrated in Figure 6-47. Another system involves the use of corner brackets and tie rods, as shown in Figure 6-48. Vertical rods running through a series of metal eyes fastened to the sheathing panels are also used to secure forms at the corners (see Figure 6-49). Another way of making a tight corner is to overlap the wales at the corner and to provide a vertical kick strip for each group of wale ends. Wedges behind each wale tighten the corner.

Openings in concrete walls are made by setting *bucks* (see Figure 6-50) into the form at the required position. Bucks are simply wooden or steel frames set between the inner and outer forms and held in place so that the forms can be tightened against them. Bucks should be braced horizontally, vertically, and diagonally so that concrete pressure will not distort their shape. In some cases, the window or door frame itself is set into the form. Care must be taken that proper anchorage is provided so that the frame will not move in the opening.

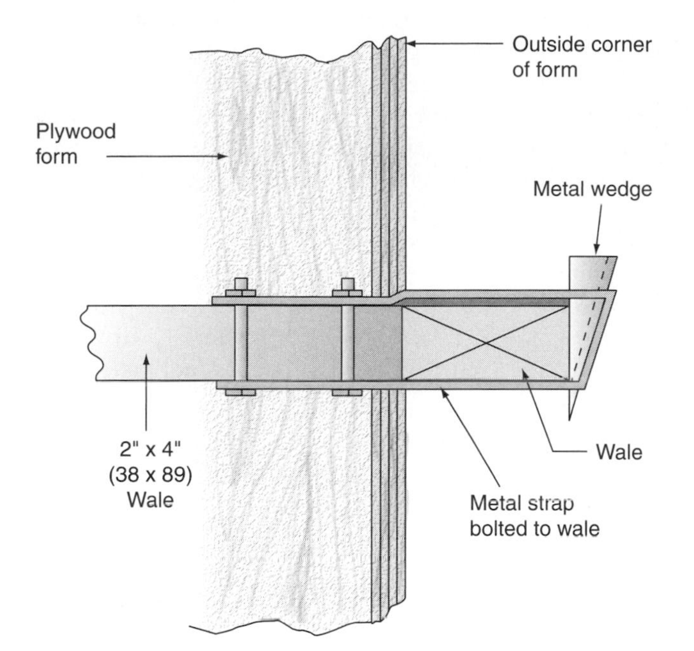


**Figure 6-47** Wedge Corner Clamps.

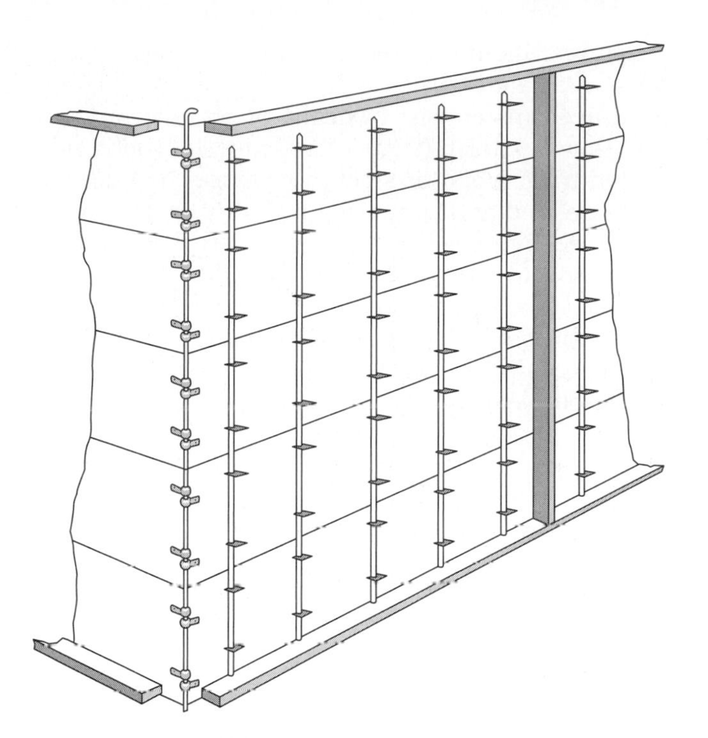

**Figure 6-49** Corner Rods and Eyes.

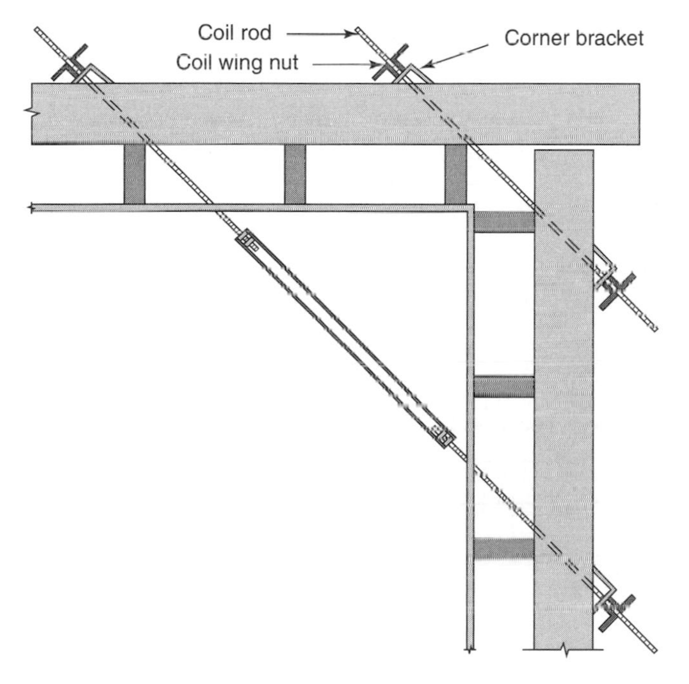


**Figure 6-48** Corner Brackets and Tie Rods.

**Figure 6-50** Simple Wood Buck.

## SLIP FORMS

*Slip forms* are a type of form that could be compared with a set of dies, originally designed for curved structures such as bins, silos, and towers for which the conventional forming systems were not suited. They are, however, increasingly used for a variety of structures, including rectangular buildings, bridge piers, and canal linings.

Slip forms consist basically of an inner and an outer form, 3 to 5 ft (900 to 1,500 mm) in height, fabricated from wood or steel to produce the building shape desired and supported by two strong, vertical *yokes* (see Figure 6-51). These yokes are tied together across the top to give the form sides the rigidity needed—to apply the pressure required—to produce reasonably smooth surfaces. The bottom ends of the form sheathing are slightly tapered to help make the forms self-clearing.

To provide a platform from which workers may look after the placing of concrete in the forms, the fabrication of the steel reinforcement, the extension of the jack rods, and the maintenance of the jacks, a working platform is attached to the inner form and rides upward with it. At the same time a finisher's scaffold is suspended from the outer form so that workers can finish the newly extruded concrete as it emerges.

Instead of remaining stationary as normal forms do, slip forms move continuously upward, drawn by *jacks* climbing on vertical steel *jack rods*.

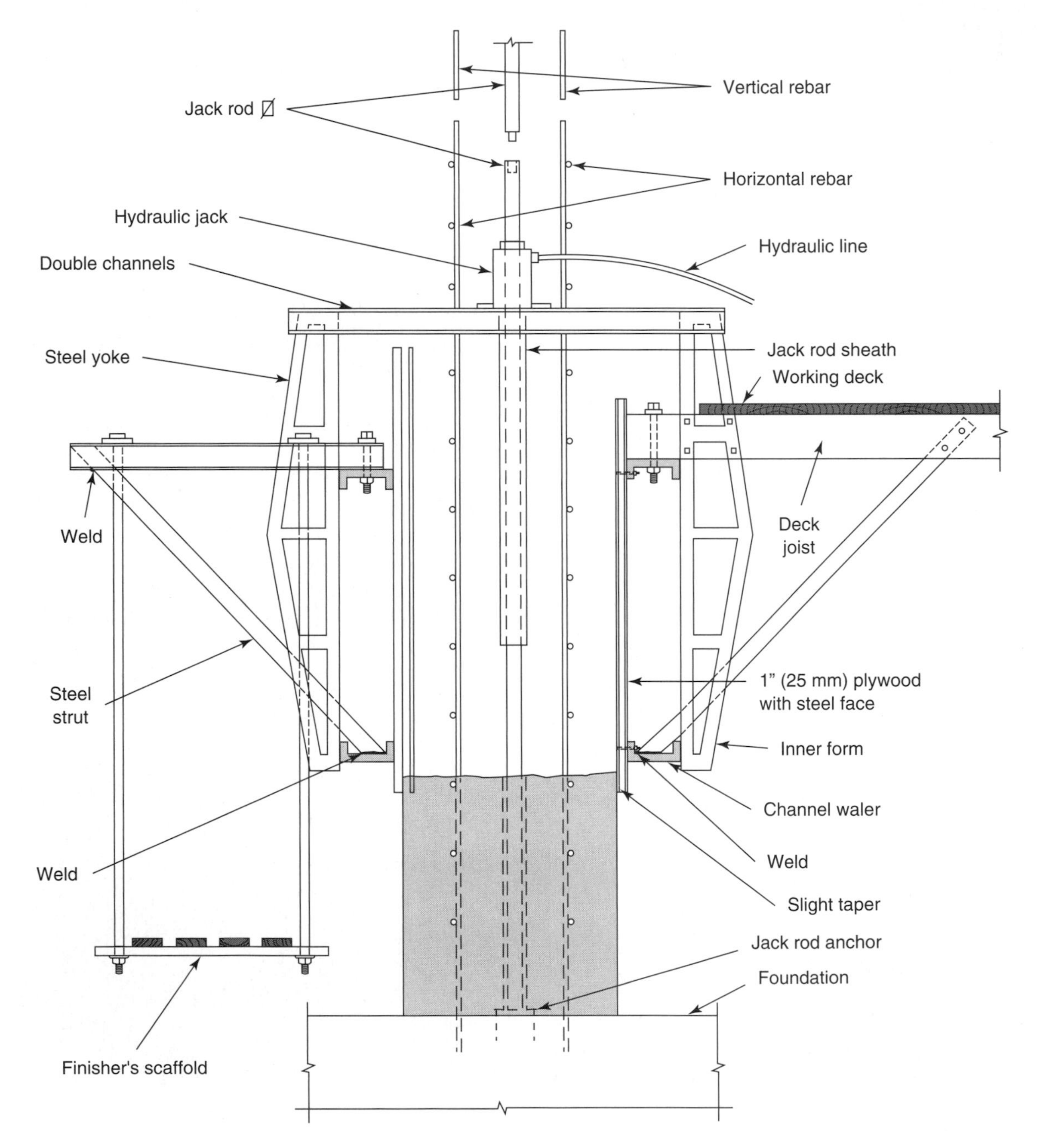


**Figure 6-51** Schematic Diagram of Slip Form.

These are anchored at the base of the structure and embedded in the concrete below the forms. The jacks may be hydraulic, electric, or pneumatic and are capable of producing form speeds of up to 20 in. (500 mm)/hr. If the jack rod is to be reused, it is withdrawn from the wall after the forming is complete. This is made possible by sheathing the rod with a thin pipe, which is attached at its top end to the jack base and moves up with the forms. The sheath prevents concrete from bonding to the jack rod and leaves it standing free within the hardened concrete. In some cases the rod is left unsheathed and remains as part of the reinforcing.

Concrete is placed into the forms at the top end, and, as they are drawn upward, they act as dies to shape the concrete so that it emerges from the bottom of the form shaped as planned and set enough to carry its own weight. Thus a continuous process is carried on, filling and moving the forms upward, often 24 hours a day until the structure is complete. However, the operation may be stopped and resumed later, resulting in the same kind of concrete joint that one would get between lifts in conventional form use.

Vertical reinforcing rods are anchored at the base of the structure and extend upward between the inner and outer form. As the form rises and reaches the top of the first set of rods, new lengths are added as concreting continues. Horizontal steel is attached to the vertical rods as work goes on (see Figure 6-51). In the same way, jack rods are extended as necessary.

Openings in the wall may be formed by introducing neatly fitting bucks into the form at the proper location. Beam pockets, key slots with dowels, and anchor slots for attaching masonry can also be placed in the same way. If a concrete projection from the wall is required, it must be added after the forming is complete. A pocket is formed in the wall with dowels bent in so as not to interfere with the operation of the forms. After the forming is complete, the dowels can be bent out, the forms for the projection built around them, and the structure cast.

The success of a slip-forming operation depends on good planning, design, and supervision so that the operation may, in fact, be as continuous as possible. Some of the major factors contributing to successful slip form construction are:

1. The proper concrete mix design and careful control of the concrete to maintain the proper slump and set, in spite of changing temperatures.
2. Adequate facilities for supplying concrete to the forms at any height and an adequate concrete supply.
3. A supply of reinforcing steel at hand and experienced workers to do the fabricating as work progresses.
4. Reliable forms, designed to stand the stresses placed on them by the constant heat of liquid concrete.
5. Supervisors who thoroughly understand the operation of slip forms.

## CONSTRUCTION JOINTS

It is often necessary to place long or high walls in sections. This results in a vertical joint (where two sections of a long wall meet) or in a horizontal joint (when high walls are placed in two or more lifts). Such joints are called *construction* joints and require special attention.

Vertical construction joints are formed by placing a *bulkhead* in the form. It may be one piece of material placed vertically in the form or a number of short pieces placed horizontally. The latter is preferable when horizontal reinforcing bars project beyond the bulkhead, because the pieces can be cut to fit around the bars. In Figure 6-52, the bulkhead is held in place by vertical strips nailed to the inside faces of the forms. Another method of securing the bulkhead is shown in Figure 6-53.

It is usually desirable to provide a key between adjoining sections to prevent lateral movement. One method of doing this is to attach a tapered strip to the inside face of the bulkhead, as shown in Figure 6-53. Another type of key consists of a 6-in. (150-mm) strip of corrugated copper or galvanized iron to bridge the joint. The strip is bent into a right angle and fastened to the inside face of the bulkhead, as

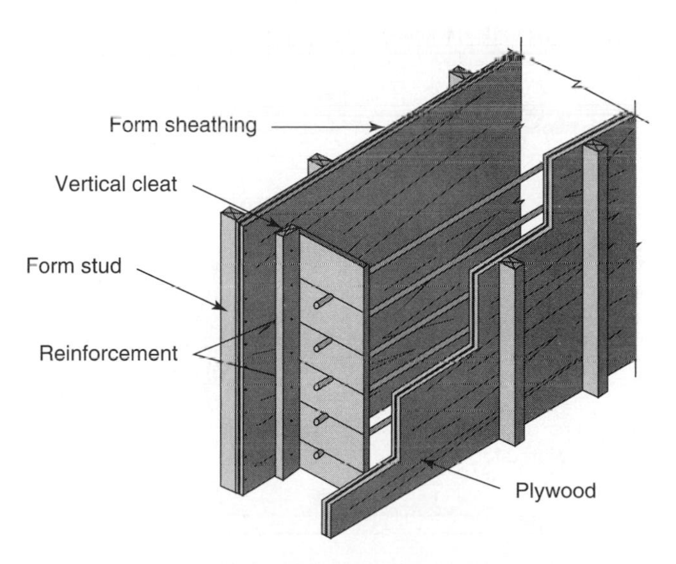


**Figure 6-52** Bulkhead Made with Horizontal Pieces.

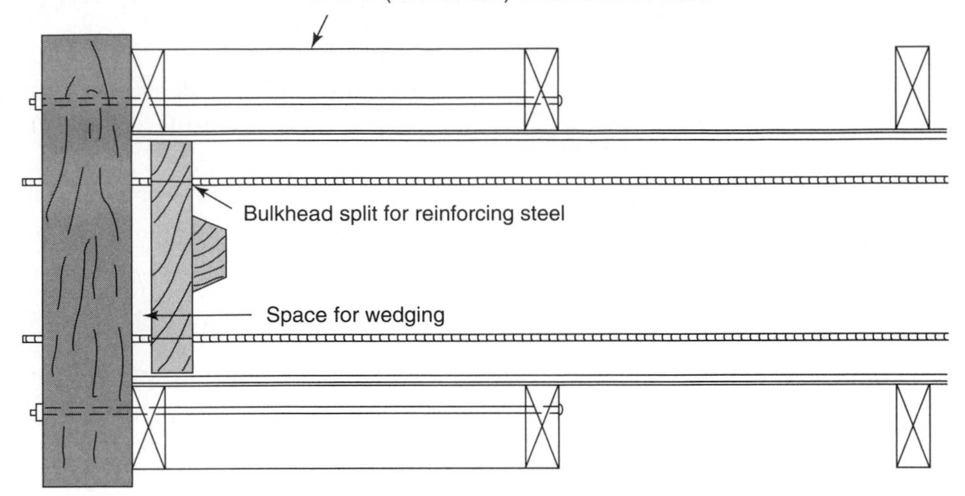


**Figure 6-53** One Method of Securing a Bulkhead.

shown in Figure 6-54. When the bulkhead is removed, the bent half is straightened and projects into the adjacent pour. The metal also acts as a *waterstop* to prevent passage of water through the joint.

Many types of waterstops are used, most of them made of rubber, neoprene, or a type of composition material. Two typical waterstops and their methods of installation are illustrated in Figures 6-55 and 6-56.

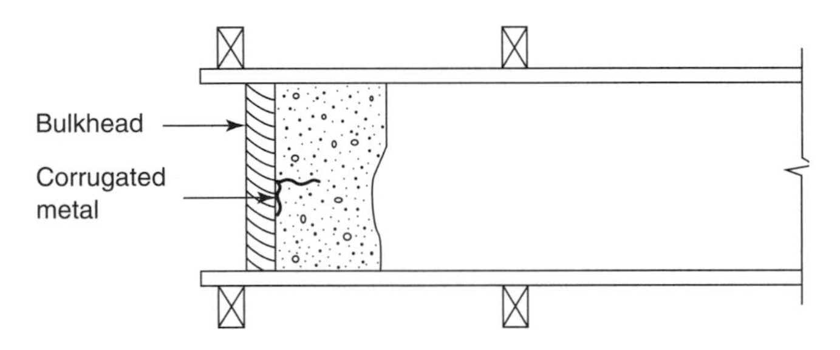

**Figure 6-54** Sheet Metal Key and Waterstop.

Two important factors are involved when making horizontal construction joints. One is the necessity of having the two concrete lifts bond at the joint, and the other, in certain cases, is the requirement that the joint be invisible or as inconspicuous as possible.

When the form for the first lift is built, a row of bolts is placed near the top of the lift. After this pour has set, the forms are stripped and raised as required for the next lift and supported on bolts placed in the top holes previously cast. A row of ties should be placed about 6 in. (150 mm) above the joint between the lifts to prevent leakage and help make the joint inconspicuous.

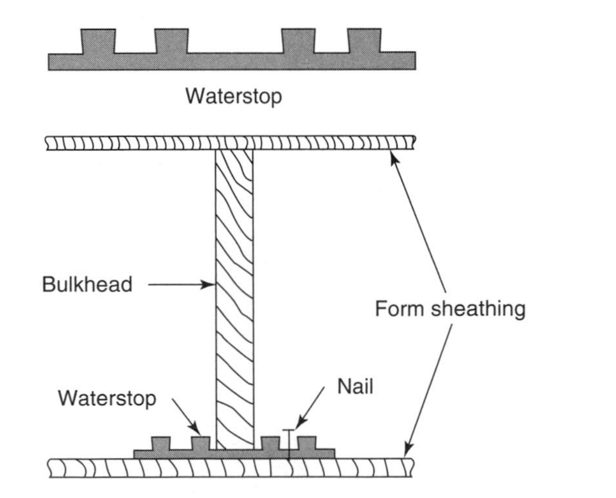

**Figure 6-55** Side Waterstop.

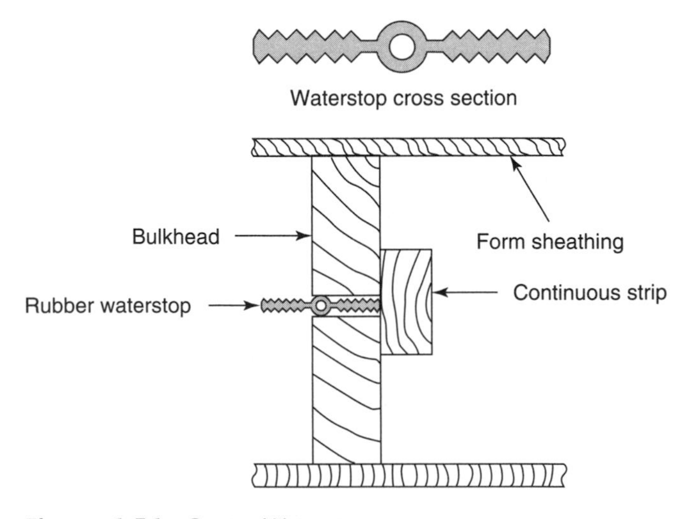

**Figure 6-56** Center Waterstop.

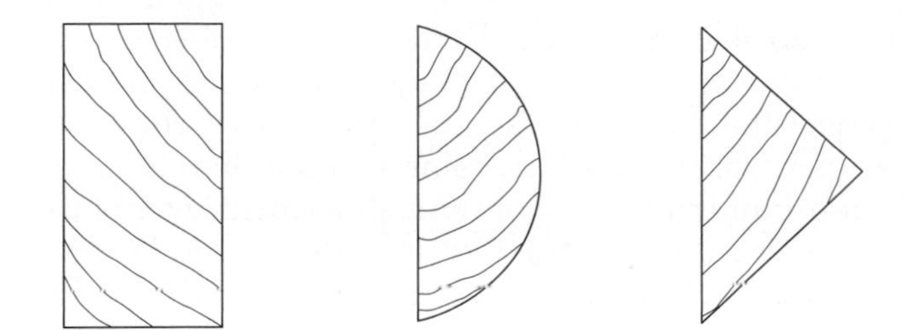

**Figure 6-57** Rustication Strips.

Hiding of horizontal construction joints is often accomplished by the use of *rustication strips.* These are narrow strips of various shapes (see Figure 6-57) nailed to the inside face of the form at the level of the joint. Concrete is placed to a height slightly above that required and then cut back to it.

## CONTROL JOINTS

Control joints in concrete are made to control cracking due to contraction. They are formed by fastening a beveled insert of wood, metal, rubber, or other material to the inside face of the form. The insert produces a groove in the concrete that will control surface cracking. After the concrete has set, the insert may be removed and the joint caulked. Rubber inserts may be left in place. Figure 6-58 illustrates the forming of control joints. When a waterstop is required at a control joint, a rubber stop like that shown in Figure 6-56 is often used.

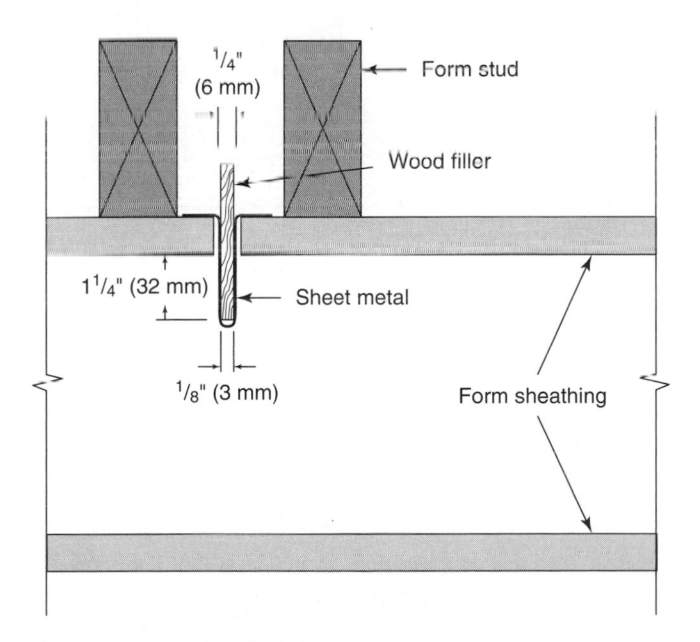

**Figure 6-58** Control Joint Form.

## COLUMN, GIRDER, AND BEAM FORMS

See Chapter 9.

## FLOOR FORMS

The design of forms for concrete floors depends a great deal on whether the floor is a *slab-on-grade* or a *structural slab* supported on a steel or concrete structural frame.

### Slab-on-Grade Forms

Forms for concrete slabs placed on grade are usually quite simple. Concrete is cast on a compacted earth or gravel base, and forms are required only for the edges. Plywood, planks, or steel edge forms are commonly used for this purpose. Wood forms may be held in place by wooden stakes or by form supports, such as the one illustrated in Figure 6-59(a). In the latter case, the subgrade stake is pushed firmly into the earth and the bolt bracket mounted on the form plank. The plank can then be adjusted for height by turning the bolt up or down. Steel edge forms (see Figure 6-60) are commonly used on larger jobs and for highway work. Provision is made for the wheels of mechanical strike-off bars and floats to ride on the top flange of steel edge forms.

To ensure a correct thickness of slab, the subgrade is normally leveled with a fine-graded granular material, compacted to its maximum density. The granular material also serves as a capillary break under the slab should moisture in the subgrade be a problem. Uniform compaction of the granular layer will minimize differential settlement in the concrete slab, and building specifications should be closely followed.

In the past, the use of vapor diffusion retarders (commonly called a vapor barrier) under all slabs on grade was considered to be good practice. Recent research into the effectiveness of these barriers has brought their use into question. Improper placing and vulnerability to puncturing during concrete pouring brings the use of vapor retarders under slabs into question and defeats their use in most instances. In some situations it has been found that these films cause stressing in the slab due to non-uniform drying and curing. It is recommended that for slabs on grade that are sealed with some type of covering such as vinyl tile, carpet, or wood, a vapor retarder should be installed. To ensure the anticipated service life of the slab on grade, each slab should be evaluated according to site conditions and its end use.

Reinforcement in slabs on grade may or may not be specified depending on the slab area and the use

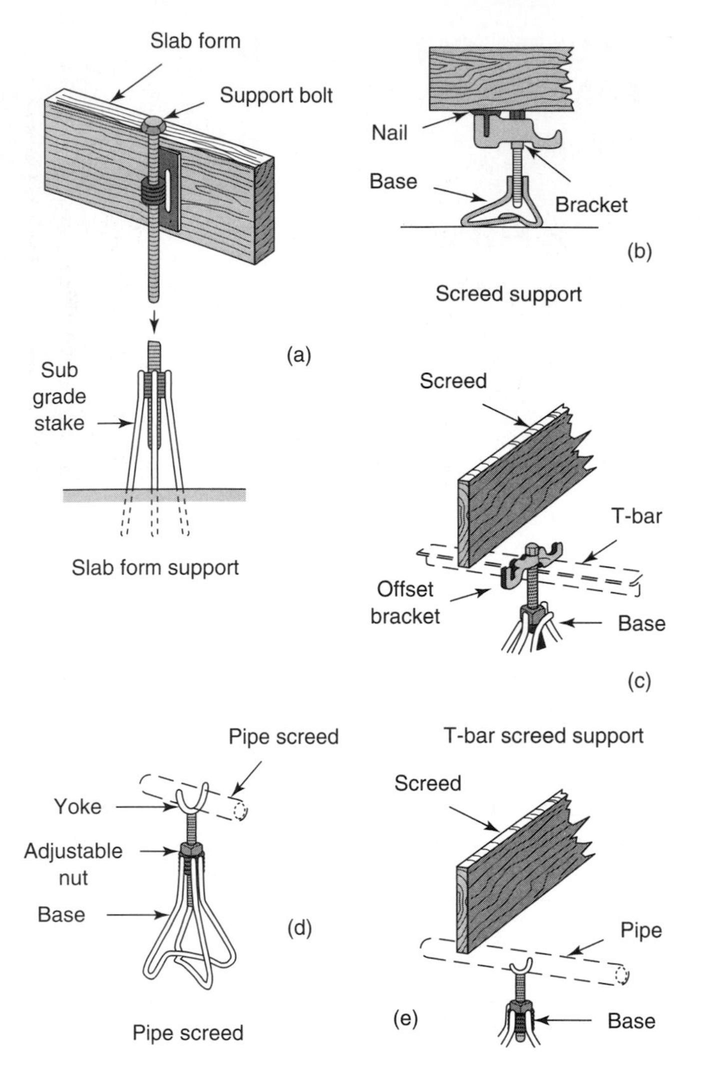


**Figure 6-59** Typical Slab Form and Screed Supports.

of control joints. Normally, the amount of steel that is specified for slabs on grade is for crack control rather than to increase the strength of the slab. Proper placement of the reinforcement in the slab is important if it is to be of value. If a single layer of reinforcement is specified, its location should be 2 in. (50 mm) below the top surface of the slab to help control cracking in the top of the slab.

Reinforcement in the form of reinforcing bars [see Figure 6-61(a)] or welded wire reinforcement, commonly called wire mesh, is used. The American Concrete Institute recommends a minimum of 0.2% steel to concrete ratio in each direction for normal applications. If this minimum is followed, reinforcing bars are usually preferred because of better placement and location control during the concrete pour. To ensure the proper location of the reinforcement above the base, *chairs, bolsters,* and *spacers* made of either metal or concrete can be used (see Figure 6-62). Bolsters do not space bars and can be used anywhere. Spacers are made to order, to provide a leg under each bar, whereas bar chairs may be used as required.

If strike-off bars cannot reach from one edge of a slab to the other, it will be necessary to set up *screeds* (temporary guides) in the slab area to be used to bring the concrete being placed to the correct grade. One method of doing this is to set a 2″ × 4″ (38 × 89) on edge, held in place by stakes, in the required position and level it to the same height as the edge forms (see Figure 6-63). Wood screeds may also be supported by *adjustable screed supports,* such as the one shown in Figure 6-59(b). Straight pipe may be used as a screed or small pipe

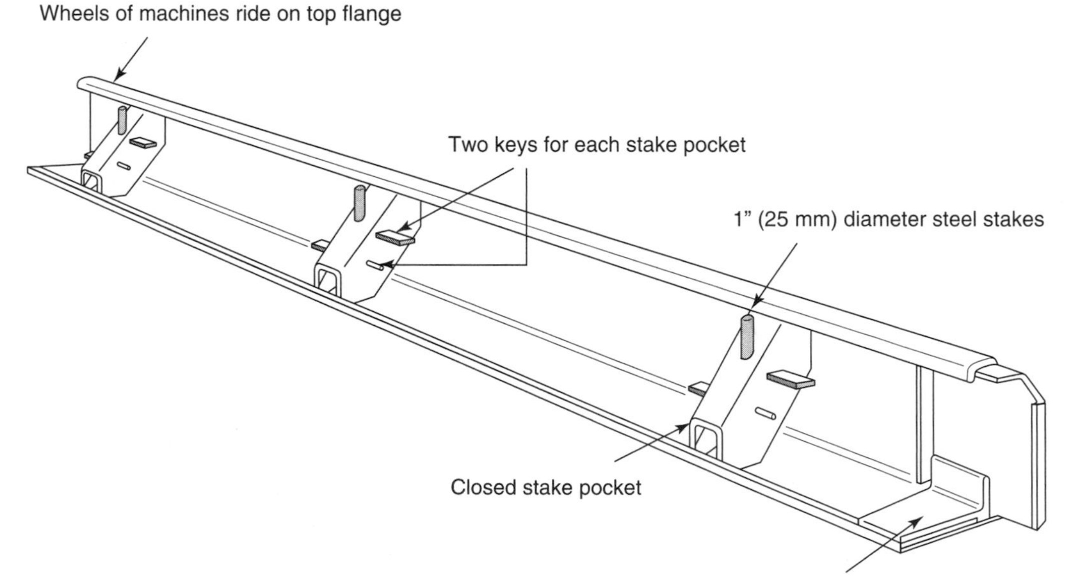


**Figure 6-60** Steel Slab Edge Form.

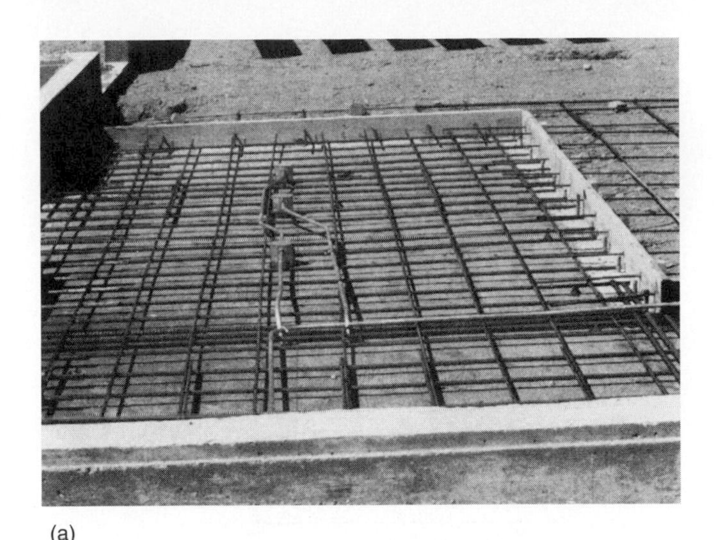

(a)

(b)

**Figure 6-61** (a) Reinforcing for Slab on Grade. (b) Welded Wire Reinforcing.

or T-bars on adjustable supports used to support wooden screeds. When concrete has been placed to the correct level, the screed is removed and the depression filled.

If the slab is to be placed in sections, *construction joints* must be made between them, which will transmit shear from one to the other. Forms for construction joints are illustrated in Figure 6-64.

Forms for *control joints,* which control the location of cracks that inevitably occur in a concrete slab as a result of the shrinking of the concrete, may be made in several ways. One common method is to insert a wood or metal strip into the slab, top, or

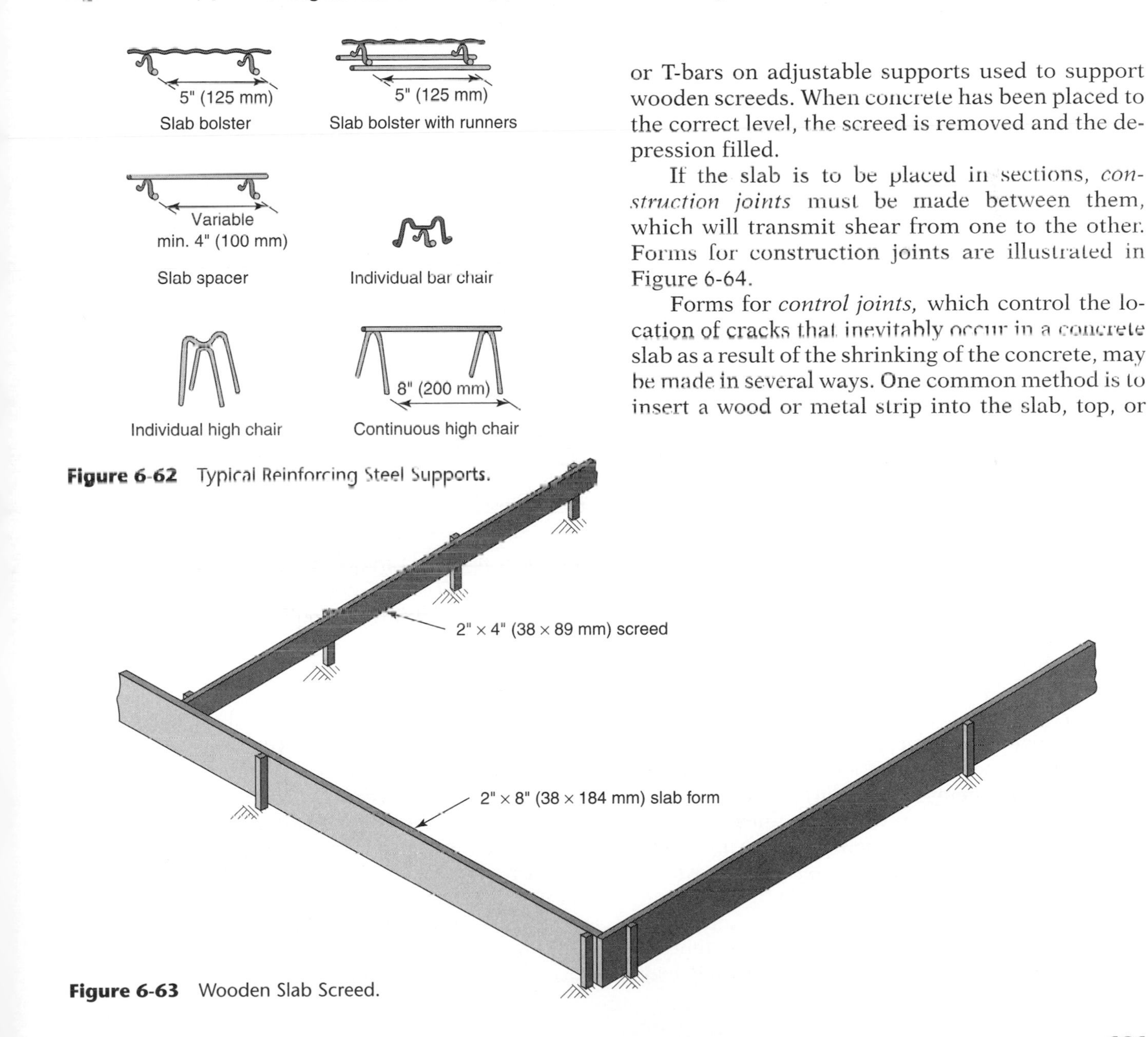


**Figure 6-62** Typical Reinforcing Steel Supports.

**Figure 6-63** Wooden Slab Screed.

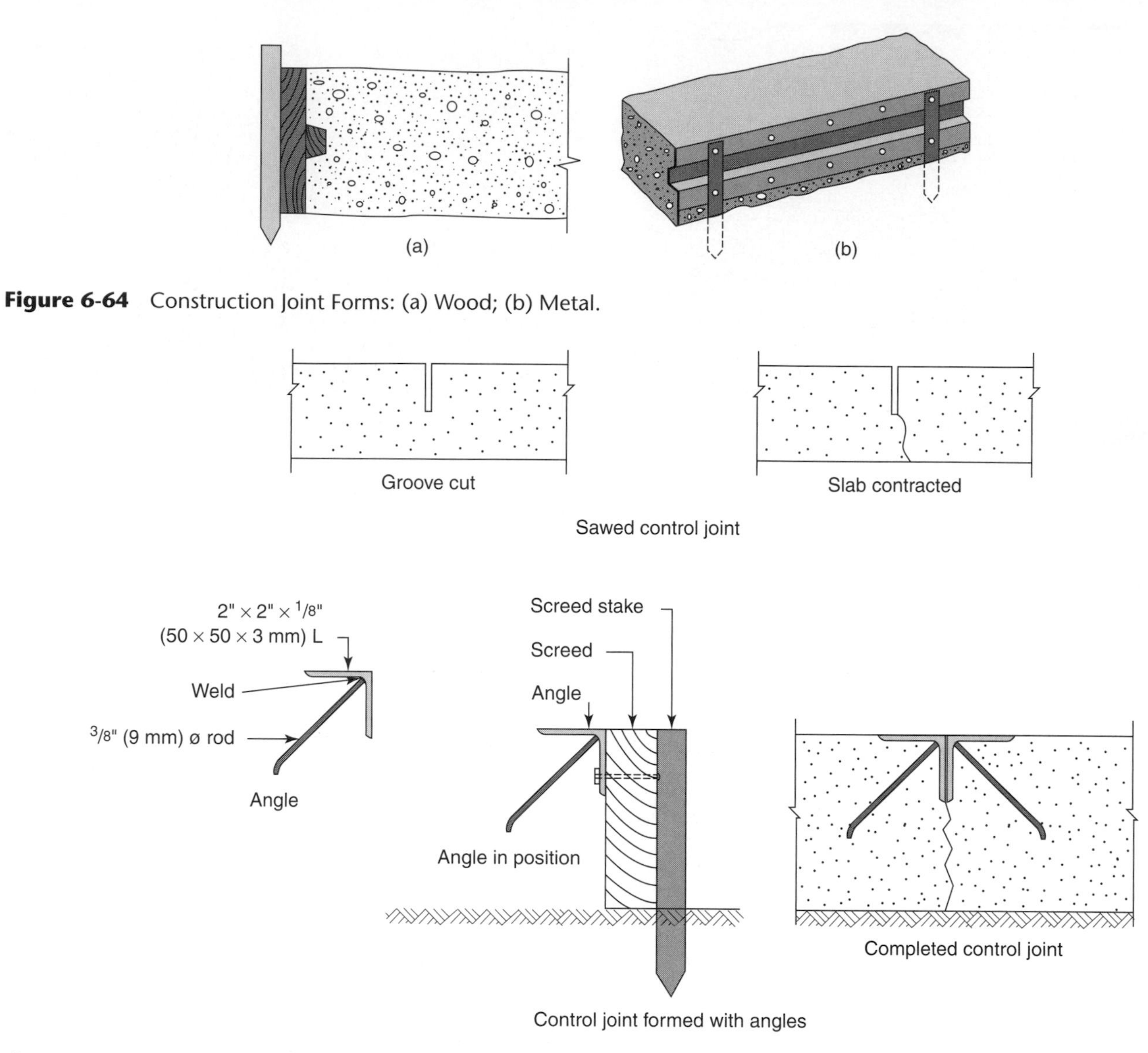


**Figure 6-64** Construction Joint Forms: (a) Wood; (b) Metal.

**Figure 6-65** Control Joints.

bottom to form a plane of weakness. Another method is to make a cut partway through the slab with a masonry blade, as shown in Figure 6-65. Still another method is to set two structural steel angles into the slab surface so that a plane of weakness is established below them.

*Expansion joints* are necessary in slab construction to provide space for the slab to expand, due to changes in temperature, without exerting damaging pressure on the member adjacent to it. They may be formed around exterior walls, columns, and machine bases by placing a tapered wood strip around the perimeter before concrete is placed, removing it after the concrete has set, and filling the void with some type of caulking material. The same result may be accomplished by using a strip of compressible material—fiberboard, for example—and leaving it in position. Joints similar to those shown in Figure 6-66(b) or (c) should be used in slabs where the expansion joint will be subjected to heavy traffic.

## Structural Slab Forms

Forms for concrete decks (floors above grade) supported by a structural frame will be as varied as the systems that support them. In some cases, such as a cast-in-place reinforced concrete frame, columns, beams, and slabs will be cast monolithically, and forms must be designed accordingly. In the case of a precast concrete frame, deck forms may be necessary, depending on the type of members used. For example, single or double tees provide the floor slab and the frame. A structural steel frame will require deck forms for the floors.

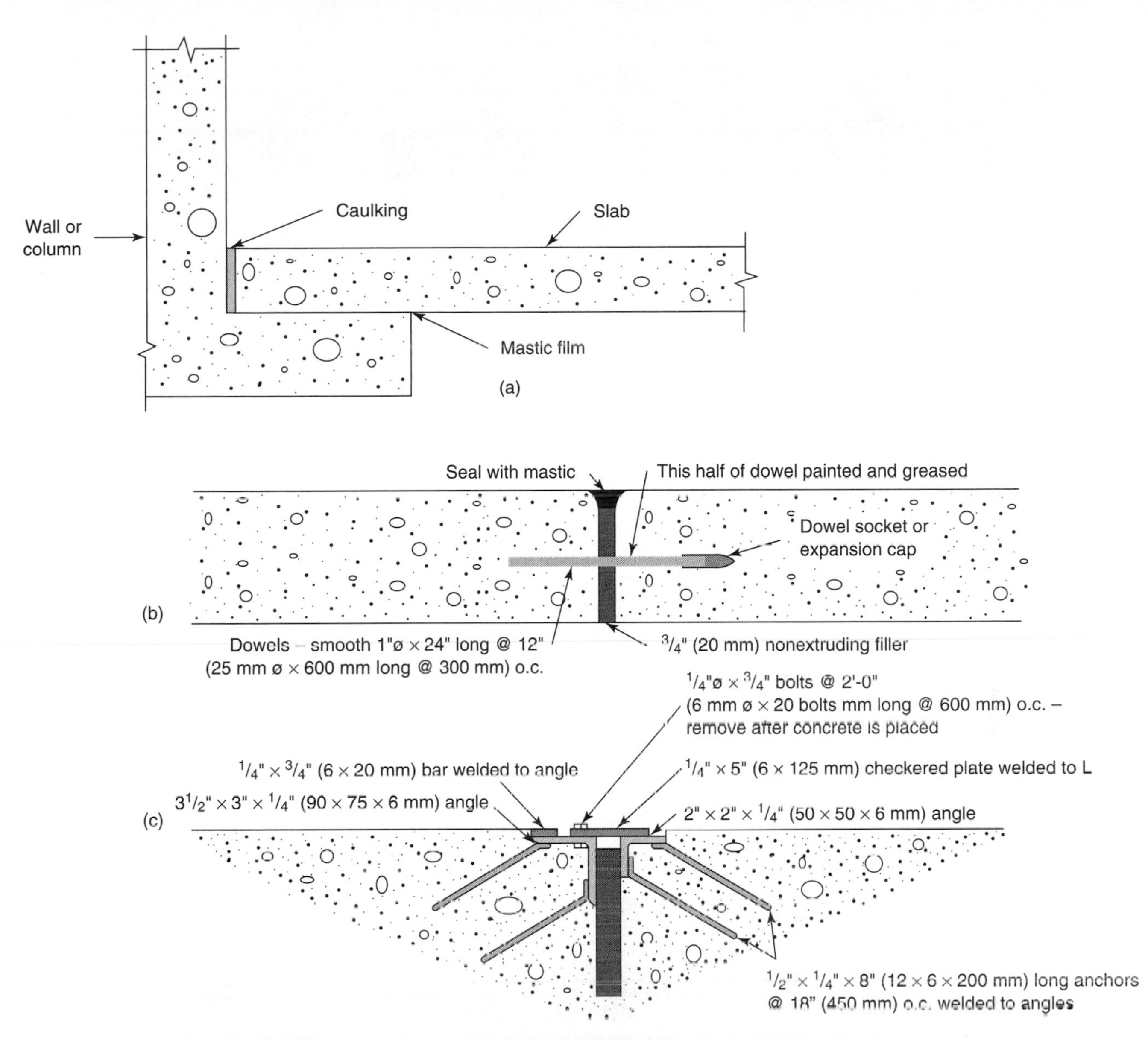


**Figure 6-66** Expansion Joints: (a) At Wall or Column; (b) Slab with Moderate Traffic; (c) Slab with Heavy Traffic.

## Beam-and-Slab Forms

Forms for beam-and-slab floors incorporate the forms for the *spandrel beam,* the regular *girder* and *beam* forms, and the *slab* forms. Support systems for these forms include wood *T-posts,* adjustable *metal posts,* and supports made with *metal scaffolding* (see Figure 6-67).

## Flat Slab Floor Form

With the flat slab floor system (see Figure 11-26), in which there are no beams and no drop panels, the first-story columns are usually cast first. Then the slab forms are set up for the floor that they will support (see Figure 6-68). When that floor is ready, the operation is repeated for successive floors.

## Ribbed Slab Floor Form

Ribbed slab floors (see Figure 6-69) are formed by the use of *metal pans* resting on a flat deck or supported by joists (see Figure 11-29). The pan system may be supported in a number of ways, one of which is shown in Figure 6-70.

## Waffle Slab Floor Form

A waffle slab floor (see Figure 6-71) is formed by using *dome pans* of metal or plastic resting on a flat deck (see Figure 11-35).

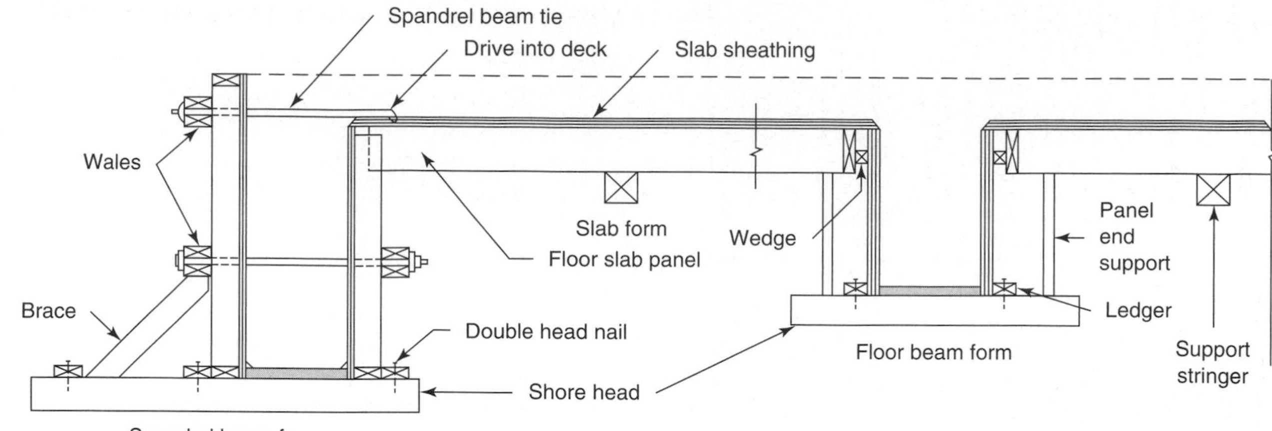

**Figure 6-67** Beam-and-Slab Floor Beam.

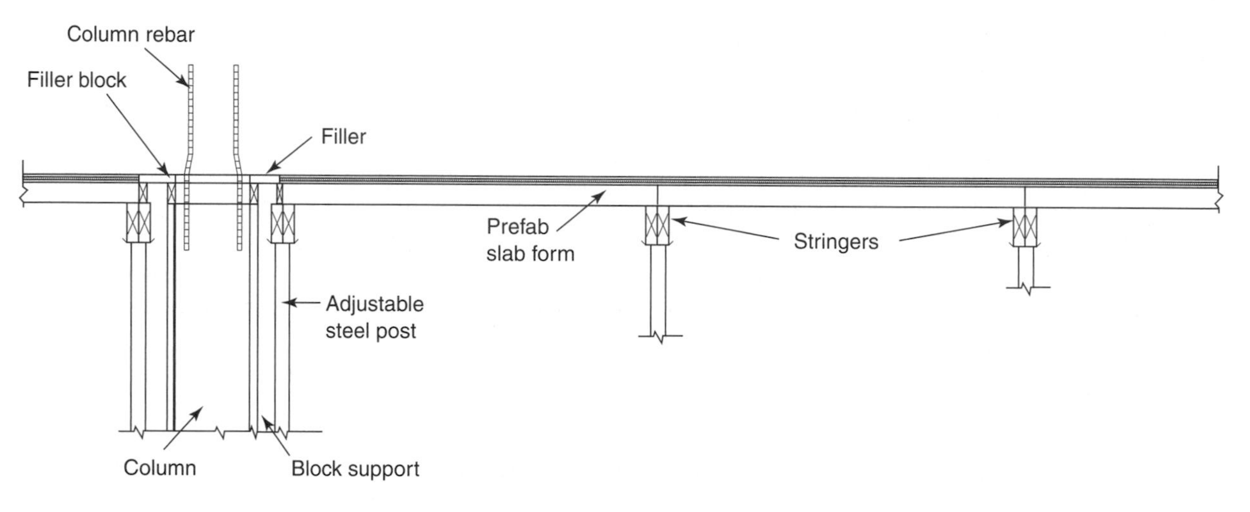


**Figure 6-68** Flat Plate Floor Form.

**Figure 6-69** Ribbed Slab Floor Forms.

**Figure 6-70** Supports for Ribbed Slab Floor.

## Prestressed Concrete Joist Floor Form

The design of a floor framed with prestressed concrete joists may specify prestressed concrete planks for the floor slab. On the other hand, the floor may be a cast-in-place concrete slab placed on a slab form suspended from the joists (see Figure 6-72).

## Steel Frame Floor Forms

Floor forms for concrete floors supported by a steel frame will vary, depending on whether or not the fireproofing for the steel beams is to be cast monolithically with the slab (see Figure 6-73). In either case the slab forms will usually be suspended from the steel frame, eliminating the necessity for a shoring system (see Figure 6-74).

## Flying Forms

In many present-day high-rise, cast-in-place concrete buildings, construction is identical from bay to bay and from floor to floor. The buildings consist essentially of either a series of columns supporting flat plate floors (see Figure 6-75) or a number of shear walls doing the same thing (also refer to Figure 6-78). A floor slab form and its support system, designed for one floor, can be reused for all the others, the main problem being the moving of the forms from one position to another. This problem has been solved by the use of *flying forms,* which are forms designed as a complete unit, large enough that the entire floor of one bay may be cast on it and capable of being moved as a unit from one position to another.

**Figure 6-71** Waffle Slab Floor Forms.

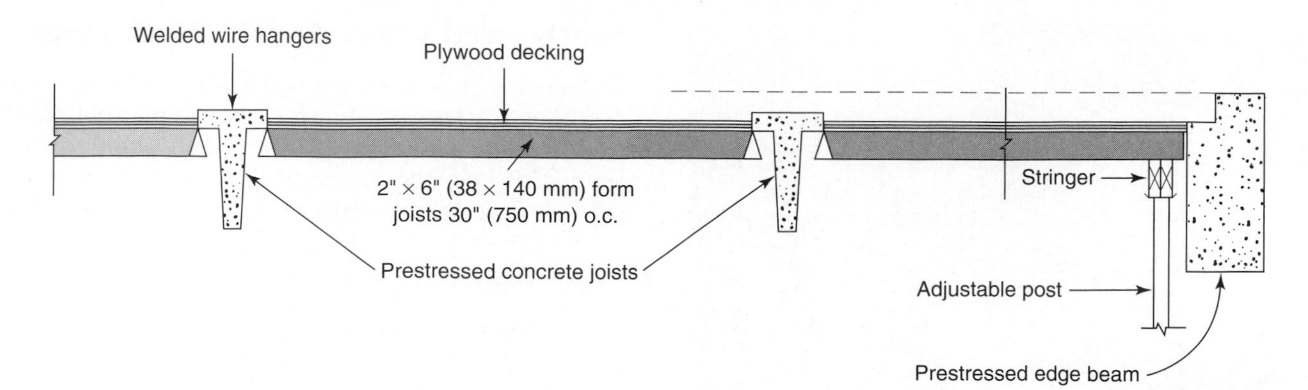


**Figure 6-72** Prestressed Concrete Joist Floor Form.

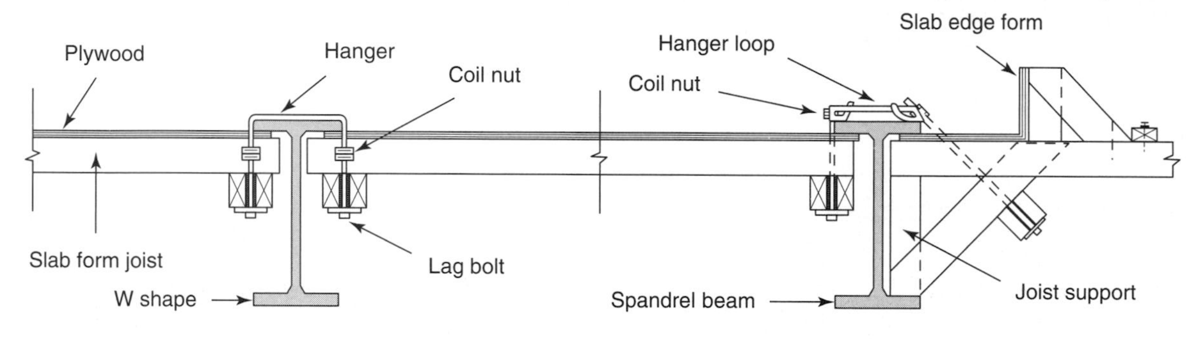


**Figure 6-73** Slab Forms with No Fireproofing.

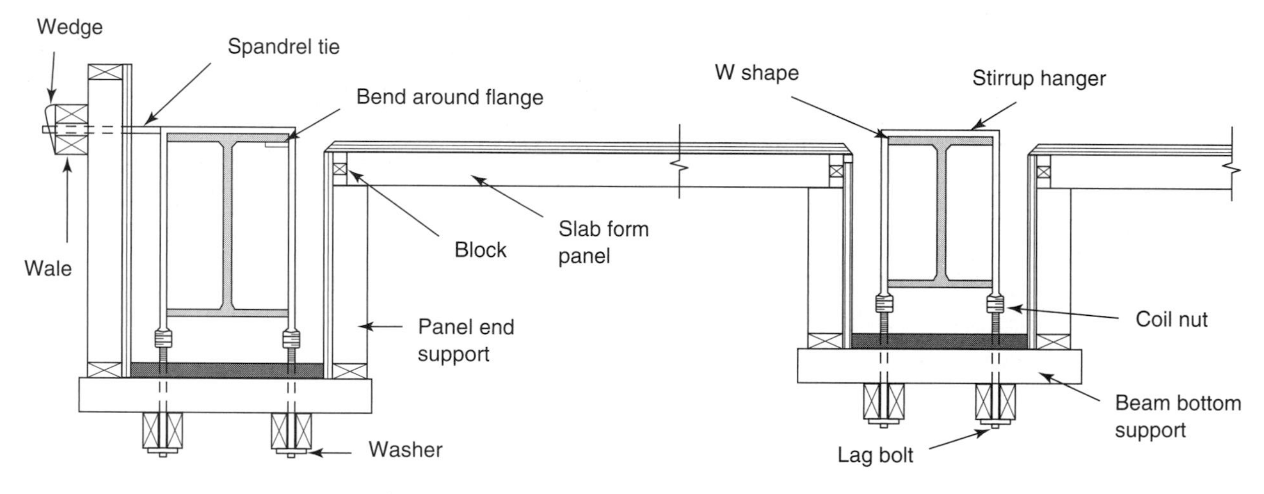


**Figure 6-74** Slab Forms with Fireproofing.

Forms are made up of a steel or aluminum metal frame supporting a wood or metal deck and resting on screw jacks, as shown in Figure 6-76. The jacks allow the form to be adjusted to exactly the right level for placing the slab, and when it has gained sufficient strength, the form can be lowered away from the slab by turning down the jacks.

The form is then lowered onto rollers (see Figure 6-77), the jacks folded up into the *flying position,* and the form rolled out of the bay to where it can be hooked to a crane, which will finish moving

**Figure 6-75** Series of Columns and Flat Plate Floors.

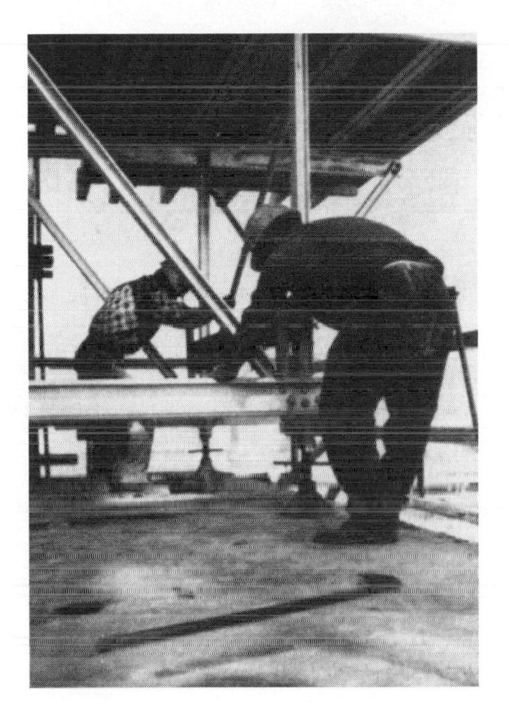

**Figure 6-76** Flying Form Frame and Jacks. (Reprinted by permission of Aluma Enterprises, Inc.)

**Figure 6-77** Section of Fly Form on Rollers, Ready to be Moved.

it out of the bay and swing it into a new position. There it will be used again and moved again, as many times as there are bays in the structure which that form will accommodate. Figure 6-78 shows a sequence of operations in *flying* a form from one location to the next.

## Shoring Systems

Shoring members are used to support concrete forms and their contents or other structural elements and may be divided into two major categories: *horizontal shoring* and *vertical shoring*. Horizontal shores range from small units, such as those that might be used to span between pairs of precast concrete joists to support the plywood deck form, to large wood or steel members with relatively few supports, used to carry much heavier loads with a minimum of interference in the work area below.

Vertical shores are those that support the horizontal ones from a firm base below, such as a concrete

(a)

(b)

(c)

(d)

**Figure 6-78** "Flying" a Form: (a) Rolling Out of Bay; (b) Suspended, Clear of Bay; (c) Closing into New Bay; (d) Being Guided to New Bay. (Reprinted by permission of Aluma Enterprises, Inc.)

slab or a *mudsill,* if there is no solid bearing on which the shore may rest [see Figure 6-81(a)]. They may be of wood or metal and take many different shapes, depending on the particular circumstances.

Vertical wood shores may be *single wood posts,* with wedges at the bottom to adjust the height, *double wood posts,* two-piece *adjustable posts,* or T-head shores. Figure 6-81(b) illustrates some typical wood shores.

Vertical metal shores may be adjustable pipe shores or shores made up of prefabricated metal scaffolding. Scaffold-type shoring is usually assembled into *towers* by combining a number of units into a single shoring structure (see Figure 6-79). A typical adjustable pipe shore is illustrated in Figure 6-80.

Regardless of the type of shoring system to be used, it should be carefully planned in advance by someone who understands the loads and stresses involved. Failure to properly take into account not only the structural loads that will occur but also the construction loads that may be imposed may result in shoring failure, which is not only very costly but potentially dangerous.

**Figure 6-79** Scaffold-Type Shoring Used to Support Structural Slab Formwork.

**Figure 6-80** Adjustable Pipe Shore.

## SHORING AND RESHORING IN MULTISTORY STRUCTURES

When building multistory structures using pour-in-place concrete, the rate at which the structure advances is controlled by the ability of the poured slabs to support load. For a contractor to wait for fresh concrete to develop its full strength before continuing to the next level would be an inefficient use of time (see Figure 6-82).

Shoring and reshoring semicured slabs allow the contractor to transfer the construction loads through the shores without affecting the freshly poured sections. The amount of shoring and reshoring depends on the strength of the partially cured slabs, the amount of load that is to be transferred to each slab, and the rate of advance that the contractor wishes to achieve. Although the operation appears to be straightforward, the loads transferred from the shoring to the slabs can be much higher than expected and, once a sequence has been decided, it should be followed for the duration of the job. Consider an example using two sets of shores and one set of reshores. The object is to determine the amount of load that will be placed on the floor slabs during the construction sequence. Beginning with the first-floor slab [Figure 6-83(a)], the shoring must support all the construction loads, dead plus live. To facilitate the construction of the next floor as quickly as possible, the shores will be left in place

(a)

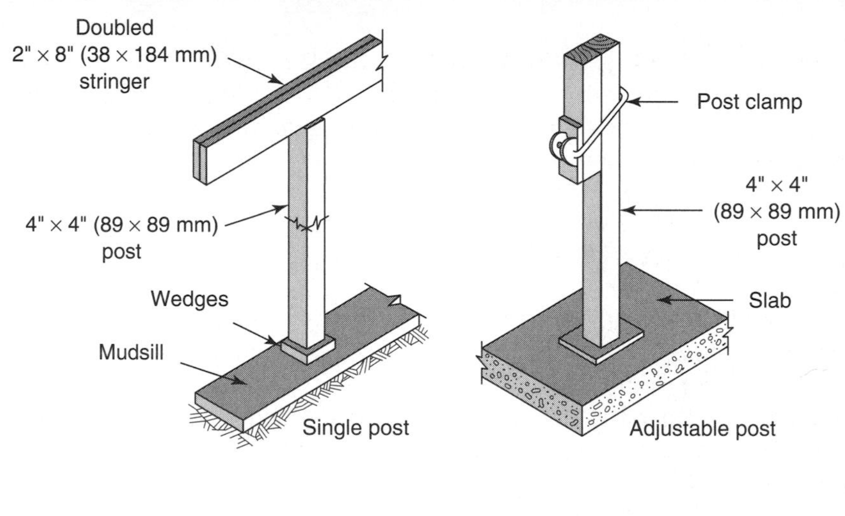



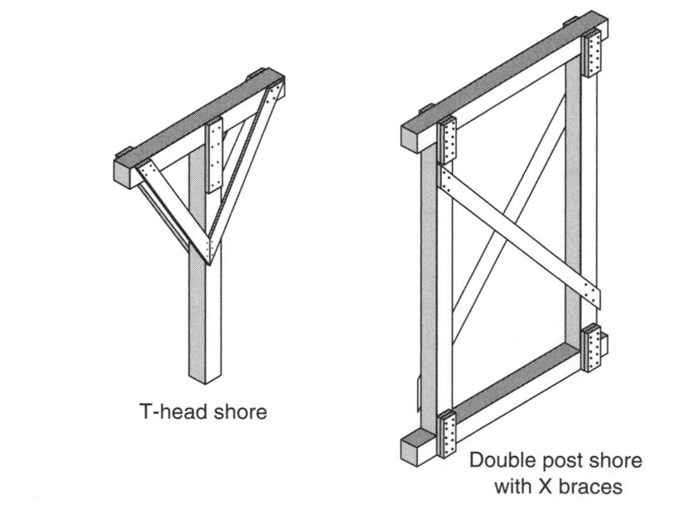



(b)

**Figure 6-81** (a) Continuous Wood Mudsills under Steel Pipe Shores; (b) Typical Wood Shores.

**Figure 6-82** Reshoring in a Multistory Structure.

under the first floor and the forming of the next floor will commence. Because the first floor is supported by its shores at this time, the second-floor loads will also be transferred to the first-floor shores.

At this point in the operation, the first-floor shores are now supporting the dead weight of the two floors [Figure 6-83(b)]. Once the second-floor slab has hardened sufficiently, the first-floor shores are removed. Each slab must now support its own weight [Figure 6-83(c)]. Before any construction loads are applied to the second floor, reshoring is placed under the first floor to ensure that no additional loads are applied to the first and second floors [Figure 6-83(d)]. With the third-floor shores in place, the load of the third floor is carried down to the first-floor reshores. No additional loads are transferred to the first and second floors; however, they continue to support their own weight [Figure 6-83(e)].

Once the third floor has set, the first-floor reshores can be removed. The load on the first-floor reshores (dead weight of the third floor) is now redistributed among the three slabs. The remaining shores must now support that load above them not supported by the slabs [Figure 6-83(f)]. The first- and second-floor slabs, as a result, support one-third of the third-floor weight each.

Removing the second-floor shores redistributes the load carried by the shores (one-third of the third-floor weight) between the two interconnected slabs [Figure 6-83(g)]. The third-floor slab now supports half its own weight, and the second-floor slab supports its own weight and half of the third floor. Placing reshores between the first and second floors does not change the loads [Figure 6-83(h)].

The concrete is now poured on the fourth floor [Figure 6-83(i)]. The load of the floor is distributed among the three slabs below. As a result, the second-floor slab must now support 1.5 + 0.33 = 1.83 times the total load, and the reshores are supporting one-third of the fourth-floor load. When the reshores are removed, their load is redistributed between the three slabs, and the second-floor slab must now support 1.83 + 0.11 = 1.94 total dead load. This type of redistribution continues as additional floors are added, each being slightly different in value.

The object here is to illustrate the effects of shoring and reshoring on the poured slabs. In this case, the second floor, at one point in the operation, must be capable of supporting almost twice its own weight. If the cycle is to be shortened, then more levels of reshoring would be required to help redistribute the weight of the additional uncured floors that would have to be supported. The most important aspect of this procedure is not to remove the reshores prematurely and not to change the sequence of events at some point throughout the operation.

To ensure that all stages of forming, shoring, pouring of concrete, and stripping of forms are executed within a minimum amount of time, careful

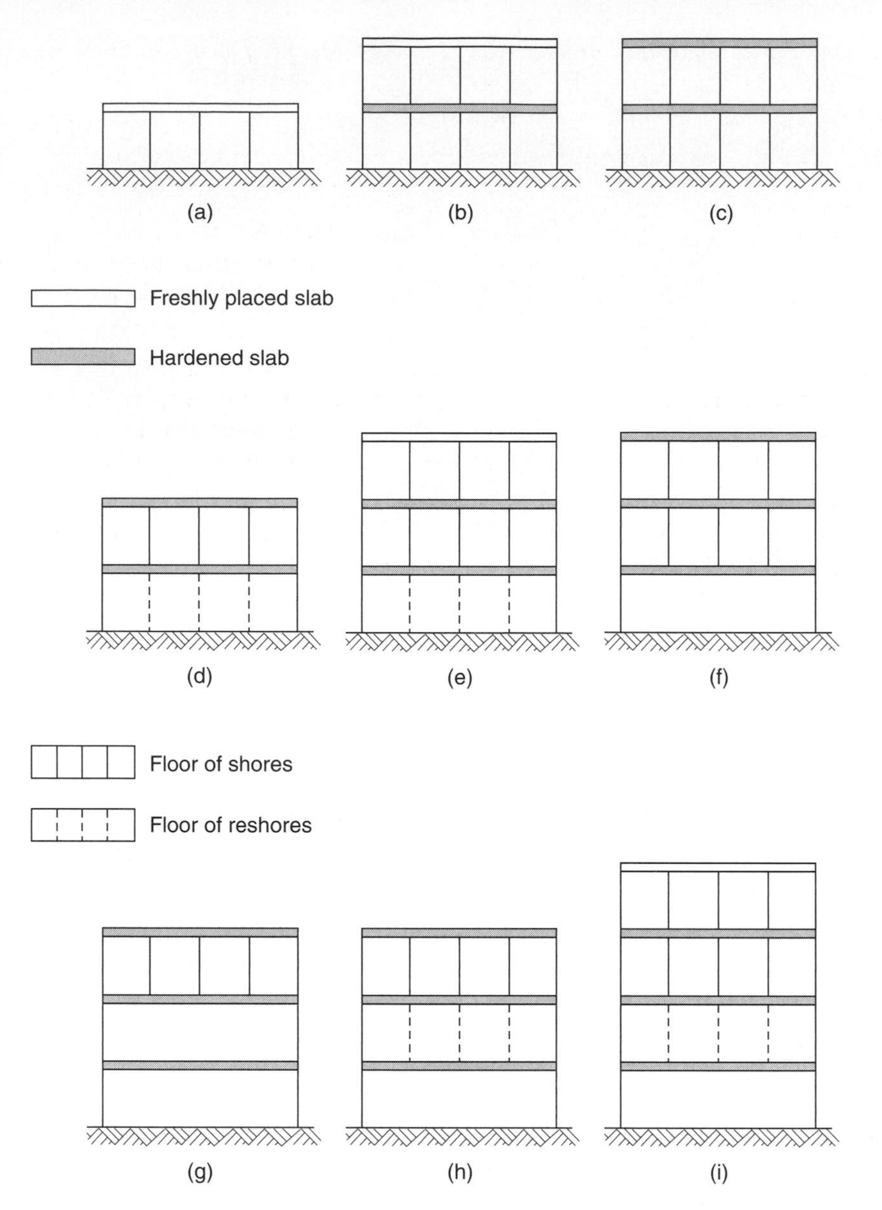


**Figure 6-83** Shoring and Reshoring Sequence in a Multistory Building.

thought must be given to personnel requirements, equipment, and materials prior to the actual commencement of work.

## FORM ACCESSORIES

A vast array of products is currently available to aid in making forms stronger and erecting them faster. These products include items that have already been mentioned in this chapter, such as ties, spreaders, wedges, corner brackets, and waler clips. They also include snap ties for special conditions, rock anchors, beam hangers, waterstops and keys, masonry ties, column clamps, shores, form rods, concrete inserts, sleeves, slab pans, and many others. Detailed information on the uses of these may be obtained from brochures and catalogs published by their manufacturers.

## FORM TREATMENT, CARE, AND REMOVAL

In nearly all types of building construction, formwork constitutes a significant part of the cost of the building. To keep this cost at a minimum, forms are often made reusable, either wholly or in part. They must therefore be designed so that removal is simple and can be accomplished without damage to the form sections. Care must be taken in handling and storing these units so that they will not be broken or damaged and will be available for reuse.

To facilitate removal, form faces must be treated to prevent concrete from adhering to them. A number of materials are available for this purpose, usually consisting of liquids that are to be brushed or sprayed on the form. Wooden forms must be treated to minimize absorption of water. Oil is one material used for this purpose. Form sealers that coat the surface of the form with an impervious film are also used for this type of treatment.

Form removal must be carried out without damaging either the forms or the structure being stripped. Levers should not be used against the concrete to pry forms away because green concrete is relatively easy to damage. If levering action is necessary, pressure should be applied against a broad, solid base. Form panels being removed from a wall section should be pulled straight off over the tie rods.

The production of concrete structures of the desired size, shape, and strength, yet pleasing to the eye, can be achieved if a number of important points are remembered. Forms must be properly designed and carefully constructed and erected. They must receive the appropriate surface treatment. Tying, bracing, and shoring must be adequate for the particular job involved. There must be proper handling, placing, consolidation, and curing of the concrete, and form removal must be carried out in such a way that no damage is incurred by the finished product.

## Review Questions

**1.** What seven basic features must forms have to fulfill their function?

**2.** Give three reasons for not including the earth as part of a form.

**3.** What is meant by *rate of pour?*

**4.** List three factors used in determining the rate of pour.

**5.** List the four main factors that determine the pressure on forms.

**6.** Give two basic reasons for treating forms before using them.

**7.** Give three common methods of providing support against lateral pressure to square or rectangular forms.

**8.** Differentiate between a *construction joint* and an *expansion joint.*

**9.** List four factors that determine when forms may be removed.

**10.** Define the following: **(a)** form clearance, **(b)** form accessories, **(c)** prefabricated forms, **(d)** slip form, **(e)** form buck, and **(f)** flying form.

**11.** Calculate the maximum pressure developed when a 10-in.-thick wall 10 ft high is to be poured at the rate of 3 ft/hr at a temperature of 70°F. Check your result with Table 6-2(a).

**12.** Calculate the resulting maximum pressure developed in a column form 20 in. × 20 in. square, 20 ft high, if the form is filled in 30 min. Assume the temperature to be 68°F.

**13.** Determine the required spacing of joists supporting ¾-in.-thick plywood that is to support an 8-in. concrete floor slab, assuming a construction live load of 50 psf and a formwork dead load of 10 psf. Use plywood with $F'_b = 1{,}930$ psi.

**14.** If the joists in Problem 13 must span a distance of 5 ft, what size of joist would be required to support the calculated loads assuming a two-span continuous situation?

**15.** Based on the loads in Problem 13 and assuming that the joists were supported by beams that were to span 10 ft, what total load would result on the shore posts supporting the beams?

**16.** If the shore posts in Problem 15 had an unsupported height of 10 ft, what size of wood post would be required, assuming an S4S section with an allowable compressive stress parallel to the grain of 1,100 psi is to be used?

**17.** For the loads calculated for the shore posts in Problem 15, what area of mudsill would be required for each shore post, assuming that the soil supporting the mudsills has an allowable bearing capacity of 1,100 lb/sq ft?

**18.** Calculate the maximum pressure developed when a 250-mm wall 3 m high is to be poured at the rate of 0.9 m/hr at a temperature of 20°C. Check your result with Table 6-2(b).

**19.** Calculate the resulting maximum pressure developed in a column form 500 × 500 mm square, 5 m high, if the form is filled in 30 min. Assume the temperature to be 18°C.

**20.** Determine the required spacing of joists supporting 18.5-mm-thick plywood that is to support a 200-mm concrete floor slab, assuming a construction live load of 2.4 kPa and a formwork dead load of 0.5 kPa. Use plywood with $F'_b = 13.3$ MPa.

**21.** If the joists in Problem 20 must span a distance of 1.2 m, what size of joist would be required to support the calculated loads assuming a two-span continuous situation?

**22.** Based on the loads in Problem 20 and assuming that the joists were supported by beams that were to span 2.4 m, what total load would result on the shore posts supporting the beams?

**23.** If the shore posts in Problem 22 had an unsupported height of 3.6 m, what size of wood post would be required assuming an S4S section with an allowable compressive stress parallel to the grain of 6.2 MPa is to be used?

**24.** For the loads calculated for the shore posts in Problem 22, what area of mudsill would be required for each shore post, assuming that the soil supporting the mudsills has an allowable bearing capacity of 55 kPa?

# 7 CONCRETE WORK

The extensive use of concrete is ample proof of its outstanding characteristics as a building material. It is such a familiar material that we take for granted the remarkable process by which cement and water, mixed with a wide range of aggregate materials, are converted into a strong and durable material of almost any desired shape. The development of modern portland cements that can set quickly, even under water, began over 100 years ago when the raw materials and processes needed for their manufacture were first recognized. Today, hundreds of scientists, engineers, and technicians are engaged in studying these materials, attempting to understand and to improve them still further.

Many of the physical and chemical reactions that take place during the setting and aging of concrete are so complicated that they are not yet fully understood. This is due in part to the wide range of chemical substances that can exist, partly by design and partly by chance, in any given concrete mix. Additional factors may be introduced in the methods of manufacture, handling, and curing at the site. The changes that take place quite rapidly in the new concrete do not cease at the end of the formal curing period. Some may continue over a long period of time, and some may only begin in the environment to which the concrete is subsequently exposed. Despite these complications, concrete of predictable properties and performance can be produced, and this does not occur by chance.

It is, fortunately, unnecessary for the designer, specification writer, job engineer, or supplier to keep in touch with the whole field of concrete technology. There are a number of guides to good practice in the form of codes, standards, and specifications from sources such as the Canadian Standards Association, Portland Cement Association, American Society for Testing and Materials, and the American Concrete Institute. Nevertheless, it is highly desirable to have some idea of the general nature of the material and its more important properties.

## AGGREGATES FOR CONCRETE

Concrete is the product of a combination of portland cement and water paste with some type of aggregate. The paste surrounds the aggregate particles and, as it sets (returns to a limestonelike state), binds them solidly together. At the same time, to form a dense concrete, the paste must fill the voids between the particles of aggregate (see Figure 7-1).

It is desirable to use a maximum amount of aggregate when making concrete of any given strength for a number of reasons. In the first place, cement is up to 10 times more costly than aggregates. Second, cement paste tends to shrink while setting and curing; consequently, the use of more paste than is necessary results in a more expensive concrete and also produces excessive shrinking.

Aggregates used for concrete mixes are generally placed into four classifications. *Sand, gravel, crushed stone* (natural aggregates), and *air-cooled blast furnace slag* are used to produce **normal-weight concrete** with a unit density ranging from 135 to 160 lb/cu ft (2,160 to 2,560 kg/m$^3$). Materials such as *expanded shale, slate, clay,* and *slag* produce structural **lightweight concrete** with a density of 85 to 115 lb/cu ft (1,360 to 1,840 kg/m$^3$). Lightweight

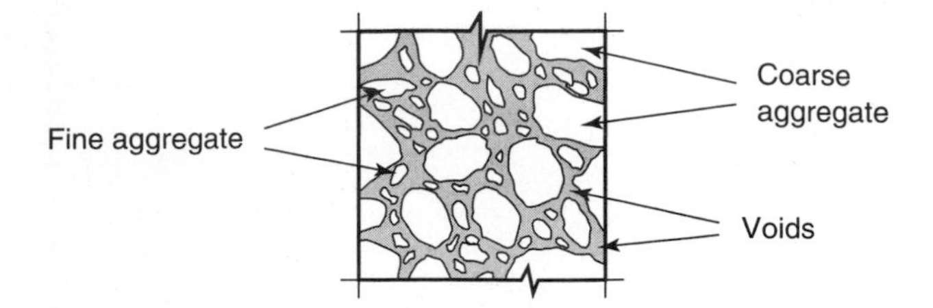


**Figure 7-1** Voids Between Aggregate Particles.

materials such as *vermiculite, perlite, pumice, scoria,* and *diatomite* are used to produce nonstructural **insulating concrete** with a density ranging from 15 to 90 lb/cu ft (240 to 440 kg/m$^3$). Heavy materials such as *hematite, barite, limonite, magnetite, steel punchings,* and *shot* produce **heavyweight concrete** with a unit density of 189 to 380 lb/cu ft (2,880 to 6,090 kg/m$^3$).

The amount of aggregates used in making concrete usually occupies 60% to 80% of the total volume; as a result, the type and quality of the aggregates have a major influence on the properties of concrete. It is therefore important to know as much as possible about the aggregates being used to be able to produce concrete to required specifications.

Natural aggregates are divided into two classes: *fine aggregate* (F.A.) and *coarse aggregate* (C.A.). Fine aggregate includes material that does not exceed ¼ (6 mm) in diameter, whereas coarse aggregate normally ranges from ⅜ to 3 in. (9.5 to 75 mm), though sizes to 6 in. (150 mm) are sometimes used. Both require study and testing to make sure that they qualify as good material for concrete. It is common practice to carry out tests on both to ensure soundness, cleanliness (the presence of excessive amounts of silt and fine organic matter), and proper gradation (distribution of sizes in a given sample) and to determine specific gravity, moisture content, particle shape, and surface texture.

## Soundness

Concrete can only be as strong as the aggregate from which it is made, and aggregate, particularly coarse aggregate, should be inspected to see that it does not contain weak or laminated particles, soft and porous particles, or bits of shale. Aggregate soundness may also be determined by its abrasion resistance, which is of particular importance when the concrete is to be used for heavy-duty floors. The abrasion resistance test is outlined in ASTM C131, in which a quantity of the material is rotated in a steel drum and then measured to ascertain the amount of material worn away during the test.

## Cleanliness

Silt clinging to the surface of coarse aggregates prevents proper bonding of the cement paste. In fine aggregate, the presence of too much fine material means a significant increase in the surface area to be covered with paste, resulting in either having to use more paste or the thinning out of that in use.

## Silt Test

The presence of silt in coarse aggregates can usually be detected visually but, in the case of fine aggregate, a *silt test* is used to determine the amount of very fine material present. To conduct a silt test, proceed as follows:

Place 2 in. (50 mm) of the fine aggregate to be tested in a standard 1-qt. (1-litre) glass jar. Fill the jar three-quarters full of water, replace the cover, and shake well. Let the jar stand for several hours until the water has cleared. Upon settling, the fine silt will form a layer on top of the sand. If this layer exceeds ⅛ (3 mm) in depth, the aggregate is unfit for making good concrete unless part of this material is removed.

## Colorimetric Test

The presence of small amounts of fine organic matter in a cement paste retards its setting action, sometimes so much so that a mix will dry out before setting properly. To test for the presence of harmful amounts of this organic matter in fine aggregate, a *colorimetric test* is used. The test is as follows:

Place 4 oz (100 g) of the aggregate to be tested into a small jar or bottle. Then fill to 7½ oz (225 ml) with a 3% solution of sodium hydroxide (NaOH). Shake well and allow to stand for 24 hr. The presence of organic matter will cause color changes in the clear solution varying from a very light straw color to a deep chocolate brown. If the color ranges from light to dark straw, the aggregate is suitable. If, on the other hand, the color is darker than this, too much organic matter is present and the aggregate is unsuitable for concrete work.

## Gradation

The proper gradation of particle size in the aggregate used to make concrete is essential for a number of reasons. It is apparent that the fewer the number of pieces required to fill a given volume, the smaller their total surface area will be. This can be demonstrated with cubes or spheres, as in Figure 7-2. Because these particles must be coated with cement paste, the smaller the surface area, the less paste is required. In other words, the greater the maximum

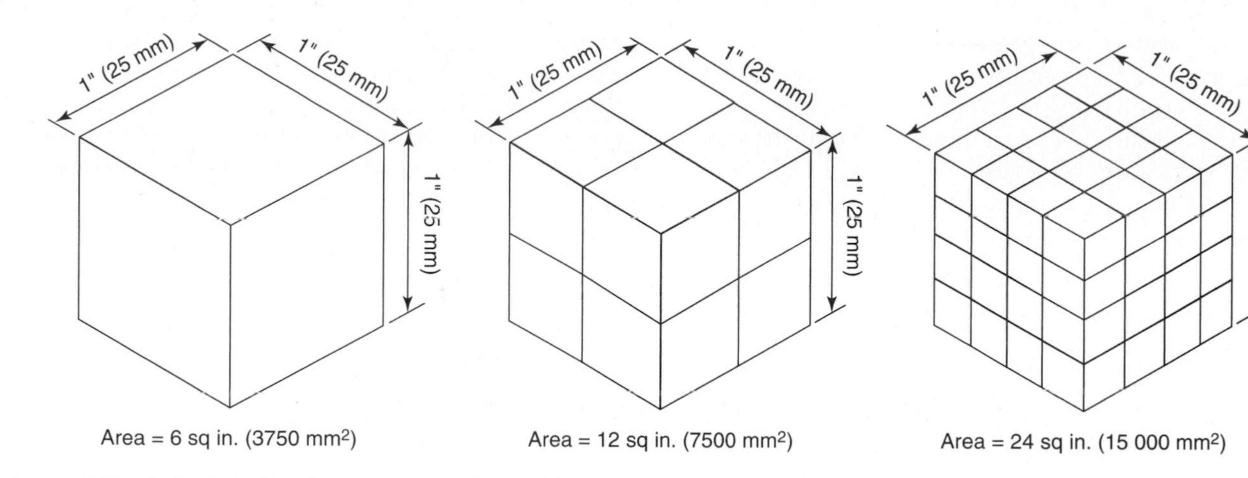


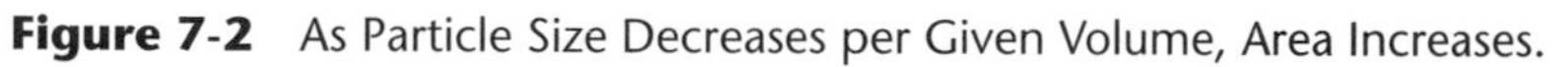

**Figure 7-2** As Particle Size Decreases per Given Volume, Area Increases.

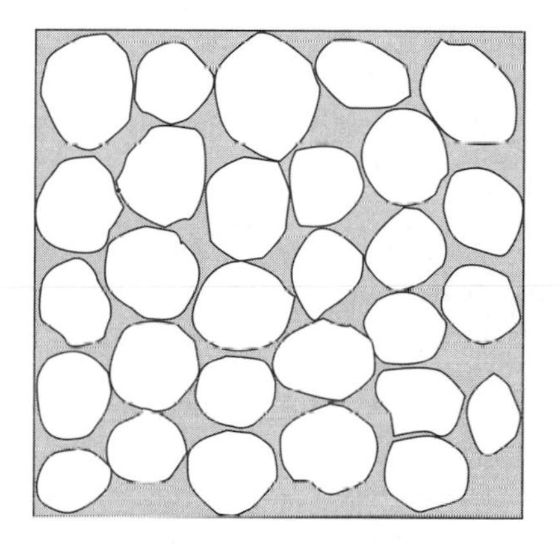

**Figure 7-3** Unavoidable Spaces when Particle Size Is Uniform.

size of the aggregate, within limits, the more economical the mix becomes.

But with a given quantity of aggregate in which all particles are the same size, it is evident that, no matter how these are arranged together, there will always be spaces between them that cannot be filled (see Figure 7-3). It can also be demonstrated that, regardless of the size of the particle chosen, the amount of voids will remain constant, as shown in Figure 7-4. If an aggregate of this type was used to make concrete, those spaces would have to be filled with cement paste, necessitating the use of a great deal more than would be required simply to coat the particles and bond them together.

If smaller pieces of material are introduced, however, as illustrated in Figure 7-4, a considerable portion of those spaces would be filled with aggregate. When still smaller particles are added, more spaces are filled. This process can be continued until all of the available particle sizes have been utilized. In other words, the proper gradation of particles, from the largest to the smallest, will effect an economy of paste (see Figure 7-5).

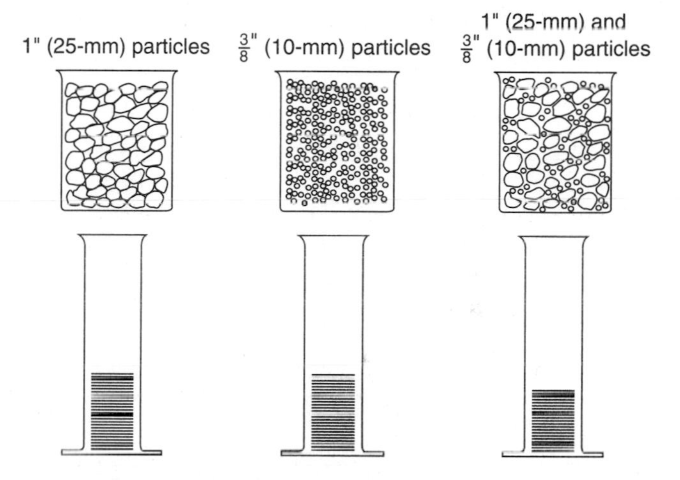


**Figure 7-4** Reducing Voids by Improving Gradation.

**Figure 7-5** Well-Graded Aggregate. (Reprinted by permission of Portland Cement Association.)

Cement paste has a tendency to shrink when it sets. If the volumes of paste are relatively large, as when the particles are all the same size, this shrinkage could result in the breaking of the bond as the paste shrinks away from the aggregates, thus weakening the concrete and also leaving passages through which water could travel. Proper gradation can therefore affect watertightness and strength of concrete. For these reasons, aggregates are tested to make sure that there is a proper gradation of sizes. This test is called a *fineness modulus* test and may be carried out on any aggregate.

## Fineness Modulus Test

When doing a fineness modulus test on fine aggregate, a set of six standard wire screens is used. The set includes a #4 sieve (four wires to the inch or 16 openings to the square inch); a #8 sieve with eight wires to the inch or 64 openings per square inch; a #16; a #30; a #50; a #100; and a pan to retain everything passing through the #100 sieve. A metric set of sieves includes a 5-mm sieve; a 2.5-mm sieve; 1.25 mm; a 630 μm; a 315 μm; a 160 μm; and a pan to retain everything passing through the 160-μm sieve. A μm is one millionth of a metre, or 0.001 mm.

These sieves are stacked as shown in Figure 7-6 so that aggregate placed in the top can pass through, to be held on any sieve with small enough mesh to retain that particular size. Very fine material passing the (#100) (160-μm) sieve will be retained in the pan at the bottom.

From a dry quantity of the fine aggregate to be tested, very carefully weigh out 500 g, and place the sample in the top sieve, (5 mm), which is the #4 sieve. The sieves may be shaken by hand or by a mechanical shaker. Shaking should continue for at least 1 min. Weigh the material retained on each sieve, as illustrated in Figure 7-7, and record the weights. Calculate the percentage of the total (500 g) that each size represents and total these percentages as cumulative percentages. This is done by beginning with the percentage retained on the #4 (5-mm) sieve and recording it as the cumulative percentage for that grade. The cumulative percentage for the next grade #8 or (2.5 mm) is found by adding its percentage to the one above. Each grade is treated the same way. Now add these cumulative percentages together and divide the result by 100. This figure represents the fineness modulus number of that aggregate and is an indication of the relative fineness or coarseness of the material. Note that any material passing the #100 (160 μm) sieve is not included in the calculations.

Fineness modulus numbers for fine aggregate should range from 2.20 to 3.20, and any fine aggregate with a fineness modulus number falling within this range will generally be a suitable concrete aggregate, from a gradation standpoint.

The smaller numbers represent fine materials, aggregates with a great proportion of fine particles, whereas larger numbers represent aggregates with a preponderance of large particles. Fine aggregate is classed as *fine, medium,* or *coarse,* depending on its fineness modulus. Fine sand ranges from 2.20 to 2.60, medium from 2.61 to 2.90, and coarse from 2.91 to 3.20. In cases when mass concrete is being designed, fine aggregate with a fineness modulus larger than 3.20 may be allowed. Specifications usually define the allowable fineness modulus.

The fineness modulus calculations may be recorded as in the following table. To better understand the procedure, consider this sample test. Suppose that a 500-g sample has been put through the standard sieves and the results are as follows:

Retained on the #4 (5-mm) sieve, 12 g; on the #8 (2.5-mm) sieve, 55 g; on the #16 (1.25-mm) sieve,

**Figure 7-6** Set of Sand Screens.

**Figure 7-7** Weighing Sand Sample.

100 g; on the #30 (630-μm) sieve, 122 g; on the #50 (315-μm) sieve, 118 g; on the #100 (160-μm) sieve, 75 g. Totaling these masses, it is found that 18 g passed through to the pan below; weigh the contents of the pan to make sure. This information completes the table. The result indicates that this particular aggregate belongs in the medium sand category.

Fineness modulus is not an indication of whether the proper amount of each grade of sand is in the sample, for a great number of gradings will give the same value for fineness modulus. It can, however, be determined from the results of the test whether the sample contains the right proportion of each grade of sand.

This determination can be made by plotting the cumulative percent passing by mass through each sieve on a graph as shown in Figure 7-8. On this graph the upper and lower boundaries are indicated by the solid lines based on CSA Standard A23.1

**FINENESS MODULUS TEST FOR SAND**

| Sieve Size | Mass Retained (g) | Percentage Retained | Cumulative Percentage Passing | Cumulative Percentage Retained |
|---|---|---|---|---|
| #4 (5 mm) | 12 | 2.4 | 97.6 | 2.4 |
| #8 (2.5 mm) | 55 | 11.0 | 86.6 | 13.4 |
| #16 (1.25 mm) | 100 | 20.0 | 66.6 | 33.4 |
| #30 (630 μm) | 122 | 24.4 | 42.2 | 57.8 |
| #50 (315 μm) | 118 | 23.6 | 18.6 | 81.4 |
| #100 (160 μm) | 75 | 15.0 | 3.6 | 96.4 |
| Pan | 18 | 3.6 | | 284.8 |
| | | 100 | | |

Fineness modulus = 284.8 ÷ 100 = 2.85

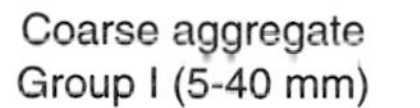

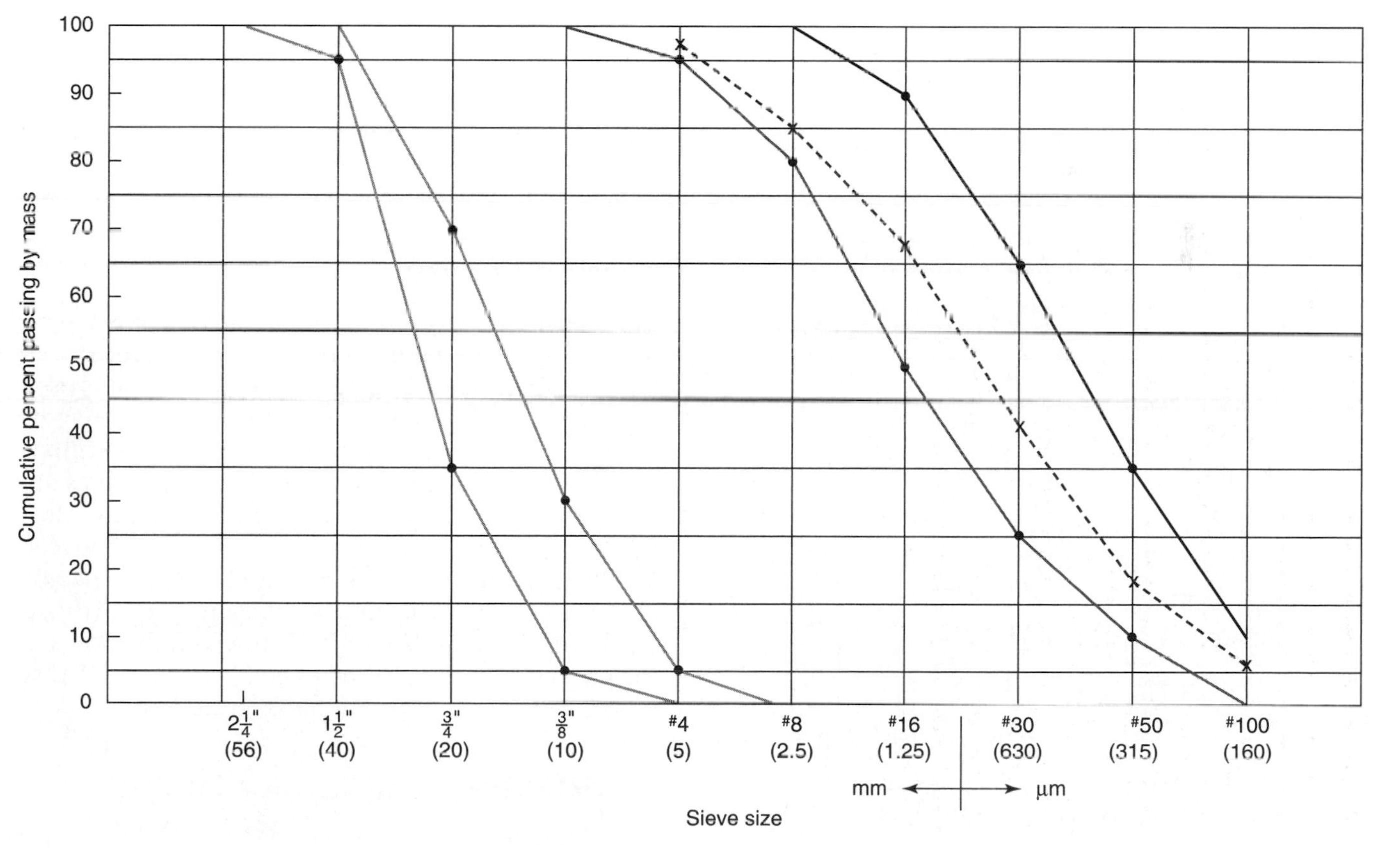


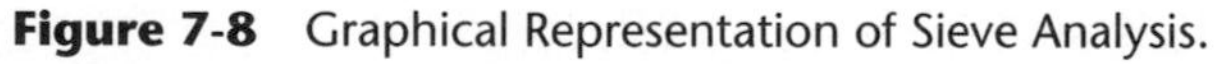
**Figure 7-8** Graphical Representation of Sieve Analysis.

requirements, and the dashed line indicates the results of the sieve analysis based on the data in the previous table. In this case, the plot of the results indicates that the sample of sand tested meets the requirements of CSA Standard A23.1 for a well-graded sand.

## Moisture Content

Moisture content tests on aggregates, particularly fine aggregates, are carried out for two reasons. One is that the moisture content in a fine aggregate may necessitate adjusting the amount of water added to the mix in order to produce a given strength. The other is that when water is present in certain proportions, a phenomenon known as *bulking* occurs. Volume increases considerably over that at a dry or saturated state because the particles are separated by a film of water. This means that, when designing concrete mixes, it is much better to specify the amount of aggregate to be used by mass rather than by volume. It takes only a small compensation to make up for the difference in mass between dry and moist sand, whereas a large compensation is sometimes required if amounts to be used are specified by volume.

The amount of bulking varies with the fineness of the sand and the amount of moisture present. By mass, 5% to 6% water will produce maximum bulking, and the amount of bulking is more pronounced in fine sand than in coarse sand. The sand bulking chart in Figure 7-9 shows the percentage of increase in volume over dry sand for amounts of water varying from 0% to 25% for fine, medium, and coarse sands.

The amount of water present in a sample is determined by carrying out a *moisture test*. Proceed as follows:

Weigh out a 500-g sample and spread it on the bottom of a flat metal pan approximately 12 in. (300 mm) in diameter and 1 in. (25 mm) deep. Set the pan over a heating unit and allow the sand to dry out, stirring continually. A change in the appearance of the sand can be noted as the moisture is driven off. When the material appears to be dry, weigh it again and record the weight mass. Dry once more and weigh again; it will probably have lost a little more weight this time. Continue this procedure until the weight remains constant. This is the dry weight. The total moisture content is found as follows:

$$\frac{\text{Loss in weight}}{\text{Dry weight}} \times 100 = \text{percentage of moisture}$$

But a portion of the moisture driven off had been present in pores and cracks in the particles (absorbed moisture) and was not available for reaction with cement. For average aggregates, it is assumed to be 1%. The free moisture, which appears on the surface of the particles, is therefore found by subtracting 1 from the previous result. In other words,

$$\frac{\text{Loss in weight}}{\text{Dry weight}} \times 100 - 1 = \text{percentage of free moisture}$$

For example, if a 500-g sample has a constant mass of 478 g after drying, the loss in weight has been 500 − 478 = 22 g. The percentage of free moisture is then

$$\frac{22}{478} \times 100 - 1 = 3.6\%$$

If this is the same sand previously tested for fineness modulus, the bulking chart shows that this medium sand, containing approximately 3½% free moisture, will increase by 24% over its dry volume.

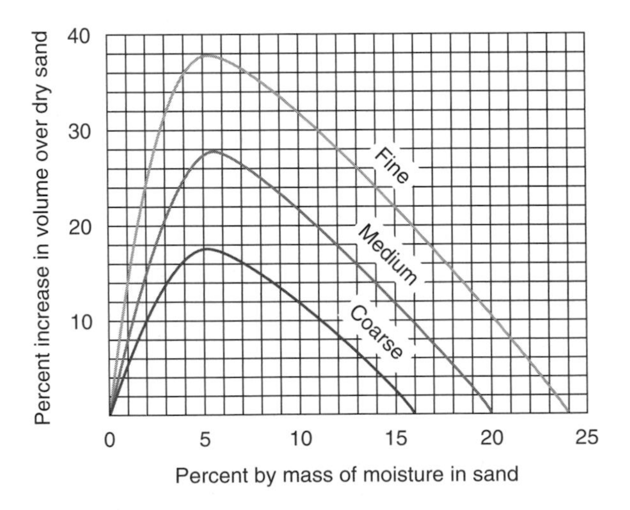


**Figure 7-9** Sand Bulking Chart.

## Specific Gravity

The specific gravity (sp. gr.) of an aggregate is the ratio of its mass to the mass of an equal volume of water. It is not a measure of aggregate quality but is used in making calculations involved in mix design, particularly those concerning *absolute volume*. Methods for determining the specific gravity of aggregates are outlined in ASTM C127 and in CSA A23.2-12A. Most natural aggregates have specific gravities between 2.4 and 2.9, and in concrete calculations the specific gravities of *saturated, surface–dry* aggregates are usually used. These are aggregates in which the pores in the particles are considered to be filled with moisture but with no excess moisture on the surface.

## Particle Shape and Surface Texture

The shape of particles of aggregate and the texture of their surface have an effect on the properties of fresh concrete. Rough-surfaced and long, thin, flat particles require more water to produce workable concrete than do rounded or cube-shaped ones. In other words, angular particles need more cement to maintain the same water:cement ratio. Long, slivery particles should be avoided if possible or at least limited to 15% by mass of the total aggregate. It is particularly important to limit such particles in aggregates used in concrete produced for floor toppings.

## PORTLAND CEMENT

The chemical reaction that is produced when cement is mixed with water is known as *hydration*. When the cement and water paste is mixed with aggregates and allowed to cure, the result is a very hard and durable substance.

The raw materials used in the manufacture of portland cement are lime, silica, alumina, and iron. These materials are readily available from natural deposits. From these basic materials, four principal compounds make up the composition of cement: *tricalcium silicate, dicalcium silicate, tricalcium aluminate,* and *tetracalcium aluminoferrite.* By varying the proportions of these basic components, the properties of the cement can be modified to best suit a particular application.

Five basic types of cement are available:

- *Type I (10)* or normal cement: Used as a general-purpose cement.
- *Type II (20)* or moderate cement: Used where protection against sulfate attack is required. It can also be used in large-mass concrete structures, as it produces less heat during the hydration process.
- *Type III (30)* or high early strength: Used where high concrete strengths are required in a very short period of time.
- *Type IV (40)* or low heat of hydration: Used where the heat of hydration must be kept to a minimum. Used in massive concrete structures such as dams.
- *Type V (50)* or sulfate resisting: Used where the concrete will be exposed to severe sulfate attack. Normally used in concrete foundations.

Cement is usually gray in color but can also be white. Coloring may be added to the white cement for special effects. Cement is transported in bulk for large projects or in bags when smaller quantities are required. In the United States, a standard bag of cement weighs 94 lb, whereas in Canada a bag of cement has a mass of 40 kg (88 lb).

## CONCRETE MIX DESIGN

Concrete is a very versatile material and has been adapted to many uses. Although the basic mixture is a remarkably durable material, its properties can be further enhanced by adding chemicals and additives and by varying the properties of the cement. A good concrete mix must include a number of different properties to ensure that it will demonstrate the quality that is expected during the length of its service life.

Every concrete mix begins with a trial mix using a relative small quantity of materials. The basic mix consists of three principal ingredients: water, cement, and aggregates. Each of these materials plays an important part in determining the quality of the mix. Although the compressive strength is the leading indicator for a concrete mix, other properties are equally important. Properties such as the consistency of the mix, durability, resistance to freeze-thaw action, and ease of pouring and finishing must be considered to ensure a successful result.

In the following paragraphs, an overview of the basic methods for calculating trial mixes are outlined to give the reader an insight into basic concrete mix design practices. The use and effects of some of the more common additives and admixtures are also discussed.

### Water:Cement Ratio

Concrete mix design is based on the theory that the strength of the concrete is regulated by the amount of water used per unit of cement, provided that the mix is plastic and workable. This is the *water:cement ratio theory,* which says that, as the amount of water used per pound (kilogram) of cement is increased, the strength of the concrete will decrease. Table 7-1 gives the probable compressive strengths of concrete at 28 days, based on various water:cement ratios.

Concrete mix design is also based on the assumption that sound, coarse aggregate is absolutely essential to the production of high-quality concrete. This being so, the best concrete is produced by using the largest size of aggregate and the greatest quantity of it, per given volume of concrete, that is compatible with job requirements. However, there is a practical limit to the amount of coarse aggregate that can be used in any given situation, depending on the maximum size of the aggregate and the fineness modulus of the fine aggregate used with it. Tables 7-2(a) and (b) give the volume of coarse aggregate of various sizes that can be used per volume of concrete, for various fineness moduli of fine aggregate.

**Table 7-1**
**COMPRESSIVE STRENGTHS FOR VARIOUS WATER: CEMENT RATIOS[a]**

| Water-cement Ratio by Weight | Probable Compressive Strength at 28 Days: Non-air-entrained Concrete (MPa) | (psi) | Air-entrained Concrete (MPa) | (psi) |
|---|---|---|---|---|
| 0.825 | 14 | 2000 | 11 | 1600 |
| 0.750 | 17 | 2500 | 14 | 2000 |
| 0.680 | 21 | 3000 | 17 | 2400 |
| 0.625 | 24 | 3500 | 19 | 2800 |
| 0.570 | 28 | 4000 | 22 | 3200 |
| 0.525 | 31 | 4500 | 25 | 3600 |
| 0.480 | 34 | 5000 | 28 | 4000 |
| 0.445 | 38 | 5500 | 30 | 4400 |
| 0.410 | 41 | 6000 | 33 | 4800 |

[a]For concrete subjected to severe exposures such as (1) being wet frequently or continuously, (2) being exposed to freezing and thawing, and (3) being exposed to seawater or sulfates, water:cement ratios should be restricted as follows: Thin sections and sections with less than 25-mm cover over the steel: (1) and (2) 0.45; (3) 0.40; all other structures: (1) and (2) 0.50; (3) 0.45.

**Table 7-2(a)**
**VOLUMES OF C.A., IN CU FT PER CU YD OF CONCRETE**

| Max. Size of C.A. (in.) | Fineness Modulus of F.A.: 2.40 | 2.60 | 2.80 | 3.00 |
|---|---|---|---|---|
| ⅜ | 13.5 | 13.0 | 12.4 | 11.9 |
| ½ | 15.9 | 15.4 | 14.8 | 14.3 |
| ¾ | 17.8 | 17.3 | 16.7 | 16.2 |
| 1 | 19.2 | 18.6 | 18.1 | 17.6 |
| 1½ | 20.2 | 19.7 | 19.2 | 18.6 |
| 2 | 21.1 | 20.5 | 20.0 | 19.4 |
| 3 | 22.1 | 21.6 | 21.1 | 20.5 |
| 6 | 23.8 | 23.2 | 22.7 | 22.1 |

**Table 7-2(b)**
**VOLUMES OF C.A., IN M$^3$ PER CUBIC METER OF CONCRETE**

| Max. Size of C.A. (mm) | Fineness Modulus of F.A.: 2.40 | 2.60 | 2.80 | 3.00 |
|---|---|---|---|---|
| 9.5 | 0.500 | 0.482 | 0.459 | 0.441 |
| 12.5 | 0.589 | 0.570 | 0.548 | 0.530 |
| 19 | 0.659 | 0.641 | 0.619 | 0.600 |
| 25 | 0.711 | 0.689 | 0.670 | 0.652 |
| 40 | 0.748 | 0.730 | 0.711 | 0.689 |
| 50 | 0.782 | 0.759 | 0.741 | 0.719 |
| 75 | 0.819 | 0.800 | 0.782 | 0.759 |
| 150 | 0.882 | 0.859 | 0.841 | 0.819 |

The largest size of coarse aggregate that can be used in any given situation depends on (1) the size of the unit being poured, (2) whether it is reinforced, and (3) the spacing of the reinforcement. Table 7-3 gives the recommended maximum sizes of coarse aggregate for various types of construction.

Another consideration in designing concrete is the desired plasticity, or flowability of the mix. This will depend on the type of construction, the size of the unit being poured, and whether the unit is reinforced. This flowability is known as *slump* and is measured by testing with a slump cone in a prescribed manner (see page 224 for the description of a slump test). Table 7-4 gives recommended slumps for various types of construction.

Variations in slump for a given strength of concrete are produced by altering the amount of aggregate used with the paste. However, experience has shown that to obtain a given slump, a fairly well-fixed amount of water must be used per cubic yard (cubic meter) of concrete. This will vary, depending on whether air entraining is used and on the maximum size of coarse aggregate used. Table 7-5(a) shows the approximate amount of water required for each cubic yard of concrete under various conditions. Table 7-5(b) provides similar information for each cubic meter of concrete.

The amount of water used per bag of cement is influenced not only by the strength of concrete required but also by the climatic conditions to which the concrete will be subjected, by the thickness of the section, and the location of concrete (on land or in water). See the note at the bottom of Table 7-1.

**Table 7-3**
**RECOMMENDED MAXIMUM SIZES OF AGGREGATE FOR VARIOUS TYPES OF CONSTRUCTION**

| Minimum Dimension of Section | | Reinforced Walls, Beams, and Columns | | Unreinforced Sections | | Heavily Reinforced Slabs | | Lightly Reinforced Slabs | |
|---|---|---|---|---|---|---|---|---|---|
| (in.) | (mm) | (in.) | (mm) | (in.) | (mm) | (in.) | (mm) | (in.) | (mm) |
| 2½–5 | 65–125 | ½–¾ | 12–20 | ¾ | 20 | ¾–1 | 20–25 | ¾–1½ | 20–40 |
| 6–11 | 150–280 | ¾–1½ | 20–40 | 1½ | 40 | 1½ | 40 | 1½–3 | 40–75 |
| 12–29 | 300–740 | 1½–3 | 40–75 | 3 | 75 | 1½–3 | 40–75 | 3 | 75 |
| 30 or more | 760 or more | 1½–3 | 40–75 | 6 | 150 | 1½–3 | 40–75 | 3–6 | 75–150 |

**Table 7-4**
**RECOMMENDED SLUMPS FOR VARIOUS TYPES OF CONSTRUCTION**

| Type of Construction | Slump (in.) | | Slump (mm) | |
|---|---|---|---|---|
| | Maximum | Minimum | Maximum | Minimum |
| Reinforced foundation walls and footings | 3 | 1 | 75 | 25 |
| Plain footings, caissons, and substructure walls | 3 | 1 | 75 | 25 |
| Slabs, beams, and reinforced walls | 4 | 1 | 100 | 25 |
| Building columns | 4 | 1 | 100 | 25 |
| Pavements | 2 | 1 | 50 | 25 |
| Bridge decks | 3 | 2 | 75 | 50 |
| Sidewalks, driveways, and slabs on ground | 4 | 2 | 100 | 50 |
| Heavy, mass construction | 2 | 1 | 50 | 25 |

**Table 7-5(a)**
**APPROXIMATE MIXING WATER REQUIREMENTS (lb/cu yd) FOR VARIOUS SLUMPS AND MAXIMUM SIZES OF C.A.**

| Slump (in.) | Water (lb/cu yd of Concrete of Indicated Maximum Sizes of C.A.) | | | | | | | |
|---|---|---|---|---|---|---|---|---|
| | ⅜ in. | ½ in. | ¾ in. | 1 in. | 1½ in. | 2 in. | 3 in. | 6 in. |
| | Non-air-entrained Concrete | | | | | | | |
| 1 to 2 | 350 | 333 | 308 | 300 | 275 | 258 | 242 | 209 |
| 3 to 4 | 383 | 367 | 342 | 325 | 300 | 284 | 267 | 234 |
| 5 to 6 | 408 | 384 | 359 | 342 | 317 | 300 | 284 | 250 |
| Approximate amount of entrapped air in % | 3 | 2.5 | 2 | 1.5 | 1 | 0.5 | 0.3 | 0.2 |
| | Air-entrained Concrete | | | | | | | |
| 1 to 2 | 308 | 300 | 275 | 258 | 242 | 225 | 209 | 184 |
| 3 to 4 | 342 | 325 | 300 | 284 | 267 | 250 | 234 | 200 |
| 5 to 6 | 359 | 342 | 317 | 300 | 284 | 267 | 250 | 217 |
| Recommended average total air content in % | 7.5 | 7.5 | 6 | 5 | 4.5 | 4 | 3.5 | 3 |

**Table 7-5(b)**
APPROXIMATE MIXING WATER REQUIREMENTS ($kg/m^3$) FOR VARIOUS SLUMPS AND MAXIMUM SIZES OF C.A.

| Slump (mm) | Water ($kg/m^3$ of Concrete of Indicated Maximum Sizes of C.A.) | | | | | | | |
|---|---|---|---|---|---|---|---|---|
| | 9.5 mm | 12.5 mm | 20 mm | 25 mm | 40 mm | 50 mm | 75 mm | 150 mm |
| | Non-air-entrained Concrete | | | | | | | |
| 25 to 50 | 208 | 198 | 183 | 178 | 163 | 153 | 144 | 124 |
| 75 to 100 | 227 | 218 | 203 | 193 | 178 | 168 | 158 | 139 |
| 125 to 150 | 242 | 228 | 213 | 203 | 188 | 178 | 168 | 148 |
| Approximate amount of entrapped air in % | 3 | 2.5 | 2 | 1.5 | 1 | 0.5 | 0.3 | 0.2 |
| | Air-entrained Concrete | | | | | | | |
| 25 to 50 | 183 | 178 | 163 | 153 | 144 | 133 | 124 | 109 |
| 75 to 100 | 203 | 193 | 178 | 168 | 158 | 148 | 139 | 119 |
| 125 to 150 | 213 | 203 | 188 | 178 | 168 | 158 | 148 | 129 |
| Recommended average total air content in % | 7.5 | 7.5 | 6 | 5 | 4.5 | 4 | 3.5 | 3 |

## Absolute Volume Method

Concrete is made by binding together aggregates with cement paste in such a way as to produce a strong, impervious material containing no voids in its bulk. In other words, a given quantity of concrete is made up of solid material; theoretically, no air spaces are involved. This product is made by mixing four basic ingredients, only one of which (water) contains no voids. The other three are made up of particles that have air spaces between them; consequently, in a given volume of any of these, only part is actually usable. The part of the given volume that is actually cement or stone is known as the *absolute volume* (volume without voids). To get a true picture of the quantities of solid materials to be used to produce concrete, it seems reasonable to calculate these materials in terms of their absolute volume and then convert the absolute volumes to weight (mass). The absolute volume of a loose, dry material is found by the formula

$$\text{Absolute volume} = \frac{\text{weight of loose, dry material}}{\text{relative density} \times \text{unit weight of water}}$$

For example, the absolute volume of 100 lb of portland cement with a relative density of 3.15 will be

$$\text{Absolute volume} = \frac{100}{3.15 \times 62.4} = 0.51 \text{ cu ft}$$

When using metric units, the procedure is the same except that mass is substituted for weight. Using 55 kg of portland cement, the absolute volume calculation is

$$\text{Absolute volume} = \frac{55}{3.15 \times 1000} = 0.1746 \text{ m}^3$$

Before calculations can be made, certain data must be known. These items include the following:

1. Relative density of F.A. and C.A.
2. Dry-rodded unit weight (mass) of C.A.
3. Fineness modulus of F.A.
4. Maximum size of C.A.
5. Cement factor or water:cement ratio for concrete to be designed
6. Percentage of air entrained or entrapped
7. Required slump of the concrete

With this information and the aid of tables or simple calculations, the quantities of cement, coarse aggregate, water, and air required in pounds per cubic yard can be determined. The absolute volumes (in cubic feet) of the four can then be calculated and totaled. Based on 1 cu yd (27 cu ft) of mix, subtracting the total of the four from 27 will provide the absolute volume of the fine aggregate required. From the absolute volume, the weight of the fine aggregate can then be calculated.

In metric units, the calculations are based on a volume of 1 $m^3$. The absolute volumes of the cement, coarse aggregate, water, and air are subtracted from 1 to obtain the absolute volume of the fine aggregate required. From the absolute volume, the mass of the fine aggregate is calculated.

Thus, the quantities of materials required for 1 cu yd (1 $m^3$) of concrete have been estimated and a trial batch based on these quantities can be made. If adjustments are necessary, further batches should be adjusted by keeping the water:cement ratio constant and adjusting the aggregates and entrained air to produce the desired slump and air content.

It is often desirable to make small trial batches using, for example, 20 lb (10 kg) of cement as a base measure or reducing the quantities of all of the ingredients to produce a trial batch of 1 cu ft or 0.1 $m^3$.

---

***Example:***

Concrete is required for a 10-in. retaining wall that will be subject to frequent wetting by fresh water. 4,000-psi concrete is specified. The sp. gr. of the fine aggregate is given as 2.75 and that of the coarse aggregate as 2.65. The dry-rodded weight of the coarse aggregate is 105 lb, with the maximum size 1½ in. Air content is to be 5 (±1)% and the slump 3 to 4 in.

Data required:

1. From Table 7-1, water:cement ratio, = 0.48
2. Max. size C.A. ½ in
3. Sp. gr. C.A. = 2.65
4. Dry-rodded wt. C.A. = 105 lb
5. Sp. gr. F.A. = 2.75
6. Slump = 3–4 in.
7. Air content = 5 (±1)%

***Solution:***

1. Weights:
   From Table 7-5(a), water required/cu yd = 267 lb
   Cement factor = 267/0.48 = 556 lb/cu yd of concrete
   From Table 7-2, vol. of C.A. required/cu yd = 19.3 cu ft
   Wt. of C.A. required/cu yd = 19.3 × 105 = 2,026 lb
2. Absolute volumes:

$$\text{Cement} = \frac{556}{3.15 \times 62.4} = 2.83 \text{ cu ft}$$

$$\text{Water} = \frac{267}{62.4} = 4.28 \text{ cu ft}$$

$$\text{Coarse aggregate} = \frac{2026}{2.65 \times 62.4} = 12.25 \text{ cu ft}$$

$$\text{Air} = 0.05 \times 27 = \underline{1.35 \text{ cu ft}}$$

$$\text{Total} \quad 20.71 \text{ cu ft}$$

$$\text{Fine aggregate} = 27 - 20.71 = 6.29 \text{ cu ft}$$

3. Wt. of F.A. = 6.29 × 2.75 × 62.4 = 1,079 lb
4. Trial batch, using 20 lb of cement:

   Cement, 20 lb.; C.A., 20/556 × 2026 = 73 lb; F.A., 20/556 × 1079 = 39 lb; water = 0.48 × 20 = 9.6 lb

---

***Example:***

Concrete is required for a 250-mm retaining wall that will be subject to frequent wetting by fresh water. The concrete specified is 30 MPa. The relative density of the fine aggregate is given as 2.75 and that of the coarse aggregate is 2.65. The fineness modulus of the fine aggregate is 2.8. The unit dry-rodded mass of the coarse aggregate is 1,682 kg/$m^3$, with a maximum size of 40 mm. Air content is to be 5 (±1)% and the slump is to be 75 to 100 mm.

***Solution:***

Data required:

1. Fineness modulus, F.A. = 2.8
2. Max. size coarse aggregate = 40 mm
3. Relative density C.A. = 2.65
4. Unit dry-rodded mass C.A. = 1,682 kg/$m^3$
5. Relative density, fine aggregate = 2.75
6. Slump = 75 to 100 mm
7. Air content = 5 (±1)%

From Table 7-1, water:cement ratio = 0.48

1. Masses:
   From Table 7-5(b), for 40-mm maximum aggregate size and 75- to 100-mm slump, required water/$m^3$ = 158 kg

   Cement content = 158 ÷ 0.48 = 329 kg/$m^3$ of concrete

   From Table 7-2, for fine aggregate having a fineness modulus of 2.8, required volume of coarse aggregate per cubic meter = 0.711 $m^3$.

   Mass of coarse aggregate/$m^3$ = 0.711 × 1,682 = 1196 kg
2. Absolute volumes:

$$\text{Cement} = \frac{329}{3.15 \times 1{,}000} = 0.104 \text{ m}^3$$

$$\text{Water} = \frac{158}{1{,}000} = 0.158 \text{ m}^3$$

$$\text{Coarse aggregate} = \frac{1196}{2.65 \times 1{,}000} = 0.451 \text{ m}^3$$

$$\text{Air} = 0.05 \times 1.000 = \underline{0.050 \text{ m}^3}$$

$$\text{Total} \quad 0.763 \text{ m}^3$$

$$\text{Fine aggregate} = 1.000 - 0.763 = 0.237 \text{ m}^3$$

3. Mass of fine aggregate = 0.237 × 2.75 × 1.000 = 652 kg
4. Trial batch using 10 kg of cement:
   Cement, 10 kg;
   Coarse aggregate, (10/329) × 1,196 = 36 kg;
   Fine aggregate, (10/329) × 655 = 20 kg;
   Water, 0.48 × 10 = 4.8 kg

---

## Trial Mix Method

An alternative method of designing a concrete mix is to use the suggested trial mixes shown in Tables 7-6 and 7-7. It is quite likely that, having made a trial batch from amounts given in one of the tables, some adjustment will have to be made to produce the desired slump. When making a trial batch, enough water should be added to produce the desired slump, whether that is the amount of water given in the table. The slump, air content, and unit mass of the fresh concrete should then be measured.

---

***Example:***

Concrete is to be designed with a water:cement ratio of 0.50, a slump of 2 to 3 in., and an air content of 5 (±1)%. Fine aggregate has a fineness modulus of 2.75 and a moisture content of 5%. Maximum size of coarse aggregate is 1 in., and its moisture content is 1%.

***Solution:***

For the preceding specifications, quantities per cubic yard may be calculated from Table 7-6(a) as follows:

Water (reduced by 3% for 1-in. slump decrease) = 276 lb
Cement (maintained at same W/C ratio) = 554 lb
Fine aggregate (interpolate for 2.75 F.M.) = 1,075 lb
Course aggregate (interpolate for 2.75 F.M.) = 1,885 lb
Total = 3,790 lb

Free moisture in F.A. = 0.05 × 1075 = 54 lb
Free moisture in C.A. = 0.01 × 1885 = 10 lb

The correct weights to allow for moisture in the aggregates are as follows:

Water = 276 − 54 − 19 = 203 lb
Cement = 554 lb
F.A. = 1,075 + 54 = 1,129 lb
C.A. = 1,885 + 19 = 1,904 lb
Total = 3,790 lb

From a trial mix made with these quantities, the following measurements were taken:
Slump = 1 in.; air content = 5.5%; unit wt. = 145 lb
20 lb of added water required to produce a 3-in. slump
Total wt. of water used = 276 + 20 = 296 lb
Total wt. of concrete produced = 3790 + 20 = 3810 lb

$$\text{Volume of concrete produced} = \frac{3810}{145} = 26.27 \text{ cu ft}$$

To produce 27 cu ft per trial batch, adjust the quantities as follows:

$$\text{Water} = \frac{27}{26.27} \times 296 = 304 \text{ lb}$$

$$\text{Cement} = \frac{304}{0.50} = 608 \text{ lb}$$

Total wt. of materials/cu yd = 145 × 27 = 3,915 lb
Percentage of F.A. (interpolated from Table 7-6(a)) = 36%

Wt. of F.A./cu yd = 0.36 × 3,003 = 1,080 lb
Wt. of C.A./cu yd = 3,003 − 1,080 = 1,923 lb

To obtain the best aggregate proportion for workability and economy of materials, further trial batches should be made, altering the ratio of fine and coarse aggregates.

---

As in the previous example that used standard units, the calculations in metric are very similar. The following example illustrates the approach taken to calculate the required proportions of the various materials for a trial mix. Additional mixes, as was suggested, should be made to ensure a workable mix.

---

***Example:***

A concrete mix is to be designed with a water:cement ratio of 0.50, a slump of 50 to 75 mm, and an air content of 5 (±1)%. Fine aggregate has a fineness modulus of 2.75 and moisture content of 5%. Maximum size of coarse aggregate is 25 mm, and its moisture content is 1%.

***Solution:***

For these provided specifications, quantities per cubic meter may be calculated from Table 7-6(b) as follows:

Water (reduced by 3% for 25-mm reduction in slump) = 164 kg
Cement (maintained at same W:C ratio) = 328 kg
Fine aggregate (interpolate for 2.75 F.M.) = 638 kg
Coarse aggregate (interpolate for 2.75 F.M.) = 1,118 kg
Total = 2,248 kg

Free moisture in F.A. = 0.05 × 638 = 32 kg
Free moisture in C.A. = 0.01 × 1,118 = 11 kg

The correct masses, to allow for moisture in the aggregates, are

Water = 164 − (32 + 11) = 121 kg
Cement = 328 kg
F.A. = 638 + 32 = 670 kg
C.A. = 1118 + 11 = 1,129 kg
Total = 2,248 kg

From a trial mix made with the foregoing quantities, the following test results were obtained:

Slump = 25 mm; air content = 5.5%; unit mass = 2323 kg/m$^3$; and 10 kg of water were added to obtain a 75-mm slump.

Total mass of water used = 121 + 10 = 131 kg
Total mass of concrete produced = 2,248 + 10 = 2,258 kg
Volume of concrete produced = 2,258 ÷ 2,323 = 0.972 m$^3$

**Table 7-6(a)**

SUGGESTED TRIAL MIXES FOR AIR-ENTRAINED CONCRETE OF MEDIUM CONSISTENCY (3- TO 4-IN. SLUMP)[a]

| | | | | | With Fine Sand, Fineness Modulus 2.50 | | | With Coarse Sand, Fineness Modulus 2.90 | | |
|---|---|---|---|---|---|---|---|---|---|---|
| Water: Cement Ratio (lb/lb) | Max. Size of Aggregate in. | Air Content (%) | Water (lb/cu yd of Concrete) | Cement (lb/cu yd of Concrete) | F.A. (% of Total Aggregate) | F.A. (lb/cu yd of Concrete) | C.A. (lb/cu yd of Concrete) | F.A. (% of Total Aggregate) | F.A. (lb/cu yd of Concrete) | C.A. (lb/cu yd of Concrete) |
| 0.40 | ⅜ | 7.5 | 340 | 850 | 50 | 1250 | 1260 | 54 | 1360 | 1150 |
| | ½ | 7.5 | 325 | 815 | 41 | 1060 | 1520 | 46 | 1180 | 1400 |
| | ¾ | 6 | 300 | 750 | 35 | 970 | 1800 | 39 | 1090 | 1680 |
| | 1 | 6 | 285 | 715 | 32 | 900 | 1940 | 36 | 1010 | 1830 |
| | 1½ | 5 | 265 | 665 | 29 | 870 | 2110 | 33 | 990 | 1990 |
| 0.45 | ⅜ | 7.5 | 340 | 755 | 51 | 1330 | 1260 | 56 | 1440 | 1150 |
| | ½ | 7.5 | 325 | 720 | 43 | 1140 | 1520 | 47 | 1260 | 1400 |
| | ¾ | 6 | 300 | 665 | 37 | 1040 | 1800 | 41 | 1160 | 1680 |
| | 1 | 6 | 285 | 635 | 33 | 970 | 1940 | 37 | 1080 | 1830 |
| | 1½ | 5 | 265 | 590 | 31 | 930 | 2110 | 35 | 1050 | 1990 |
| 0.50 | ⅜ | 7.5 | 340 | 680 | 53 | 1400 | 1260 | 57 | 1510 | 1150 |
| | ½ | 7.5 | 325 | 650 | 44 | 1200 | 1520 | 49 | 1320 | 1400 |
| | ¾ | 6 | 300 | 600 | 38 | 1100 | 1800 | 42 | 1220 | 1680 |
| | 1 | 6 | 285 | 570 | 34 | 1020 | 1940 | 38 | 1130 | 1830 |
| | 1½ | 5 | 265 | 530 | 32 | 980 | 2110 | 36 | 1100 | 1990 |
| 0.55 | ⅜ | 7.5 | 340 | 620 | 54 | 1450 | 1260 | 58 | 1560 | 1150 |
| | ½ | 7.5 | 325 | 590 | 45 | 1250 | 1520 | 49 | 1370 | 1400 |
| | ¾ | 6 | 300 | 545 | 39 | 1140 | 1800 | 43 | 1260 | 1680 |
| | 1 | 6 | 285 | 520 | 35 | 1060 | 1940 | 39 | 1170 | 1830 |
| | 1½ | 5 | 265 | 480 | 33 | 1030 | 2110 | 37 | 1150 | 1990 |
| 0.60 | ⅜ | 7.5 | 340 | 565 | 54 | 1490 | 1260 | 58 | 1600 | 1150 |
| | ½ | 7.5 | 325 | 540 | 46 | 1290 | 1520 | 50 | 1410 | 1400 |
| | ¾ | 6 | 300 | 500 | 40 | 1180 | 1880 | 44 | 1300 | 1680 |
| | 1 | 6 | 285 | 475 | 36 | 1100 | 1940 | 40 | 1210 | 1830 |
| | 1½ | 5 | 265 | 440 | 33 | 1060 | 2110 | 37 | 1180 | 1990 |
| 0.65 | ⅜ | 7.5 | 340 | 525 | 55 | 1530 | 1260 | 59 | 1640 | 1150 |
| | ½ | 7.5 | 325 | 500 | 47 | 1330 | 1520 | 51 | 1450 | 1400 |
| | ¾ | 6 | 300 | 460 | 40 | 1210 | 1880 | 44 | 1330 | 1680 |
| | 1 | 6 | 285 | 440 | 37 | 1130 | 1940 | 40 | 1240 | 1830 |
| | 1½ | 5 | 265 | 410 | 34 | 1090 | 2110 | 38 | 1210 | 1990 |
| 0.70 | ⅜ | 7.5 | 340 | 485 | 55 | 1560 | 1260 | 59 | 1670 | 1150 |
| | ½ | 7.5 | 325 | 465 | 47 | 1360 | 1520 | 51 | 1480 | 1400 |
| | ¾ | 6 | 300 | 430 | 41 | 1240 | 1800 | 45 | 1360 | 1680 |
| | 1 | 6 | 285 | 405 | 37 | 1160 | 1940 | 41 | 1270 | 1830 |
| | 1½ | 5 | 265 | 380 | 34 | 1110 | 2110 | 38 | 1230 | 1990 |

[a]Increase or decrease water per cubic yard by 3% for each increase or decrease of 1 in. in slump; then calculate quantities by absolute volume method. For manufactured fine aggregate, increase percentage of fine aggregate by 3 and water by 17 lb/cu yd of concrete. For less workable concrete, as in pavements, decrease percentage of fine aggregate by 3 and water by 8 lb/cu yd of concrete.

*Source:* Reprinted by permission of Portland Cement Association.

**Table 7-6(b)**

**SUGGESTED TRIAL MIXES FOR AIR-ENTRAINED CONCRETE OF MEDIUM CONSISTENCY (75- TO 100-MM SLUMP)[a]**

| | | | | | With Fine Sand Fineness Modulus 2.50 | | | With Coarse Sand Fineness Modulus 2.9 | | |
|---|---|---|---|---|---|---|---|---|---|---|
| Water: Cement Ratio (kg/kg) | Max. Size of Aggregate (mm) | Air Content (%) | Water ($kg/m^3$ of Concrete) | Cement ($kg/m^3$ of Concrete) | F.A. (% of Total Aggregate) | F.A. ($kg/m^3$ of Concrete) | C.A. ($kg/m^3$ of Concrete) | F.A. (% of Total Aggregate) | F.A. ($kg/m^3$ of Concrete) | C.A. ($kg/m^3$ of Concrete) |
| 0.4 | 9.5 | 7.5 | 202 | 504 | 50 | 742 | 748 | 54 | 807 | 682 |
| | 12.5 | 7.5 | 193 | 484 | 41 | 629 | 902 | 46 | 700 | 831 |
| | 20 | 6 | 178 | 445 | 35 | 576 | 1068 | 39 | 647 | 997 |
| | 25 | 6 | 169 | 424 | 32 | 534 | 1151 | 36 | 599 | 1086 |
| | 38 | 5 | 157 | 395 | 29 | 516 | 1252 | 33 | 587 | 1181 |
| 0.45 | 9.5 | 7.5 | 202 | 448 | 51 | 789 | 748 | 56 | 854 | 682 |
| | 12.5 | 7.5 | 193 | 427 | 43 | 676 | 902 | 47 | 748 | 831 |
| | 20 | 6 | 178 | 395 | 37 | 617 | 1068 | 41 | 688 | 997 |
| | 25 | 6 | 169 | 377 | 33 | 576 | 1151 | 37 | 641 | 1086 |
| | 38 | 5 | 157 | 350 | 31 | 552 | 1252 | 35 | 623 | 1181 |
| 0.5 | 9.5 | 7.5 | 202 | 403 | 53 | 831 | 748 | 57 | 896 | 682 |
| | 12.5 | 7.5 | 193 | 386 | 44 | 712 | 902 | 49 | 783 | 831 |
| | 20 | 6 | 178 | 356 | 38 | 653 | 1068 | 42 | 724 | 997 |
| | 25 | 6 | 169 | 338 | 34 | 605 | 1151 | 38 | 670 | 1086 |
| | 38 | 5 | 157 | 314 | 32 | 581 | 1252 | 36 | 653 | 1181 |
| 0.55 | 9.5 | 7.5 | 202 | 368 | 54 | 860 | 748 | 58 | 926 | 682 |
| | 12.5 | 7.5 | 193 | 350 | 45 | 742 | 902 | 49 | 813 | 831 |
| | 20 | 6 | 178 | 323 | 39 | 676 | 1068 | 43 | 748 | 997 |
| | 25 | 6 | 169 | 309 | 35 | 629 | 1151 | 39 | 694 | 1086 |
| | 38 | 5 | 157 | 285 | 33 | 611 | 1252 | 37 | 682 | 1181 |
| 0.6 | 9.5 | 7.5 | 202 | 335 | 54 | 884 | 748 | 58 | 949 | 682 |
| | 12.5 | 7.5 | 193 | 320 | 46 | 765 | 902 | 50 | 837 | 831 |
| | 20 | 6 | 178 | 297 | 40 | 700 | 1068 | 44 | 771 | 997 |
| | 25 | 6 | 169 | 282 | 36 | 653 | 1151 | 40 | 718 | 1086 |
| | 38 | 5 | 157 | 261 | 33 | 629 | 1252 | 37 | 700 | 1181 |
| 0.65 | 9.5 | 7.5 | 202 | 311 | 55 | 908 | 748 | 59 | 973 | 682 |
| | 12.5 | 7.5 | 193 | 297 | 47 | 789 | 902 | 51 | 860 | 831 |
| | 20 | 6 | 178 | 273 | 40 | 718 | 1068 | 44 | 789 | 997 |
| | 25 | 6 | 169 | 261 | 37 | 670 | 1151 | 40 | 736 | 1086 |
| | 38 | 5 | 157 | 243 | 34 | 647 | 1252 | 38 | 718 | 1181 |
| 0.7 | 9.5 | 7.5 | 202 | 288 | 55 | 926 | 748 | 59 | 991 | 682 |
| | 12.5 | 7.5 | 193 | 276 | 47 | 807 | 902 | 51 | 878 | 831 |
| | 20 | 6 | 178 | 255 | 41 | 736 | 1068 | 45 | 807 | 997 |
| | 25 | 6 | 169 | 240 | 37 | 688 | 1151 | 41 | 753 | 1086 |
| | 38 | 5 | 157 | 225 | 34 | 659 | 1252 | 38 | 730 | 1181 |

[a]Increase or decrease water per cubic meter by 3% for each increase or decrease of 25 mm in slump; then calculate quantities by absolute volume. For manufactured fine aggregate, increase percentage of fine aggregate by 1.8 and water by 10 $kg/m^3$ of concrete. For less workable concrete, as in pavements, decrease percentage of fine aggregate by 1.8 and water by 4.8 $kg/m^3$ of concrete.

*Source:* Original data reprinted by permission of Portland Cement Association; metric conversion by author.

**Table 7-7(a)**

**SUGGESTED TRIAL MIXES FOR NON-AIR-ENTRAINED CONCRETE OF MEDIUM CONSISTENCY (3- TO 4-IN. SLUMP)[a]**

| | | | | | With Fine Sand, Fineness Modulus 2.50 | | | With Coarse Sand, Fineness Modulus 2.90 | | |
|---|---|---|---|---|---|---|---|---|---|---|
| Water: Cement Ratio (lb/lb) | Max. Size of Aggregate (in.) | Entrapped Air Content (%) | Water (lb/cu yd of Concrete) | Cement (lb/cu yd of Concrete) | F.A. (% of Total Aggregate) | F.A. (lb/cu yd of Concrete) | C.A. (lb/cu yd of Concrete) | F.A. (% of Total Aggregate) | F.A. (lb/cu yd of Concrete) | C.A. (lb/cu yd of Concrete) |
| 0.40 | ⅜ | 3 | 385 | 965 | 50 | 1240 | 1260 | 54 | 1350 | 1150 |
| | ½ | 2.5 | 365 | 915 | 42 | 1100 | 1520 | 47 | 1220 | 1400 |
| | ¾ | 2 | 340 | 850 | 35 | 960 | 1800 | 39 | 1080 | 1680 |
| | 1 | 1.5 | 325 | 815 | 32 | 910 | 1940 | 36 | 1020 | 1830 |
| | 1½ | 1 | 300 | 750 | 29 | 880 | 2110 | 33 | 1000 | 1990 |
| 0.45 | ⅜ | 3 | 385 | 855 | 51 | 1330 | 1260 | 56 | 1440 | 1150 |
| | ½ | 2.5 | 365 | 810 | 44 | 1180 | 1520 | 48 | 1300 | 1400 |
| | ¾ | 2 | 340 | 755 | 37 | 1040 | 1800 | 41 | 1160 | 1680 |
| | 1 | 1.5 | 325 | 720 | 34 | 990 | 1940 | 38 | 1100 | 1830 |
| | 1½ | 1 | 300 | 665 | 31 | 960 | 2110 | 35 | 1080 | 1990 |
| 0.50 | ⅜ | 3 | 385 | 770 | 53 | 1400 | 1260 | 57 | 1510 | 1150 |
| | ½ | 2.5 | 365 | 730 | 45 | 1250 | 1520 | 49 | 1370 | 1400 |
| | ¾ | 2 | 340 | 680 | 38 | 1100 | 1800 | 42 | 1220 | 1680 |
| | 1 | 1.5 | 325 | 650 | 35 | 1050 | 1940 | 39 | 1160 | 1830 |
| | 1½ | 1 | 300 | 600 | 32 | 1010 | 2110 | 36 | 1130 | 1990 |
| 0.55 | ⅜ | 3 | 385 | 700 | 54 | 1460 | 1260 | 58 | 1570 | 1150 |
| | ½ | 2.5 | 365 | 665 | 46 | 1310 | 1520 | 51 | 1430 | 1400 |
| | ¾ | 2 | 340 | 620 | 39 | 1150 | 1800 | 43 | 1270 | 1680 |
| | 1 | 1.5 | 325 | 590 | 36 | 1100 | 1940 | 40 | 1210 | 1830 |
| | 1½ | 1 | 300 | 545 | 33 | 1060 | 2110 | 37 | 1180 | 1990 |
| 0.60 | ⅜ | 3 | 385 | 640 | 55 | 1510 | 1260 | 58 | 1620 | 1150 |
| | ½ | 2.5 | 365 | 610 | 47 | 1350 | 1520 | 51 | 1470 | 1400 |
| | ¾ | 2 | 340 | 565 | 40 | 1200 | 1800 | 44 | 1320 | 1680 |
| | 1 | 1.5 | 325 | 540 | 37 | 1140 | 1940 | 41 | 1250 | 1830 |
| | 1½ | 1 | 300 | 500 | 34 | 1090 | 2110 | 38 | 1210 | 1990 |
| 0.65 | ⅜ | 3 | 385 | 590 | 55 | 1550 | 1260 | 59 | 1660 | 1150 |
| | ½ | 2.5 | 365 | 560 | 48 | 1390 | 1520 | 52 | 1510 | 1400 |
| | ¾ | 2 | 340 | 525 | 41 | 1230 | 1800 | 44 | 1350 | 1680 |
| | 1 | 1.5 | 325 | 500 | 38 | 1180 | 1940 | 41 | 1290 | 1830 |
| | 1½ | 1 | 300 | 460 | 35 | 1130 | 2110 | 39 | 1250 | 1990 |
| 0.70 | ⅜ | 3 | 385 | 550 | 56 | 1590 | 1260 | 60 | 1700 | 1150 |
| | ½ | 2.5 | 365 | 520 | 48 | 1430 | 1520 | 53 | 1550 | 1400 |
| | ¾ | 2 | 340 | 485 | 41 | 1270 | 1800 | 45 | 1390 | 1680 |
| | 1 | 1.5 | 325 | 465 | 38 | 1210 | 1940 | 42 | 1320 | 1830 |
| | 1½ | 1 | 300 | 430 | 35 | 1150 | 2110 | 39 | 1270 | 1990 |

[a]See Table 7-6 footnote.

*Source:* Reprinted by permission of Portland Cement Association.

**Table 7-7(b)**

SUGGESTED TRIAL MIXES FOR NON-AIR-ENTRAINED CONCRETE OF MEDIUM CONSISTENCY (75- TO 100-MM SLUMP)[a]

| | | | | | With Fine Sand Fineness Modulus 2.50 | | | With Coarse Sand Fineness Modulus 2.9 | | |
|---|---|---|---|---|---|---|---|---|---|---|
| Water: Cement Ratio (kg/kg) | Max. Size of Aggregate (mm) | Entrapped Air Content (%) | Water (kg/m³ of Concrete) | Cement (kg/m³ of Concrete) | F.A. (% of Total Aggregate) | F.A. (kg/m³ of Concrete) | C.A. (kg/m³ of Concrete) | F.A. (% of Total Aggregate) | F.A. (kg/m³ of Concrete) | C.A. (kg/m³ of Concrete) |
| 0.4 | 9.5 | 3 | 228 | 573 | 50 | 736 | 748 | 54 | 801 | 682 |
| | 12.5 | 2.5 | 217 | 543 | 42 | 653 | 902 | 47 | 724 | 831 |
| | 20 | 2 | 202 | 504 | 35 | 570 | 1068 | 39 | 641 | 997 |
| | 25 | 1.5 | 193 | 484 | 32 | 540 | 1151 | 36 | 605 | 1086 |
| | 38 | 1 | 178 | 445 | 29 | 522 | 1252 | 33 | 593 | 1181 |
| 0.45 | 9.5 | 3 | 228 | 507 | 51 | 789 | 748 | 56 | 854 | 682 |
| | 12.5 | 2.5 | 217 | 481 | 44 | 700 | 902 | 48 | 771 | 831 |
| | 20 | 2 | 202 | 448 | 37 | 617 | 1068 | 41 | 688 | 997 |
| | 25 | 1.5 | 193 | 427 | 34 | 587 | 1151 | 38 | 653 | 1086 |
| | 38 | 1 | 178 | 395 | 31 | 570 | 1252 | 35 | 641 | 1181 |
| 0.5 | 9.5 | 3 | 228 | 457 | 53 | 831 | 748 | 57 | 896 | 682 |
| | 12.5 | 2.5 | 217 | 433 | 45 | 742 | 902 | 49 | 813 | 831 |
| | 20 | 2 | 202 | 403 | 38 | 653 | 1068 | 42 | 724 | 997 |
| | 25 | 1.5 | 193 | 386 | 35 | 623 | 1151 | 39 | 688 | 1086 |
| | 38 | 1 | 178 | 356 | 32 | 599 | 1252 | 36 | 670 | 1181 |
| 0.55 | 9.5 | 3 | 228 | 415 | 54 | 866 | 748 | 58 | 931 | 682 |
| | 12.5 | 2.5 | 217 | 395 | 46 | 777 | 902 | 51 | 848 | 831 |
| | 20 | 2 | 202 | 368 | 39 | 682 | 1068 | 43 | 753 | 997 |
| | 25 | 1.5 | 193 | 350 | 36 | 653 | 1151 | 40 | 718 | 1086 |
| | 38 | 1 | 178 | 323 | 33 | 629 | 1252 | 37 | 700 | 1181 |
| 0.6 | 9.5 | 3 | 228 | 380 | 55 | 896 | 748 | 58 | 961 | 682 |
| | 12.5 | 2.5 | 217 | 362 | 47 | 801 | 902 | 51 | 872 | 831 |
| | 20 | 2 | 202 | 335 | 40 | 712 | 1068 | 44 | 783 | 997 |
| | 25 | 1.5 | 193 | 320 | 37 | 676 | 1151 | 41 | 742 | 1086 |
| | 38 | 1 | 178 | 297 | 34 | 647 | 1252 | 38 | 718 | 1181 |
| 0.65 | 9.5 | 3 | 228 | 350 | 55 | 920 | 748 | 59 | 985 | 682 |
| | 12.5 | 2.5 | 217 | 332 | 48 | 825 | 902 | 52 | 896 | 831 |
| | 20 | 2 | 202 | 311 | 41 | 730 | 1068 | 45 | 801 | 997 |
| | 25 | 1.5 | 193 | 297 | 38 | 700 | 1151 | 41 | 765 | 1086 |
| | 38 | 1 | 178 | 273 | 35 | 670 | 1252 | 39 | 742 | 1181 |
| 0.7 | 9.5 | 3 | 228 | 326 | 56 | 943 | 748 | 60 | 1009 | 682 |
| | 12.5 | 2.5 | 217 | 309 | 48 | 848 | 902 | 53 | 920 | 831 |
| | 20 | 2 | 202 | 288 | 41 | 753 | 1068 | 45 | 825 | 997 |
| | 25 | 1.5 | 193 | 276 | 38 | 718 | 1151 | 42 | 783 | 1086 |
| | 38 | 1 | 178 | 255 | 35 | 682 | 1252 | 39 | 753 | 1181 |

[a]See Table 7-6 footnote.

*Source:* Original data reprinted by permission of Portland Cement Association; metric conversion by author.

To produce 1 $m^3$ per trial batch, adjust the quantities as follows:

$$\text{Water} = 131/0.972 = 135 \text{ kg}$$
$$\text{Cement} = 135/0.5 = 270 \text{ kg}$$

Because total mass of materials per cubic meter = 2,323 kg, and the percentage of fine aggregate (interpolated from Table 7-6(b)) = 36%

$$\text{Mass of fine aggregate/m}^3 = 0.36 \times (2{,}323 - 135 - 270) = 690 \text{ kg}$$

$$\text{Mass of course aggregate/m}^3 = (2{,}323 - 135 - 270) - 690 = 1{,}228 \text{ kg}$$

## Concrete Admixtures

Ordinary concrete consisting of Portland cement, water, and aggregates, although very useful, was found to be lacking in durability when exposed to cycles of freezing and thawing. Efforts to increase its useful life span led researchers to discover that a concrete mix containing a small percentage of entrained air had much better resistance to the effects of freeze-thaw cycles than an ordinary concrete mix. Very quickly, air-entrained concrete became the standard for all structural applications where durability and strength were a concern. In addition to an increase in durability, it was found that the addition of a small amount of air-entraining agent produced a more fluid concrete mix, thereby reducing the amount of water that had to be used in the mix design. Thus the new age of concrete mix design began.

Since that initial development, many additives are now available to modify and enhance concrete properties. Organizations such as the Portland Cement Association (PCA) and the American Concrete Institute (ACI) have contributed extensively to the knowledge of proper use of admixtures in concrete mix design. The American Society for Testing and Materials (ASTM) provides standard tests for the evaluation of concrete mix properties to ensure consistency in the quality of the concrete throughout the industry. Some of the more common admixtures and their effects are as follows:

1. Accelerators, chlorine type and non-chlorine type, improve the setting time and early strength gain.
2. Air-entraining admixtures improve the durability and workability of the concrete mix. ASTM standard test C233 is the test method used for evaluating air-entraining admixtures for concrete. The Canadian Standards Association (CSA) provides CSA Standard A266.1 for the same purpose. ASTM standard C260 provides specifications for air-entraining admixtures for concrete.
3. Coloring agents provide different colors for concrete. ASTM C979 provides a specification for pigments for integrally colored concrete.
4. Pumping additives improve the pumpability of concrete. Organic polymers, organic flocculents, bentonite, natural pozzolans, fly ash, and hydrated lime are typical admixtures.
5. Retarders increase the setting time of a concrete mix by slowing the initial set of the cement. This slowing provides a more homogenous integration of concrete poured in lifts (layers), reducing the formation of cold joints within structures such as a wall or deep girder.
6. Superplasticizers increase concrete flow and retard the setting time. Because flow is increased, the quantity of the water in the mix can be reduced.
7. Water reducers reduce the quantity of water required in the mix.

Bonding admixtures to increase bond strength, fungicides, hardeners, and dampproofing admixtures are also available where circumstances may warrant their use. Comprehensive information on admixtures available to the concrete industry may be obtained from organizations such as the PCA. It is also good practice to contact the appropriate manufacturer and obtain full details on the proper use of these products prior to their application.

## High Strength Concrete

The term high strength concrete is applied to mixes that achieve strengths of 6,000 psi (40 MPa) or higher. Strengths as high as 20,000 psi (140 MPa) can be achieved with the proper selection of materials and with strict quality control. To obtain these high strengths, adherence to all design specifications must occur at every stage of the operation—from the mix design to the final curing of the concrete. The concrete producer plays a key role in the operation by providing uniform batches of concrete to the site.

During the mix design stage, the final selection and proportioning of all the materials to be used in the mix will depend on the required design strength of the concrete. Cement selection is based on cement cube sample strength at seven days rounded out by compression tests on standard concrete samples. Cement contents ranging between 25 to 35 lb/cu ft (400 to 550 kg/m$^3$) are recommended for initial trial mixes. To affirm the quality of the cement, tests on cement cube test samples should have a seven day minimum strength of approximately 4,000 psi (30 MPa) to ensure the desired results. Cement quality can be further verified by supplementing standard twenty-eight day testing of concrete samples with additional tests done at 56 and 90 days.

Adding cementing materials to the concrete mix such as silica fume or fly ash enhances the strength of the cement paste. However, this must be done with care to ensure that actual production volumes of concrete maintain the desired strength, durability, and volume stability. Usual amounts of these additives range between 5% to 20% by weight of Portland cement. ASTM Standard C618 provides guidelines for the addition of fly ash and natural pozzolans. CSA Standard A23.1 provides limits on quantities of silica fume that may be used in concrete mixes. The PCA and the ACI also have a wealth of information on the appropriate use of these materials. Trial mixes should maintain a slump between 3–4 inches (75–100 mm).

The selection of aggregates for high strength concrete is as critical as the selection of the cementing materials. It has been found that crushed rock aggregates provide better results than gravel aggregates due to their rough angular shapes. Coarse aggregates may need to be washed to ensure a maximum bond between the cementitious paste and the aggregates. It has been found that maximum aggregate sizes ranging from ⅜″ to 9/16″ (10 to 14 mm) provide excellent strength characteristics to the mix. Aggregates up to 1⅛ inches (28 mm) in size, however, have been used. Because of the high amounts of cement and other additives, good workability can be obtained with coarser sand than is required for conventional concrete mixes. A fineness modulus of 3 for the sand has proven to work well. Finer sands may produce a sticky mix.

Admixtures such as superplasticizers to improve workability and keep the water-cementitious materials ratio to a practical minimum are highly recommended. Consultation with the concrete mix supplier will ensure the best results. Air entraining of high strength concrete for protected sections is not necessary (air entraining decreases concrete strength); however, entrained air is a must for exposed structures and structures exposed to freeze-thaw cycles.

Placing of high strength concrete on the construction site must be done as quickly as possible. Adjustments made to the concrete mix at the site must be done under the supervision of qualified personnel fully knowledgeable of the properties of high strength concrete. The arbitrary changing of water:cement ratios can produce dire consequences. Delays between lifts during the placing of the concrete must be avoided and quantities delivered to the site must match the rate of placement. Consolidation of the concrete must be done quickly and thoroughly. Because high strength concrete is usually relatively stiff, under consolidation rather than over consolidation can be a problem. Proper curing cannot be overemphasized. Adequate moisture and a moderate temperature must be maintained for an appropriate length of time to ensure the desired strength. This is particularly important when 56-day and 96-day strengths are specified.

## Use of Fast-Setting Cements

It is sometimes necessary to use concrete in situations when it is impossible to wait the required period for concrete made with normal portland cement to set and strengthen. In these situations it is possible to use a high-early-strength portland cement. The difference in early strength is illustrated in Figure 7-10.

It is also possible to use aluminous cement, which gains its strength in a very short time compared with portland cement. Concrete made with portland cement will gain about 10% of its 28-day strength in 1 day at a temperature of approximately 65°F (18°C). On the other hand, concrete made with aluminous cement attains the 28-day strength of that made from portland cement in approximately 24 hr at the same temperature. At 36°F (2°C), concrete made with normal portland cement will have gained very little if any of its 28-day strength, whereas under similar conditions, concrete made with aluminous cement will have gained about 50% of the 28-day strength of portland cement concrete (see Table 7-8). This fast-setting feature makes aluminous cement particularly valuable when concrete must be placed in cold weather. A considerable amount of heat is generated during the hydration process, and this heat is usually sufficient to ensure the continuation of the curing process if the concrete is protected from freezing for the first six hours.

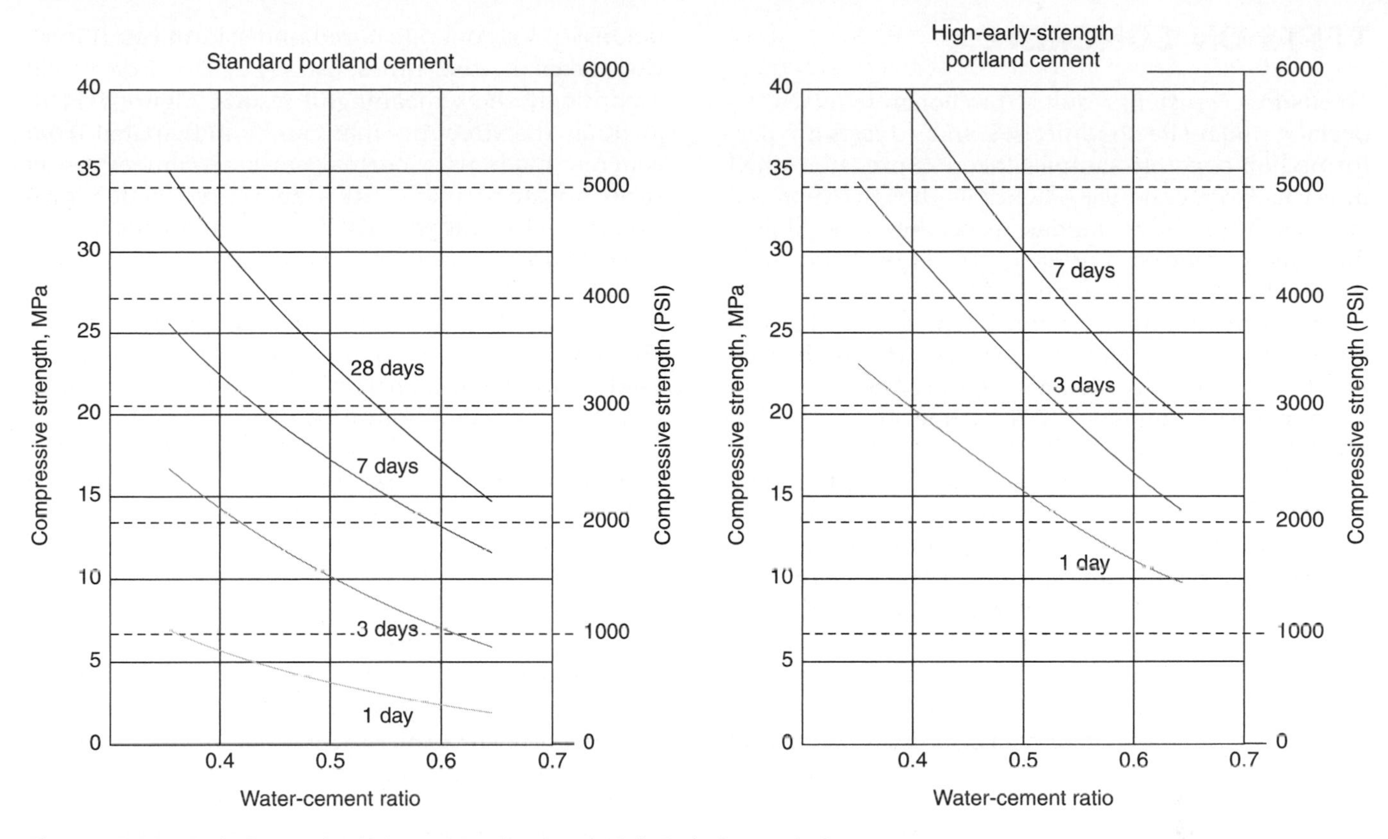


**Figure 7-10** Early Strength of Normal Portland and High-Early-Strength Cements.

**Table 7-8**
**EFFECT OF TEMPERATURE ON THE STRENGTH OF CONCRETE**

**Crushing strength of 1:2:4 concrete expressed as a percentage of 28-day strength of normal portland cement concrete.**

| Type of Cement | Age (Days) | Low Temperature 36°F (2°C) | Normal Temperature 64°F (18°C) |
|---|---|---|---|
| Normal portland cement | 1 | 0 | 10 |
| | 3 | 10 | 37 |
| | 7 | 25 | 64 |
| | 14 | 55 | 80 |
| | 28 | 92 | 100 |
| | 56 | 108 | 110 |
| Rapid-hardening portland cement | 1 | 0 | 20 |
| | 3 | 12 | 59 |
| | 7 | 37 | 93 |
| | 14 | 70 | 107 |
| | 28 | 118 | 126 |
| | 56 | 135 | 135 |
| High alumina cement | 1 | 54 | 146 |
| | 3 | 164 | 188 |
| | 7 | 188 | 197 |
| | 14 | 197 | 212 |
| | 28 | 207 | 218 |
| | 56 | 222 | 229 |

## TESTS ON CONCRETE

To ensure consistent results from concrete mixes, especially under site conditions, standard tests are performed on concrete samples taken at predetermined intervals from concrete batches as they arrive at the site. ASTM provides standard procedures for all testing that is required. For each project, specifications set out by structural engineers list the number and type of tests that must be done. Typically, as the concrete arrives on the site, the fluid concrete is checked for slump and the percentage of entrained air within the mix. ASTM standard C231 outlines the procedure for determining the amount of entrained air in fresh concrete using the pressure method. It is favoured by technicians in that it can be done quickly on the site and it provides reliable results. Additional samples of concrete are taken from the batch in the form of standard test cylinders. The number of cylinders and the method of curing are outlined in the project specifications. These samples are then tested at predetermined intervals to ensure that the concrete develops its required design strength. On occasion, flexural tests are done on concrete used in pavements and slabs on grade.

A *slump test* (ASTM C143) is made to ensure that the concrete has the flowability required for placing. A slump or slump range will usually be stipulated in specifications. The test is carried out as soon as a batch of concrete is mixed, and a standard slump cone and tamping rod are also required. The cone is made of sheet metal, 4 in. (100 mm) in diameter at the top, 8 in. (200 mm) in diameter at the bottom, and 12 in. (300 mm) high. The rod is a 5/8-in. (16-mm) bullet-nosed rod approximately 24 in. (600 mm) long. The cone is filled in three equal layers, each tamped twenty-five times with the rod. After the third layer is in place and has been tamped, the concrete is struck off level, the cone is lifted carefully and set down beside the slumped concrete, and the rod is laid across the top. The distance from the underside of the rod to the average height of the concrete is measured and registered as the amount of slump in millimeters.

A *compression test* (ASTM C39) is made to determine the quality of the cured concrete and to ensure that it meets the design specifications after being cured for a certain length of time. The tested samples will also provide proof of the quality of the cement paste and its ability to bind with the aggregates. When tested in compression, good concrete should fall through the aggregates, not around them.

In order that concrete may demonstrate its quality, it must be allowed time to gain strength—to *cure*. Concrete is assumed to have gained its full working strength within 28 days under ideal curing conditions, and that period is taken as a standard for testing purposes. But experience has shown an age–strength relationship for concrete cured under ideal conditions. Tests taken at other times, usually 1, 3, or 7 days, will accordingly show meaningful results. Figure 7-11 illustrates the strengths that can be anticipated from concrete made with normal portland cement, using various water:cement ratios and cured under ideal conditions for test periods of 1, 3, 7, and 28 days.

A standard size of test specimen (ASTM C192) is used to make the test. It is molded in a cylindrical container, 6 in. (150 mm) in diameter and 12 in. (300 mm) high (see Figure 7-11). One commonly used type of container is made of stiff waxed cardboard with a metal bottom and cardboard top.

The cylinder is filled in three equal layers, each layer having been tamped twenty-five times with the standard tamping rod. The top is struck off level, the cap is placed on, and the specimen is weighed (see Figure 7-12). It is then set aside for 24 hr. Several specimens (at least four) should be taken of each mix to be tested (see Figure 7-13). Two test cylinders can be set under actual conditions and two can be taken to a laboratory for curing under ideal conditions.

After 24 hr, those cylinders to be cured in the lab will be removed from their containers and placed in warm, moist conditions, preferably a curing cabinet, to cure for the designated period. Those cylinders left to cure under job conditions may be left in their containers but should preferably be removed when the forms are taken off or, in the case of slabs, after 24 hr to simulate job conditions as closely as possible.

At the end of the curing period the cylinders are weighed again and capped with a thin layer of capping compound (see Figure 7-14) usually made of a sulfur-based material. Caps should be planed to within 0.002 in. (0.05 mm).

Allowing the caps to harden for a 2-hour period, each cylinder is placed in a universal testing machine and loaded until failure, as illustrated in Figure 7-15. The maximum compressive load is registered on the machine consul in pounds (kilograms). The ultimate compressive strength of the concrete is calculated in pounds per square inch (MPa) by dividing the maximum load on the test sample by its cross-sectional area. Tests done at 28 days become the reference strength for use in many design calculations.

*Flexure tests* (ASTM C78 & C293) are done to provide information on the flexural or bending strength and are used to calculate the modulus of rupture of the concrete. Forms used must provide a beam with a length of at least 2 in. (50 mm) greater than 3 times its depth as tested. This minimum cross-sectional dimension must be at least 3 times the maximum nominal size of the coarse aggregate used, but in no case less than 6 by 6 in. (150 by 150 mm). After proper curing, the beam is tested for flexural strength by third-point or center-point loading.

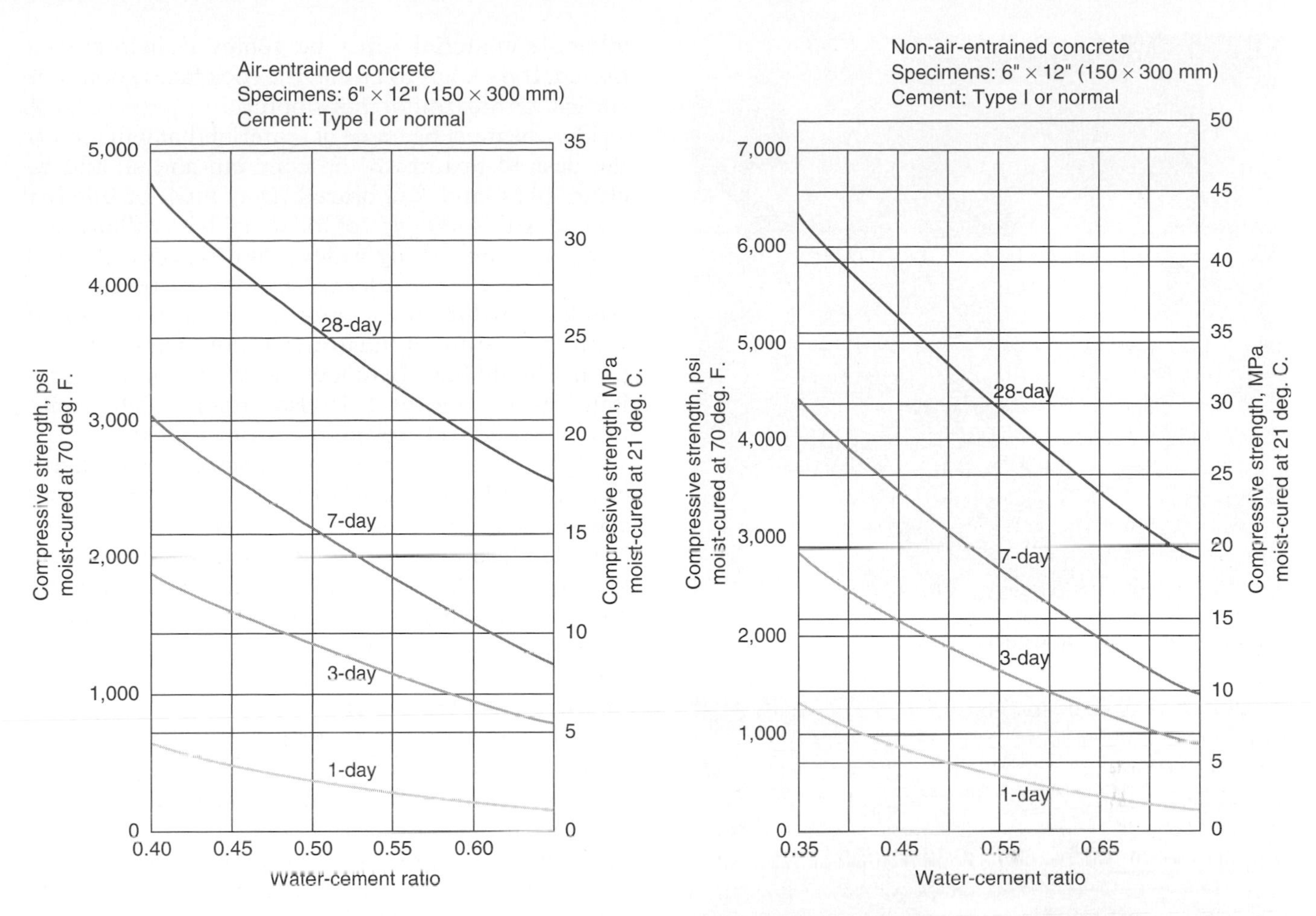


**Figure 7-11** Typical Strength-Age Relationships in Concrete, Based on Compression Tests. (Reprinted by permission of Portland Cement Association.)

**Figure 7-12** Filling Test Cylinder.

**Figure 7-13** Test Cylinders Filled and Covered. (Reprinted by permission of ELE International, Soiltest Products Division.)

**Figure 7-14** Capping Concrete Test Samples.

**Figure 7-15** Concrete Test Cylinder Ready for Loading.

## PREPARATION FOR PLACING CONCRETE

Before placing concrete, the subgrade must be properly prepared, and forms and reinforcing must be erected according to specifications. Subgrades must be trimmed to specific elevation and thoroughly compacted (see Figure 7-16) and should be moist when the concrete is placed. A moist subgrade is particularly important to prevent overly rapid extraction of water from concrete when flat work, such as floors and pavements, is being placed in hot weather. When concrete is being placed on a rock foundation, all loose material must be removed before it is placed. If rock has to be cut, the rock faces should be vertical or horizontal, not sloping.

Forms must be made of material that will impart the desired texture to the concrete and should be clean, tight, and well braced. They must be oiled or treated with some type of form seal that will prevent them from absorbing water from the concrete and will facilitate form removal. Care must be taken to see that sawdust, nails, or other debris are removed from the bottom of the forms. Column and tall wall forms should have "windows" at the bottom to facilitate debris removal. It is also important that reinforcing steel be clean and free from scales or rust.

When fresh concrete is to be placed on hardened concrete, it is important to secure a good bond and a watertight joint. The hardened concrete must be fairly level, rough, clean, and moist. Some of the aggregate particles must be exposed by cutting away part of the existing surface by sandblasting, by cleaning with hydrochloric acid, or by using a wire brush. Any laitance or soft layer of mortar must be removed from the surface.

When concrete is to be placed on a hardened concrete surface or on rock, a layer of mortar must be placed on the hard surface first. This provides a cushion on which the new concrete can be placed and stops aggregate from bouncing on the hard surface and forming stone pockets. The mortar should be approximately 2 in. (50 mm) deep, with the same water content as the concrete being placed, and should have a slump of 5 to 6 in. (125 to 150 mm).

Preparations must also include the building of adequate runways for wheelbarrows or buggies. Build runways to be reasonably smooth and rigid enough to prevent the wheelbarrows or buggies from bouncing or jarring as they travel. Some means of transferring concrete from the wheelbarrow, buggy, or bucket into the forms must also be provided. This may consist of a baffle board, hopper, downspout, or chute. These will be discussed later in the chapter.

## MIXING CONCRETE

Concrete should be mixed until it is uniform in appearance and consistency. The time required for mixing depends on the volume and stiffness of the mix, to some extent on the size of the C.A. and the fineness of the sand, and on the type of mixer being used.

### Drum Mixer

A *drum mixer,* one size of which is shown in Figure 7-17, consists of a rotating drum with stationary blades fixed on the inside. Mixing is accom-

**Figure 7-16** Compacting Subgrade.

**Figure 7-17** Six-Cubic-Foot (0.17 m$^3$) Concrete Mixer. (Courtesy of Terex Light Construction.)

plished by the action of the blades passing through the fluid concrete. For this type of mixer, specifications usually require a minimum of 1 min for mixers of up to 1 cu yd (0.75 m$^3$) capacity, with an increase of at least 15 sec for every ½ cu yd (0.38 m$^3$) (or fraction thereof) of additional capacity. If mixes are stiff, additional time will be required, and the same is true for mixes containing fine sand or small C.A. The mixing time begins when all materials are in the mixer drum.

Batch mixers are available in sizes varying from 1½-cu ft to 4-cu yd (0.04 to 3 m$^3$) capacity. For general construction work, standard mixers are rated at 3½, 6, 11, 16, and 28 cu ft (0.10, 0.17, 0.30, 0.45, and 0.80 m$^3$) of mixed concrete. For larger projects and central mixing plants, mixers of 2 to 8 cu yd (1.5 to 6 m$^3$) capacity are used.

Mixers may be tilting or nontilting with either high or low loading skips. High loading skips are suitable when the skip is being charged from a bin batcher, but the low loading type is preferable if the skip has to be charged by shovel or wheelbarrow. Nontilting mixers may be equipped with a swinging discharge chute. Many mixers are provided with timing devices so that they can be set for a given mixing period and cannot be opened until the designated mixing time is up.

Mixers should not be loaded beyond their rated capacity and should not be operated at speeds other than those for which they were designed. Overloading or running either too quickly or too slowly prevents the proper mixing action from taking place. When increased output is required, a larger mixer or additional ones should be used.

Under usual conditions, no more than approximately 10% of the mixing water should be placed in the drum before the dry materials are added. Water should then be added uniformly with the dry materials, leaving approximately 10% to be added after the dry materials are in the drum. If heated water is used during cold weather, this sequence may have to be changed. In this case, addition of cement should be delayed until most of the aggregate and water are mixed in the drum. This is done to prevent a flash set of the paste, a set brought about by too much heat when the water and cement combine.

### Countercurrent Mixer

A *countercurrent mixer* is so called because of the mixing system employed. It consists of a horizontally revolving pan, into which is suspended one or more vertical, three-paddle mixing tools, offset from the center of the pan. The pan rotates in a clockwise direction, while the mixing tools turn counterclockwise, thus producing a high degree of agitation in the materials being mixed. Additional agitators are optional equipment that intensify the movement of the particles in the mix and bring about rapid homogenization of otherwise difficult mixes. Such a mixer is particularly efficient at mixing additives, colors, and the like, into a concrete mix (see Figure 7-18).

### Temperature and Slump Control

It is important that the maximum and minimum temperatures of the concrete in the mixer be controlled in all seasons. Significant temperature changes may bring about changes in water requirement, slump, or air content. (See the sections on hot and cold weather concreting, pages 239 and 240.)

Compensation for the loss of slump due to a change in concrete temperature or to delays in delivery to the forms is sometimes brought about by the addition of some water to the mix (retempering). This practice should be carefully controlled and should be allowed only to the degree that the specified water:cement ratio is not exceeded.

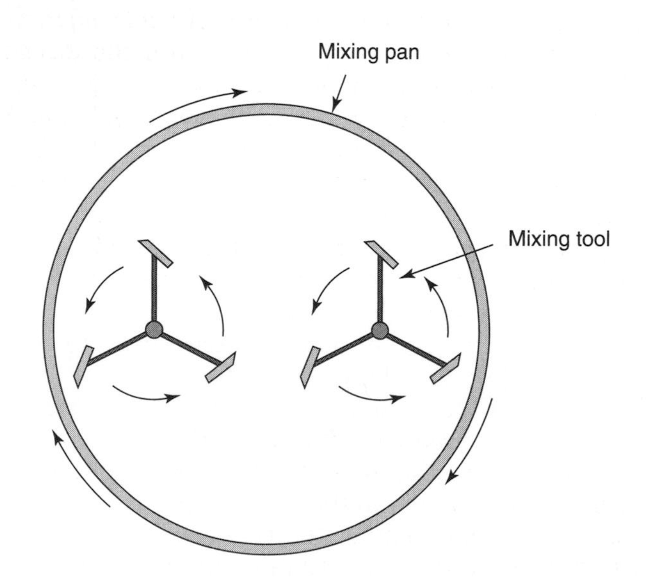


**Figure 7-18** Diagram of Countercurrent Mixer.

It is also important that the mixer be of a type that is capable of mixing and discharging concrete at the slump designated for the job at hand. In other words, the mixer must be suitable for the job, not the concrete made suitable for the mixer.

## MACHINERY FOR PLACING CONCRETE

Concrete may be mixed on the job site or brought to the site by truck from a concrete batching plant. In either case, various methods are available to the contractor to move the concrete from the mixer location or concrete truck to the pour location. Although this distance may be quite short, care must be taken to prevent excessive segregation within the concrete mix—the separation of the coarse aggregates from the cement paste and the fine aggregates. The site conditions and the type of pour will determine how the concrete will be transported and placed. Careful planning and co-ordination of the crew and the equipment on site before the concrete arrives will ensure a safe and efficient operation.

Various methods are used to move concrete on the construction site; however, the concrete pump has become the method of choice for many contractors. Although traditional methods such as wheelbarrows, push buggies, power buggies, chutes, belt conveyors, and crane-hoisted buckets have their place on the construction site as viable methods for moving concrete, the versatility and speed of the concrete pump has relegated them to secondary roles.

### Wheelbarrows and Buggies

Wheelbarrows are used where concrete volumes are small and the distance from the supply location to the forms is short. Smooth runways are required to minimize segregation of the concrete mix. Because wheelbarrows require manual labor to move them, runways must also have gradual slopes and provide a non-slip surface to ensure safe footing for the operator.

Pushcarts and power buggies are made in a variety of sizes, from approximately 32/1 to 11 cu ft (0.10 to 0.3 m$^3$) in capacity, and are equipped with pneumatic tires for smoother operation (See Figure 7-19). Although they can carry more concrete, they require robust runways that must be of sufficient width to ensure safe operating conditions. When used in high-rise construction, the loaded buggy is raised to the appropriate floor by means of a materials hoist such as the one illustrated in Figure 7-20. Moving and placing concrete in this fashion requires smooth operation of the hoist to ensure that segregation of the concrete mix is kept to a minimum.

**Figure 7-19** Concrete Power Buggy. (Reprinted by permission of Multiquip, Inc.)

**Figure 7-20** Materials Hoist for High-Rise Building.

## Chutes

Chutes may be used to carry concrete directly from mixer to forms or from a hopper conveniently situated to allow chuting. Chutes should be metal or metal lined, with rounded bottoms, and of sufficient size to guard against overflow. They should be designed so that concrete will travel fast enough to keep the chute clean but not fast enough to cause segregation. It is generally recommended that the slope of chutes be between 3 to 1 and 2 to 1, although steeper chutes may be used to carry stiff mixes.

## Buckets

Buckets lifted and moved about by crane or cable are commonly used where concrete has to be placed at a considerable height above ground level or where forms are in otherwise inaccessible locations (see Figure 7-21).

Buckets vary in size from approximately ½ to 8 cu yd (0.4 to 6 m$^3$) of capacity and may be circular or rectangular in cross section (see Figure 7-22). The load is released by opening a gate that forms the bottom of the bucket. Gates that can regulate the flow and close when only part of the load has been discharged are preferred where sections are small and it is not desirable to place a full load in one location [see Figure 7-22(b)].

Care should be taken to prevent jarring or shaking of buckets while they are in transit. This movement will cause segregation, particularly if the concrete has a relatively high rate of slump.

## Belt Conveyors

Belt conveyors used to transfer concrete may be classified into three types: portable (see Figure 7-23), series, and side discharge conveyors. Series (feeder) conveyors operate at fairly high belt speeds 500 ft per min (150 m/min, or better), whereas the other two conveyors operate best at considerably slower speeds.

Concrete should be fed onto a conveyor belt from a hopper to get an even distribution of material along the belt, and it should be well enough supported that it will not vibrate and cause segregation of the concrete. The slope used will vary with the concrete mix and with the type of belt used. Those with straight ribs on their surface work best on steep slopes. Conveyors should be covered to prevent climatic conditions (sun, rain, or wind) from affecting the concrete during its transfer.

## Pumps

Concrete pumps are used to transport concrete under pressure through some type of piping system. Pumps are of three types: *piston pump, pneumatic pump,* and *squeeze pressure pump.* A piston pump consists of an inlet and valve, a piston, and a cylinder connected to a hopper on the intake end and to

**Figure 7-21** Placing Concrete into a Column Form Using a Bucket Equipped with a Chute.

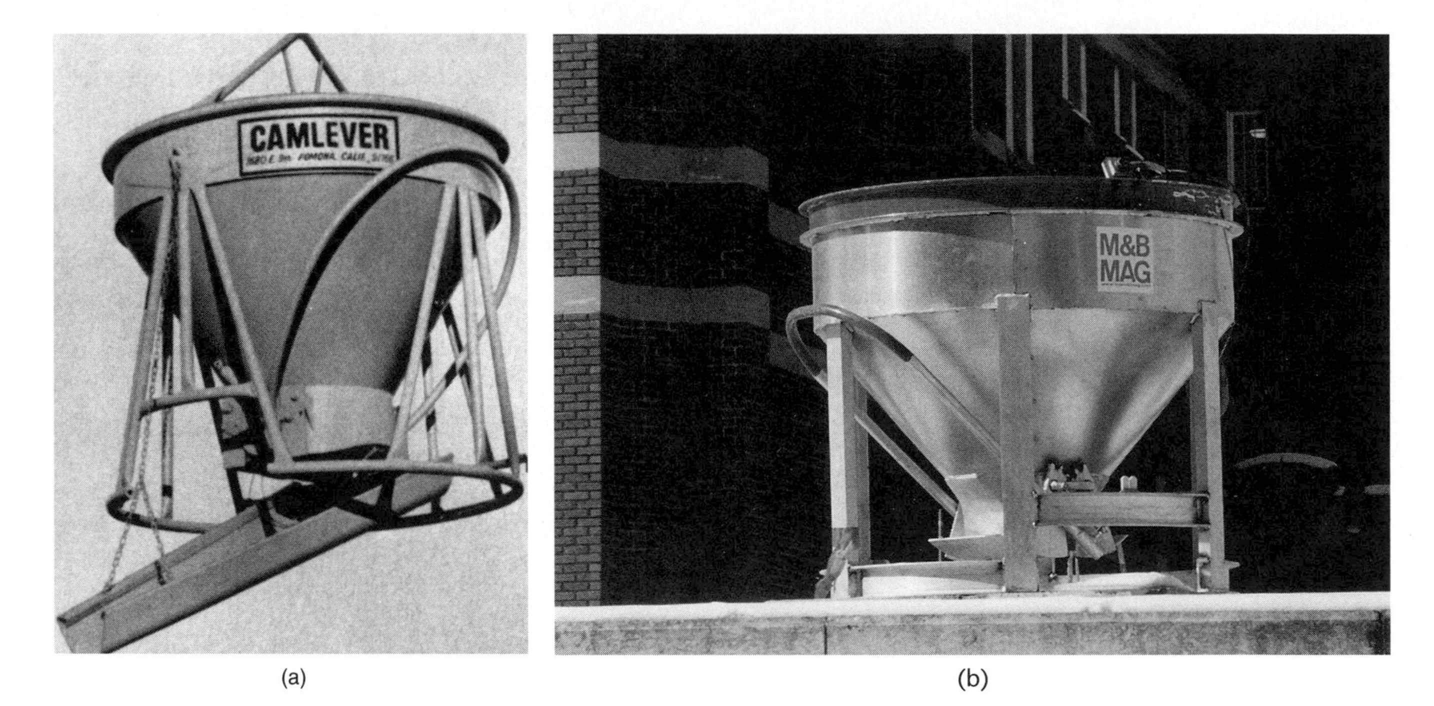


(a) (b)

**Figure 7-22** Concrete Buckets: (a) Bucket with Metal Chute (Reprinted by permission of Camlever Inc.); (b) Bucket with Regulated Gate.

a hose or pipe on the discharge end. The cylinder receives concrete from the hopper, and the piston forces it out into the hose and, by continuous action, eventually to the form.

A pneumatic pump consists of a pressure vessel and equipment for supplying compressed air. Concrete is taken into the pressure vessel, the intake valve is closed, and compressed air is supplied into the top end of the vessel. The pressure forces the concrete out through a pipe at the bottom and into the delivery system (see Figure 7-24).

A squeeze pressure pump consists of a steel drum maintained under high vacuum, inside which hydraulically powered rollers operate. A flexible hose runs from a hopper, enters the bottom of the drum, and runs around the inside surface and out

**Figure 7-23** Mobile Series Conveyor with Drop Chute.

**Figure 7-24** Truck-Mounted Concrete Pump.

the top. The vacuum maintains a supply of concrete from the hopper in the hose, while the rollers, rotating on the hose inside the drum, force the concrete out at the top. A delivery system carries the concrete to the job.

The delivery line may be either rigid pipe or flexible hose. Depending on the equipment, a concrete pump will deliver from 10 to 90 cu yd/hr (7.5 to 70 $m^3$/hr) through lines that can range up to 2,000 ft (600 m) horizontally and up to 300 ft (90 m) vertically. Three feet (1 meter) of vertical lift is considered to equal 24 ft (8 m) of horizontal run, a 90° bend in rigid pipe is equivalent to 40 ft (12 m) of horizontal run, and a 45° bend is equal to 20 ft (6 m).

Pumping may be used in nearly all types of concrete construction, but it is especially useful where space for construction equipment is at a premium. The pump can be set in an out-of-the-way location where concrete can readily be delivered to it (see Figure 7-25), while hoists and cranes are used to deliver other material to the job. Pumping will generally be restricted to concrete that has a maximum coarse aggregate size of 1½ in. (40 mm) and ultimate strength of 2,500 psi (17 MPa) or better.

As discussed, there is a limit to the amount of pressure that can be applied to the concrete, which will vary, depending on slump, water:cement ratio, gradation, and size and type of aggregates. All concretes, however, have their *bleed-out point*—the pressure point beyond which the cement and water will be forced (bled) out of the concrete at the pump by applied pressure into the concrete ahead of it. The concrete that has bled out becomes dry and becomes what is called a *slug* in the line, which makes pumping very difficult. Concretes that have poor gradation of aggregates will bleed at low pressures, and such concrete should be avoided if pumping is to be the method of transfer to be used.

**Figure 7-25** Hydraulic Boom Allows Concrete Placement on a Difficult Site.

## CONCRETE TRANSPORTATION

When concrete is brought to the job from a mixing area some distance away, any of a number of methods may be used. They include dump trucks, transit mix trucks, agitator trucks, and rail cars.

Dump trucks should have a body of special shape, with rounded and sloping front and rounded bottom. The rear end should taper to a discharge gate to facilitate delivery of the load.

A transit mix truck is essentially a heavy-duty truck chassis on which is mounted a large, drum-type concrete mixer. It is equipped with a water tank and auxiliary engine that operates the mixer (see Figure 7-26). The dry ingredients for a concrete mix are charged into the truck from a batching plant (see Figure 7-27). If the distance to be traveled from batching plant to job site can be covered within the initial setting time of the cement paste, water may also be added at the batcher. The concrete is then mixed and agitated en route. But when long distances are involved, the driver must add the water at the appropriate time along the way.

Transit mix trucks are available in sizes ranging from 1 to 12 cu yd (0.75 to 9 $m^3$) of capacity. Each batch of concrete should be mixed not less than 50 or more than 100 revolutions of the drum or blades at the prescribed rate of rotation. Any additional mixing should be done at the designated agitating speed. Concrete should be delivered and discharged from the truck mixer within 1½ hr after the introduction of water to cement and aggregate.

An agitator truck is similar to a transit mix truck, except that it carries no water tank. This means that the complete mix is made at the batching plant and charged into the truck drum. The truck simply keeps the concrete agitated until it is delivered. As a result, the distance that may be traveled is limited to that which can be covered within the initial setting time of the paste. In extremely hot weather, it may be necessary to use ice to keep the temperature down. This is done to prevent the initial set of concrete from taking place before it can be delivered.

Rail cars specially designed for transporting concrete are used only on large projects. Some are tilted to discharge through side or end gates, whereas others discharge through bottom gates. Concrete is normally dumped into a large hopper from which short chutes or downspouts direct it to the forms. It is essential to supervise this operation closely to prevent segregation. Figure 7-28 illustrates correct and incorrect procedures in handling and transporting concrete.

**Figure 7-26** Transit Mix Truck.

**Figure 7-27** Concrete Batching Plant.

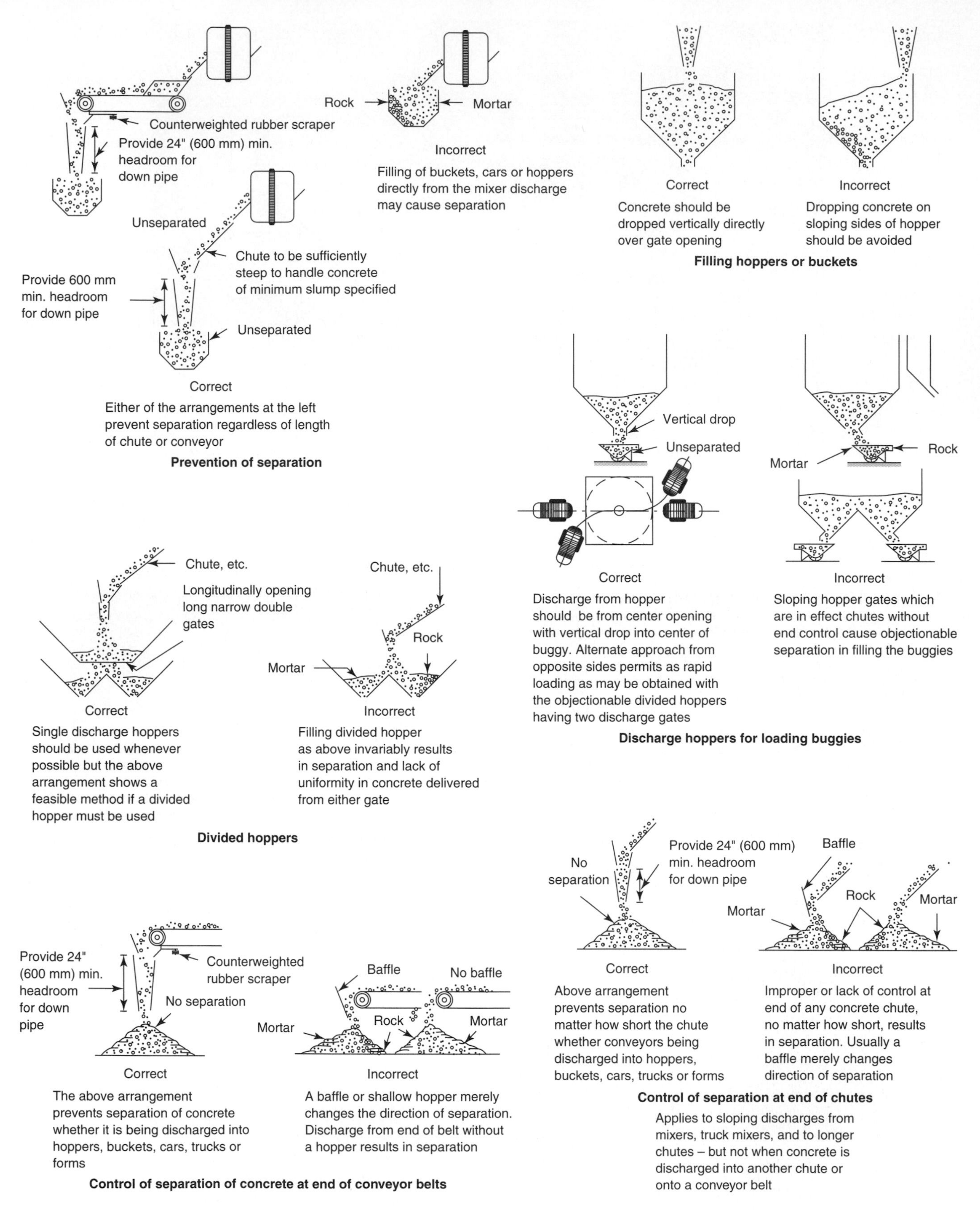

**Counterweighted rubber scraper**

Provide 24" (600 mm) min. headroom for down pipe

Unseparated

Provide 600 mm min. headroom for down pipe

Chute to be sufficiently steep to handle concrete of minimum slump specified

Unseparated

Correct

Either of the arrangements at the left prevent separation regardless of length of chute or conveyor

**Prevention of separation**

Rock — → — Mortar

Incorrect

Filling of buckets, cars or hoppers directly from the mixer discharge may cause separation

Correct

Concrete should be dropped vertically directly over gate opening

Incorrect

Dropping concrete on sloping sides of hopper should be avoided

**Filling hoppers or buckets**

Chute, etc.

Longitudinally opening long narrow double gates

Chute, etc.

Rock

Mortar

Correct

Single discharge hoppers should be used whenever possible but the above arrangement shows a feasible method if a divided hopper must be used

Incorrect

Filling divided hopper as above invariably results in separation and lack of uniformity in concrete delivered from either gate

**Divided hoppers**

Vertical drop

Unseparated

Mortar — Rock

Correct

Discharge from hopper should be from center opening with vertical drop into center of buggy. Alternate approach from opposite sides permits as rapid loading as may be obtained with the objectionable divided hoppers having two discharge gates

Incorrect

Sloping hopper gates which are in effect chutes without end control cause objectionable separation in filling the buggies

**Discharge hoppers for loading buggies**

Provide 24" (600 mm) min. headroom for down pipe

Counterweighted rubber scraper

No separation

Mortar

Correct

The above arrangement prevents separation of concrete whether it is being discharged into hoppers, buckets, cars, trucks or forms

Baffle — No baffle

Rock — Mortar

Incorrect

A baffle or shallow hopper merely changes the direction of separation. Discharge from end of belt without a hopper results in separation

**Control of separation of concrete at end of conveyor belts**

No separation

Provide 24" (600 mm) min. headroom for down pipe

Baffle

Rock

Mortar — Mortar

Correct

Above arrangement prevents separation no matter how short the chute whether conveyors being discharged into hoppers, buckets, cars, trucks or forms

Incorrect

Improper or lack of control at end of any concrete chute, no matter how short, results in separation. Usually a baffle merely changes direction of separation

**Control of separation at end of chutes**

Applies to sloping discharges from mixers, truck mixers, and to longer chutes – but not when concrete is discharged into another chute or onto a conveyor belt

**Figure 7-28** Correct and Incorrect Methods of Handling Concrete. (Reprinted by permission of Portland Cement Association.)

## PLACING CONCRETE

Concrete must be placed as nearly as possible in its final position. It should not be placed in large quantities in one position and allowed to flow or be worked over a long distance in the form. The mortar will tend to flow ahead of the coarser materials, thus causing stone pockets and sloping work planes.

Concrete placed in forms must *never* be allowed to drop freely more than 1 m. When concrete must be placed into deep forms, rubber or metal chutes must be lowered into the formwork to control the rate of fall of the concrete. Figure 7-29 illustrates the use of a flexible hose to help minimize the height that the concrete will be dropped. In some instances, when it may not be possible to use chutes inside the formwork, concrete may be deposited through openings in the side of the form, known as *windows*. Figure 7-30(b) illustrates the construction of such a window, which also provides an outside pocket from which the concrete can flow into the form at a controlled rate, rather than allowing it to enter directly into the form at a high velocity.

**Figure 7-29** Using a Flexible Hose to Minimize the Free Fall of Concrete.

When concrete is placed on a sloping surface, placing should begin at the bottom of the slope. This system will not only improve the compaction of the concrete as placing progresses, but will prevent the flowing out of mortar, which would occur if pouring had begun at the top of the slope.

Placing of concrete for a slab must begin against a wall or against previously placed concrete. Dumping away from previously placed concrete allows the coarse aggregate to separate from the mortar. Figure 7-30(d) illustrates this problem.

Concrete should be placed in wall forms in relatively thin layers, or *lifts*, 12 to 18 in. (300 to 450 mm) deep, each layer placed the full length of the form before the next lift is begun. This operation must take place so that each lift is placed in time to integrate easily and completely with the one below. This time factor will be determined by the kind of cement used in making the concrete, the richness of the mix, the presence or absence of accelerators, the temperature of the concrete, and atmospheric conditions at the time of placing. In addition, the first batches of each lift must be placed at the ends of the form section or in corners, and placing should then proceed toward the center. This positioning is done to prevent the trapping of water at ends of the sections, in corners, and along form faces. The integration of each lift with the one below is done with the aid of puddling spades or vibrators.

Concrete being placed in columns and walls must be allowed to stand for approximately 2 hr before placing the concrete for monolithic girders, beams, and slabs. This time allows the concrete in the walls or columns to settle and thus prevents cracking due to settlement, which would occur if all members were placed at one time.

The correct placing of concrete includes proper consolidation to ensure that no pockets or spaces remain unfilled and that the face of the formed concrete has been made as smooth as required by specifications. The compacting or consolidation is done by means of *puddling spades* or *vibrators*. A puddling spade is simply a flat piece of metal attached to a handle, which is worked up and down in the freshly placed concrete. Vibrators may be either *internal* (see Figures 7-31 and 7-32) or *external*. Internal vibrators must always be inserted vertically into the concrete (see Figure 7-33) and should be used for consolidation only, not to move concrete from one place to another within a form. External vibrators are used against the outside of forms and are most effective in producing smooth surfaces against the form faces.

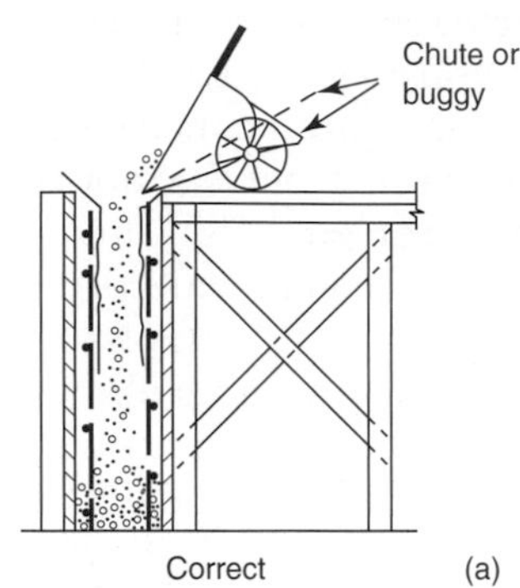


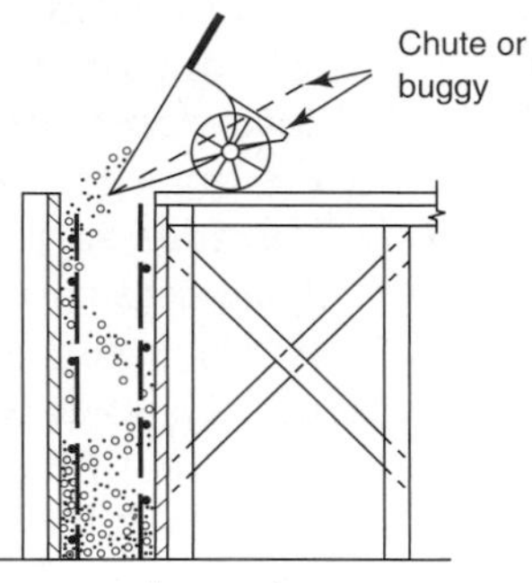


(a)

Correct

Separation is avoided by discharging concrete into hopper feeding into drop chute. This arrangement also keeps forms and steel clean until concrete covers them.

Incorrect

Permitting concrete from chute or buggy to strike against form and ricochet on bars and form faces causes separation and honeycomb at the bottom.

**Placing in top of narrow form**

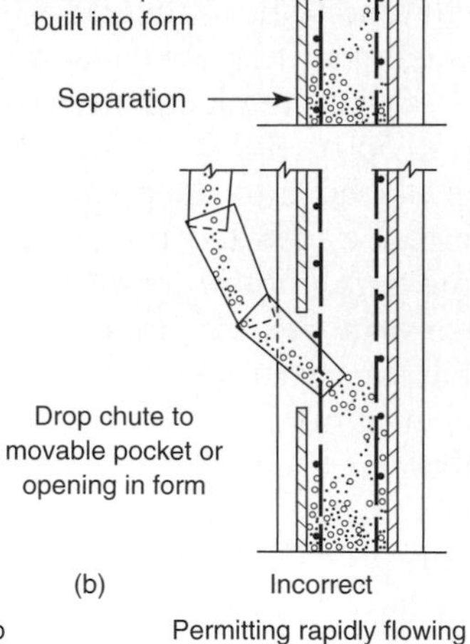


(b)

Correct

Drop concrete vertically into outside pocket under each form opening so as to let concrete stop and flow easily over into form without separation.

Incorrect

Permitting rapidly flowing concrete to enter forms on an angle invariably results in separation.

**Placing in deep narrow wall through port in form**

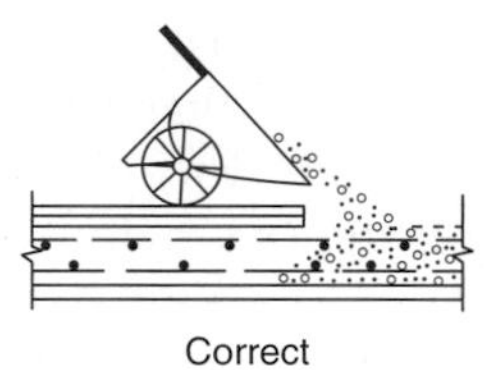

(c)

Correct

Concrete should be dumped into face of previously placed concrete.

Incorrect

Dumping concrete away from previously placed concrete causes separation.

**Placing slab concrete from buggies**

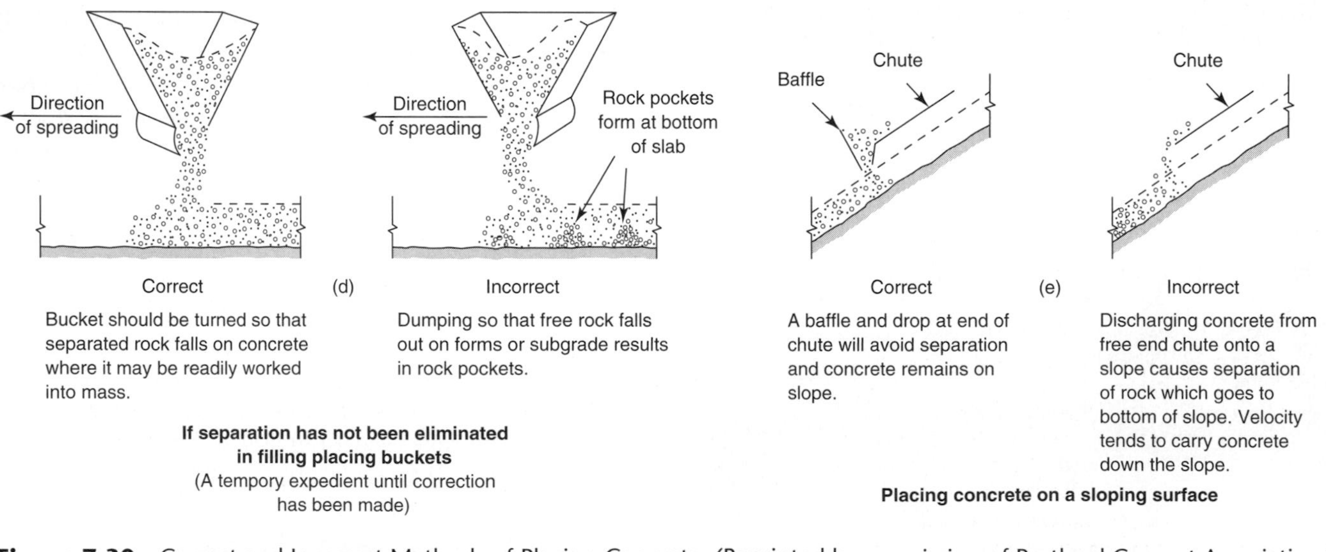


(d)

Correct

Bucket should be turned so that separated rock falls on concrete where it may be readily worked into mass.

Incorrect

Dumping so that free rock falls out on forms or subgrade results in rock pockets.

**If separation has not been eliminated in filling placing buckets**

(A tempory expedient until correction has been made)

(e)

Correct

A baffle and drop at end of chute will avoid separation and concrete remains on slope.

Incorrect

Discharging concrete from free end chute onto a slope causes separation of rock which goes to bottom of slope. Velocity tends to carry concrete down the slope.

**Placing concrete on a sloping surface**

**Figure 7-30** Correct and Incorrect Methods of Placing Concrete. (Reprinted by permission of Portland Cement Association.)

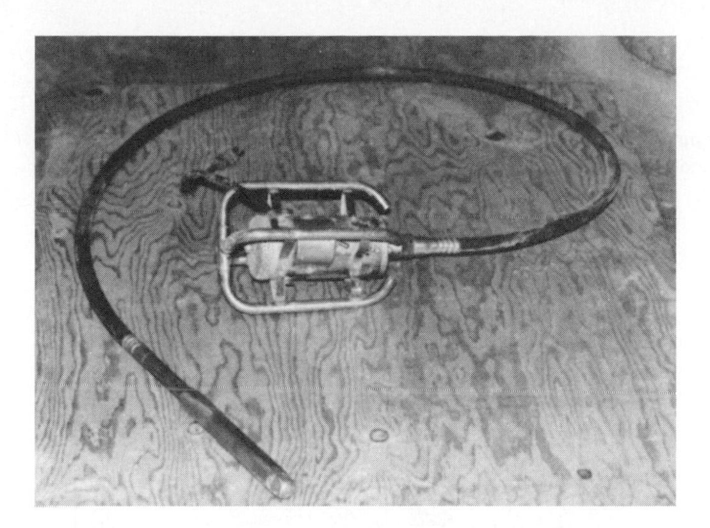

**Figure 7-31** Electrically Driven Internal Vibrator.

**Figure 7-32** Gasoline-Powered Internal Vibrator. (Reprinted by permission of Multiquip, Inc.)

Vibrating eliminates stone pockets and air bubbles in freshly placed concrete and consolidates each successive layer of concrete with the one below. This activity must also bring enough mortar to the concrete surface or the form face to ensure a smooth finish. Excessive vibration should be avoided as it causes segregation by forcing the coarse aggregates away from the vibrator, resulting in pockets of cement mortar lacking in coarse aggregates. It should also be remembered that vibrating of the concrete mix increases the pressure on the formwork, so special care must be taken to ensure that the formwork is designed and constructed to withstand the additional pressure.

## Underwater Concreting

Placing concrete under water requires special techniques. Of course, it cannot be dropped through the water but must be conducted to its final position. A number of methods are used to place concrete for underwater structures, among them the use of a *tremie*, an *underwater bucket*, a *concrete pump*, or *preplaced aggregate*.

**Tremie.** One of the most widely used methods of placing concrete under water is by the use of a tremie. This is a pipe with a normal diameter eight times the maximum size of the coarse aggregate in the concrete being used. It is commonly made in 10- and 12-in (250- and 300-mm) diameters in 10-ft (3-m) lengths. A funnel-shaped hopper is bolted to the top and sections are bolted together with a gasket at each joint, to make a pipe long enough to reach from the water surface to the bottom of the underwater structure.

The bottom end is fitted with a watertight valve or a plug of some kind, and the tremie is lowered into position by hoisting equipment. Placement is begun by filling the tremie with concrete while the bottom valve is closed, and then opening it to allow concrete

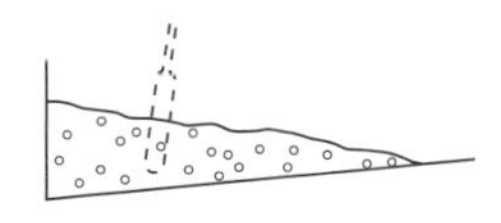

Correct
(a)
Incorrect
Start placing at bottom of slope so that compaction is increased by mass of newly added concrete. Vibration consolidates the concrete.
When placing is begun at top of slope the upper concrete tends to pull apart especially when vibrated below as this starts flow and removes support from concrete above.
When concrete must be placed in a sloping lift



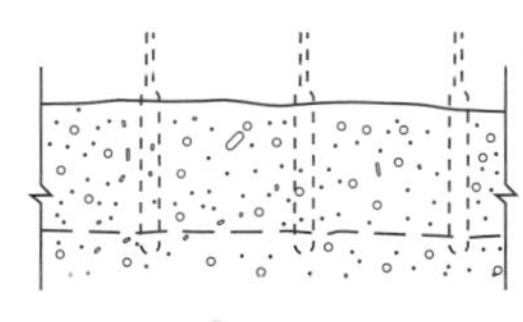

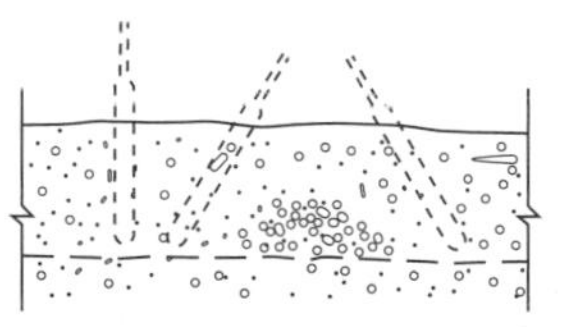

Correct
(b)
Incorrect
Vertical penetration of vibrator a few inches into previous lift (which should not yet be rigid) at systematic regular intervals will give adequate consolidation.
Haphazard random penetration of the vibrator at all angles and spacings without sufficient depth will not ensure intimate combination of the two layers.
Systematic vibration of each new lift



**Figure 7-33** Correct and Incorrect Methods of Consolidating Concrete. (Reprinted by permission of Portland Cement Association.)

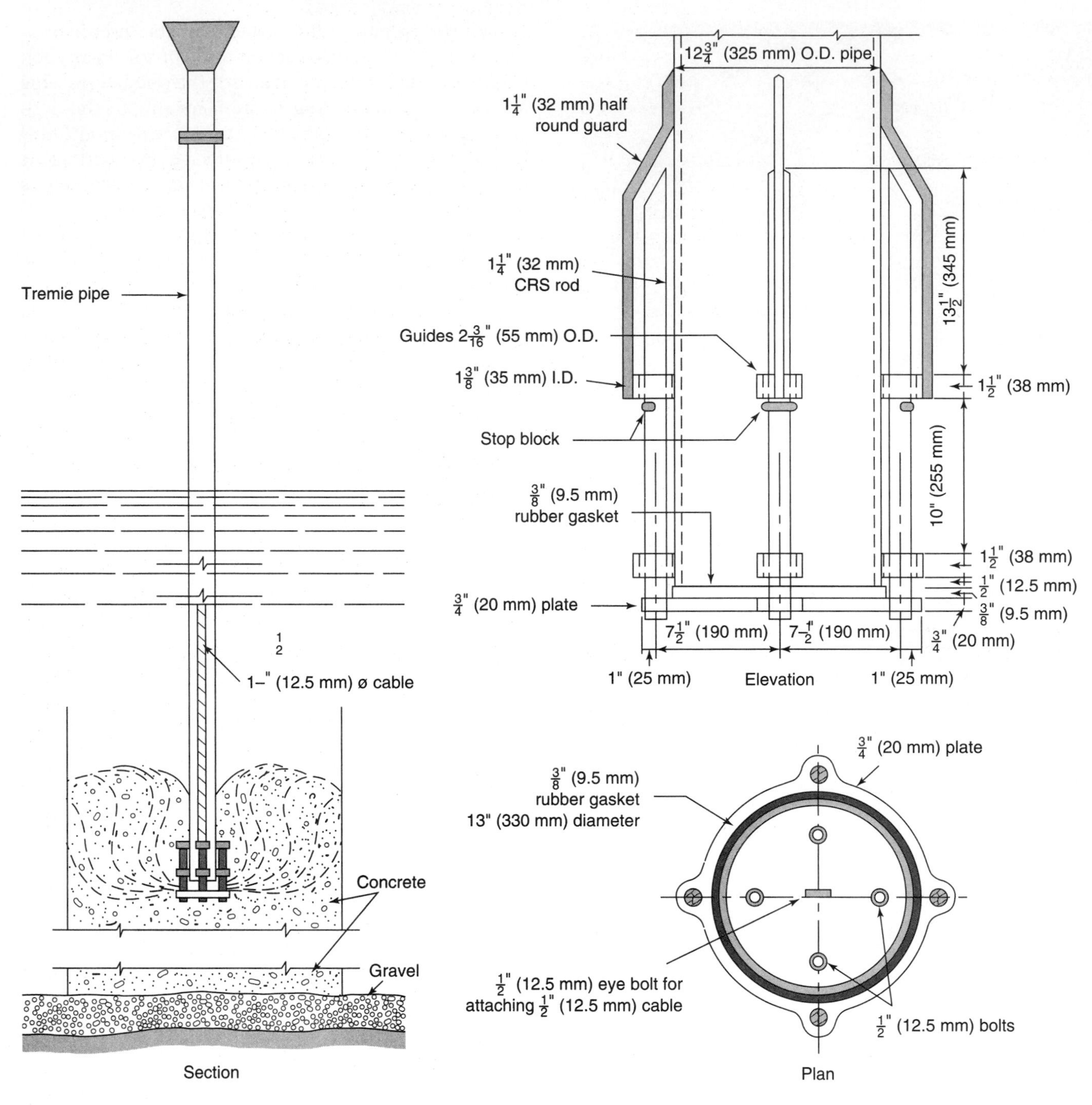


**Figure 7-34** Diagram of Tremie.

to flow out. It flows outward from the bottom of the pipe, pushing the existing concrete surface outward and upward. Concrete is fed to the tremie continuously as long as the valve is open, and, as long as the flow is smooth so that the concrete surface next to the water is not physically disturbed, high-quality concrete will result. The tremie is raised slowly as concreting progresses, but the end is kept submerged in concrete at all times, as illustrated in Figure 7-34. If a plug is used at the bottom of the pipe, it will be pushed out as concrete is introduced into the tremie and the process will continue as described.

Where large areas are involved, several tremies are used together, the concrete from each flowing into a common mass. In most cases, tremies should not be spaced more than 20 to 25 ft (6 to 7.5 m) apart.

The mix proportions for tremie concrete will be somewhat different from normal structural mixes because the concrete must flow into place and consolidate without any mechanical help. The slump should be from 6 to 9 in. (150 to 225 mm), and generally rounded coarse aggregate is preferable to crushed stone, because it will aid flowability. The fine aggregate proportions will usually be from 40%

to 50% of the total aggregates, and the water:cement ratio is kept to 0.44 or less by mass. Air-entraining agents and pozzolans are often used to improve the concrete flow characteristics.

**Underwater Bucket.** An underwater bucket has bottom gates that cannot be opened until the bucket is resting on the bottom or on previously placed concrete. This operation allows depositing the concrete and removing the bucket without disturbing the concrete or unduly agitating the water. A canvas cover prevents the swirling action of water from disturbing the concrete while the bucket is being lowered.

**Concrete Pump.** Using a concrete pump to place concrete under water is similar to using a tremie. The lower end of the line must be kept submerged in concrete and is then slowly raised to prevent the pressure from overcoming the pump as placing progresses.

**Preplaced Aggregate.** Grouted aggregate concrete involves placing a series of small vertical pipes in the form, reaching from the bottom to a level above that which the concrete is to reach. The forms are then filled with the required coarse aggregate. Finally, structural grout is pumped through the pipes, forcing the water out of the forms and filling the spaces between the particles of aggregate. The pipes are slowly withdrawn as grouting proceeds, but their lower ends must be kept below the level of grout already in the form.

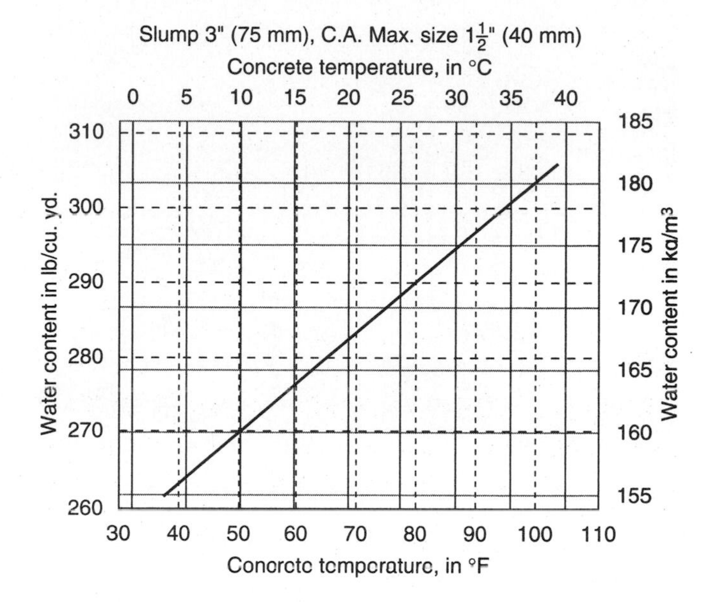


**Figure 7-35** Relationship of Water Content to Concrete Temperature. (Reprinted by permission of Portland Cement Association.)

## Hot-Weather Concreting

Placing concrete during hot weather presents some special problems, and the success of a concreting project under such conditions depends on their solution.

High temperatures accelerate the set of concrete, and only a short time is available in which finishing may be done. In addition, more mixing water is usually required for the same amount of slump, and if more cement is not added also, to maintain the water:cement ratio, the concrete strength will be reduced. Higher water content will also result in greater drying shrinkage. Figure 7-35 indicates the relationship between concrete temperature and the amount of water required per cubic yard of concrete.

During hot weather, a concrete temperature of 50° to 60°F (10° to 16°C) is desirable, and for massive projects it should be from 40° to 50°F (5° to 10°C). The most practical way to achieve these temperatures is to control the temperature of the concrete ingredients.

Aggregate stockpiles should be shaded from the sun and kept moist by sprinkling to take advantage of the cooling brought about by evaporation. Coarse aggregate may be sprayed with cold water or chilled by blowing cold air over it while it is on the way to the batching plant or mixer.

Water may be cooled by insulating and burying the supply line, by refrigeration, or by adding crushed or flaked ice to the mixing water, provided that the ice is all melted by the time mixing is completed.

The subgrade, forms, and steel should be sprinkled with cool water just before concrete is placed during hot weather. It is also helpful to wet down the entire area because the evaporation will help to cool the surrounding air and increase the relative humidity.

Transporting and placing should be carried out as rapidly as possible. The depth of lifts in walls may have to be reduced to ensure consolidation with the concrete below and the area of slab pours reduced to avoid cold joints. All protection possible should be given from the sun and wind.

Finishing steps should be attended to at the earliest possible moment and the curing and protection procedures carried out promptly. Wood forms should be sprayed with water while still in place and loosened as soon as practicable so that water can be run down inside the forms. On flat surfaces, the curing water should not be much cooler than the concrete, to minimize cracking due to stresses caused by temperature change. Preferably, after 24 hr of moist curing, curing paper, heat-reflecting plastic sheets, or a curing compound should be applied without delay.

**Figure 7-36** Reinforced Polyethylene Film Encloses Building.

**Figure 7-37** Oil-Burning Heater. (Reprinted by permission of Portland Cement Association.)

## Cold-Weather Concreting

Experiments in cold weather construction have been carried out by contractors and subcontractors for many years, and much valuable experience has been gained. This experience, together with new techniques and materials that have recently appeared on the market, have led to a marked change in the attitude of owners, builders, architects, and engineers toward cold weather construction.

Most industrial building projects are not interrupted during the winter months, though it has been considered desirable to have the building "closed in" before the advent of cold weather. But closing in meant completing exterior walls and roofs, a long job in the case of large buildings. So it has become common practice to enclose a structural frame with large tarpaulins or polyethylene film, as shown in Figure 7-36. Polyethylene has the added advantage of letting light into the enclosed space.

Within the enclosure, the use of equipment such as space heaters or automatic oil furnaces (see Figure 7-37) permits temperatures that allow most types of construction work to continue throughout the winter.

In multistory construction—buildings too large to be completely covered—individual floors are enclosed. This takes the form of a hanging scaffold enclosed in tarpaulin or plastic film, which protects masons and masonry while closure walls are being built (Figure 7-38). When heat is required, warm air may be piped into the enclosure.

In recent years it has become standard practice for major projects to be initiated during the winter. In most regions except the northern states, the Prairie Provinces, and northwestern Ontario, frost penetration under snow cover seldom exceeds 12 in. (300 mm). This depth presents no problem for modern excavating machinery and, in fact, sometimes facilitates excavation by providing firmer footing for equipment and trucks.

**Figure 7-38** Enclosed Hanging Scaffold. (Reprinted by permission of Portland Cement Association.)

When it is known in advance that excavating is to be done in soil having considerable frost penetration, the depth of frost can be kept to a minimum by leaving the snow undisturbed over the area or by covering the site with a thick layer of straw before the first snowfall. As soon as an excavation has been completed, steps must be taken to prevent frost from getting into the ground. Straw may again be used, or a shelter such as that shown in Figure 7-39 may be provided for even better protection.

Protection for concrete during placing and curing operations is essential whenever below-freezing temperatures are to be expected. Concrete should be warm when placed and should remain at a reasonable temperature until it has gained sufficient strength to prevent frost damage. Concrete requires approximately 24 hrs, at normal temperatures, to obtain a strength of 500 psi (3.5 MPa). Once concrete has developed a strength of 500 psi (3.5 MPa) it is usually

**Figure 7-39** Tarpaulin-Enclosed Foundation. (Reprinted by permission of Portland Cement Association.)

considered past the danger stage, although it cannot withstand repeated cycles of freezing and thawing.

In cold weather, it is sometimes necessary to heat the water and aggregate to produce concrete that will have a placing temperature of between 50 and 70°F (10° and 21°C). Water is the most practical component to heat, because more heat units can be stored in water than in other materials. However, water should not be heated above 175°F (80°C) because of the danger of causing a *flash* set. If the bulk of the aggregate has a temperature appreciably below 45°F (7°C), it must also be heated. Aggregate is usually heated by either steam coils or by the injection of live steam.

A formula has been devised to calculate the temperatures to which water (and aggregate, if necessary) must be heated to produce concrete of a given temperature.

$$D = \frac{Wt + 0.22Wt'}{W + 0.22W'}$$

where $D$ = desired concrete temperature (°F)
$W$ = weight of water (LB)
$W'$ = weight of cement and aggregate (LB)
$t$ = temperature of water (°F)
$t'$ = temperature of solids (°F)

Remember that the moisture in the aggregate will have the same temperature as the aggregate. Remember also that, if it is necessary to calculate a temperature to which aggregate must be heated, it is assumed that the added water will also have been heated to a given temperature.

Large exposed areas such as floor slabs should be protected on both sides. It is difficult to supply heat to the upper surface of the slab if it is not enclosed, but it should be covered with insulation, as shown in Figure 7-40. Heat can be supplied to the underside of the slab by heaters, as in Figure 7-41.

Enclosures such as that shown in Figure 7-42 may be used while erecting concrete or masonry walls. As building progresses, protection may be removed from one location and reused in another (see Figure 7-43).

It may at times be difficult to supply heat to isolated structures during their curing period. An alternative to doing so is to build insulated forms (Figure 7-44) and to use concrete with the maximum allowable placing temperature. The insulation will allow the concrete to retain its heat and to cure properly despite low temperatures.

Two particular precautions should be taken when using heaters in enclosed spaces. Heaters in spaces containing freshly placed concrete should be adequately vented. Failure to do so may result in an accumulation of carbon monoxide gas, which presents a hazard to health and acts as a retarder to the setting of concrete. The second precaution concerns the use of oil-burning heaters in rooms that have received their base plaster coat. An oily film may be deposited on the plaster surface and may seriously affect the adhesion of the finish plaster coat.

In general, direct-fired portable air heaters are regulated by fire codes and are not permitted within closed areas. All heaters used on a construction site must be kept clear from all flammable materials and monitored by qualified personnel. Appropriate fire

**Figure 7-40** Surface of Slab Insulated Against Heat Loss. (Reprinted by permission of Portland Cement Association.)

**Figure 7-41** Supplying Heat to Underside of Slab. (Reprinted by permission of Portland Cement Association.)

**Figure 7-42** Wall Enclosure for Winter Construction. (Reprinted by permission of Portland Cement Association.)

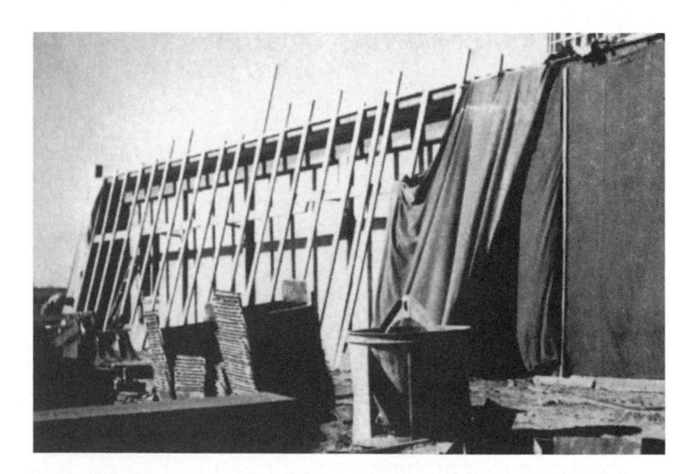

**Figure 7-43** Removing Part of Protective Enclosure. (Reprinted by permission of Portland Cement Association.)

**Figure 7-44** Insulated Forms. (Reprinted by permission of Portland Cement Association.)

fighting equipment must be kept on hand in case of an emergency.

Shelters used in winter construction can be divided into two general groups. The first is the self-supporting type. One example is a shelter made from light laminated arches covered with plastic film, now in fairly common use. Another is an air-supported unit made from plastic-coated fabric and often large enough to cover an area the size of a football field. The second type of shelter may use the existing frame of the building for support and encompass the shelters discussed earlier in this chapter.

Cost is a prime factor in deciding whether to build in the winter. It is generally conceded that the average increase in costs for winter construction will vary from 5% to 10%, which includes the cost of providing tarpaulins, heaters, insulation, fuel, and snow removal. Indirect savings may result from higher productivity, uninterrupted schedules, and greater control of temperatures on the job, thus offsetting direct costs.

## Shotcrete

The American Concrete Institute has defined *shotcrete* as "mortar or concrete conveyed through a hose and pneumatically projected at high velocity onto a surface." Placing concrete in this manner is considerably different from the conventional approach in that the concrete can be placed in a relatively thin, high-density layer over any surface. Two approaches have been developed for placing concrete in this manner: the *dry process* and the *wet process*.

In the dry process, cement and slightly damp sand (in some cases aggregates are used) are mixed without water, fed into a pneumatic gun, and forced through a small hose, 1 to 2½ in. (25 to 62.5 mm) in diameter, with compressed air to a nozzle where, with additional water for the hydration process, the mix is projected on the surface being covered. The dry ingredients may be mixed on site (see Figure 7-45), or they may come prebagged and be introduced to the pneumatic gun directly.

The basic equipment for the dry process consists of a mixer for mixing the dry ingredients, the

**Figure 7-45** Two-Component Proportioning Rig with Rotary Gun. (Reprinted by permission of Allentown Equipment.)

pneumatic gun, a water supply, and a source of compressed air. The purpose of the pneumatic gun is to feed the dry ingredients smoothly into the compressed air stream and move them through the air hose. Compressed air is provided by an air compressor, and water can be supplied from a pressurized tank or from the local water supply system.

The wet process uses a low-slump premixed mortar or pea-sized concrete mix that is forced into a relatively small diameter hose 1½ to 2½ in. (40 to 62.5 mm) in diameter, by a high-capacity concrete pump (Figure 7-46) to a nozzle, where compressed air is introduced to project the mix onto the required surface. In this process the consistency of supply and the design mix specifications are important to ensure that the mix is suitable for pumping through the hose without inducing blockages.

**Figure 7-46** Heavy-Duty Shotcrete/Concrete Pump. (Reprinted by permission of Allentown Equipment.)

In both processes, the placing of the mix depends on the experience of the person operating the nozzle (Figure 7-47). In the dry process the nozzle operator must add the appropriate amount of water, maintain a correct nozzle angle, and control the amount of rebound. In addition, the operator must have an appreciation for basic concrete principles such as minimum cover for reinforcement, finished thickness, and surface finish. As in the dry process, material application in the wet process requires much the same techniques, except that no additional water is required at the nozzle.

Concrete applied in this manner has several positive advantages over concrete placed in the traditional manner:

- Concrete can be placed in relatively thin layers.
- The density of the mix is enhanced by the high velocities that are used to place the concrete mix.
- No formwork is necessary. Excellent adhesion to other materials, even in an overhead application, is achieved.
- High compressive strength is achieved because of the low water:cement ratio, resulting in extreme hardness, high abrasion resistance, and excellent water impermeability.

**Figure 7-47** Shotcrete Application. The Rocks in the Background Were Also Formed Using Shotcrete. (Reprinted by permission of Allentown Equipment.)

## CONCRETE JOINTS

See Chapter 6.

## FINISHING AND CURING CONCRETE

The treatment of exposed concrete surfaces to produce the desired appearance, texture, or wearing qualities is known as *finishing*. The procedure to follow depends primarily on whether the surface is horizontal or vertical. Horizontal surfaces are usually exposed and must be finished before the concrete has hardened. Timing is an important factor in this operation. Finishing must be done when the concrete is neither too hard to be worked nor so soft that it will fail to retain the desired finish.

The concrete must first be struck off level as soon as it has been placed, which is done by the use of strike-off bars worked against the top edge of screeds previously set to the proper height (see Figure 7-48). These bars are operated either by hand or by power, the latter acting as vibrators as well as strike-off bars. Striking off removes all humps and hollows, leaving a true and even surface.

Further finishing is then delayed until the surface is quite stiff but still workable. Any water appearing on the surface should be removed with a broom, burlap, or other convenient means. Do not sprinkle cement or a mixture of cement and fine aggregate on the surface to take up excess water. This forms a layer that will dust and wear easily or develop many fine cracks due to shrinkage. A short wood *float* (see Figure 7-49) produces an even surface, with all larger particles of aggregate embedded in mortar. This should be followed by a long float (see Figure 7-50), which removes the marks left by the short one and produces a gritty nonskid surface.

If a scored surface is required, it may be produced by drawing a stiff, coarse broom across the surface, as shown in Figure 7-51.

**Figure 7-48** Power-Operated Strike-Off Bars. (Reprinted by permission of Multiquip, Inc.)

**Figure 7-49** Short Wood Float.

**Figure 7-50** Magnesium Float. (Reprinted by permission of Portland Cement Association.)

**Figure 7-51** Broomed Surface. (Reprinted by permission of Portland Cement Association.)

When a smooth surface is required, it may be produced by using a steel trowel, shown in Figure 7-52. Finishing can also be done by the use of power floats and power trowels. These are commonly used on large projects and are operated as shown in Figure 7-53. Smooth surfaces may also be produced by the use of power grinders.

Trimmed edges and joints are made by using edging tools such as those shown in Figure 7-54. Special hard surfaces are applied and finished as described in Chapter 11.

When vertical surfaces have to be given a smooth finish, a different technique is used. First, the forms must be removed after the concrete is sufficiently strong. The tie holes must then be filled with mortar, or the snap ties must be broken off, and all depressions must be patched. All rough fins and protrusions must be removed, which can be done by rubbing the surface with a flat emery stone. Next, the surface is washed to remove all loosened material and at the same time the concrete is dampened.

**Figure 7-52** Steel Trowel.

**Figure 7-53** Power Trowel in Operation. (Reprinted by permission of Multiquip, Inc.)

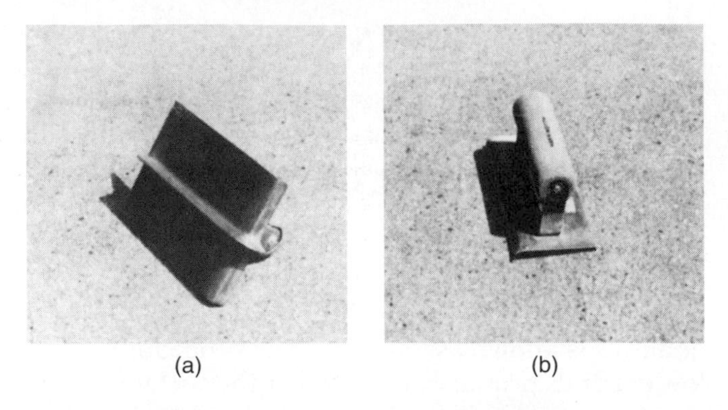

(a) (b)

**Figure 7-54** (a) Jointing Tool; (b) Edging Tool.

Grout is then applied to the surface with a stiff brush and rubbed in with a piece of burlap or a cork float, which will fill in the pores and small holes. After the grout has set sufficiently so that it will not rub out of the holes easily, the surface must be rubbed down with clean burlap to remove all excess material.

It has been shown that the strength, watertightness, and durability of concrete improve with age as long as conditions are favorable for continued hydration of the cement. The improvement in strength is rapid at the early stages, but continues more slowly for an indefinite period. The conditions required are sufficient moisture and temperatures preferably in the 65° to 70°F (18° to 21°C) range.

Fresh concrete contains more than enough water for complete hydration, but under most job conditions, much of it will be lost through evaporation unless certain precautions are taken. In addition, hydration proceeds at a much slower rate when temperatures are low and practically ceases when they are near freezing. Thus, for concrete to develop its full potential, it must be protected against loss of moisture and low temperatures, particularly during the first 7 days.

Concrete can be kept moist in several ways, such as by leaving forms in place, sprinkling with water, covering with water or a waterproof tarp, and sealing the surface with a wax coating.

When forms are left on, the top surfaces should be covered to prevent evaporation. In hot weather, it may be necessary to sprinkle water on wooden forms to prevent them from drying out and taking moisture from the concrete.

When concrete is to be kept moist by sprinkling, care should be taken not to allow the surfaces to dry out between applications. One method of doing this is to cover the concrete with burlap or canvas.

Flat surfaces can be covered with water, damp sand, or waterproof material such as waxed paper or polyethylene film. Sprinkling may also be used, but it must be constant enough to prevent the concrete from drying out. A wax film applied as a spray is commonly used to seal a concrete surface.

## HOT CONCRETE

Previously, a good deal of emphasis has been placed on the necessity for keeping fresh concrete temperatures in the range of 50° to 90°F (10° to 32°C). For concrete that is to be placed in the field, the maintenance of such temperatures is necessary to allow time for consolidation, finishing, and the like.

But in situations in which concrete is to be mixed in a central plant for use in producing reinforced precast products or for concrete block, it has been found that, under the proper conditions, fresh concrete can be placed successfully at temperatures up to 175°F (80°C).

The primary reason for using hot concrete is that it sets and gains early strength much more rapidly than concrete placed at conventional temperatures. For example, it has been shown that curing times of 3 hr can produce approximately 60% to 70% of the 28-day design strength. Therefore, more intensive use can be made of expensive plant facilities. Molds can be used two or sometimes three times in an 8-hr day, rather than once, as is the case with normal temperature concrete.

Moderate concrete temperatures of up to 90°F (32°C) can be achieved economically by heating the mixing water and the aggregates. When hotter concrete is required, injection of steam into the mixer is used to produce temperatures of 120° to 175°F (50° to 80°C). Temperature is controlled by regulating the amount of steam entering the mixer. Steam pressure of 30 to 75 psi (200 to 480 kPa) is used, with steam temperature from 300° to 320°F (150° to 160°C).

To achieve the best results, the steam should be injected when mixing intensity is high, and for this reason, countercurrent mixers, which produce high-intensity mixing, are often used in plants producing hot concrete (see Figure 7-18). Normally, steam is injected into the mixer at the same time the aggregates are dumped, but just prior to adding cement and water. Water produced in the mix due to the condensation of steam is included in the total amount of water required for strength and slump control. If more water is required for slump adjustment, it should be the same temperature as the concrete.

A fast, efficient transportation system is necessary to convey fresh concrete from the mixer to the forms. In some cases, monorail hoppers, traveling at approximately 200 ft/min (60 m/min), are used. It has been found that external form vibrators have usually been more effective than internal vibrators.

Cleanup of mixing and casting equipment is important because hot concrete sets up quickly. The mixers will have to be cleaned twice a day, and conveyors and buckets up to six times a day.

## DESIGN OF STRUCTURAL LIGHTWEIGHT CONCRETE

*Structural lightweight concrete* is defined as concrete that will have a 28-day compressive strength of more than 2,500 psi (15 MPa) and an air-dry unit mass of less than 115 lb/cu ft (1,850 kg/m$^3$), but normally more than 85 lb/cu ft (1,360 kg/m$^3$).

### Properties of Structural Lightweight Aggregates

Structural lightweight aggregates, usually made from expanded *clay, shale, slate,* or *blast furnace slag* or from *pelletized fly ash,* normally are produced in two sizes: *fine,* with a maximum size of ⅜ in. (9.5 mm), and *coarse,* with a usual maximum of ¾ in. (19 mm). The unit density of lightweight aggregate is considerably less than that of normal density aggregate and can vary from 35 to 70 lb/cu ft (560 to 1,120 kg/m$^3$). ASTM Standard C330 limits the dry, loose unit weight (mass) of lightweight aggregate for structural concrete to a maximum of 70 lb/cu ft (1,120 kg/m$^3$) for fine aggregates; 55 lb/cu ft (880 kg/m$^3$) for coarse aggregates; and 65 lb/cu ft (1,040 kg/m$^3$) for a combination of fine and coarse aggregates.

In contrast to normal density aggregates, which usually absorb 1% to 2% water by weight (mass) of dry aggregates, lightweight aggregates may absorb from 5% to 20% water by weight (mass) of dry material, based on 24 hr absorption tests. This can amount to as much as 250 lb of water per cu yd (150 kg of water per cubic meter) of concrete, if total absorption is reached. As a result, to help control the uniformity of lightweight mixtures, prewetted but not saturated aggregates are generally used. Prewetting should be done 24 hr in advance of mixing to allow the moisture to distribute itself evenly through the aggregate.

The bulk specific gravity for lightweight aggregates is generally between 1.0 and 2.4, and for any specific aggregate, the bulk specific gravity will usually increase as the maximum particle size decreases. It is often difficult to determine accurately the bulk specific gravity of many lightweight aggregates; consequently, the mix design method for structural lightweight concrete is based on the concept of *specific gravity factors,* which take into account the actual moisture conditions of the aggregate, rather than on bulk specific gravity.

In some cases, normal-density fine aggregate may be substituted for part or all of the lightweight fines in lightweight concrete mixtures. This substitution may be done for economy or to improve strength, workability, durability, modulus of elasticity, or finishing qualities. It generally decreases the water required for a given slump. Such a combination, however, will increase unit weight (mass) from 10 to 18 lb/cu ft (160 to 290 kg/m$^3$) and have a detrimental effect on such properties as fire resistance and thermal insulation.

**Table 7-9**
**APPROXIMATE RELATIONSHIP BETWEEN STRENGTH AND CEMENT CONTENT**

| Compressive Strength | | Cement Content | |
|---|---|---|---|
| (psi) | (MPa) | (lb/cu yd) | (kg/m$^3$) |
| 2500 | 15 | 425–700 | 252–415 |
| 3000 | 20 | 475–750 | 282–445 |
| 4000 | 30 | 550–850 | 326–504 |
| 5000 | 35 | 650–950 | 386–564 |

## Proportioning

Proper proportioning of lightweight concrete mixtures may sometimes be achieved with the conventional procedures used for normal density mixes in a few specific instances. Generally, however, the variations in rate of absorption and total absorption of most lightweight aggregates make it difficult to establish a water:cement ratio accurately enough to be used as a basis for proportioning. Instead, it is found to be more practical to design lightweight aggregate mixes by a series of trial mixes proportioned on a *cement content* basis. Tests are made at various ages to determine what cement content is required to produce a given strength.

Lightweight structural concrete intended for watertight structures should be designed for at least 3,750 psi (25 MPa) if exposed to fresh water or 4,000 psi (30 MPa) for saltwater exposure. The latter strength should also be used for concrete exposed to sulfate concentrations.

## Water Content

Concrete made from normal-density aggregates with ¾″-in. (20-mm) maximum size will require from 310 to 360 lb of water per cubic yard (185 to 215 kg of water per cubic meter). On the other hand, if lightweight aggregates with the same maximum size are used, water requirements may range from 290 to 500 lb/cu yd (170 to 300 kg/m$^3$) of concrete. When lightweight fines are replaced by normal-weight fine aggregate, the amount of water required will be reduced, and entrained air in the mix will usually mean a reduction in water of 2% to 3% for each 1% of entrained air.

## Cement Content

The cement content required to produce a given strength will vary widely, depending on the characteristics of the lightweight aggregates. In Table 7-9 an approximation is given of the cement contents to be used for establishing trial mix proportions to obtain a specified strength.

## Proportion of Fine to Coarse Aggregate

Lightweight concrete mixes require a greater percentage of fines than does normal-density concrete because there are significant differences between the bulk specific gravities of fine and coarse lightweight particles and because most lightweight aggregates are rough surfaced and angular in shape. Workable mixes of lightweight concrete require 40% to 60% of fines by volume. This percentage can be reduced when normal-density fines are substituted for the lightweight.

Usually, between 28 and 32 cu ft (1.04 and 1.20 m$^3$) of combined dry, loose, lightweight aggregates are required to make 1 cu yd (1 m$^3$) of concrete. The total volume of aggregates and the proportion of fine to coarse, which depends on gradation, shape, size, and surface texture, have a great influence on the workability and finishing qualities of the fresh concrete.

## Entrained Air

Most lightweight concretes contain 2% to 4% entrapped air, but this has little effect on workability and durability, and it is desirable to add entrained air to most lightweight mixes. When the concrete is subject to freezing and thawing cycles, air should not be less than 6 (±1½)% total when maximum aggregate size is ¾ in. (20 mm) and 7½ (±1½)% when maximum aggregate size is ⅜ in. (10 mm). Even when freeze–thaw resistance is not required, 4% to 8% air is recommended for workability.

## Trial Mixes

It is common practice to determine the cement content required for a specific strength by the use of trial mixes, and these should be made with at least three different cement contents. Each mix should have sufficient entrained air to ensure workability and durability. Each trial batch should produce at least 1½ cu ft (0.1 $m^3$) of concrete, from which measurements of air content and unit weight will be made, and four or five test cylinders molded, on which compressive strength tests will be made. The test results will determine the cement content that produces the required strength. Based on the mix proportions of the trial mixes, the mix proportions for the selected cement content can be accurately estimated. Small trial batches made in the lab may require some further adjustment when enlarged to job mixes.

The following examples illustrate the procedure to be followed when determining the quantities of materials required in the design of lightweight concrete mixes. Examples are provided in both standard and metric units.

---

***Example:***

Design a lightweight aggregate concrete mix that is to test 3,000 psi in 28 days. Concrete is to have 6 (±1½)% air entrainment and 4 (±1) in. of slump. Type 1 or normal portland cement and an air-entraining agent are to be used.

Proceed as follows:

**1.** Determine the loose, dry unit weights of aggregates and their moisture content at time of mixing.

Loose, dry unit wt. of C.A. = 48 lb/cu ft
Loose, dry unit wt. of F.A. = 62 lb/cu ft
Moisture content of C.A. = 4%
Moisture content of F.A. = 7%

**2.** Determine the specific gravity factors (the relationship of aggregate weight to displaced volume) of the fine and coarse aggregates by the pycnometer method (see PCA literature for procedure).

Specific gravity factor for C.A. = 1.40
Specific gravity factor for F.A. = 1.98

**3.** Estimate quantities. From Table 7-9, the cement content should range from 475 to 750 lb/cu yd. Design trial mixes with cement contents of 500, 600, and 700 lb/cu yd, respectively. Assume the total dry, loose volume aggregates required to be 30 cu ft, with equal volumes of fine and coarse for the first trial.

**4.** Calculate the loose weights of the damp fine and coarse aggregate required per cubic yard.

Loose wt. of damp C.A. = 15 × 48 × 1.04 = 749 lb
Loose wt. of damp F.A. = 15 × 62 × 1.07 = 995 lb

**5.** Calculate the displaced volumes of the damp aggregates, using the specific gravity factors at hand.

$$\text{Displaced vol. of C.A.} = \frac{749}{62.4 \times 1.40} = 8.57 \text{ cu ft}$$

$$\text{Displaced vol. of F.A.} = \frac{995}{62.4 \times 1.98} = 8.05 \text{ cu ft}$$

**6.** Calculate the displaced volumes of cement and air.

$$\text{Displaced vol. of cement} = \frac{500}{62.4 \times 3.15} = 2.54 \text{ cu ft}$$

$$\text{Vol. of air} = 27 \times 0.06 = 1.62 \text{ cu ft}$$

**7.** Calculate the volume and weight of water required. The volume of water will be 27 combined volumes of cement, air, and aggregates.

$$\text{Vol. of water} = 27 - (8.57 + 8.05 + 2.54 + 1.62) = 6.22 \text{ cu ft}$$

$$\text{Wt. of water} = 6.22 \times 6.24 = 388 \text{ lb}$$

**8.** Calculate the weights of materials required for a 1½-cu ft trial batch. Divide the weights required per cubic yard by 18 (27/1.5).

Wt. of cement = 500 ÷ 18 = 28 lb
Wt. of water = 388 ÷ 18 = 21.5 lb
Wt. of C.A. = 749 ÷ 18 = 41.5 lb
Wt. of F.A. = 995 ÷ 18 = 55 lb

**9.** Make a chart similar to that shown in Table 7-10 and record all calculations made to this point for ready reference.

**10.** Mix the materials and test the mixture for *air content, slump, unit weight,* and *workability.* If all are satisfactory, take cylinders for compression tests. If a correction in the amount of aggregate or water is required, adjust mix proportions as shown in steps 11 through 14.

**11.** If more coarse aggregate may be used and more water is required to produce the desired slump, add the amounts required. Assume 4.0 lb of C.A. and 4.5 lb of water.

**12.** Calculate the correction required per cubic yard for each.

Correction for C.A. = 4 × 18 = 72 lb
Correction for water = 4.5 × 18 = 81 lb

**13.** Calculate the corrected batch proportions.

**(a)** Corrected weights.

Cement (no change) = 500 lb
F.A. (no change) = 995 lb
C.A. = 749 + 72 = 821 lb
Water = 388 + 81 = 469 lb

**(b)** Corrected displaced volumes:

Cement (no change) = 2.54 cu ft
F.A. (no change) = 8.05 cu ft
Air (no change) = 1.62 cu ft

$$\text{C.A.} = \frac{821}{62.4 \times 1.40} = 9.39 \text{ cu ft}$$

$$\text{Water} = 469 \div 62.4 = 7.51 \text{ cu ft}$$

Total = 29.11 cu ft

**14.** Multiply the weights and displaced volumes of materials from step 13 by 0.925 (27/29.11) to adjust the quantities to a yield of 27 cu ft.

**Table 7-10**
INITIAL BATCH PROPORTIONS AND TRIAL BATCH ADJUSTMENTS

| | 1 | 2 | 3 | 4 | 5 | 6 | 7 | 8 | 9 |
|---|---|---|---|---|---|---|---|---|---|
| | Estimated Batch Proportions (Damp Aggregates) | | | | | Corrected Batch, Proportions (Damp Aggregates) | | Adjusted Batch Proportions (Damp Aggregates) | |
| | Weight (lb) | Displaced volume (cu ft) | Weights for 1½-cu ft Trial Batch (lb) | Correction for 1½ cu ft (lb) | Correction for 1 cu yd (lb) | Weight (lb) | Displaced Volume (cu ft) | Weight (lb) | Displaced Volume (cu ft) |
| Cement | 500 | $\frac{500}{62.4 \times 3.15} = 2.54$ | 28.0 | | | 500 | 2.54 | $0.925 \times 500 = 462$ | $0.925 \times 2.54 = 2.35$ |
| Air | | $27 \times 0.06 = 1.62$ | | | | | 1.62 | | 1.62 |
| C.A. | 749 | $\frac{749}{62.4 \times 1.40} = 8.57$ | 41.5 | +4.0 | +72 | $749 + 72 = 821$ | $\frac{8.21}{62.4 \times 1.40} = 9.39$ | $0.925 \times 821 = 759$ | $0.925 \times 9.39 = 8.67$ |
| F.A. | 995 | $\frac{995}{62.4 \times 1.98} = 8.05$ | 55.0 | | | 995 | 8.05 | $0.925 \times 995 = 920$ | $0.925 \times 8.05 = 7.44$ |
| Water | $6.22 \times 6.24 = 338$ | $27 - 20.78 = 6.22$ | 21.5 | +4.5 | +81 | $338 + 81 = 469$ | $469 \div 62.4 = 7.51$ | $0.925 \times 469 = 434$ | $0.925 \times 7.51 = 6.92$ |
| Total | | 27.00 | | | | | | | 27.00 |

**(a)** Adjusted weights:

$$\begin{aligned} \text{Cement} &= 500 \times 0.925 = 462 \text{ lb} \\ \text{F.A.} &= 995 \times 0.925 = 920 \text{ lb} \\ \text{C.A.} &= 821 \times 0.925 = 759 \text{ lb} \\ \text{Water} &= 469 \times 0.925 = 434 \text{ lb} \end{aligned}$$

**(b)** Adjusted displaced volumes:

$$\begin{aligned} \text{Cement} &= 2.54 \times 0.925 = 2.35 \text{ cu ft} \\ \text{Air} &= 1.62 \text{ cu ft} \\ \text{F.A.} &= 8.05 \times 0.925 = 7.44 \text{ cu ft} \\ \text{C.A.} &= 9.39 \times 0.925 = 8.67 \text{ cu ft} \\ \text{Water} &= 7.51 \times 0.925 = 6.92 \text{ cu ft} \\ \text{Total} &= 27.00 \text{ cu ft} \end{aligned}$$

Record all of these calculations on the chart.

The aforementioned adjusted quantities are for damp aggregates. For future adjustments of batch proportions, it is necessary to convert the batch weights of materials to dry weights. These will then serve as a basis for the calculation of adjustments for change in *aggregate moisture conditions, cement content, proportions of aggregates, slump,* or *air content,* in order to maintain a yield of 27 cu ft. The procedure for converting batch weights of damp aggregates to batch weights for dry aggregates is as follows:

1. Calculate the dry weights of C.A. and F.A. from the batch weights for damp aggregates. The weight of cement is to remain constant.

$$\text{Dry wt. of C.A.} = 759 \div 1.04 = 730 \text{ lb}$$
$$\text{Dry wt. of F.A.} = 920 \div 1.07 = 860 \text{ lb}$$

2. Calculate the specific gravity factors for the dry state. Assume that the specific gravity factor for C.A. is 1.37 and F.A., 1.99.

3. Calculate the displaced volumes of dry aggregates, using specific gravity factors.

$$\text{Displaced vol. of C.A.} = \frac{730}{62.4 \times 1.37} = 8.54 \text{ cu ft}$$

$$\text{Displaced vol. of F.A.} = \frac{860}{62.4 \times 1.99} = 6.92 \text{ cu ft}$$

4. Maintaining the displaced volumes of cement and air constant, calculate the required displaced volume of adding water and the required weight of added water.

Displaced vol. of added water
$= 27 - (2.35 + 1.62 + 8.54 + 6.92) = 7.57$ cu ft

Req'd wt. of added water $= 7.57 \times 62.4 = 472$ lb

To record these calculations, make a chart similar to that shown in Table 7-11. Into the first two columns, transfer the figures from columns 8 and 9 of Table 7-10. Into columns 3 and 4 enter the calculations from above.

Aggregates may have different moisture contents from time to time, and it will be necessary, under these circumstances, to adjust the batch weights of materials to maintain a yield of 27 cu ft.

Assume new moisture conditions of 2% for C.A. and 4% for F.A. Proceed as follows:

1. Calculate the specific gravity factors for the new moisture contents. Assume that of the C.A. to be 1.35 and of the F.A., 1.97.
2. Maintain the weight of cement and the displaced volumes of cement and air constant.
3. Calculate the weights of aggregates for the new moisture conditions.

$$\text{Wt. of moist C.A.} = 730 \times 1.02 = 745 \text{ lb}$$
$$\text{Wt. of moist F.A.} = 860 \times 1.04 = 894 \text{ lb}$$

**Table 7-11**
**ADJUSTMENTS FOR CHANGES IN AGGREGATE MOISTURE CONDITIONS**

| | 1 | 2 | 3 | 4 | 5 | 6 |
|---|---|---|---|---|---|---|
| | Batch Proportions (Damp Aggregates) | | Converted Batch Proportions (Dry Aggregates) | | Adjusted Proportions (Aggregate in New Moisture Condition) | |
| | Weight (lb) | Displaced Volume (cu ft) | Weight (lb) | Displaced Volume (cu ft) | Weight (lb) | Displaced Volume (cu ft) |
| Cement | 462 | 2.35 | 462 | 2.35 | 462 | 2.35 |
| Air | | 1.62 | | 1.62 | | 1.62 |
| C.A. | 759 | 8.67 | $\frac{759}{1.04} = 730$ | $\frac{730}{62.4 \times 1.37} = 8.54$ | $730 \times 1.02 = 7.45$ | $\frac{745}{62.4 \times 1.35} = 8.84$ |
| F.A. | 920 | 7.44 | $\frac{920}{1.07} = 860$ | $\frac{860}{62.4 \times 1.99} = 6.92$ | $860 \times 1.04 = 894$ | $\frac{894}{62.4 \times 1.97} = 7.27$ |
| Water | 434 | 6.92 | $7.57 \times 62.4 = 472$ | $27 - 19.43 = 7.57$ | $6.92 \times 62.4 = 432$ | $27 - 20.08 = 6.92$ |
| Total | | 27.00 | | | | 27.00 |

4. Calculate the displaced volume of aggregates. using new specific gravity factors.

$$\text{Displaced vol. of C. A.} = \frac{745}{62.4 \times 1.35} = 8.84 \text{ cu ft}$$

$$\text{Displaced vol. of F. A.} = \frac{894}{62.4 \times 1.97} = 7.27 \text{ cu ft}$$

5. Calculate the displaced volume of added water and, from this, the required weight of added water.

   Displaced volume of added water
   = 27 − (2.35 + 1.62 + 8.84 + 7.27) = 6.92 cu ft

   Wt. of added water = 6.92 × 62.4 = 432 lb

6. Enter these results into Table 7-11.

As indicated earlier, two more trial mixes will be made, using cement contents of 600 and 700 lb/cu yd. Much of the trial-and-error work may be simplified by beginning with proportions obtained from the first trial mix. They will have to be adjusted to compensate for the increase in cement content by a corresponding decrease in fine aggregate. Proceed as follows:

1. Maintain the weights and volumes displaced by dry, coarse aggregate, air, and water, shown in columns 3 and 4 of Table 7-11, constant. Transfer these values to columns 1 and 2 of Table 7-12.
2. Calculate the volume displaced by the new cement content, 600 lb/cu yd.

$$\text{Displaced vol. of cement} = \frac{600}{62.4 \times 3.15} = 3.05 \text{ cu ft}$$

3. Calculate the required displaced volume of dry fine aggregate as the difference between 27 cu ft and the sum of the displaced volumes of air, cement, coarse aggregate, and water.

   Req'd vol. of dry F.A.
   = 27 − (3.05 + 1.62 + 8.54 + 7.57) = 6.22 cu ft

4. Calculate the required weight of dry F.A., using the specific gravity factor for the dry state.

   Req'd wt. of dry F.A. = 6.22 × 62.4 × 1.99 = 772 lb

5. Convert the weight of dry F.A. to the weight at the original moisture condition.

   Wt. of F.A. at original moisture condition
   = 772 × 1.07 = 826 lb

6. Calculate the weight of the C.A. at the original moisture condition.

   Wt. of C.A. at original moisture condition
   = 730 × 1.04 = 759 lb

7. Calculate the displaced volumes of damp F.A. and C.A.

$$\text{Displaced vol. of damp C.A.} = \frac{759}{62.4 \times 1.35} = 9.01 \text{ cu ft}$$

$$\text{Displaced vol. of damp F.A.} = \frac{826}{62.4 \times 1.97} = 6.72 \text{ cu ft}$$

8. Calculate the required volume of added water.

   Req'd vol. of added water
   = 27 − (3.05 + 1.62 + 9.01 + 6.72) = 6.60 cu ft

9. Calculate the required weight of added water.

   Req'd wt. of added water = 6.60 × 62.4 = 412 lb

Minor changes in slump, air content, or proportion of fine to total aggregate may be desirable at times. These changes will necessitate corresponding adjustments in mix proportions to maintain yield and other characteristics. The following general rules may be used as a guide.

**Table 7-12**
**ADJUSTMENTS OF PROPORTIONS FOR CHANGE IN CEMENT CONTENT**

| | 1 | 2 | 3 | 4 | 5 | 6 |
|---|---|---|---|---|---|---|
| | Values from Cols. 3 and 4 of Table 7-11 | | Adjusted Proportions (Dry Aggregates) | | Converted Proportions (Damp Aggregates) | |
| | Weight (lb) | Displaced Volume (cu ft) | Weight (lb) | Displaced Volume (cu ft) | Weight (lb) | Displaced Volume (cu ft) |
| Cement | 462 | 2.35 | 600 | $\frac{600}{62.4 \times 3.15} = 3.05$ | 600 | 3.05 |
| Air | | 1.62 | | 1.62 | | 1.62 |
| C.A. | 730 | 8.54 | 730 | 8.54 | 730 × 1.04 = 759 | $\frac{759}{62.4 \times 1.35} = 9.01$ |
| F.A. | 860 | 6.92 | 6.22 × 62.4 × 1.99 = 772 | 27 − 20.78 = 6.22 | 772 × 1.07 = 826 | $\frac{826}{62.4 \times 1.97} = 6.72$ |
| Water | 472 | 7.57 | 472 | 7.57 | 6.6 × 62.4 = 412 | 27 − 20.4 = 6.60 |
| Total | | | | 27.00 | | 27.00 |

1. For each 1% increase or decrease in the percentage of fine to total aggregate, increase or decrease water by approximately 3 lb and cement content by 1%/cu yd.
2. For each 1-in. increase or decrease in slump, increase or decrease water by approximately 10 lb and cement content by 3%/cu yd.
3. For each 1% increase or decrease in air content, decrease or increase water by approximately 5 lb/cu yd and increase or decrease cement content by 2%/cu yd.

To calculate the adjustments, use the mix proportions that are based on dry aggregates. Maintaining the coarse aggregate constant, the quantities of cement, air, or water are increased or decreased according to the rules just outlined. These adjustments may result in an increase or decrease in the total displaced volume. To maintain the original volume, the proportion of fine aggregates should be decreased or increased accordingly.

***Example:***

Assume that the proportions given in columns 3 and 4 of Table 7-12 should be adjusted to decrease the slump by 2 in. and increase air content by 2%. Proceed as follows:

1. Transfer the proportions given in columns 3 and 4 of Table 7-12 to columns 1 and 2 of Table 7-13.
2. Decrease of 2 in. of slump requires $2 \times 10 = 20$-lb decrease in water. Decrease in cement $= 2 \times 0.03 \times 600 =$ 36 lb/cu yd.
3. Two percent increase in air content requires $2 \times 5 = 10$-lb decrease in water. Increase in cement $= 2 \times 0.02 \times 600 = 24$ lb/cu yd.
4. Total decrease in water $= 20 + 10 = 30$ lb
   Total decrease in cement $= 36 - 24 = 12$ lb
   Increase in air content $= 0.02 \times 27 = 0.54$ cu ft
5. Enter these calculations in Table 7-13 and complete in accordance with previous similar steps.

***Example:***

Design a lightweight aggregate concrete mix that is to test 20 MPa at 28 days. Concrete is to have 6 (±1.5)% air entrainment and 100 (±25) mm of slump. Type 10 normal portland cement and an air-entraining agent are to be used.

***Solution:***

The procedure is as follows:

1. Determine the loose, dry unit mass of the aggregates and their moisture content at time of mixing.

   Loose, dry unit mass of C.A. = 770 kg/m$^3$
   Loose, dry unit mass of F.A. = 990 kg/m$^3$

   Moisture content of C.A. = 4%
   Moisture content of F.A. = 7%

2. Determine the relative density factors (the relationship of aggregate mass to displaced volume) of the coarse and fine aggregates (use procedures as provided in ASTM C127 and C128 respectively).

   Relative density factor for C.A. = 1.40
   Relative density factor for F.A. = 1.98

3. Estimate quantities. From Table 7-9, the cement content should range from 280 to 445 kg/m$^3$. Design trial mixes with cement contents of 300, 350,

**Table 7-13**
**ADJUSTMENT OF PROPORTIONS FOR CHANGE IN AIR CONTENT AND SLUMP**

| | 1 | 2 | 3 | 4 | 5 | 6 |
|---|---|---|---|---|---|---|
| | Values from Cols. 3 and 4 of Table 7-12 | | Adjusted Proportions (Dry Aggregates) | | Converted Proportions (Damp Aggregates) | |
| | Weight (lb) | Displaced Volume (cu ft) | Weight (lb) | Displaced Volume (cu ft) | Weight (lb) | Displaced Volume (cu ft) |
| Cement | 600 | 3.05 | $600 - 12 = 588$ | $\frac{588}{62.4 \times 3.15} = 2.99$ | 588 | 2.99 |
| Air | | 1.62 | | $1.62 + 0.54 = 2.16$ | | 2.16 |
| C.A. | 730 | 8.54 | 730 | 8.54 | $730 \times 1.04 = 759$ | $\frac{759}{62.4 \times 1.35} = 9.01$ |
| F.A. | 772 | 6.22 | $6.23 \times 62.4 \times 1.99 = 774$ | $27 - 20.77 = 6.23$ | $774 \times 1.07 = 828$ | $\frac{828}{62.4 \times 1.97} = 6.74$ |
| Water | 472 | 7.57 | $472 - 30 = 442$ | $\frac{442}{62.4} = 7.08$ | $6.10 \times 62.4 = 381$ | $27 - 20.90 = 6.10$ |
| Total | | 27.00 | | 27.00 | | 27.00 |

and 400 kg/m$^3$, respectively. Assume the total dry, loose volume of aggregates required to be 1.10 m$^3$ with equal volumes of coarse and fine for the first trial.

4. Calculate the loose weights of the damp coarse and fine aggregate required per cubic meter.

Loose mass of damp C.A. = 0.55 × 770 × 1.04 = 440 kg
Loose mass of damp F.A. = 0.55 × 990 × 1.07 = 583 kg

5. Calculate the displaced volumes of the damp aggregates, using the previously determined relative density factors.

$$\text{Displaced volume of C.A.} = \frac{440}{1{,}000 \times 1.4} = 0.314 \text{ m}^3$$

$$\text{Displaced volume of F.A.} = \frac{583}{1{,}000 \times 1.98} = 0.294 \text{ m}^3$$

6. Calculate the displaced volumes of the cement and air.

$$\text{Displaced volume of cement} = \frac{300}{1{,}000 \times 3.15} = 0.095 \text{ m}^3$$

Volume of air = 1.000 × 0.06 = 0.060 m$^3$

7. Calculate the volume and mass of water required. The volume of water will be 1.000 m$^3$ minus the combined volumes of cement, air, and aggregates.

Volume of water

= 1.000 − (0.314 + 0.294 + 0.095 + 0.060) = 0.237 m$^3$

Mass of water = 1000 kg/m$^3$ × 0.237 m$^3$ = 237 kg (mass density for water is assumed to be 1,000 kg/m$^3$)

8. Calculate the masses of the materials (assuming a 0.10 m$^3$ trial batch) by multiplying the quantities per cubic meter by 0.10.

Mass of cement = 300 × 0.10 = 30.0 kg
Mass of water = 237 × 0.10 = 23.7 kg
Mass of C.A. = 440 × 0.10 = 44.0 kg
Mass of F.A. = 583 × 0.10 = 58.3 kg

9. Make a chart similar to that shown in Table 7-14 and record all calculations made to this point for ready reference.
10. Mix the materials and test the mixture for *air content, slump, unit mass,* and *workability.* If all are satisfactory, take cylinders for compression tests. If a correction in the amount of aggregate or water is required, adjust mix proportions as shown in Steps 11 through 14.
11. If more coarse aggregate is to be used, and more water is needed to produce the required slump, add the amounts needed. Assume 1.8 kg of C.A. and 2.0 kg of water.
12. Calculate the correction required per cubic meter for each.

Correction for C.A. = 18 kg
Correction for water = 2.0 × 10 = 20 kg

13. Calculate the corrected batch proportions.
(a) Corrected masses:

Cement (no change) = 300 kg
F.A. (no change) = 583 kg
C.A. = 440 + 18 = 458 kg
Water = 237 + 20 = 257 kg

(b) Corrected displaced volumes:

Cement (no change) = 0.095 m$^3$
F.A. (no change) = 0.294 m$^3$
Air (no change) = 0.060 m$^3$

$$\text{C.A.} = \frac{458}{1{,}000 \times 1.40} = 0.327 \text{ m}^3$$

Water = 257 ÷ 1,000 = 0.257 m$^3$
Total = 1.033 m$^3$

14. Multiply the masses and displaced volume of the materials in Step 13 by 0.968 (1/1.033) to adjust the quantities to yield 1 m$^3$.

(a) Adjusted masses:

Cement = 300 × 0.968 = 290.40 kg
F.A. = 583 × 0.968 = 564.30 kg
C.A. = 458 × 0.968 = 443.30 kg
Water = 257 × 0.968 = 248.80 kg

(b) Adjusted volumes:

Cement = 0.095 × 0.968 = 0.090 m$^3$
Air = 0.060 m$^3$
F.A. = 0.294 × 0.968 = 0.280 m$^3$
C.A. = 0.327 × 0.968 = 0.320 m$^3$
Water = 0.257 × 0.96 = 0.250 m$^3$
Total = 1.000 m$^3$

Record these calculations on the chart.

The foregoing adjusted quantities are for damp aggregates. For future adjustments of batch proportions, it is necessary to convert the batch mass of the materials to a dry mass. This will then serve as a basis for the calculation of adjustments for change in *aggregate moisture conditions, cement content, proportions of aggregates, slump,* or *air content,* to maintain a yield of 1 m$^3$. The procedure for converting the batch mass of damp aggregates to a batch mass for dry aggregates is as follows:

1. Calculate the dry mass of the C.A. and the F.A. from the batch mass for damp aggregates. The mass of cement is to remain constant.

Dry mass of C.A. = 443.3 ÷ 1.04 = 426.3 kg
Dry mass of F.A. = 564.3 ÷ 1.07 = 527.4 kg

**Table 7-14**
INITIAL BATCH PROPORTIONS AND TRIAL BATCH ADJUSTMENTS

| | 1 | 2 | 3 | 4 | 5 | 6 | 7 | 8 | 9 |
|---|---|---|---|---|---|---|---|---|---|
| | Estimated Batch Proportions (Damp Aggregates) | | | | | Corrected Batch Proportions (Damp Aggregates) | | Adjusted Batch Proportions (Damp Aggregates) | |
| | Mass (kg) | Displaced Volume ($m^3$) | Masses for 1/10 $m^3$ Trial Batch (kg) | Correction for 1/10 $m^3$ Trial Batch (kg) | Correction for 1 $m^3$ (kg) | Mass (kg) | Displaced Volume ($m^3$) | Mass (kg) | Displaced Volume ($m^3$) |
| Cement | 300 | $\frac{300}{1000 \times 3.15} = 0.095$ | 30.0 | | | 300 | 0.095 | $0.968 \times 300 = 290.40$ | $0.968 \times 0.095 = 0.090$ |
| Air | | $1.000 \times 0.06 = 0.060$ | | | | | 0.060 | | 0.060 |
| C.A. | 440 | $\frac{440}{1000 \times 1.4} = 0.314$ | 44.0 | +1.8 | +18 | $440 + 18 = 458$ | $\frac{458}{1000 \times 1.40} = 0.327$ | $0.968 \times 458 = 443.30$ | $0.968 \times 0.327 = 0.320$ |
| F.A. | 583 | $\frac{583}{1000 \times 1.98} = 0.294$ | 58.3 | | | 583 | 0.294 | $0.968 \times 583 = 564.30$ | $0.968 \times 0.294 = 0.280$ |
| Water | $0.237 \times 1000 = 237$ | $1.000 - 0.763 = 0.237$ | 23.7 | +2 | +20 | $237 + 20 = 257$ | $\frac{257}{1000} = 0.257$ | $0.968 \times 257 = 248.80$ | $0.968 \times 0.257 = 0.250$ |
| Total | | 1.000 | | | | | | | 1.000 |

2. Calculate the relative density factors for the dry state. Assume that the relative density factor for C.A. is 1.37 and for F.A. is 1.99.
3. Calculate the displaced volumes of the dry aggregates, using the relative density factors.

$$\text{Displaced volume of C.A.} = \frac{426.3}{1{,}000 \times 1.37} = 0.311 \text{ m}^3$$

$$\text{Displaced volume of F.A.} = \frac{527.4}{1{,}000 \times 1.99} = 0.265 \text{ m}^3$$

4. Maintaining the displaced volumes of cement and air constant, calculate the required displaced volume of adding water and the required mass of added water.

   Displaced volume of added water
   $= 1.000 - (0.090 + 0.060 + 0.311 + 0.265) = 0.274 \text{ m}^3$

   Required mass of added water
   $= 0.274 \times 1000 = 274 \text{ kg}$

To record these calculations, make a chart similar to that shown in Table 7-15. Into the first two columns, transfer the figures from columns 8 and 9 of Table 7-14. Into columns 3 and 4 enter the calculations from Steps 1 to 4.

Aggregates may have different moisture contents from time to time, and it will be necessary, under these circumstances, to adjust the batch masses of the materials to maintain a yield of 1 m$^3$.

Assume new moisture conditions of 2% for C.A. and 4% for F.A. Proceed as follows:

1. Determine the relative density factors for the new moisture contents. Assume that of the C.A. to be 1.35 and of the F.A., 1.97.
2. Maintain the mass of the cement constant and the displaced volumes of cement and air constant.
3. Calculate the mass of the aggregates for the new moisture conditions.

   Mass of moist C.A. $= 426.3 \times 1.02 = 434.83 \text{ kg}$
   Mass of moist F.A. $= 527.4 \times 1.04 = 548.50 \text{ kg}$

4. Calculate the displaced volumes of aggregates, using new relative mass density factors.

$$\text{Displaced volume of C.A.} = \frac{434.83}{1{,}000 \times 1.35} = 0.322 \text{ m}^3$$

$$\text{Displaced volume of F.A.} = \frac{548.50}{1{,}000 \times 1.97} = 0.278 \text{ m}^3$$

5. Calculate the required displaced volume of added water and, from this result, the required mass of added water.

   Displaced volume of added water $= 1.000 - (0.090 + 0.060 + 0.322 + 0.278) = 0.250 \text{ m}^3$

6. Enter these results into Table 7-15.

As indicated earlier, two more trial mixes will be made, using cement contents of 350 and 400 kg/m$^3$. Much of the trial-and-error work may be simplified by beginning with proportions obtained from the first trial mix. They will have to be adjusted to compensate for the increase in cement content by a corresponding decrease in fine aggregate. Proceed as follows:

1. Maintain the masses and volumes displaced by the dry, coarse aggregates, air, and water, shown in columns 3 and 4 of Table 7-15, constant. Transfer these values to columns 1 and 2 of Table 7-16.
2. Calculate the volume displaced by the new cement content, 350 kg/m$^3$.

$$\text{Displaced volume of cement} = \frac{350}{1{,}000 \times 3.15} = 0.111 \text{ m}^3$$

3. Calculate the required volume of dry fine aggregate as the difference between 1 m$^3$ and the sum of the displaced volumes of air, cement, coarse aggregate, and water.

   Required volume of dry F.A.
   $= 1.000 - (0.111 + 0.060 + 0.311 + 0.274) = 0.244 \text{ m}^3$

4. Calculate the required mass of dry F.A., using the relative density factor for the dry state.

   Required mass of F.A. $= 0.244 \times 1{,}000 \times 1.99$
   $= 485.56 \text{ kg}$

5. Convert the mass of dry F.A. to the mass at the original moisture condition.

   Mass of F.A. at original moisture condition
   $= 485.56 \times 1.07 = 519.55 \text{ kg}$

6. Calculate the mass of the C.A. at the original moisture condition.

   Mass of C.A. at original moisture condition
   $= 426.3 \times 1.04 = 443.35 \text{ kg}$

7. Calculate the displaced volumes of damp F.A. and C.A.

$$\text{Displaced volume of damp C.A.} = \frac{443.35}{1{,}000 \times 1.35} = 0.328 \text{ m}^3$$

$$\text{Displaced volume of damp F.A.} = \frac{519.55}{1{,}000 \times 1.97} = 0.264 \text{ m}^3$$

8. Calculate the required volume of added water.

   Required volume of added water
   $= 1.000 - (0.111 + 0.060 + 0.328 + 0.264) = 0.237 \text{ m}^3$

9. Calculate the required mass of added water.

   Required mass of added water $= 0.237 \times 1{,}000 = 237 \text{ kg}$

Minor changes in slump, air content, or proportion of fine to total aggregate may be desirable at times. These changes will necessitate corresponding adjustments in mix proportions to maintain yield and other characteristics. The following general rules may be used as a guide.

1. For each 1% increase or decrease in the percentage of fine to total aggregate, increase or decrease water content by approximately 1.5 kg and cement content by 1% per cubic meter.

**Table 7-15**
**ADJUSTMENTS FOR CHANGES IN AGGREGATE MOISTURE CONDITIONS**

| | 1 | 2 | 3 | 4 | 5 | 6 |
|---|---|---|---|---|---|---|
| | Batch Proportions (Damp Aggregates) | | Converted Batch Proportions (Dry Aggregates) | | Adjusted Proportions (Aggregates in New Moisture Condition) | |
| | Mass (kg) | Displaced Volume ($m^3$) | Mass (kg) | Displaced Volume ($m^3$) | Mass (kg) | Displaced Volume ($m^3$) |
| Cement | 290.4 | 0.090 | 290.4 | 0.090 | 290.4 | 0.090 |
| Air | | 0.060 | | 0.060 | | 0.060 |
| C.A. | 443.3 | 0.320 | $\frac{443.3}{1.04} = 426.3$ | $\frac{426.3}{1000 \times 1.37} = 0.311$ | $426.3 \times 1.02 = 434.8$ | $\frac{434.8}{1000 \times 1.35} = 0.322$ |
| F.A. | 564.3 | 0.280 | $\frac{564.3}{1.07} = 527.4$ | $\frac{527.4}{1000 \times 1.99} = 0.265$ | $527.4 \times 1.04 = 548.5$ | $\frac{548.5}{1000 \times 1.97} = 0.278$ |
| Water | 248.8 | 0.250 | $0.274 \times 1000 = 274$ | $1.000 - 0.726 = 0.274$ | $0.250 \times 1000 = 250$ | $1000 - 0.750 = 0.250$ |
| Total | | 1.000 | | | | 1.000 |

**Table 7-16**
**ADJUSTMENTS OF PROPORTIONS FOR CHANGE IN CEMENT CONTENT**

| | 1 | 2 | 3 | 4 | 5 | 6 |
|---|---|---|---|---|---|---|
| | Values from Cols. 3 and 4 of Table 7-11 | | Adjusted Proportions (Dry Aggregates) | | Converted Proportions (Damp Aggregates) | |
| | Mass (kg) | Displaced Volume ($m^3$) | Mass (kg) | Displaced Volume ($m^3$) | Mass (kg) | Displaced Volume ($m^3$) |
| Cement | 290.40 | 0.090 | 350.00 | $\frac{350}{1000 \times 3.15} = 0.111$ | 350.0 | 0.111 |
| Air | | 0.060 | | 0.060 | | 0.060 |
| C.A. | 426.30 | 0.311 | 426.3 | 0.311 | $426.3 \times 1.04 = 443.4$ | $\frac{443.4}{1000 \times 1.35} = 0.328$ |
| F.A. | 527.40 | 0.265 | $0.244 \times 1000 \times 1.99 = 485.6$ | $1000 - 0.756 = 0.244$ | $485.6 \times 1.07 = 519.6$ | $\frac{519.6}{1000 \times 1.97} = 0.264$ |
| Water | 274.00 | 0.274 | 274.0 | 0.274 | $0.237 \times 1000 = 237.0$ | $1000 - 0.763 = 0.237$ |
| Total | | | | 1.000 | | 1.000 |

2. For each 25-mm increase or decrease in slump, increase or decrease water by approximately 6 kg and cement content by 3% per cubic meter.
3. For each 1% increase or decrease in air content, decrease or increase water content by approximately 3 $kg/m^3$ and increase or decrease cement content by 2% per cubic meter.

---

To calculate the adjustments, use the mix proportions that are based on dry aggregates. Maintaining the coarse aggregate constant, the quantities of cement, air, or water are increased or decreased according to the rules just outlined. These adjustments may result in an increase or decrease in the total displaced volume. To maintain the original volume, the proportion of fine aggregates should be decreased or increased accordingly.

---

***Example:***

Assume that the proportions given in columns 3 and 4 of Table 7-16 should be adjusted to decrease the slump by 50 mm and increase air content by 2%. Proceed as follows:

1. Transfer the proportions given in columns 3 and 4 of Table 7-16 to columns 1 and 2 of Table 7-13.
2. Decrease of 50 mm in slump requires $2 \times 6 = 12$-kg decrease in water.

$$\text{Decrease in cement} = 2 \times 0.03 \times 350 = 21 \text{ kg/m}^3$$

3. A 2% increase in air content requires $2 \times 3 = 6$-kg decrease in water.

$$\text{Increase in cement content} = 2 \times 0.02 \times 350 = 14 \text{ kg/m}^3$$

4. Total decrease in water $= 12 + 6 = 18$ kg

$$\text{Total decrease in cement} = 21 - 14 = 7 \text{ kg}$$
$$\text{Increase in air content} = 0.02 \times 1.000 = 0.020 \text{ m}^3$$

5. Enter these calculations in Table 7-17 and complete in accordance with previous similar steps.

## REVIEW QUESTIONS

**1.** Give two reasons for using as much aggregate as possible when designing a concrete mix of any given strength.

**2.** Explain the purpose of each of the following aggregate tests: **(a)** silt test, **(b)** specific gravity test, **(c)** fineness modulus test, and **(d)** abrasion test.

**3.** Concrete is required for interior reinforced building columns 12 in. by 12 in., 10 ft high. It is to test 4,000 psi at 28 days and is to have 3 to 4 in. slump, with 5 (±1)% air. The relative density of the coarse aggregate is 2.65, its dry-rodded unit weight is 106 lb per cubic foot, and its maximum size is 1 in. The relative density of the fine aggregate is 2.78 and its fineness modulus is 2.60. **(a)** Design a trial mix, using 30 lb cement. **(b)** If there are 24 columns, what quantity of cement is required to do the job?

**4.** Concrete is required for interior reinforced building columns 300 × 300 mm, 3.0 m high. It is to test 30 MPa at 28 days and have 75 to 100 mm of slump, with 5 (±1)% air. The relative density of the coarse aggregate is 2.65, its dry-rodded unit mass is 1,700 $kg/m^3$ and its maximum size is 25 mm. The relative density of the fine aggregate is 2.78 and its fineness modulus is 2.60. **(a)** Design a trial mix, using 10 kg of cement. **(b)** If there are 24 columns, what quantity of cement is required to do the job?

**5.** Prepare a seven-point checklist to be used when preparing to place concrete.

**6.** Describe the purpose of each of the following when finishing a concrete slab: **(a)** screed, **(b)** strike-off bar, **(c)** float, **(d)** trowel, and **(e)** drag.

**7.** Explain briefly **(a)** what is meant by a *flash set* of cement, **(b)** why it may be necessary to heat both water and aggregates for concrete in very cold weather, **(c)** why the placing temperature of concrete is lower for massive structures than it is for ordinary reinforced structures, and **(d)** how rounded coarse aggregate usually produces better flowability in concrete than crushed stone.

**8.** A lightweight aggregate concrete mix is required to test 4,000 psi at 28 days. It is to have 5 (±1)% air entrainment and 4 (±1) in. of slump. The coarse aggregate has a dry unit weight of 50 lb/cu ft, moisture content of 3%, maximum size of ¾ in., and a relative density of 1.35. Fine aggregate has a dry unit weight of 65 lb/cu ft, free moisture of 6%, and a relative density factor of 1.90. Calculate the weight of materials required for a 2-cu-ft trial batch, using middle range cement content.

**9.** A lightweight aggregate concrete mix is required to test 30 MPa at 28 days. It is to have 5 (±1)% air entrainment and 100 (±25) mm of slump. The coarse aggregate has a dry unit mass of 800 $kg/m^3$, moisture content of 3%, maximum size of 20 mm, and a relative density of 1.35. Fine aggregate has a dry unit mass of 1,040 $kg/m^3$, free moisture of 6%, and a relative density factor of 1.90. Calculate the mass of materials required for a 0.1 $m^3$ trial batch, using middle-range cement content.

**10.** Outline four steps that you would take to ensure that 4,000 psi (30 MPa) concrete mixed and placed in cold weather would in fact achieve its design strength.

**Table 7-17**
ADJUSTMENTS OF PROPORTIONS FOR CHANGE IN AIR CONTENT AND SLUMP

| | 1 | 2 | 3 | 4 | 5 | 6 |
|---|---|---|---|---|---|---|
| | Values from Cols. 3 and 4 of Table 7-12 | | Adjusted Proportions (Dry Aggregates) | | Converted Proportions (Damp Aggregates) | |
| | Mass (kg) | Displaced Volume ($m^3$) | Mass (kg) | Displaced Volume ($m^3$) | Mass (kg) | Displaced Volume ($m^3$) |
| Cement | 350.0 | 0.111 | 350 − 7 = 343.0 | $\frac{343}{1000 \times 3.15} = 0.109$ | 343.0 | 0.109 |
| Air | | 0.060 | | 0.060 + 0.02 = 0.080 | | 0.080 |
| C.A. | 426.3 | 0.311 | 426.3 | 0.311 | 426.3 × 1.04 = 443.4 | $\frac{443.4}{1000 \times 1.35} = 0.328$ |
| F.A. | 485.6 | 0.244 | 0.244 × 1000 × 1.99 = 485.6 | 1000 − (0.756) = 0.244 | 485.6 × 1.07 = 519.6 | $\frac{519.6}{1000 \times 1.97} = 0.264$ |
| Water | 274.0 | 0.274 | 274 − 18 = 256.0 | $\frac{256}{1000} = 0.256$ | 0.219 × 1000 = 219.0 | 1000 − 0.781 = 0.219 |
| Total | | 1.000 | | 1.000 | | 1.000 |

# 8 Structural Timber Frame

Timber, as a building frame material, has been used successfully in many forms throughout the world. Timber is one of the most versatile materials used in the construction industry because it is relatively light in weight, has good tension and compression strength, and is very durable when protected from decay and insects, yet it can be easily shaped to accommodate almost any situation. Timber is used for pile foundations, beams, columns, bracing, and trusses. Many old wood-frame buildings are still in good structural condition and are being used today.

Originally, buildings made with a heavy timber frame were structures of two or more stories intended primarily for industrial and storage purposes. The earliest of these buildings used whole logs of various diameters as structural members. The next step was the use of sawn timbers for the structural frame and, more recently, timber members have been built from a number of small pieces, nailed, bolted, or glued together.

Heavy timber framing has now been expanded from the original industrial purpose to include schools, churches, auditoriums, gymnasiums, apartment buildings, supermarkets, and the like.

A number of important factors have contributed to the more efficient use of timber in modern building construction. One is the development and refinement of stress-graded timber, both solid and glue laminated. These have made it possible to apply precise structural design procedures to heavy timber framing, resulting in a completely engineered structure.

The production of modern types of timber connectors has made it possible to develop the full strength of wood when stresses are being transferred from one member to another. It is no longer necessary to provide much larger sections to accommodate the fastenings formerly used to transfer stresses from member to member.

Full recognition of the fire resistivity of large timber sections has helped in the growth of timber building. Extensive tests have been successfully carried out to determine the ability of timber to withstand fire. The use of smooth surfaces and rounded edges has improved the ability of timber to oppose the inroads of fire. The use of fire-retardant chemicals has also proved successful in increasing the fire resistance of timber elements. Chemicals such as water-soluble ammonium salts are impregnated into the wood under pressure and, when exposed to flame, release a gas that impedes the spread of the fire.

Another important development in wood technology has been the use of preservative materials to help timber withstand the deteriorating effects of moisture, disease, and insects. Pressure treatment with these preservatives, including creosote, Wolman salts, and solutions of copper compounds, has given deep penetration and long-lasting effects.

Probably the most significant development in the timber industry is the production of heavy timber sections by glue laminating and the production of structural sections using random-length veneers and strips. Their applications in timber frame construction are dealt with later in the chapter.

Heavy timber construction implies that the section sizes used in the frame are of sufficient size to provide the necessary degree of fire resistance required by local fire codes. All sections that must support loads are solid masses, have smooth flat surfaces without any sharp projections, and do not have any concealed or inaccessible voids. Structural components that do not meet fire code requirements must be protected to ensure that prescribed fire ratings are met.

## THE STRUCTURAL FRAME

A structural frame is an interrelated arrangement of load-bearing sections that, when put under the influence of external loads, will ensure the structural integrity of the building. That is, the structural members must be capable of resisting the stresses imposed on them without showing signs of failure or without producing failure in related members of the frame.

The structural frame must be capable of resisting applied loads in various combinations as determined by the structural designers. Typically, the loads fall into two categories: dead loads and live loads. Included in the dead load category are the materials used in the construction of the building, including the weight of the framing members. Stationary equipment and permanent partitions are also considered as dead loads. Live loads include gravity loads such as occupancy, weight of stored materials, snow, rain, and movable equipment. Other live loads such as wind or earthquake, pressure due to backfill, hydrostatic pressure due to water, and forces due to temperature changes must also be included in various combinations with the gravity loads.

In the design of a structural frame, the objective of the structural designer is to provide an economical structure using the appropriate materials without compromising the performance of the frame. Performance of the frame is its ability to support loads without exceeding specific deflection limits. These limits are usually dictated by the type of materials used in the nonstructural elements of the building and the type of occupancy.

Nonstructural elements depend on the structural frame for support. Glass panels, masonry partitions, and drywall ceilings are typical nonstructural elements that require fairly stringent deflection limits to ensure that they do not become overstressed by excessive deflections. Deflections of floor beams are also dictated by the need of the occupants to feel comfortable within a building; that is, vibrations and deflections produced by walking and other movements must be kept below certain values so that they are not noticed by the occupants. Roof deflections, especially for flat roofs, must be limited so that excessive ponding does not occur during heavy rains.

Structural frames can be classified as braced frames and unbraced frames. A frame can be defined as a braced frame when lateral displacement has been prevented and any lateral movement that occurs is small. In a braced frame, the connections between the structural members are normally designed to resist gravity loads. Cross-bracing or other structural components such as shear walls, stairwells, and elevator shafts are used to provide lateral stability to the frame. In an unbraced frame, lateral stability is developed through the rigidity of the frame members and the ability of the connections to resist rotation.

Because wood is relatively soft compared with other materials, timber connections using bolts and other steel components are usually designed as pin-type connections. These connections use thin plates in conjunction with bolts or other timber connectors that allow some rotation between the connected members. To ensure lateral stability of the frame, some type of bracing must be provided.

Wood frames that have large open areas, such as gymnasiums and hangars, depend primarily on the diaphragm action of the wall sheathing and roof decking to transfer lateral loads into the foundation. Diaphragm action is developed by nailing plywood sheathing or plank decking to the frame (see Figure 8-1). Bracing in the form of steel

**Figure 8-1** Diagonal Decking Adds Rigidity to Structural Frame. (Reprinted by permission of Canadian Wood Council.)

rod cross-bracing placed in the plane of the roof and in the walls is also used to transfer lateral loads from the superstructure to the foundation and provide added stability to the building. Walls built of other materials, such as masonry or reinforced concrete, and integrated into the timber frame also provide excellent lateral support for the structure (Figure 8-2).

Basic components of structural timber frame buildings may be classified as columns, beams, purlins, trusses, the connectors that are required for them, and decking or planking. The size of each used in any specific case will depend on the load; the span of girders, beams, and planks; the unsupported height of columns; and the species and grade of timber being used. Although timber sections may be cut to almost any size, most sections used in construction applications are based on recognized standard sizes (see Table 8-1). When specifying timber sections in metric units, the actual size of the section is used rather than the nominal size as is customary for standard units of measure.

Complete data on allowable working stresses for species of timber used in construction and grades within a species are available from such sources as the Canadian Institute of Timber Construction, the American Institute of Timber Construction, the Southern Pine Association, the National Lumber Manufacturers Association, or the West Coast Lumberman's Association.

## TIMBER CONNECTORS

Various types of connections have been developed for the joining of members in the timber frame and for the anchoring of the timber frame to its supporting foundation (Figure 8-3).

Timber connectors are required to transfer load from one member to another without producing any adverse effects on the connected members; that is, they must be able to distribute the load over a large enough area to prevent member cracking and material deformation in the vicinity of the connection. T-straps, U-straps, wood or steel side plates, angles and saddles in conjunction with bolts, lag screws, glulam rivets, shear plates, and split-ring connectors are used in various combinations to achieve this transfer of load.

*Glulam rivets* are special heavy-duty nails made of high-strength steel for use in glue-laminated member connections. At present these rivets are limited to Douglas fir sections only. Further testing of this connection is still being done to extend its use to other species. Glulam rivets are used in conjunction with prepunched steel side plates to achieve load transfer from one section to another. The rivets are driven with a pneumatic hammer; no predrilling of the timber sections is required.

*Shear plates* (Figure 8-4) are intended to prevent the deformation of the timber due to the forces in the bolt by distributing the forces over a much larger

**Figure 8-2** Masonry Walls Provide Lateral Stability for Timber Frame Structure. (Reprinted by permission of Canadian Wood Council.)

**Table 8-1**
**SIZES OF SAWN TIMBER SECTIONS**

| Standard Dimensions | | Metric Dimensions |
|---|---|---|
| Nominal Size (in.) | Actual Size (in.) | Actual Size (mm) |
| 2 × 2 | 1½ × 1½ | 38 × 38 |
| × 3 | × 2½ | × 64 |
| × 4 | × 3½ | × 89 |
| × 5 | × 4½ | × 114 |
| × 6 | × 5½ | × 140 |
| × 8 | × 7¼ | × 184 |
| × 10 | × 9¼ | × 235 |
| × 12 | × 11¼ | × 286 |
| × 14 | × 13¼ | × 337 |
| 3 × 4 | 2½ × 3½ | 64 × 89 |
| × 5 | × 4½ | × 114 |
| × 6 | × 5½ | × 140 |
| × 8 | × 7¼ | × 184 |
| × 10 | × 9¼ | × 235 |
| × 12 | × 11¼ | × 286 |
| × 14 | × 13¼ | × 337 |
| 4 × 4 | 3½ × 3½ | 89 × 89 |
| × 5 | × 4½ | × 114 |
| × 6 | × 5½ | × 140 |
| × 8 | × 7¼ | × 184 |
| × 10 | × 9¼ | × 235 |
| × 12 | × 11¼ | × 286 |
| × 14 | × 13¼ | × 337 |
| × 16 | × 15¼ | × 387 |
| 6 × 6 | 5½ × 5½ | 140 × 140 |
| × 8 | × 7¼ | × 191 |
| × 10 | × 9¼ | × 241 |
| × 12 | × 11¼ | × 292 |
| × 14 | × 13¼ | × 343 |
| × 16 | × 15¼ | × 394 |
| × 18 | × 17½ | × 445 |
| 8 × 8 | 7¼ × 7¼ | 191 × 191 |
| × 10 | × 9¼ | × 241 |
| × 12 | × 11¼ | × 292 |
| × 14 | × 13¼ | × 343 |
| × 16 | × 15¼ | × 394 |
| × 18 | × 17 | × 445 |
| 10 × 10 | 9¼ × 9¼ | 241 × 241 |
| × 12 | × 11¼ | × 292 |
| × 14 | × 13¼ | × 343 |
| × 16 | × 15¼ | × 394 |
| × 18 | × 17½ | × 445 |
| 12 × 12 | 11¼ × 11¼ | 292 × 292 |
| × 14 | × 13¼ | × 343 |
| × 16 | × 15¼ | × 394 |

area of wood. They can be used for steel-to-wood or wood-to-wood connections.

*Split-ring connectors* (Figure 8-5) use the same principle as do shear plates except that the bolt is not used in the transfer of load from one member to another. The load is passed from one member into the split ring and on into the other member (Figure 8-6). The bolt serves only as a clamp to keep the whole connection together. Split rings are used in wood-to-wood connections only. Like shear plates, split rings come in two sizes, 2½ in. (64 mm) and 4 in. (100 mm) in diameter.

*Nails* are used extensively in light truss construction, light framing, application of sheathing and decking, formwork, built-up beams, and stressed-skin panels. They come in many sizes, shapes, finishes, and coatings. When being considered for a particular application, they must be selected to ensure that the resulting connection will have the required strength to resist all anticipated loads without encountering deformation of the connection or splitting of the wood.

The capacity of a nailed connection is based on the lateral resistance developed by the nail when driven into the side grain of the wood section. The nail capacity depends on nail type (smooth or spiral), coating, length, and the species of wood being nailed. The lateral load resistance of a nail is greatly reduced when driven into the end grain and is not recommended. Toe nailing (nailing at an angle to the side grain) also results in reduced capacity and should be applied with care. Building code specifications and manufacturers' recommendations can be followed to ensure that the appropriate size and nail type are used to obtain satisfactory results.

*Truss plates* (Figure 8-7) are used in the manufacture of trusses. This type of connector replaces the need for gusset plates and provides a quick method for connecting the web members to the chord sections. The plates are pressed into the sections of the truss to be connected using a portable hydraulic press. Various sizes and types are available, providing a good selection of capacities for the truss manufacturer.

## WOOD COLUMNS

Wood columns in a structural timber frame are normally solid sawn timber or glue-laminated sections. To achieve a minimum ¾-hr fire rating, columns supporting floors must not be smaller than 8″ × 8″ nominal (191 × 191 mm actual) in cross section and columns supporting roofs must not be less than 6″ × 8″ nominal (140 × 191 mm actual). Columns may be continuous throughout the entire building height or spliced at various levels by means of dowels, wood or metal splice plates, or metal straps (Figure 8-8). In some cases, upper-story columns may bear on the floor beam of the lower story (also see Figure 8-13(d)).

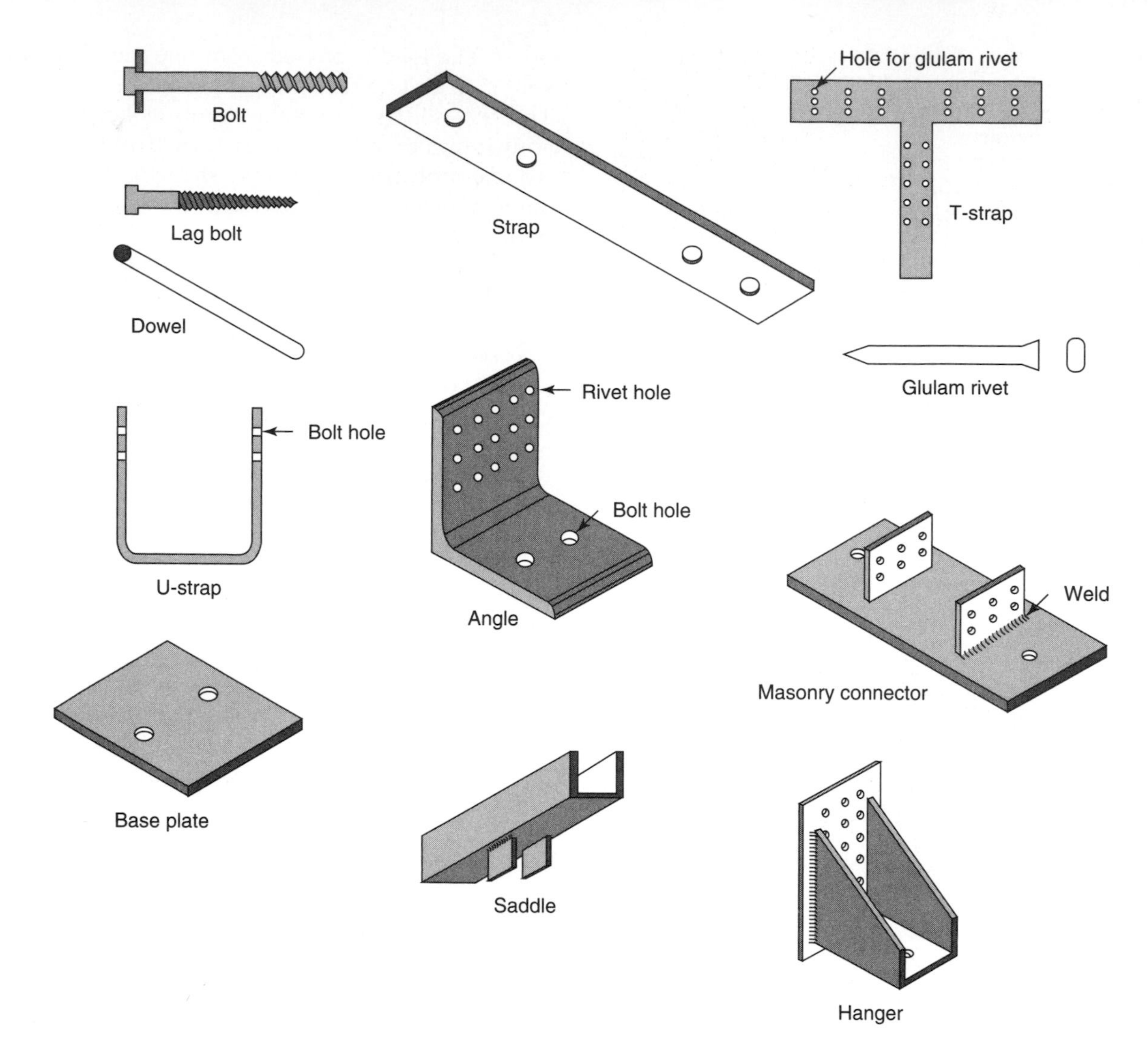


**Figure 8-3** Typical Timber Connectors.

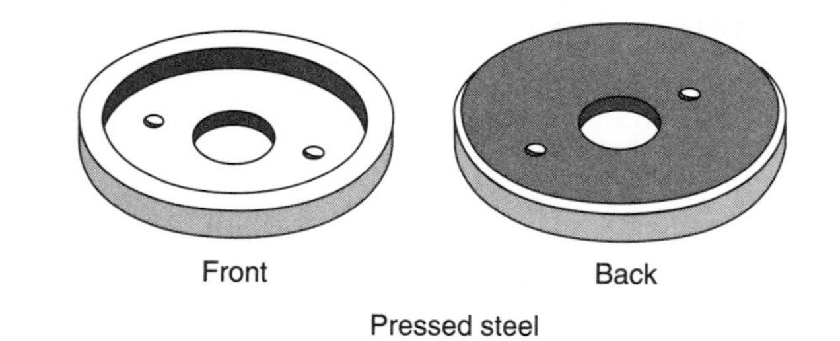


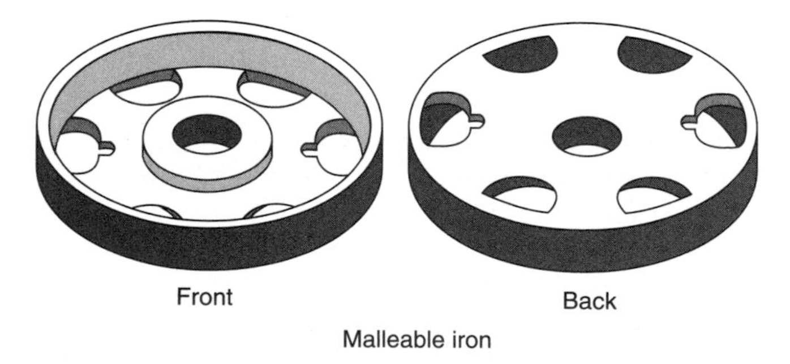


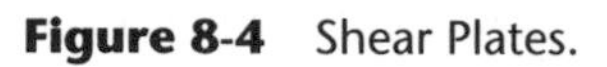

**Figure 8-4** Shear Plates.

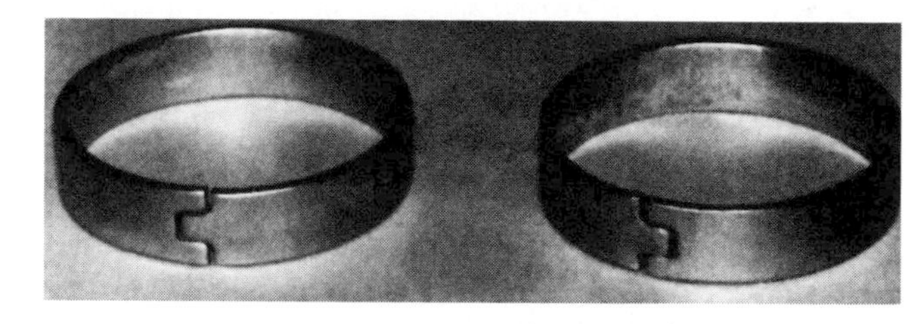

**Figure 8-5** Split-Ring Connectors.

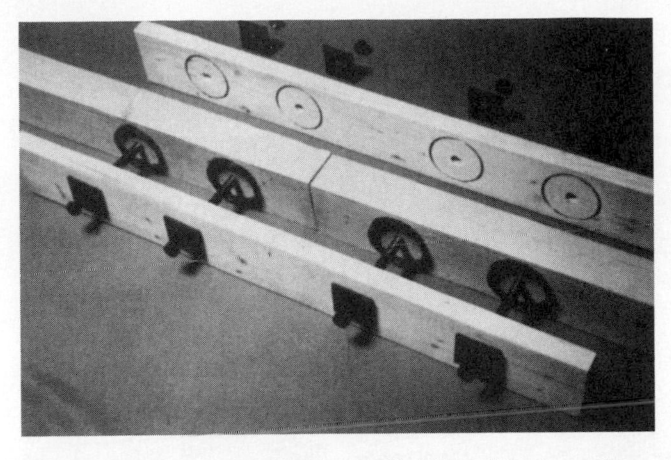

**Figure 8-6** Connection Using Split Rings. (Reprinted by permission of Canadian Wood Council.)

**Figure 8-7** Truss Plate.

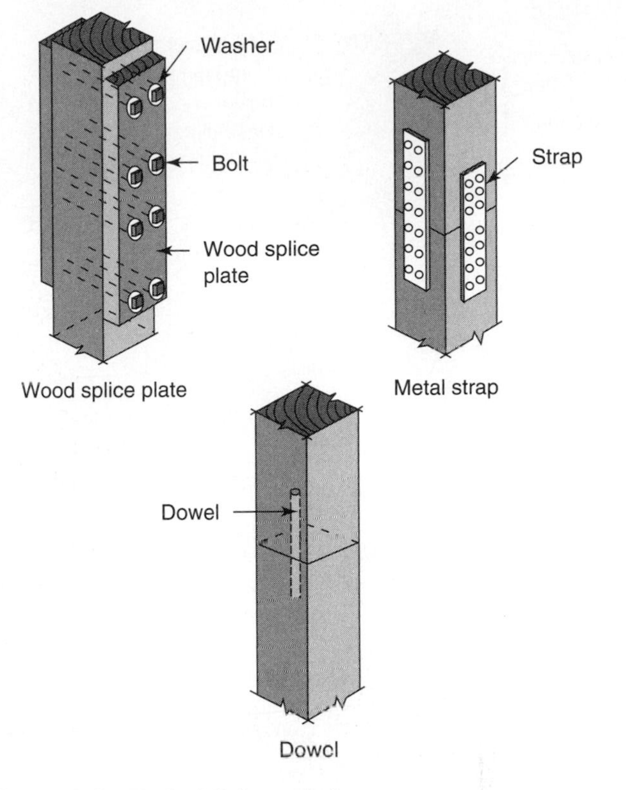


**Figure 8-8** Typical Column Splices.

## Column-to-Base Connections

The first consideration in erecting a structural timber frame is the method to be employed in anchoring the columns to the foundation. This may be done in several ways, some of which are illustrated in Figure 8-9. Notice in Figure 8-9(f) that *shear plates* are set into the column behind the straps. This may be done in any of the connections when conditions require them.

The dowel connection shown in Figure 8-9(a) is the simplest type of column-to-base connection and may be used where uplift and horizontal forces do not have to be considered, but there must be some means of preventing rotation of the column. The dowel is normally ¾ in. (20 mm) in diameter and the base plate a minimum of ⅛ in. (3 mm) thick. The connections in Figure 8-9(b) and (f) are used where uplift and horizontal forces must be considered. Shear plates may be used if necessary to transmit horizontal forces. The connection in Figure 8-9(c) is used where the column load must be distributed over a larger area than the end of the column. Plate thickness should not be less than ¼ in. (6 mm). The connection in Figure 8-9(d) is common in industrial buildings and warehouses. The connection in Figure 8-9(e) is intended for use with a raised concrete pedestal of limited dimensions, when the column cross section is large enough to distribute the load adequately. The end of the column is countersunk to accommodate the anchor bolt, nut, and washer. The base plate should be ¼ in. (6 mm) in minimum thickness (see Figure 8-10).

## Stud Walls

Structural components used in a timber frame do not always consist of heavy timber sections. Wall framing consisting of vertical sections or studs spaced at 12 in. to 24 in. (300 mm to 600 mm) o.c. and nailed top and bottom to horizontal plates (see Figure 8-11) is a very common method of wall framing. Studs and plates normally consist of sawn lumber having a nominal thickness of 2 in. (38 mm actual) and a nominal depth of 4 in. (89 mm) or 6 in. (140 mm actual). This framing is then covered with plywood sheathing, oriented strand board, or drywall to form a wall section.

Wall sections can be placed into three categories—load-bearing walls, non-load-bearing walls or partitions, and shear walls. Partition walls serve to divide large areas into smaller areas and, in general, do not support other major components of the structural frame. They can be moved or removed entirely without jeopardizing the stability of the overall building frame. However, these types of walls should not be confused with shear walls, which resist horizontal loads such as wind loads and act as a

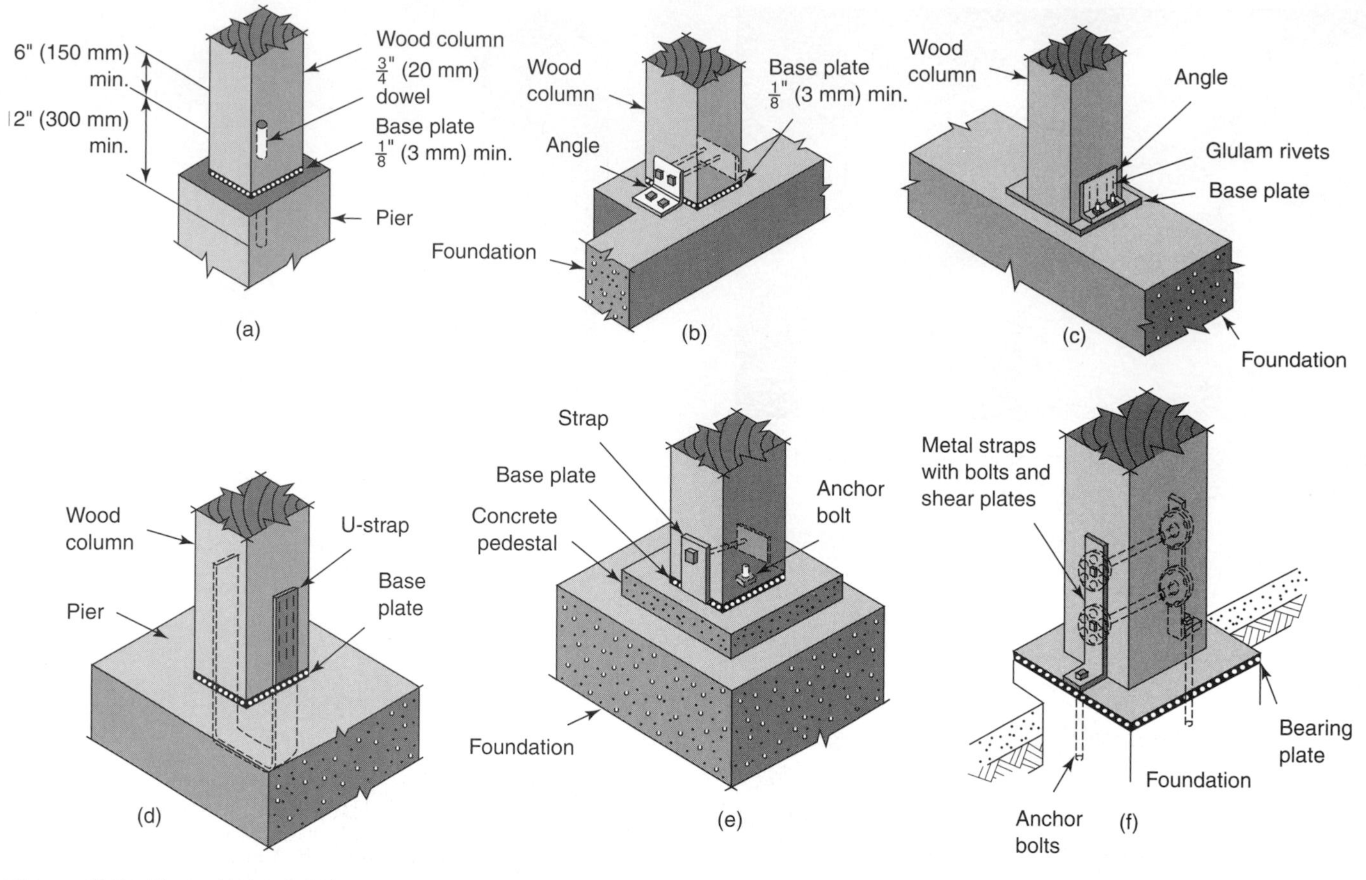


**Figure 8-9** Typical Wood Column-to-Base Connections.

**Figure 8-10** Column Base Plate Detail for Glulam Column.

method of bracing within a building frame. Load-bearing walls normally support gravity loads due to floors and roofs; however, they may also serve as shear walls.

Exterior walls may be load-bearing or serve as secondary framing between load-bearing columns and beams. When used as secondary framing between columns and beams, the stud wall framing must be robust enough to resist horizontal loads due to wind normal to the plane of the wall (see Figure 8-12). Exterior load-bearing walls that support gravity loads may be required to resist horizontal shear loads in the plane of the wall as well as bending that is normal to the plane of the wall due to wind loads.

The slenderness of the studs limits the maximum unsupported height of the wall. In general, the greater the unsupported height of the wall, the greater the depth of the stud. Structural codes provide practical limits on the slenderness to ensure that walls do not become unstable or too flexible. The amount of axial compressive load that a wall can support is based on the cross sectional area of the studs as well as their slenderness. Blocking and sheathing are used to stiffen the studs with respect to their weak axis and the overall stability of the wall is based on the depth of the studs used. Because most plank sections are sawn to a length of 16 ft (4.8 m), the practical unbraced height of a stud wall is usually 16 ft (4.8 m) or less. However, greater wall heights can be achieved if the lumber lengths are available. For heavy axial compressive loads, the stud thickness may be increased to 3 in. nominal (64 mm actual) and the depth of the section can be increased to provide additional stability.

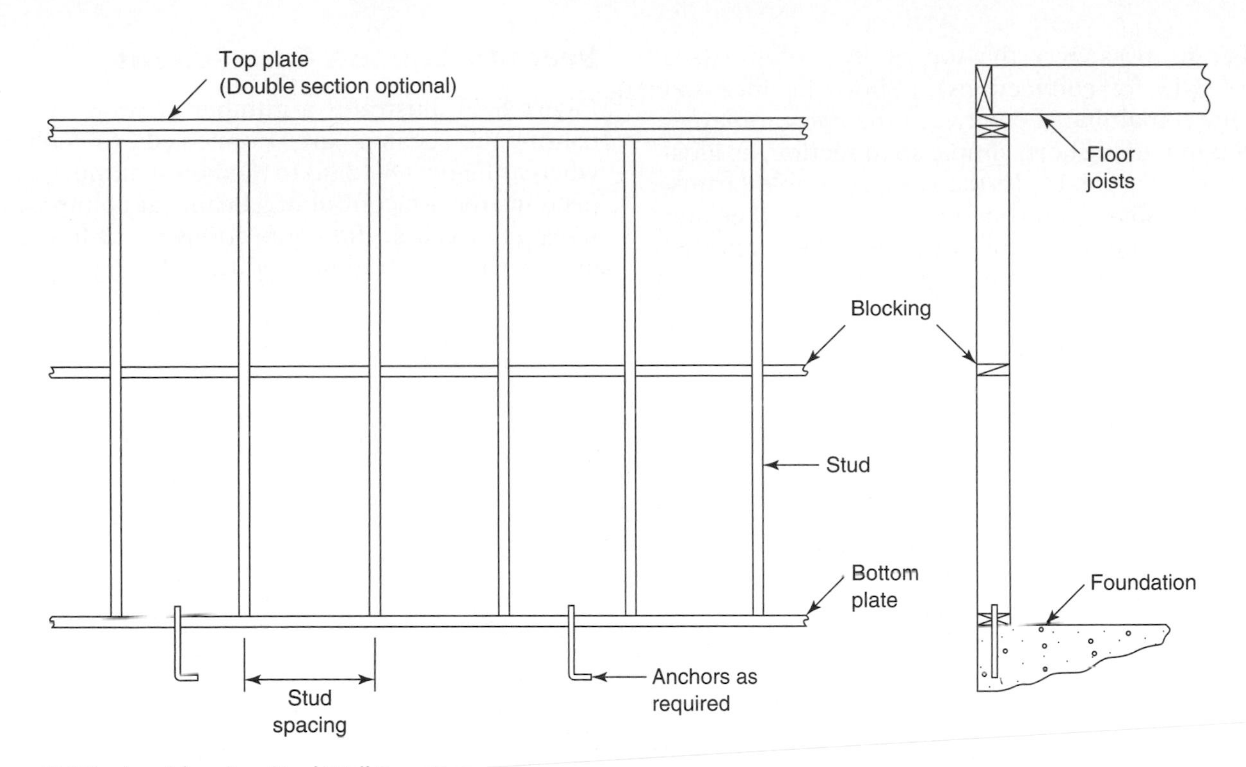


**Figure 8-11** Load-bearing Stud Wall Framing.

**Figure 8-12** Stud Framing Between Load-bearing Columns.

## WOOD BEAMS

Sections that support floors or roofs in a structural timber frame may be glue-laminated members or sawn timber. In either case, surfaces should be smooth and edges rounded to retard combustion. To achieve a ¾-hr fire rating, floor beams must not be smaller than 6″ × 10″ nominal (140 × 241 mm actual) in size, and beams supporting roofs must not be smaller than 4″ × 6″ nominal (89 × 140 mm actual) in cross section.

Beams used in a timber frame are either simple span or are extended over the column in what is known as *cantilever construction*. In cantilever construction, the beams pass over the top of the column (see Figure 8-13 for connections) a short distance and then the remaining gap between the two cantilevers is filled in with a short, simple span section, as illustrated in Figure 8-14. In the case of simple beams, the beams span from one column to another and may rest on top of the column or may be connected to the sides of the column (see Figure 8-15).

## Beam-to-Column Connections

Figure 8-13 illustrates a number of ways by which continuous beams are connected to columns, whereas Figures 8-15 (a) to (f) show a number of connections for noncontinuous beams at columns. In a number of cases, the connections are interchangeable, perhaps with some modification. Figure 8-15(g)

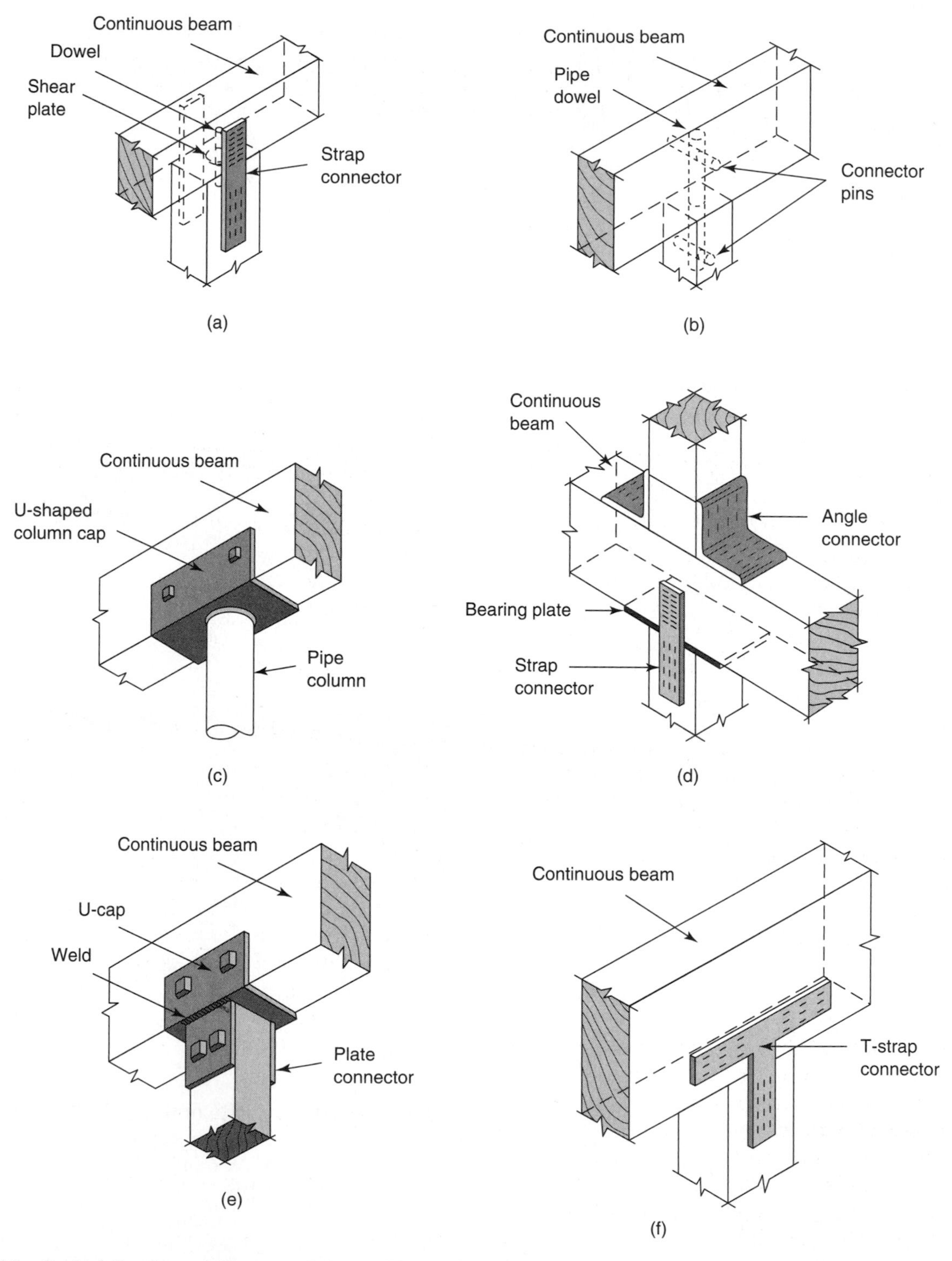


**Figure 8-13** Typical Continuous Beam-to-Column Connections.

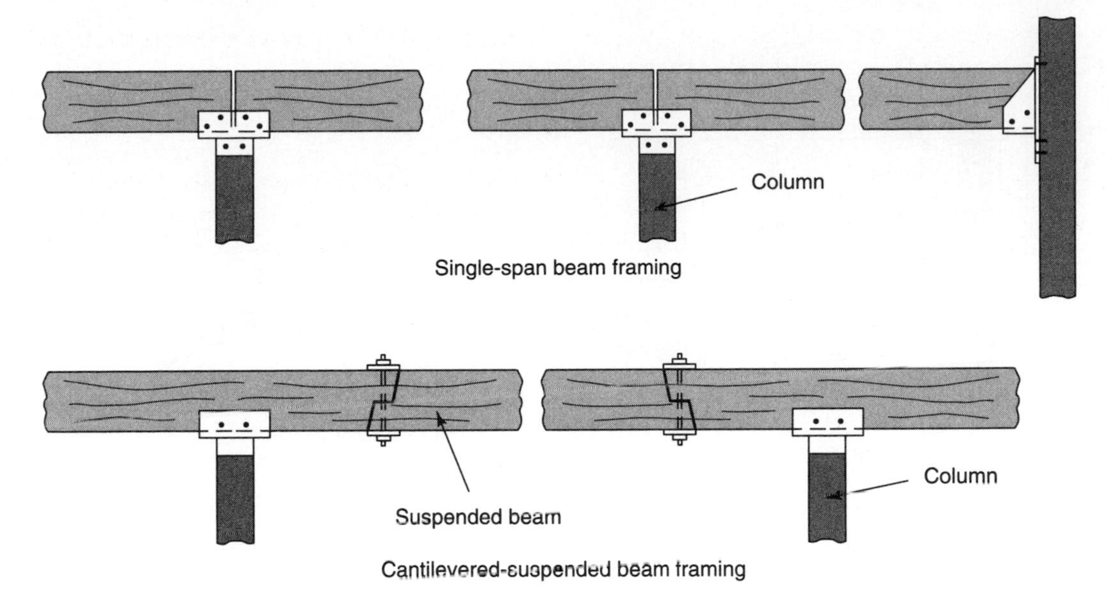


**Figure 8-14** Timber Beam Framing Methods.

illustrates an actual glulam roof beam consisting of single-span beam sections resting in a steel saddle on top of a square hollow steel column. The length of the saddle under each beam end is determined by the compressive stress perpendicular to the grain of the beam sections.

Figure 8-13(a) is a standard beam to-column connection providing for uplift. The shear plate and dowel are used only when lateral forces make it necessary. A loose bearing plate may be added if the cross section of the column does not provide enough bearing area for the beam in compression perpendicular to the grain. Concealed connections are shown in Figure 8-13(b) and in Figure 8-15(a). In the latter connection, the bolt heads may also be hidden by countersinking and plugging the holes. Figure 8-13(c) is designed specifically for pipe columns, whereas Figures 8-13(e) and 8-15(e) are designed for situations in which the beam and the column are not the same width.

## Beam-to-Masonry Wall Connections

Buildings may be designed with masonry exterior walls and a structural timber floor and roof frame. In such cases, the framing members must be connected to the masonry wall or to pilasters built into it. Figure 8-16 illustrates several methods for connecting timber beams to a masonry wall.

Figure 8-16(a) is a standard anchorage to a masonry wall, providing for uplift and resisting horizontal forces. The bearing plate should be a minimum of ¼ in (6 mm) thick and the clearance between the end of the beam and the wall at least 1 in. (25 mm). Figure 8-16(d) is used where the pilaster is not wide enough to permit outside anchor bolts. The beam must be countersunk to accommodate the anchor nut and washer. Figure 8-16(e) is intended for beams not exceeding 24 in. (600 mm) in depth. Figure 8-16(f) is used with curved or pitched beams, the slotted hole in the angles allowing for slight horizontal movement resulting from vertical deflection.

## Roof Beams

The method used to connect roof beams to exterior masonry walls depends on whether there is a parapet extending above the roof or if the roof is a *flush deck*, which covers the top of the wall [see Figure 8-17(c)]. In the case of a flush deck, a fascia is required along the top outside edge of the wall to provide solid backing for the roof flashing. Notice in Figure 8-17(c) that the beam pocket is large enough to allow at least ¾-in. (20-mm) clearance at the sides and end of the beam.

When a party wall or fire wall is involved in structural timber framing, the framing members tie into it from both sides, as shown in Figure 8-18.

## Intermediate Beams and Purlins

In many timber designs, smaller, *intermediate* beams are framed into the main ones, at right angles to them, and it is to these that the decking is fastened (see Figure 8-19). In the case of the roof, purlins may span from beam to beam and carry

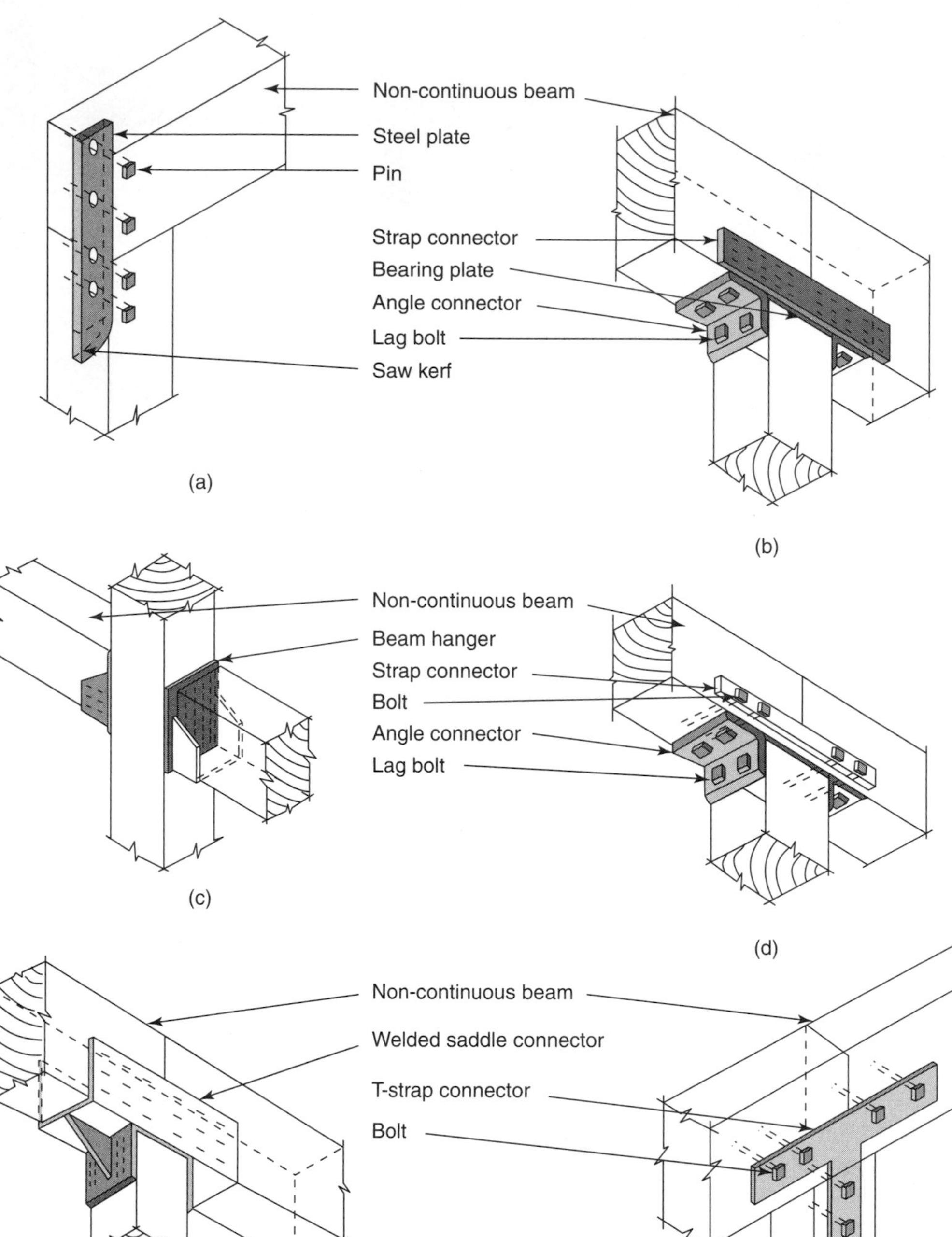


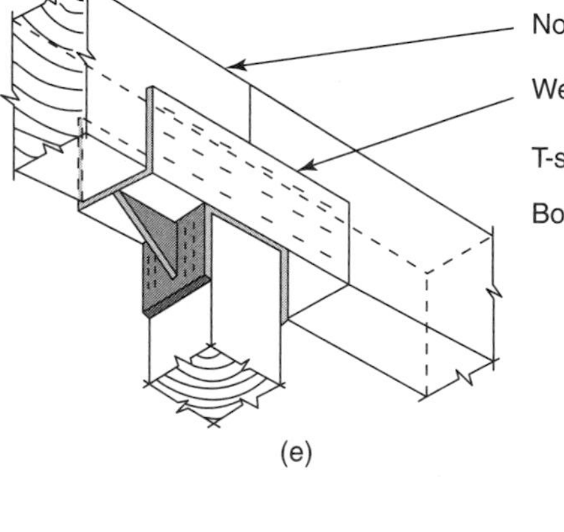

**Figure 8-15** (a to f) Typical Noncontinuous Beam-to-Column Connections. (g) Single Span Glulam Beams Supported by Steel Saddle and Column.

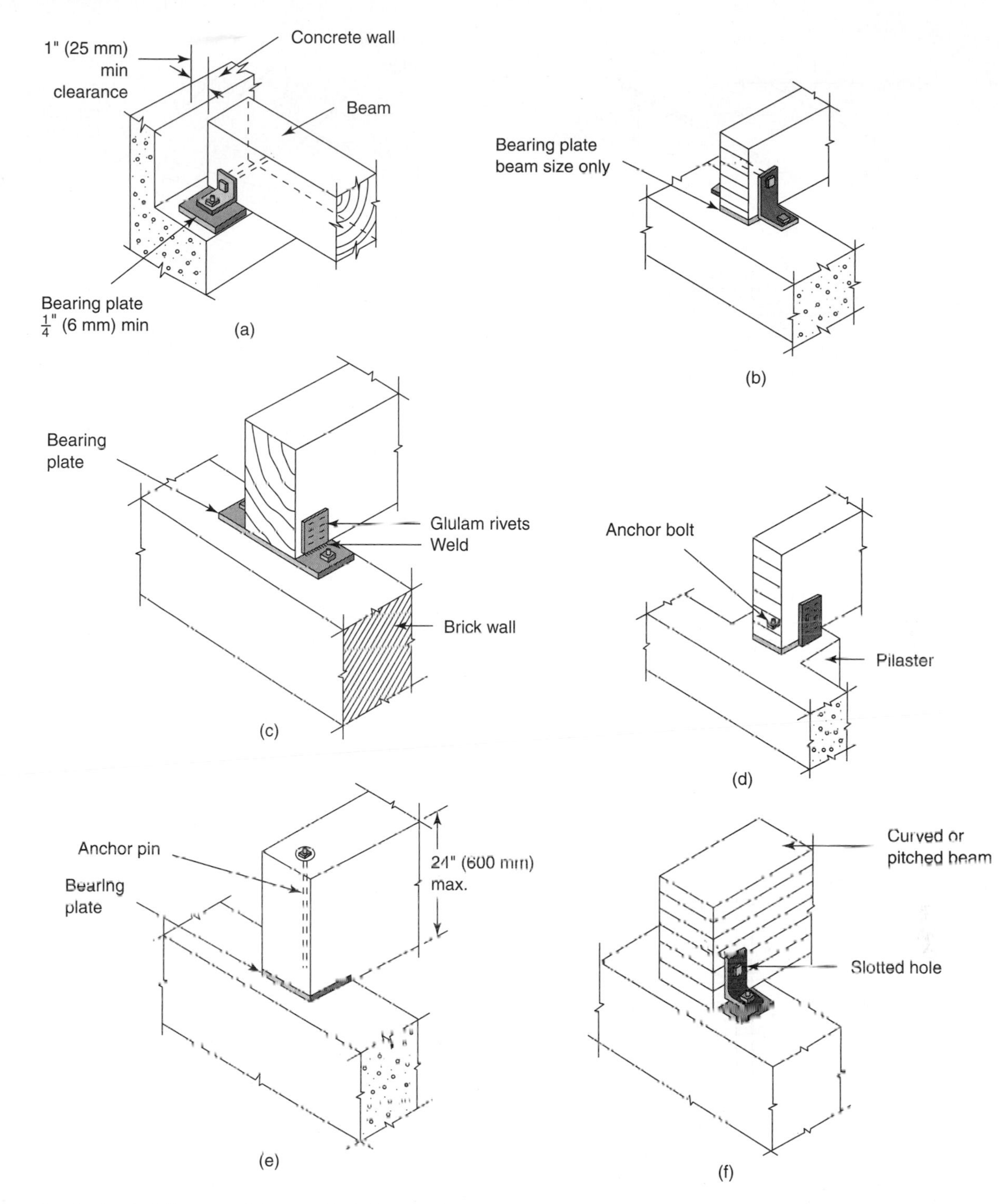


**Figure 8-16** Typical Beam-to-Masonry Wall Connections.

the roof deck. These smaller members may be carried on top of the main beams, supported on ledgers fastened to the sides of the beams, or suspended by hangers anchored to the beam sides (see Figure 8-20). Other designs will call for the decking to be applied directly to the main beams (see Figure 8-21).

## FLOOR AND ROOF DECKS

Timber floor and roof decks can be constructed of *plank decking* or *laminated decking.* Plank decking is used in most applications where extreme wearing of the exposed surface is not a concern. Laminated decking is used for bridge decks, loading

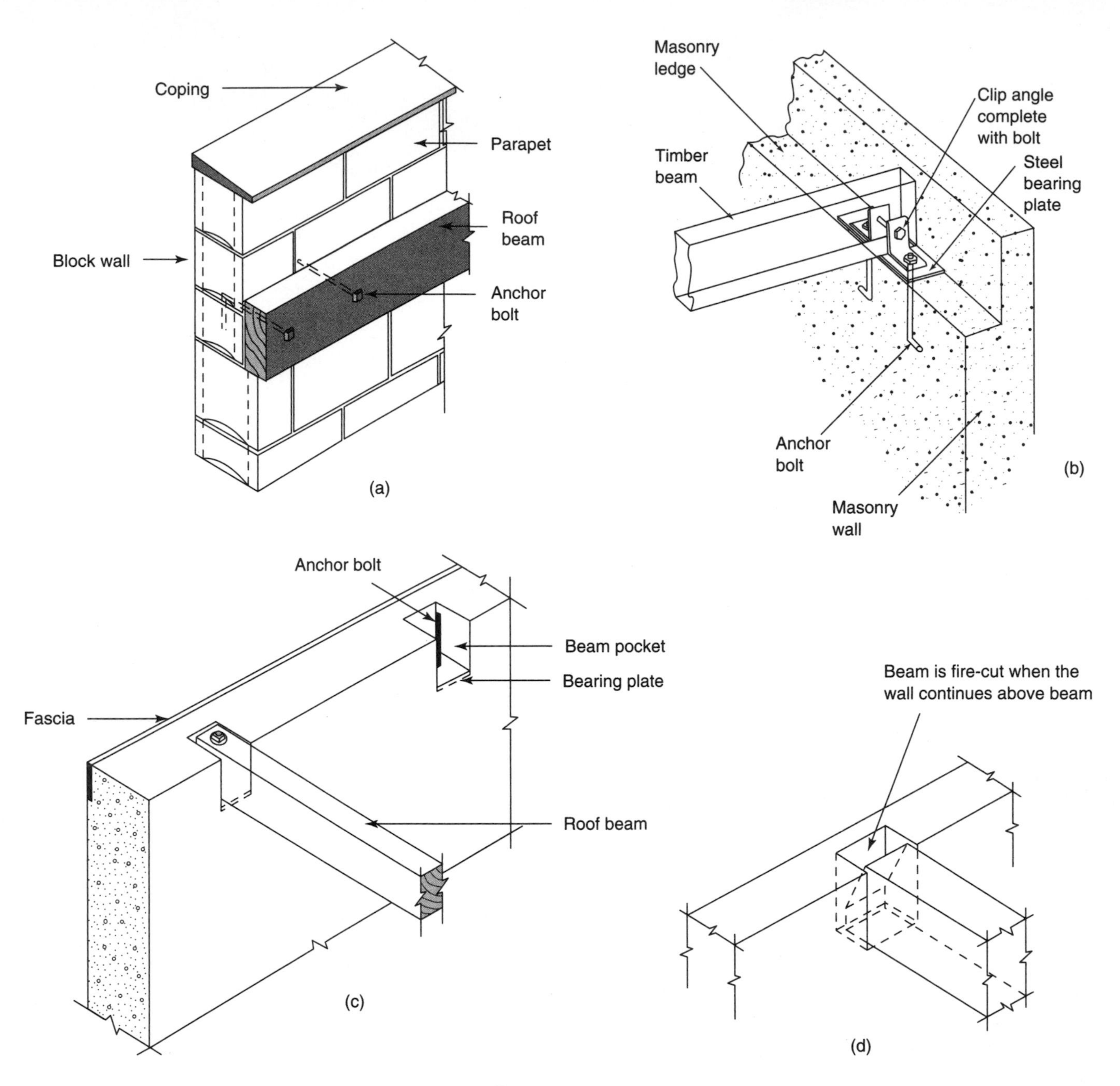


**Figure 8-17** Roof Beam Connections to Masonry Wall.

ramps, and warehouse floors where high wear is anticipated.

Plank decking is tongue-and-groove material that is laid on its wide side over supporting members (Figure 8-22). Plank decking is normally available in three thicknesses: 2″, 3″, and 4″ nominal (38 mm, 64 mm, and 89 mm actual). The 2-in. (38-mm) decking has a single tongue and groove, while the 3-in. and 4-in. (64- and 89-mm) thicknesses have a double tongue and groove (Figure 8-22). The 3-in. and 4-in. (64- and 89-mm) sections are predrilled horizontally with ¼-in. (6-mm) diameter holes spaced at 30 in. (750 mm) o.c. for 8-in. (200-mm) spikes that are driven from plank to plank to ensure a snug tongue and groove fit. The 2-in. (38-mm) decking may come in 6-in. and 8-in. nominal (127- and 178-mm actual) widths, while the 3-in. and 4-in. (64- and 89-mm) material comes in 6-in. nominal (127-mm actual) width only. Plank decking may be plain or have decorative grooves cut into the exposed side for acoustical and aesthetic purposes. Floor decking is usually *blind-nailed;* that is, the nails are driven through the

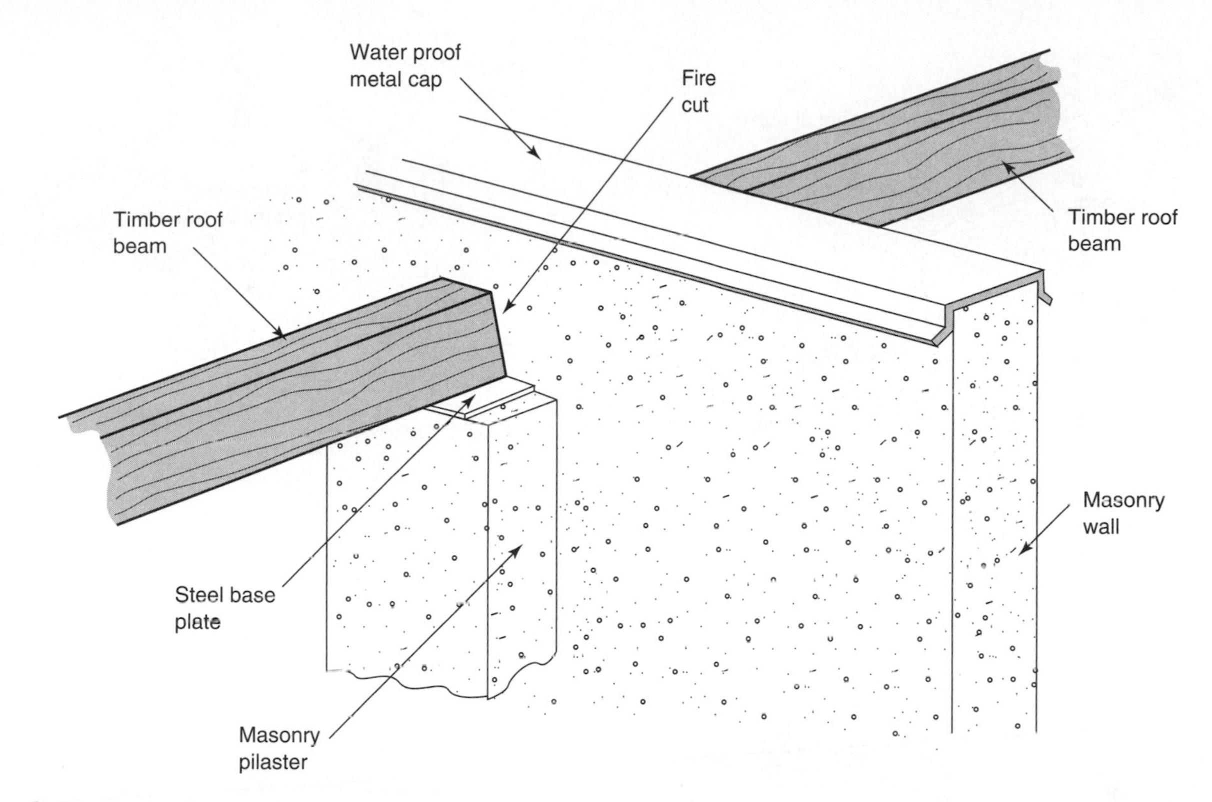


**Figure 8-18** Timber Frame to Masonry Fire Wall.

**Figure 8-19** Intermediate Beams Framed with Hangers. (Reprinted by permission of Canadian Wood Council.)

tongue of the decking into the supporting member. The nail head is concealed by the groove of the next plank (Figure 8-23).

Laminated decking is usually 2-in. thick nominal (38-mm-thick actual) material that can vary in width from 4 to 12 in. nominal (89 to 286 mm actual). Laminated decking is placed on edge over supporting members and spiked into place to provide a load-carrying surface (Figure 8-24).

Wood species used for plank decking include red cedar, white pine, Douglas fir, western hemlock, and the spruce–pine–fir group. Douglas fir, spruce, or jack pine are recommended for floor decks because they wear well. Cedar is used in exposed areas because of its durability against decay and its attractive appearance. The two grades of plank decking are *select* and *commercial*. Select grade is used when strength and appearance are both important (churches, chapels, and the like) and commercial grade is used where appearance is not a major consideration. Commercial grade has a lower strength; however, it is less costly than the select grade. Almost any species of wood can be used for laminated decking. The strength and durability required for its anticipated use will determine the species chosen.

Three basic patterns are used for laying decking: simple span, two-span continuous, and controlled random spliced (see Figure 8-25). Because most decking is supplied in random lengths, the controlled random pattern is the most popular as it makes the best use of the material supplied. A deck using this arrangement must extend over at least three spans and have the end joints staggered at least 24 in. (600 mm) in adjacent courses and 6 in. (150 mm) or more in every other course. Each plank must bear on at least one support, and end joints in the first half of the end spans are not allowed. To obtain continuity, nailing requirements must be followed carefully.

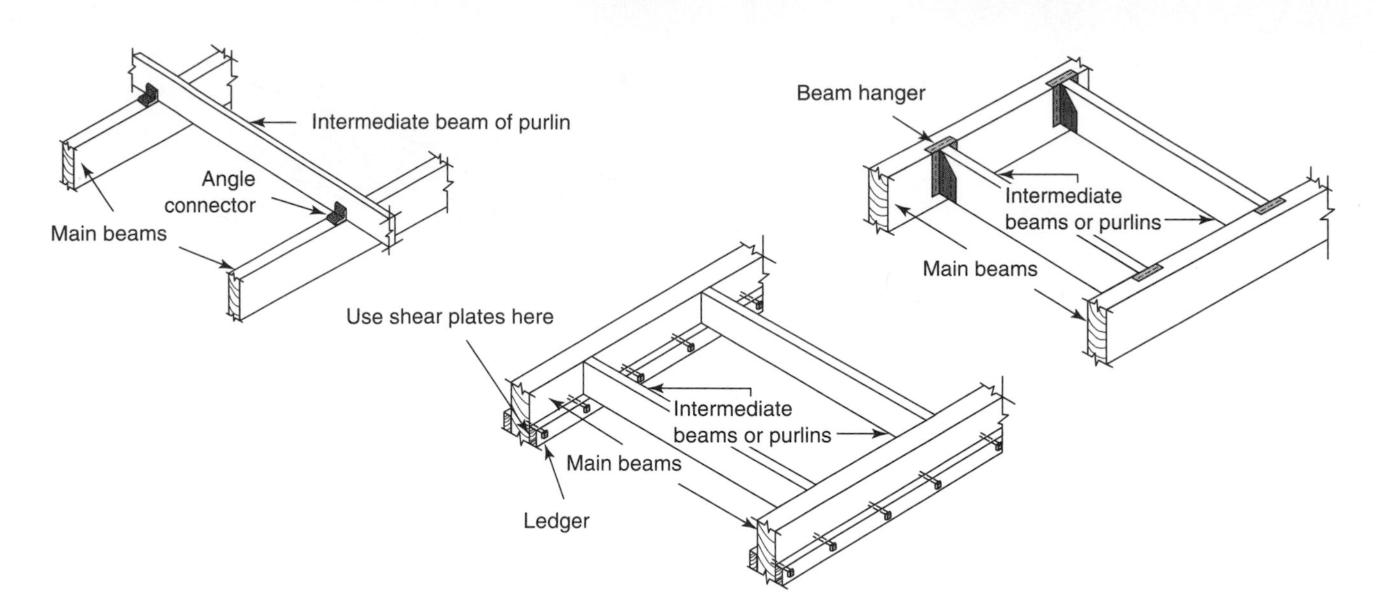


**Figure 8-20** Intermediate Beam Supports.

**Figure 8-21** Decking on Main Beams.

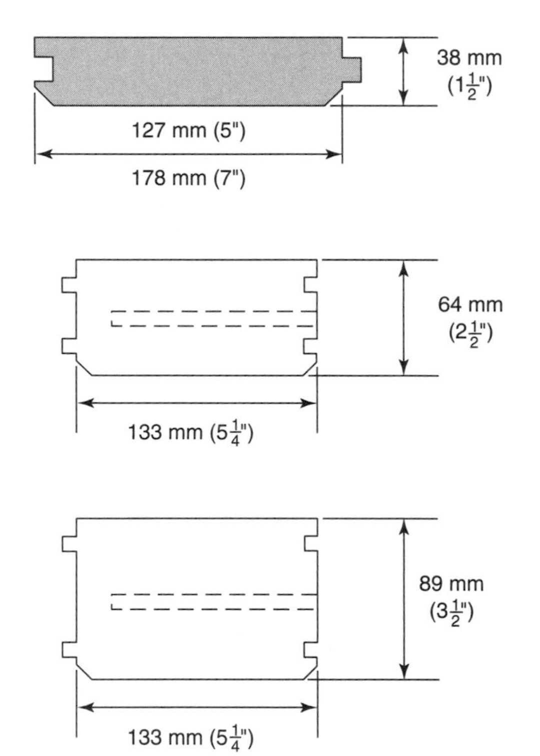


**Figure 8-22** Types of Plank Decking.

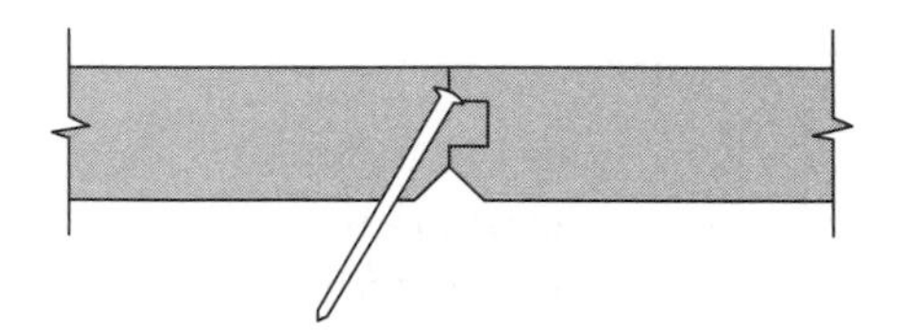

**Figure 8-23** Blind Nailing.

## GLUED-LAMINATED TIMBER

Bonding wood with adhesives is an old art; examples of it have been found in Egyptian tombs. It was used in Europe for construction purposes as early as the twentieth century, where it gained ready acceptance, but not until the early 1930s was glued-laminated structural timber introduced in the United States. World War II and its heavy demand for military and industrial construction provided the impetus that made glued-laminated construction commercially important.

Architects and builders are finding glued-laminated timber the answer to some of their most pressing problems. Such timbers have proved to be extremely versatile and have allowed architects to develop shapes and sizes that would be difficult to obtain with solid timber, concrete, or steel.

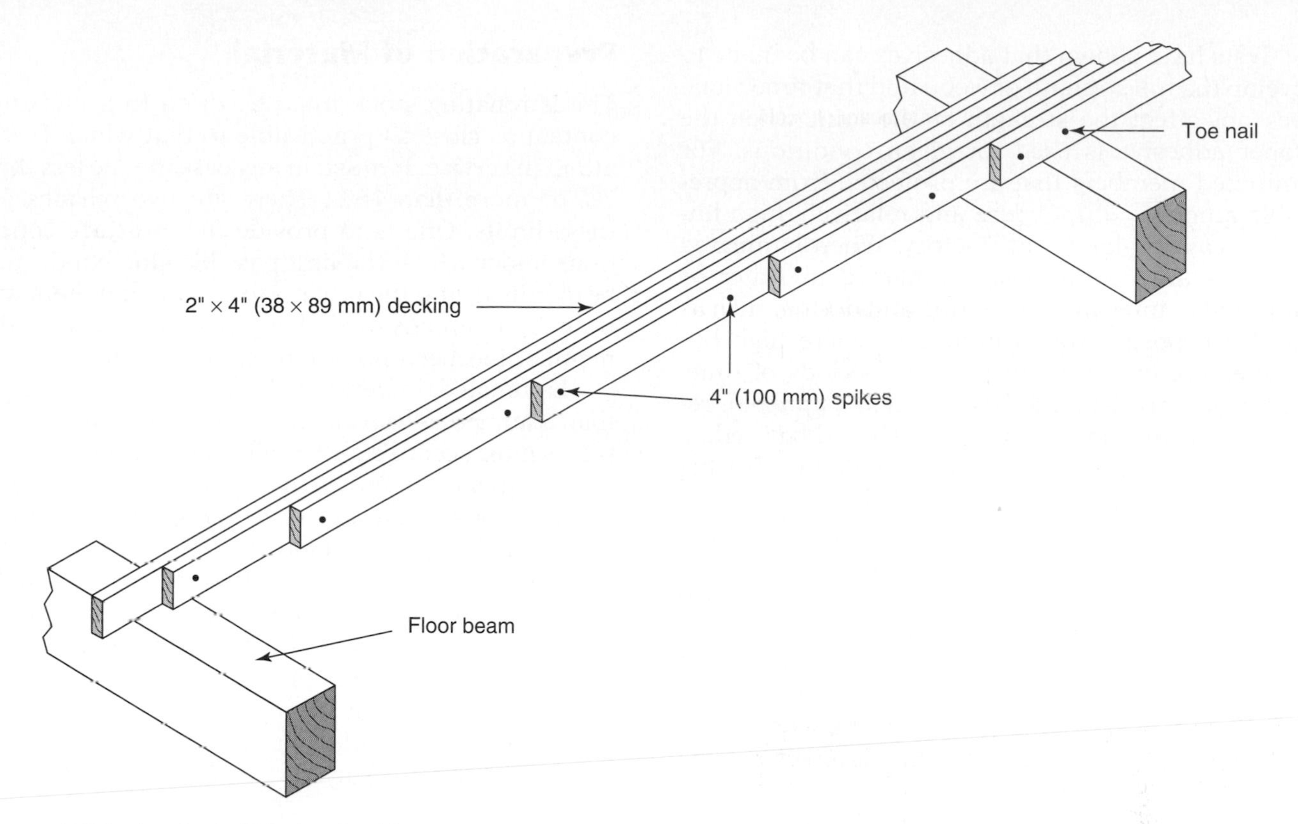


**Figure 8-24** Laminated Timber Decking.

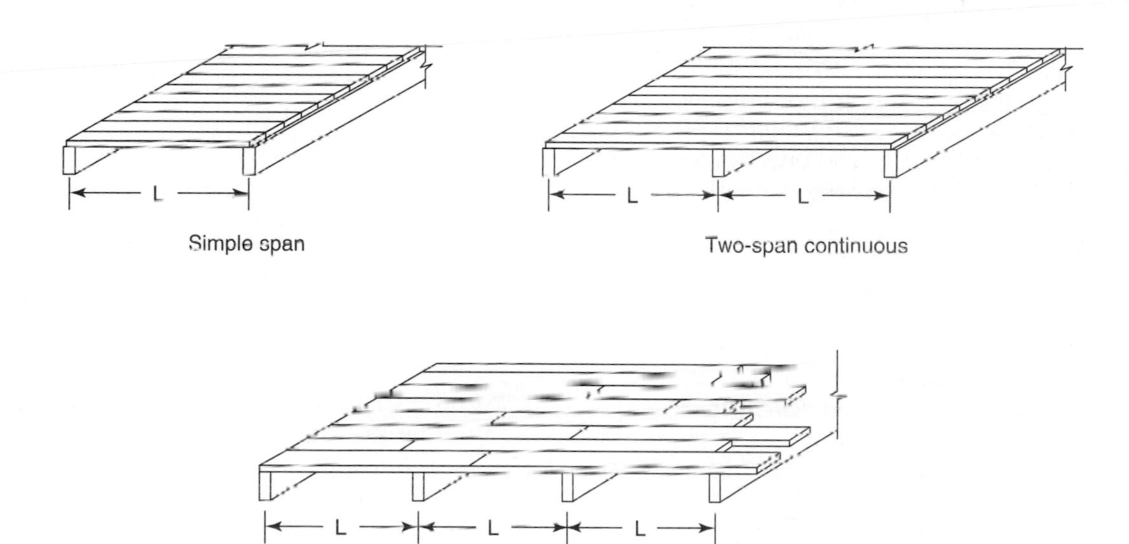


**Figure 8-25** Deck-Laying Patterns.

## Materials Used

Laminated timbers are widely recognized as being among the finest permanent structural materials. They are made from kiln-dried stock, mostly Douglas fir and southern pine, with some western larch and western hemlock. Both fir and pine are extremely strong, straight-grained, tough, resilient, and durable woods. The larch and hemlock are not quite as strong but match the fir and pine closely in color and texture and are used in areas of the laminated beam where flexural stresses are less of a concern, that is, in the middle third of the beam cross section.

Tests have shown that adhesives can be made to develop the full strength of wood and that time alone does not affect the strength of the joint when the proper adhesive is used for given conditions. For laminated members that are protected from appreciable amounts of moisture and relatively high humidity, casein glue is satisfactory. When laminated members are continuously immersed in water or subjected to intermittent wetting and drying, such as exterior exposures or in buildings where high humidities are encountered for long periods of time, highly water-resistant adhesives such as phenol, resorcinol, or melamine resin glues should be used.

The manufacture of glued-laminated timber begins with the selection and grading of the laminating lumber. The finished product will be only as good as the pieces that went into it, so the laminating stock must be of good quality. Of the materials available to the laminator, the high-quality laminations are almost free of all natural defects that may restrict strength and compromise appearance. These laminations are placed in the top and bottom of the glued beam section where bending stresses are the greatest; and if appearance is a concern, they are used on the side faces of the beam section as well.

The lower grades of laminating stock are almost equivalent to select structural, construction, and standard grades of joists and planks, with special additional limitations that qualify them for laminating purposes. One of these limitations relates to cross grain. Limitations in this aspect of material selection are more severe than those commonly imposed by grading rules. They apply to only the outer 10% of the depth of a member stressed principally in bending, but apply to all laminations of members under tension or compression. Working unit stresses must be modified to the slope of grain, as specified in Table 8-2.

**Table 8-2**
**STRENGTH RATIOS CORRESPONDING TO SLOPE OF GRAIN**

| | Maximum Strength Ratio | |
|---|---|---|
| Slope of Grain | Fiber Stress in Bending or Tension | Stress in Compression Parallel to Grain |
| 1 in 6 | — | 0.56 |
| 1 in 8 | 0.53 | 0.66 |
| 1 in 10 | 0.61 | 0.74 |
| 1 in 12 | 0.69 | 0.82 |
| 1 in 14 | 0.74 | 0.87 |
| 1 in 15 | 0.76 | 1.00 |
| 1 in 16 | 0.80 | — |
| 1 in 18 | 0.85 | — |
| 1 in 20 | 1.00 | — |

## Preparation of Material

The laminating stock must be dried to a moisture content as close as practicable to that which it will attain in service. It must, in any case, not be less than 7% or more than 16%. There are two reasons for these limits. One is to provide the moisture conditions under which the best possible glue bonds may be obtained, and the other is to ensure that there will be as little dimensional change as possible once the member has been placed in service (Figure 8-26).

Laminated timbers are often required to be longer than commercially available stock, and the individual pieces must accordingly be spliced end to end to make full-length laminations. This is done in several ways, including *plain scarf joints, stepped scarf joints, finger joints,* and *butt joints.* Figure 8-27 illustrates all four.

Plain scarf joints are relatively easy to make and are used extensively in important structural members. When they occur in tension members or in the tension portion of a member subjected to bending, a reduction in allowable stress is necessary. The scarf is made by a machine that produces a smooth, sloping surface at one or both ends of each board.

A stepped scarf joint employs a mechanical as well as a glue bond, but is more difficult to make. In addition, when stepped scarf joints are used, the portion of the thickness of the lamination occupied by the step is disregarded in calculating the moment of inertia and the net effective area.

Finger joints are cut with special saws and are often used where the laminating stock is ¾ in. (19 mm) material. Notice in Figure 8-27 that each finger has a square end and that the bottom of each V has a flat surface. This feature eliminates the likelihood of the fingers splitting the end of the matching piece as the two halves of the joint are fitted together.

Butt joints are not recommended, especially in important structural members, but are sometimes used in members or portions of members subjected to compression only. If butt joints are used in the compression portion of a member subjected to bending, the cross-sectional area of all laminations containing butt joints at a particular cross section should be disregarded in computing the effective moment of inertia for that cross section. Butt joints in adjacent laminations should not be spaced closer than 10 times the thickness of the lamination.

Every piece comprising a single lamination (and the completed end joint) must be of equal thickness to avoid gaps in the glue lines, which will affect strength. For practical reasons, laminating plants maintain a constant thickness of all laminating stock. The standard thicknesses used by most manufacturers are ¾ and 1½ in. (19 and 38 mm), with a

**Figure 8-26** Measuring Moisture Content of Laminating Stock with Electric Moisture Meter. (Reprinted by permission of Canadian Wood Council.)

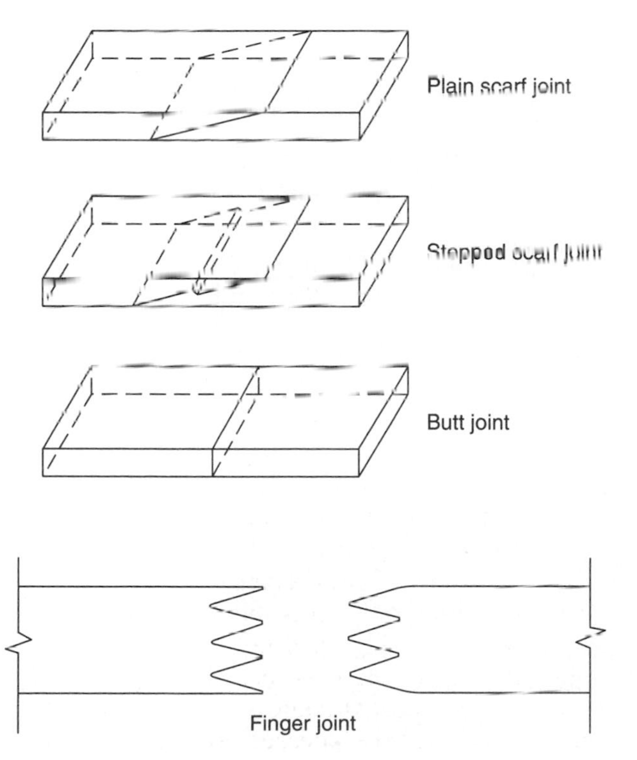


**Figure 8-27** End Joints for Laminating Stock.

permissible tolerance of ±0.0075 in. (±0.2 mm). Figure 8-28 shows laminating stock being checked for thickness as it emerges from a surfacer.

## Assembly of Material

The first step in the assembly of a glued laminated timber is the positioning and gluing of end joints. This must be done in such a way that the completed joint, after bonding, will be the same thickness as the lamination. End jointing may be a separate operation, but many manufacturers have devised systems of indexing scarf joints for gluing during the assembly operation. One is to bore a small hole at the time the scarf is made so that a dowel can be inserted during assembly to hold the two pieces in the right position. When used, the stepped scarf helps to position the two pieces at a joint. Other methods of marking meeting points automatically, including hand pinning, are also employed. Finger joints do not require indexing prior to assembly.

Laminations are coated with adhesive by passing them through a mechanical glue spreader that controls the amount of adhesive applied. Scarf joints are coated and pinned during this operation when the single operation assembly system is used. Finger joints are sprayed with adhesive, fitted together, and

**Figure 8-28** Checking Thickness of Laminating Stock with a "Go-No-Go" Gauge as It Comes from Surfacer. (Reprinted by permission of Canadian Wood Council.)

**Figure 8-30** Tightening Clamp Bolts with an Air-Operated Impact Wrench. (Reprinted by permission of Canadian Wood Council.)

set electronically in a single operation. The entire lamination is then laid on edge in a jig or press made of steel frames, according to the shape of the member being formed. The position of each lamination is previously determined according to grade, either by assembling in a trial run or by showing the position of every piece on a drawing. Successive laminations are laid on edge next to the first until the full depth of the member has been reached (see Figure 8-29).

Clamping begins as soon as the final laminations have been placed in the jig. Steel bolts with wood or metal clamp blocks are commonly used to press the laminations together. Clamping begins at the middle and progresses toward the ends, tightening the nuts with an impact wrench set to produce a pressure of from 10 to 200 psi (70 to 1380 kPa) on the glue lines (Figure 8-30). A torque wrench is used to check the tension on the clamp nuts (Figure 8-31).

**Figure 8-29** Assembling a Large Glued-Laminated Beam. (Reprinted by permission of Canadian Wood Council.)

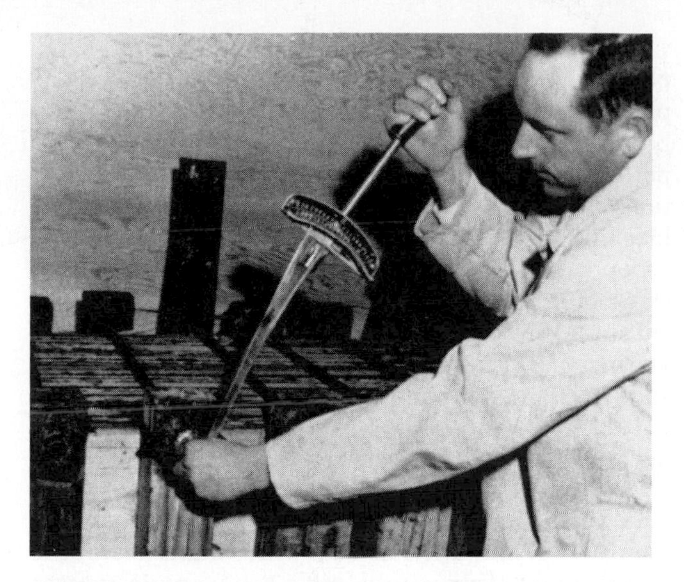

**Figure 8-31** Checking Tension on Clamp Bolts with a Calibrated Torque Wrench. (Reprinted by permission of Canadian Wood Council.)

Some adhesives require a minimum glue line temperature of approximately 75°F (20°C), and the temperature of the laminations may be checked prior to assembly. When the adhesive has set, the clamps are removed and the member is passed through a large double surfacer to remove irregularities (such as squeezed-out beads of glue) and to bring the member to its finished width. Irregular buildups are sawed to the required shape, ends are trimmed, and daps or other fabrication completed. Surface irregularities are corrected as necessary for the required appearance grade, and identification marks are placed on the ends of the member. Finally, it is given a coat of transparent moisture seal.

If necessary, finished members are wrapped before shipment to protect their appearance. Some manufacturers use a wrap of double-creped, impregnated paper, whereas others use an inner wrap of polyethylene covered with jute.

When dictated by service conditions, glued-laminated timbers may be pressure treated for protection against decay or insect damage. Straight, or almost straight, members to be accommodated in a pressure-treating cylinder may be treated after laminating. Creosote or Wolman salts are generally used for this purpose. If the shape of the member prevents it from being placed in the cylinder after fabrication, the laminations may be pressure treated beforehand. Wolman salts are used in this case, because this preservative does not affect subsequent glue bonding.

## Properties of Glued-Laminated Sections

The Laminated Timber Institute of Canada has adopted the SI metric system for the sizing of glued-laminated timber sections. Table 8-3 indicates the standard widths adopted. Nominal sizes are not used in the metric system, so all member sizes are designated by their actual dimensions. Lamination widths greater than 10¾ in. (275 mm) are composed of two boards with joints staggered in successive layers (Figure 8-32). The interior laminations normally are not edge glued unless loading is to be applied parallel to the width of the laminations. The American Institute of Timber Construction uses similar standards for dimensioning.

Douglas fir laminations, 1½ in. (38 mm) in thickness, may be bent to a minimum radius of 27 ft 6 in. (8.4 m) when the ends are tangent, and 35 ft 6 in. (10.8 m) when the member is curved throughout its length. Limiting radii for ¾-in. (19-mm) laminations are 9 ft 4 in. and 12 ft 6 in. (2.8 m and 3.8 m), respectively. When a smaller radius is necessary, the laminations must be reduced in thickness, resulting in higher costs due to additional material handling and manufacturing costs.

Having the flexibility to select laminations according to their strength, the manufacturer can produce beams with specific strength characteristics. Balanced beams, as they are sometimes called, have

**Table 8-3**
**NOMINAL AND FINISHED WIDTHS**

| Initial Width of Glulam Stock | | Finished Width of Glulam Member | |
|---|---|---|---|
| Nominal Width (in.) | Actual Width (mm) | Width (in.) | Width (mm) |
| 4 | 89 | 3 | 80 |
| 6 | 140 | 5 | 130 |
| 8 | 184 | 6¾ | 175 |
| 10 | 235 | 8¾ | 215 |
| 12 | 286 | 10¾ | 275 |
| 4 + 10, 6 + 8 | 89 + 235, 140 + 184 | 12¼ | 315 |
| 6 + 10, 8 + 8 | 140 + 285, 184 + 184 | 14¼ | 365 |

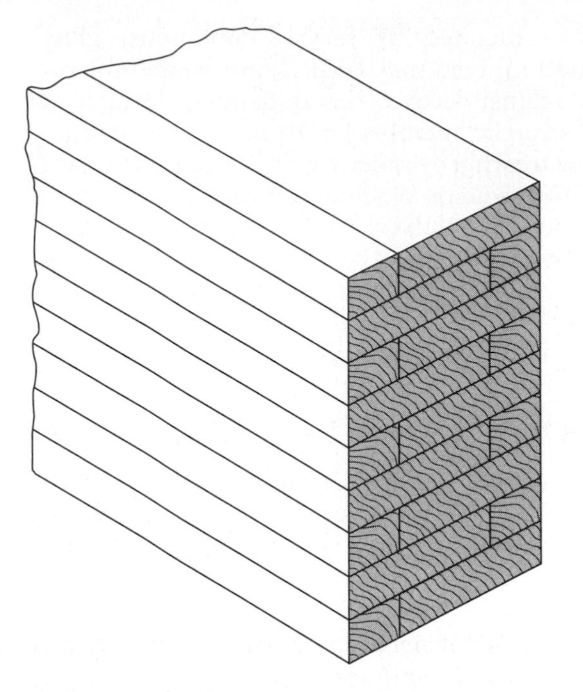

**Figure 8-32** Edge Joints Staggered.

a different value for flexural compression than for flexural tension and must have the top of the beam marked "TOP" so that they will not be installed upside down in the field. Beams that may have tension in both the top and bottom, such as beams that have an overhang, must have the same strength in the top as well as the bottom. Glulam sections are available in three appearance grades: industrial, architectural, and premium or quality grade. Industrial grade glulams are used where appearance and irregularities in the exposed surfaces are not a concern. Usual applications are in industrial buildings. Architectural grade glulams have fewer variations and are usually painted or finished with flat varnish. Premium or quality grade glulams have very few irregularities, have very few knots, and are sanded smooth. The finish applied is usually a high gloss transparent finish to accentuate the natural beauty of the wood.

Glulam sections are wrapped with a protective paper covering that should be left wrapped as long as possible. Ideally it is recommended that the wrapping be left in place until just before the glulam sections are permanently protected from the weather. If the sections must be stored on site for any interval of time, blocking must be provided to keep them off the ground and spacers are recommended between stacked sections. The protective covering should be slit along the bottom to prevent condensation buildup. During the installation of glued laminated sections, sudden changes in temperature should be avoided.

## Types of Units

Glued-laminated members may be used independently or in conjunction with other materials in the construction of structural units. For example, trusses may be built with glued-laminated chords and sawn timber webs. Units such as folded plate or other stressed skin panels may be combinations of glued-laminated timber and plywood (see Figure 8-33).

The greatest use of glued-laminated lumber is in the fabrication of beams and arches. Three basic types of beams are made: *straight, tapered,* and *curved* (see Figure 8-34). Straight and tapered beams (Figure 8-35) are usually made with a slight camber to allow for normal deflection and to improve appearance. Straight beams are common in the type of construction shown in Figure 8-36.

The *three-hinged arch* (most commonly used) may have legs that are straight, circular, or of varying curvature. Straight-legged arches are normally either *high V* or *low V* (see Figure 8-37), as the V arch allows maximum utilization of enclosed space. The building shown in Figure 8-38 is an example of high V arches.

Circular arches are actually segments of a circle. Notice the purlins, spanning from arch to arch, which will be used to carry the sheathing. Arches of varying curvature may be parabolic or gothic in shape or may be a combination of curve

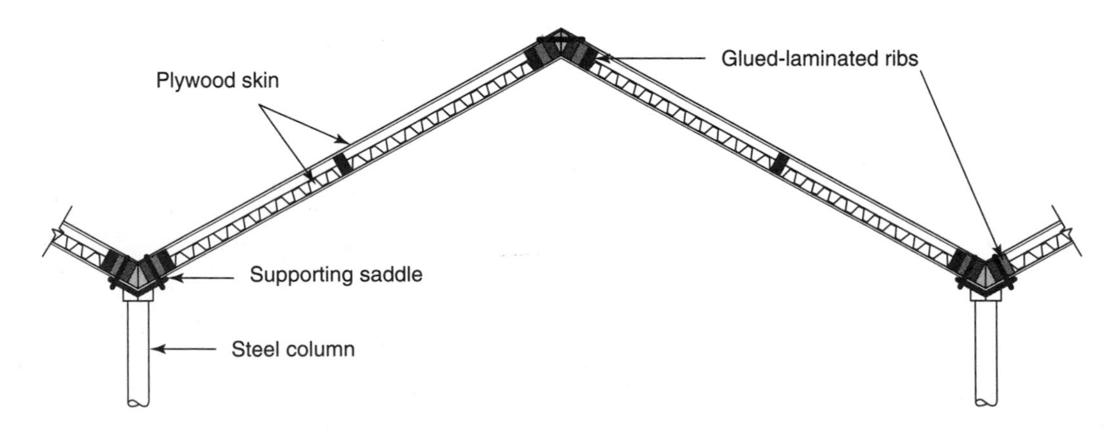


**Figure 8-33** Folded Plate Roof Detail.

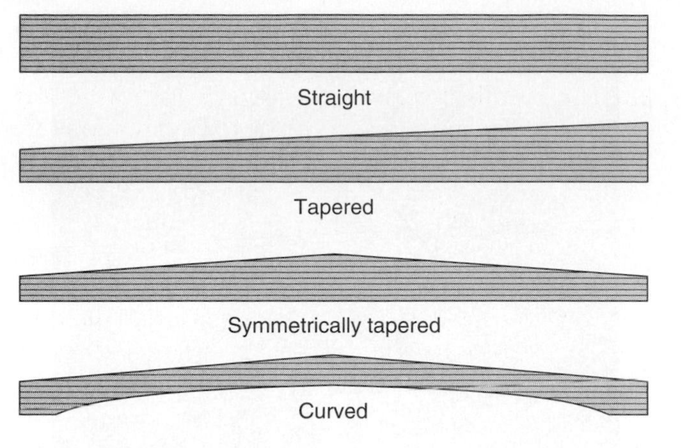

**Figure 8-34** Basic Beam Shapes.

**Figure 8-35** Symmetrically Tapered Beam.

**Figure 8-36** Straight Glued-Laminated Beams. (Reprinted by permission of Canadian Wood Council.)

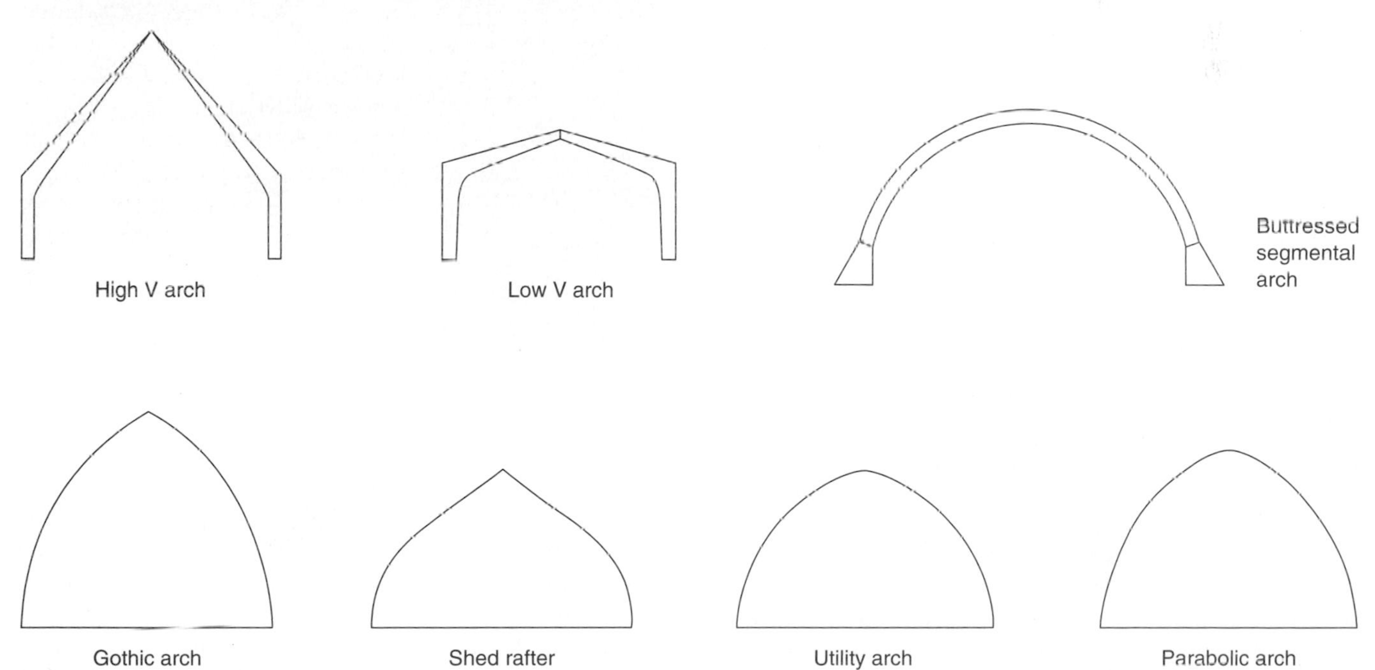

**Figure 8-37** Arch Shapes.

**Figure 8-38** High V Arches.

**Figure 8-39** Glued-Laminated Gothic Arches. (Reprinted by permission of Canadian Wood Council.)

and straight line. In Figure 8-39, gothic arches form the main supporting structure, but additional framing has been added to produce a straight line, peaked roof. Figures 8-40 and 8-41 illustrate the variation in design possible using glulam structural members.

Figure 8-42 illustrates an example of a shed rafter arch. The shed rafter is partly curved and partly straight, with the straight portion tangent to the curve.

## CONNECTIONS FOR GLULAM BEAMS AND ARCHES

### Glulam Beams

The only connection that will be different from those described in Figures 8-9, 8-13, 8-15, and 8-16 is the one required when a beam is cantilevered over a support. To produce the required framing in such a case, the beams are connected between supports. Three such connections are illustrated in Figure 8-43. Figure 8-43(a) is commonly used for this purpose when large beams are involved, designed so that the bolt is in tension, with the plates providing the required bearing perpendicular to the grain. To design the shape of cut so that the full depth of the beam is available for shear, draw lines at 60° from the ends of both top and bottom plates. The horizontal cut should be at the center of the beam and the sloping cuts within the boundaries of the 60° lines.

**Figure 8-40** Arches with Straight Top End. (Reprinted by permission of American Institute of Timber Construction.)

Figure 8-43(b) and (c) show *saddle* connections, with square-cut beam ends. Figure 8-43(b) may be used for moderate loads and Figure 8-43(c) for light loads.

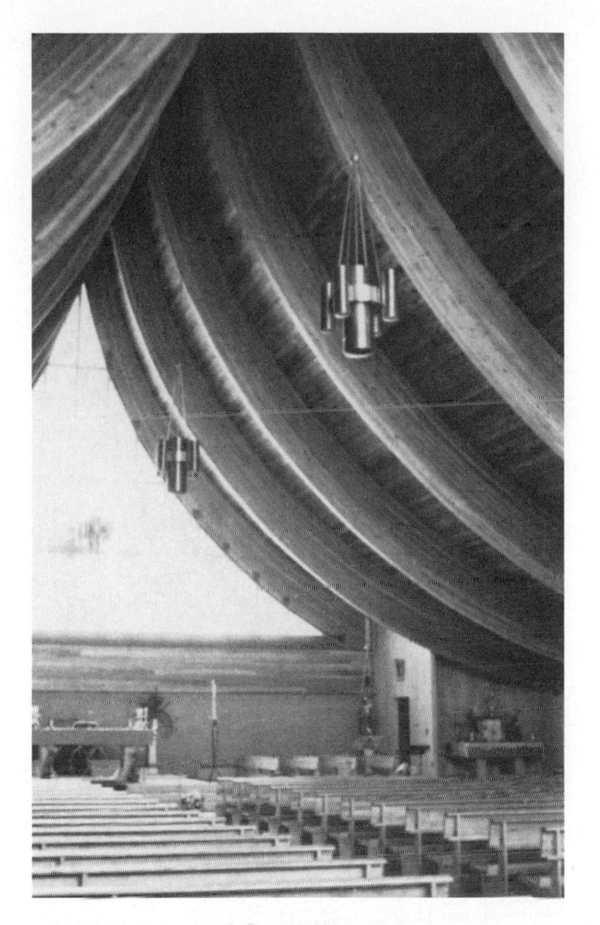

**Figure 8-41** Reversed Curve Glulam Arches. (Reprinted by permission of Canadian Wood Council.)

For beams framing into girders, hanger connections are normally used. Glulam rivets are used to secure the hanger to the girder (Figure 8-44) and the saddle is made wide enough to provide the necessary bearing area for the beam being supported. Lag screws or glulam rivets are used to secure the beam in the saddle.

## Glulam Arches

Arch connections are of two general types: *upper hinge* and *lower hinge.* Upper hinge connections may be made by gussets, bolts, or bolted plates (see Figure 8-45). Light arch legs are usually connected at the top by gussets, whereas the other two methods are used with heavier arches. The bolted connection is used for arches with a steep pitch, with double shear plates at the center. When the slope of the arch is such that a bolt length becomes excessive, the bolted plate connection is used. In both, a dowel and double shear plates are added for vertical shear.

The bottom or lower hinge of light arches is generally one of the connections shown in Figure 8-46. Heavier arches may be connected at the base by any of the methods illustrated in Figure 8-47. Figure 8-47(a) shows the most common connection used for this purpose, where the foundation has been designed to carry the horizontal thrust. If the foundation is not so designed, the type of connection shown in Figure 8-47(b) will be used, which includes a tie rod. Figure

**Figure 8-42** A Shed Rafter Arch Frame. (Reprinted by permission of Canadian Wood Council.)

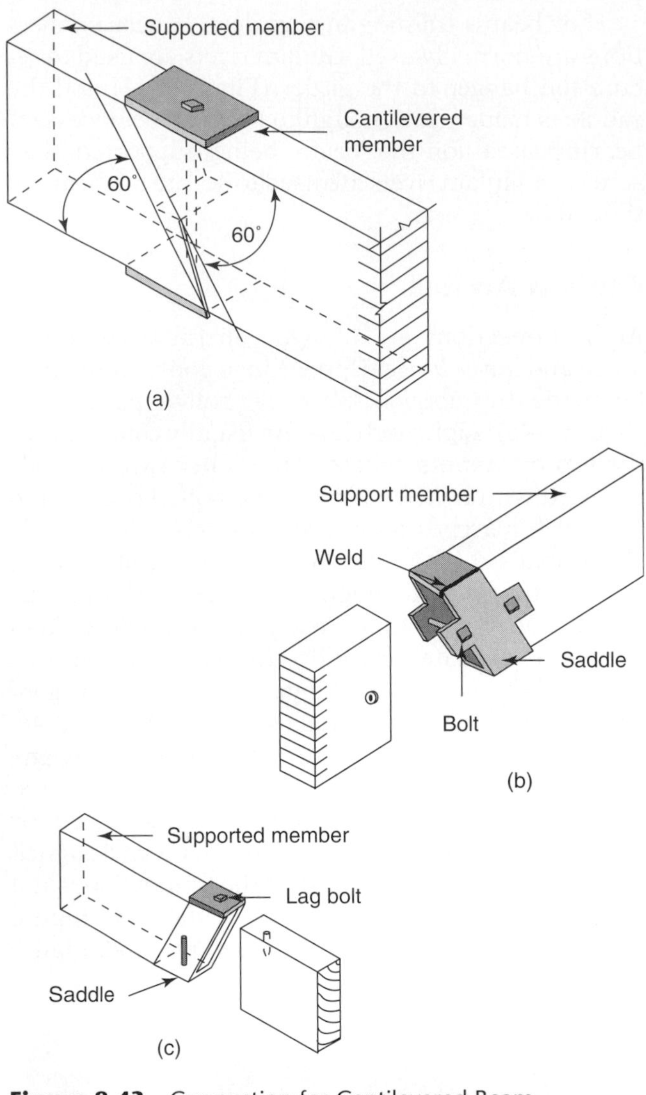

Figure 8-43 Connection for Cantilevered Beam.

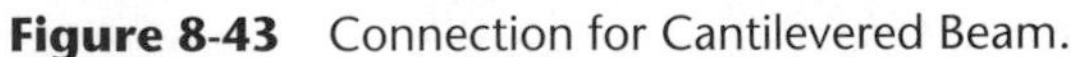

**Figure 8-44** Intermediate Glulam Beam Supported on Metal Hanger Fastened to Main Girder with Glulam Rivets. (Reprinted by permission of Canadian Wood Council.)

8-47(c) and (d) show connections designed for wood and steel bases, respectively, whereas the connection in Figure 8-47(e) is intended for smaller arches that may not require a true hinge. A true hinge connection,

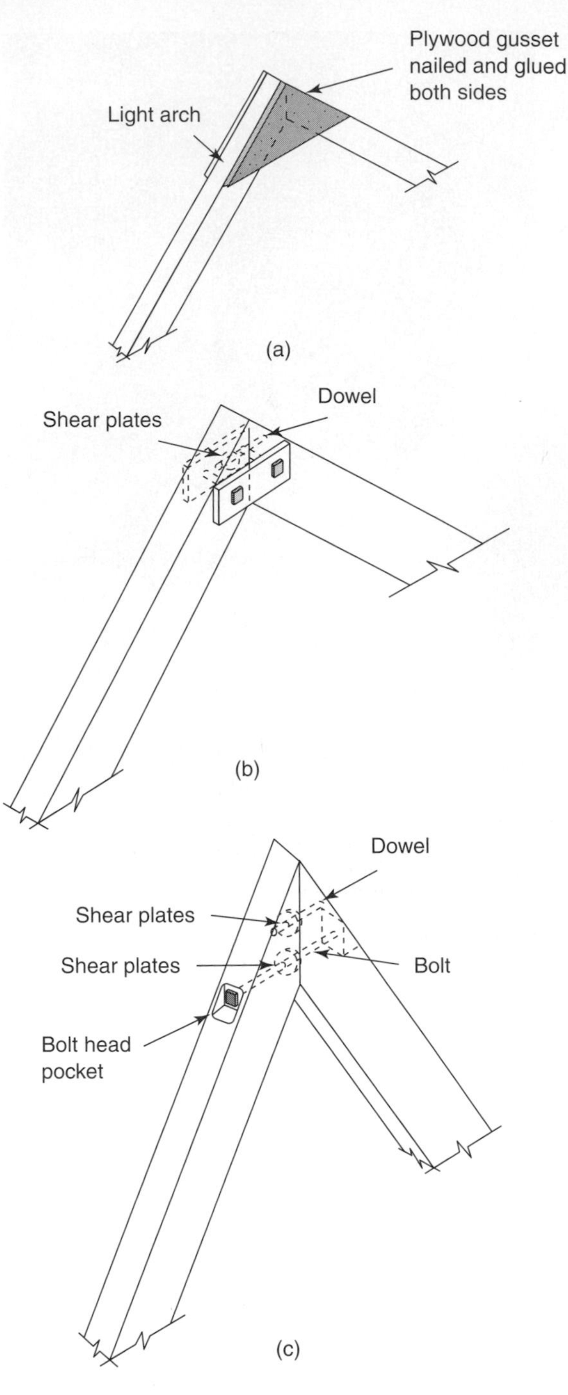


**Figure 8-45** Arch Upper Hinge Connections.

shown in Figure 8-48, is used for long-span, deep-section arches.

Moment resisting connections are sometimes required for splicing long beams or arches, and Figure 8-49 illustrates such a connection. In Figure 8-49(a), the bending moment is resisted by steel straps top and bottom, and in Figure 8-49(b) the moment connection is provided by straps and shear plates bolted to the sides of the members. In both connections a pressure plate is included, and shear transfer is provided by short dowels and shear plates in the end faces.

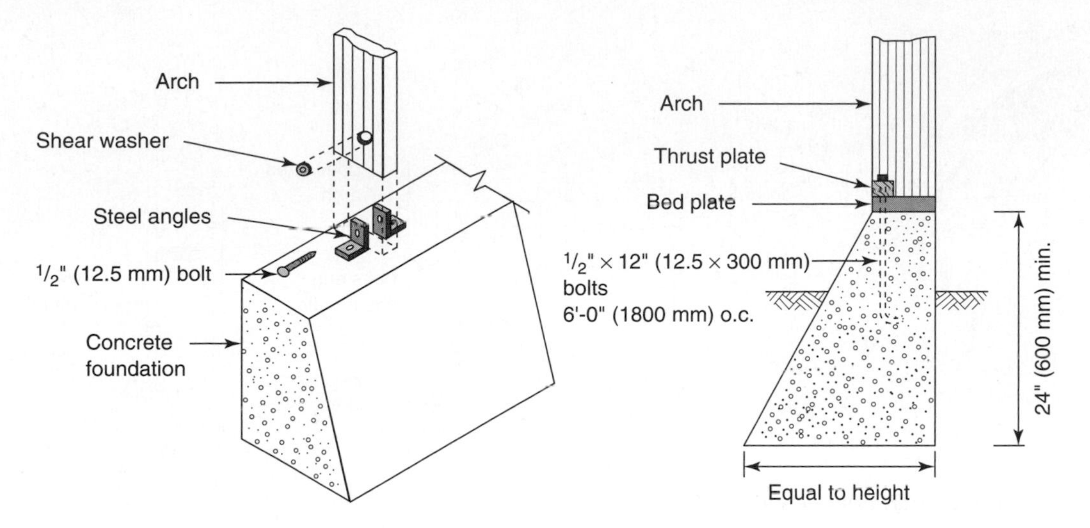


**Figure 8-46** Bottom Connections for Light Arches.

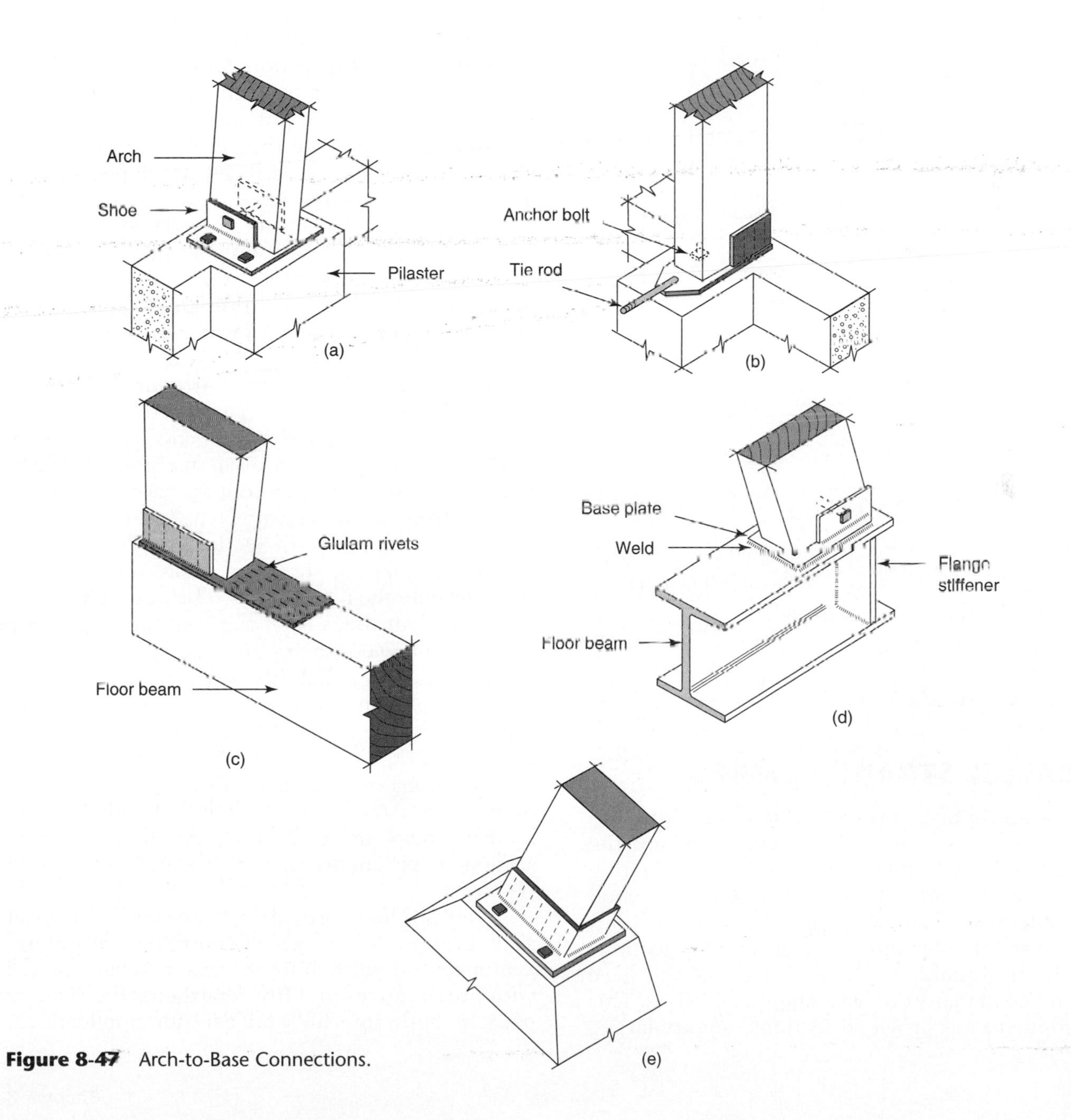


**Figure 8-47** Arch-to-Base Connections.

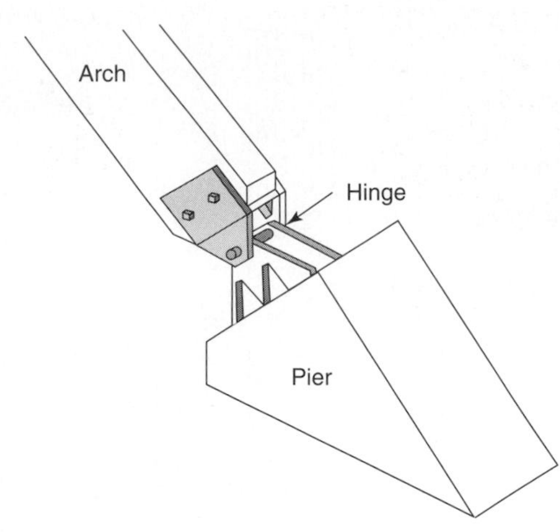


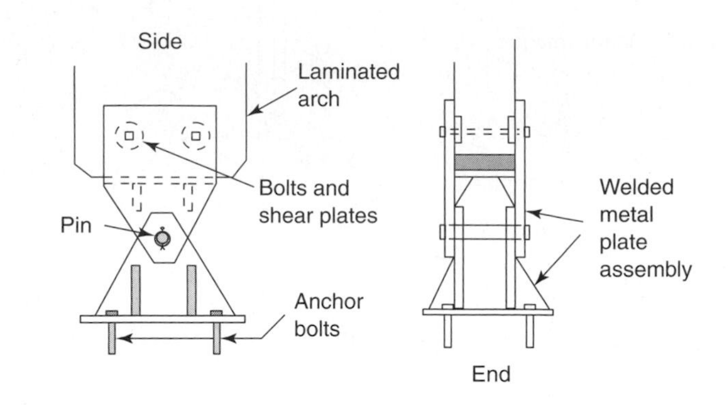


**Figure 8-48** True Hinge Arch Connection.

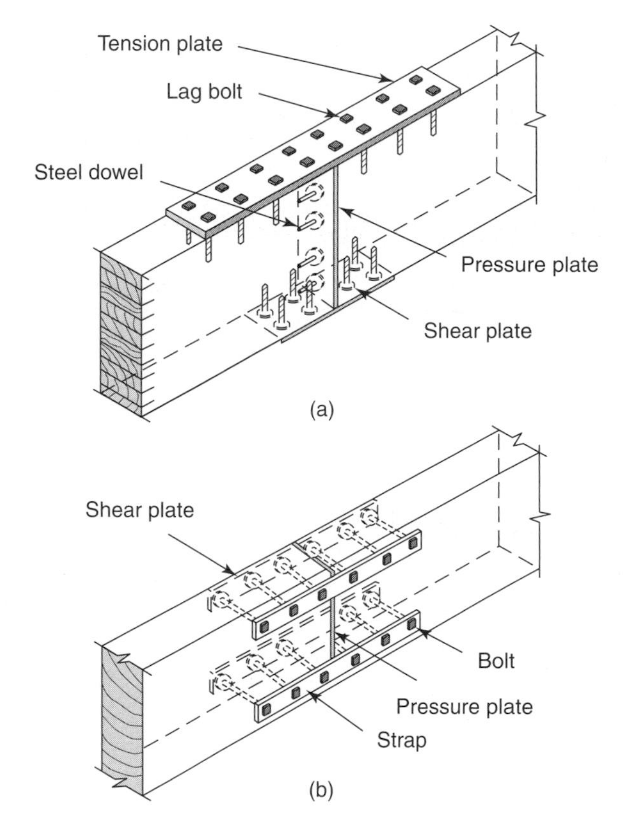


**Figure 8-49** Moment Connections.

## PARALLEL STRAND LUMBER

With large timbers now being at a premium, the wood industry is continually striving to expand the use of wood as a building material by investigating new methods and products that will use wood more effectively. Glued-laminated timber sections made from smaller dimensioned lumber sections have been available for several decades and have been made in many different shapes and sizes. This development has provided sections that are larger, stronger, and more uniform in their physical properties than that of a section cut from a single tree. However, even this approach requires sawn sections that must be straight and free of defects.

More recently, with the development of better glues, new methods have been developed in producing relatively large wood sections from material that at one time was considered as waste. Rather than using dimensioned lumber, as was the case in the glulam sections, thin veneers are now being glued together under pressure to produce usable structural sections known as *parallel strand lumber.*

Two different approaches are currently being taken in the production of parallel strand sections. One is to use thin veneer panels covered with glue, stacked so that the ends overlap in a staggered fashion, and pressed to form a continuous slab of wood. The resulting product is known as *laminated veneer* lumber (see Figure 8-50). Sections produced in this manner are denser, straighter, more uniform, and stronger than the original wood from which the veneers were cut. Because it is a continuous process, sections can be cut to any size and length.

The other approach is to use random lengths of veneer strips ½ in. (12.5 mm wide) that are aligned, coated with glue, pressed into a compact slab, and cooked with microwaves. Again, the resulting section is denser, stronger, and more uniform than the original tree. As previously noted, parallel strand lumber is made in a continuous length, allowing the section to be cut to size and length as required (Figure 8-51).

Both products provide sections with improved structural properties, yet maintain the natural appearance and workability of wood. When treated with preservatives and fire retardants, they can be used in both interior and exterior applications.

**Figure 8-50** Laminated Veneer Built-Up Girder Supporting Dimensioned Lumber Beam. (Reprinted by permission of Canadian Wood Council.)

**Figure 8-51** Parallel Strand Sections are Straight and Can Be Cut to Any Length. (Reprinted by permission of Trus Joist a Weyerhaeuser business.)

Sections are now being used in various structural applications such as beams (Figure 8-52), headers, purlins, columns (Figure 8-53), joists, and trusses.

## PLYWOOD STRUCTURES

Plywood is recognized as one of the most versatile and widely used products in the building construction

**Figure 8-52** Parallel Strand Beam Fitted with Steel Joist Hangers Being Placed into Position. (Reprinted by permission of Trus Joist a Weyerhaeuser business.)

**Figure 8-53** Parallel Strand Spaced Column Sections Supporting Heavy Beams. (Reprinted by permission of Trus Joist a Weyerhaeuser business.)

industry. In addition to the ordinary uses to which it is put, such as for sheathing, decking, and formwork, it is also used in the construction of such structural units as *box beams, web beams,* and *stressed-skin panels.* Just as glue plays an important part in the makeup of glued-laminated members, so it has an important role not only in the manufacture of plywood itself but also in the fabrication of plywood beams and panels. In all of these, plywood is combined with sawn lumber or glued-laminated framing members to produce a variety of structural units that are strong, relatively light, and usually simple to install.

Plywood sheathing is manufactured by placing three or more wood plies or veneers with their grains at right angles to each other to produce a built-up section of some predetermined thickness. The plies are bonded together with waterproof glue that is set with heat and pressure. The result is a piece of material that is uniform in thickness and very strong in both directions.

Plywood sheathing is available in various grades depending on the intended use. For construction purposes, exterior quality softwood plywood is the standard. Two types of softwood plywood are available; these are labeled as Douglas fir or softwood depending on the species of tree used for the veneers. Plywood labeled as softwood is manufactured using common softwoods other than Douglas fir, such as western hemlock, southern pine, ponderosa pine, larch, and white fir.

Standard sheet size for both types is 4 ft × 8 ft (1220 mm × 2440 mm). Each type of plywood is available in various thicknesses, both in standard units and metric units. Thicknesses range from 5/15 in. to 1¼ in. (7.5 mm to 31.5 mm), either sanded or unsanded, in various grades. Good two sides and good one side are sanded, while select-tight face, select, and sheathing grades are unsanded. For example, plywood used for formwork is exterior grade, either sanded or unsanded and is usually available in three thicknesses: ⅝, 11/16, and ¾ in. if unsanded; and two thicknesses, 11/16 and ¾ in. if sanded. In metric units, the three thicknesses for unsanded plywood are 15.5 mm, 18.5 mm, and 20.5 mm. For sanded, two thicknesses are provided: 17 mm and 19 mm. A special Douglas fir plywood (providing extra stiffness) either sanded, unsanded, or overlaid, is also available.

## Box Beams

In wood frame buildings in which long, deep beams are required, the weight of a solid glulam beam might be excessive and a properly designed box beam may be substituted. This beam is made by sheathing a framework of sawn or glulam members on both sides with plywood (see Figure 8-54), the thickness of which depends on the load to which the beam will be subjected.

The strength of such a beam largely depends on the glue bond between plywood and frame but is also improved by using plywood sheets that are as long as possible. Long sheets can be produced by *end scarfing* (joining end to end) regular sheets, as shown in Figure 8-55.

Glue may be applied with a brush or by gun (see Figure 8-56) at the temperature recommended by the manufacturer. Facilities for applying pressure to the glue joints until the glue has set are also important.

## Plywood Web Beams

A plywood web beam consists of one or more plywood webs to which sawn lumber or glue-laminated timber flanges are attached along the top and bottom edges. Notice that in the beam shown in Figure 8-56 the top and bottom flanges are laminated and the webs are sawn lumber. At intervals along the beam, lumber stiffeners are used to separate the flanges and control web buckling. Such beams may be fabricated with nails alone, with nails and glue, or by pressure gluing. The strength of the beam will vary somewhat, depending on which method of fabrication is used. Figure 8-57 illustrates typical plywood web beam designs.

There is no definite limit to the span that may be bridged with plywood web beams. Shallow beams spanning from 10 to 50 ft (3 to 15 m) and deep beams spanning to 100 ft (30 m) have been found to

**Figure 8-54** Large Box Beam under Construction. (Reprinted with permission of Council of Forest Industries of British Columbia.)

**Figure 8-56** Large Plywood Web Beam under Construction. (Reprinted with permission of Council of Forest Industries of British Columbia.)

**Figure 8-55** Joining Plywood Sheets by End Scarfing. (Reprinted with permission of Council of Forest Industries of British Columbia.)

be economical. Usually the limiting factor is the 48 in. (1,220 mm) width, which is the standard used in plywood manufacture. If sheets are edge spliced to increase width, greater spans can be achieved.

## Stressed-Skin Panels

A stressed-skin panel consists of longitudinal framing members of wood, covered on one or both sides by plywood skins (see Figure 8-58). The plywood skin, in addition to providing a surface covering, acts with the framing members to contribute to the strength of the unit as a whole. The plywood skins must be attached to the frame with glue to develop the full potential strength of the panel.

Spacing of the longitudinal members depends on the thickness of the plywood skin and the direction of the face grain in relation to the long dimension of the longitudinal members. Table 8-4 gives basic spacing with various plywood grades and thicknesses.

Stressed-skin panels are used for roof and floor decking (see Figure 12-32) in the standard rectangular shapes but may be fabricated in almost any desired shape to fit a particular design. For example, in Figure 8-59, rectangular panels are used to form a folded plate roof, and in Figure 8-60, triangular panels make up segments of a roof for an octagonal-shaped building.

A special kind of stressed-skin panel is involved when plywood is applied to a curved frame to make skin-stressed panel segments for a vaulted roof. Figure 8-61 shows a curved, skin-stressed panel, and Figure 8-62 shows a completed plywood vaulted roof.

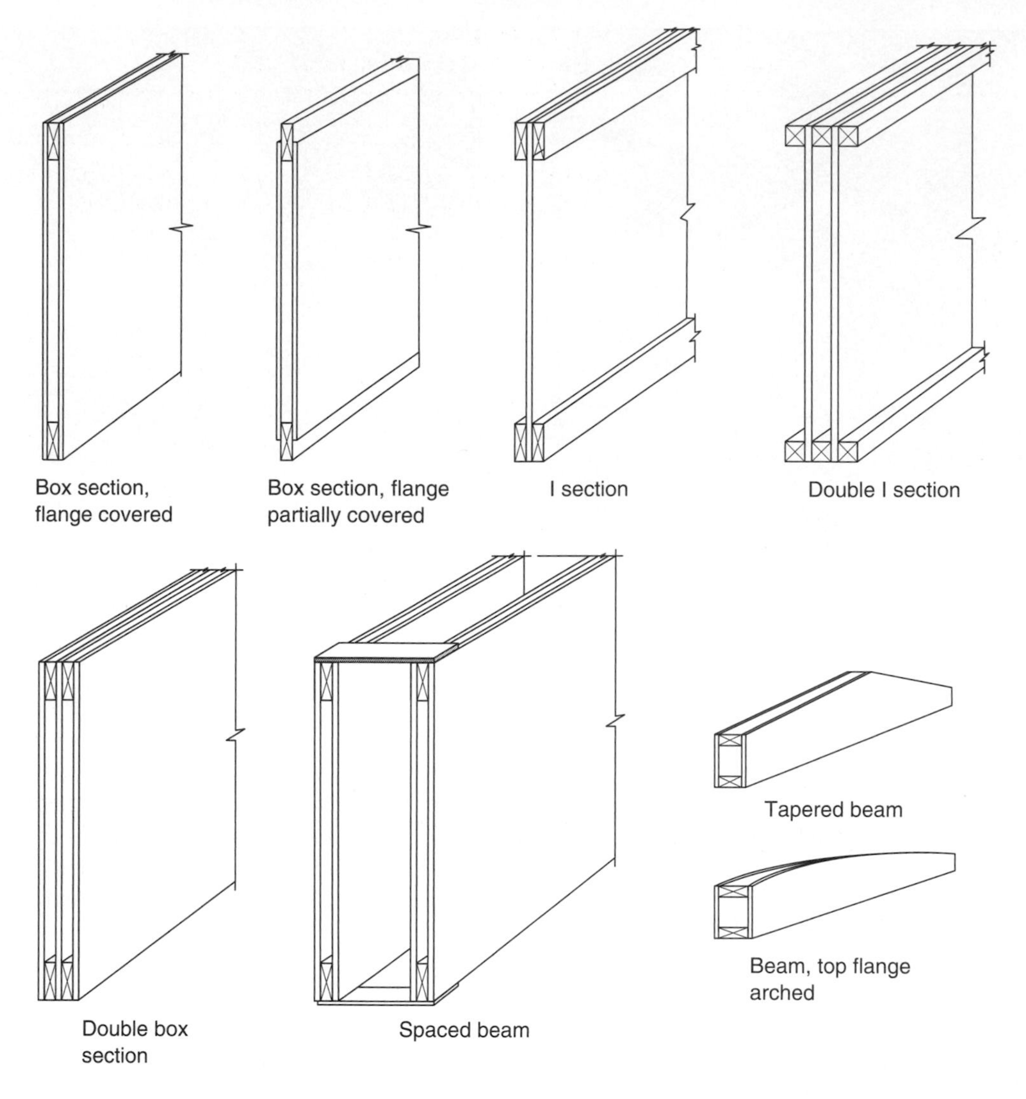

**Figure 8-57** Typical Plywood Web Beam Designs.

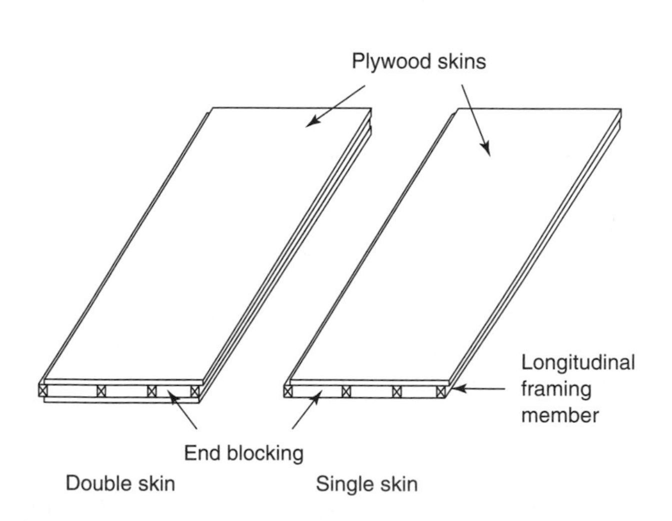

**Figure 8-58** Stressed-Skin Panels.

**Figure 8-59** Folded Plate Roof of Stressed-Skin Panels.

**Table 8-4**
**BASIC SPACING OF PANEL FRAMING MEMBERS**

| Plywood Thickness | | Basic Spacing | | | |
|---|---|---|---|---|---|
| | | Face Grain Parallel to Longitudinal Members | | Face Grain Perpendicular to Longitudinal Members | |
| (in.) | (mm) | (in.) | (mm) | (in.) | (mm) |
| ¼ (sanded) | 6 (sanded) | 10.1 | 255 | 12.1 | 305 |
| 5/16 (unsanded) | 7.5 (unsanded) | 11 | 280 | 15.6 | 395 |
| ⅜ (sanded) | 9.5 (sanded) | 17.5 | 445 | 15.5 | 395 |
| ⅜ (unsanded) | 9.5 (unsanded) | 15.3 | 390 | 16.6 | 420 |
| ½ (sanded) | 12.5 (sanded) | 25.6 | 650 | 25.3 | 645 |
| ½ (unsanded) | 12.5 (unsanded) | 23 | 585 | 28.2 | 715 |
| ⅝ (sanded) | 15.5 (sanded) | 33.1 | 840 | 30.7 | 780 |
| ⅝ (unsanded) | 15.5 (unsanded) | 26.8 | 680 | 38.8 | 985 |
| ¾ (sanded) | 20.5 (sanded) | 40.3 | 1025 | 36.3 | 920 |
| ¾ (unsanded) | 20.5 (unsanded) | 36.3 | 920 | 40.2 | 1020 |

**Figure 8-60** Stressed-Skin Roof Panels. (Reprinted with permission of Council of Forest Industries of British Columbia.)

## TERMITE CONTROL FOR WOOD BUILDINGS

In some areas and climates, wood that is in contact with or close to the ground is subject to attack by insects called *termites*. They bore into wood and, working from inside, may almost completely destroy it, leaving only an outer shell. Where wooden buildings or portions of them are subject to termite attack, protection against such infestation must be provided. Different methods of protection, both mechanical and chemical, have been tried with various degrees of success. Mechanical systems such as termite shields (Figure 8-63) have proven to be ineffective; however, if used in conjunction with chemical treatment of the soil around the foundation their use may be of some value.

One common method is to use pressure treated material for all framing and structural members that are near or in contact with the ground. Wood treated with waterborne preservatives such as chromated copper arsenate and ammoniacal copper arsenate is resistant to decay as well as insects. Chemical treatment of the soil around the building site is also an alternative; however, this option may present environmental problems in many areas.

A new method that appears to be quite promising is the Trap-Treat-Release method. Once an insect problem is identified, a number of the termites are trapped, treated with a surface toxicant, and then released to spread the toxicant throughout the colony. Combined with proper construction practices such as minimizing humidity levels in basement and crawlspace areas, avoiding soil-lumber contact, and burning or disposing of all scrap lumber during construction, will usually alleviate the problem. The most rational approach to an insect problem is to evaluate each situation on its own merits and apply the appropriate method or combination of methods.

**Figure 8-61** Curved Panel for Vaulted Roof. (Reprinted with permission of Council of Forest Industries of British Columbia.)

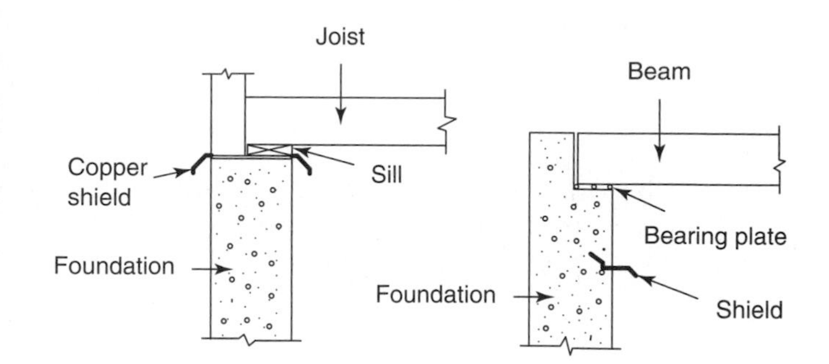


**Figure 8-63** Termite Shields.

**Figure 8-62** Plywood Vaulted Roof. (Reprinted with permission of Council of Forest Industries of British Columbia.)

## Review Questions

**1.** What is the prime function of a structural frame?

**2.** What four loads would normally be considered in the design of a timber frame?

**3.** What is meant by the term *pin-type connection?*

**4.** What three methods can be used to brace frames with pin-type connections?

**5.** What three methods can be used for connecting structural timber sections?

**6.** Outline five factors that have contributed to the more efficient use of timber in modern building construction.

**7.** List six advantages of using glued-laminated members for timber construction.

**8.** Answer briefly: **(a)** Why are there severe limitations to the use of cross-grained material in laminated members? **(b)** Why is the moisture content of laminating stock so closely controlled? **(c)** Why is there relatively small tolerance in the thickness of laminating stock?

**9.** Explain briefly the purpose of **(a)** a bearing plate, **(b)** a shear plate, and **(c)** a lateral tie.

**10.** Explain the difference in action between a *shear plate* and a *split ring.*

**11.** Explain what is meant by **(a)** a three-hinged arch, **(b)** an arch rib, and **(c)** a true hinge connection.

**12.** Answer the following: **(a)** What are the two types of wood decks and what are the main differences between them? **(b)** List the three types of arrangements used in plank decking placement. **(c)** Which of the three placement patterns is the most efficient and why?

**13.** Show by a scale diagram how you would suspend an object such as an overhead heater from a wooden roof beam.

**14.** What three advantages does parallel strand lumber have over regular sawn lumber when considered for use as structural framing?

# 9 REINFORCED CONCRETE FRAME

We have seen that concrete is a basic material in the construction of footings, foundations, and bearing walls. It is also widely used for making the structural frames of large buildings—the columns, the girders, and the beams. Because these members are subjected to tensile stresses and concrete is weak in resisting tension, steel is added to the member; hence the term *reinforced concrete.*

A reinforced concrete building frame may be erected by either of two methods. The members may be *cast in place* or the frame may be assembled from *precast* members. In the first case, forms are built and erected to form the shape of the frame and concrete is placed on the site. Precast members, on the other hand, are formed, cast, and cured in a plant and are subsequently brought to the building site ready for assembly.

## PRINCIPLES OF REINFORCED CONCRETE

It is a well-known fact that plain concrete sections are very susceptible to cracking and failure if loaded in bending or in axial tension (Figure 9-1). Plain concrete is a brittle material, and when loaded in tension, fails suddenly and with very little warning. To overcome this lack of tensile strength, steel reinforcing is added to the concrete to provide the necessary tensile strength. Steel has good tensile strength, is ductile, and can be rolled into rods, which makes it an ideal material for use as reinforcement.

The number of bars and their location in the concrete section are based on the frame analysis done by the structural designer. The proper location of the bars and the bond that develops between the steel bars and the concrete make the concept of reinforced concrete a viable structural material. By limiting the amount of reinforcing steel that is placed into the concrete section and ensuring that the bars are properly located and distributed, the reinforcing will provide the necessary tensile properties to complement the compressive strength of the plain concrete.

In the analysis of reinforced concrete sections, the tensile strength of the bars and the 28-day compressive strength of the concrete mix determine the section size and its load resistance. The proportion of steel area to concrete area is usually such that the steel will yield in tension before the concrete fails in compression. This approach ensures a ductile response to failure rather than a brittle failure, which would occur if the concrete was to fail in compression before the reinforcing steel yielded.

Codes provide the necessary guidelines for the proper placement and distribution of the steel, while designers establish the amount of steel required within the section by calculating the loads that must be supported. Because it takes relatively small amounts of steel to reinforce a concrete section, many variations in section size and shape are easily attained.

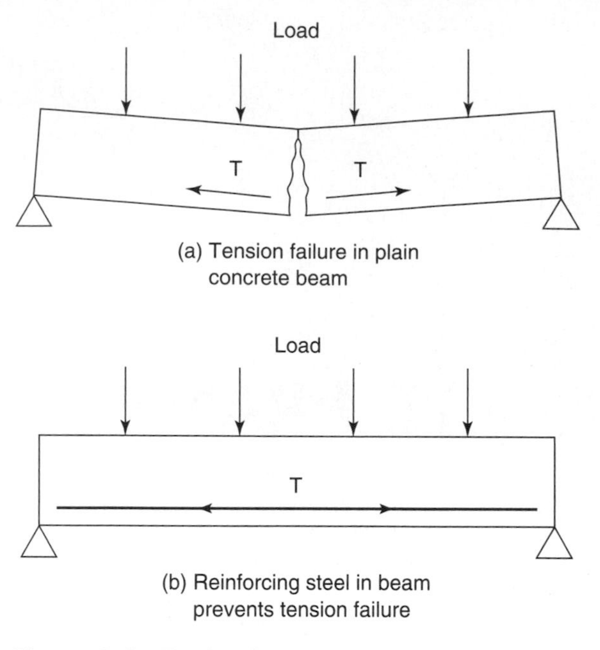


**Figure 9-1** Tension Stresses in Beams.

## THE REINFORCED CONCRETE FRAME

The components of a reinforced concrete frame are similar to those of the timber frame (see Figure 9-2). The frame consists of columns that support beams, which in turn support the slabs. In many instances, the beams can be eliminated and the slab supported directly by the columns.

Two alternatives are available to designers when considering a concrete frame: a cast-in-place concrete frame or a precast concrete frame. If the frame is made of precast sections, the connections are normally considered as pin type, and lateral stability of the frame must be transferred through the connections into portions of the frame that will act as bracing. Shear walls and diaphragm action of roof and floor decks are normally used in a frame of this type to transfer the horizontal loads to the foundation and ensure lateral stability. In some instances, structural steel sections can be used in conjunction with the precast sections to provide lateral stability (see Figure 9-3).

In cast-in-place concrete frames, the connections can be designed to develop resistance to rotation and, although the structure requires additional lateral support, the interaction between the structural sections adds to the stability of the frame. This added rigidity developed between the interacting sections results from the positioning of the reinforcing bars in areas of tension stress and pouring the sections monolithically; that is, without any joints or breaks between the interacting sections. This process allows for a more efficient design, resulting in smaller structural sections and thereby decreasing the weight of the overall structure. To further increase the efficiency of the frame, posttensioning can be incorporated into the design to provide additional stability and strength.

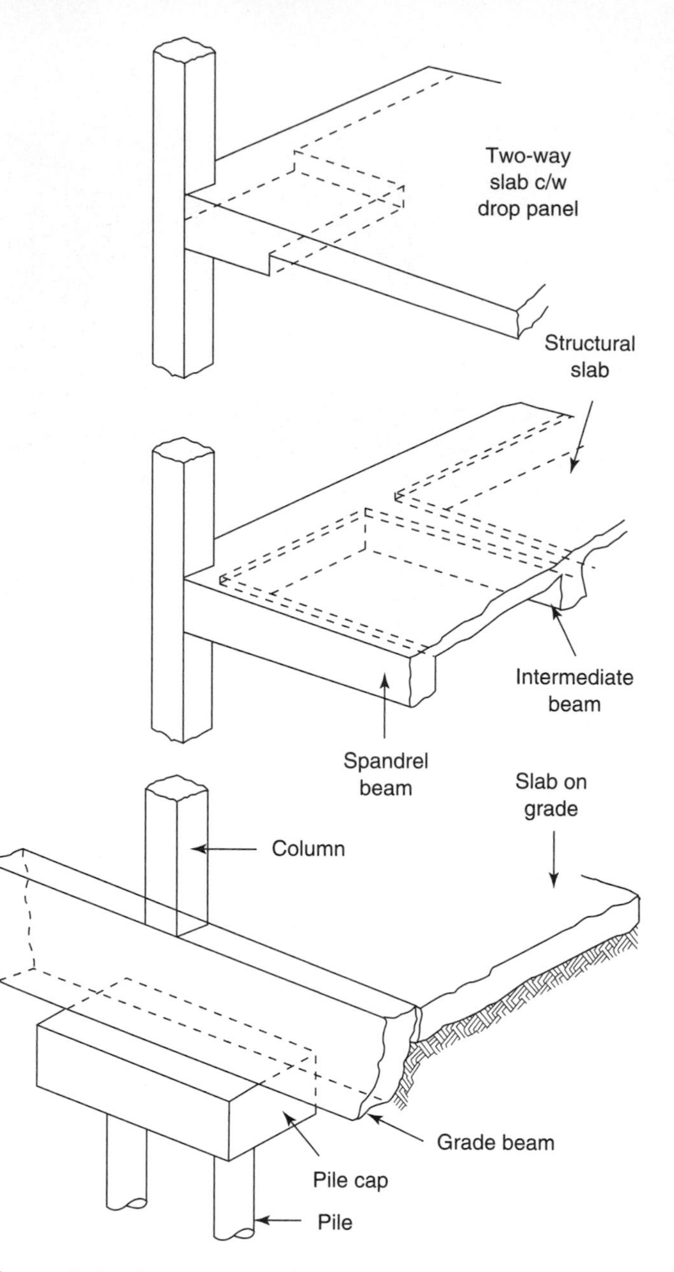


**Figure 9-2** Components of a Reinforced Concrete Frame.

**Figure 9-3** Structural Steel Cross-Bracing Provides Lateral Stability for Precast Concrete Structure.

**Table 9-1**
**REINFORCING BAR PROPERTIES**

| Imperial | | | | Metric | | | |
|---|---|---|---|---|---|---|---|
| Size | Dia. (in.) | Area ($in^2$) | Weight (lb/ft) | Size | Dia. (mm) | Area ($mm^2$) | Mass (kg/m) |
| #3 | 0.375 | 0.11 | 0.376 | 10M | 11.3 | 100 | 0.785 |
| #4 | 0.500 | 0.20 | 0.668 | 15M | 16.0 | 200 | 1.570 |
| #5 | 0.625 | 0.31 | 1.043 | 20M | 19.5 | 300 | 2.355 |
| #6 | 0.750 | 0.44 | 1.502 | 25M | 25.2 | 500 | 3.925 |
| #7 | 0.875 | 0.60 | 2.044 | 30M | 29.9 | 700 | 5.495 |
| #8 | 1.000 | 0.79 | 2.670 | 35M | 35.7 | 1000 | 7.850 |
| #9 | 1.128 | 1.00 | 3.400 | 45M | 43.7 | 1500 | 11.775 |
| #10 | 1.270 | 1.27 | 4.303 | 55M | 56.4 | 2500 | 19.625 |
| #11 | 1.410 | 1.56 | 5.313 | | | | |
| #14 | 1.693 | 2.25 | 7.650 | | | | |
| #18 | 2.257 | 4.00 | 13.600 | | | | |

## REINFORCING STEEL

Reinforcing bars are available in various sizes and grades. The grade designates the yield strength of the material used in the manufacture of the bars. The three most common grades are 40, 50, and 60 (300, 350, and 400). For example, a grade 40 (300) bar has a yield strength of 40,000 psi (300 MPa). Higher grades are available but are subject to a premium.

The size of a bar is designated by a number that represents the number of eighths of an inch that make up the bar diameter (see Table 9-1). A #4 bar has a diameter of $4 \times \frac{1}{8} = \frac{1}{2}$ in. In Canada, metric bar sizes have been adopted using a similar approach for bar designation (Table 9-1). In this instance, the number of the bar approximates the bar diameter in millimetres.

Reinforcing bars may be smooth or deformed; the deformations are standard indentations rolled into the surface of the bar during the manufacturing process.

## Welded Wire Reinforcing

Another form of concrete reinforcing consists of wire or round rods, either smooth or deformed, welded into grids (see Figure 6-61(b)). It is commonly known as welded wire mesh or welded wire fabric because the spacing between wires is relatively small. The welded grids may come as sheets or rolls depending on the diameter of the wire or rod used. Diameters of smooth wire used range from 1⁄16 to ½ in. (2 to 13 mm), while deformed rods can range from ¼ to ½ in. (6 mm to 13 mm or larger). Spacings between wires range from 2 to 12 in. (50 mm to 300 mm).

A typical designation of welded wire reinforcing is written as 6 × 6 − 10⁄10, where the first two numbers indicate the spacing of the mesh in inches and the next two numbers indicate the gauge of the wires in each direction. In this instance, the wires are spaced at 6 in. o.c. in each direction and the size of the wire in each direction is ten gauge.

Using metric units, a typical designation appears as

$$150 \times 100 \text{ MW9.1} \times \text{MW13.3}$$

where 150 represents longitudinal wire spacing, mm

100 represents transverse wire spacing, mm

MW9.1 = longitudinal wire size, $mm^2$

MW13.3 = transverse wire size, $mm^2$

This type of reinforcement is used primarily in slabs because of ease of placement. Straightening of the welded wire mesh is necessary when it is delivered on site in a roll to ensure its correct location in the concrete slab.

## Spacing of Reinforcing Bars

Two important criteria must be considered in the location of reinforcing bars in a concrete section: (1) sufficient cover must be provided to protect the reinforcing steel from the environment, and (2) proper spacing must be provided between the bars to ensure that the concrete flows around the bars and develops the necessary bond with the bars.

The amount of cover provided depends on the type of environment in which the concrete section must perform. The American Concrete Institute and the Canadian Standards Association provide guidelines for minimum cover for reinforced concrete sections in a noncorrosive environment. If conditions are more extreme or if additional fire protection is required, the amount of cover can be increased.

The American Concrete Institute, in standard ACI 318-95, lists its clear cover requirements for standard bar sizes in noncorrosive environments as follows:

*Concrete* cast against earth—3 in.

*Concrete exposed* to earth or weather

#6 through #18 bars—2 in.

#5 and smaller—1½ in.

*Concrete not exposed* to earth or weather

*Slabs, walls, joists:* #14 and #18 bars—1½ in.

#11 bars and smaller—¾ in.

*Beams and columns:* primary reinforcement, ties, stirrups, spirals—1½ in.

*Shells and folded plates:* #6 bars and larger—¾ in.

#5 bars and smaller—½ in.

The following guidelines are provided in CSA Standard A23.1-94 for minimum clear cover using metric bar sizes.

- *Beams and columns:* If not exposed to weather or not in contact with the ground, minimum cover is 40 mm for principal reinforcing consisting of 35M bars and smaller; ties, stirrups, and spirals, 30 mm. For sections exposed to weather or earth, minimum cover is 50 mm for principal reinforcing consisting of 35M bars and smaller; ties, stirrups, and spirals, 40 mm.
- *Structural slabs, walls, and joists:* If not exposed to earth or weather, minimum cover for 20M bars and smaller, 20 mm; exposed to earth or weather, 30 mm.
- *Structural shells and folded plates:* Shells and folded plates with bars 15M and smaller, not exposed to earth and weather, minimum cover is 15 mm; exposed to earth and weather, 30 mm.
- *Footings:* For footings or sections poured against the ground, minimum clear cover is 75 mm.

For bars having a diameter larger than those listed, the clear cover must be at least 1.5 bar diameters but need not be more than 2½ in. (60 mm).

Minimum cover is also dependent on the maximum aggregate size used in the concrete mix; 1.5 times the maximum aggregate size for exposed conditions and 1.0 times the maximum aggregate size for nonexposed conditions. The spacing of the bars depends primarily on the maximum size of the aggregates in the concrete mix and on the size of the reinforcing bars in question. Figure 9-4 provides some typical spacings for beam and column sections.

When designers establish the size of a section, they must allow for the necessary cover and spacing

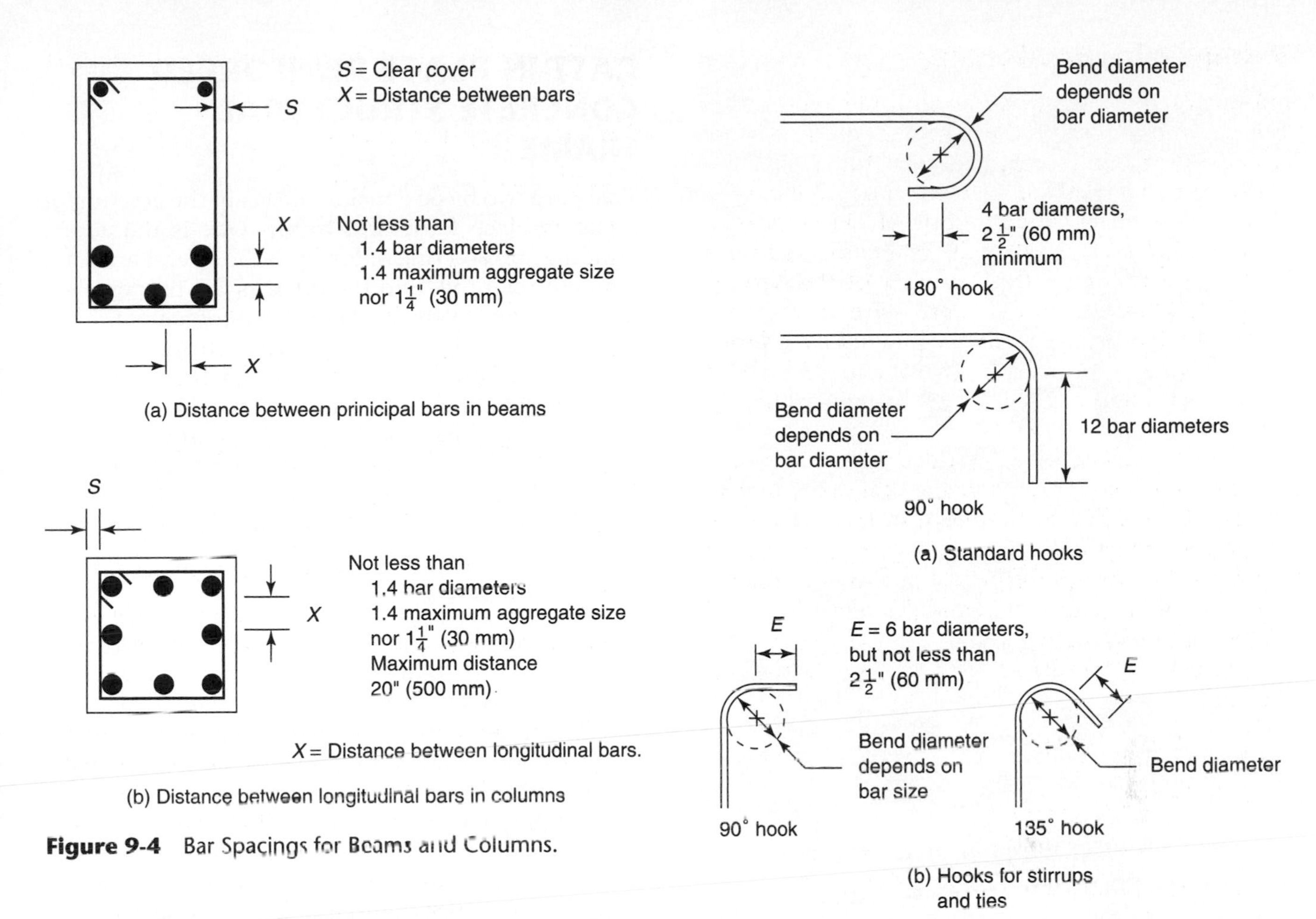


**Figure 9-4** Bar Spacings for Beams and Columns.

**Figure 9-5a** Standard Hooks for Reinforcement.

of bars. The field crew must also be aware of these code requirements to ensure that the actual placement of the bars in the formwork conforms to the original design calculations. Tolerances for placing vary with the type of section being formed and the depth of the overall section. These tolerances must be adhered to by the field crew and are enforced by the site inspector.

## Reinforcing Bar Details

Reinforcing bars are cut and bent to the required shapes based on the requirements shown in the structural drawings. Typical details have been developed by the various interested parties within the industry. The American Concrete Institute provides standards for typical details for bending of reinforcing bars. Figure 9-5(a) illustrates some of the standard hook and stirrup details that are recognized throughout the industry for use in reinforced concrete applications. The contractor responsible for the supply and fabrication of the reinforcing steel ensures that all bar details are based on these standards and provides placement drawings for the field crew to ensure that all reinforcing is placed according to the requirements of the structural drawings.

## Splicing of Reinforcing Steel

Because reinforcing bars can be fabricated and shipped only in predetermined, manageable lengths, there are times when splicing of the reinforcing steel is necessary to ensure continuity within the structural members in the building frame.

Three types of splices are available to the designer for the splicing of reinforcing steel: *welded* splices, *mechanical* splices, and *lap* splices. Each method has its limitations, and one method should not be substituted for another without approval from the structural engineer.

Normally, welded splices are not recommended when splicing normal reinforcing bars. Because of the large amount of heat required in the welding process, the properties of the bar will be affected in the area of the weld. Cracking of the joint can occur if the weld is not done with the proper preparation of the bar. If bars are to be welded, special weldable reinforcing steel is specified by the engineer.

Mechanical splices are used where butt-type splices are needed because of space limitations [see Figure 9-5(b)]. A mechanical splice usually consists

**Figure 9-5b** Mechanical Sleeve Splices for Large Column Reinforcing Steel.

of some type of metal sleeve that is held securely in place by a wedge driven over the sleeve ends. In some applications the sleeve can be filled with a grout or a molten metal. Mechanical splices can also consist of couplers that are threaded over the bar ends. Mechanical methods of splicing are normally used for large bars for butt-type splices where welding, if allowed, would be time consuming and expensive. Again, they must be used only where specified.

The lap splice is the most common type of splice primarily because of its simplicity, especially for the smaller bar sizes. Lap splices are considered as either tension lap splices or compression lap splices depending on the stresses that are encountered at the point of splicing. Their lengths are established by the structural designer according to code requirements and must be indicated on the placing drawing to ensure that sufficient lap lengths are provided by the placing crew. They are popular because they are cost efficient and do not need any special hardware to make them work.

## CAST-IN-PLACE REINFORCED CONCRETE STRUCTURAL FRAME

There are two basic considerations in the erection of a cast-in-place structural frame. One is the actual building and erecting of forms of the size, shape, and strength required, and the other is the proper placing of the steel rods needed for reinforcement.

The size and shape of the members to be formed are indicated on the building plans, whereas the size and amount of material necessary to make the forms strong enough are based on the loads and pressures involved. The size, shape, and amount of steel required are also indicated on the plans in the form of column, girder, and beam *schedules*.

### Column Forms

Column forms are often subjected to much greater lateral pressure than wall forms because of their comparatively small cross section and relatively high rates of placement. It is therefore necessary to provide tight joints and strong tie support. Some means of accurately locating column forms, anchoring them at their base, and keeping them in a vertical position are also prime considerations. Wherever possible, a *cleanout* opening should be provided at the bottom of the form so that debris may be removed before placing begins (see Figure 9-6). *Windows* are often built into one side of tall column forms to allow the placing of concrete in the bottom half of the form without having to drop it from the top.

Columns may be square, rectangular, round, or irregular and forms may be of wood, steel, or fiberboard. Wood (boards or plywood) is commonly used for square or rectangular forms and may also be used for irregular or round ones. Most round columns, however, are formed by the use of steel or fiberboard tubes of various diameters (see Figure 9-7). Steel forms are also available for square or rectangular columns.

There are many ways to build square or rectangular wood column forms, depending on column size and height and on the type of tying equipment to be used. Figure 9-8 illustrates a simple method of making a form for a light column, to approximately 12 in $\times$ 12 in. (300 mm by 300 mm). The sheathing is generally plywood, with wood battens and steel rods used as a tying system.

As column sizes increase, either the thickness of the sheathing must be increased or vertical stiffeners must be added to prevent sheathing deflection. In Figure 9-6, vertical stiffening is used. Also notice the adjustable metal clamps that tie the form together. Several methods of tying column forms are available,

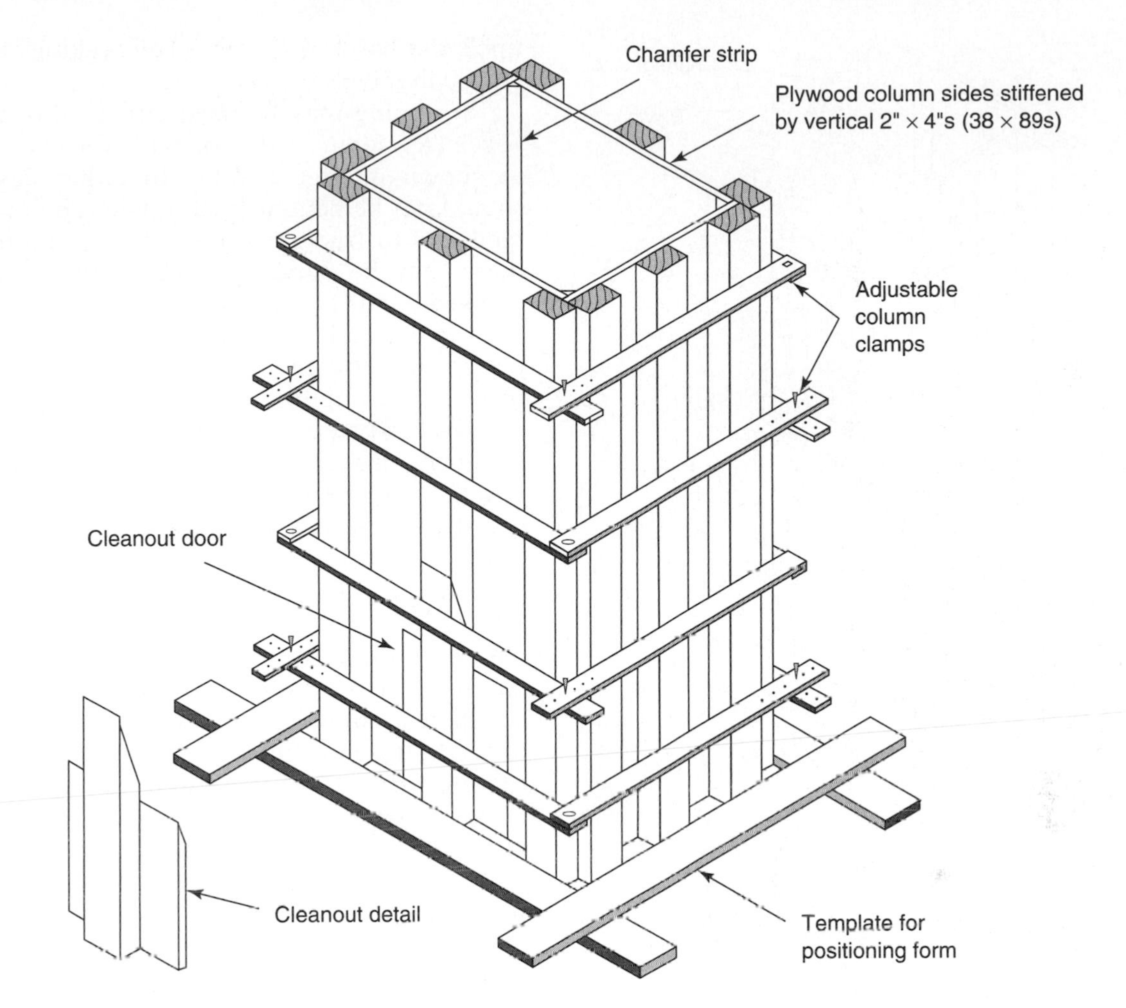


**Figure 9-6** Heavier Column Form.

**Figure 9-7** Fiberboard Column Forms: (a) In Place; (b) Removed.

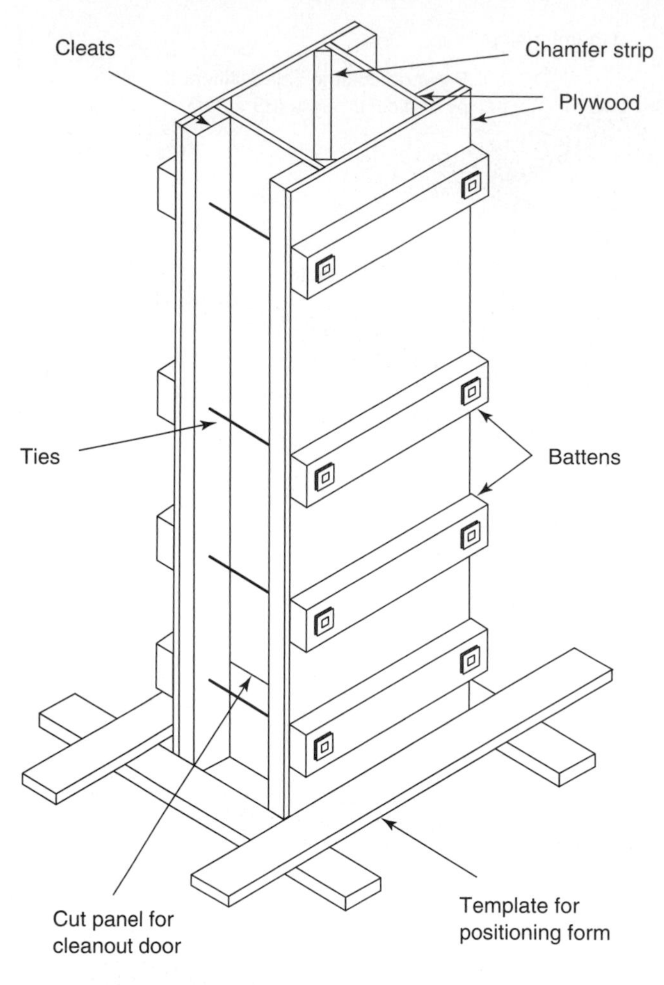


**Figure 9-8** Light Column Form.

and some are illustrated in Figure 9-9. Ties of this type are generally referred to as *yokes*.

Large columns require heavy yokes and a strong tying system. In Figure 9-10(a), the yokes are made from 2″ × 4″s (38 × 89s) on edge, and the tighteners are patented U clamps and wedges. In Figure 9-10(b), vertical stiffeners consisting of 2″ × 4″s have been used in conjunction with horizontal steel yokes to ensure that the formwork sheathing is prevented from deflecting.

Another method of tying column forms involves the use of steel strapping, normally available in three widths: ¾, 1¼, and 2 in. (20, 30, and 50 mm). The ¾-in. (20-mm) band is available in two thicknesses: 0.028 and 0.035 in. (0.7 and 0.9 mm), with approximate breaking strengths of 2,300 and 3,100 lb (10 and 14 kN), respectively. The 1¼-in. (30-mm) width is available in thicknesses of 0.035 and 0.050 in. (0.9 and 1.3 mm), with approximate breaking strengths of 5,000 and 7,000 lb (22 and 31 kN). The 2-in. (50-mm) wide band is 0.050 in. (1.3 mm) thick and has an approximate breaking strength of 11,000 lb (50 kN).

Strapping may be used either with horizontal yokes (see Figure 9-11) or with vertical stiffeners, as shown in Figure 9-12. In either case, bands should not be bent at a 90° angle. The metal has a tendency to fracture under those conditions, and there may be danger of form failure caused by cracked bands (see Figure 9-11). The band is tightened by a special tool, after which a metal clip is placed over the ends of the band and clamped in place by a crimping tool.

The first step in building a column form is to select the proper type and thickness of sheathing. Boards may be used, but plywood is more frequently employed because of its greater resistance to splitting and warping. It may be ½, ⅝, ¾, or 1 in. (12.5, 15.5, 19, or 25.5 mm) thick, depending on the size and height of the column. Table 6-4 may be used to determine the spacing of yokes, based on lateral pressures indicated in Table 6-3. It will be assumed that the first tie will be as close to the bottom of the form as is practical, within 6 to 8 in. (150 to 200 mm), and that the top tie will be at or near the top. Yokes or bands must be checked to ensure that they withstand bending and shear and that deflection will not exceed 1⁄16 in. (1.5 mm).

One pair of form sides can be cut the exact width of one column dimension. The other pair must be made wider, depending on the method of form assembly. In Figure 9-6, for example, the second pair of sides will be wider by twice the thickness of the sheathing. On the other hand, in Figure 9-8, extra allowance must be made for the width of the cleats.

Lay out, on the wide sides, the position of chamfer strips by drawing lines parallel to the edge of the board. The distance between these lines will be the second dimension of the column (see Figure 9-13). Fasten the chamfer strips in place with nails and glue.

If yokes or bands are to be used to tie the column, mark the location of each on the side of the form. If vertical stiffeners are used, tack them in place on the column sides and mark the position of the clamps or bands on a stiffener. When bands are used to tie the column, the yoke pieces can be tacked in place on each side before assembly. With other tying systems, all possible parts should be tacked in place before the form is assembled.

The form can now be assembled on a temporary basis. All four sides can sometimes be put together before erection, whereas in other situations it is preferable to put three sides together (see

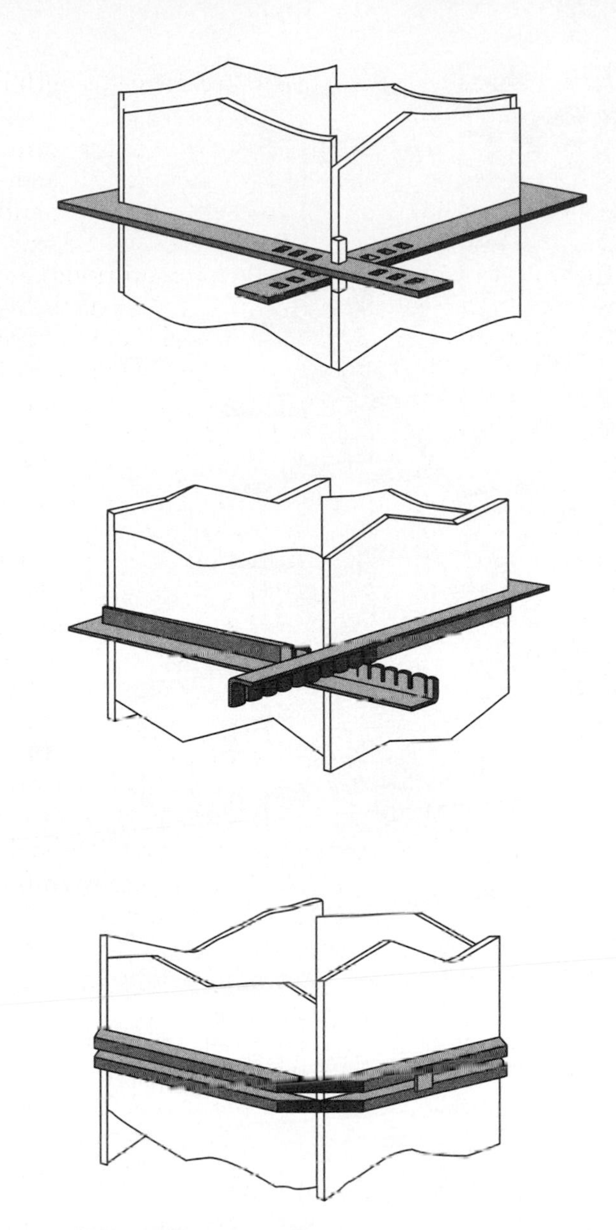
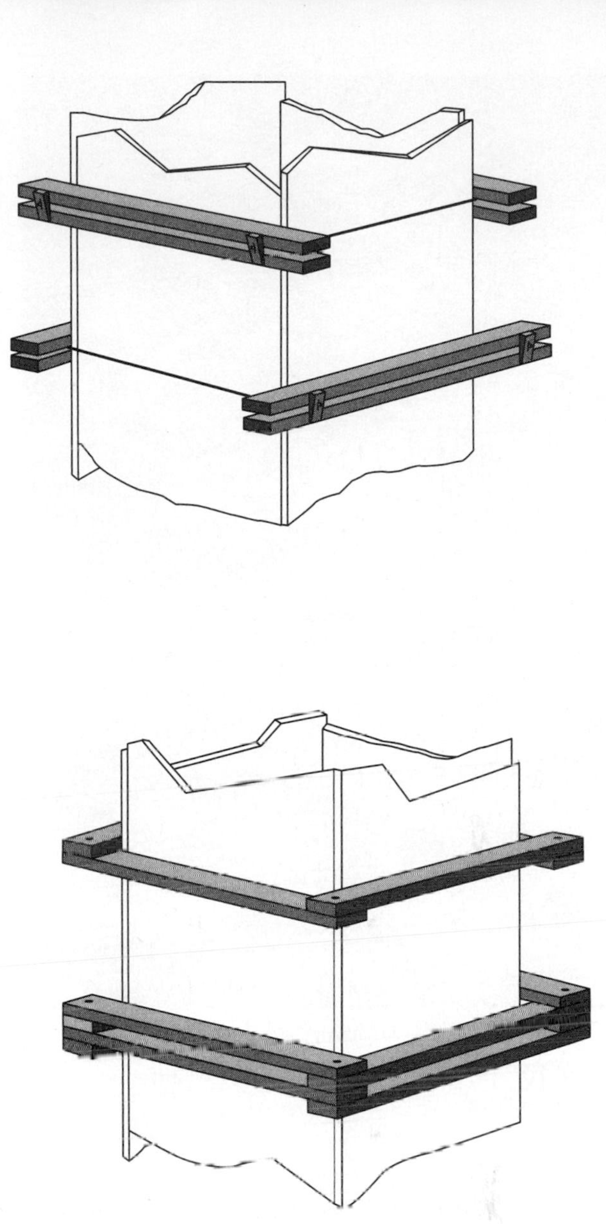

**Figure 9-9** Light Column Ties.

Figure 9-14), set the partially completed form in place, and add the fourth side later. This would probably be done in setting column forms for a job such as that shown in Figure 9-16, where the reinforcement is already in position.

Round wood columns are usually made of 2-in. (38-mm) *cribbing,* as shown in Figure 9-15. The edges of each piece of cribbing must be beveled so that they will form a circle when fitted together. The amount of bevel on each will depend on the diameter of the column and the number of pieces used to make the form. Wood form construction is used only in special instances because of the time and cost involved in the construction of the form. Round column forms are usually constructed of metal or reinforced plastic. Where the reuse of formwork is not a requirement, round fiberboard forms, available in standard diameters, is an inexpensive alternative (see Fig. 9-7).

To locate column forms accurately, templates such as the one illustrated in Figure 9-8 must be made. These are carefully located by chalk line or other convenient means and anchored in position [see Figures 9-16(a) and (b)].

The length of the column form is determined by subtracting the thickness of the bottom of the girder form that the column is to carry from the column height indicated on the plans or in the column

**Figure 9-10** (a) Horizontal Column Form Stiffeners. (b) Horizontal Steel Yokes and Vertical Form Stiffeners on a Column Form.

schedule [see Tables 9-2(a) and (b)]. Once the column forms are in place, they must be plumbed, braced, and made ready to support the ends of the girder and beam forms that will be built to them.

## Column Reinforcement

The reinforcing steel in columns consists of two parts: the longitudinal bars that run the length of the columns, and the lateral reinforcing that encloses the longitudinal steel.

When load is applied to a reinforced concrete column, the longitudinal bars absorb a portion of the load. Consequently, each bar acts as a small, slender column susceptible to buckling within the concrete shell. The lateral reinforcing, which is in the form of horizontal ties or a spiral, provides lateral support for and maintains proper spacing and alignment of the longitudinal bars.

The type, number of bars, spacing of bars, and size of reinforcing to be used in each column will be indicated in a column schedule on the structural drawings along with column sizes and heights. A typical column schedule is outlined in Table 9-2.

The column reinforcing is usually assembled as a unit on the job site (see Figure 9-17), based on the material listed in the column schedule. Tie wire is used to hold the vertical and lateral bars in position, forming a cage. The unit is placed in position and fastened to the dowels that project from the base surface [see Figure 9-16(b)], forming a lap splice. Notice that the longitudinal bars are cut long enough for their upper ends to form a lap splice with the column steel on the floor above.

Columns that must resist substantial lateral loads such as those encountered during an earthquake require lateral reinforcing to provide additional ductility to the overall frame. This can be achieved by using spiral reinforcing (see Figure 9-18), which is a continuous rod formed into a spiral of a given diameter.

**Figure 9-11** Banding a Plywood Column Form. (Reprinted with permission of Council of Forest Industries of British Columbia.)

**Figure 9-12** Steel-Banded Column Form. (Courtesy Acme Steel Co.)

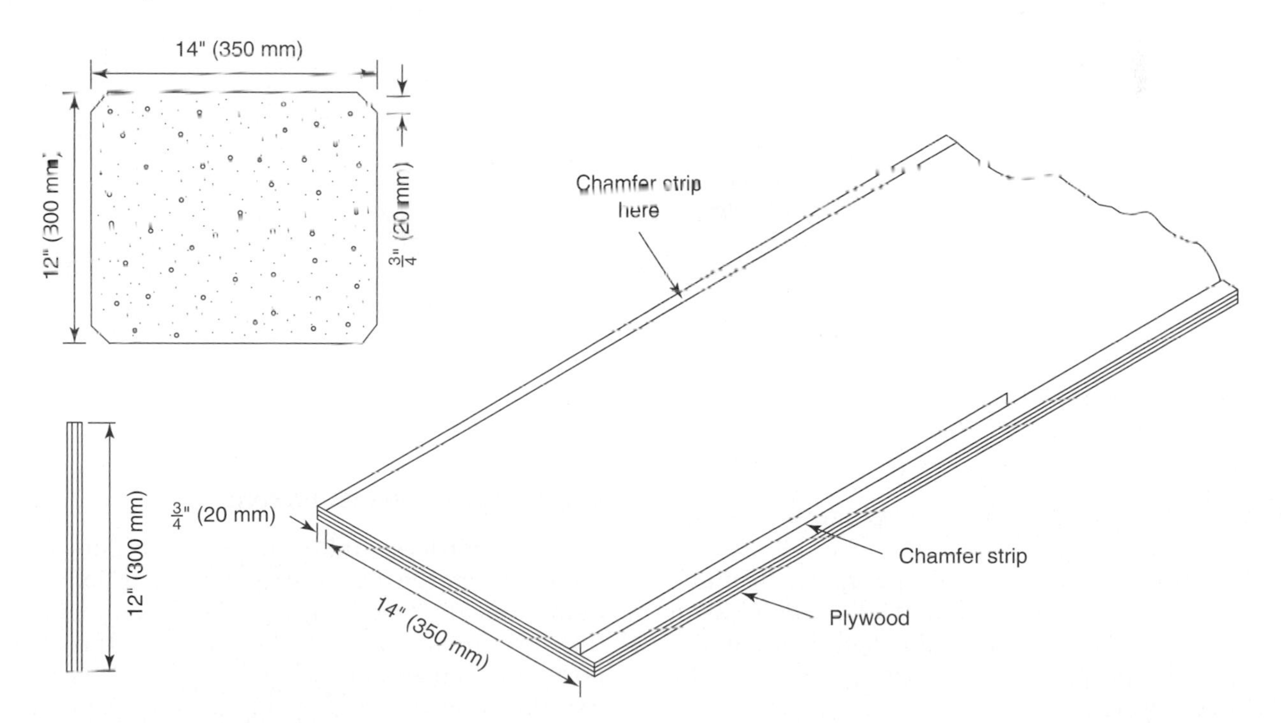


**Figure 9-13** Column Form Sides.

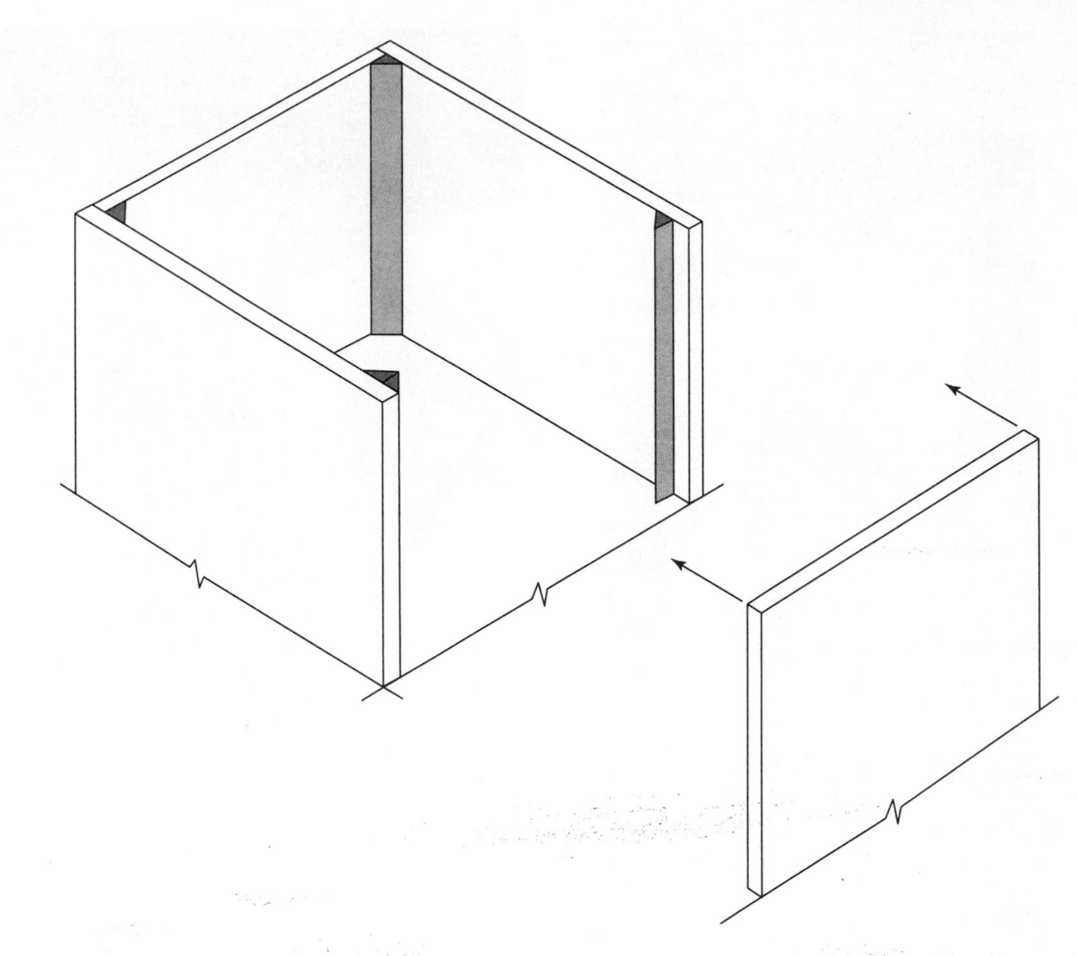

**Figure 9-14** Three Sides Assembled.

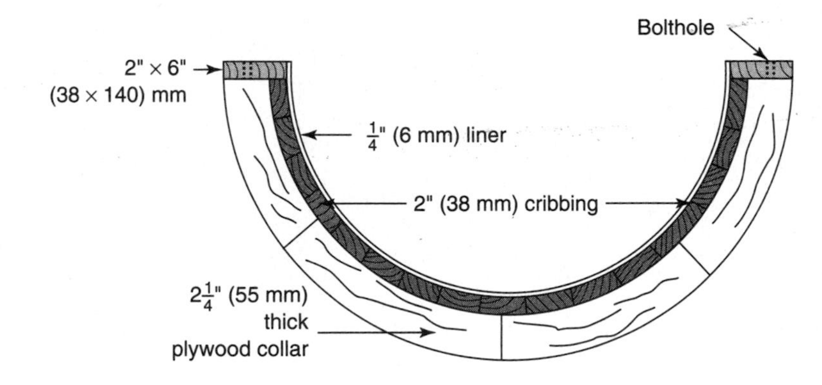


**Figure 9-15** Wood Form for Round Column.

Spiral reinforcement also provides additional support to the longitudinal bars because the spacing (*pitch*) between the loops of the spiral is much smaller than the spacing that is allowed between ordinary ties. The loops of the spiral are maintained at the proper pitch by spacer bars, usually three per spiral.

## Girder and Beam Forms

Columns, girders, and beams on each floor of a reinforced concrete frame structure are usually placed monolithically (see Figure 9-19), which means that the forms must frame into one another. It is also important to remember that some parts of these forms may be removed before others. For example, the

(a)

(b)

**Figure 9-16** (a) Column Pile Cap Marked for Column Formwork Template Location. (b) Column Form Templates and Column Reinforcement in Place. (Reprinted by permission of Bethlehem Steel Co.)

beam and girder sides may be removed first, followed later by the column forms, and finally by the beam and girder bottoms. It is therefore necessary to construct the girder and beam forms so that each part can be removed without disturbing the remainder of the form. It is critical that proper lateral bracing is provided for the column formwork to ensure that the columns remain plumb during the construction of beam and girder forms and while the concrete is being placed.

The first step in building a girder or beam form is to set the form bottom. Make its width exactly that

**Table 9-2(a)**
**TYPICAL COLUMN SCHEDULE, STANDARD UNITS**

| | | Reinforced Concrete Column Schedule | | | |
|---|---|---|---|---|---|
| Column No. | Column Size (in.) | Vertical Reinforcement | Lateral Ties (in. o.c.) | Top Elevation (ft) | Bottom Elevation (ft) |
| 1, 5, 20 | 20 × 18 | 4–#6 | #3 @ 11 | 98.67 | 87.33 |
| 2 | 34 × 18 | 4–#8 | #3 @ 12 | 97.50 | 87.33 |
| 3 | 26 × 18 | 4–#7 | #3 @ 12 | 97.50 | 87.33 |
| 4 | 24 × 16 | 4–#7 | #3 @ 12 | 98.00 | 87.33 |
| 6 | 18 × 17 | 4–#6 | #3 @ 11 | 98.67 | 87.33 |
| 7, 8, 9 | 18 × 18 | 4–#6 | #3 @ 12 | 96.50 | 87.67 |
| 10 | 20 × 20 | 4–#7 | #3 @ 12 | 96.00 | 87.33 |
| 11, 12, 13 | 36 × 18 | 4–#8 | #3 @ 12 | 97.00 | 87.33 |
| 14, 15, 16, 17 | 20 × 20 | 4–#7 | #3 @ 12 | 98.67 | 87.33 |
| 18 | 36 × 18 | 6–#6 | #3 @ 12 | 97.67 | 87.67 |
| 19 | 36 × 18 | 6–#6 | #3 @ 12 | 97.67 | 85.50 |
| 20 | 20 × 18 | 4–#6 | #3 @ 11 | 98.00 | 85.50 |

**Table 9-2(b)**
**TYPICAL COLUMN SCHEDULE, METRIC UNITS**

| | | Reinforced Concrete Column Schedule | | | |
|---|---|---|---|---|---|
| Column No. | Column Size (mm) | Vertical Reinforcement | Lateral Ties (mm o.c.) | Top Elevation (m) | Bottom Elevation (m) |
| 1, 5, 20 | 500 × 450 | 4–20M | 10M @ 320 | 30.000 | 26.620 |
| 2 | 850 × 450 | 4–25M | 10M @ 400 | 29.720 | 26.620 |
| 3 | 650 × 450 | 8–15M | 10M @ 240 | 29.720 | 26.620 |
| 4 | 600 × 400 | 8–15M | 10M @ 240 | 29.870 | 26.620 |
| 6 | 450 × 425 | 4–20M | 10M @ 320 | 30.000 | 26.620 |
| 7, 8, 9 | 450 × 450 | 4–20M | 10M @ 320 | 29.410 | 26.720 |
| 10 | 500 × 500 | 8–15M | 10M @ 240 | 29.260 | 26.620 |
| 11, 12, 13 | 900 × 450 | 4–25M | 10M @ 400 | 29.570 | 26.620 |
| 14, 15, 16, 17 | 500 × 500 | 8–15M | 10M @ 240 | 30.000 | 26.620 |
| 18 | 900 × 450 | 6–20M | 10M @ 320 | 29.770 | 26.720 |
| 19 | 900 × 450 | 6–20M | 10M @ 320 | 29.770 | 26.000 |
| 20 | 500 × 450 | 4–20M | 10M @ 320 | 29.870 | 26.000 |

of the member being formed and its length equal to the distance between columns. The two ends of the form bottom are cut at a 45° angle, as shown in Figure 9-20, to produce a chamfer at the junction of the girder bottom and column. Chamfer strips with beveled ends are nailed to the girder bottom, flush with the outside edge. Notice that the girder bottom rests on the narrow side of the column form.

The girder must be supported between columns by some type of shoring. T-head shores are a frequent choice (see Figure 9-21); their number depends on the load and carrying capacity of each shore. Stringers supported by adjustable metal posts or metal scaffolding (see Figure 9-22) may also be used to support girder bottoms.

Girder sides overlap the bottom and rest on the shore heads and the sides of the column form (see Figure 9-20). They are held in place by ledger strips nailed to the shore heads with double-headed nails. By removing these ledger strips and the *kickers* (see Figure 9-23), girder form sides may be removed without disturbing the bottom. The bottom and its sup-

**Figure 9-17** Column Cages Being Assembled on Site.

**Figure 9-18** Spiral Reinforcement.

port must remain in place until the beam concrete has gained a major portion of its design strength. Full concrete strength will not be attained for 28 days. For larger girders, the side forms will have to be provided with vertical stiffeners to prevent buckling.

Beam forms are constructed in the same manner as girder forms. They may frame into either columns or girders (see Figure 9-19), and they must be framed in such a way that form removal is as simple as possible.

A beam pocket is cut into the girder or column form (see Figure 9-20) of sufficient size to receive the end of the beam form (see Figure 9-24). Cut the ends of the side forms for the beam at a 45° angle (see Figure 9-26). The beam bottom length should be such that its end is flush with the outside of the girder or column form. The end is supported by a block resting on girder ledger strips (see Figures 9-25 and 9-26). Cut 45° beveled ends on a length of chamfer strip and nail it to the bottom edge of the beam pocket so that the ends fit against the beveled edges of the beam sides.

Forms for girders and beams that frame into round columns require special attention. Fiberboard column forms have relatively thin shells, and a yoke or collar is needed around the top of the form to support the girder bottom and sides (see Figure 9-27). The girder bottom is cut the width of the column diameter, and a half-circle of the same diameter is cut in the ends. The girder bottom rests on the column collar, as shown in Figure 9-28, and is supported by shores. Beam sides rest on the shores and on the column collar and are held in place by ledgers.

## Spandrel Beam Forms

Forms for spandrel beams (deep beams that span openings in outer walls) need to be carefully formed. Form alignment must be accurate to produce an

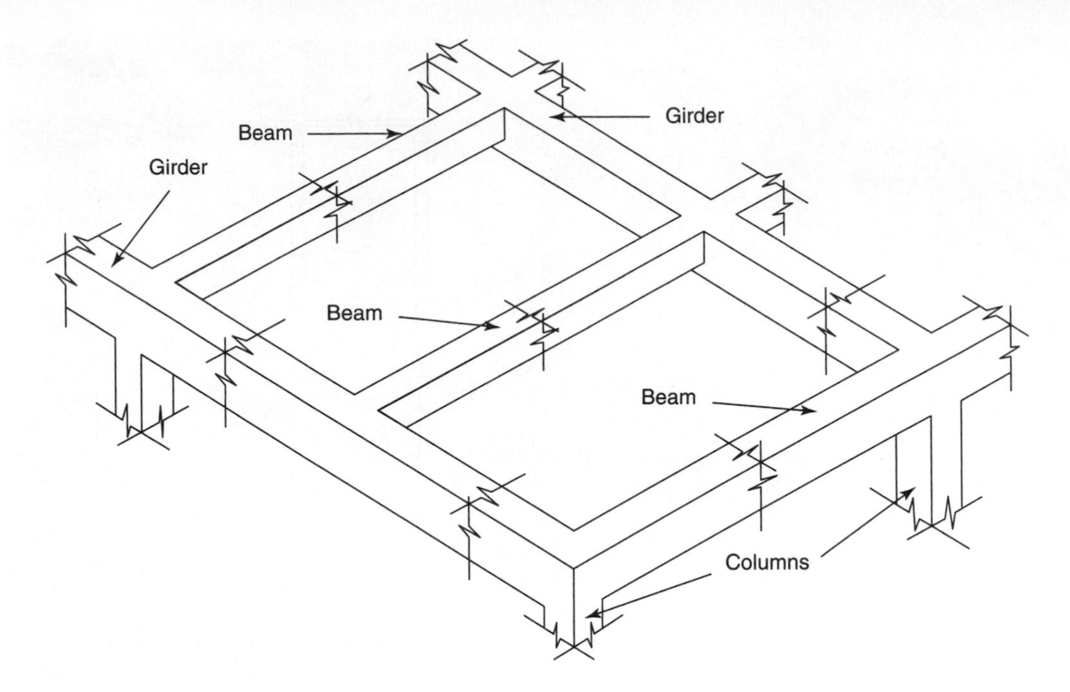


**Figure 9-19** Concrete Structural Frame.

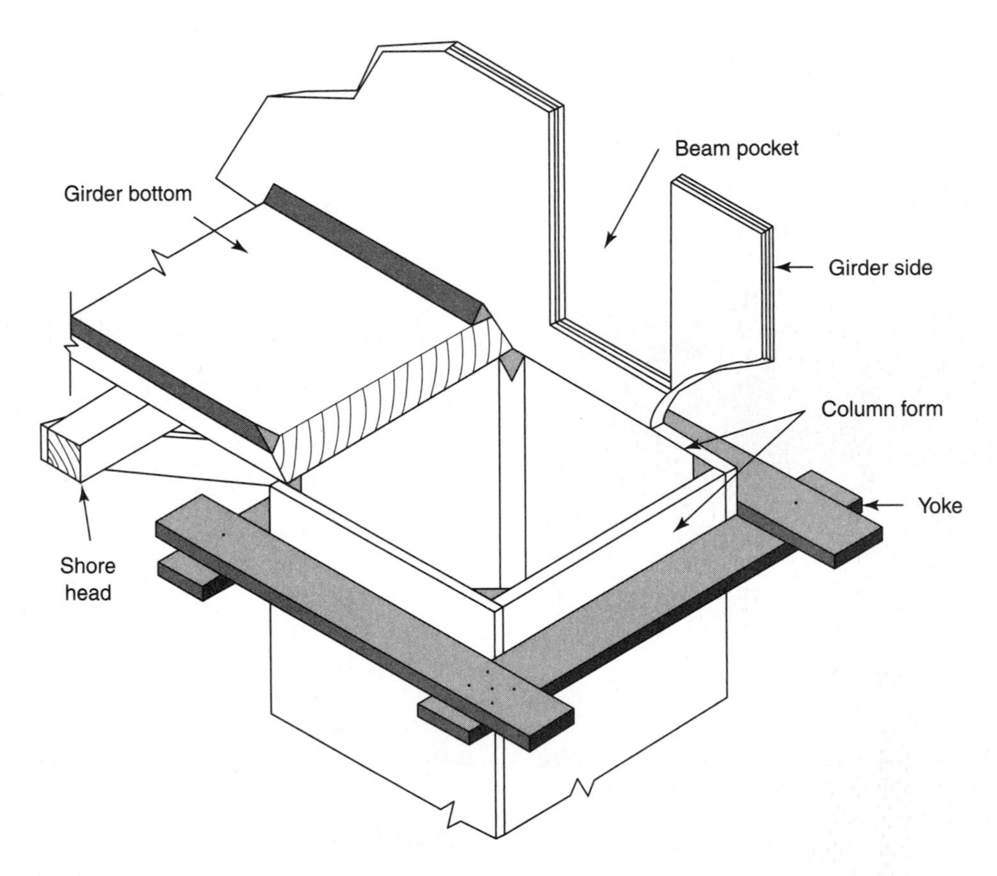

**Figure 9-20** Girder Form at Column.

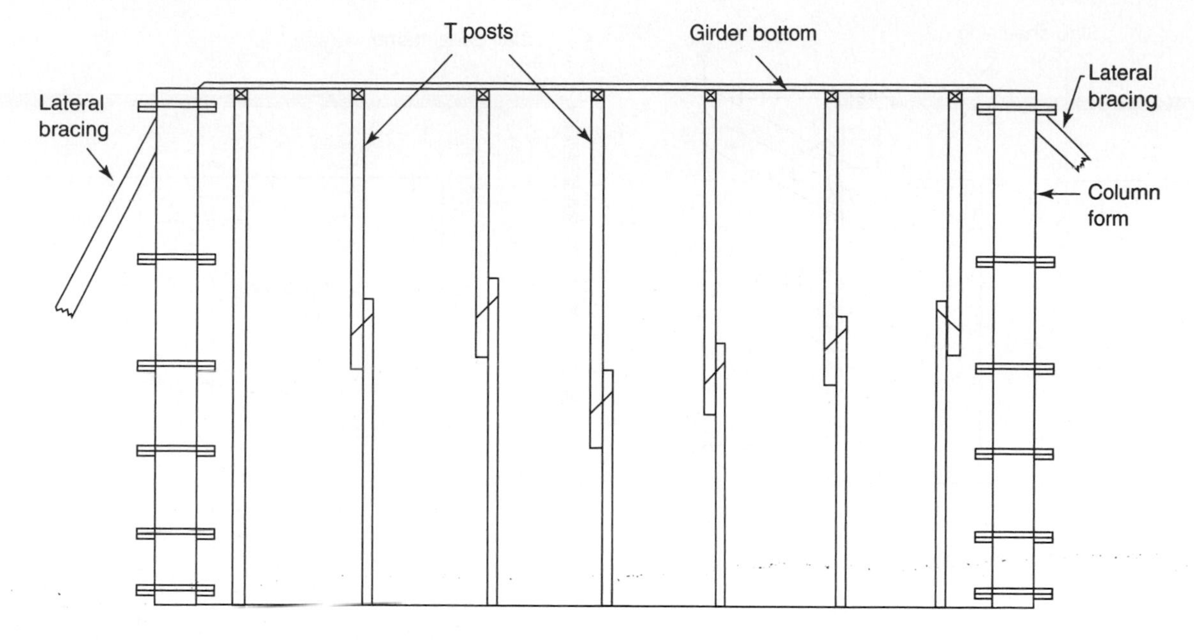


**Figure 9-21** Girder Bottom Supported by T Posts.

**Figure 9-22** Scaffolding Supports Girder Form.

attractive wall. Shore heads are often extended on the outside to accommodate the knee braces used to keep the forms in alignment. The extended shore head also frequently supports a catwalk for workers (see Figure 9-29).

## Beam Reinforcement

Reinforcing steel in beams serves three purposes: (1) to provide tensile strength to that part of the beam cross section that undergoes tension due to an applied load, (2) to add compression strength to the portion of the beam cross section that is in compression, and (3) to provide additional shear strength to the beam cross section. Tension and compression bars may be straight or bent and are held in alignment by the vertical bars that also resist shear. The whole assembly is supported on *chairs* set on the form bottom. To ensure that the reinforcement is not moved during the concrete pour, the bars are held in position inside the formwork by securing them to the form ties.

Consider the reinforcing steel for the beam in Figure 9-30. Note the concentration of steel in the top of the beam above the column. The beam is designed to be continuous over the support, resulting in high-tension stresses in the top portion of the beam. Similar stresses occur in the bottom of the beam between the supports, and appropriate reinforcing must be placed along the bottom of the beam to absorb them.

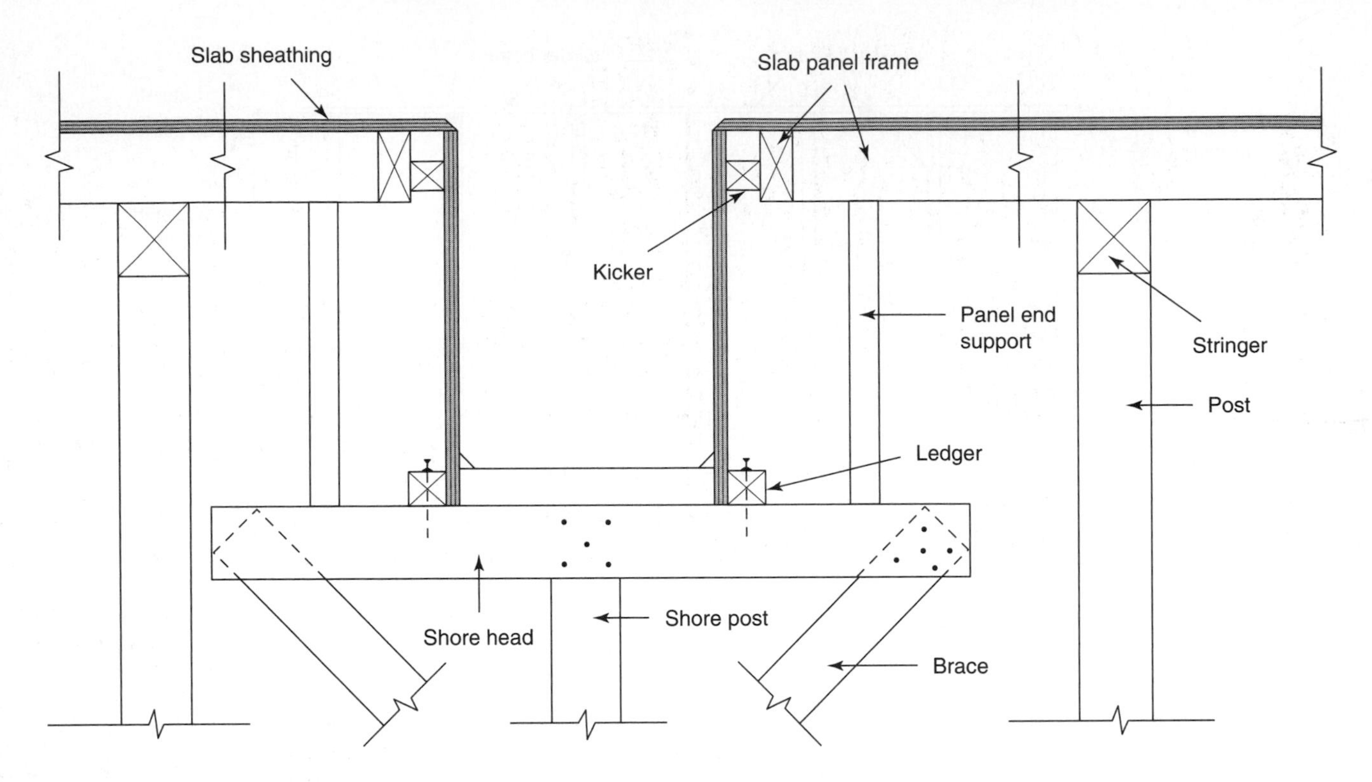

**Figure 9-23** Girder Form Details.

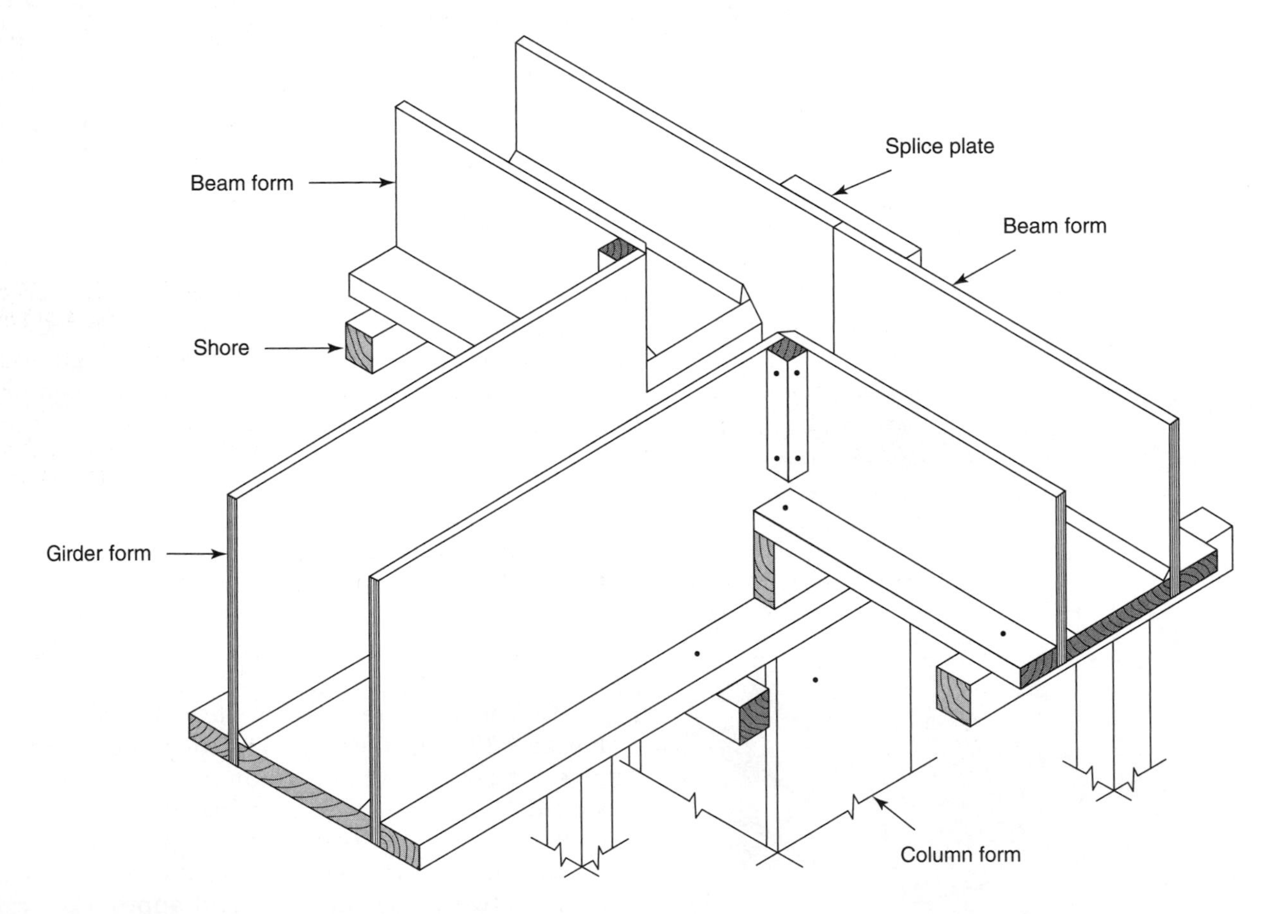


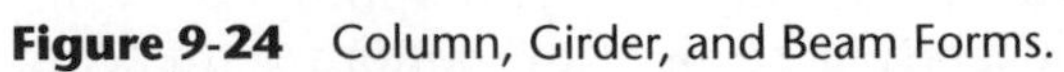

**Figure 9-24** Column, Girder, and Beam Forms.

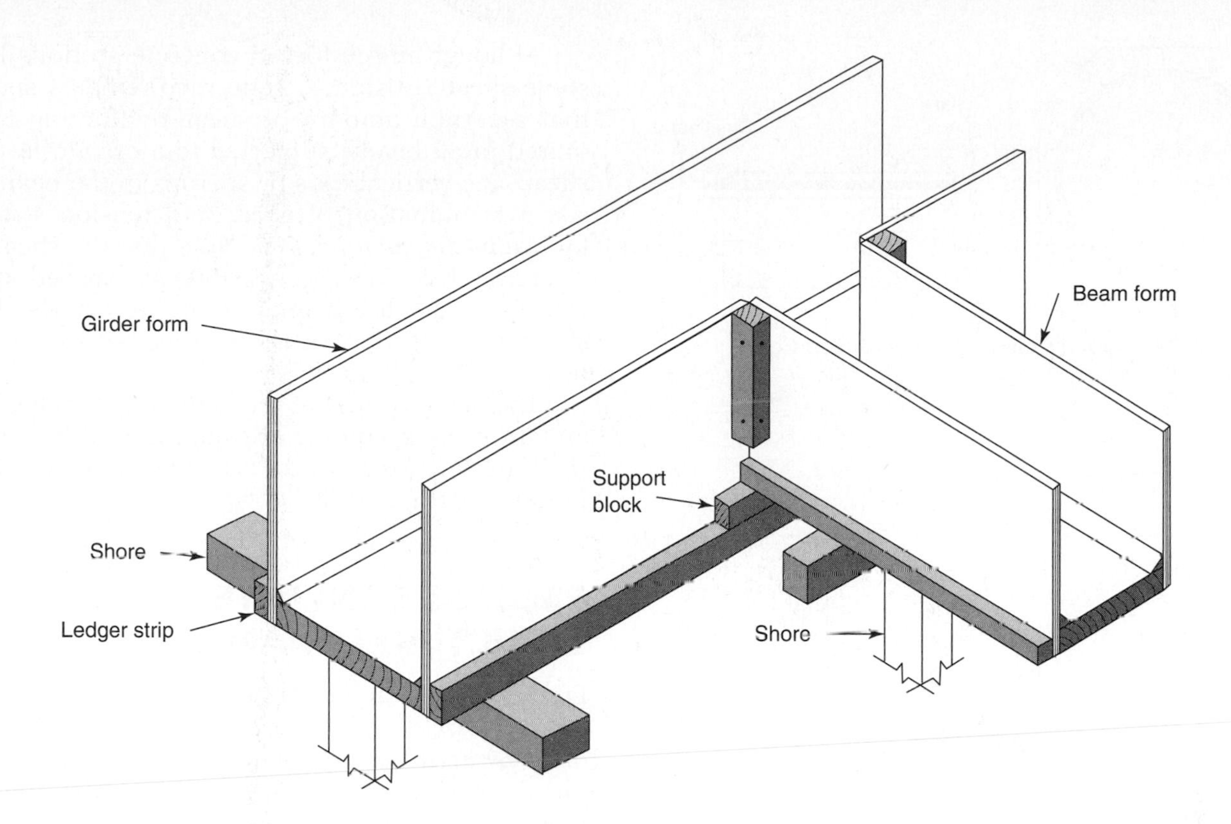

**Figure 9-25** Girder and Beam Form.

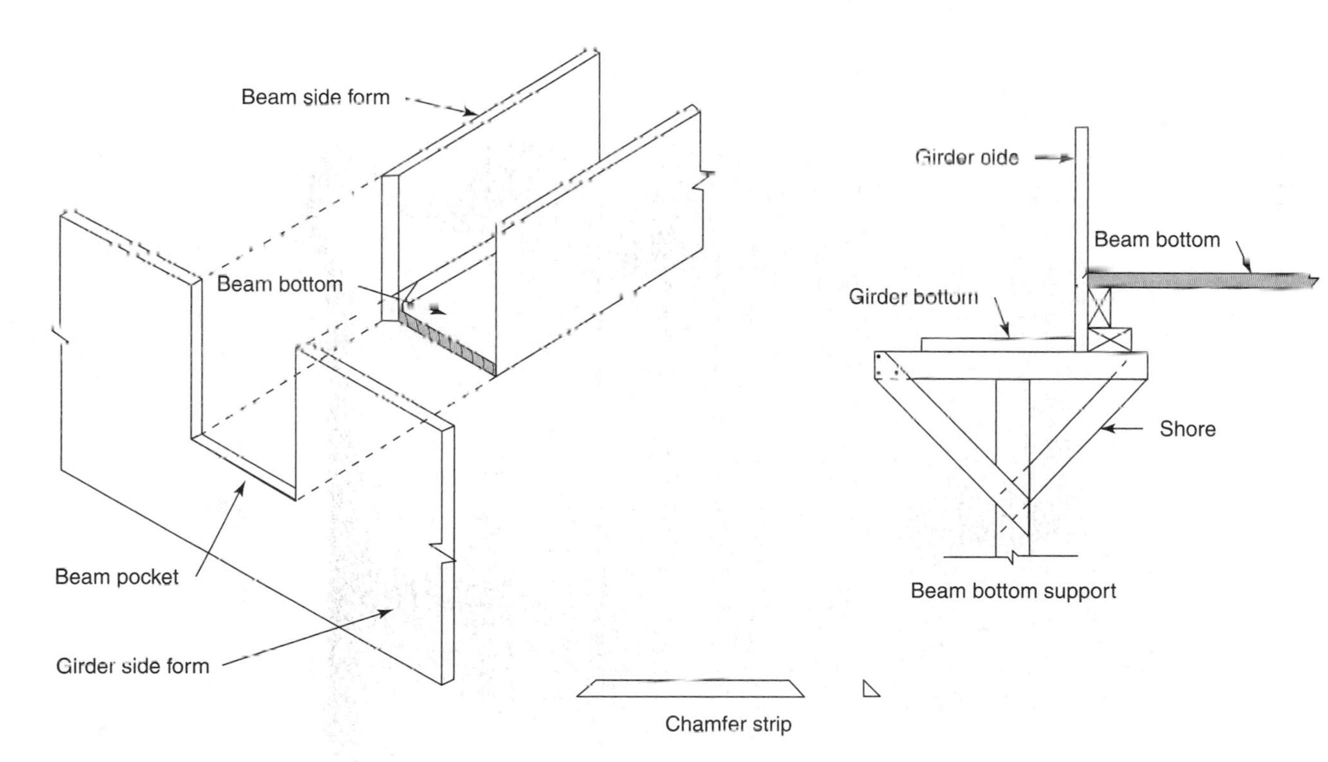


**Figure 9-26** Beam to Girder Form.

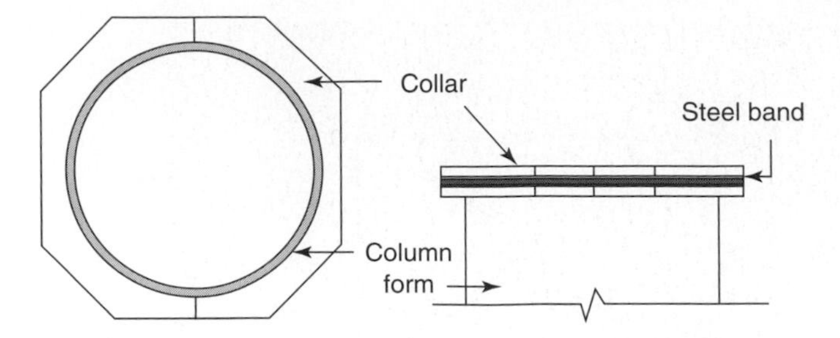


**Figure 9-27** Round Column Form with Top Collar.

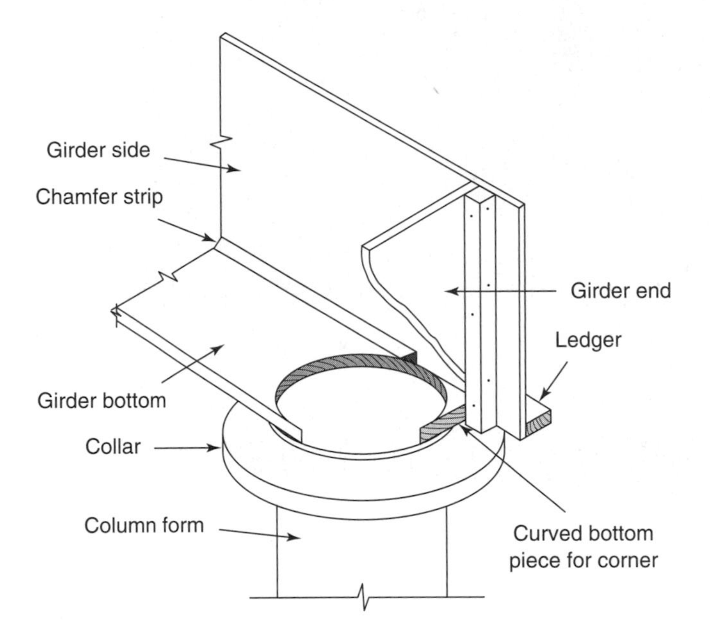


**Figure 9-28** Girder Form to Round Column.

Although unreinforced concrete sections have some shear resistance, code requirements specify that a certain amount of shear reinforcing is required in all beams subjected to a certain level of shear. The vertical bars or stirrups in the beam resist a combination of shear and tension stresses known as *diagonal tension.* Note that the shear reinforcing bars in Figure 9-30 are spaced quite closely at the beam end to counteract the high shear stresses that occur in this portion of the beam span.

The details of the reinforcement for each girder and beam in a building are summarized in a *beam schedule.* Part of a typical schedule is illustrated in Figures 9-31(a) and (b).

## PRECAST CONCRETE STRUCTURAL FRAMES

Precast structural members are of three types. They may be *normally reinforced, prestressed pretensioned,* or *prestressed posttensioned.* Reinforced precast members are normally designed according to common reinforced concrete practice. As a result of factory control, however, high-strength concrete is produced whether or not it is required, for it is actually more economical to design reinforced con-

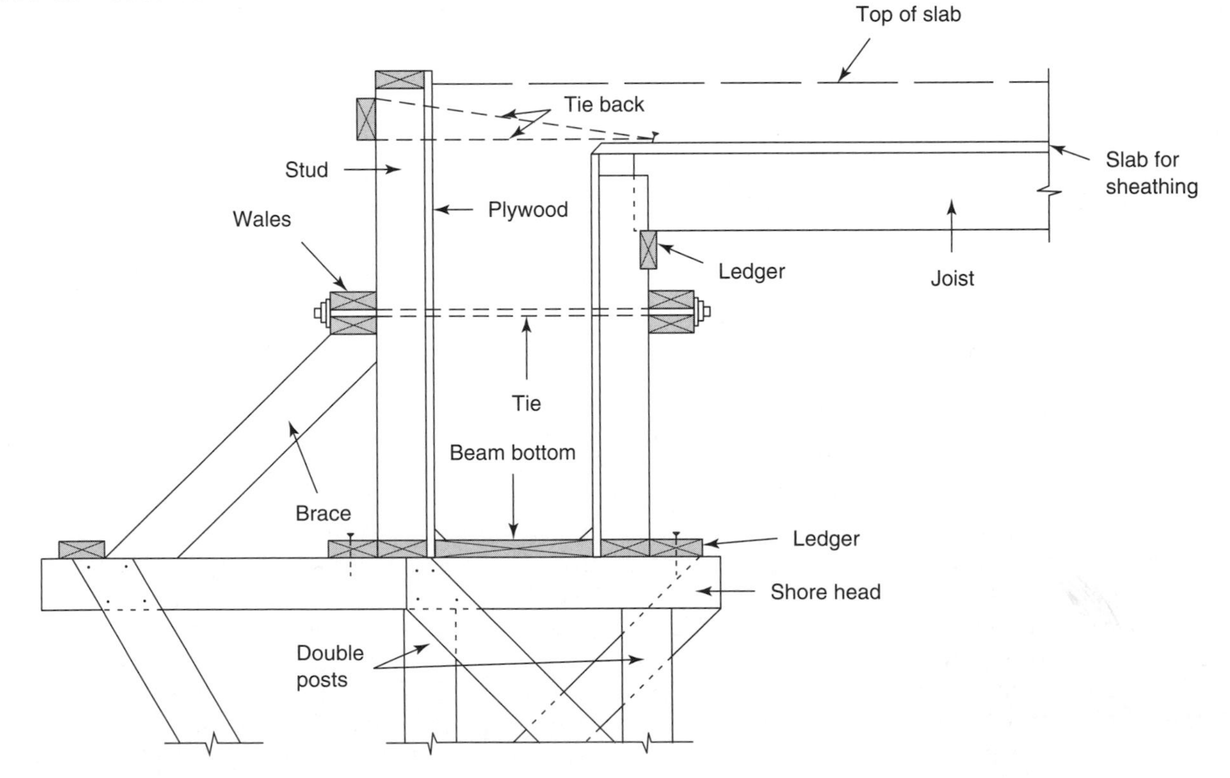


**Figure 9-29** Spandrel Beam Form.

**Figure 9-30** Beam Reinforcement for Continuous Beam.

crete for higher strengths in the range of 5,000 psi (35 MPa) or better.

The members of a precast structural frame are often prestressed; that is, the tensile reinforcement has been placed under tension before the member is erected. Pretensioning is a technique in which the reinforcement is placed in the forms and stretched against fixed abutments; concrete is then poured around the wires or cables. Pretensioned members are made several at a time in a long bed (see Figure 9-32), and the major requirements are therefore simplicity and quantity.

Posttensioning involves first forming and casting the member with ducts through its length. After the concrete is cured, reinforcement is placed in the ducts and anchored at one end. It is stretched against the concrete by means of jacks such as the one shown in Figure 9-33. The *tendons* are gripped and tension is maintained by various patented gripping devices that fit into the end of the member. Posttensioning is economical if the member is long or large. Both weight and cost may generally be reduced by posttensioning units longer than 45 ft (15 m) or having a weight over 7 tons (6,500 kg).

Posttensioning may be applied to the individual sections before they are placed in the building frame or, as in the case of slabs, posttensioning is applied after the concrete has been poured and set (see Figure 9-34). In many structures, posttensioning is used to add stability to the overall frame by providing a fixed connection between the beams and columns.

Columns are normally reinforced with regular reinforcing steel and usually do not contain any prestressing steel. Because they are cast individually, column sections can be produced to suit any framing and load requirements. Figure 9-35 shows columns that have seats, or *corbels,* to support the beams, whereas in Figure 9-36, beams simply bear on top of the column section.

Precast girder and beam shapes have become standardized to a considerable degree, but it is possible to produce a special shape if required. In Figure 9-37, a number of beam shapes in common use are shown. One great advantage in the use of these precast members is the possibility of erecting long free spans. In Figure 9-38, for example, the single T girders shown have a span of 75 ft (23 m), and greater spans are quite common.

Precast concrete joists are widely used to support concrete floor and roof slabs in many types of buildings. They are small, prestressed units, which are described in Chapter 12 and illustrated in Figure 12-11.

| Beam No | Bars | | | | | Stirrups | | | | | Section | Elevation |
|---|---|---|---|---|---|---|---|---|---|---|---|---|
| | No | Size | Length Ft | Length Ins | Mark | No | Size | Length Ft | Length Ins | Mark | | |
| 1 | 2 | 6ø | 25 | 0 | 1A | 12 | 3 | 5 | 6 | S1A | | |
| | 2 | 6ø | 20 | 0 | Str | | | | | | | |
| 2 | 2 | 7ø | 19 | 6 | 2A | 8 | 4 | 5 | 6 | S2A | | |
| | 2 | 6ø | 12 | 6 | 2B | | | | | | | |

(a)

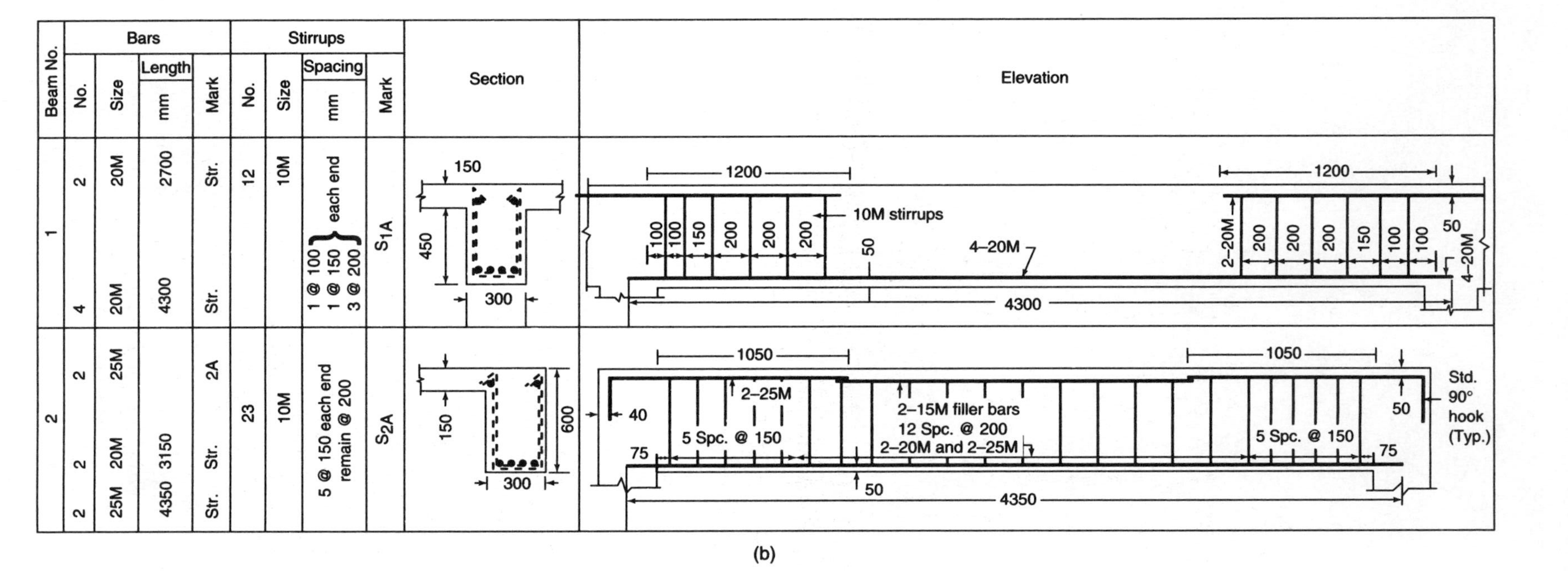

| Beam No. | Bars | | | | Stirrups | | | | Section | Elevation |
|---|---|---|---|---|---|---|---|---|---|---|
| | No. | Size | Length mm | Mark | No. | Size | Spacing mm | Mark | | |
| 1 | 2 | 20M | 2700 | Str. | 12 | 10M | 1 @ 100, 1 @ 150, 3 @ 200 each end | S1A | | |
| | 4 | 20M | 4300 | Str. | | | | | | |
| 2 | 2 | 25M | | 2A | 23 | 10M | 5 @ 150 each end remain @ 200 | S2A | | |
| | 2 | 20M | 3150 | Str. | | | | | | |
| | 2 | 25M | 4350 | Str. | | | | | | |



(b)

**Figure 9-31** (a) Beam Schedule, Standard Units. (b) Beam Schedule, Metric Units.

**Figure 9-32** Pretensioned, Precast Beams. (Reprinted with permission of Portland Cement Association.)

**Figure 9-33** Hydraulic Tensioning Jack. (Reprinted with permission of Portland Cement Association.)

**Figure 9-34** Posttensioning Cables in Poured Concrete Slab Beam.

**Figure 9-35** Prestressed Precast Concrete Structural Frame. (Reprinted with permission of Portland Cement Association.)

**Figure 9-36** Prestressed Precast Concrete Beams Bearing on Top of Columns.

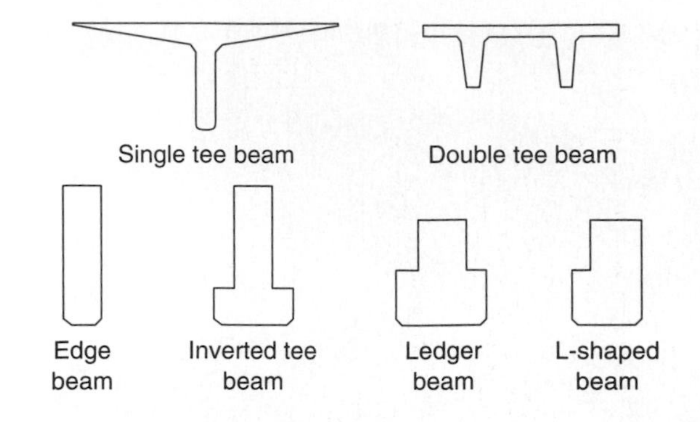


**Figure 9-37** Precast Beam Shapes.

**Figure 9-38** Single-T Prestressed Girders.

## Column Erection

The first step in the erection of a precast structural frame is the preparation of the column footings. Because the columns must be mechanically attached to them, anchor bolts have to be set accurately into the footing (see Chapter 6, dowel templates). Each column foot has anchor plates [Figure 9-39(a)] cast into it so that the column can be bolted to the footing. Large columns require special steel pads such as those shown in Figure 9-39(b) to transfer the column load to the foundation. Note the large splicing nuts on top of each anchor bolt.

A nut is turned onto each anchor bolt before the column is set over them. After erection, the column is plumbed, using guy wires if necessary (see Figure 9-35), and the nuts are adjusted until the column rests firmly over them. Top nuts are then put on and tightened to hold the column in position. Finally, the space between the top of the footing and the column is filled with grout, often made with metallic aggregate, to provide even pressure over the entire area.

## Column-to-Base Connections

Figure 9-40 illustrates five common arrangements for connecting precast columns to a foundation pier, a pile cap, a wall footing, or a spread footing. In each, the *double-nut system* with a 2- to 2½-in. (50- to 65-mm) space between the top of the foundation and the bottom of the base plate for *nonshrink grout* is used. At the column locations, it is necessary that sufficient ties be placed in the top of the pier or wall to secure the anchor bolts. Each connection type shown has unique features, as follows:

1. The base plate is larger in dimension than the cross section of the column, and the anchor bolts may be either at the corners or at the middle of the sides of the plate, depending on erection requirements. Column reinforcement is welded to the base plate.
2. The *internal* base plate is either the same size or smaller than the column cross section, so there must be anchor bolt pockets in the base of the column. After erection, the pockets are usually grouted at the same time as the underbase grout is placed.
3. The base plate is restricted to not more than column cross-section size, and a section of ½-in. (12.5-mm) thick angle, with a cap on one end, is cast into the corners of the column. The base plate is welded to the angles. Reinforcement may be welded to the base plate or to the angles.
4. A full base plate is not used. Instead, angles are attached to the bottom of the column by welding to the main reinforcement or to *dowel bars,* which lap the main reinforcement. Often a plate is inserted at the base of the column between the angles. To prevent rotation, welded studs can be attached to the vertical legs of the angles.
5. The column reinforcing bars project from the bottom of the column and are inserted into grout-filled metallic conduit embedded in the foundation. Temporary bracing is required for the column until the grout has gained the desired strength. The angles shown are one means of providing temporary bracing.

## Beam Erection

Beams are raised into position by crane, and some type of temporary connection is usually provided until the permanent ones have been completed. In the structure shown in Figure 9-35, pins cast into the column seats hold the girders in place during erection. Angle iron clips bolted through holes in the columns restrain the girders until the final connections can be made.

(a)

(b)

**Figure 9-39** (a) Column-to-Footing Connection. (Reprinted with permission of Portland Cement Association.) (b) Steel Bearing Support for Precast Concrete Column.

Connections may be either *moment resisting* or *pin type*. Moment-resisting connections prevent rotation between the column and the beam and contribute to the lateral stability of the frame. For moment connections to resist the tension force resulting from the force couple that is developed, some method to produce continuity at the connection is required. Welding, shear anchors, inserts, additional reinforcing bars, posttensioning, or some combination of these can be used to provide this continuity. Care must be taken to provide temporary bracing during the erection of beams with these types of connections until the connections have been completed.

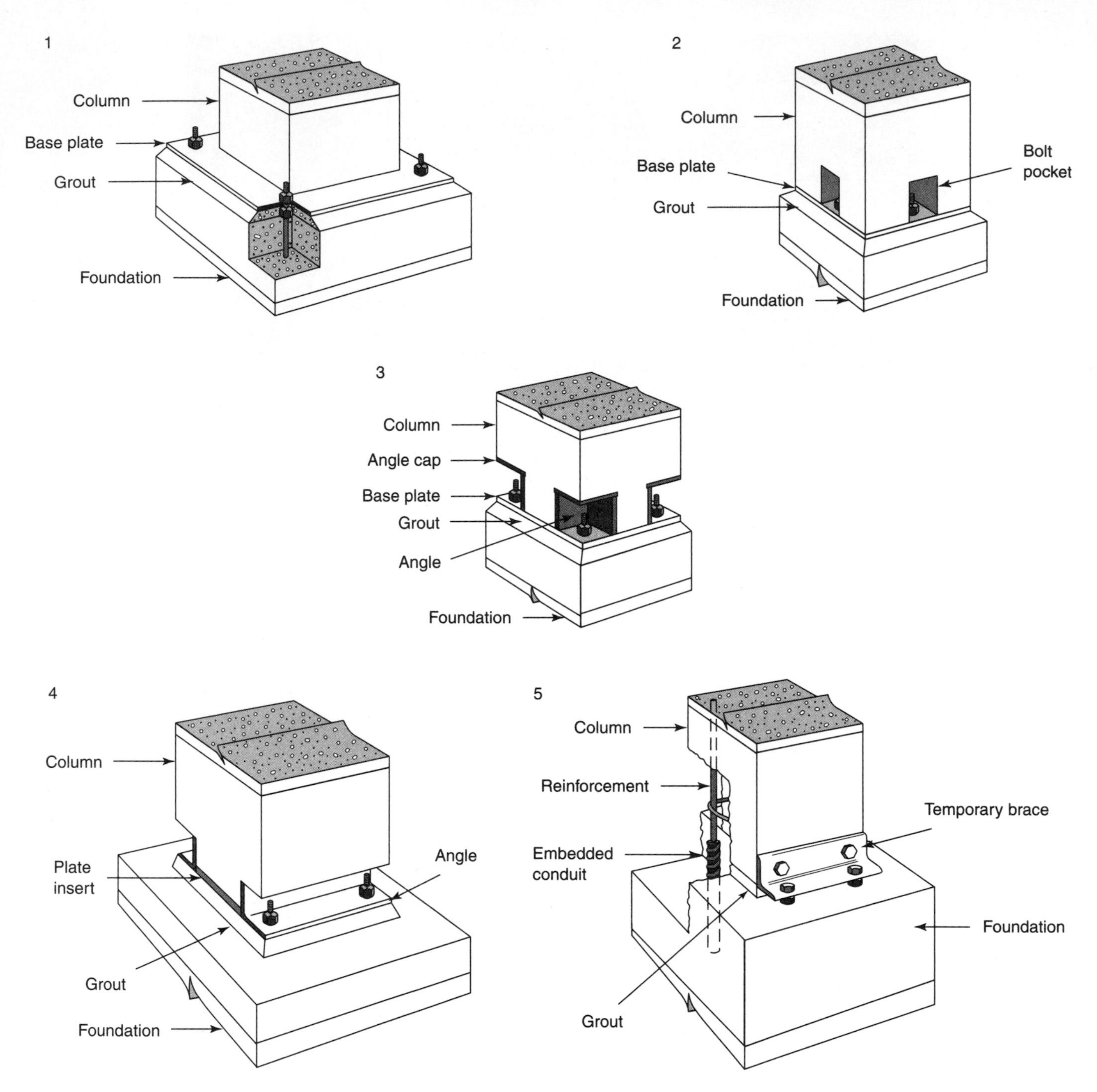


**Figure 9-40** Column-to-Foundation Connections. (Reprinted with permission of Portland Cement Association.)

Pin-type connections, on the other hand, are designed primarily for gravity loads, but are not expected to resist lateral loads as such. Connections of this nature usually consist of plates or angles embedded in the beam ends that bear on similar plates embedded in the supporting section. They are normally welded together to prevent the beam from slipping off its support. Pins cast into column seats that project into holes in the beam end are another form of pin-type connection.

## Beam-to-Column Connections

In Figure 9-41, seven different methods are shown for connecting beams to columns. These are by no means the only types of connections used for this purpose, but they illustrate some of the basic techniques. Although the beams shown are rectangular in cross section, any other type of beam commonly used could be connected by the same methods. The seven variations of beam-to-column

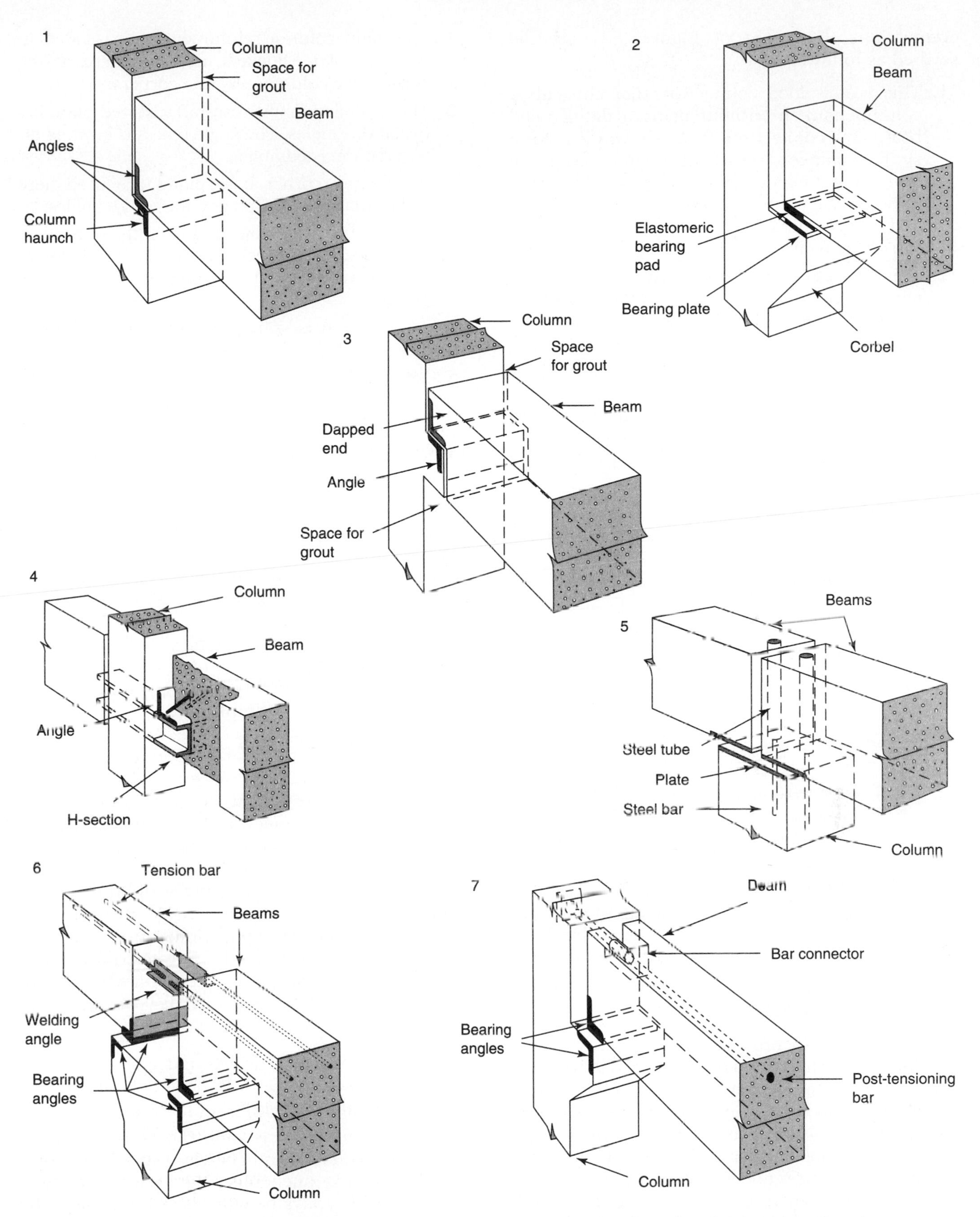


**Figure 9-41** Beam-to-Column Connections. (Reprinted by permission of Precast/Prestressed Concrete Institute.)

connections illustrated in Figure 9-41 are described as follows:

1. This is a welded connection that conceals a column haunch without using a dapped-end beam (see connection 3). As shown, it is for a simple-span condition, but it can be used as a moment connection by using nonshrink grout between the end of the beam and the column and by providing for transfer of tension at the top of the beam.
2. This is a welded connection, with the beam resting on a corbel projecting from the side of the column (see Figure 9-35). It is also for a simple-span condition but can be converted into a moment connection. Elastomeric bearing pads are shown, but are optional.
3. This is called a dapped-end connection, also welded. Its development to a moment connection requires grout at two interfaces, which is a difficult field procedure.
4. This welded connection is often used when it is required that it be hidden for architectural reasons. An H-shape section is shown projecting from the column, but other structural steel shapes may be used instead.
5. This is a doweled connection, with bars projecting from the column into steel tubes cast into the beam. Tubes are later filled with grout. The connection can be made continuous by welding tension reinforcement as shown in Connection 6. Bearing pads are used at the bearing surfaces.
6. This is a welded connection in which the tension bars are welded to small angles. If future extension of the column is required, anchor bolts can be set into the cast-in-place concrete between the beam ends.
7. In this type of connection, a posttensioning bar is tensioned after the nonshrink grout has been placed between the column and beam end.

## Erection of Additional Stories

Subsequent stories are added to a precast frame building by splicing upper floor columns to those already erected. In many ways they are similar to column-to-beam connections. Most column splices require nonshrink grout between the sections to provide for variations in their dimensions. To allow for alignment tolerances during erection, methods such as the double-nut method are employed. Figure 9-42 illustrates five basic methods for splicing columns and they are described as follows:

1. The main column reinforcement is welded to the base plate. Pockets may be placed at the sides of the column or in the corners.
2. This connection also uses an inserted plate between the angles. The angle recess is grouted after erection is complete.
3. Four small corner base plates are used here rather than a full-sized one under the angles.
4. Reinforcing bars projecting from the upper column extend into steel tubes cast into the lower one. Provisions for temporary bracing must be provided.
5. The column is spliced through a continuous beam and reinforcement is required within the beam to transfer the load from column to column.

## Floor and Roof Slab Erection

Finally, in most precast structural frame designs, prestressed floor and roof slabs span from beam to beam, from beam to wall, or from wall to wall to complete the structure. In most cases the slabs will be set edge to edge, but they may be spaced and the space between them filled with cast-in-place, reinforced concrete.

Slab-to-beam connections should take into consideration the transfer of horizontal forces from slab to beam if it is assumed that the floor or roof is acting as a diaphragm. Where cast-in-place topping is used on floors, additional reinforcement in the form of welded wire mesh is placed *across* the beam to minimize cracking. Figure 9-43 illustrates two typical slab-to-beam connections. In Connection 1, the ends of tee legs may be dapped to accommodate units with a greater depth.

Slabs supported by bearing walls utilize connections similar to those used in beam-to-slab connections. In most applications, some degree of continuity is obtained at the slab-to-wall connection, but generally a fully fixed connection is not required. Figure 9-44 illustrates six typical connections, which are described as follows:

1. A bond beam is provided under the slab ends, and the joint between the meeting ends of the slabs is fully grouted.
2. *Hairpin anchors* cast into the bond beam are embedded in the grout between slab ends to provide positive anchorage. Bars may be grouted into the keyways between adjacent slabs to tie the slabs together lengthwise and help to prevent roofing problems at the joint.

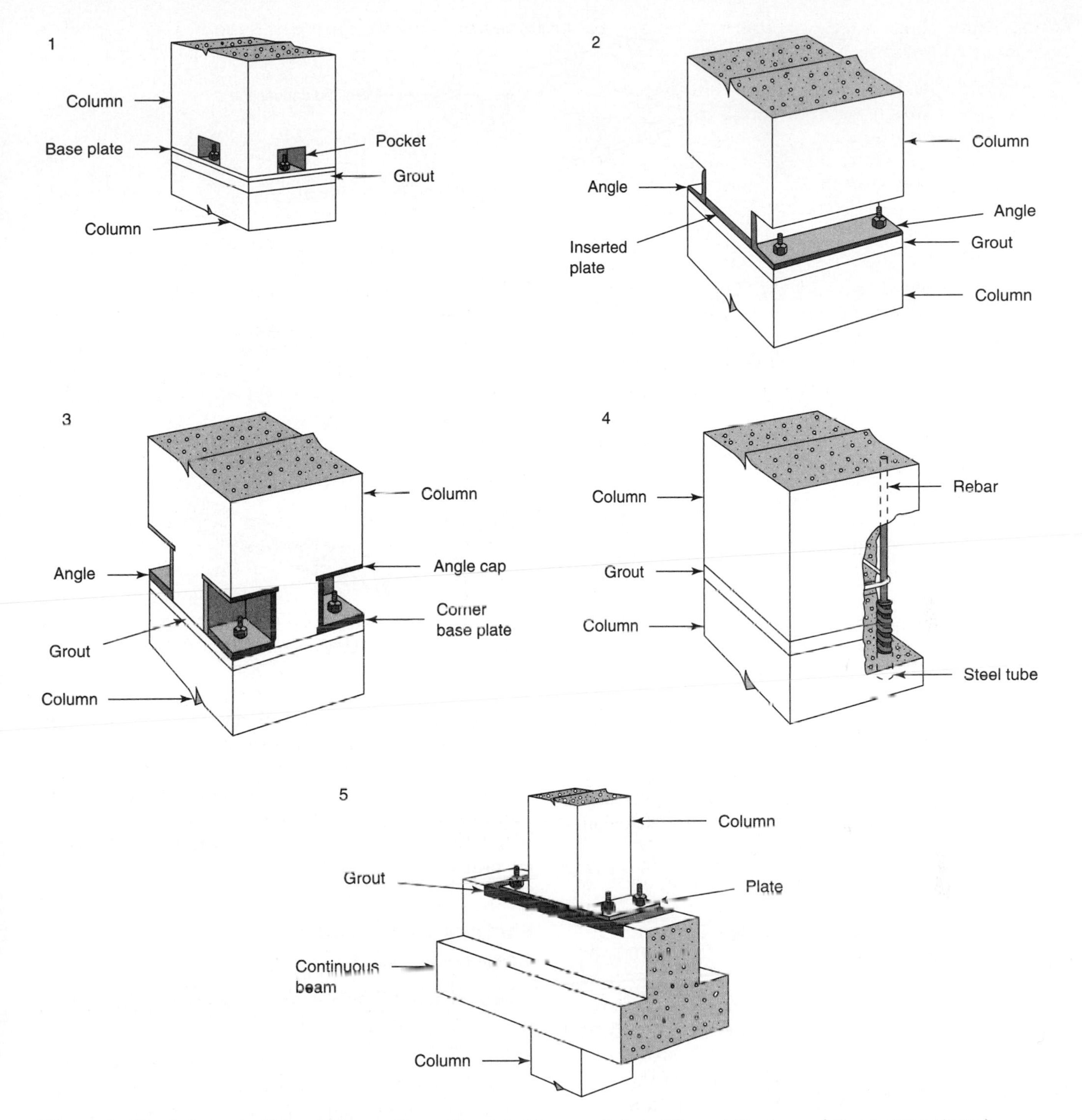


**Figure 9-42** Column-to-Column Connections. (Reprinted by permission of Precast/Prestressed Concrete Institute.)

3. The pocket in the wall must be sufficiently large to receive the extended stem of the single tee slab without difficulty. Elastomeric bearings pads may be used if required.
4. Double tee slabs rest on a bond beam cast on a block wall. Precast fill-ins between the tee stems may be used, as forms for concrete between the slab ends.
5. Single or double tee units used as wall panels may have corbels cast with them to support floor or roof slabs. Vertically slotted bolted connections may be used to tie the slabs to the

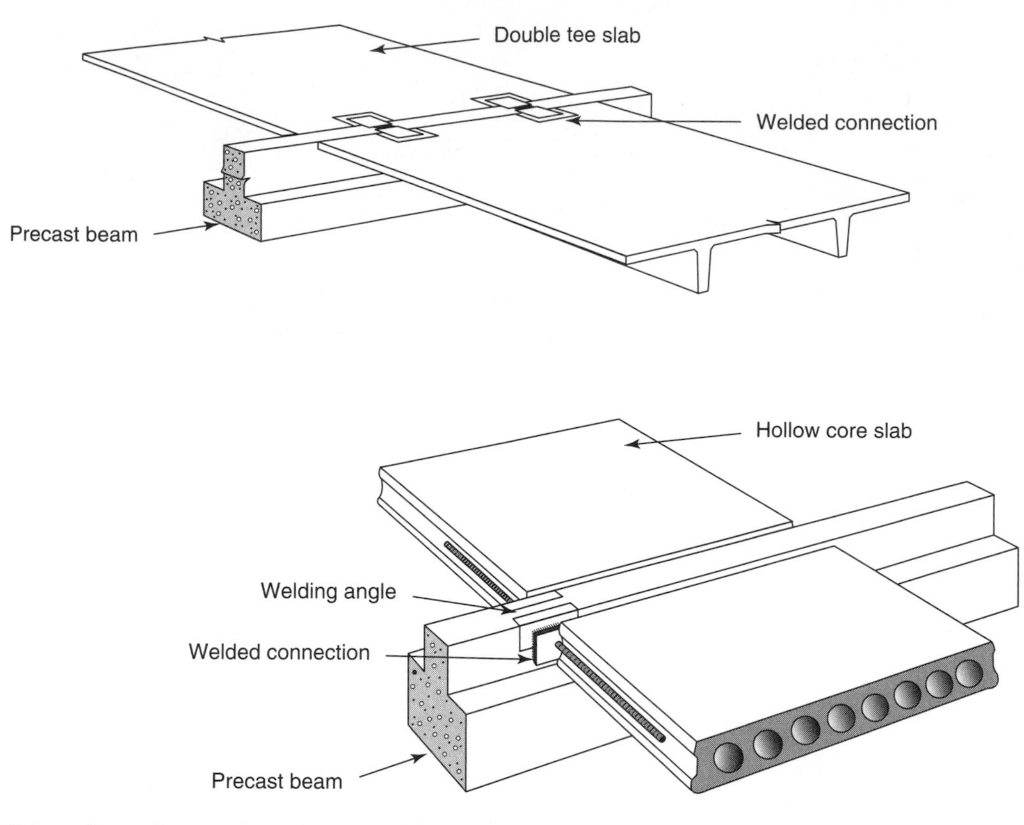


**Figure 9-43** Slab-to-Beam Connections. (Reprinted by permission of Precast/Prestressed Concrete Institute.)

load-bearing wall panels or the welded connections shown may be used.

6. Connecting a slab to a parallel wall requires a connection that can accommodate vertical movements. Slotted angles with low-friction washers permit this movement.

## ARCHITECTURAL PRECAST CONCRETE

A relatively new advance in the precast concrete industry has been the development of architectural precast concrete. Originally, precast structural units such as those described in the foregoing section were intended, in the main, to form a structural frame in the same way wood or steel might be used. More recently, advances in design, control, and manufacturing techniques have made possible and economical the production of precast units that are themselves finished components of a building.

There are two principal methods by which this type of unit is produced. One is known as *tilt-up construction,* in which the units are cast on the job site and tilted up into position. The other is an independent, *factory-type* production, geared to making a variety of precast products, in which all the facilities and economies of a production line type of operation may be utilized.

In most cases, the units produced by the tilt-up method are intended strictly as curtain wall components and, as such, are described in Chapter 14 on curtain wall construction.

### Factory-Produced Precast Concrete Units

There are a number of distinct advantages in the factory method of producing precast units. They include the following:

1. A freedom of design is possible. The styles and designs of buildings may be as diversified as the skill and ingenuity of those in charge of production permit (see Figures 9-45 and 9-46).
2. A high degree of quality control of the ingredients in the units is possible.
3. A wide variety of shapes, sizes, and configurations for any one building is possible.
4. There is the possibility of including in the precast units some or all of the services that are normally installed after erection (see Figures 9-47 and 9-48). This can result in considerable saving

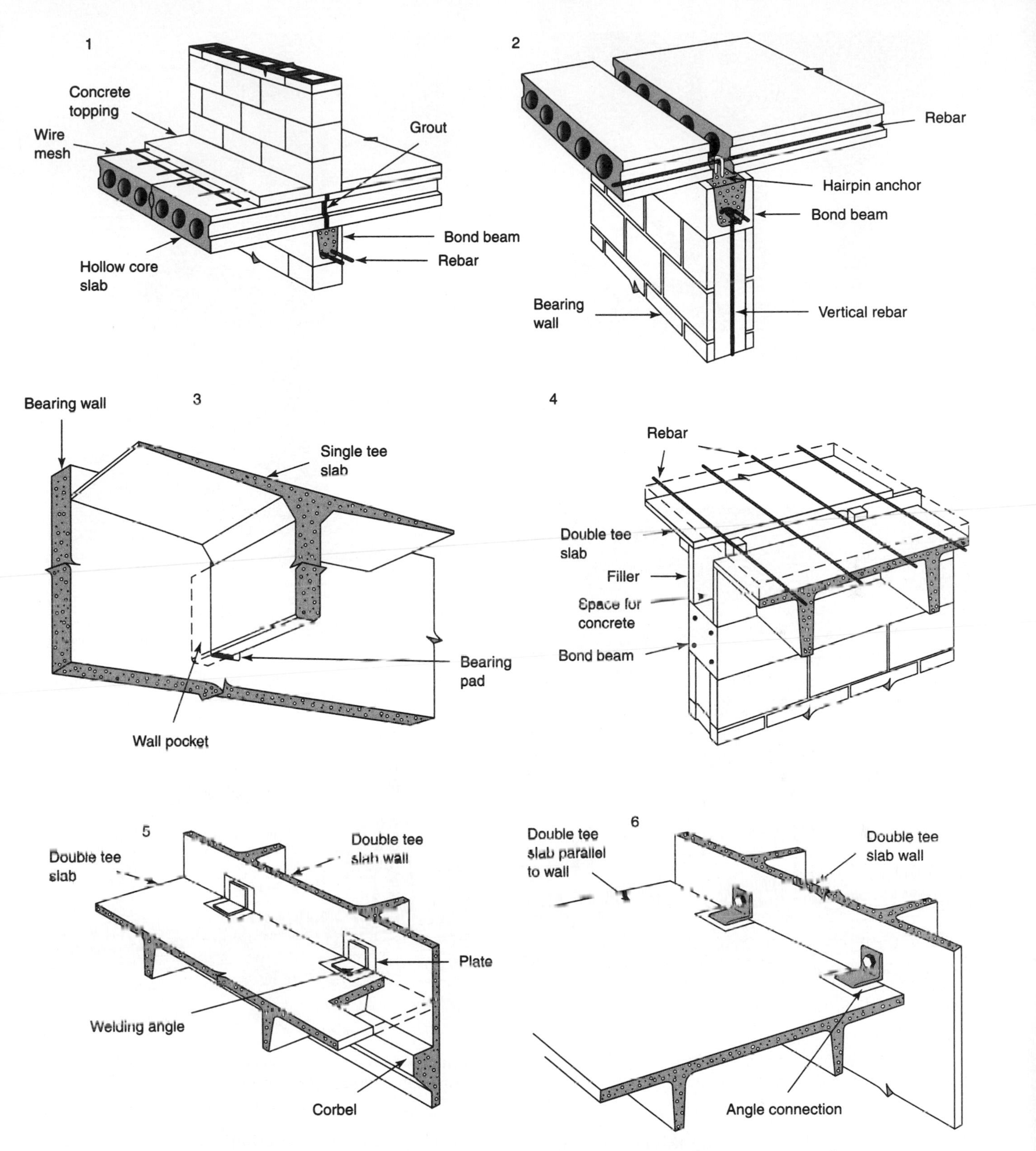


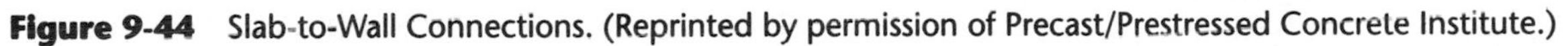
**Figure 9-44** Slab-to-Wall Connections. (Reprinted by permission of Precast/Prestressed Concrete Institute.)

**Figure 9-45** Toronto City Hall in Architectural Precast Concrete. (Reprinted by permission of Precast/Prestressed Concrete Institute.)

**Figure 9-47** Panels May Come Complete with Glazing, Mechanical Services, and Interior Finish. (Reprinted by permission of Precast/Prestressed Concrete Institute.)

**Figure 9-46** Fine Example of Precast Concrete Architecture. (Reprinted by permission of Precast/Prestressed Concrete Institute.)

**Figure 9-48** Preglazed, Precast Wall Panel. (Reprinted by permission of Precast/Prestressed Concrete Institute.)

of time and labor following the erection of the building. The precast wall panels for the building shown in Figure 9-47 were shipped to the job site complete with glazing, mechanical services, and interior finish. Only a minimum of mechanical connections and interior painting were required following panel installation.

5. The use of precast components makes possible the rapid closure of the building. This allows for earlier access by the service trades.
6. Precast panels can be fabricated as sandwich units with insulation laminated between the inner and outer layers, thus providing an energy-efficient wall section.

**Figure 9-49** Large Architectural Precast Wall Panel. (Reprinted by permission of Precast/Prestressed Concrete Institute.)

**Figure 9-50a** Erecting Load-Bearing Wall Panel. (Reprinted by permission of Precast/Prestressed Concrete Institute.)

Among the types of units produced by this method, perhaps the major one is the *precast wall panel* (see Figure 9-49). Such panels may be used as curtain walls, load-bearing members, wall-supporting units, formwork for exterior walls, or shear walls.

Curtain wall panels are units that carry no loads except wind load. They are set into or hung from the structural frame to act as closure units and to provide the architectural effect desired.

Load-bearing panels carry the loads of floor and roof. In Figure 9-50(a), for example, the load-bearing panels were designed in widths to suit the standard width of floor and roof slabs that they support. In Figure 9-50(b), the load-bearing panels have built-in pockets at predetermined spacing to provide bearing locations for steel roof joists. Another way in

**Figure 9-50b** Load-Bearing Wall Panel Supporting Open Web Steel Joists.

Figure 9-51 Frame Made from Posttensioned Precast Segments. (Reprinted by permission of Precast/Prestressed Concrete Institute.)

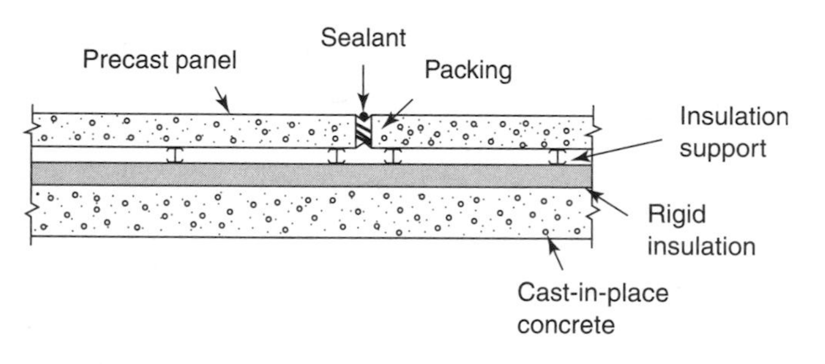


Figure 9-52 Precast Facing Panels Used as Formwork.

which load-bearing units are utilized is in the segmental construction of a precast frame or of large wall panels. This technique involves the posttensioning together of a number of segments to form a complete unit. In the building shown in Figure 9-51, precast segments are posttensioned together to form a complete building frame.

Precast wall-supporting units are those that support a part of the wall but do not carry any loads from floor or roof. Such units are stacked one above the other so that they carry the weight of the units above them.

In some cases, precast panels may be intended for exterior finish only; in such situations, they may be used as forms for the cast-in-place structural wall

Figure 9-53 Precast Facing Panels Used as Formwork for Cast-in-Place Structural Wall. (Reprinted by permission of Precast/Prestressed Concrete Institute.)

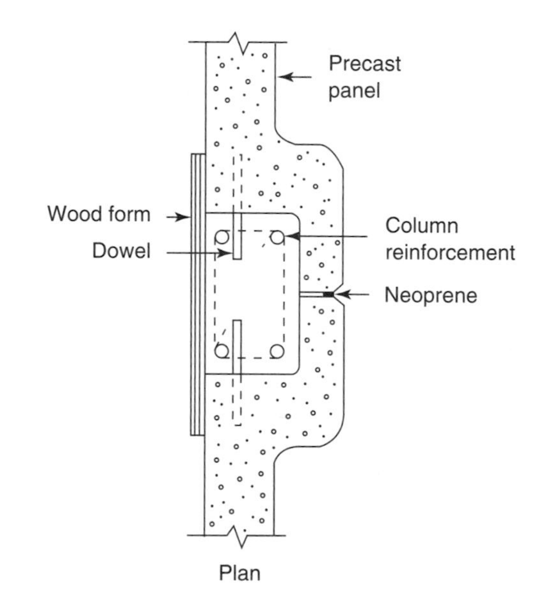


Figure 9-54 Precast Panels as Formwork for Column.

behind them. Figure 9-52 indicates how this is done, and Figure 9-53 shows an example of this technique in operation. Panels may also be used as formwork for cast-in-place columns (see Figure 9-54). Other types of precast units may also be used for this purpose (see Figure 9-55).

Walls utilized for the transfer of lateral forces are known as shear walls (see Figure 9-56), and precast wall panels are ideally suited for such a purpose.

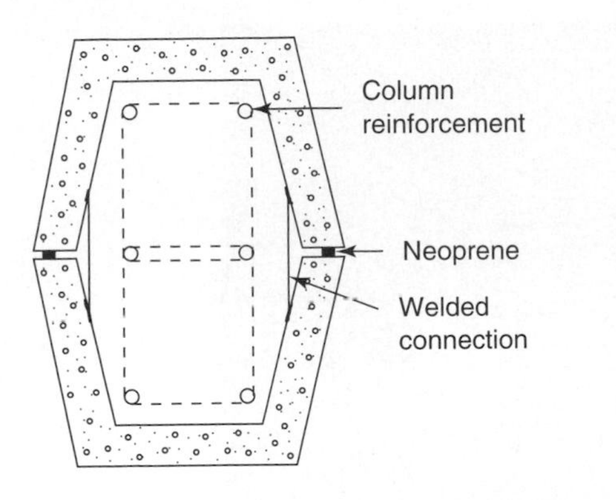


**Figure 9-55** Precast Column Covers Used as Formwork. (Reprinted by permission of Precast/Prestressed Concrete Institute.)

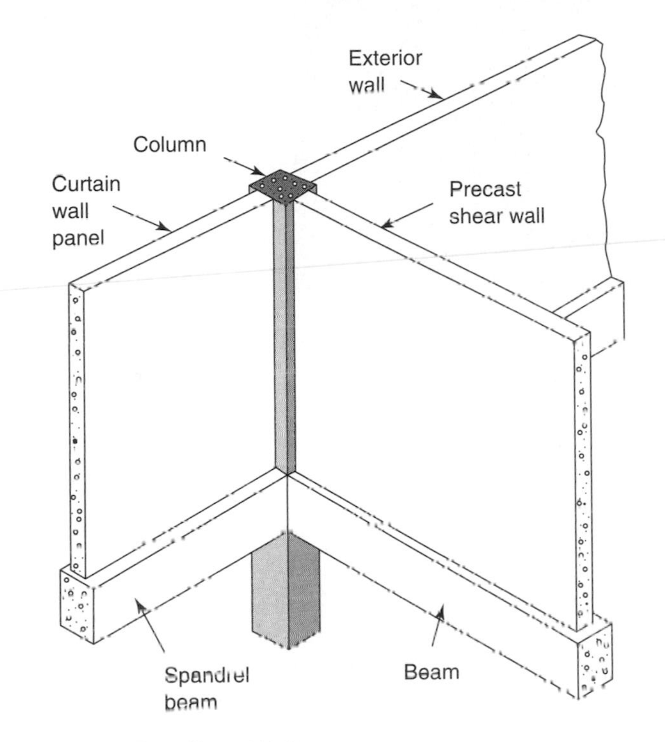


**Figure 9-56** Shear Wall.

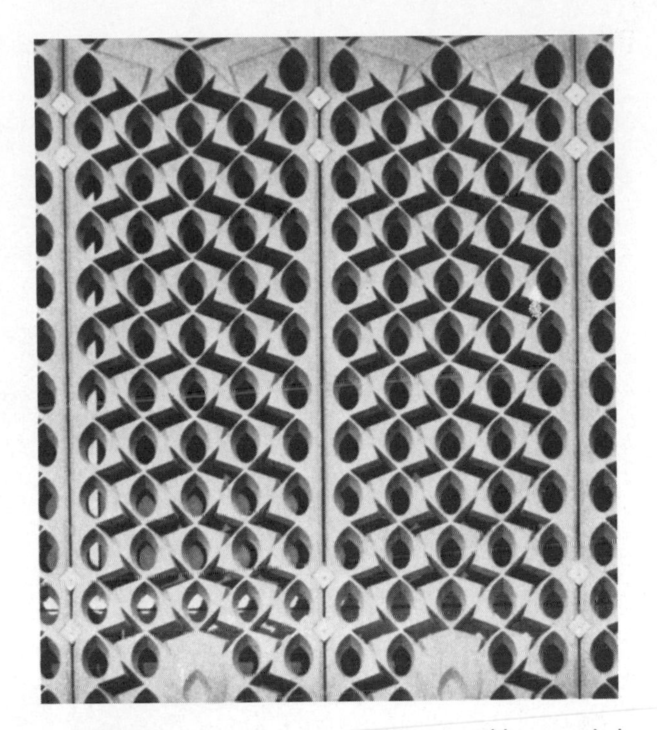

**Figure 9-57** Precast Sun Screen. (Reprinted by permission of Precast/Prestressed Concrete Institute.)

In addition to its use in panels and other structural members, architectural precast concrete has a number of other uses, such as for fences and screens (see Figure 9-57), planters, bases, light standards, street furniture (Figure 9-58), and sculpture.

The key to the success of architectural precast concrete projects probably rests in the planning that goes into them. One major item in that planning should be the concept that the cost of making the molds should be spread over as many units as possible. In other words, the more units that can be made in one mold, the less the unit cost will be. This leads to the concept of a *master mold* from which a maximum number of units for one project may be made, some of them perhaps with minor modifications in the original mold. Figure 9-59 shows an office building in which all of the panels were cast from one master mold.

Molds are usually one of two general types: *complete envelope* or *conventional*. A complete envelope mold is one in which all sides remain in place during the entire casting and stripping operation. A conventional mold has one or more removable side or end bulkheads (see Figure 9-60). Figure 9-61 shows a complete envelope mold that was used to cast the panels shown in Figure 9-49. Panels produced by these molds may be *closed*, *open*, or a *combination of open and closed* units (see Figure 9-62).

Precast units are transported to the job site by truck, and erection is carried out by the precaster or by the general contractor, with the aid of hoists or mobile cranes. In any case, one of the most common methods of connecting to the unit in order to lift it is by means of inserts and bolts (see Figure 9-63). Inserts are properly placed during the casting process and should be protected by plastic caps to prevent dirt or ice from getting into them until they are ready to be used.

Units are held in position in the building by connection hardware, part of which is cast into the units at the plant (see Figure 9-64), while the corresponding parts are attached to the structural frame. Connections are usually either bolted or welded and are designed to suit the particular project. Figure 9-65 illustrates a few typical connections, but a number of

**Figure 9-58** Precast Street Furniture. (Reprinted by permission of Precast/Prestressed Concrete Institute.)

**Figure 9-59** Walls Built from One Master Mold. (Reprinted by permission of Precast/Prestressed Concrete Institute.)

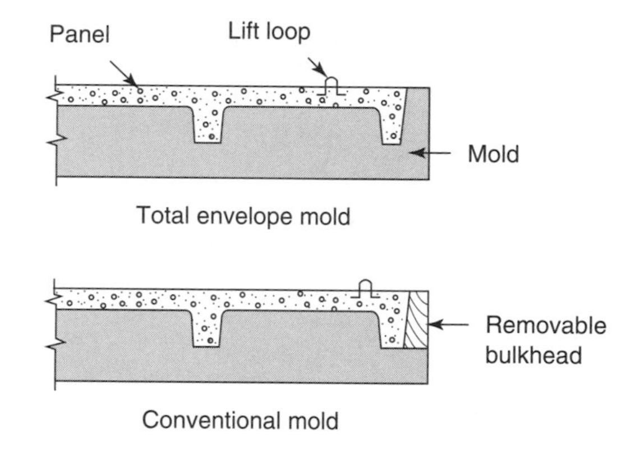


**Figure 9-60** Mold Types. (Reprinted by permission of Precast/Prestressed Concrete Institute.)

others may also be used, depending on the circumstances in any particular case.

The joints between precast concrete panels are an important part of the overall project and will have been designed taking into account the structural requirements, degree of watertightness required, appearance, and cost. They will usually be one of two basic types: *one-stage* or *two-stage* joints.

One-stage joints are simply sealed against water penetration by the use of a sealant near the

**Figure 9-61** Total Envelope Panel Mold. (Reprinted by permission of Precast/Prestressed Concrete Institute.)

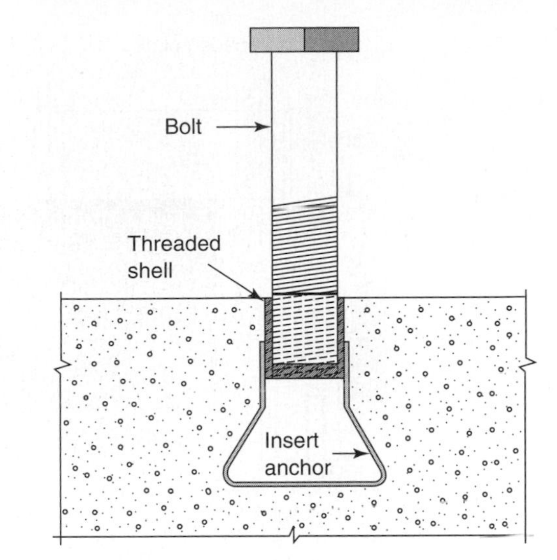


**Figure 9-63** Concrete Insert.

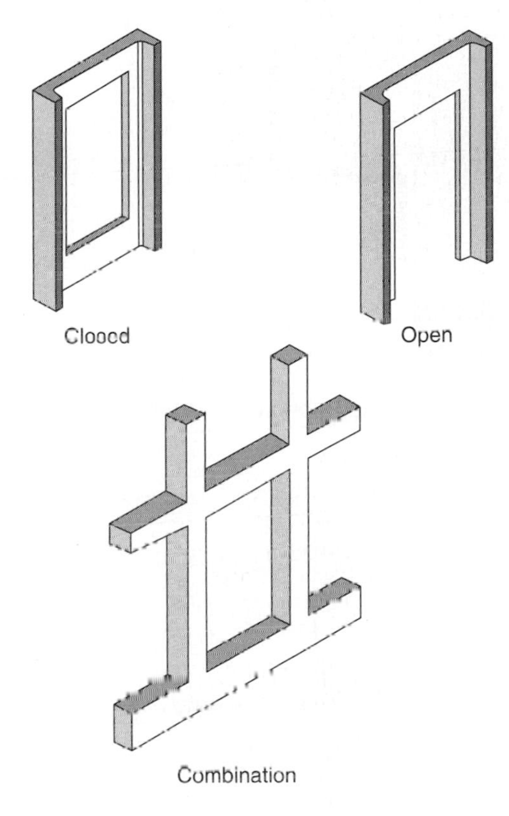


**Figure 9-62** Precast Panel Shapes.

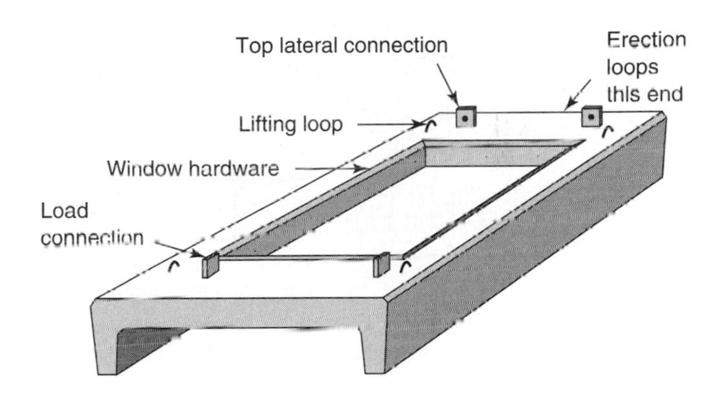


**Figure 9-64** Typical Wall Panel Hardware. (Reprinted by permission of Precast/Prestressed Concrete Institute.)

outer surface (see Figure 9-66). Two-stage joints have a sealant near the outer surface and an air seal that is usually close to the inner face of the panels. Between the two is a chamber that must be vented and drained to the outside. For example, the outer sealant in a two-stage horizontal joint must be broken at intervals to allow any moisture that collects behind it to drain to the outside (see Figure 9-67).

One major advantage of architectural precast concrete panels lies in the great choice of surface finishes available, with regard to both color and surface texture. Colors are varied by using white cement, color pigments, and colored aggregates. A variety of surface textures is produced by casting on a smooth surface, by using a ribbed or textured form face, by casting on a sand-molded surface, by sand blasting, by exposing the aggregate to various degrees on the panel surface, and by grinding. Color or surface texture variations on a single panel are often accentuated by the use of ribs and recesses. Figure 9-68 illustrates a panel with a highly ground and polished surface, whereas in Figure 9-69 the opposite type of surface—an effective change in texture in a panel—is brought about by changing the size of aggregate in one of the mixes and using a recess to outline the change.

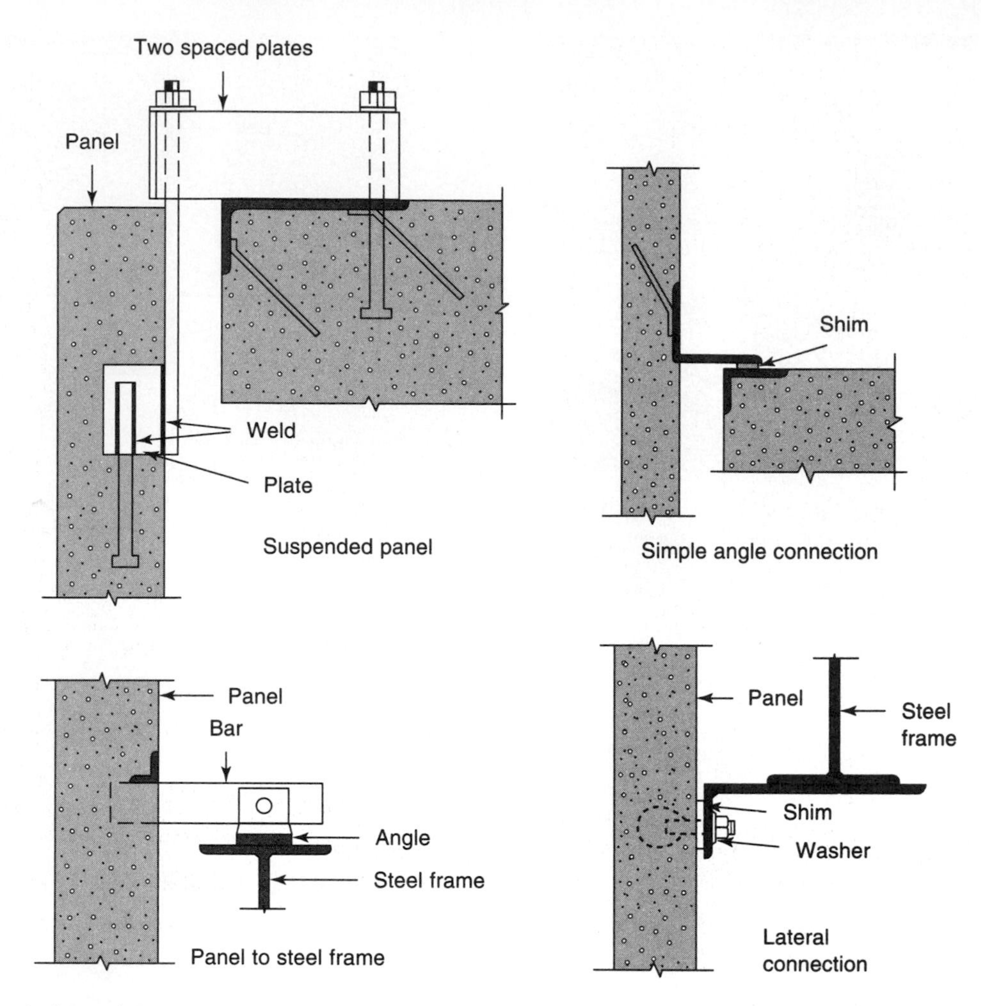


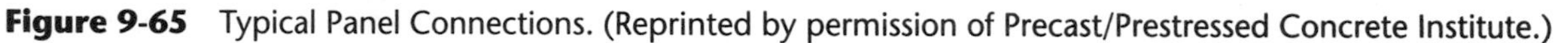

**Figure 9-65** Typical Panel Connections. (Reprinted by permission of Precast/Prestressed Concrete Institute.)

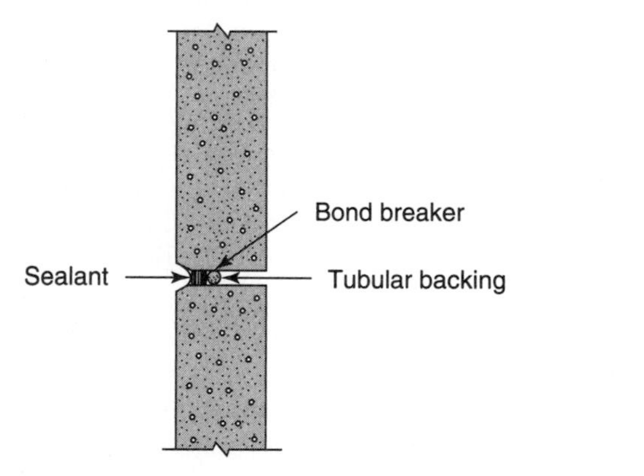


**Figure 9-66** One-Stage Joint. (Reprinted by permission of Precast/Prestressed Concrete Institute.)

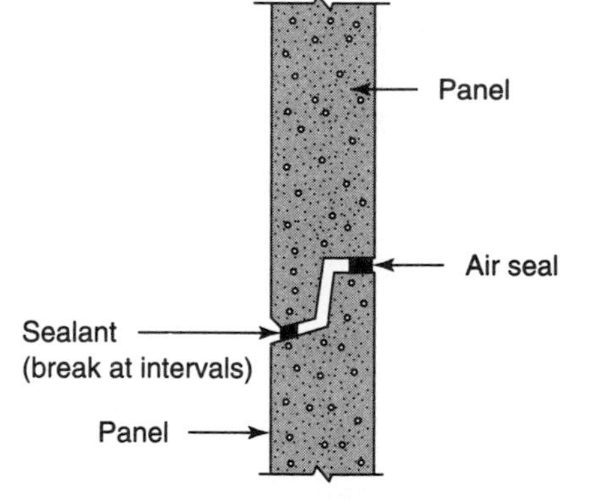


**Figure 9-67** Horizontal Two-Stage Joint. (Reprinted by permission of Precast/Prestressed Concrete Institute.)

The final step in the erection of an architectural precast concrete structure is the cleaning of the exposed surfaces. They should be thoroughly wet before cleaning to prevent cleaning solutions from being absorbed into the concrete. Then cleaning can usually be done with detergents, followed by a thorough rinsing. In some cases an alkali or acid cleaning agent must be used, but before this is done, tests should be made to determine the result of using such an agent.

For a comprehensive understanding of the subject of architectural precast concrete, refer to the ex-

**Figure 9-68** Ground Panel Surface.

**Figure 9-69** Changing Panel Surface Texture.

cellent literature published by the Precast/Prestressed Concrete Institute.

## TILT-UP LOAD-BEARING WALL PANELS

Tilt up construction is a technique for producing precast wall panels for one- or two-story buildings at the job site, rather than in a precast plant. Originally developed for economic reasons, improvements in technique now make it possible to produce panels that are also architecturally pleasing.

Load-bearing panels may be produced by this system and function in exactly the same way as architectural load-bearing members. There are three common types of panels used as load-bearing members, each intended to support a different type of roof system. Panels intended to carry prestressed roof slabs are cast with a continuous ledger near their top edge for this purpose (see Figure 9-70). Panels that will support a steel beam roof frame are cast with weld plates in their top inner surface (see Figure 9-71), whereas panels intended to support open web joists are cast with seated weld plates and shelf angles (see Figure 9-72). If they are two-story

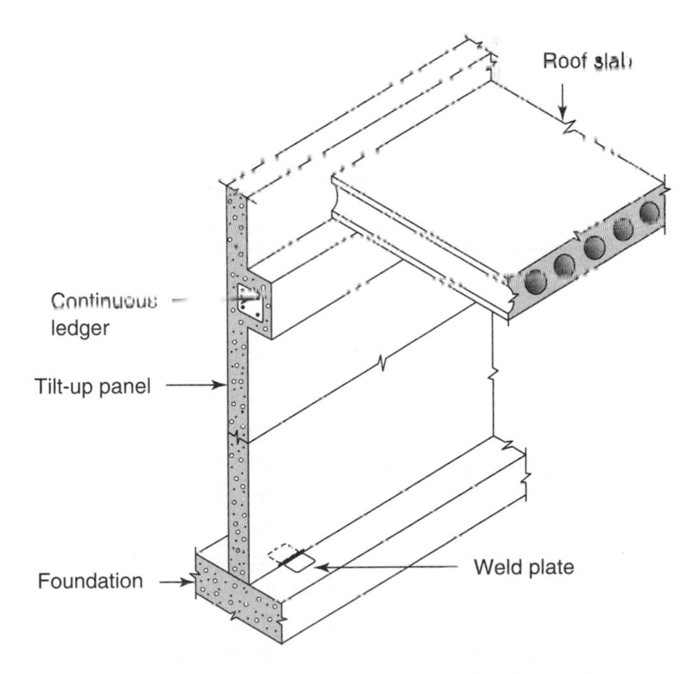


**Figure 9-70** Load-Bearing Tilt-Up Panel with Continuous Ledger.

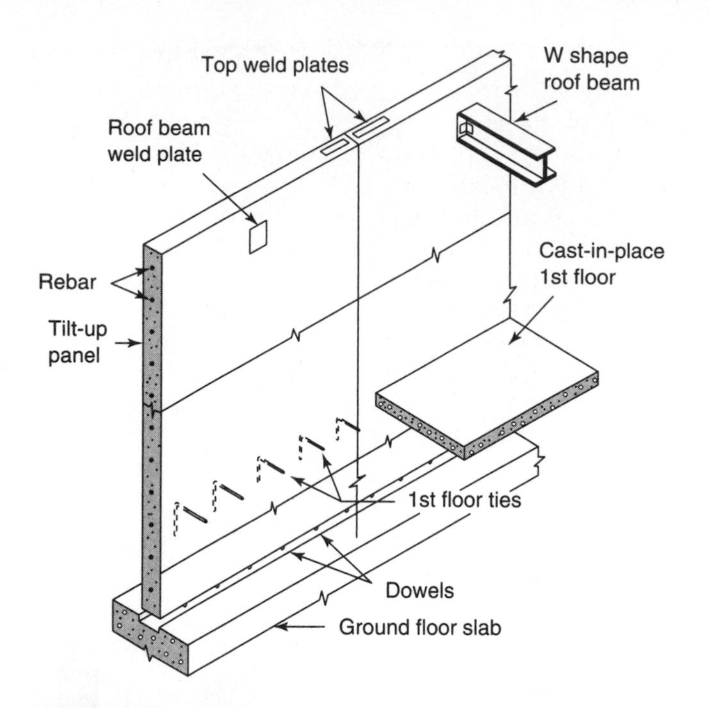

**Figure 9-71** Load-Bearing Tilt-Up Panel with Weld Plates for Roof Beams.

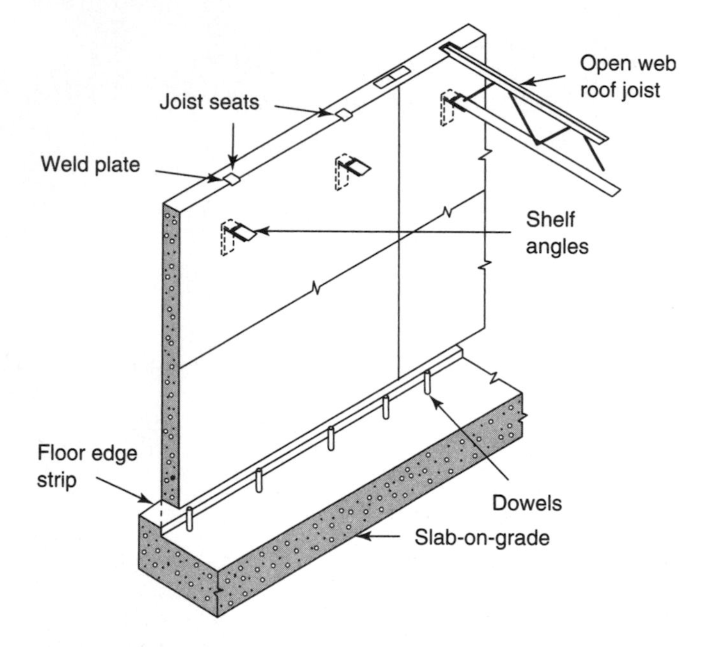


**Figure 9-72** Load-Bearing Tilt-Up Panel with Weld Plates and Shelf Angles.

panels, they may have ties cast into their inner face at the first-floor level to tie into a cast-in-place concrete floor (see Figure 9-71).

At the base, panels may be attached to the foundation or slab by weld plates or dowels or with a floor edge strip and dowels.

The concrete floor slab for the building is often the most convenient place on which to cast tilt-up panels (see Figure 9-73), but other casting beds

**Figure 9-73** Casting Panels on Floor Slab. (Reprinted with permission of Portland Cement Association.)

**Figure 9-74** Hold-Downs for Panel Forms. (Reprinted with permission of Portland Cement Association.)

may be set up nearby if it is not convenient to use the floor.

When the floor slab is used, it must be level, smoothly troweled, and strong enough to carry the materials, trucks or mobile cranes that may be used during casting and erection. It must also be surfaced with a good bond-breaking agent to get a clean, trouble-free lift from the floor.

Panel edge forms must be bolted, pinned, weighted, or otherwise held down (see Figure 9-74). If they have to have holes for rebars, they will usually be split to facilitate removal (see Figure 9-75).

Window and door openings are formed by casting steel window or door frames into the panel or by setting a form buck into the panel form (see Figure 9-76). In either case, such opening forms must be held down securely and braced internally to hold the square. Wood forms and exposed sash should be given a coat of bond-breaking compound.

Reinforcement may be in the form of #4 (10M) bars or welded wire fabric. This size bar is often used. A common requirement for steel area is 0.0015 to 0.0025 of the panel cross-sectional area in each direction. Two #5 (15M) bars are added around openings to strengthen edges and corners. If panels and columns are to be tied together and wire fabric is being used as reinforcement, dowel bars must be added for this purpose. Reinforcing mats are placed on chairs, which will hold them near the center of the panel. This position, in most cases, will provide the greatest strength for the panel (see Figure 9-75).

Concrete is placed and finished in the same way as for a floor slab. Finishing techniques will depend on the texture desired. Curing is begun as soon as possible and may be carried out by any of the accepted methods that are feasible in such a situation.

The method of tilting up depends on whether the crane can operate on the floor slab and on whether the panel has been cast face up or face down (see Figure 9-77). Lifting will be done by the use of

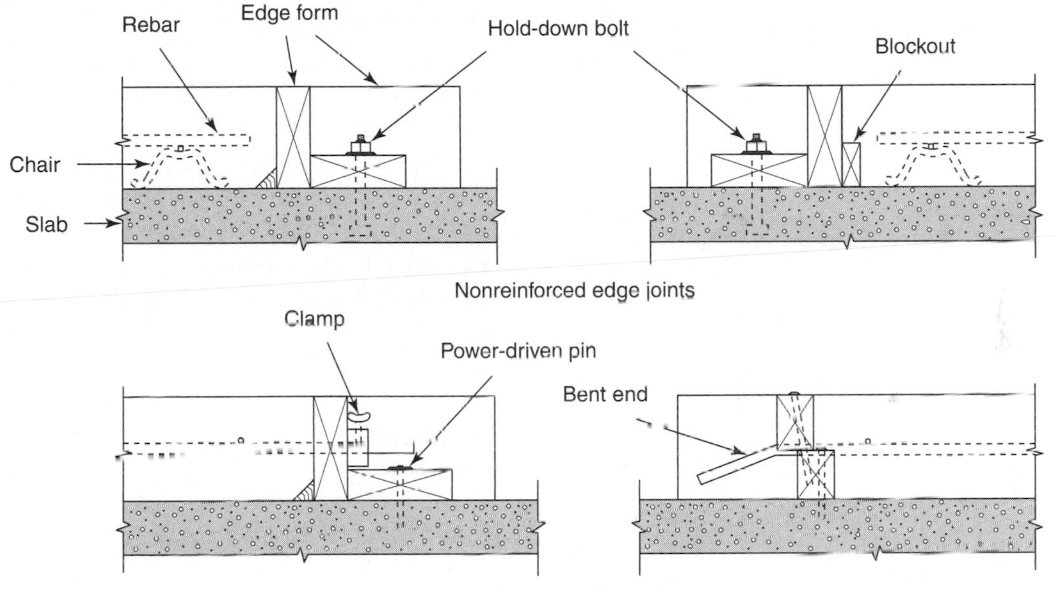


**Figure 9-75** Panel Edge Forms

**Figure 9-76** Openings Cast in Tilt-Up Panels. (Reprinted with permission of Portland Cement Association.)

**Figure 9-77** Lifting Tilt-Up Panel. (Reprinted with permission of Portland Cement Association.)

inserts and bolts (see Figure 9-63) or, in some cases, by the use of vacuum lifters.

Surface treatments and colors for tilt-up panels may be almost as varied as those found with architectural precast panels. Color variations may be achieved by using color pigments or white cement, and decorative designs are produced by using wood strips, form liners of foamed plastic, or rubber matting and by exposing aggregates. See Chapter 14 for details of surface treatments. For further details on tilt-up construction, consult the literature published by the Portland Cement Association.

## Review Questions

**1.** What is the primary purpose of reinforcement in **(a)** concrete girders and beams? **(b)** concrete columns?

**2.** What property of concrete is used when analyzing reinforced concrete sections?

**3.** What advantages can be achieved by using rigid connections between members in a structural frame?

**4.** What two considerations must be made when placing bars in a reinforced concrete section?

**5.** What three methods are used in splicing reinforcing bars?

**6.** What is the reason for including a window in a column form?

**7.** Twenty-four pieces of 2″ × 6″ (38 × 140 mm) planking are to be used as cribbing to make a circular wooden form. How much bevel should there be on the edges of the cribbing pieces?

**8.** What is the purpose of the following items used in beams and girder forming: **(a)** T-head shore, **(b)** ledger strip, **(c)** kicker, and **(d)** vertical stiffener?

**9.** What is a spandrel beam?

**10.** What two types of reinforcing are used in reinforced concrete beams? Explain the reason for each.

**11.** What is the purpose of the following items used to reinforce beams or columns: **(a)** stirrups, **(b)** chairs, and **(c)** lateral ties?

**12.** Differentiate between **(a)** normal reinforcing and prestressing, and **(b)** pretensioning and posttensioning.

**13.** What factors are usually considered to determine whether a prestressed member will be pretensioned or posttensioned?

**14.** Outline three major advantages of using precast structural members instead of a cast-in-place concrete frame.

**15.** Differentiate between rigid and simply supported connections when joining precast structural members.

# 10 STRUCTURAL STEEL FRAME

The use of structural steel in the construction of building frames was the beginning of a new concept in building design, the high-rise or multistory building. Because of its versatility and strength as a building material, steel has become one of the dominant materials in the construction industry.

The properties of steel make it one of the most efficient materials available for use in building frames. It has a high strength to weight ratio, it is ductile and equally strong in compression and in tension, and it can be formed and fabricated into any shape without experiencing any detrimental effects to its physical properties. Even materials such as concrete and timber rely on steel components to enhance their performance as structural materials.

Constant improvements in methods of production and fabrication have rendered its use virtually limitless, as is evident from the diversity of steel buildings being constructed (Figures 10-1 and 10-2).

## STRUCTURAL STEEL SHAPES

The most common steel section shapes used in the fabrication and construction of structural steel frames are W (wide flange), HSS (hollow structural sections), WT (tees cut from wide flange sections), S (standard I sections), angle (equal leg and unequal leg), C (channels), HP (I-shaped pile sections), and plate. All shapes are available in a great variety of sizes (in the case of W, S, and C shapes, the nominal size represents the approximate depth of the section) (see Figure 10-3), and each size is manufactured in several different masses per lineal meter. Differences in mass are produced by altering the thickness of the flanges and the thickness of the webs in the case of S shapes, the width and thickness of flanges and the thickness of webs in the case of W shapes, and the thickness of channel webs, the thickness of the legs of angles, and the thickness of the shell of HSS. The American Institute of Steel Construction (standard units) and the Canadian

**Figure 10-1** Steel Frame High-Rise Office Building. (Reprinted by permission of Bethlehem Steel Co.)

**Figure 10-2** Steel Framework for Factory Building. (Reprinted by permission of Bethlehem Steel Co.)

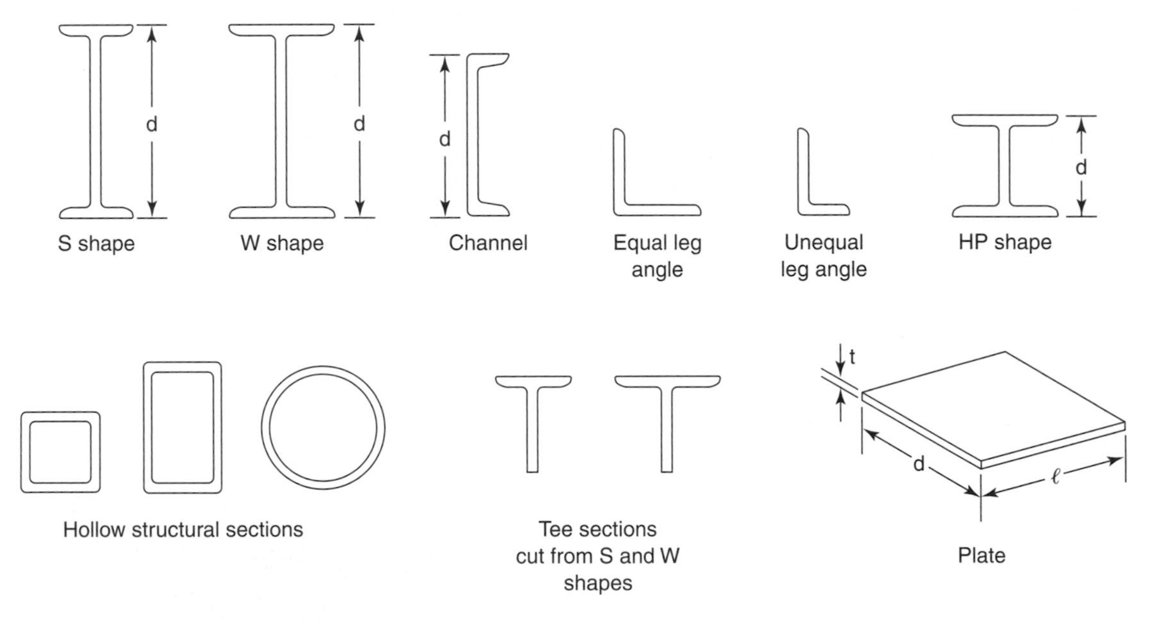


**Figure 10-3** Standard Rolled Structural Shapes.

Institute of Steel Construction (metric units) publish manuals dealing with all aspects of steel construction. Tables 10-1(a) and (b) and 10-2 illustrate the type of information that is provided. Estimating data, detailing information, section properties of hot rolled and cold rolled sections, and fabrication and erection guidelines are all thoroughly covered.

The arrival of structural steel on the construction site is the beginning of the second phase of a two-part process. The first phase consists of the design, detailing, and fabrication of structural shapes into beams, columns, trusses, and bracing. Upon its arrival at the site, each piece of structural steel must be the right length, have the necessary bolt holes, and have all the detail pieces attached (see Figure 10-4). During fabrication, each piece is given an *erection mark* that carries it through the fabrication process and serves to

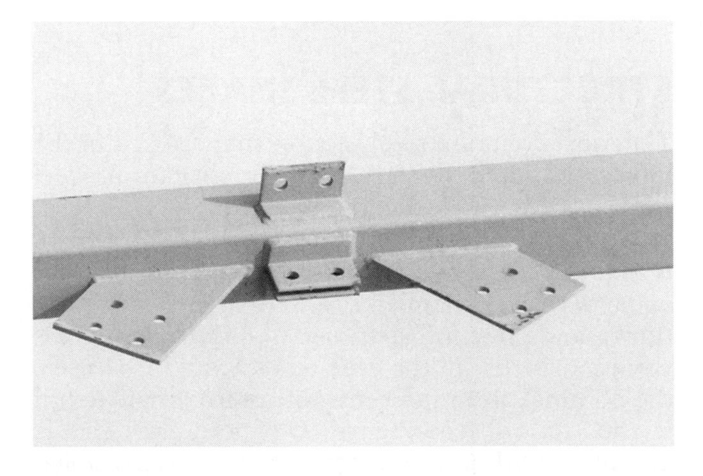

**Figure 10-4** Connection Gusset Plates on a Bracing Member.

**Table 10-1(a)**
**DESIGN PROPERTIES OF TYPICAL W SHAPES, STANDARD UNITS.**

W Shapes Dimensions | W Shapes Properties

| Designation | Area $A$ | Depth $d$ | | Web Thickness $t_w$ | | Web $\frac{t_w}{2}$ | Flange Width $b_f$ | | Flange Thickness $t_f$ | | Distance $T$ | Distance $k$ | Distance $k_1$ | Nominal Wt. per ft | Compact Section Criteria $\frac{b_f}{2t_f}$ | $F_y'$ | $\frac{d}{t_w}$ | $F_y'''$ | $r_T$ | $\frac{d}{A_f}$ | Elastic Properties Axis X-X $I$ | $S$ | $r$ | Axis Y-Y $I$ | $S$ | $r$ | Plastic Modulus $Z_x$ | $Z_y$ |
|---|---|---|---|---|---|---|---|---|---|---|---|---|---|---|---|---|---|---|---|---|---|---|---|---|---|---|---|---|
| | in.$^2$ | in. | | in. | | in. | in. | | in. | | in. | in. | in. | lb. | | Ksi | | Ksi | in. | | in.$^4$ | in.$^3$ | in. | in.$^4$ | in.$^3$ | in. | in.$^3$ | in.$^3$ |
| W 14 × 132 | 38.8 | 14.66 | 14⅝ | 0.645 | ⅝ | 5/16 | 14.725 | 14¾ | 1.030 | 1 | 11¼ | 1 11/16 | 15/16 | 132 | 7.1 | — | 22.7 | — | 4.05 | 0.97 | 1530 | 209 | 6.28 | 548 | 74.5 | 3.76 | 234 | 113 |
| × 120 | 35.3 | 14.48 | 14½ | 0.590 | 9/16 | 5/16 | 14.670 | 14⅝ | 0.940 | 15/16 | 11½ | 1⅝ | 15/16 | 120 | 7.8 | — | 24.5 | — | 4.04 | 1.05 | 1380 | 190 | 6.24 | 495 | 67.5 | 3.74 | 212 | 102 |
| × 109 | 32.0 | 14.32 | 14⅜ | 0.525 | ½ | ¼ | 14.605 | 14⅝ | 0.860 | ⅞ | 11¼ | 1 9/16 | ⅞ | 109 | 8.5 | 58.6 | 27.3 | — | 4.02 | 1.14 | 1240 | 173 | 6.22 | 447 | 61.2 | 3.73 | 192 | 92.7 |
| × 99 | 29.1 | 14.16 | 14⅛ | 0.485 | ½ | ¼ | 14.565 | 14⅝ | 0.780 | ¾ | 11¼ | 1 7/16 | ⅞ | 99 | 9.3 | 48.5 | 29.2 | — | 4.00 | 1.25 | 1110 | 157 | 6.17 | 402 | 55.2 | 3.71 | 173 | 83.6 |
| × 90 | 26.5 | 14.02 | 14 | 0.440 | 7/16 | ¼ | 14.520 | 14½ | 0.710 | 11/16 | 11½ | 1⅜ | ⅞ | 90 | 10.2 | 40.4 | 31.9 | — | 3.99 | 1.36 | 999 | 143 | 6.14 | 362 | 49.9 | 3.70 | 157 | 75.6 |
| W 14 × 82 | 24.1 | 14.31 | 14¼ | 0.510 | ½ | ¼ | 10.130 | 10⅛ | 0.855 | ⅞ | 11 | 1⅜ | 1 | 82 | 5.9 | — | 28.1 | — | 2.74 | 1.65 | 882 | 123 | 6.05 | 148 | 29.3 | 2.48 | 139 | 44.8 |
| × 74 | 21.8 | 14.17 | 14⅛ | 0.450 | 7/16 | ¼ | 10.070 | 10⅛ | 0.785 | 13/16 | 11 | 1 5/16 | 15/16 | 74 | 6.4 | — | 31.5 | — | 2.72 | 1.79 | 796 | 112 | 6.04 | 134 | 26.6 | 2.48 | 126 | 40.6 |
| × 68 | 20.0 | 14.04 | 14 | 0.415 | 7/16 | ¼ | 10.035 | 10 | 0.720 | ¾ | 11 | 1½ | 15/16 | 68 | 7.0 | — | 33.8 | 57.7 | 2.71 | 1.94 | 723 | 103 | 6.01 | 121 | 24.2 | 2.46 | 115 | 36.9 |
| × 61 | 17.9 | 13.89 | 13⅞ | 0.375 | ⅜ | 3/16 | 9.995 | 10 | 0.645 | ⅝ | 11 | 1 7/16 | 15/16 | 61 | 7.7 | — | 37.0 | 48.1 | 2.70 | 2.15 | 640 | 92.2 | 5.98 | 107 | 21.5 | 2.45 | 102 | 32.8 |
| W 14 × 53 | 15.6 | 13.92 | 13⅞ | 0.370 | ⅜ | 3/16 | 8.060 | 8 | 0.660 | 11/16 | 11 | 1 7/16 | 15/16 | 53 | 6.1 | — | 37.6 | 46.7 | 2.15 | 2.62 | 541 | 77.8 | 5.89 | 57.7 | 14.3 | 1.92 | 87.1 | 22.0 |
| × 48 | 14.1 | 13.79 | 13¾ | 0.340 | 5/16 | 3/16 | 8.030 | 8 | 0.595 | ⅝ | 11 | 1⅜ | ⅞ | 48 | 6.7 | — | 40.6 | 40.2 | 2.13 | 2.89 | 485 | 70.3 | 5.85 | 51.4 | 12.8 | 1.91 | 78.4 | 19.6 |
| × 43 | 12.6 | 13.66 | 13⅝ | 0.305 | 5/16 | 3/16 | 7.995 | 8 | 0.530 | ½ | 11 | 1 5/16 | ⅞ | 43 | 7.5 | — | 44.8 | 32.9 | 2.12 | 3.22 | 428 | 62.7 | 5.82 | 45.2 | 11.3 | 1.89 | 69.6 | 17.3 |
| W 14 × 38 | 11.2 | 14.10 | 14⅛ | 0.310 | 5/16 | 3/16 | 6.770 | 6¾ | 0.515 | ½ | 12 | 1 1/16 | ⅝ | 38 | 6.6 | — | 45.5 | 31.9 | 1.77 | 4.04 | 385 | 54.6 | 5.87 | 26.7 | 7.88 | 1.55 | 61.5 | 12.1 |
| × 34 | 10.0 | 13.98 | 14 | 0.285 | 5/16 | 3/16 | 6.745 | 6¾ | 0.455 | 7/16 | 12 | 1 | ⅝ | 34 | 7.4 | — | 49.1 | 27.4 | 1.76 | 4.56 | 340 | 48.6 | 5.83 | 23.3 | 6.91 | 1.53 | 54.6 | 10.6 |
| × 30 | 8.85 | 13.84 | 13⅞ | 0.270 | ¼ | ⅛ | 6.730 | 6¾ | 0.335 | ⅜ | 12 | 15/16 | ⅝ | 30 | 8.7 | 55.3 | 51.3 | 25.1 | 1.74 | 5.34 | 291 | 42.0 | 573 | 19.6 | 5.82 | 1.49 | 47.3 | 8.99 |
| W 14 × 26 | 7.69 | 13.91 | 13⅞ | 0.255 | ¼ | ⅛ | 5.025 | 5 | 0.420 | 7/16 | 12 | 15/16 | 9/16 | 26 | 6.0 | — | 54.5 | 22.2 | 1.28 | 6.59 | 245 | 35.3 | 5.65 | 8.91 | 3.54 | 1.08 | 40.2 | 5.54 |
| × 22 | 6.49 | 13.74 | 13¾ | 0.230 | ¼ | ⅛ | 5.000 | 5 | 0.335 | 5/16 | 12 | ⅞ | 9/16 | 22 | 7.5 | — | 59.7 | 18.5 | 1.25 | 8.20 | 199 | 29.0 | 5.54 | 7.00 | 2.80 | 1.04 | 33.2 | 4.39 |

*Source:* American Institute of Steel Construction, Inc. Reprinted with permission. 

## Table 10-1(b)
## DESIGN PROPERTIES OF TYPICAL W SHAPES, METRIC UNITS

W shapes
W610-W530

Properties

| Designation‡ | Dead Load | Total Area | Axis X-X | | | | Axis Y-Y | | | | Torsional Constant | Warping Constant |
|---|---|---|---|---|---|---|---|---|---|---|---|---|
| | | | $I_x$ | $S_x$ | $r_x$ | $Z_x$ | $I_y$ | $S_y$ | $r_y$ | $Z_y$ | $J$ | $C_w$ |
| | kN/m | mm$^2$ | 10$^6$mm$^4$ | 10$^3$mm$^3$ | mm | 10$^3$mm$^3$ | 10$^6$mm$^4$ | 10$^3$mm$^3$ | mm | 10$^3$mm$^3$ | 10$^3$mm$^4$ | 10$^9$mm$^6$ |
| **W610** | | | | | | | | | | | | |
| ×140 | 1.37 | 17 900 | 1 120 | 3 630 | 250 | 4 150 | 45.1 | 392 | 50.2 | 613 | 2 180 | 3 990 |
| ×125 | 1.22 | 15 900 | 985 | 3 220 | 249 | 3 670 | 39.3 | 343 | 49.7 | 535 | 1 540 | 3 450 |
| ×113 | 1.11 | 14 400 | 875 | 2 880 | 247 | 3 290 | 34.3 | 300 | 48.8 | 469 | 1 120 | 2 990 |
| ×101 | .997 | 13 000 | 764 | 2 530 | 242 | 2 900 | 29.5 | 259 | 47.6 | 404 | 781 | 2 550 |
| ×91† | .892 | 11 600 | 667 | 2 230 | 240 | 2 560 | 24.8 | 219 | 46.2 | 343 | 577 | 2 130 |
| ×84† | .824 | 10 700 | 613 | 2 060 | 239 | 2 360 | 22.6 | 200 | 46.0 | 313 | 462 | 1 930 |
| **W610** | | | | | | | | | | | | |
| ×92* | .905 | 11 800 | 646 | 2 140 | 234 | 2 510 | 14.4 | 161 | 34.9 | 258 | 710 | 1 250 |
| ×82* | .803 | 10 400 | 560 | 1 870 | 232 | 2 200 | 12.1 | 136 | 34.1 | 218 | 488 | 1 040 |
| **W530** | | | | | | | | | | | | |
| ×599* | 5.87 | 76 300 | 5 070 | 15 300 | 258 | 18 500 | 524 | 3 080 | 82.9 | 4 840 | 122 000 | 44 300 |
| ×543* | 5.33 | 69 300 | 4 490 | 13 900 | 255 | 16 600 | 465 | 2 760 | 81.9 | 4 320 | 92 300 | 38 400 |
| ×496* | 4.86 | 63 200 | 4 010 | 12 600 | 252 | 15 000 | 415 | 2 490 | 81.0 | 3 890 | 71 400 | 33 500 |
| ×447* | 4.38 | 56 900 | 3 530 | 11 300 | 249 | 13 400 | 364 | 2 210 | 80.0 | 3 440 | 53 400 | 28 800 |
| ×409* | 4.01 | 52 100 | 3 170 | 10 300 | 247 | 12 100 | 325 | 1 990 | 79.0 | 3 100 | 41 500 | 25 300 |
| ×370* | 3.50 | 45 500 | 2 780 | 9 220 | 247 | 10 700 | 287 | 1 770 | 79.4 | 2 730 | 29 700 | 21 900 |
| ×331* | 3.25 | 42 200 | 2 480 | 8 360 | 242 | 9 660 | 254 | 1 580 | 77.6 | 2 440 | 22 700 | 19 000 |
| ×300* | 2.93 | 38 200 | 2 210 | 7 550 | 241 | 8 680 | 225 | 1 410 | 76.7 | 2 180 | 17 100 | 16 600 |
| ×272* | 2.66 | 34 600 | 1 970 | 6 830 | 239 | 7 800 | 200 | 1 260 | 76.0 | 1 950 | 12 800 | 14 600 |
| ×248* | 2.42 | 31 400 | 1 780 | 6 220 | 238 | 7 070 | 180 | 1 140 | 75.7 | 1 760 | 9 830 | 13 000 |
| ×219* | 2.14 | 27 900 | 1 510 | 5 390 | 233 | 6 110 | 157 | 986 | 75.0 | 1 520 | 6 420 | 11 000 |
| ×196* | 1.92 | 25 000 | 1 340 | 4 840 | 232 | 5 460 | 139 | 877 | 74.6 | 1 350 | 4 700 | 9 640 |
| ×182* | 1.78 | 23 100 | 1 240 | 4 480 | 232 | 5 040 | 127 | 808 | 74.1 | 1 240 | 3 740 | 8 820 |
| ×165* | 1.62 | 21 100 | 1 110 | 4 060 | 229 | 4 550 | 114 | 726 | 73.5 | 1 110 | 2 830 | 7 790 |
| ×150* | 1.47 | 19 200 | 1 010 | 3 710 | 229 | 4 150 | 103 | 659 | 73.2 | 1 010 | 2 160 | 7 030 |
| **W530** | | | | | | | | | | | | |
| ×138 | 1.35 | 17 600 | 861 | 3 140 | 221 | 3 610 | 38.7 | 362 | 46.9 | 569 | 2 500 | 2 670 |
| ×123 | 1.20 | 15 700 | 761 | 2 800 | 220 | 3 210 | 33.8 | 319 | 46.4 | 499 | 1 800 | 2 310 |
| ×109 | 1.06 | 13 900 | 667 | 2 480 | 219 | 2 830 | 29.5 | 280 | 46.1 | 437 | 1 260 | 2 000 |
| ×101 | .995 | 12 900 | 617 | 2 300 | 219 | 2 620 | 26.9 | 256 | 45.7 | 400 | 1 020 | 1 820 |
| ×92 | .907 | 11 800 | 552 | 2 070 | 216 | 2 360 | 23.8 | 228 | 44.9 | 355 | 762 | 1 590 |
| ×82† | .808 | 10 500 | 479 | 1 810 | 214 | 2 070 | 20.3 | 194 | 44.0 | 303 | 530 | 1 340 |
| ×72† | .705 | 9 160 | 402 | 1 530 | 209 | 1 760 | 16.2 | 156 | 42.1 | 245 | 344 | 1 060 |
| **W530** | | | | | | | | | | | | |
| ×85* | .830 | 10 800 | 485 | 1 810 | 212 | 2 100 | 12.6 | 152 | 34.2 | 242 | 737 | 849 |
| ×74* | .733 | 9 520 | 411 | 1 550 | 208 | 1 810 | 10.4 | 125 | 33.1 | 200 | 480 | 692 |
| ×66* | .644 | 8 370 | 351 | 1 340 | 205 | 1 560 | 8.57 | 104 | 32.0 | 166 | 320 | 565 |

‡Nominal depth in millimeters and mass in kilograms per meter.
*Not available from Canadian mills.
†Produced exclusively by Algoma Steel.
*Source:* Reproduced by permission of the Canadian Institute of Steel Construction.

**Table 10-2**
**TYPICAL GAUGE DISTANCES USED IN DETAILING BOLTED CONNECTIONS, METRIC UNITS**

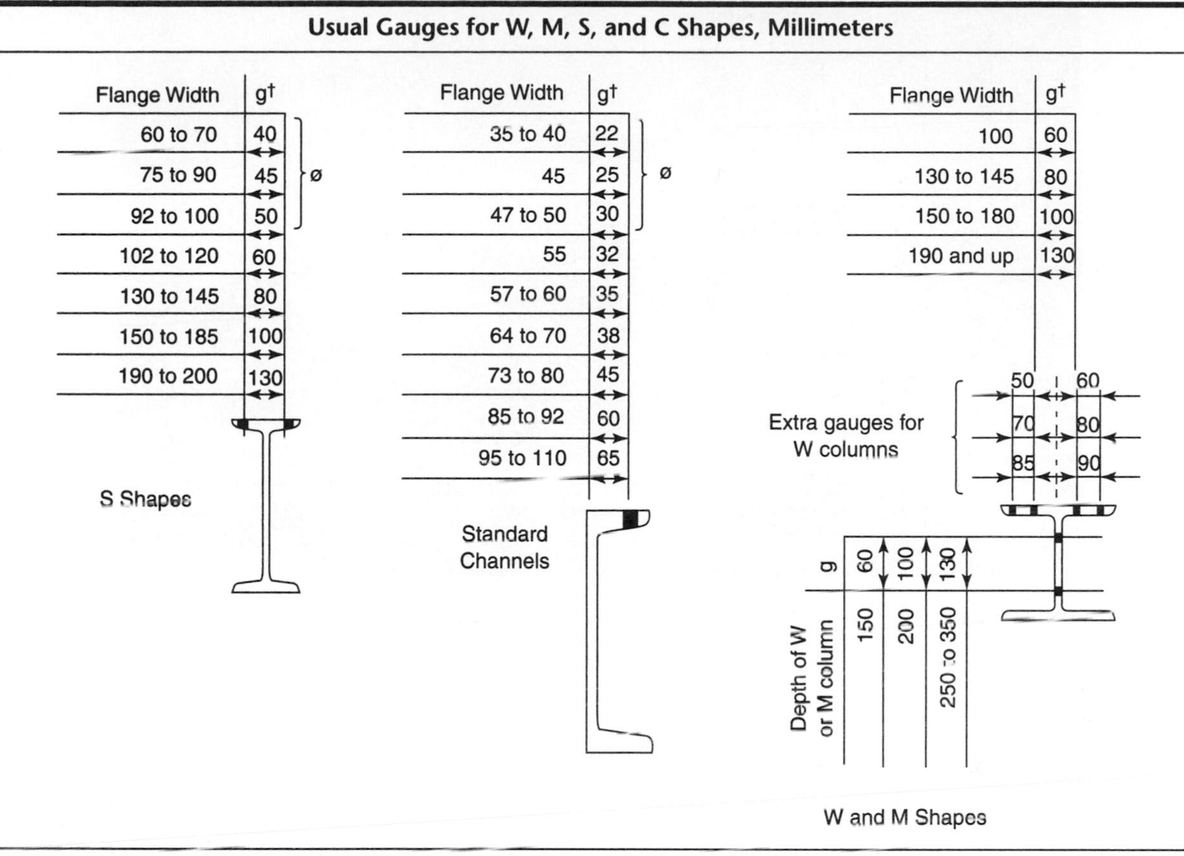

Usual Gauges for W, M, S, and C Shapes, Millimeters

S Shapes

| Flange Width | g† |
|---|---|
| 60 to 70 | 40 ø |
| 75 to 90 | 45 ø |
| 92 to 100 | 50 ø |
| 102 to 120 | 60 |
| 130 to 145 | 80 |
| 150 to 185 | 100 |
| 190 to 200 | 130 |

Standard Channels

| Flange Width | g† |
|---|---|
| 35 to 40 | 22 ø |
| 45 | 25 ø |
| 47 to 50 | 30 ø |
| 55 | 32 |
| 57 to 60 | 35 |
| 64 to 70 | 38 |
| 73 to 80 | 45 |
| 85 to 92 | 60 |
| 95 to 110 | 65 |

W and M Shapes

| Flange Width | g† |
|---|---|
| 100 | 60 |
| 130 to 145 | 80 |
| 150 to 180 | 100 |
| 190 and up | 130 |



ϕ Holes usually drilled due to size of punch die block.
† Dependent on edge distance.
*Source:* Reproduced by permission of the Canadian Institute of Steel Construction.

identify the member on the construction site. The erection mark is usually placed on the left end of the horizontal members to eliminate the possibility of placing the member end for end or upside down. Erection drawings, complete with erection marks, are issued to the erection crew for the location and proper positioning of each piece in the frame.

During the detailing stage, detailers must be aware of the erection sequences used in the field and provide the appropriate connections on beams and columns. Girders or beams can be swung into place from the side or lowered from up above. The type of connection used on one end of the beam must not conflict with the type used at the other end.

## Cranes and Derricks

All of the historic high rise buildings of structural steel were erected with the use of derricks, either stiff-leg or guyed. Although derricks could lift relatively heavy loads, they were cumbersome to set up and move from one place to another, and their reach was limited. With the development of tower cranes and high capacity mobile cranes, guyed derricks and stiff-leg derricks have been relegated to secondary roles such as removing tower cranes from the tops of buildings (see Figure 10-5).

On a historical note, a guyed derrick (see Figure 10-6) consists of a vertical mast and a swinging boom attached to a rotating bull wheel, with the whole assembly fastened to the building frame. Guy lines, running from the guy cap collar at the top of the mast, are anchored at designated points to the foundation or the building superstructure. The stiff-leg derrick (see Figure 10-7) has its mast held erect by a stiff leg, attached to the mast at one end and to a horizontal sill at the other, forming a right triangle. A boom is attached to the bottom of the mast and the whole assembly is mounted on a bull wheel, much like the guyed derrick.

With large capacity tower cranes now available to the contractor, the tower crane has become the workhorse of the construction site. A tower crane (see Figure 10-8) is made up of a vertical tower with a horizontal boom that can rotate 360°. Figure 10-9 illustrates the basic parts of a tower crane. It can be set up within the building or beside the building, depending on material storage requirements and truck access to the construction site. With the operator's

**Figure 10-5** Guyed Derrick used for Removing Tower Crane from Top of High Rise Building.

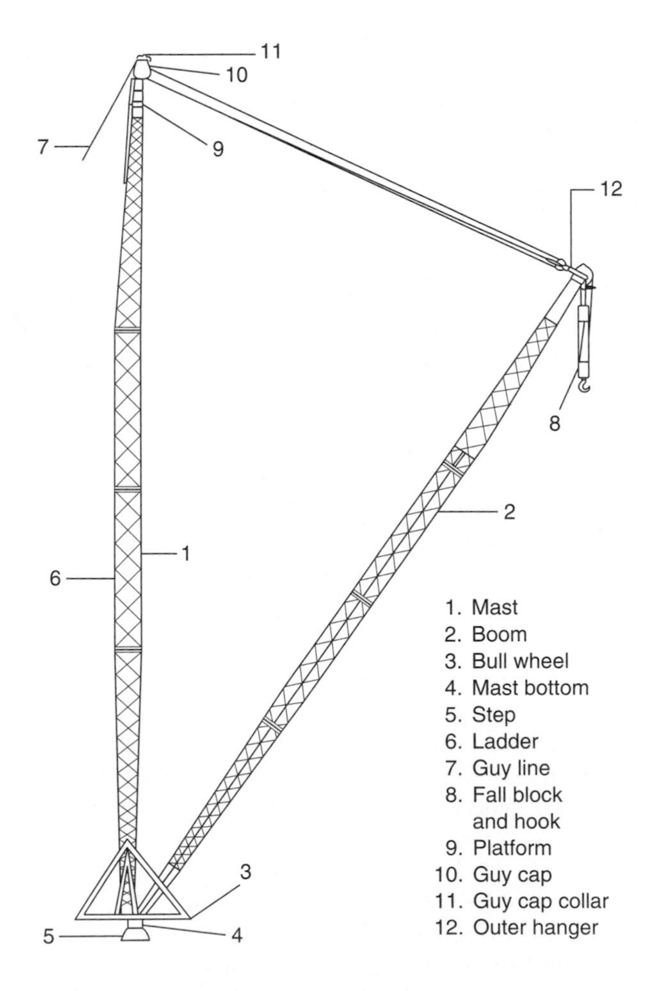


**Figure 10-6** Diagram of Guyed Derrick. (Reprinted by permission of Hydralift AmClyde, Inc.)

cab mounted on the boom, it also provides excellent visibility for the operator.

Mobile cranes, although limited in the height that they can reach by the length of their boom, are most useful because of their versatility on the site (see Figure 10-10). Equipped with hydraulic outriggers and booms, they can be set up and moved quickly, and they can work in relatively small areas. Mobile cranes are preferred for low-rise buildings up to ten stories and on industrial and commercial sites where buildings cover large areas.

## PRINCIPLES OF STRUCTURAL STEEL FRAMES

As with any frame, the structural steel frame must be considered from two points of view: (1) the strength of the individual sections that make up the frame, and (2) the overall stability of the frame. In short, the stability of any structure is contingent on the arrangement of the framing members and the type of connections used to fasten them together.

Structural steel frames can be categorized into two basic types: braced frames and unbraced frames. The concept of a braced frame implies that some type of lateral bracing is used to prevent horizontal movement of the frame. Conversely, in an unbraced frame, the stiffness of the load-carrying sections in the assembled frame must resist the horizontal movement that is anticipated.

Similar to the reinforced concrete frame, the connections between the framing members can be

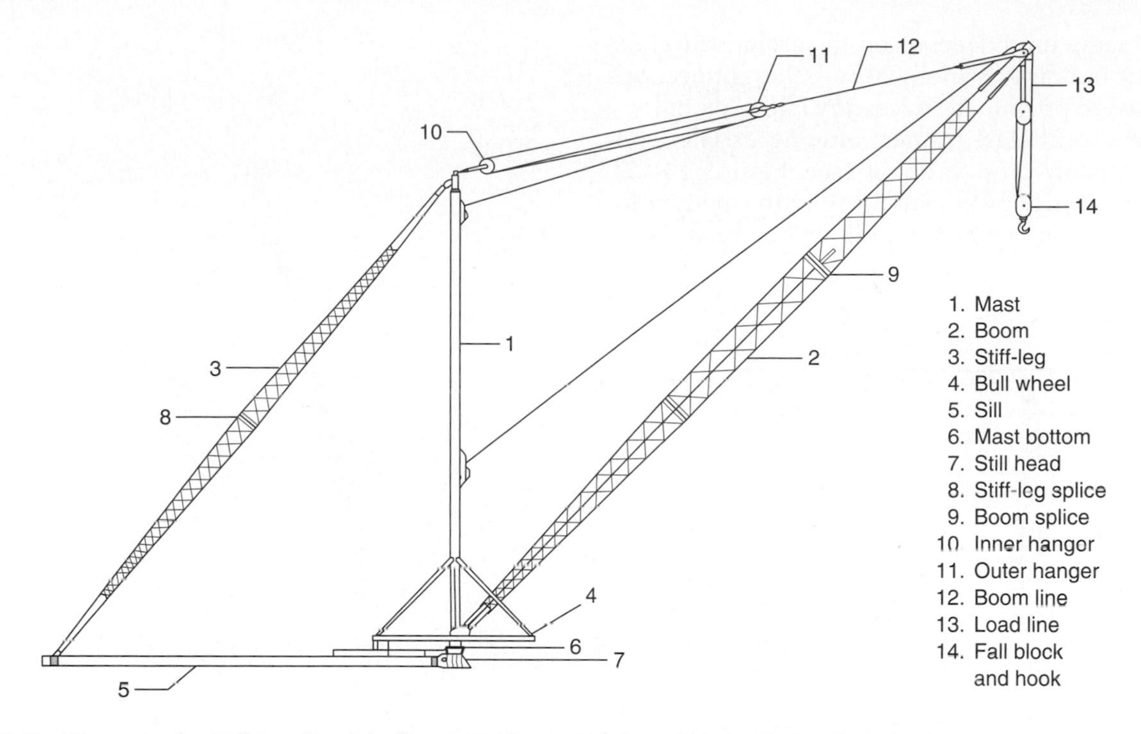


**Figure 10-7** Diagram of a Stiff-Leg Derrick. (Reprinted by permission of Hydralift AmClyde Inc.)

**Figure 10-8** Tower Cranes on Top of a High Rise Building.

considered to be either pin type or moment resisting. Pin-type connections are flexible and allow rotation to occur between the connected members when deflections occur due to applied loads. Rigid or moment-resisting connections add stability to the frame by their ability to resist the rotation that occurs between the frame sections when the frame is under the influence of lateral loads. The amount of lateral movement experienced by the unbraced frame now depends on the relative stiffness of the frame members and the rigidity of the connections.

Consider the structural steel frame in Figure 10-11. Notice that no bracing is apparent, other than some temporary bracing used to ensure that the building frame remains plumb during construction. If the beam and column connections allow rotation, the frame would certainly lack stability. To stabilize such a frame, it is imperative to provide some type of bracing. The type of bracing used depends on the architectural details for the building, or moment-resisting connections can be used to provide the necessary lateral stability.

In structural steel frames, bracing usually consists of tension cross-bracing (see Figure 10-12), provided that it does not interfere with opening locations. This type of bracing allows for load rever-

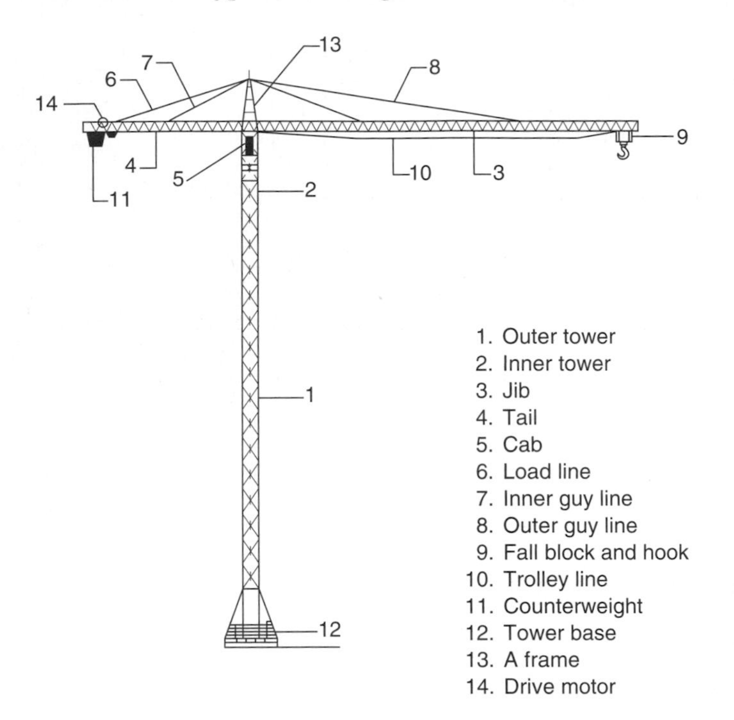


**Figure 10-9** Diagram of Tower Crane.

**Figure 10-10** Mobile Crane with Hydraulic Boom is used for Erecting Tower Crane.

**Figure 10-11** Unbraced Structural Steel Frame.

sals and ensures that stability is provided in both directions. The removal of one of these braces will produce instability in the frame because tension braces are not designed to resist compression loads.

Where moment-resisting connections are used, bracing can be eliminated. A structural steel building using rigid bents as shown in Figure 10-13 is an example of an unbraced frame that develops its stability through its rigid connection at the haunch. However, the connections used to join the bents to one another are considered to be pin type, and bracing is required along the length of the building to ensure stability in that direction. A building using rigid frames such as these embraces the principles applied to a braced frame and an unbraced frame.

In a three-dimensional structure, to ensure stability in all directions, the bracing must be provided in both planes of the structure. In some cases, masonry walls or reinforced concrete stairwells are used to provide lateral stability. Many structural steel frame buildings, especially multistory buildings, use a reinforced concrete core to provide the necessary resistance to lateral forces. Lateral forces are transferred through the floor slabs by diaphragm action to the core, which in turn transfers these loads into the foundations.

Although buildings are stable after the completion of construction, it is during the construction process, when all the connections are not yet fully tightened and all of the required bracing is not yet installed, that particular attention must be given to the stability of the frame. During the construction phase, the frame is subjected to construction forces that can be more severe than those experienced after construction is complete. Concrete pouring, lateral forces due to equipment, and forces due to temperature changes produce, in many cases, unusual and abnormal strains on the partially completed frame. To keep the frame square and plumb during construction, all temporary bracing must be kept in place until removal is allowed by the engineer in charge.

## STRUCTURAL STEEL CONNECTIONS

The two most common methods used for connecting steel members in structural steel frames are *bolting* and *welding*. Experience has shown that welded connections are usually less expensive when done in the shop under controlled conditions, and

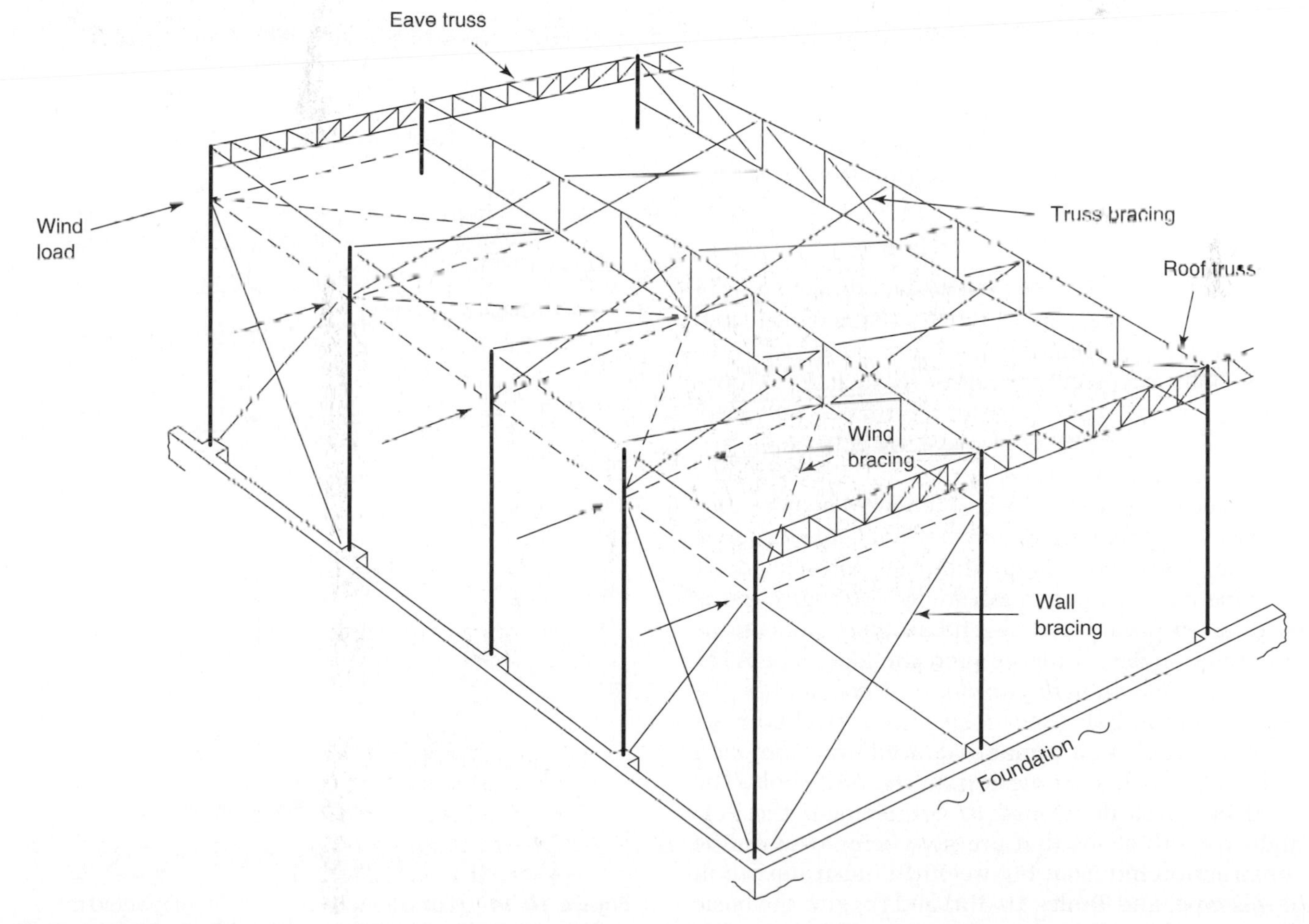


**Figure 10-12** Tension Cross Bracing in a Braced Frame.

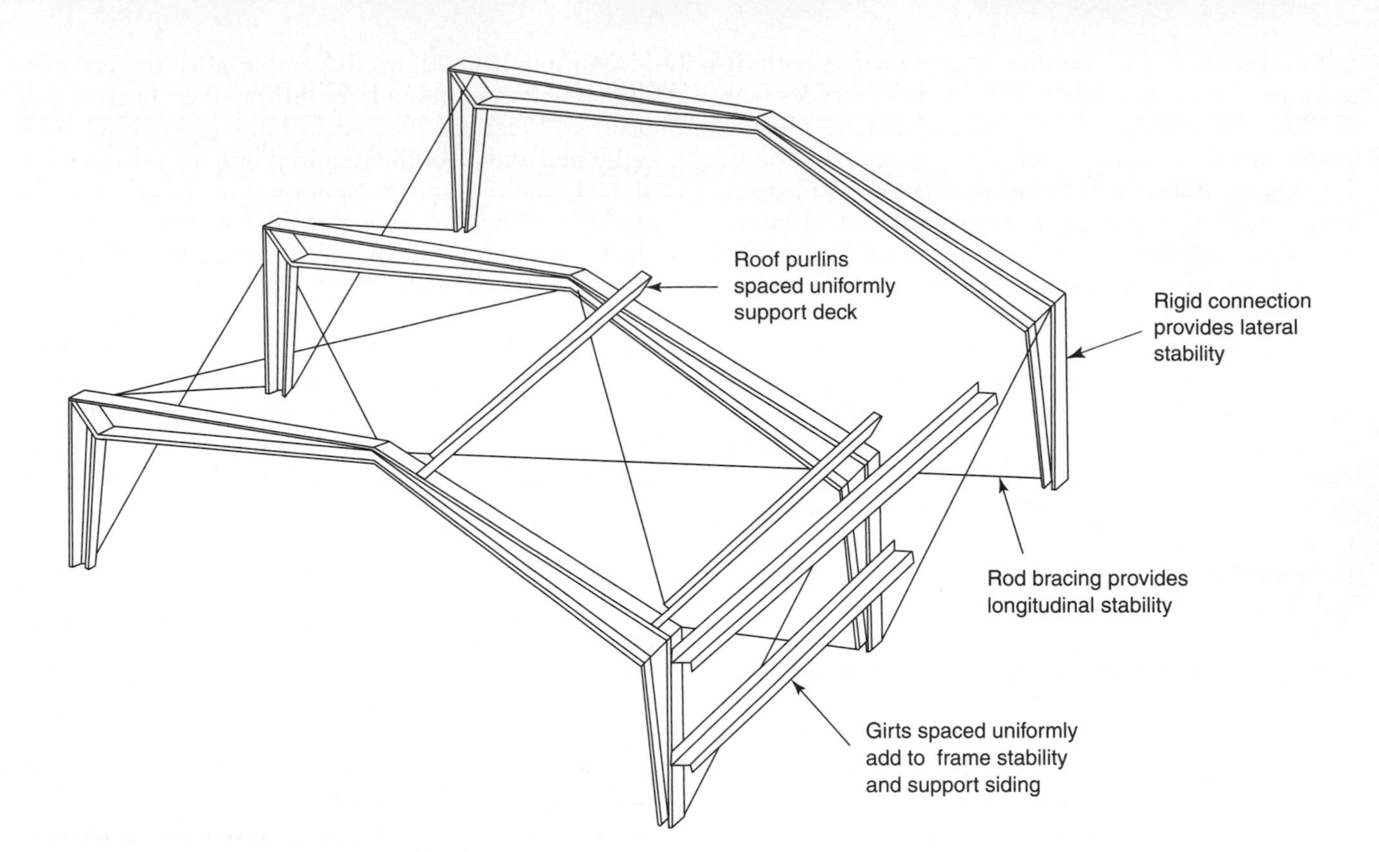


**Figure 10-13** Rigid-Frame Structural Steel Building Frame.

bolted connections are usually better suited to field conditions. In most structural steel buildings, welds and bolts are used in combinations to produce the most practical and economical connections possible.

## Bolts

Structural bolts may be classified according to type of thread, type of steel, steel strength, shape of head and nut, and type of shank, but for structural steel purposes, bolts are usually specified by their ASTM designation. Three types of bolts are usually specified: A307, A325, and A490. They may have either square or hexagonal heads and nuts; hexagonal heads and nuts are preferred for the A325 and the A490 bolts as they are easier to grip with a wrench when being torqued.

The A307 bolt is a standard unfinished or *machine bolt*. It is of relatively low strength and is used in temporary and secondary-type connections only (clip angles, bridging, and the like). The A325 bolt is a *high-strength carbon steel bolt* and is the most common bolt used in structural steel connections. The A490 bolt is manufactured from *alloy steel* and is of a higher strength than the A325 bolt. The A490 bolt was developed to complement the new high-strength steels that are now being used in the construction industry. Figure 10-14 illustrates a bolt of this type, and Tables 10-3(a) and (b) give the basic dimensions for high-strength structural bolts.

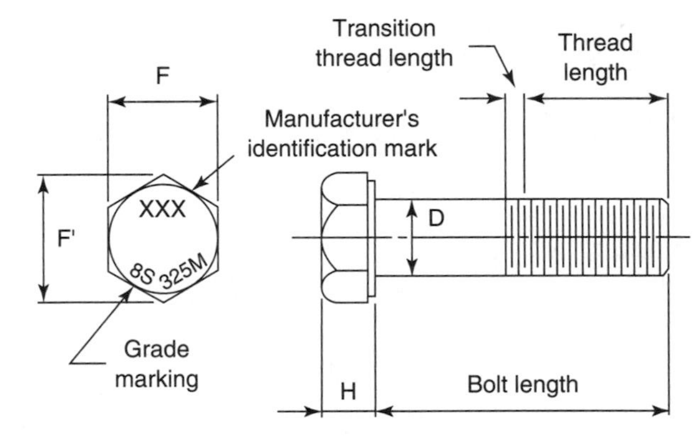

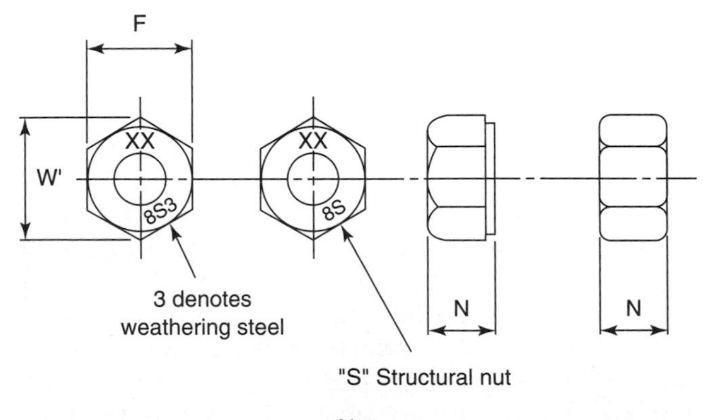


**Figure 10-14** High-Strength Bolt Details. (Reprinted by permission of Canadian Institute of Steel Construction.)

**Table 10-3(a)**
**BASIC DIMENSIONS FOR HIGH-STRENGTH BOLTS, ALL TYPES (INCHES)**

| Nominal Bolt Diameter | Heavy Hex Structural Bolts | | | Heavy Hex Nut[a] | | To Determine Required Bolt Length Add to Grip[b] (in.) | | |
|---|---|---|---|---|---|---|---|---|
| | Thread Length | Width Across Flats Heavy | Height | Width Across Flats | Height | No Washers | 1 Washer | 2 Washers |
| 1/2 | 1 | 7/8 | 5/16 | 7/8 | 31/64 | 11/64 | 27/32 | 1 |
| 5/8 | 1 1/4 | 1 1/16 | 25/64 | 1 1/16 | 39/64 | 7/8 | 1 1/32 | 1 3/16 |
| 3/4 | 1 3/8 | 1 1/4 | 15/32 | 1 1/4 | 47/64 | 1 | 1 5/32 | 1 5/16 |
| 7/8 | 1 1/2 | 1 7/16 | 35/64 | 1 7/16 | 55/64 | 1 1/8 | 1 9/32 | 1 7/16 |
| 1 | 1 3/4 | 1 5/8 | 39/64 | 1 5/8 | 63/64 | 1 1/4 | 1 13/32 | 1 9/16 |
| 1 1/8 | 2 | 1 13/16 | 11/16 | 1 13/16 | 1 7/64 | 1 1/2 | 1 21/32 | 1 13/16 |
| 1 1/4 | 2 | 2 | 25/32 | 2 | 1 7/32 | 1 5/8 | 1 25/32 | 1 15/16 |
| 1 3/8 | 2 1/4 | 2 3/16 | 27/32 | 2 3/16 | 1 11/32 | 1 3/4 | 1 29/32 | 2 1/16 |
| 1 1/2 | 2 1/4 | 2 3/8 | 15/16 | 2 3/8 | 1 15/32 | 1 7/8 | 2 1/32 | 2 3/16 |

[a]May be either washer faced or double chamfered.
[b]To get bolt length required if beveled washers are used, add to grip length the amount in "no washer" column plus 5/16 in. for each beveled washer. The length determined by the use of this table should be adjusted to the next longer 1/4 in.

**Table 10-3(b)**
**BASIC DIMENSIONS FOR HIGH-STRENGTH BOLTS (MILLIMETERS)**

| Nominal Bolt Size | Heavy Hex Bolt or Nut Dimension | | | | Heavy Hex Nut Max. Height N | Heavy Hex Structural Bolt | | | Max. Transition Thread Length |
|---|---|---|---|---|---|---|---|---|---|
| | Across Flats F or W | | Across Corners F′ or W′ | | | Max. Head Height H | Thread Length* | | |
| | Max. | Min. | Max. | Min. | | | Bolt Lengths < 100 | Bolt Lengths > 100 | |
| mm | mm | mm | mm | mm | mm | mm | mm | mm | mm |
| M16 × 2 | 27.00 | 26.16 | 31.18 | 29.56 | 17.1 | 10.75 | 31 | 38 | 6.0 |
| M20 × 2.5 | 34.00 | 33.00 | 39.26 | 37.29 | 20.7 | 13.40 | 36 | 43 | 7.5 |
| M22 × 2.5 | 36.00 | 35.00 | 41.57 | 39.55 | 23.6 | 14.90 | 38 | 45 | 7.5 |
| M24 × 3 | 41.00 | 40.00 | 47.34 | 45.20 | 24.2 | 15.90 | 41 | 48 | 9.0 |
| M27 × 3 | 46.00 | 45.00 | 53.12 | 50.85 | 27.6 | 17.90 | 44 | 51 | 9.0 |
| M30 × 3.5 | 50.00 | 49.00 | 57.74 | 55.37 | 30.7 | 19.75 | 49 | 56 | 10.5 |
| M36 × 4 | 60.00 | 58.80 | 69.28 | 66.44 | 36.6 | 23.55 | 56 | 63 | 12.0 |

*Does not include transition thread length.
Bolt dimensions conform to those listed in ANSI B18.2.3.7M-1979 "Metric Heavy Hex Structural Bolts," and the nut dimensions conform to those listed in CSA Standard B18.2.4.6-M1980 "Metric Heavy Hex Nuts."

Two types of bolted connections are used in structural steel connections: *bearing* type and *friction* type. In the bearing-type connection, the bolted parts bear on the shank of the bolt, thus using the bolt as a direct means of transferring load from one part to the other. Bearing-type connections are used in locations where the applied load acts primarily in one direction without much variation in magnitude. Because the connection depends on the connected parts to bear on the bolt, and the bolt hole being larger than the bolt, changes in direction of the load could produce movement in the connection. This movement could eventually cause the bolted connection to work loose, producing an undesirable situation.

Friction-type connections transfer load from one member to the other by means of the friction developed between the connected surfaces. In this type of connection, the bolt acts as a clamp only and is not used in the transfer of load from one member to the other as in the bearing-type connection. Because the friction between the connected parts determines the capacity of the connection, no movement in the connection is expected. This type of connection can be used where the load direction varies without producing any undesirable effects on the connection.

The bolts in all bolted connections must be properly tightened to produce good results. In a friction-type connection especially, proper bolt tension is of utmost importance because the friction developed between the connected parts depends directly on the amount of bolt tension provided. This bolt tension is achieved by applying the necessary force to the nuts as they are turned down. The required force may be applied by using a *pneumatic impact wrench,* with a clutch set to apply the required tension (see Figure 10-15). Nuts may also be tightened by means of a *torque wrench,* such as that used in Figure 10-16, and the same type of wrench may be used to test the tension on nuts tightened with an impact wrench.

Another method of checking the tension on bolts is by the use of *load indicator washers.* These are hardened steel washers with a series of protrusions on one face that bear against the underside of the bolt head when assembled (see Figure 10-17). When the nut is tightened down, the protrusions are flattened and the gap between the bolt head and the washer reduced. At a specified average gap, measured by a feeler gauge, the bolt tension will not be less than that specified by various standards. Washers are produced for both ASTM A325 and A490 bolts.

Bolt holes for bolted connections are punched or drilled 1/16 in. (2 mm) larger in diameter than the diameter of the bolts to be used, in set patterns normally determined by the size of the connectors and the type of member being connected. The distance between rows of bolts is usually known as the *gauge,* and the center-to-center distance between holes in a row is called the *pitch* (Figure 10-18). Steel manuals generally provide standard gauge and pitch distances for various structural shapes. Examples of the type of information available are indicated in Tables 10-4(a) and (b), in which the usual gauges for angles are given.

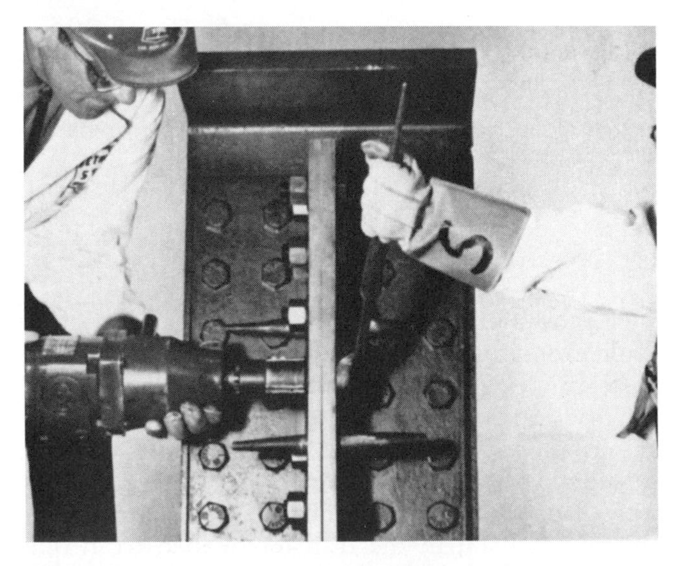

**Figure 10-15** Pneumatic Impact Wrench. (Reprinted by permission of Bethlehem Steel Co.)

**Figure 10-16** Tightening High-Strength Bolts with Torque Wrench. (Reprinted by permission of Bethlehem Steel Co.)

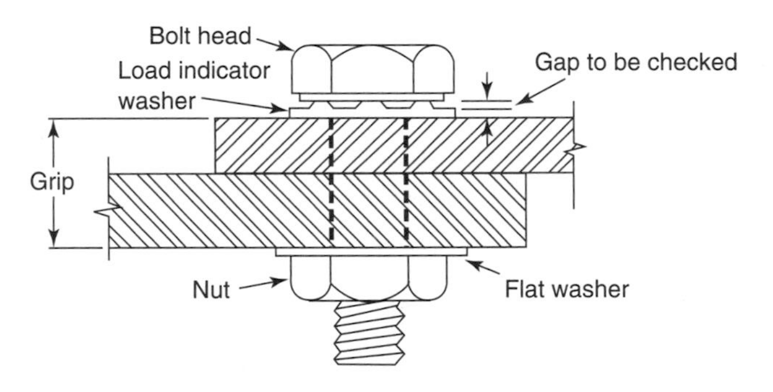


**Figure 10-17** Load Indicator Washer. (Reprinted by permission of Bethlehem Steel Co.)

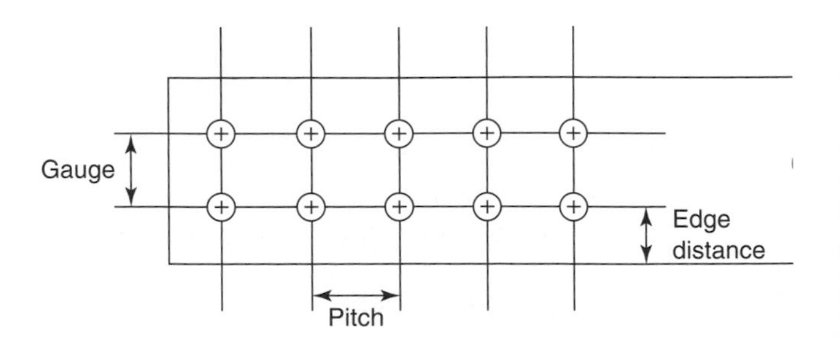


**Figure 10-18** Bolt Hole Gauge and Pitch.

Table 10-4(a)
**USUAL GAUGES FOR ANGLES (in.)**

| | Leg (in.) | | | | | | | | | | | | | |
|---|---|---|---|---|---|---|---|---|---|---|---|---|---|---|
| | **8** | **7** | **6** | **5** | **4** | **3½** | **3** | **2½** | **2** | **1¾** | **1½** | **1⅜** | **1¼** | **1** |
| g | 4½ | 4 | 3½ | 3 | 2½ | 2¼ | 1¾ | 1⅜ | 1⅛ | 1 | ⅞ | ⅞ | ¾ | ⅝ |
| $g_1$ | 3 | 2½ | 2¼ | 2 | | | | | | | | | | |
| $g_2$ | 3 | 3 | 2½ | 1¾ | | | | | | | | | | |

Table 10-4(b)
**USUAL GAUGES FOR ANGLES (mm)**

Notes:
Those values shown above the dashed line allow for full socket wrench clearance requirements.

The bolt sizes shown in italics to the left of g and $g_1$ are the maximum bolt sizes permissible for the dimensions shown.

$g_2 \geq$ 2-2/3 bolt diameters.

| Leg \ Gauge | g | | $g_1$ | | $g_2$ | |
|---|---|---|---|---|---|---|
| 200 | M36 | 115 | M30 | 80 | | 80 |
| 150 | M36 | 90 | M24 | 55 | | 65 |
| 125 | M30 | 80 | M20 | 45 | | 54 |
| 100 | M27 | 65 | | | | |
| 90 | M24 | 60 | | | | |
| 80 | M24 | 50 | | | | |
| 75 | M24 | 45 | | | | |
| 65 | M24 | 35 | | | | |
| 60 | M24 | 30 | | | | |
| 55 | M22 | 27 | | | | |
| 50 | M16 | 28 | | | | |
| 45 | M16 | 23 | | | | |

*Source:* Reproduced by permission of the Canadian Institute of Steel Construction.

## Welds

With the development of good welding practices and procedures, the process of welding has given designers and fabricators an alternative method of connecting structural steel sections.

Two types of welds are most often used in welded connections: the *fillet* weld and the *groove* or *butt* weld. Figure 10-19 illustrates some of the more common groove and fillet welds that are used. Fillet welds are used wherever possible, because no edge preparation is required on the material to be welded. When the full strength of the connected parts is to be developed, a groove weld is usually used. Welds that are used in critical areas must be inspected thoroughly to ensure that no defects are hidden below the surface. X-rays or ultrasonic testing are two methods that can indicate whether the weld is sound or defective.

As much welding as possible is done in the fabricator's shop, where better quality control can be achieved for less cost than when the welding is done on site. High-speed automatic and semiautomatic welding equipment used in the shop is much faster than the manual equipment used in the field. However, extensive welding may be done in the field when necessary. Figure 10-20 shows beam-to-column moment connections that were field welded. In this application the connections were required to develop the full moment capacity of the beams. By using a full-penetration groove weld to connect the beam flanges to the columns, this requirement was achieved without adding additional gussets or stiffeners to the beam. The final result is a neat and efficient connection.

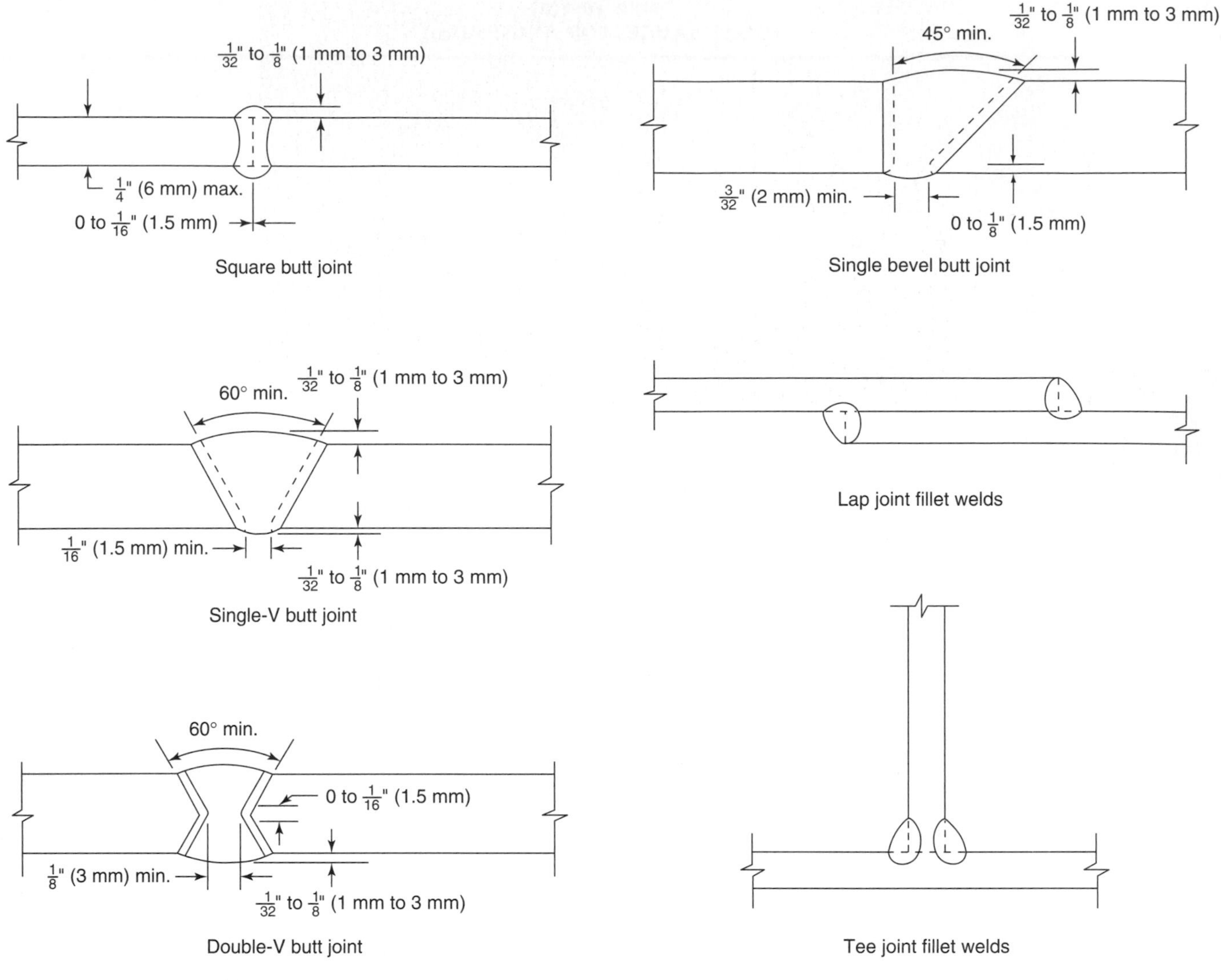


**Figure 10-19** Typical Welded Joints.

**Figure 10-20** Field-Welded Connections. (Reprinted by permission of Bethlehem Steel Co.)

## ERECTION OF STEEL FRAME

Erection of the steel frame normally begins with the completion of the building foundation. Because supporting foundations are usually poured concrete, sufficient time must be allowed for the concrete to cure. This curing period can be used to advantage by the structural steel contractor to check the locations of embedded plates and anchor bolts to which the structural steel will be anchored. Much time can be saved during the erection of the structural steel when foundation elevations and distances between anchor bolts are confirmed.

### Anchor Bolts

The first step in the successful erection of a steel frame depends upon the correct placement of the *anchor bolts* (see Figure 10-21(a)) that tie the steel

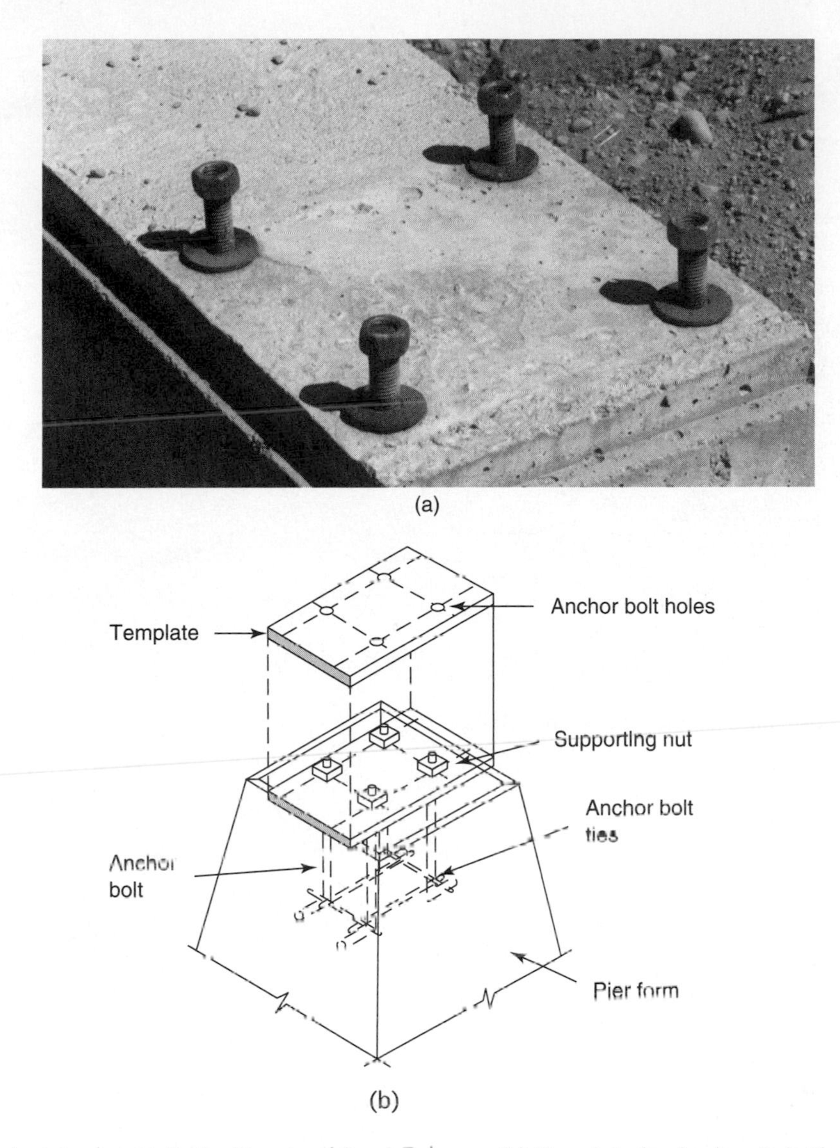


**Figure 10-21** (a) Typical Anchor Bolts for Structural Steel Column, (b) Template for Anchor Bolt Placement.

frame to the concrete foundation. Normally, the general contractor positions the anchor bolts in conjunction with the placement of the reinforcing bars in the concrete formwork. The anchor bolts must be set according to the building grid lines at the correct elevation and must have the correct orientation. To ensure that the anchor bolts are not displaced during the concrete pour, templates are used to position the anchor bolts in the formwork [see Figure 10-21(b)]. The anchor bolts must extend above the top of the finished concrete to allow for minor elevation adjustments to the base plate and provide enough room for the placement of grout between the base plate and the foundation (see Figure 10-22). It is an astute steel erector who checks all anchor bolt settings before the erection crew and heavy equipment arrive on the site.

Anchor bolts vary in diameter and length and their size will depend on the forces that they must resist. For light loads, the anchor bolt is usually ¾ in. (20 mm) in diameter and 16 to 24 in. (400 to 600 mm) in length. Where anchor bolts must resist horizontal loads or resist loads due to uplift, lengths and diameters can be 5 ft (1.5 m) and 3 in. (75 mm) respectively.

## Bearing Plates

The column bearing plates are steel plates of various thicknesses in which holes have been drilled to receive the anchor bolts. The holes are slightly larger than the bolts so that some lateral adjustment of the bearing plate is possible. The angle connections by which the columns will be attached to the bearing plates are bolted or welded on them. Plates are marked on the axis lines of their four edges to allow positioning by a survey instrument used to line them up accurately (see Figure 10-23).

**Figure 10-22** Column Base Plate Secured by Anchor Bolts.

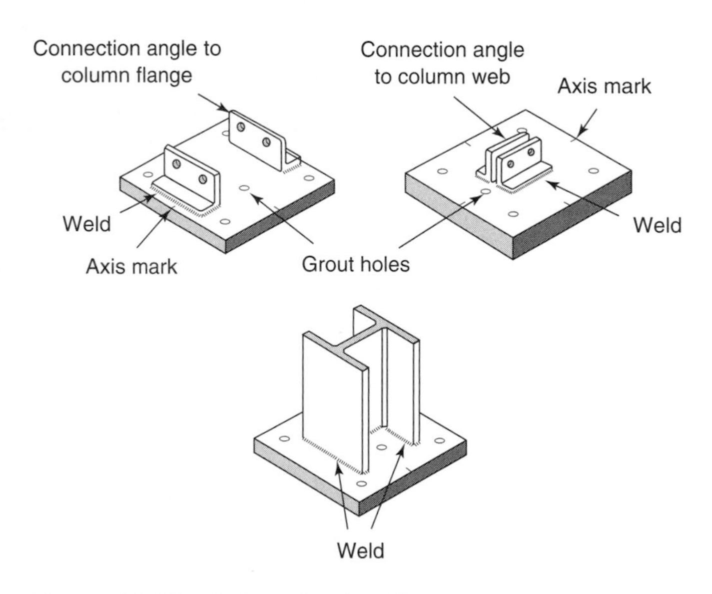


**Figure 10-23** Column Bearing Plates.

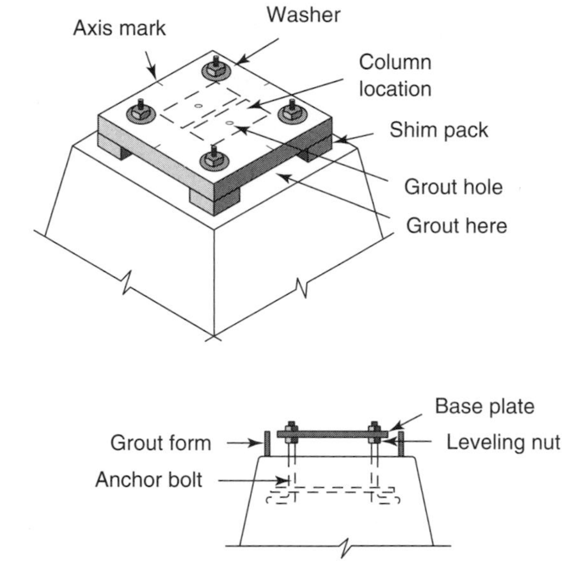


**Figure 10-24** Leveling Base Plates.

*Shim packs* may be set under the four corners of each bearing plate as it is placed. These are 3- to 4-in. (75- to 100-mm) metal squares of thicknesses ranging from 1⁄16 to 3⁄4 in. (1.5 to 20 mm) that are used to bring the bearing plates to the correct grade level and to level each on its own base. The bearing plates are first leveled individually by adjusting the thickness of the shim packs. They are then brought to the proper grade by raising or lowering each by the same amount at all four corners. Finally, they are lined up in both directions by adjusting each on its anchor bolts as necessary.

An alternative to shim packs as a means of leveling base plates is the use of leveling nuts (see Figure 10-24). They are placed on the anchor bolts under the base plate and can be adjusted to level the plate. After it is level, the plate is adjusted horizontally and the top nuts turned down to hold it in place.

After all bearing plates have been set and aligned, they must be *grouted*. This is done by completely filling the space between the bearing plate and the top of the concrete with mortar, preferably a nonshrinking type made with metallic aggregate.

A simple wood form is placed around the bearing plate approximately 2 in. (50 mm) away from its edges. Grout is poured or pumped into the space until it is completely filled. Large bearing plates have grout holes drilled near the center to make sure that no air becomes trapped under them. When grout can be seen in these holes, it indicates that the space has been filled.

## Columns

First-tier columns are the next group of members to be erected. They are often two stories long or, in any case, the same size for at least two stories. Column lengths are such that splices will come 1½ to 2 ft (450 to 600 mm) above the floor levels, which is done to prevent splice connections from interfering with girder or beam-to-column connections. Column ends are milled to exactly the right length and to make sure that column loads will be evenly distributed over the entire bearing area.

HP shapes, as nearly square in cross section as possible (see Figure 10-3), are often chosen for columns. Round or square HSS sections may also be used, though the round column may present some connecting problems. Columns may also be built up by welding or bolting a number of other rolled shapes into a single unit (see Figure 10-25).

## Column Splices

Column sections are joined together by *splice plates* that are bolted or welded to the column flanges, and in special cases, to the webs. If the column flanges match reasonably well, it is common to butt the column ends directly and join the column sections with splice plates, as illustrated in Figure 10-26(a). When the columns are of different sizes, a plate is used between the two column ends to provide bearing for the smaller column. The smaller column is bolted in place with positioning bolts and then welded to the bearing plate to complete the splice (see Figure 10-26(b)).

## Girders

Girders, the primary horizontal members of a frame, span from column to column (see Figure 10-27) and support the intermediate floor beams. They carry wall and partition loads and the point loads transmitted to them by the beams. Rolled W shapes or

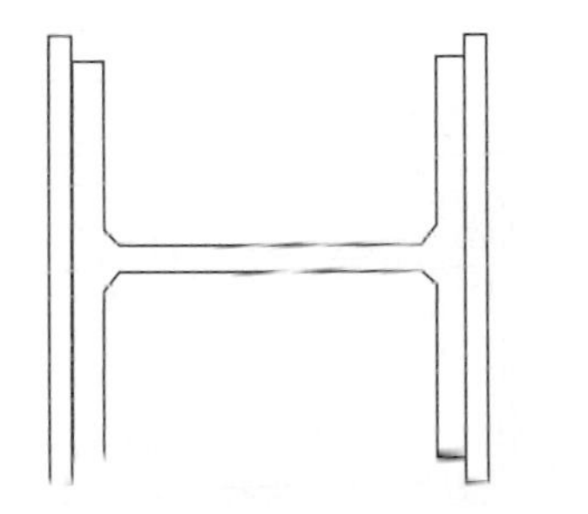


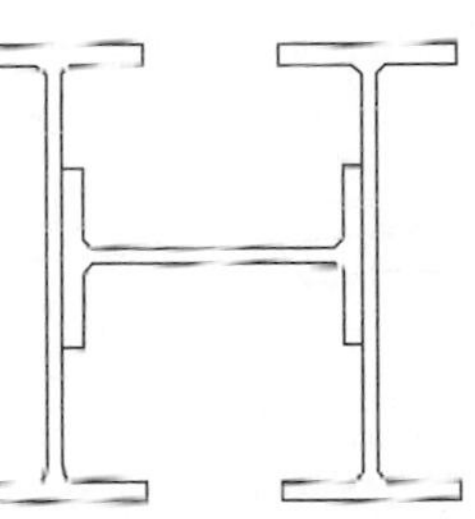
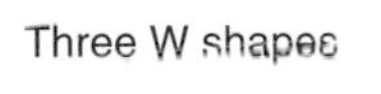


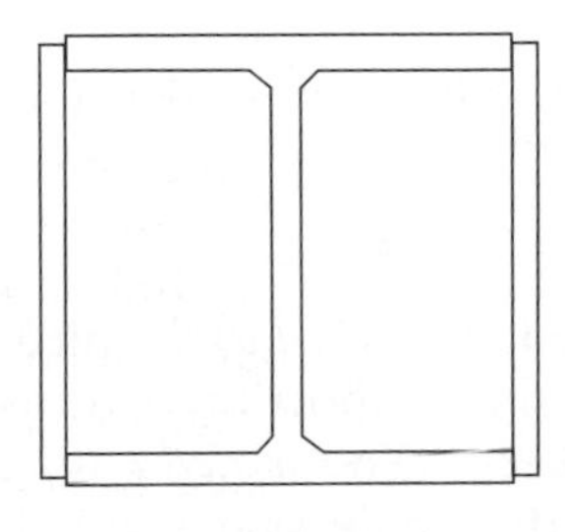


**Figure 10-25** Built-Up Column Sections.

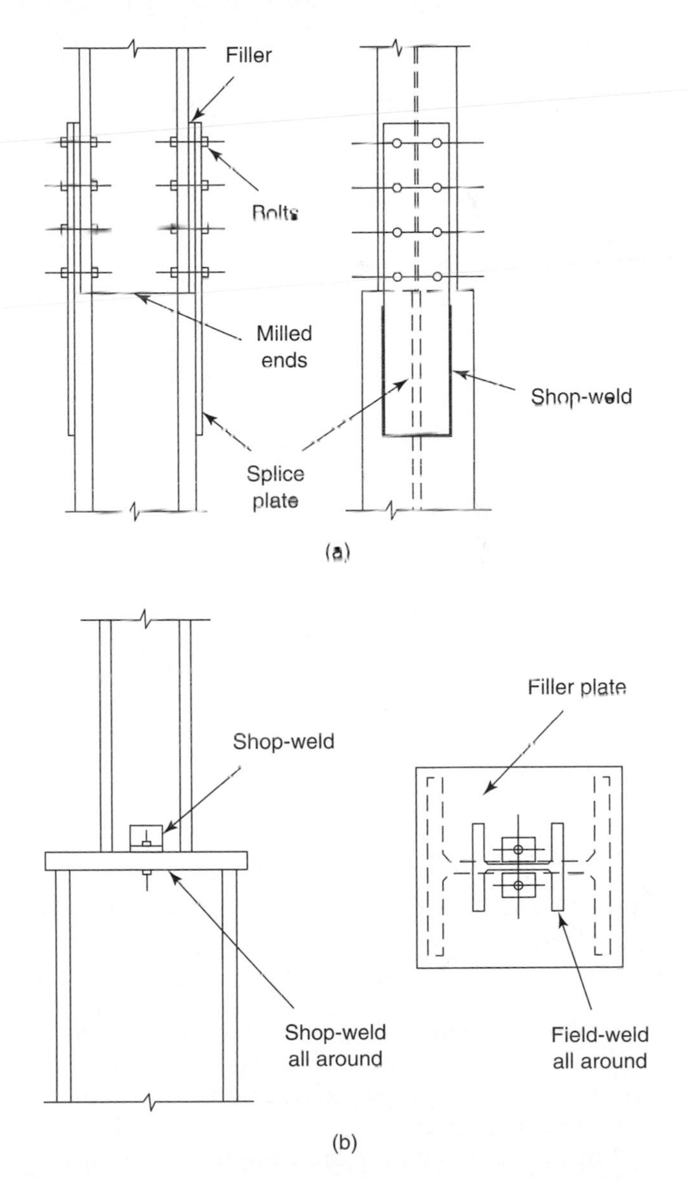


**Figure 10-26** Column Splices.

**Figure 10-27** Typical Structural Steel Frame.

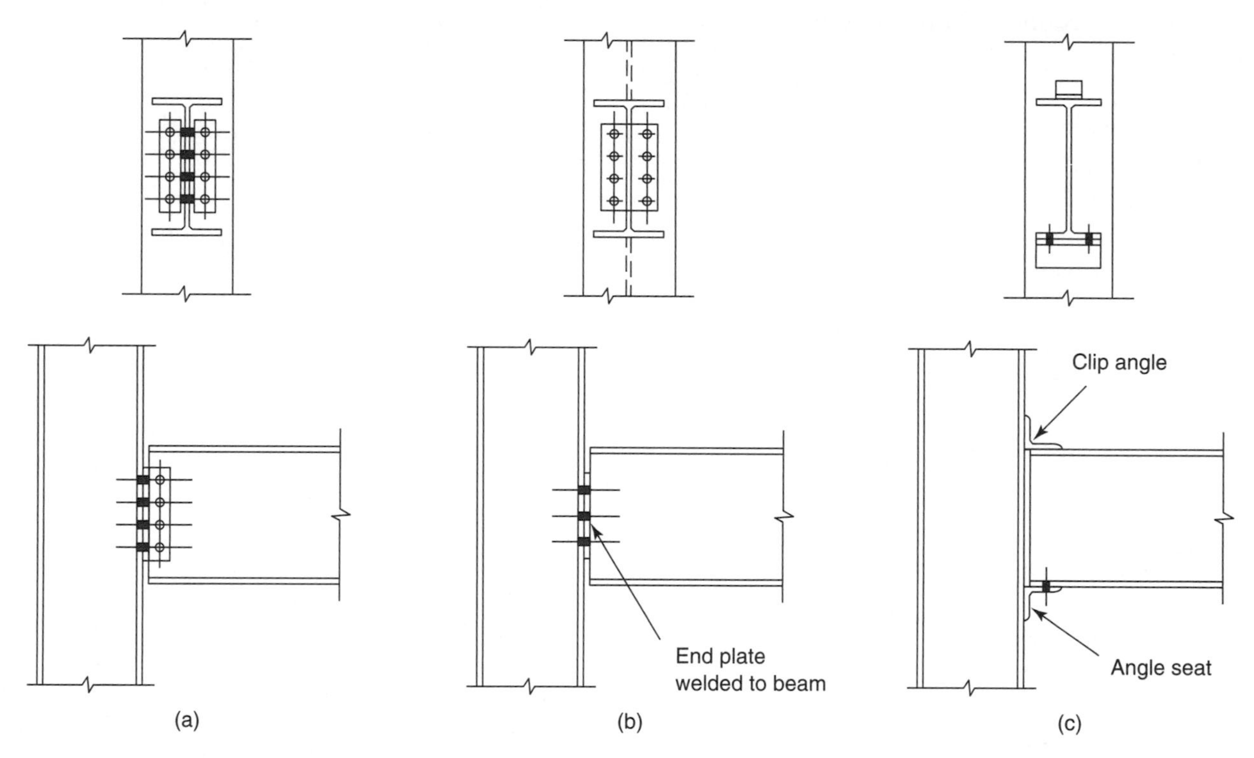


**Figure 10-28** Pin-Type Connections: (a) Double-Angle Bolted Connections; (b) End-Plate Connection; (c) Seated Connection.

three-plate welded I shapes are normally used as girders. Connections between girders and columns can be either pin-type connections or moment-resisting connections.

Pin-type connections are designed to resist vertical loads only. These connections consist of angles or plates bolted or welded to the girder web and usually bolted to the supporting column flange or web (see Figure 10-28). The *seated connection* in Figure 10-28(c) has the angle seat welded to the column and the girder may be bolted or welded to the seat. A clip angle, usually 4″ × 4″ × ¼″ (100 × 100 × 6),

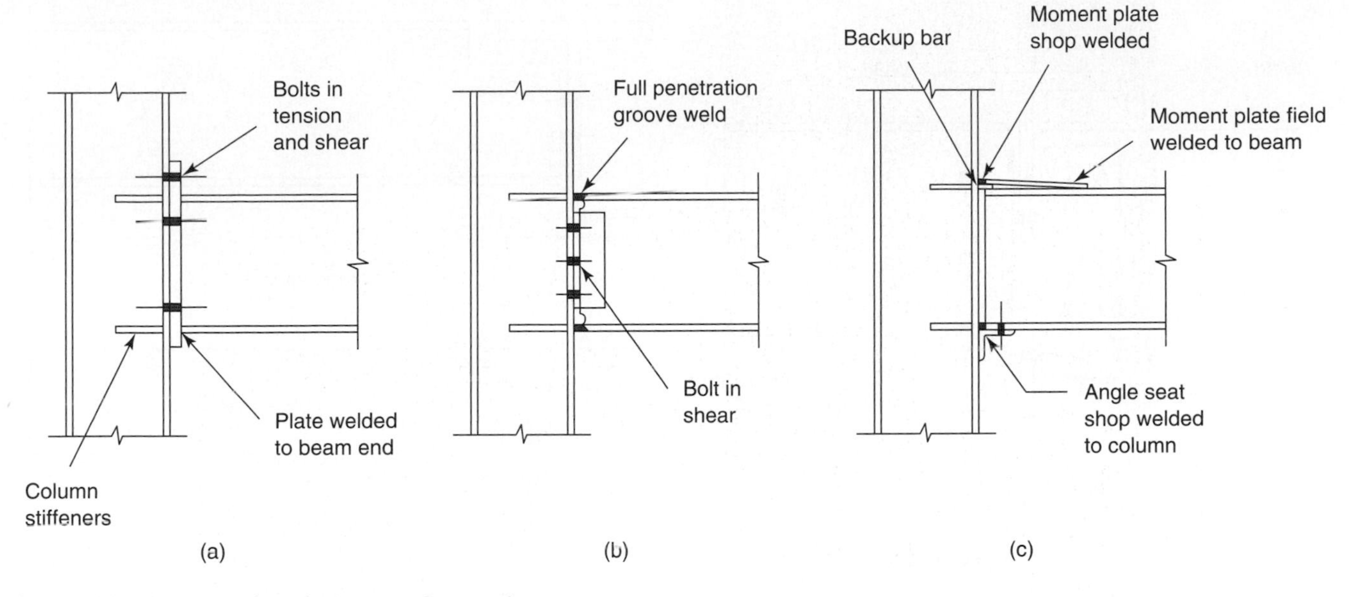


**Figure 10-29** Examples of Moment Connections.

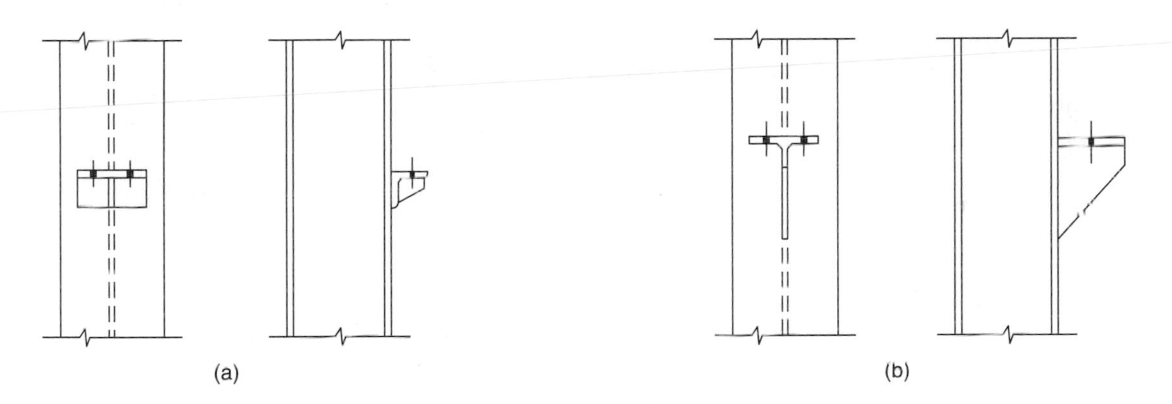


**Figure 10-30** Reinforced Seated Connections: (a) Angle Seat with Stiffener; (b) Tee Seat.

is welded to the column and the girder top flange to provide lateral support for the girder. Because of their flexibility, pin-type connections cannot be reliable to resist lateral loads. They therefore provide no lateral stability to the structural frame. Cross-bracing, shear walls, or a concrete core must be used to absorb the effects of lateral forces and to provide the necessary stability.

Building frames that are designed as *rigid frames* that depend on the stiffness of the column sections to provide lateral stability use moment-resisting connections (see Figure 10-29). Moment connections between the girders and the columns provide lateral stability in the structural frame by preventing rotation between the girders and the columns. Moment connections may be welded, bolted, or a combination of bolts and welds. Figure 10-29 shows three different arrangements of bolts and welds that may be used. Moment connections use a combination of bolts and welds; the bolts are designed to carry the shear or vertical load, whereas the welds are used to resist the effects of bending. In Figure 10-29(c), the seat aids in the positioning of the girder during erection and in the resisting of the vertical load due to the beam reaction.

Seated connections that must support very heavy loads are reinforced to prevent the seat from deforming. Figure 10-30 illustrates two common types of reinforced seats: an angle seat with a stiffener added and a tee section cut from a rolled W shape.

Many connections have been standardized to allow for easier detailing and fabrication. Most structural steel design manuals have tables to aid in the selection and design of good structural connections.

## Beams

Beams, generally smaller than girders, may be connected to either a column or a girder. Beam connections at a column are similar to the girder-to-column

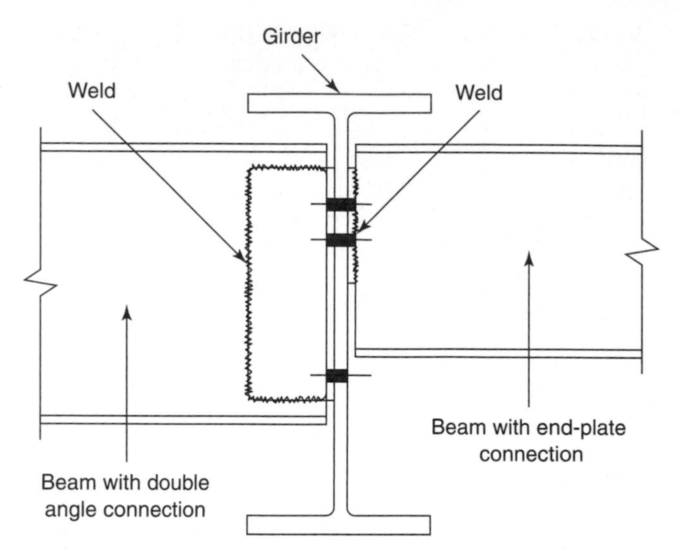


**Figure 10-31** Floor Beams Framed Between Girder Flanges.

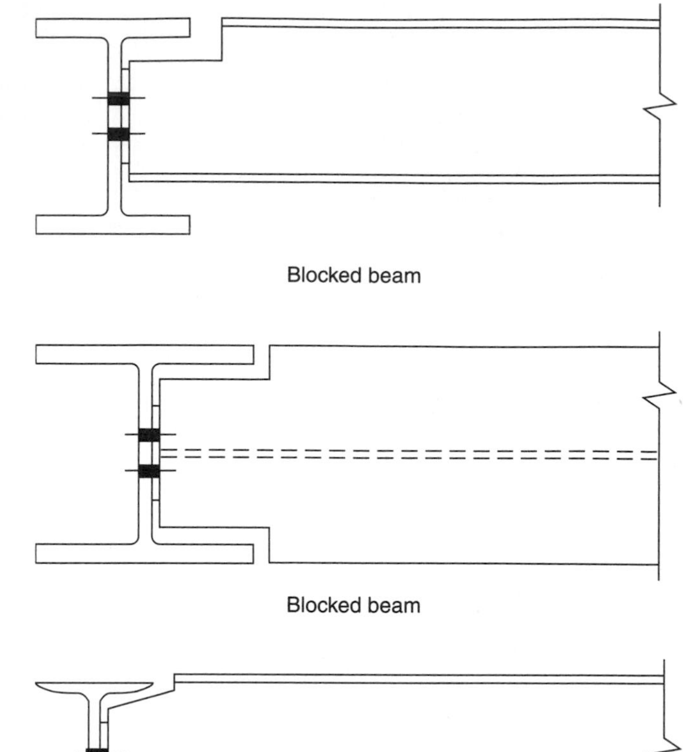


**Figure 10-32** Coped and Blocked Beam Ends.

connections. In beam-to-girder connections, the main consideration is the shear connection, because the beams are considered to be carrying the floor loads and transferring them to girders as vertical loads.

Because beams are usually not as deep as girders, there are several alternative methods of framing one into the other. The simplest is to frame the beam between the top and bottom flanges of the girder, as shown in Figure 10-31. If it is required that the top or bottom flanges of girders and beams be flush, it becomes necessary to cut away a portion of the upper or lower beam flange, as illustrated in Figure 10-32. If the girder is an S shape, the end of the beam is *coped,* whereas if it is a W shape, the end is *blocked.* In many cases, the shear connection angles are bolted or welded to the beam ends in the shop, and the member comes to the job ready for connection to the girder web.

## Channels and Angles

Channels and angles, because of their shape, light weight, and wide range of sizes, have many different uses in the structural frame. Channels are usually used as secondary framing members when loads and spans are not too great. They are used as *wall girts* and *roof purlins,* girts being the horizontal members attached to the columns to support siding (see Figure 10-33) and purlins being the roof framing members spanning between the roof beams and supporting the roof deck. Other uses for channels are as framing for doors and windows, as stair stringers, and as web and chord members in trusses; as built-up sections, they are used as eave struts and spandrel beams.

Angles are used as the connecting pieces for beams, as chord and web members in light trusses and joists, and as bracing (see Figure 10-34). Angles are also used as reinforcing around openings and as supports for mechanical equipment.

## Hollow Structural Sections

Many buildings are designed using the structural frame as an architectural feature. The use of various colors on different parts of the frame can help create an air of spaciousness within the building. Because of their clean appearance, high strength, and good stability, hollow structural sections are popular with designers for use as column sections, trusses, and space frames [see Figures 10-35(a) and (b)].

## Open-Web Steel Joists

*Open-web steel joists* are lightweight trusses that are used for supporting roofs and floors. They are normally fabricated in depths that range from 10 to

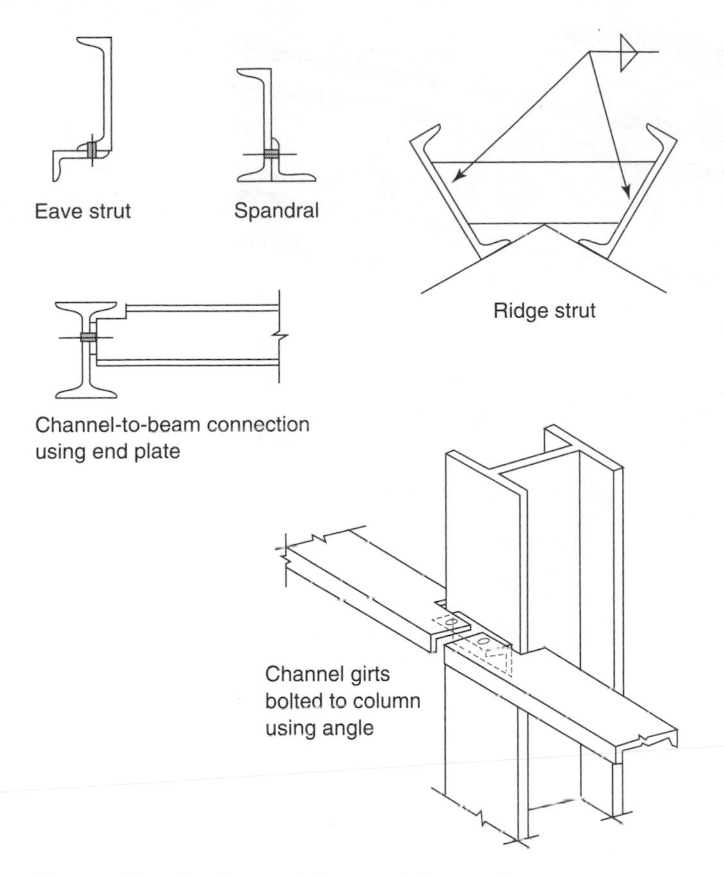


**Figure 10-33** Channel Uses and Connections.

(a)

**Figure 10-34** Double-Angle Cross Bracing.

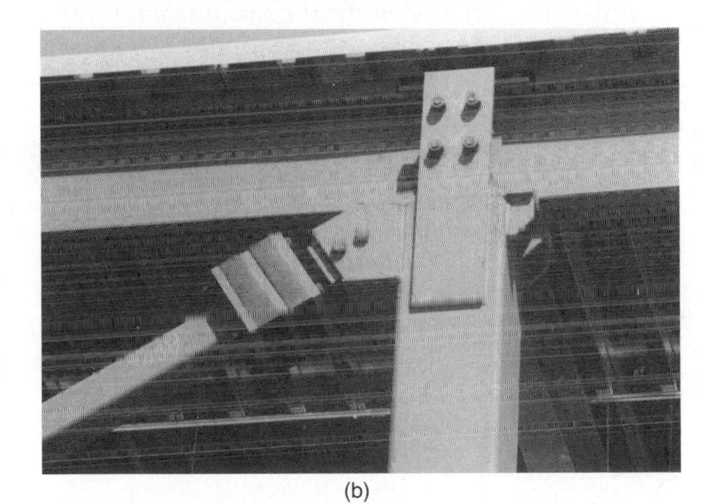

(b)

**Figure 10-35** (a) Space Frame. (b) Hollow Structural Sections Used as Column and Brace. Note the Rubber Damper in the Diagonal Brace. This is to Absorb Loads Due to Earthquakes.

48 in. (250 to 1200 mm) and in lengths that range from 10 to 100 ft (3 to 30 m). Many manufacturers use an assembly-line approach for the fabrication of their joists to ensure uniformity, quality, and economy in the final product.

The makeup of the joist can vary from one manufacturer to another but, in general, the *webs* are usually made of bent rod for joists in the range from 10 to 24 in. (250 to 600 mm) and a lightweight hollow tube is substituted for the rod in the deeper joists. The *chords* may be small angles (Figure 10-36) or some special shape designed specifically for use as a chord section. Open-web steel joists are normally designed to support gravity loads only. Joists that must resist horizontal

**Figure 10-36** Open-Web Steel Joists with Double-Angle Chords.

loads require special seats or cantilever ends, which can be obtained from the manufacturer upon request.

Joists used in a steel frame are normally field welded to the supporting beams to maintain proper spacings. Continuous horizontal or cross bridging is required at regular intervals along the span of the joists to provide vertical stability. Joists can be spaced anywhere from 2 ft to 6 ft o.c. (600 mm to 2,000 mm o.c.). or farther, depending on the depth and thickness of the steel decking, and they provide long clear spans between supports. Because of their slender web and chord sections, they provide ample room between the chords and web members for the placement of electrical and mechanical systems.

## Special Girders

In particular situations when regular rolled shapes are not suitable because they are either not deep or wide enough, special girders may be fabricated. There are two general types: *plate* girders, used to provide extra depth, and *built-up* girders, used to provide extra width.

Plate girders are made by using a plate of appropriate width and thickness for the web and angles, riveted or welded at the top and bottom edges for flanges. Deep plate girders often require vertical stiffeners in the form of angles welded to the web. In Figure 10-37, plate girder made from plate and channels and with vertical stiffeners is being hoisted into place as a spandrel beam.

Built-up girders are made by joining two or more regular rolled shapes side by side by means of plates or pipes known as *separators*.

**Figure 10-37** Built-Up Girder Being Hoisted into Position.

## PREENGINEERED STRUCTURAL STEEL FRAMES

Another concept used in structural steel frames is that of the preengineered building. Although it is fabricated and erected in much the same way as a regular steel frame, its design is based on standard modular sizes that are prefabricated and supplied as complete units. Frames are designed to accommodate different loading requirements over different spans, thus giving the client a wide range of options.

Preengineered frames are a combination of rigid frame construction and pin-type construction. Usually placed on relatively simple foundations (Figure 10-38) with a slab-on-grade floor, they are ready for occupancy in minimum time.

Lateral stability across the width of the structure (Figure 10-13) is provided by the frame, which is designed primarily as a three-pinned arch. Pin-type connections are used at the base and at the ridge. The lateral resistance across the width of the structure is provided by the rigid connection at the haunch between the column section and the roof beam section.

Horizontal sections on the walls (girts) support the siding, and purlins support the roof (Figure 10-39). These sections are normally cold-formed sections (Figure 10-40), usually of gauge thickness, and once covered with siding and roofing they provide additional stability to the overall structure. Additional bracing, usually rod bracing between frames, prevents racking of the building

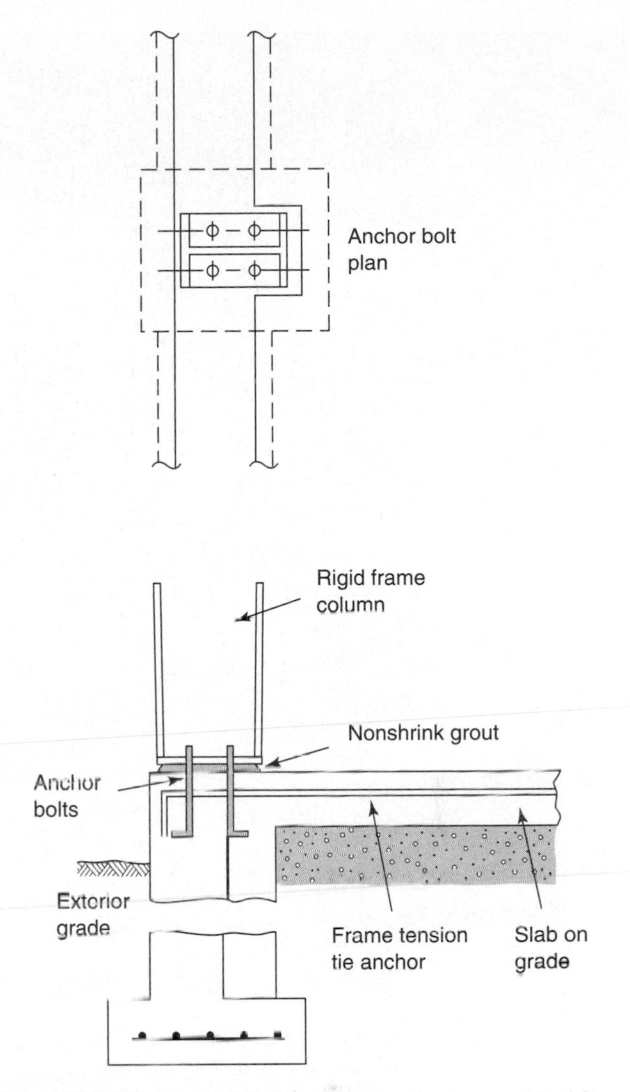


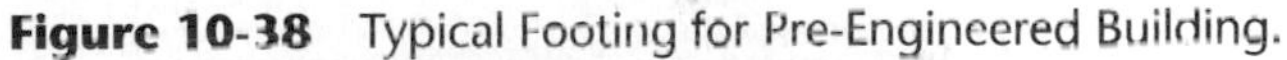
**Figure 10-38** Typical Footing for Pre-Engineered Building.

**Figure 10-39** Cold-Formed Girts and Roof Purlins.

**Figure 10-40** Cold-Formed Gauge Material C Shapes.

and maintains frame stability during the construction period.

The advantages of a preengineered frame over a fabricated steel frame are as follows:

1. Engineering costs are held to a minimum. Because the building is available in standard units, the cost of design is very moderate.
2. Unless special details are required by the client, delivery time of the frame components is short, because components are standard sizes.
3. A complete package including insulation, siding, and roofing can be obtained from one supplier, thereby decreasing the number of contractors on the site.
4. Additions can be made with relative ease by adding additional bays to the existing building.
5. A wide range of sizes is available. Frames come in many widths, heights, and load capacities.

When a building is to be an odd shape or must support unusually high loads, the preengineered building may prove to be inappropriate. However, where a straightforward structure with large open areas is required, this type of frame is most competitive.

## CONSTRUCTION PROCEDURES AND PRACTICES

There are two distinct stages in the construction of a steel frame. The first stage is performed in the shop, where the individual pieces are cut to length, coped or blocked, and punched or drilled for bolts. Detail material such as gusset plates, clip angles, and end plates for beams are attached so that the connections in the field can be made quickly.

The second stage begins when the fabricated steel arrives on the site. The field crew, with the aid of a crane (Figure 10-41), assembles the individual pieces according to the erection drawings. The connections are aligned with *drift pins,* and temporary bolts are used to hold the sections together. The building is plumbed using a building transit and held in alignment with temporary cable bracing equipped with *turnbuckles* (to allow for adjustments) until permanent bracing has been put in place and final bolting and welding have been completed. Figure 10-42 shows the final stages of completing a bolted connection—removal of a drift pin that was used to align the bolt holes—so that the final structural bolt can be put in place.

The conventional methods of steel erection, beginning at the bottom and working up, do not always apply. Unusual designs sometimes require unorthodox erection schemes to be used, as illustrated in Figure 10-43. Two concrete support towers have steel tension hangers draped over reinforced concrete saddles from which the steel frame floors were suspended. The cantilever steel frames were preassembled on the ground, raised to the proper elevation, and suspended from the steel hangers. Nine of the thirteen floors were erected in this manner, providing large, column-free areas on each floor.

Under certain conditions, abnormally high structures require innovative construction methods to be applied. Figure 10-44 shows a Sikorsky skycrane placing the steel sections of an antenna on top of a communications tower. The erection of steel requires skill, thoughtfulness, and constant attention to rules of safety. Some general safety rules to observe are as follows:

1. Do not attempt steel erection on rainy days. It is dangerous to work on wet steel.
2. Wear shoes with sewn leather or rubber soles.
3. Do not allow anyone to stand beneath a loaded boom.

**Figure 10-41** Iron Workers Erecting Steel Frame. (Reprinted by permission of Canadian Institute of Steel Construction.)

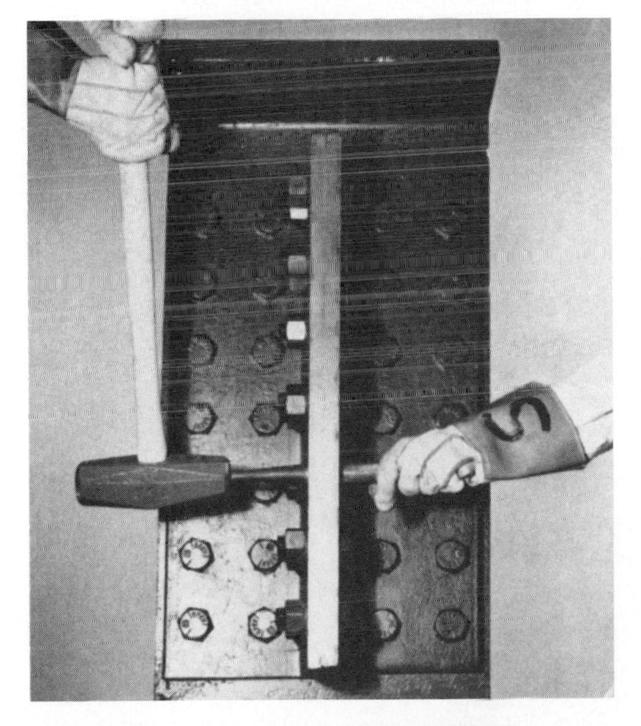

**Figure 10-42** Final Stage of Completing Bolted Connection. (Reprinted by permission of Bethlehem Steel Co.)

**Figure 10-43** Structural Steel Floor Frames Hang from Straps. (Reprinted by permission of Bethlehem Steel Co.)

**Figure 10-44** Sikorsky Sky-Crane Placing Steel Antenna Sections Atop a Tower. (Reprinted by permission of Canadian Institute of Steel Construction.)

4. Do not allow anyone to ride a load.
5. Wear heavy gloves that do not have loose cuffs.
6. Do not wear loose clothing that could easily catch on projections or swinging objects.
7. Use shackles instead of hooks.
8. Use cable slings instead of chains for lifting steel.
9. Make sure that the brakes on hoisting drums are inspected daily.
10. Keep a constant lookout for frayed or broken strands in cables and repair or replace them if necessary.

In addition to the foregoing, the safety of the crew must be ensured at all times. Safety is everyone's responsibility, and the quality of the job done relates directly to the level of safety that is provided at the site. Some essential site requirements, which a good contractor will provide and insist upon to ensure safe working conditions, are as follows:

1. Proper ladders are provided for access from one level to another.
2. Handrails are provided around elevator shafts and the building perimeter once floor decking is in place.
3. Safety nets are provided (Figure 10-45), not only for the safety of the workers above, but also to protect personnel working below from falling objects.
4. All equipment must be maintained and have the proper warning devices when in operation.
5. Conduct weekly safety meetings to ensure that all personnel are aware of all safety rules.

In many areas, minimum safety requirements are spelled out by the local authorities, and every contractor must abide by them or be asked to close the site until remedial steps have been taken to the satisfaction of the safety officer.

**Figure 10-45** Safety Nets Help Prevent Accidents.

## REVIEW QUESTIONS

**1.** What are the two main concerns in the design of structural steel frames?

**2.** What are the two types of structural steel frames? How do they differ in their ability to resist lateral loads?

**3.** What is the purpose of temporary bracing on a structural steel frame?

**4.** What are two advantages of using a tower crane over using a mobile crane in the erection of a structural steel building frame?

**5.** Where might a light guyed derrick be used to advantage on a construction site?

**6.** When connecting a beam to a column, give two situations on the construction site where welding would be more suitable than bolting.

**7.** Explain the purpose of **(a)** erection marks on structural steel members, and **(b)** the precise location of the erection mark on a beam.

**8.** Outline two alternative methods of leveling bearing plates.

**9.** Explain why **(a)** column splices are made above floor level, **(b)** column ends are milled in the fabricating shop, **(c)** it is sometimes necessary to use a plate between the ends of two columns being connected, and **(d)** HP shapes are often chosen for columns.

**10.** With the aid of diagrams, illustrate how you would try to overcome the connecting problems involved if you were required to connect a W shape beam to a round column.

**11.** Explain the difference between **(a)** *coping* and *blocking,* as applied to steel beam ends; **(b)** *plate girders* and *build-up girders;* and **(c)** *shear* connections and *moment* connections.

**12.** Using a scale drawing, illustrate how a bolted shear connection will be made between an S shape beam and an HP shape column. Use specific sizes for both.

**13.** An electronics company wishes to open a plant to assemble electronic equipment. What type of structural frame may be considered? Give reasons for your selection.

**14.** A mail order company wishes to build an automated warehouse that will have conveyor belts suspended from the roof framing throughout the building at various heights and in different directions. How might a building of this type be braced, assuming that metal insulated panels are being used on the exterior walls?

**15.** Where might a safety net be used to advantage on a construction site?

# 11 FLOOR SYSTEMS AND INDUSTRIAL FLOORING

The erection of the structural frame marks the completion of a major phase in the construction of a building. As a major component of the building frame, the floor system will determine, to a great extent, the construction sequence for the entire building. In floor systems suspended above grade, the *subfloor* is considered part of the structural frame and it must be capable of supporting all superimposed design loads. Dead loads such as additional floor toppings, floor finishes, and partition walls, as well as occupancy live loads, must be considered when designing the subfloor.

For the purpose of discussion, floor systems in buildings can be divided into two parts: (1) the base or *subfloor,* and (2) the application of the *floor finish.* Subfloors are constructed of wood, concrete, or a combination of concrete and steel. In wood frame buildings, the subfloor is supported by floor beams or floor joists. It may consist of wood sheathing or metal decking supporting a lightweight concrete slab. In steel frame buildings, a subfloor may be a reinforced concrete slab supported by the steel floor beams. An alternative approach to subfloor construction in a steel frame building is to use metal decking covered with a poured concrete slab. Precast concrete floor sections can be used to advantage in both steel and concrete frame buildings, allowing for fast erection times and relatively long spans between supports.

The construction of the subfloor is related to the material used for the structural frame. In a reinforced concrete frame building, for example, the concrete slab can be poured monolithically with the beams and columns of the frame. In timber and structural steel building frames, the supporting beams and columns are erected prior to the start of all floor construction.

## TIMBER SUBFLOORS

Wood subfloors are common in timber framed buildings (see Figure 11-1), but are also used in steel framed structures. Wood subfloors are constructed in two ways. For heavy loading conditions, 2″ × 4″s or 2″ × 6″s (38 × 89s or 38 × 140s) are placed on edge at right angles to the floor beams. The sections are laminated together side by side, using spikes, usually in a controlled random arrangement. This allows for the use of sections of variable lengths resulting in a minimum of waste. Rules for the location of end splices are provided in wood design codes.

The more common approach to the construction of timber subfloors is to use heavy planking, laid flat as shown in Figure 11-2. To ensure tolerable deflections, the thickness of the plank subfloor is based on the distance between the supporting beams and the loads that must be carried. Planks are usually tongue and grooved or splined for greater rigidity and are placed in much the same manner as the laminated floor sections.

When the supporting floor beams are of timber, the subfloor is nailed directly to the beam sections. In the case of steel floor beams, a wood nailer is first attached to the top flange of the beam (Figure 11-2) and the subfloor is then nailed to the nailer.

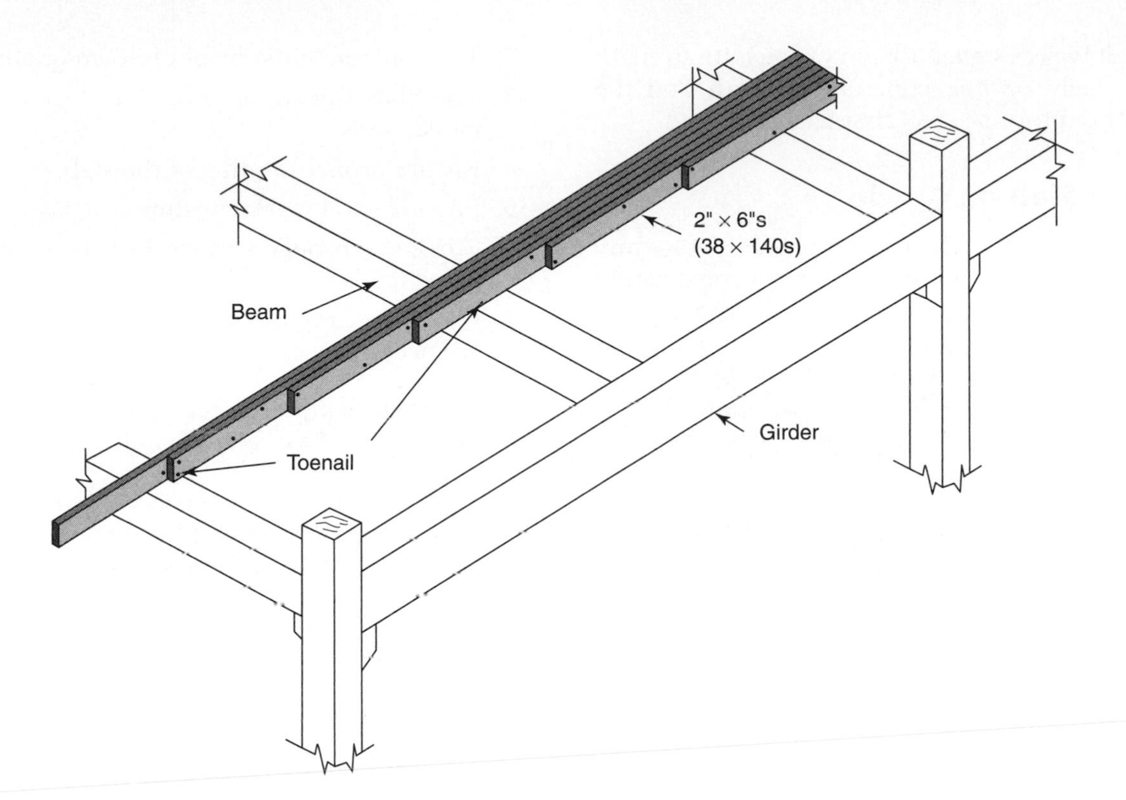


**Figure 11-1** Laminated Wood Deck on Timber Frame.

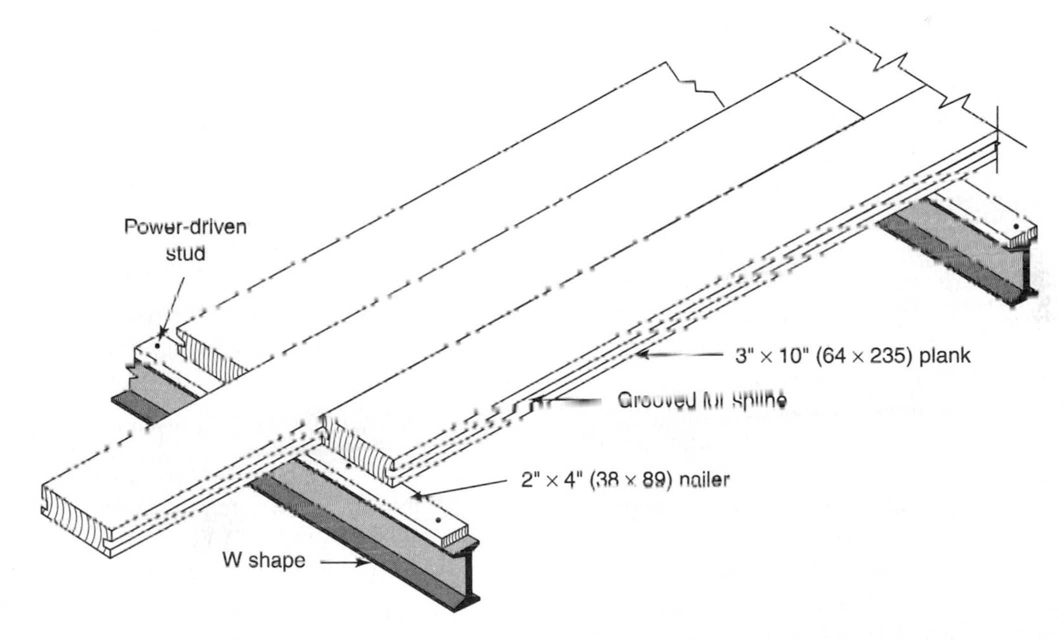


**Figure 11-2** Plank Deck on Steel Frame.

## CONCRETE SUBFLOORS

There are two distinct types of concrete floors: *slabs on grade* and *structural slabs*. The slab on grade is a floor poured directly on a prepared subgrade to provide a clean and durable surface for the building occupants. The durability of the slab depends on the concrete strength and finish, but the strength of the slab depends on the strength of the supporting subgrade. The structural slab, which must span between supporting walls, columns, or beams, must be able to support its own dead weight and the superimposed load that it was designed to carry. Thus, its strength depends solely on the concrete strength and the amount of reinforcing steel that has been used.

## Concrete Slab on Grade

Concrete slabs on grade may be placed before any other portion of the building has been constructed (Figure 11-3), on a prepared subgrade to provide a level surface on which the structural frame is erected. When the frame is supported by footings and grade beams, the floor may be poured after the building has been closed in. In the first instance, side forms of wood or metal are placed, leveled, and staked around the perimeter of the building, and screed strips are placed at convenient intervals to provide guides for the leveling of the concrete. In Figure 11-3, a vibrating straightedge and drag are used to level the slab. A slab placed on grade after the walls and roof have been erected is common practice in commercial and industrial buildings, where large floor areas poured in sections require longer periods of protection from the weather.

Slabs on grade often provide less than adequate results when put into service due to excessive cracking, dusting, and spalling. To ensure that the slab will provide a long-lasting and serviceable surface, the Portland Cement Association puts forward the following recommendations:

1. The subgrade must be uniform and have adequate bearing capacity.
2. The concrete must be of uniform quality.
3. The slab thickness should reflect the anticipated loads.
4. Ensure proper jointing of the slab.
5. Provide good workmanship.
6. Provide a proper surface finish for the anticipated loads.
7. Use proper repair procedures if and when they are necessary.

**Figure 11-3** Power-Screeding a Slab-on-Grade Concrete Floor.

A standard procedure to produce a good concrete slab on grade may be as follows:

1. Backfill all ditches and trenches under the floor with material similar to the surrounding subgrade, ensuring similar density and moisture content. Uniform compaction with some type of mechanical equipment (Figure 11-4) is most important.
2. Isolate all columns from the floor slab by boxing them with square wood or metal forms or with round fiberboard forms. Forms should be set to the level of the top of the slab (Figure 11-5).
3. Isolate the walls from the slab by fastening strips of asphalt-impregnated fiberboard or other joint material not more than ½ in. (12 mm) thick around the walls, level with top of the slab.
4. Prepare any changes in slab thickness, as at doorways, to be as gradual as possible and at slopes of not more than 1 in 10.
5. Use a template with legs equal to the slab thickness to check the grade (see Figure 11-5).
6. Provide a vapor diffusion barrier should water vapor migration through the slab be a concern.
7. Place the reinforcement (mesh or rod) as specified in plans.
8. Set screed strips at the same elevation at convenient intervals throughout the area to be concreted (see Figure 11-6).
9. Oil the screed strips.
10. Place the concrete as close to its final position as possible. Consolidate with an internal vibrator, especially at corners, walls, and bulkheads.
11. Straightedge the concrete to the level of the screed strips.
12. Smooth the surface with a long-handled float or a darby (Figure 11-7) to remove the high and low spots. Cover with damp burlap until ready for the next operation.

(a)

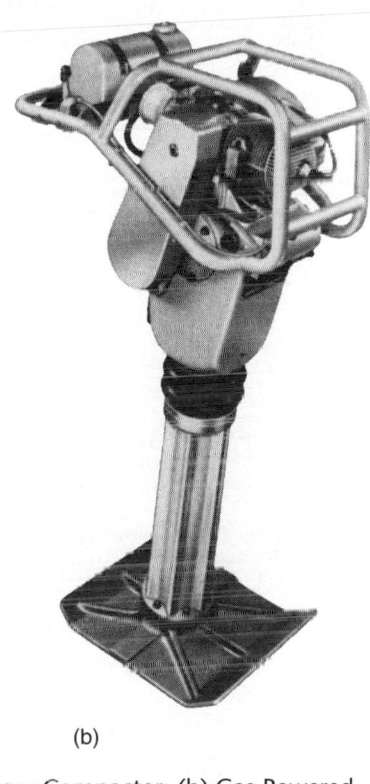

(b)

**Figure 11-4** (a) Vibratory Compactor; (b) Gas-Powered Compactor.

13. Float the surface with hand or mechanical floats as soon as the concrete supports the weight of a worker.
14. If specified, apply metallic aggregate hardener (see directions for applying hardener, page 385).

**Figure 11-5** Leveling Subgrade.

15. Trowel to a hard, dense surface with hand or power trowels (Figure 11-8).
16. Cure by covering with (a) waterproof curing paper, (b) two coats of curing compound, (c) burlap kept moist at all times, or (d) a layer of damp sand (see Figure 11-9).
17. Remove the forms around columns and attach joint material to the vertical faces of the slab and the base of the columns. Fill with concrete, edge, and finish.
18. Cut control joints to a depth of at least one-fifth the slab thickness with a power saw every 20 to 25 ft (6 to 8 m) in both directions (Figure 11-10).
19. Caulk the joints with mastic joint filler.
20. Cure for at least 7 days before allowing regular traffic on the floor.

## Structural Concrete Floor Slabs

A concrete floor supported by a structural frame may be one of a number of floor types now available. The one to be used for any particular building depends on a number of factors, including (1) the intended use of the building, (2) the type and magnitude of the loads to which the floor will be subjected, (3) the length of the span between supports, (4) the type of building frame, and (5) the number of stories in the building. Concrete structural floor slab types include the following:

1. *One-way solid slabs:* One-way solid slabs (Figure 11-11) are structural slabs that have the load-carrying steel running in the direction of the span, with temperature and shrinkage steel—steel to control cracking—running perpendicular to the span. One-way slabs are used where the span of the slab is relatively short, and they are designed as single span or

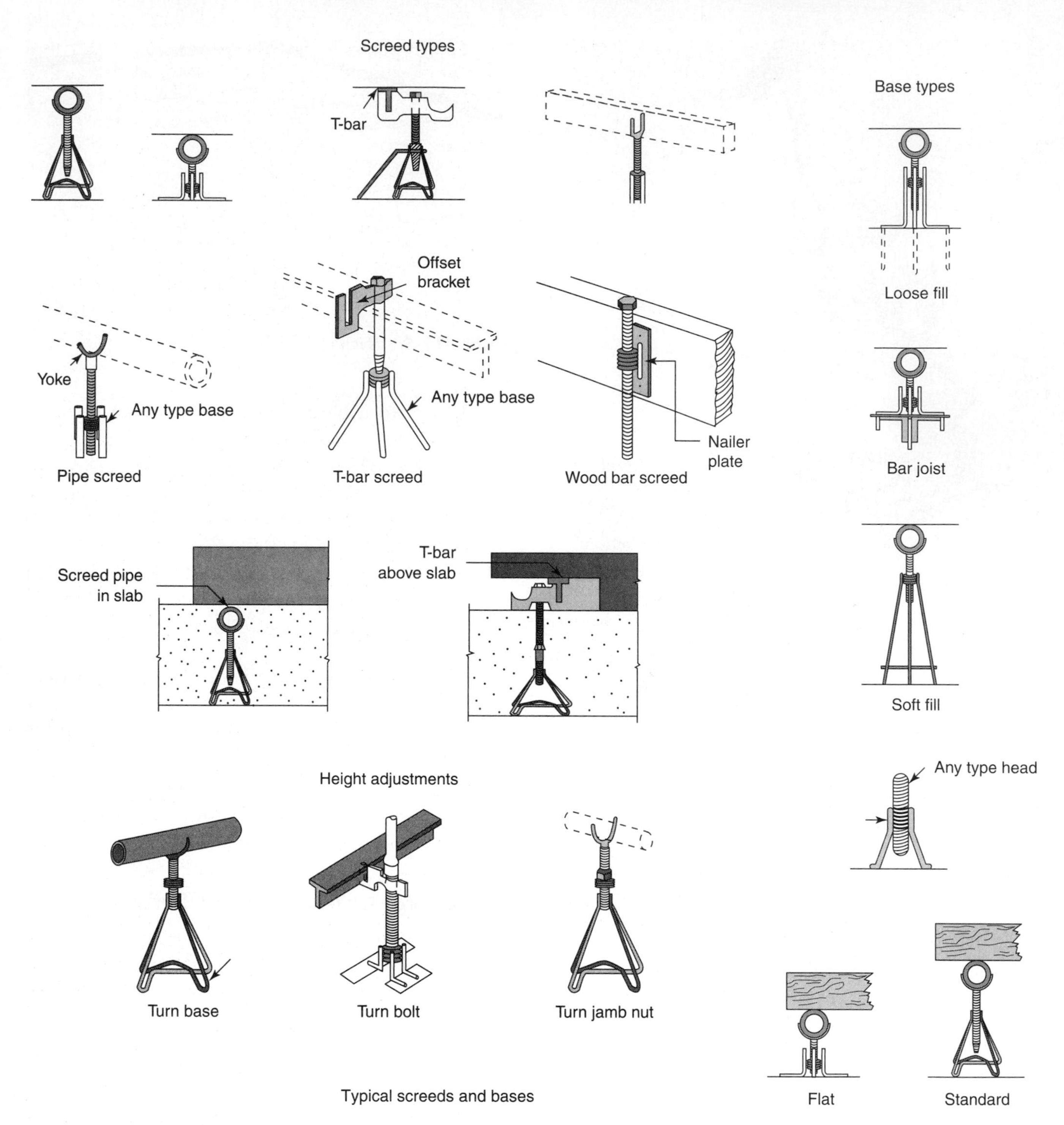


**Figure 11-6** Typical Screeds and Bases.

multispan (Figure 11-12). One-way slab formwork and the placement of the reinforcing steel are straightforward, allowing for the integration of the slab with structural steel framing (Figure 11-13) and masonry, as well as with reinforced concrete frames. Over short spans, they can support reasonably high loads relative to their thickness, but become uneconomical on spans over 20 ft (6 m) because of their mass.

2. *Two-way solid flat slabs:* Two-way flat slabs have load-supporting reinforcing steel run-

**Figure 11-7** Darbying a Concrete Slab. (Reprinted with permission of Portland Cement Association.)

**Figure 11-8** Power Trowel in Action. (Reprinted with permission of Portland Cement Association.)

ning in two directions. Beams or walls are not necessary in the support of this type of slab, as it can be placed directly on supporting concrete columns. Thickened sections over the columns known as *drop panels* (Figure 11-14) ensure that sufficient shear capacity is available in the slab. In the past, *column capitals* (Figure 11-15) have been used with or without drop panels. These have been eliminated to a great extent because of higher forming costs. The slab thickness depends on the span between columns; the minimum thickness between drop panels is 4 in. (100 mm) and the maximum is 12 in. (300 mm). The minimum size of drop panel used is one-third of the span, and the total drop panel thickness is usually 1¼ times the slab thickness. The two-way slab system is used for heavier applications such as industrial storage buildings and parking garages. Maximum practical spans for the two-way slab are approximately 34 ft (10 m).

3. *Two-way flat plate slab:* The two-way flat plate slab is similar to the two-way solid slab with the exception that it does not have a drop panel. Flat plates are very efficient from the point of forming, steel layout, and total thickness. The slab thickness varies from 5 to 10 in. (125 to 250 mm) and maximum practical spans reach approximately 32 ft (10 m). Because of its simplicity and minimum structural depth, the flat plate is popular in high-rise offices and apartments (Figure 11-16).
4. *One-way joist slab:* The one-way structural system consists of a series of concrete ribs or joists that contain the reinforcing steel, cast monolithically with a relatively thin slab. The ribs, in turn, frame into supporting beams (see Figures 11-17 and 11-18). The main purpose of using such a system is to reduce the dead load by concentrating the reinforcing steel in the ribs and eliminating the concrete between them. The slab thickness varies between 2½ and 4½ in. (65 and 115 mm) and the joist depth varies from 6 to 20 in. (150 mm to 500 mm). Standard pan forms produce between-rib distances of 20 and 30 in. (500 and 900 mm). Maximum joist spans are about 45 ft (15 m).
5. *Waffle (two-way joist) slab:* In this type of floor system, ribs run at right angles to one another. The ribs are produced by the use of dome pans, resulting in a two-way structural system (see Figures 11-19 and 11-20). The domes are available in various sizes, with two sizes being the most common: 19 × 19 and 30 × 30 in. (475 × 475 and 900 × 900 mm). Dome heights vary from 8 to 20 in. (200 mm to 500 mm) in 2-in. (50-mm) increments to produce rib depths appropriate for the anticipated spans of the floor system. The slab depth can vary from 3 to 5 in. (75 to 125 mm), and the rib depth can vary from 7 to 20 in. (175 to 500 mm). The normal rib widths are 5 and 6 in. (125 and 150 mm) and the distance between rib centers can vary from

(a)

(b)

**Figure 11-9** Concrete Curing Preparations: (a) Plastic Film Covering; (b) Wax Compound Covering. (Reprinted with permission of Portland Cement Association.)

24 to 36 in. (600 to 900 mm). This slab system is found to be economical for spans ranging from 25 to 50 ft (8 to 6 m).

## Posttensioned Concrete Slabs

Adding posttensioning strands to concrete slabs produces sections that can span over greater distances than nonprestressed sections and yet are able to support greater loads with less deflection and virtually no cracking.

High-strength steel strands, or *tendons,* enclosed in protective plastic sheathing are distributed throughout the slab at predetermined intervals. Figure 11-21 illustrates a typical tendon with slab anchor, wedge anchor, and protective sheathing. The sheathing ensures that the concrete does not bond to the posttensioning tendon. Once the concrete has cured sufficiently, the tendons are tensioned to a predetermined stress and anchored. Figure 11-22 illustrates strands being placed for a two-way posttensioned flat plate slab. Note that the tendons are raised higher above the form in the vicinity of the supporting column than at the midspan of the slab. This arrangement ensures that the compressive forces induced in the concrete slab by the tendons counteract the tension stresses produced in the slab by anticipated loads. Figure 11-23 illustrates a completed arrangement of posttensioning tendons used in conjunction with regular reinforcing steel.

Any concrete slab may be posttensioned, including slabs on grade, where it has been found that the slab performance improved even when subgrade conditions were less than ideal. However, the appli-

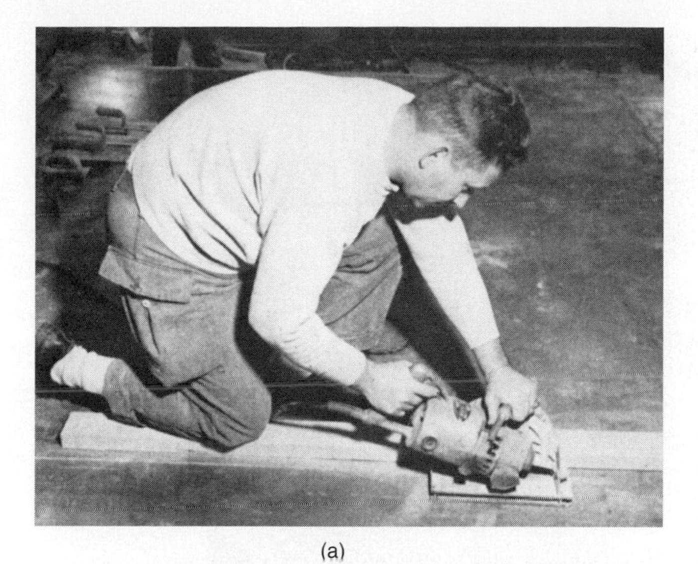

(a)

(b)

(c)

**Figure 11-10** (a) and (b) Cutting Control Joints; (c) Abrasive Wheel for Cutting Concrete. (Reprinted with permission of Portland Cement Association.)

cation of posttensioning has become most popular in solid one- and two-way slabs (see Figure 11-24) and in two-way flat plate slabs. Posttensioning can be used to advantage in all types of buildings, ranging from single-story car parks to multistory apartment blocks and office buildings.

## FLOOR SLAB FORMS & WORK

Forms for a beam-and-slab floor consist of slab forms and the shores that support them, beam forms and the T shores that support them, *kickers* between slab and beam forms (see Figure 9-23), and supports for the edges of the slab form over the T shores.

The slab forms should be made first. The frames are of 2″ × 4″s (38 × 89s) in any required size, but one edge of the sheathing should be beveled at 45° and extended approximately 2¼ in. (55 mm) beyond the frame on panels that will meet the beam form (see Figure 11-25).

The column forms are erected next, plumbed, and braced. The T shores are set between each pair of columns (see Figure 9-21), using a chalk line from one column top to another as a guide. If a camber is required in the beam, raise the center shore the required amount, run the chalk line over it, and set the remainder of the shores to the line.

The beam form may now be built on the shore heads, and the slab form panels can be raised and supported. It is important to see that the kickers are in place between the beam sides and the edges of the slab forms. These kickers prevent the beam sides from being pushed out by the pressure of the concrete and allow easy removal of beam sides when the concrete has cured sufficiently.

**Figure 11-11** Beam-and-Slab Floor.

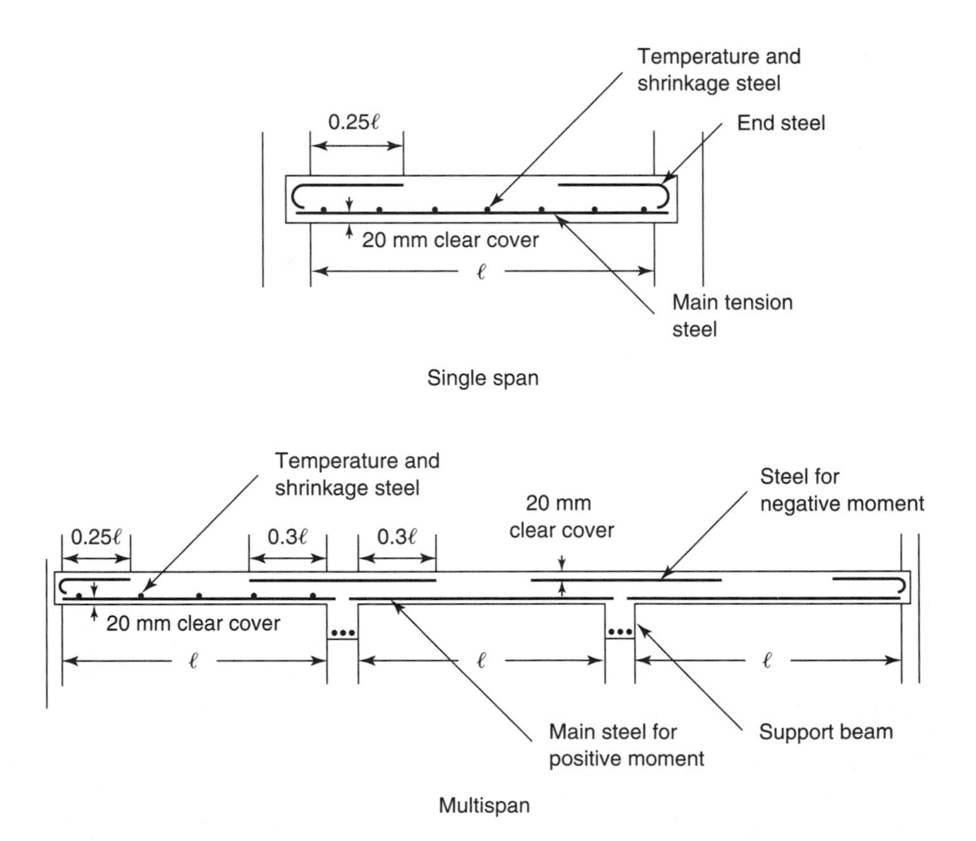


**Figure 11-12** Reinforcing Steel in Solid One-Way Slabs.

**Figure 11-13** One-Way Slab Supported by Steel Frame.

**Figure 11-14** Flat Slab Floor with Drop Panel. (Reprinted with permission of Portland Cement Association.)

**Figure 11-15** Column Capital. (Reprinted with permission of Portland Cement Association.)

**Figure 11-16** Flat Plate Floors.

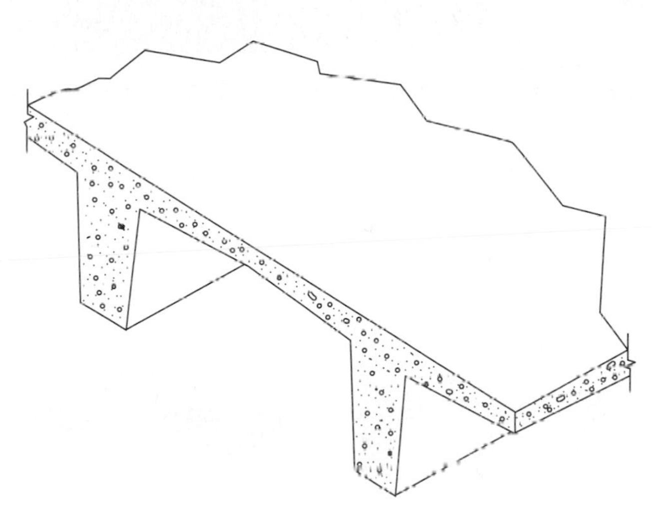

**Figure 11-17** Ribbed Slab.

**Figure 11-18** One-Way Joist Floor.

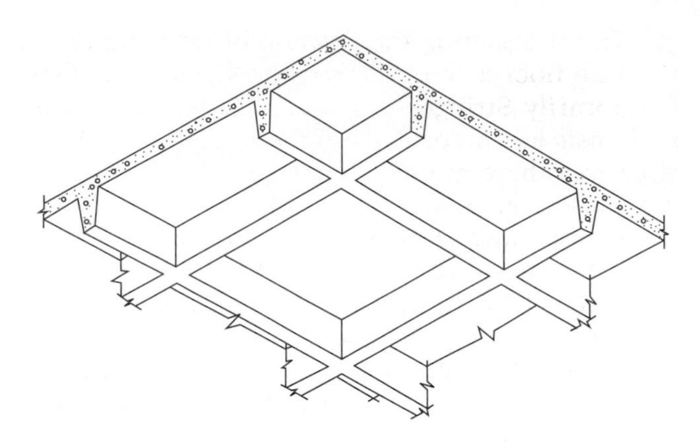

**Figure 11-19** Two-Way Ribbed Slab.

**Figure 11-22** Posttensioning Cables Being Positioned on Chairs.

**Figure 11-20** Under Side of Waffle Slab.

**Figure 11-23** Two-Way Slab Ready for Concrete.

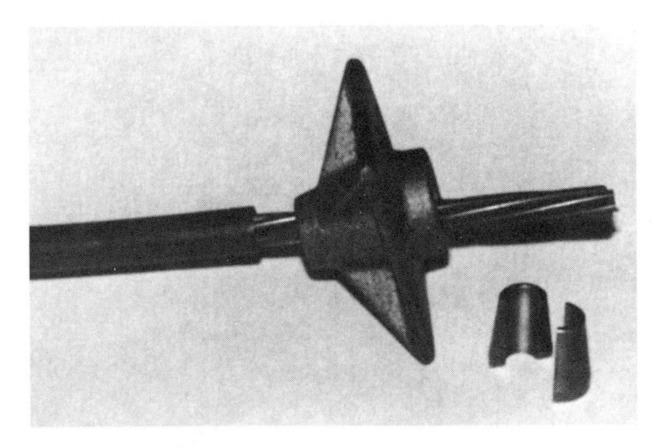

**Figure 11-21** Posttensioning Cable with Slab Anchor, Wedge Anchor, and Sheathing.

**Figure 11-24** Posttensioned Solid Slab with Posttensioned Beam Ribbons.

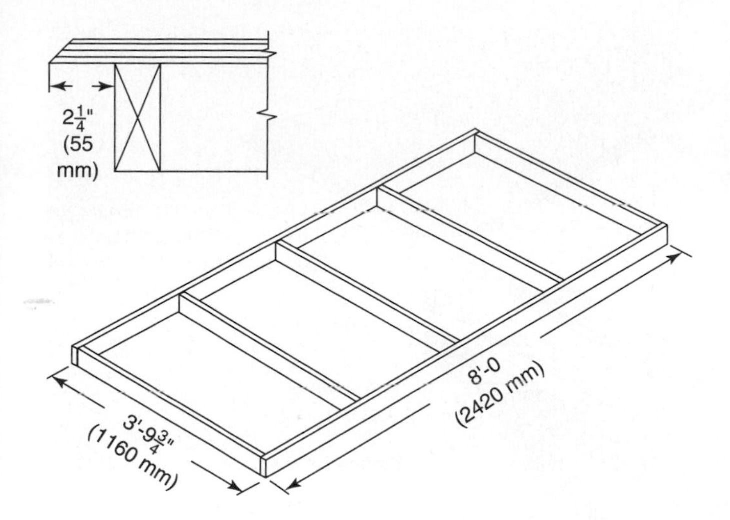


**Figure 11-25** Frame for Slab Form Panel.

The first step in the erection of formwork for a flat slab floor is to erect the shores and brace them temporarily. Stringers are then placed on the shores, and joists laid across the stringers, leaving openings wherever there are to be drop panels. This framework is leveled by wedging or adjusting the length of the shores, and sheathing is finally laid over the joists. Where sheathing meets a drop panel, the edge is beveled and extended over the top of the drop panel in the same way that slab form sheathing extends over beam side forms (see Figure 11-26).

Drop panels are built separately and are supported on separate shores or on sections of steel scaffolding set up to form a square supporting structure. Figure 11-27 illustrates one method of forming the drop panel. The sheathing must be cut to fit the column capital that will be part of the column form.

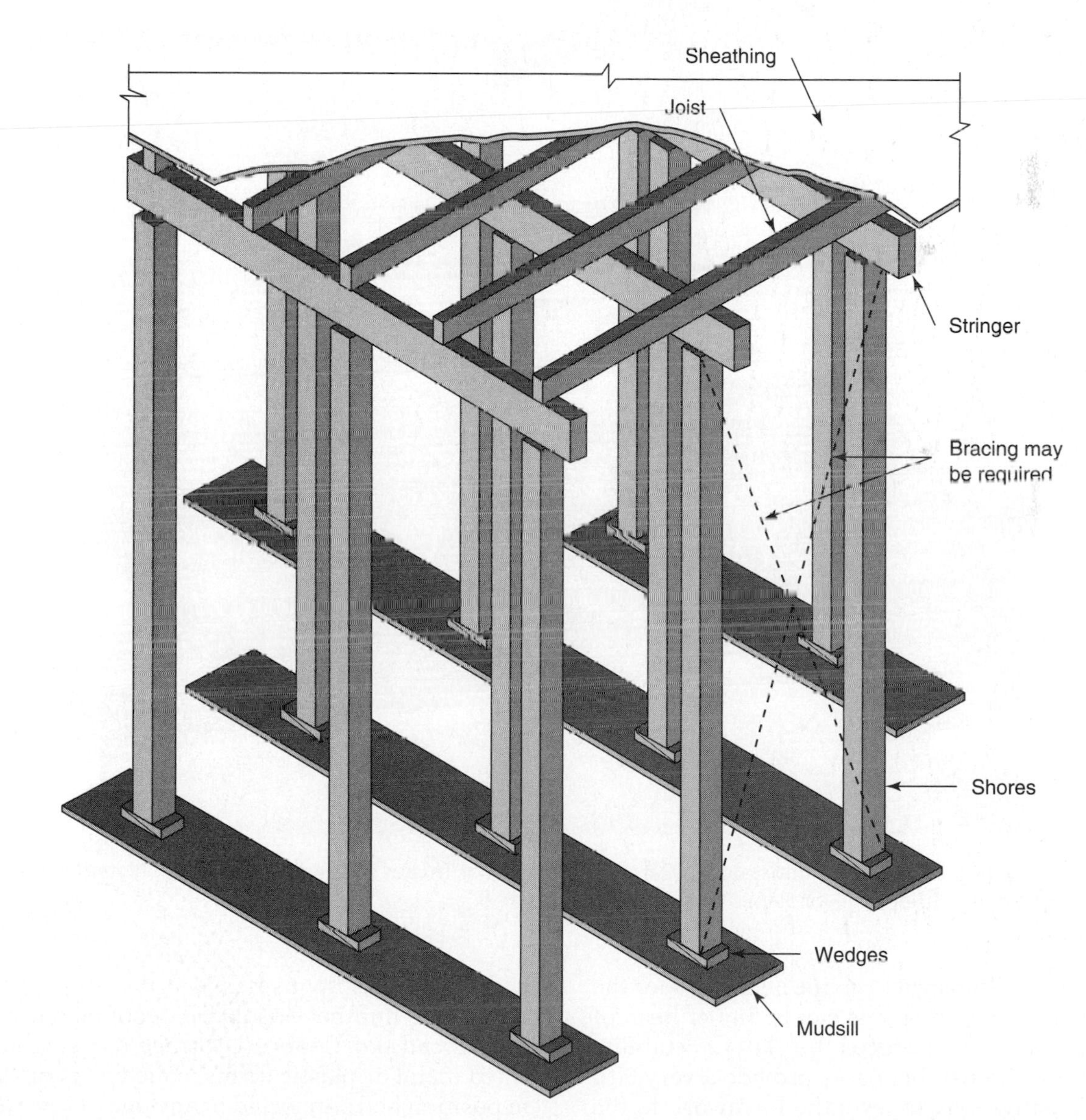


**Figure 11-26** Typical Flat Slab Formwork.

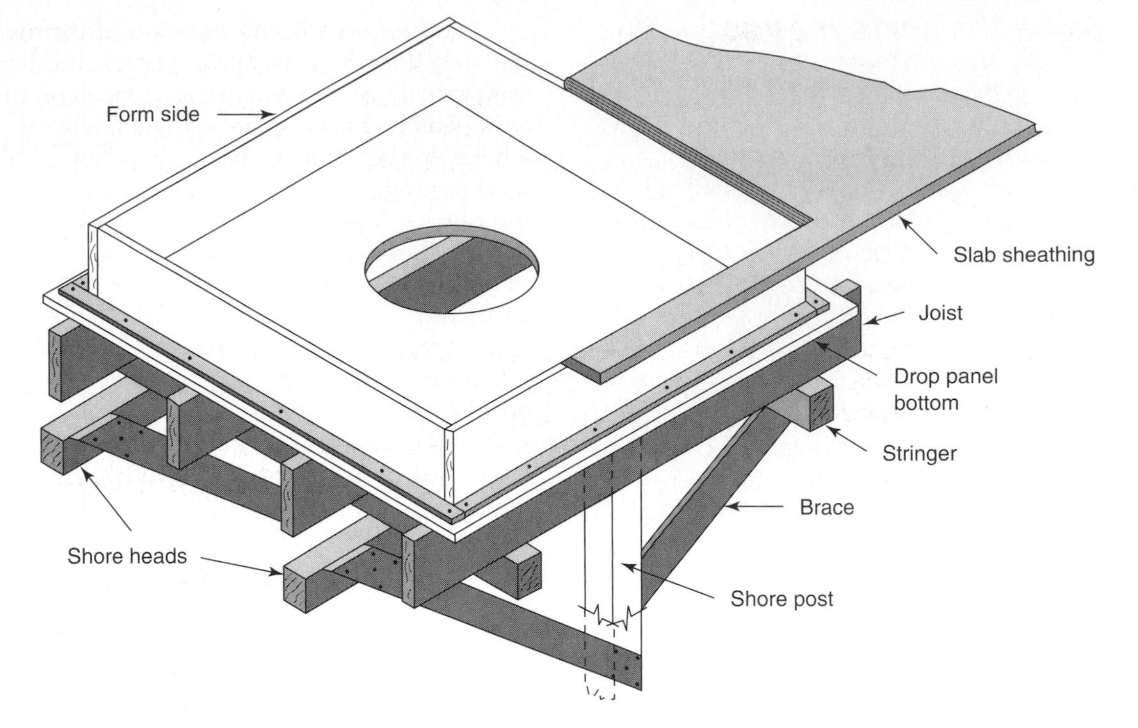


**Figure 11-27** Drop Panel Form.

**Figure 11-28** Shoring in Multistory Construction Can Take the Form of Trusses with Adjustable Bases. Note the Metal Shores under the Freshly Stripped Floor Slab.

In multistory construction, the shores under the flat slab sheathing and beams can be in the form of trusses that are cross-braced for lateral stability (Figure 11-28). Adjustable bases provide a very efficient and positive way to level the formwork to the proper elevation for the slab. The aluminum beams allow for larger spans between the trusses, thus reducing the number of support sections required.

Ribbed slab floors are formed by using prefabricated metal or plastic forms, called *pans* or *tile,* held in position and supported in any one of several ways. Pans for one-way ribs are usually made from sheet metal (see Figure 11-29), but those used to form two-way ribs—a waffle floor—may be either sheet metal or plastic.

Pans for one-way ribs are made in two ways. One type has the ends of its legs turned out flat, as illustrated in Figure 11-30, so that the pan may rest on a flat surface, whereas the other has straight legs (also see Figure 11-33) that must be fastened to the sides of supports. The *bent-leg* type is made in several widths and depths, whereas the *straight-leg* type is usually made in 20-in. (500-mm) widths. Both are made in 30-in. (900-mm) sections, with tapered end sections to provide greater width of rib at the beam. Both types have sheet metal end caps.

**Figure 11-29** Pans for One-Way Joist System in Place.

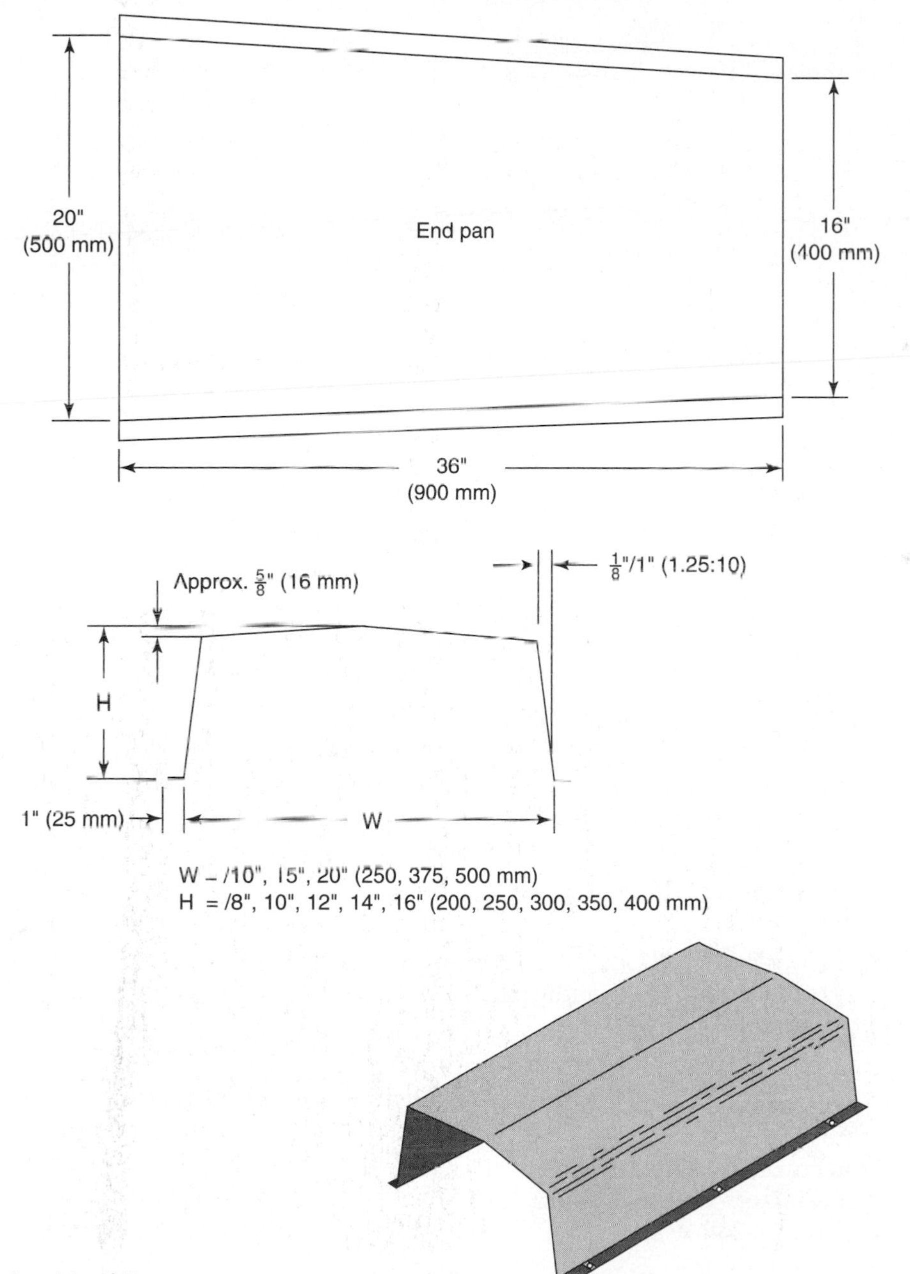


**Figure 11-30** Bent-Leg Metal Pan.

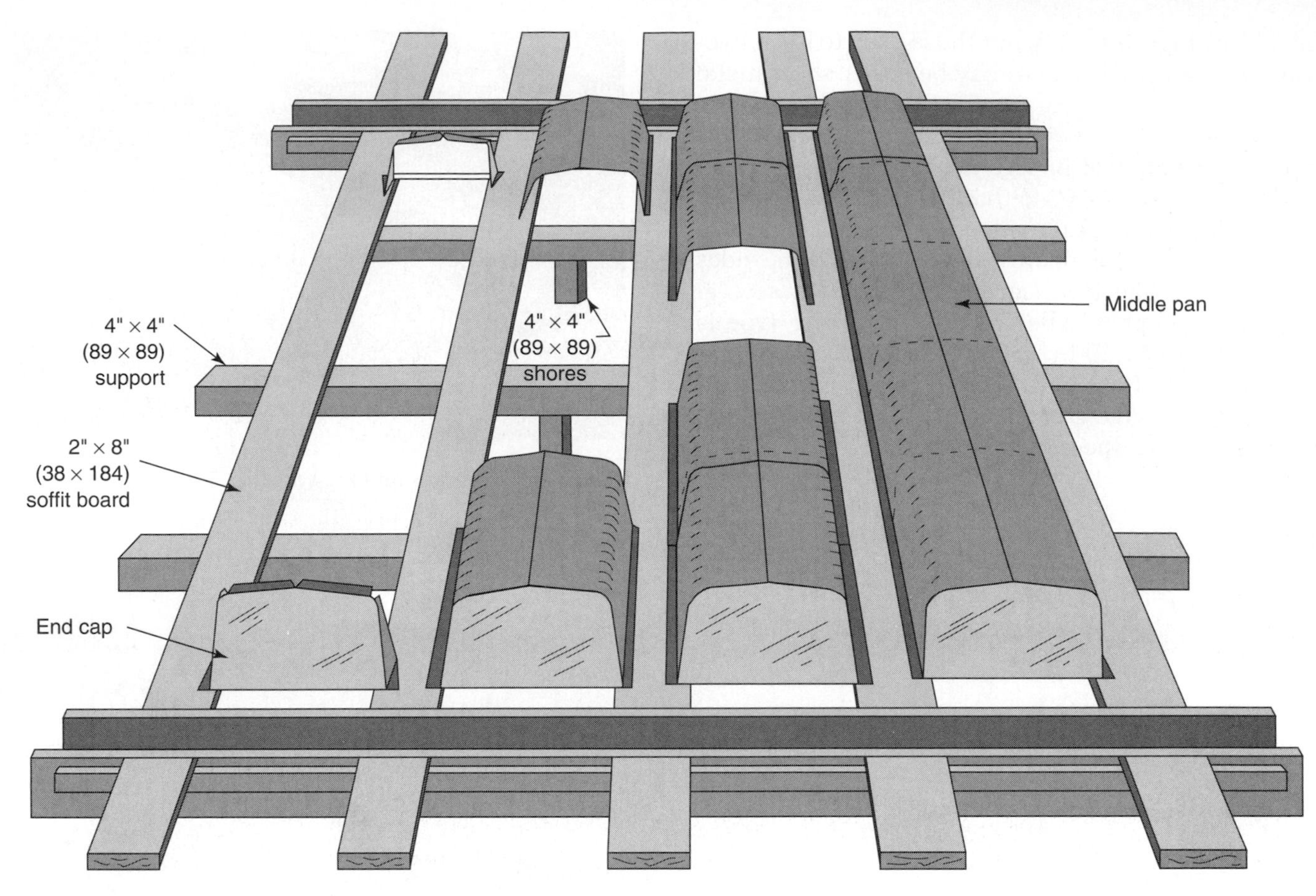


**Figure 11-31** Metal Pans in Nail-Down System.

The bent-leg pans may be used in two ways. In the nail-down system, the pans are placed and nailed to a series of *soffit boards* supported on stringers and shores (Figure 11-31). The depth of the ribs is determined by that of the pans. When this system is used, the procedure for setting up pans is as follows:

1. Erect and support soffit boards as required.
2. Nail the end caps to the soffit boards at each end of a bay.
3. Place the first pan, usually a tapered one, over the end cap.
4. Place the next pan over the first and lap at least 2 in. (50 mm).
5. Continue setting pans with at least 2-in. (50-mm) laps until they meet at the center. The middle pan must always be placed last.

The other system of using bent-leg pans, known as the *adjustable system,* is illustrated in Figure 11-32. Here the depth of ribs may be altered, within limits, according to the position of the ledger strips on the joists. The procedure for setting up pans this way is as follows:

1. Set up and level stringers as required.
2. Nail ledger strips to the joists at the specified level.
3. Set joists on the stringers on the required centers.
4. Cut spreaders to length (face to face of joists) and separators 2¼ in. (55 mm) shorter, with ends cut to the slope of the pans. Tack separators to spreaders.
5. Set the spreaders on the ledger strips, two per pan section.
6. Place end caps on spreaders at each end of bay.
7. Place metal pans on spreaders, using the same sequence as previously outlined.
8. Cut soffit pieces from ¾-in. (19-mm) material, wide enough to fit between pans and rest on the top of joists. The ends of these soffits must be widened to compensate for the taper on the end pans.

Straight-leg pans are supported by nailing them to the sides of the joists with double-headed nails (Figure 11-33). Three or more sets of nail holes are drilled in the legs, and the depth of ribs is altered by adjusting the position of the legs against the joists and using the appropriate set of holes. Joists to be used with this type of pan are usually built to form a standard rib width. The ends of the joists are widened to compensate for the tapered end pans (Figure 11-34).

Pans for a two-way joist (waffle slab) floor system may be set on a plywood deck or on soffit boards

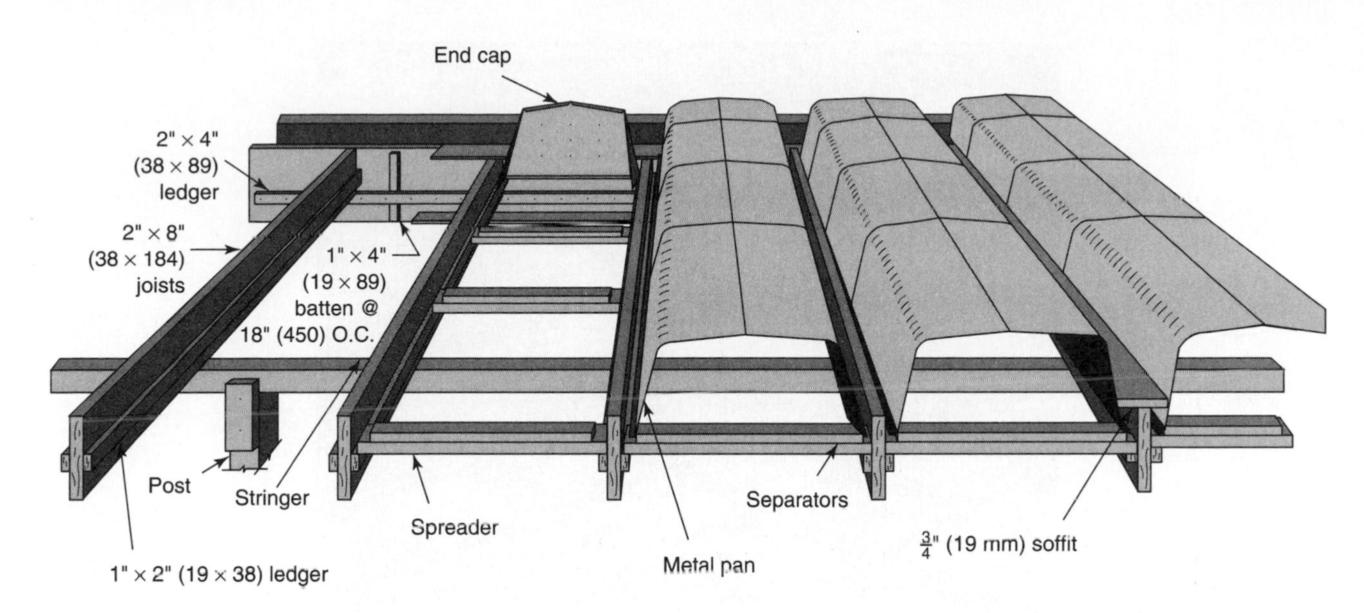


**Figure 11-32** Metal Pans in Adjustable System.

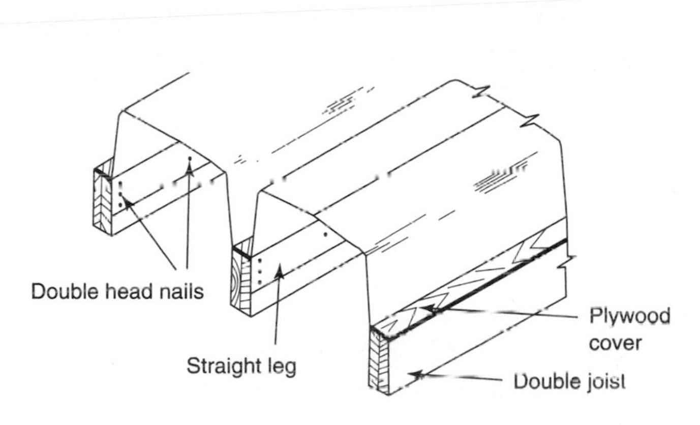


**Figure 11-33** Straight-Leg Pans on Double Joists.

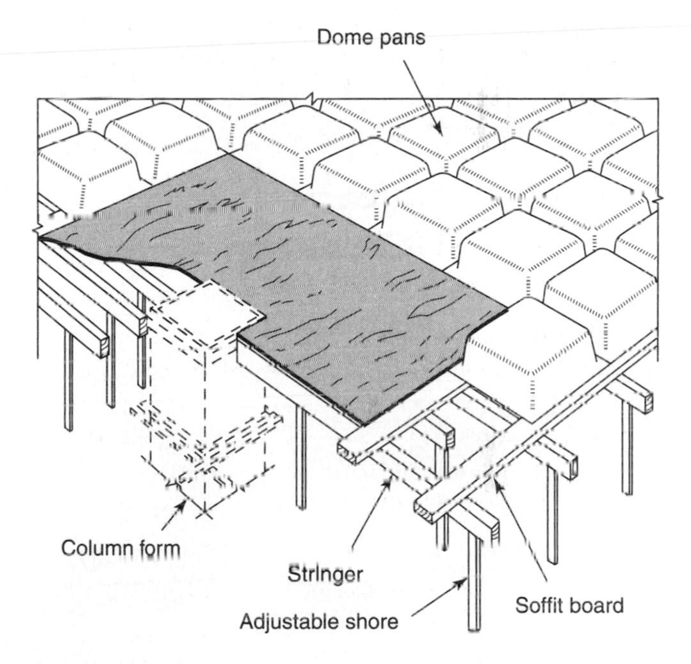


**Figure 11-35** Dome Pans for Waffle Floor.

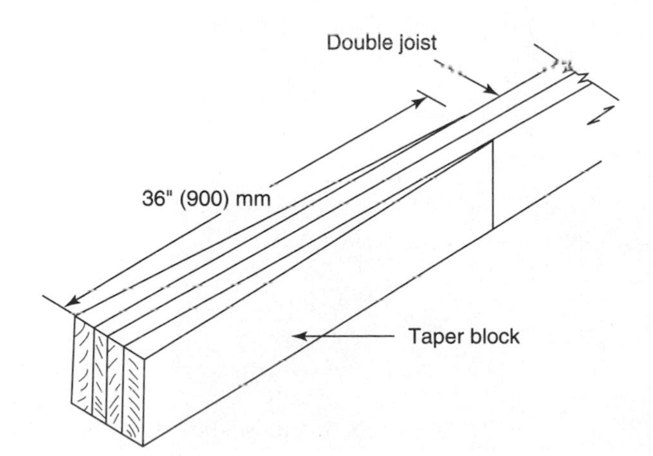

**Figure 11-34** Joist Ends Widened.

much like the one-way joist system illustrated in Figure 11-31. In Figure 11-35, soffit boards are used to support the dome pans and plywood decking is used to support the solid section of the slab (column head) over the supporting column. Figure 11-36 illustrates a two-way joist system ready for concrete. Positive reinforcement in ribbed floors is concentrated in the ribs, and negative reinforcement as well as temperature steel is placed within the slab thickness, as shown in Figure 11-37.

Another approach to the two-way slab system is to use larger beams in one direction, spanning from

**Figure 11-36** Dome Pans for Waffle Floor in Place. (Reprinted with permission of Portland Cement Association.)

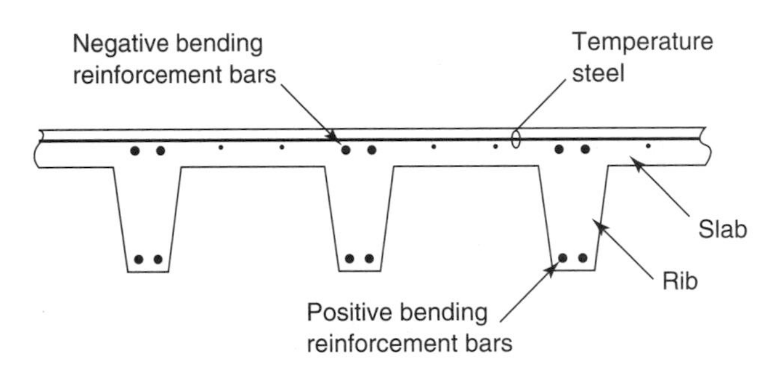


**Figure 11-37** Ribbed Slab Reinforcing

column to column, with short joists between the beams. This configuration can be achieved by using larger pans (Figure 11-38) and providing a larger space between the pans in the direction of the beams to allow for the beam reinforcing and a smaller space for the short rib or joist sections in the other direction. This type of form provides for deeper beam sections, thus allowing for greater spans than those obtained with normal-size pans.

**Figure 11-38** Pan Forms for a Beam and Joist Two-Way Ribbed Slab. Note the Heavy Reinforcing for the Beams.

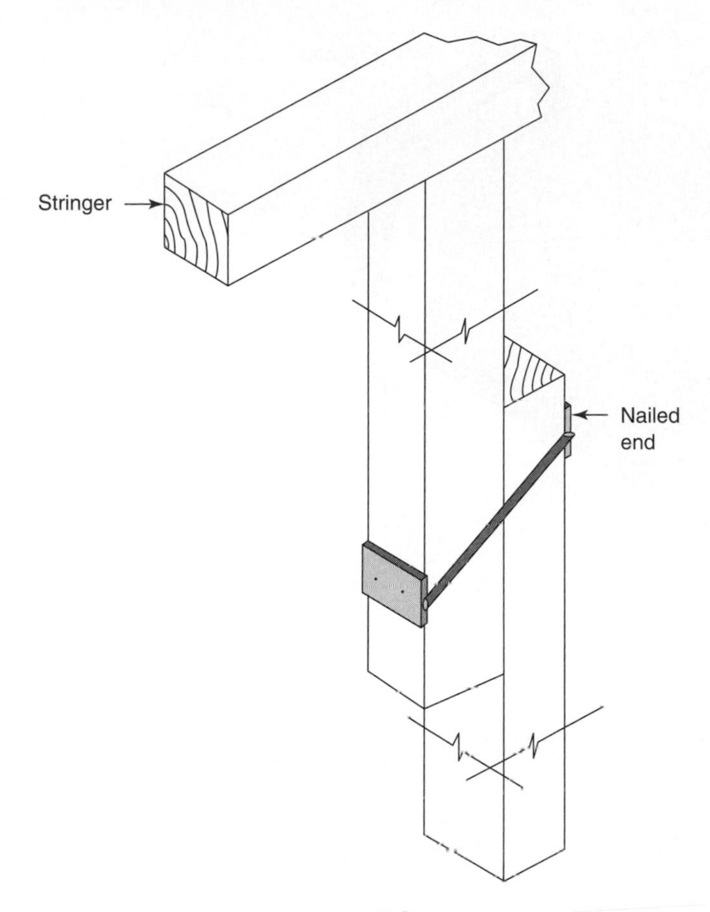


Figure 11-39 Adjustable Wood Post.

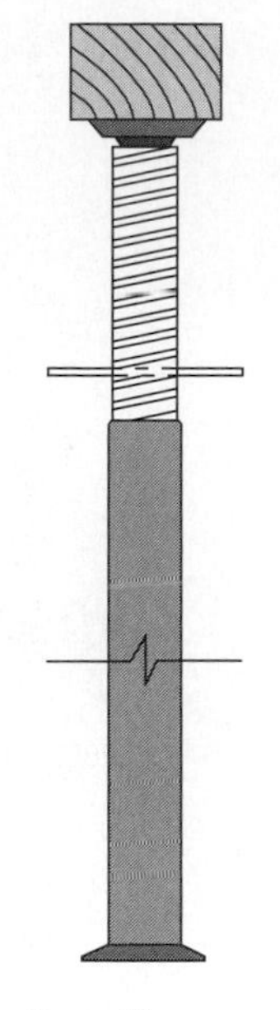
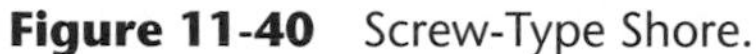

Figure 11-40 Screw-Type Shore.

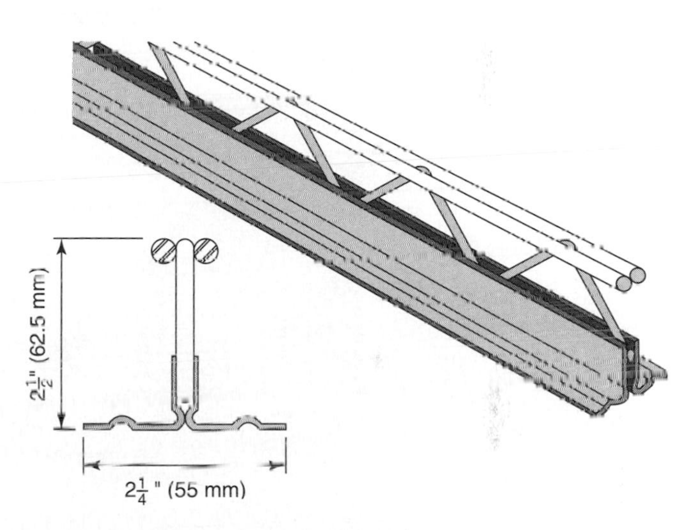


Figure 11-41 Steel Truss T. (Reprinted with permission of Portland Cement Association.)

Forms for these concrete floors must have some type of adjustable support, which is usually provided by posts or by sections of steel scaffolding. One-piece wood posts are adjusted by wedges under the bottom end, as shown in Figure 11-26. Two-piece posts are made and used as illustrated in Figure 11-39. Several kinds of adjustable metal posts are available, one of which is shown in Figure 11-40. In Figure 9-22, steel scaffolding is being used as form support.

## Integrated-Distribution Floor System

The integrated-distribution (I/D) floor (and roof) system has been developed by the Portland Cement Association and is specially designed for use in low-rise apartments, light commercial and industrial buildings, and residences.

The I/D concept includes three basic components: (1) steel truss Ts, (2) forms to be left in place, and (3) cast-in-place concrete; it incorporates heating, plumbing, and electrical utilities in the system.

The truss Ts (see Figure 11-41) do double duty as supports for the I/D forms and as the only reinforcement normally required for a simple-span system. They are supported on and anchored to load-bearing walls and intermediate shoring on 24-in. (600-mm) centers. Additional top reinforcement may be installed when required (also see Figure 11-46).

*I/D forms* (see Figure 11-42) are lightweight units having a weight from 10 to 12 lb (5 to 6 kg)/form, intended to remain in place when the floor is completed. They are easily cut by hand or saber saw for length adjustments or to form cross channels for utility connections.

Shoring should be spaced not more than 6 ft (1,800 mm) o.c. and should be left in place for a minimum of

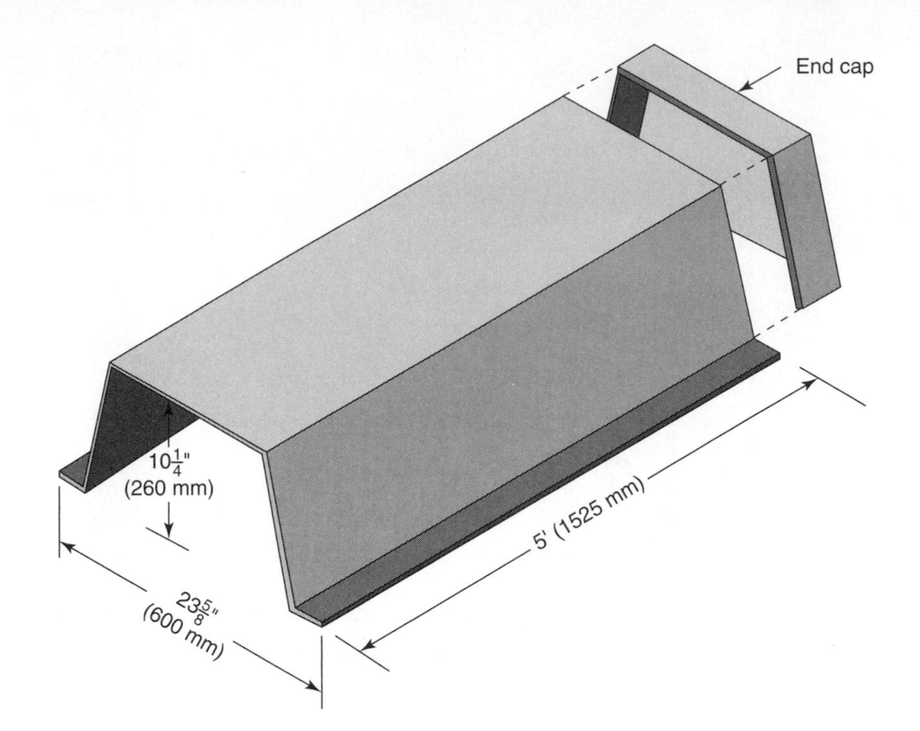


**Figure 11-42** I/D Form. (Reprinted with permission of Portland Cement Association.)

**Figure 11-43** Scaffold-Supported Shoring for I/D System. (Reprinted with permission of Portland Cement Association.)

7 days, depending on the temperature. Shores may be single jacks, properly spaced; stringers with scaffolding supports (see Figure 11-43); or non-load-bearing interior partitions. Where the last are used, the truss Ts should not rest directly on the partition, but should be supported by removable shims to provide ¾-in. (20-mm) clearance (see Figure 11-44).

Concrete, which should normally be designed for 4,000 psi (30 MPa), may be placed by any of the commonly used methods and consolidated with a spud vibrator or a vibrating screed. Care must be taken not to touch the sides of the forms with the vibrator. The resulting floor (see Figure 11-45) will have a mass of approximately 53 lb/sq ft (260 kg/m$^2$), using normal-weight concrete or 40 lb/sq ft (200 kg/m$^2$) using lightweight concrete. It provides significant acoustical control, is fire resistant, is free from squeaks, and allows for the installation of mechanical services between ribs. For more complete details on the system, consult PCA literature.

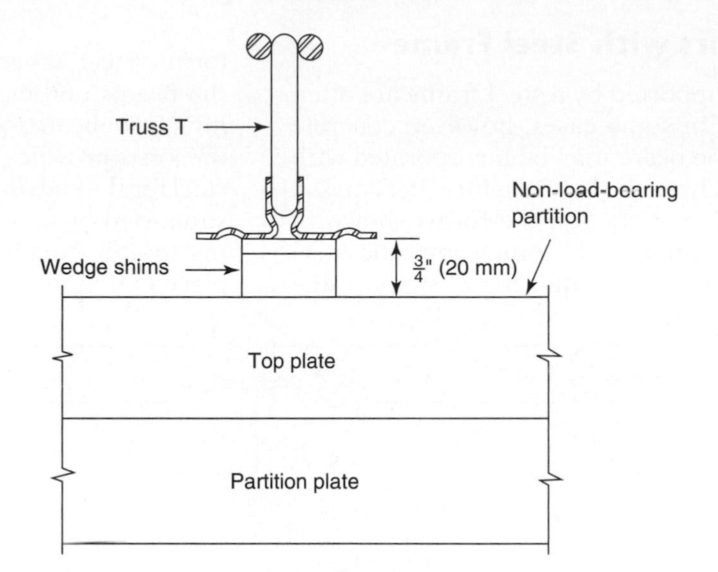


**Figure 11-44** Shims for Truss T's Supported on Non-Load-Bearing Partition. (Reprinted with permission of Portland Cement Association.)

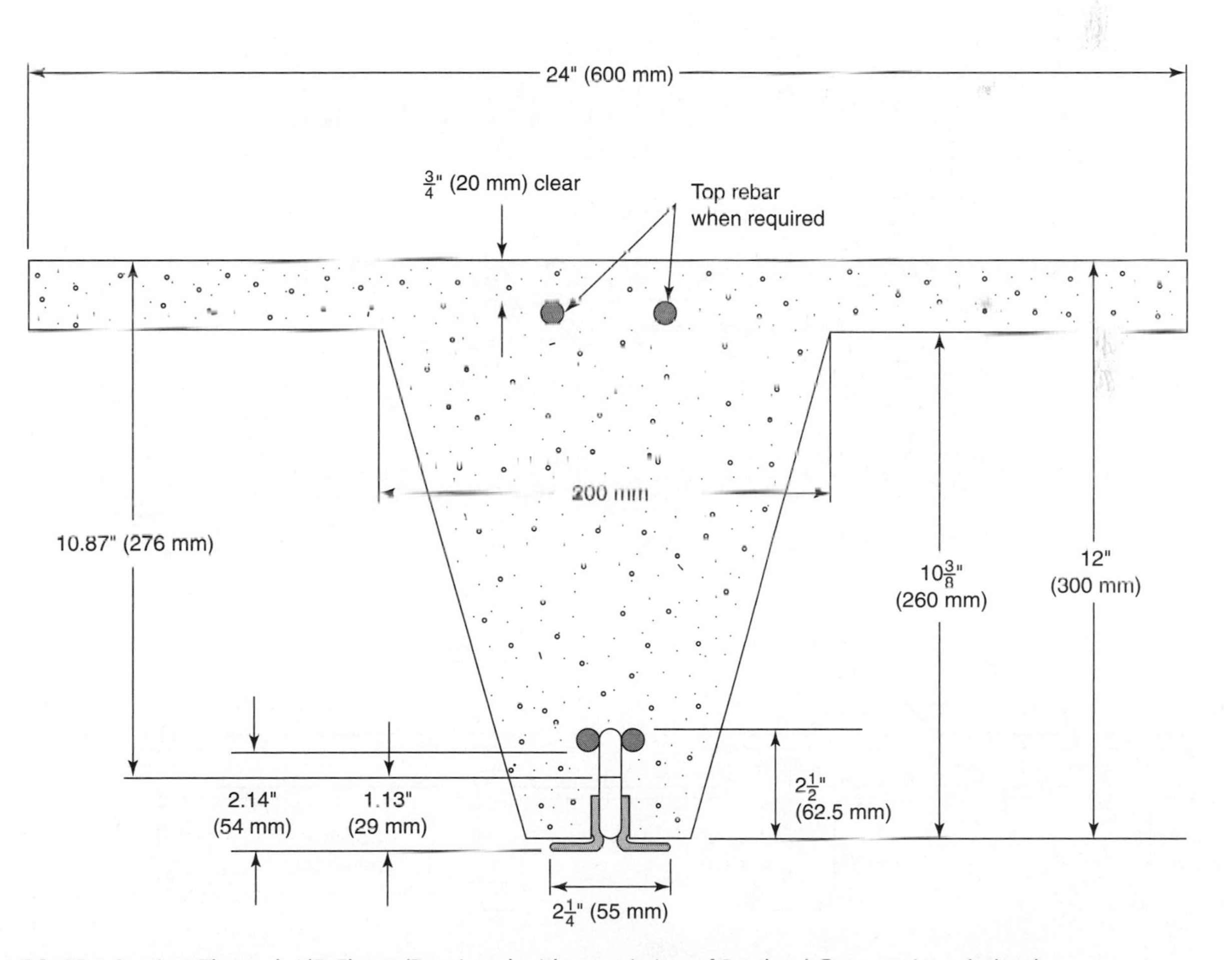


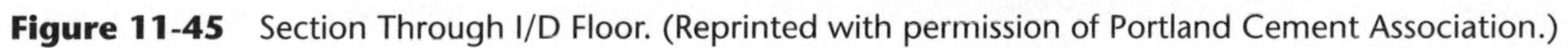

**Figure 11-45** Section Through I/D Floor. (Reprinted with permission of Portland Cement Association.)

## Concrete Floors with Steel Frame

Concrete floors supported by a steel frame are often simple flat slabs. In some cases, however, concrete fireproofing for the beam may be incorporated with the slab, in which case the structure becomes a beam-and-slab floor. Although the forms shown in Figure 11-46 are supported from below, the steel frame itself is frequently utilized to support the forms. Steel hangers are placed over the top web of the beams and carry the ends of wood joists spanning from beam to beam. Plywood sheathing laid on the joists provides the base for the slab. Reinforcing steel for the slab is similar to a one-way concrete slab supported by concrete beams. Reinforcing steel using straight bars for positive and negative bending is placed as shown in Figure 11-47. Temperature steel

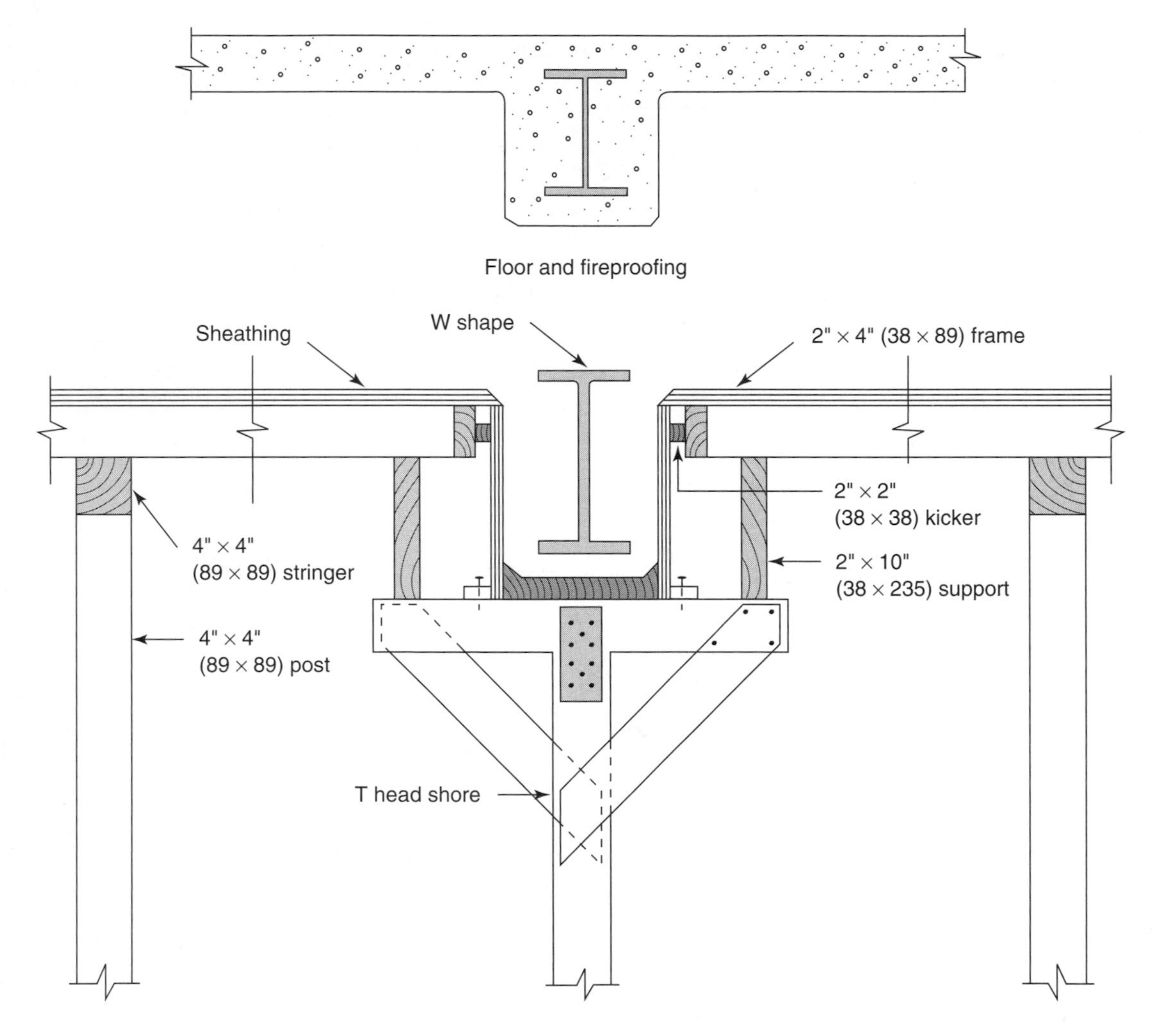

**Figure 11-46** Forming for Slab and Fireproofing.

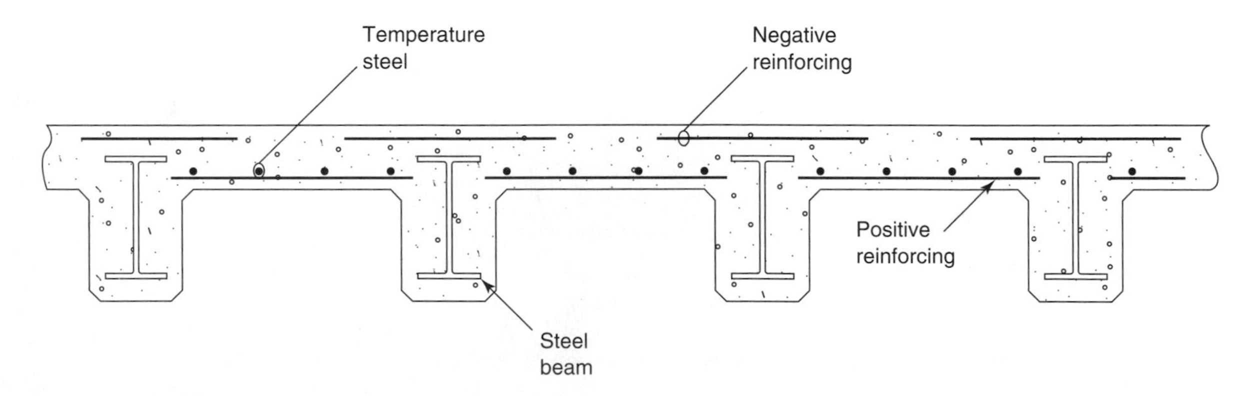


**Figure 11-47** Reinforcing Bars in Concrete Slab Supported by Steel Beams.

may be placed in one or two layers, depending on the slab thickness.

Hangers are available in a number of styles and lengths. Long *loop* hangers, such as those used in Figure 11-48, allow the top flange and the upper portion of the beam webs to be encased by the slab. *Coil* hangers (Figure 11-49) make it possible to keep the sheathing tight under the top flanges so that the entire slab is above the beams.

## Composite Floor Slabs

Composite floor slabs combine the tensile strength of structural steel with the compressive strength of concrete to produce a monolithic section that is more efficient from a load-supporting point of view than using the materials individually.

Composite slabs are designed in two ways. In the first approach, the structural steel is designed to support the weight of the concrete and the construction loads without any support from temporary shoring. Metal decking welded to the steel frame or temporary plywood forming, as discussed in the foregoing section, can be used to support the concrete slab until it cures sufficiently.

If corrugated or cellular steel decking is used, it must be securely fastened to the steel frame (see Figure 11-50) because it becomes an integral part of the floor system. Several different styles of steel decking, ranging in thickness from 22 to 14 gauge (0.76 to 1.52 mm), are available with different cell heights; the most common are 1½ and 3 in. (38 and 75 mm). In some cases, decking is supplied with closed cells, as illustrated in Figure 11-51, for additional strength and for enclosing service wiring. The decking is normally fastened to the steel frame by spot welding (Figure 11-52), by plug welding, or by using weld studs as shown in Figure 11-50. The studs provide an additional means of anchorage between the concrete slab and the steel.

In the second approach, the structural steel sections require temporary support until the concrete has cured sufficiently to provide the necessary bond between the concrete and the steel to produce a true monolithic section (see Figure 11-53). Sections such as these can be developed by bonding the concrete slab to the steel beams with shear stud connectors, by encasing the steel beam completely with concrete, or by using a steel deck anchored to the beams with stud anchors covered with concrete.

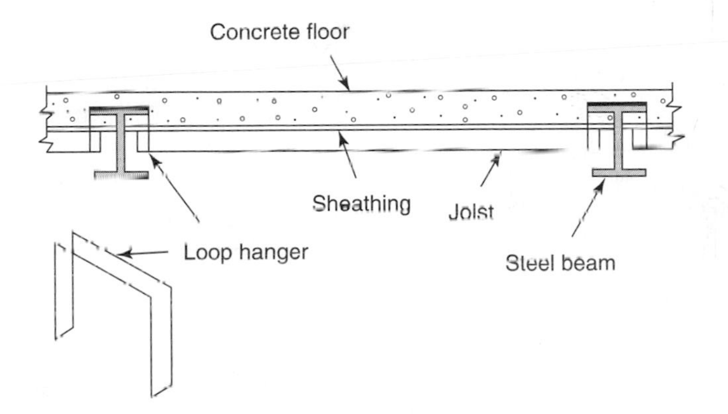


**Figure 11-48** Floor Form Suspended by Loop Hangers.

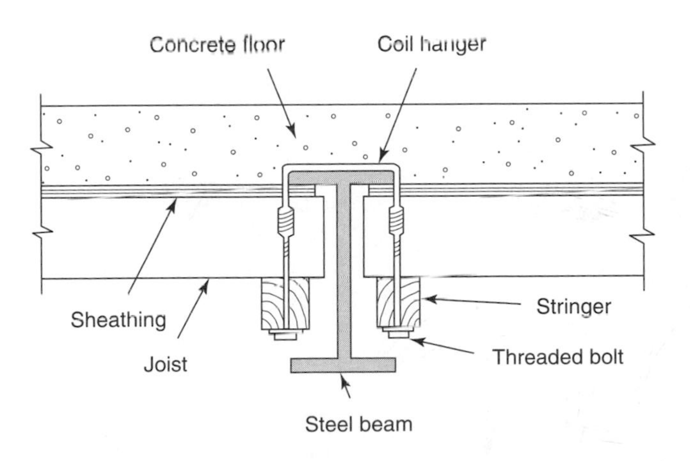


**Figure 11-49** Floor Form Suspended by Coil Hangers.

**Figure 11-50** Corrugated Sheet Steel Decking.

**Figure 11-51** Cellular Steel Decking. (Reprinted by permission of Bethlehem Steel Co.)

**Figure 11-52** Spot-Welding Steel Deck. (Reprinted by permission of Bethlehem Steel Co.)

The combination of welded stud anchors and steel decking is the most preferred method for composite floors because the decking serves as formwork for the concrete slab and provides additional strength to the assembly. The studs ensure positive anchorage of the decking to the steel frame, while providing the necessary resistance against the shear stresses that are developed at the plane between the steel and the concrete. In some applications, additional reinforcing bars are placed in the slab to control cracking and deflections. Decking that has additional ribs on the flutes is also used to further increase the bonding action between the concrete slab and the steel frame.

Another variation of a concrete floor slab supported by a steel frame is to use open web steel joists as the structural supports for the concrete slab. Joists are normally spaced 24 in. to 48 in. o.c. and covered with 1½-in. (38-mm) deep fluted steel decking. The steel decking is available in material thickness ranging from 16 to 22 gauge (to 0.76 to 1.52 mm) and comes in lengths ranging from 6.5 ft (2 m) to 40 ft (12.2 m). The steel deck is spot welded to the top chord of the steel joist, resulting in additional stability in the floor system. The poured concrete subfloor is a minimum of 2 in. (50 mm) thick over and above the depth of the deck flutes and is reinforced with welded wire mesh to control temperature and shrinkage cracking. This approach to poured concrete subfloors eliminates the need for any type of formwork and provides for shorter construction times.

## INDUSTRIAL FLOOR FINISHES

The structural floors described previously generally require some type of *floor finish*. It may be wood (block or strip flooring), concrete, terra-cotta tile, mastic flooring, resilient tile, access flooring, or terrazzo.

### Wood Floors

Strip flooring for industrial use is made from both hardwood and softwood in thicknesses of from 2 to 4 in. nominal (38 to 89 mm). Widths of strips vary from 1½ in. (38 mm) for hardwood flooring to 12 in. (286 mm) for heavy softwood flooring. Strips may be single or double tongue and grooved or grooved on

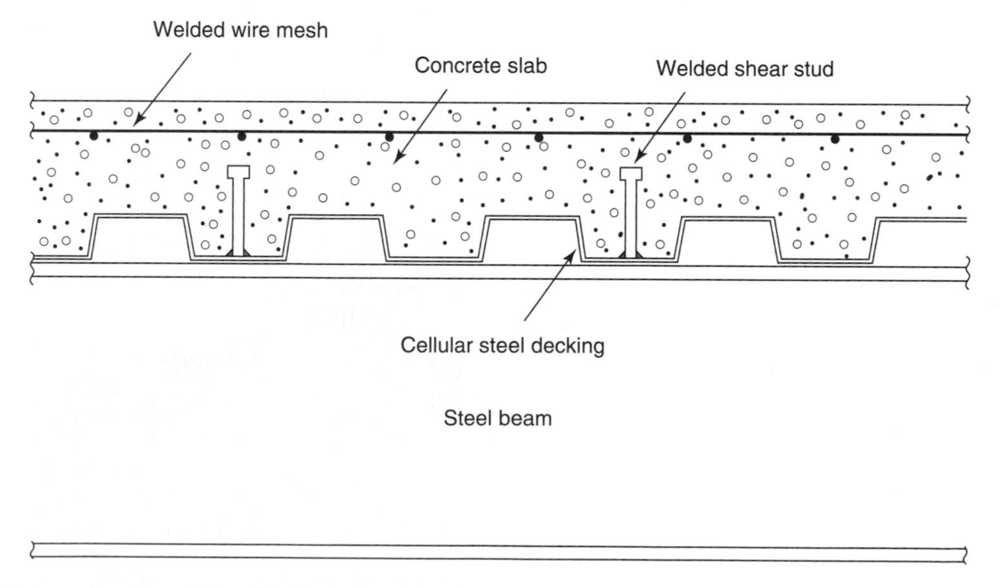


**Figure 11-53** Composite Floor Slab with Shear Studs and Steel Decking.

both edges for splines. Strip flooring may be laid over plank or laminated wood subfloors by direct nailing (see Figure 11-54) or over a concrete subfloor by nailing to sleepers that are laid in asphalt or partially cast into concrete (see Figure 11-55).

## Concrete Floor Finishes

Concrete floor finishes are used widely in industrial and commercial buildings and may consist of a single, finished slab or a base and topping slab.

A two-course slab is sometimes specified to produce a harder, denser, more durable surface. If this is the case, the base slab should be left approximately 1 in. (25 mm) below the finish grade. Just before the concrete sets, the surface should be roughened with a stiff broom to provide a better bond with the topping.

The topping mix should be made from the hardest, densest aggregates available, with a high design strength and a very low slump not exceeding 1 in. (25 mm). Just prior to the application of the topping, the base slab should be washed and a coat of cement paste applied with a stiff brush. The topping mix is spread over the surface, raked, leveled, and tamped (preferably with a power tamper). Notice in Figure 11-56 that, in spite of the dry mix, the machine is

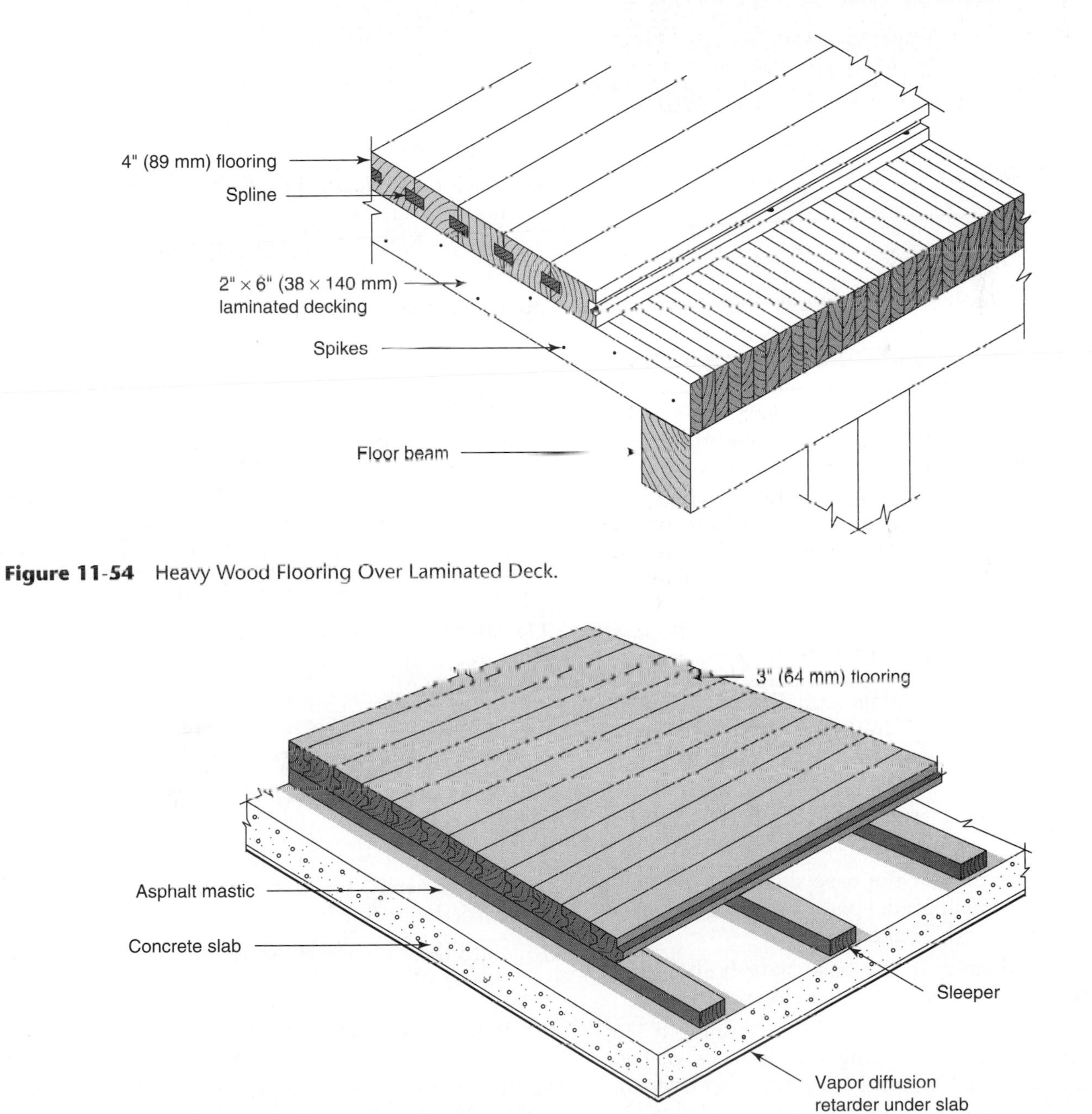


**Figure 11-54** Heavy Wood Flooring Over Laminated Deck.

**Figure 11-55** Heavy Wood Flooring over Concrete Base.

**Figure 11-56** Power Tamper and Float. (Reprinted with permission of Portland Cement Association.)

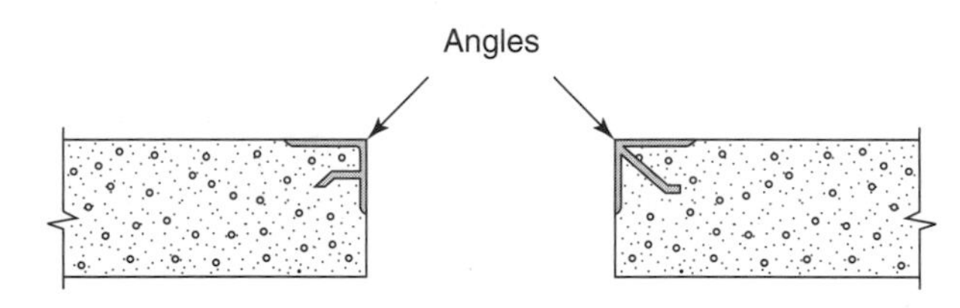


**Figure 11-57** Reinforcing for Expansion Joints.

bringing enough moisture to the top to produce a smooth surface. Final floating and troweling are best accomplished by power machinery.

A durable surface may be produced on a concrete floor slab by introducing metallic aggregate into the topping. Metallic aggregate is made up of iron particles that have been size graded to within the range of #4–#100 (5 mm–160 μm) mesh sieves, specially processed for ductility, and mixed with a cement-dispersing agent—calcium lignosulfonite.

A series of operations for producing a floor of this type follows:

1. Provide expansion joints around columns and at least every 50 ft (15 m) in both directions. Reinforce the top of the joints with angle iron (see Figure 11-57).
2. Design the base concrete for at least 3,500 psi (25 MPa).
3. Set the screeds for the base slab 1 in. (25 mm) below the finish grade, place the slab, and leave the surface with a raked finish.
4. Clean and saturate the base slab. It should be saturated 4 to 6 hr before placing the topping.
5. Set the *screed level pats* for the topping a few hours before the topping slab is to be placed. These are small mounds of mortar set over the surface at frequent intervals with their tops leveled to the finish grade.
6. For old or hard base coats, a slush bond coat should be brushed into the surface just before placing the topping. This may be made up of 100 lb (45 kg) of expanding metallic aggregate to one bag of cement, mixed with water to form a slurry. About 16 lb (7.5 kg) of the metallic aggregate will cover 100 sq ft (10 $m^2$).
7. Design the topping mix. A good design mix consists of one sack of cement with enough water to produce the specified strength of concrete. To this, add a sufficient quantity of clean, well-graded fine aggregate and ¾-in. (20-mm) maximum size coarse aggregate in equal amounts, resulting in not more than 1 in. (25 mm) of slump. This will usually require about 1½ cu ft (0.04 cubic metres) or 150 lb (70 kg) each, of fine and coarse aggregates. However, enough moisture must be available to place the metallic aggregate.
8. Place the topping mix. Rake, screed, level, and tamp. A grill-type tamper is preferred (Figure 11-58). Remove the level pats as placing progresses, and fill the holes.
9. Immediately after tamping, float the surface to bring moisture to the top.
10. First metallic aggregate shake: Mix one sack of portland cement and two sacks of metallic aggregate together dry and shake evenly over the surface at the following rates: for light-duty floors, 45 lb/100 sq ft (22 kg/10 sq m); for moderate-duty floors, 75 lb/100 sq ft (36.5 kg/10 sq m); for heavy-duty floors, 90 lb/100 sq ft (44 kg/10 sq m). If the floor is to be finished with mechanical equipment, these amounts may be increased.
11. After the shake has absorbed the surface moisture, tamp the surface. Then follow with a wood float.
12. Apply a second metallic aggregate shake as in Step 10, tamp, and float.
13. Immediately after floating, trowel the floor with a steel trowel.
14. As soon as the surface becomes hard enough to ring under it, burnish with a steel trowel to a hard, dense finish.
15. Cure the finished floor as required (see Figure 11-59).

When a nonslip concrete floor is specified, it may be produced by the addition of an aggregate composed of ceramically bonded aluminum oxide abrasive to the leveled surface. The following procedure is recommended:

**Figure 11-58** Consolidating and Leveling a Concrete Floor. (Reprinted with permission of Portland Cement Association.)

**Figure 11-59** Well-Finished Metallic Aggregate Floor. (Reprinted by permission of Master Builders Inc.)

1. Allow the leveled concrete surface to set until it will bear some weight.
2. Soak the abrasive aggregate in water for 10 min before applying.
3. Shake the aggregate evenly over the surface at the rate of ¼ lb/sq ft (1.25 kg/m$^2$).
4. Float the aggregate into the surface but do not trowel smooth.

A colored surface may be produced by the addition of a prepared coloring material. It consists of a synthetic mineral oxide of a given color, mixed with fine silica sand or silicon carbide. The coloring procedure is as follows:

1. As soon as the slab is placed, float the surface with a wood float.
2. Shake the coloring material evenly over the surface at the rate of ¼ lb/sq ft (1.25 kg/m$^2$).
3. Float the surface until the color is evenly distributed and worked into the topping.
4. Immediately apply another shake at the same rate.
5. Float again until the color has been worked into the surface to a depth of approximately ¼ in. (6 mm).
6. When the concrete has set sufficiently, trowel smooth as required.

## Terra-Cotta Tile

Tile flooring consists of a layer of some type of ceramic tile, from ¼ to 1 in. (6 to 25 mm) thick, laid over a concrete base slab. The tile is laid in a mortar bed spread evenly over the surface, and the spaces between tiles are filled with grout.

The procedure for laying terra-cotta tile is as follows:

1. Wash and saturate the base slab. If the concrete is old or hard, apply a slush bond coat of cement and water with a stiff brush.
2. Stretch two lines across the floor at right angles to one another so that the area is divided into four equal parts.
3. Mix the bedding mortar in the proportion of 1 part cement to 3 parts sand, with enough water to make a plastic, workable mix.
4. Start at the intersection of the two lines, apply a layer of mortar, and lay the first row of tiles to

the line. Tap each one firmly so that it is well bedded in mortar.

5. Lay a second row at right angles to the first, along the second line.
6. Lay the remainder of the tiles in that quarter, keeping them as level as possible and maintaining an even spacing between tiles.
7. Repeat in the opposite quarter and then lay the other two quarters.
8. When the tiles are set, prepare a grouting mix using the same proportions of cement and sand, but make the mortar a little more fluid. Pour some over a section of tiled surface and rub it into the spaces between tiles with a piece of heavy burlap.
9. Rub off the excess grout, and when it has set sufficiently so as to not pull out of the spaces, clean the surface of the tiles thoroughly.

## Mastic Floors

Mastic flooring materials are applied to the base floor in a stiff plastic state by spreading, rolling, and troweling. Three types of material are commonly used: *magnesite, asphalt,* and *epoxy resin* compounds.

Magnesite flooring is composed of calcined magnesium oxide and magnesium chloride, mixed with water to form a stiff plaster. The material is applied in two coats, each approximately ¼ in. (6 mm) thick. It may be installed over either wood or concrete; when used over a wood subfloor, metal lath is laid first to provide a better bond.

The first coat of magnesite should contain some fibrous material such as asbestos or nylon to give it greater strength. The second coat, which may contain color, fine marble chips, or an abrasive aggregate, is placed as soon as the base coat has set. It is screeded, compacted, and troweled in much the same way as concrete topping.

Asphalt flooring consists of emulsified asphalt containing asbestos fiber mixed with portland cement and stone aggregates to form a stiff plastic material. It is spread over the base floor and rolled smooth and level. Asphalt flooring may be laid over either wood or concrete bases, though the application is somewhat different in each case. For a wood base, a layer of 15-lb (No. 15) asphalt-saturated felt is laid first with 4-in. (100-mm) laps. Stucco wire is laid over the paper and nailed down, followed by a coat of asphalt primer.

A fill mix is then applied to level the surface and cover the stucco wire. It consists of 1 part portland cement, 1 part emulsified fibrated asphalt, and 5 parts clean, coarse sand. The final coat is the same as that used over a concrete floor.

The concrete base slab should first be coated with asphalt primer thinned with equal parts of mineral spirits. When it is dry, a slurry coat consisting of 1 gal (3.8 litre) of portland cement, 1 gal (3.8 litre) of emulsified asphalt primer, and 15 gal (57 litres) of water is painted over the surface. The flooring mix consists of 1 part portland cement, 2 parts emulsified fibrated asphalt, 2½ parts clean coarse sand, 4½ parts ⅛- to ⅜-in. (3- to 10-mm) pea gravel, and not more than 1 part water. This mixture is spread, leveled, compacted, and rolled to a depth of from ⅝ to ¾ in. (16 to 20 mm) to a smooth finish.

Epoxy resin is a synthetic that is used as a bonding agent in various construction applications. It may be used as a basic ingredient of floor topping. It is made of two components, a liquid resin and a curing agent. These must be mixed in prescribed proportions when the topping mixture is made.

Two types of floor topping can be made with this epoxy resin. Kiln-dry, salt-free sand and portland cement may be mixed with it to produce a smooth floor topping. The mix is composed of 182½ lb (83 kg) of liquid resin, 17½ lb (8 kg) of curing agent, 160 lb (72.5 kg) of portland cement, and 640 lb (290 kg) of 30–50 (630–315 μm) mesh sand. The topping may be colored by adding pigments to the liquid portion. The surface is first primed with an epoxy prime coat, and the mix is then spread over the floor, floated, and troweled smooth to a depth of from ⅛ to ¼ in. (3 to 6 mm). Mixes should be limited to 100 lb. (45-kg) batches because of the limited pot life of the mixed topping.

A terrazzo topping may be prepared by mixing marble chips with liquid plastic. Coloring pigment may also be added. These ingredients are mixed in the following ratios: 160 lb (68.5 kg) of liquid plastic, 87 lb (37 kg) of calcium silicate, 35 lb (15 kg) of rutile titanium dioxide paste, and 18 lb (8 kg) of hardener per 700 lb (300 kg) of marble chips. The surface must again be primed, and the mix must be applied and troweled to a depth of from 3/16 to ¼ in. (4 to 6 mm). After 24 hr, the surface is ready for grinding and polishing.

Although magnesite and asphalt floorings provide a durable finish, the mixing process is quite involved and their use has been replaced by flooring that requires less work on location. Also, the use of asbestos fibres is prohibited in many areas and some other fibrous material would be required as the binder if these types of flooring are to be used.

## Resilient Tile

Resilient tile flooring includes a number of products such as asphalt, linoleum, cork, vinyl or vinyl-asbestos, and rubber. All require a smooth surface, either wood or concrete, and a special mastic. Asphalt and vinyl-asbestos tiles are particularly suited for concrete slabs on grade, because both are laid in moistureproof asphalt mastic. Laying should in all cases begin at a line stretched across the center of the floor.

## Access Flooring

With the development of sophisticated communications systems, building owners have found that tenants require additional flexibility in the placing of electrical outlets and telecommunications wiring throughout their office space.

To deal with these requirements, especially in older buildings, the concept of access flooring was developed. Access flooring is sectional flooring that is placed on top of existing floors with sufficient space for the running of wires and conduits between the two. The flooring sections are supported by preformed metal frames (Figure 11-60), which in turn are supported by adjustable stools. The height between the two floors varies, depending on the needs of the client, and usually ranges from 6 to 12 in. (150 to 300 mm).

The flooring panels are finished with a durable material such as vinyl and fit snugly into the frames. Electrical outlets, computer network lines, and telephone lines can be installed where necessary without major expense. The floor panels can be covered with carpet to deaden noise. The advantage of this type of floor system is that its installation does not affect the existing structure in any way. If a client's needs change, the flooring can be removed and used elsewhere.

## Terrazzo Floors

*Terrazzo* is defined as a composition material, cast in place or precast, that is used for floor topping and wall treatments. It consists of marble chips, seeded or unseeded, with a binder or *matrix,* which may be cementitious, noncementitious, or a combination of both. The terrazzo is cast, cured, and then ground and polished or otherwise finished. *Rustic terrazzo* is a variation in which quartz, quartzite, onyx, or granite chips may be substituted for the marble chips and, in place of grinding and polishing, the surface is washed with water or otherwise treated to expose the stone chips.

There are a number of types of terrazzo installations, including sand cushion, bonded, monolithic, and structural, in which portland cement paste is used as the binder for the topping (see Figure 11-61). In addition, there are a number in which emulsions of plastic resins such as epoxy, polyester, and polyacrylate or combinations of these with portland cement paste are used as the binder.

In any of these types of installations, *divider strips* are used to control cracking of the terrazzo surface due to expansion and contraction of the base floor. Divider strips are of two basic types: those that are grouted into openings cut in the base slab and those that are cemented or nailed to the surface (tees

**Figure 11-60** Installation of Access Flooring.

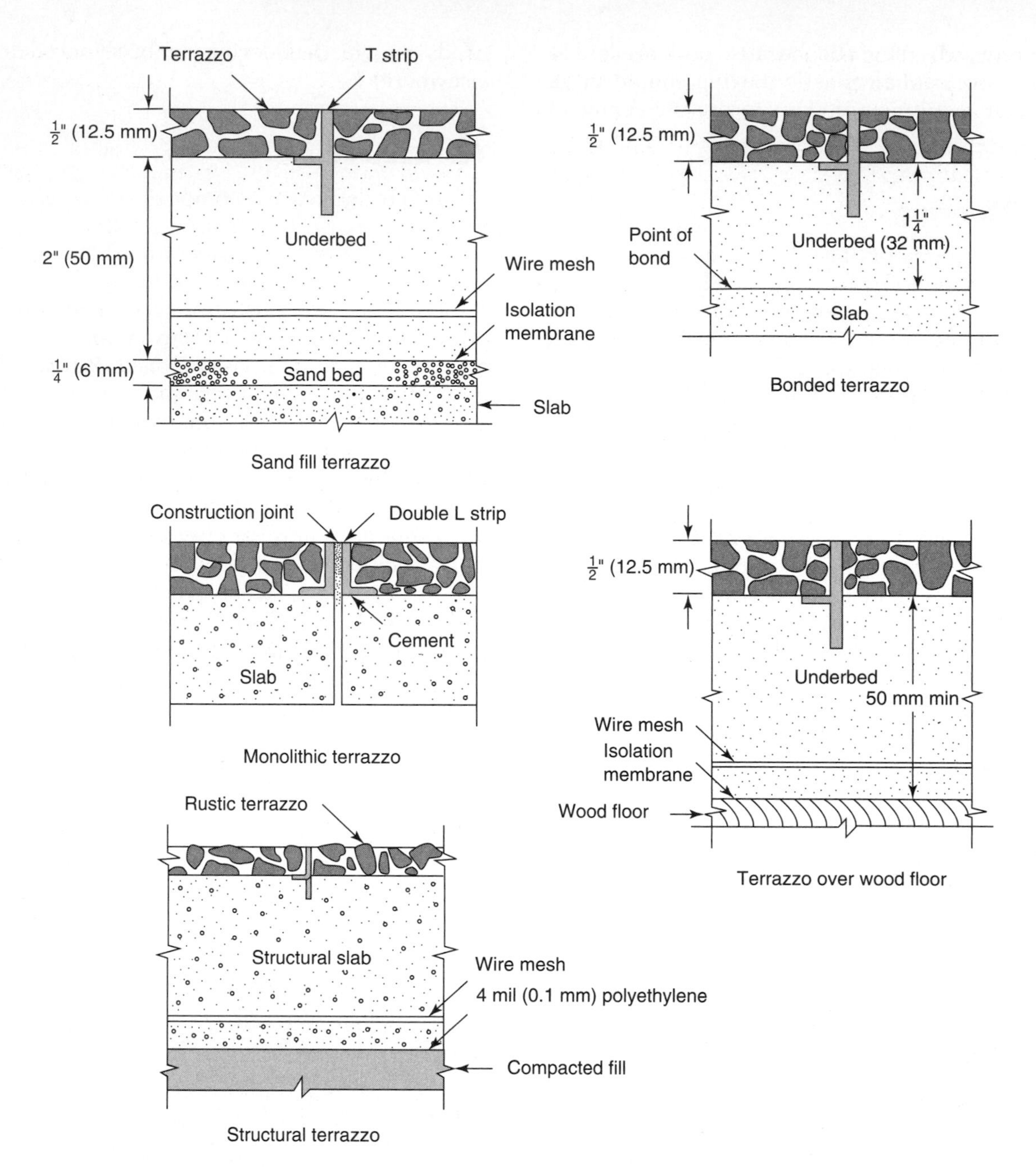


**Figure 11-61** Divider Strips. (Reprinted by permission of National Terrazzo and Mosaic Association.)

and angles) (see Figure 11-62). They are available in half-hard brass, white alloy zinc, and plastic. In addition, an expansion joint strip is available, which is recommended for use over all expansion joints in the base slab.

Divider strips should be installed over all control joints, construction joints, and isolation joints in the base slab and, in addition, may be set on the surface when there are no joints or cuts, to divide the floor area into squares or rectangles containing 200 to 300 sq ft (18 to 25 m$^2$). Rectangles should be no more than 50% longer than wide and squares no more than 16 × 16 ft (4.8 by 4.8 m).

## Terrazzo Installation

Terrazzo installation procedures will vary somewhat, depending on the type of installation (see Figure 11-61). For example, monolithic terrazzo, as the name implies, is placed at the same time as the base slab and bonded directly to it (see Figure 11-63). As a result, any structural movement occurring in the slab is likely to cause damage to the finished floor. To minimize any such damage, the following steps should be taken when placing the slab:

1. For on-grade slabs (structural terrazzo), the fill should be thoroughly compacted.

2. For on-grade slabs particularly, a power screed should be used to bring some cement paste to the surface to seal off surface pores.
3. All plumbing and electrical conduits should be placed neatly and rise at right angles through the slab.

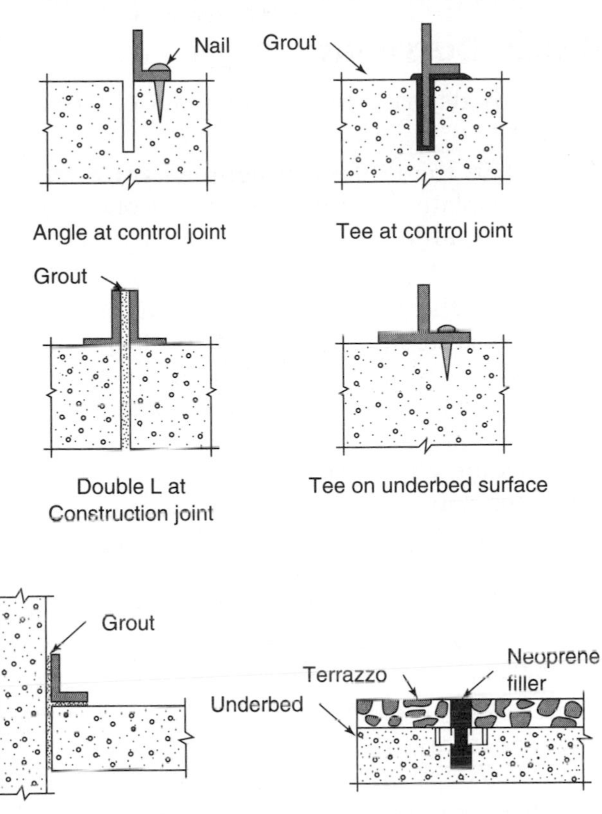


**Figure 11-62** Terrazzo installation types. (Reprinted by permission of National Terrazzo and Mosaic Association.)

4. When the slab has set enough to walk on, it should be broomed to roughen it to ensure a good bond.
5. In large areas, the concrete should be placed in alternating slabs, to extend more control over expansion and contraction.

For any type of terrazzo installation, great care must be taken in placing the subfloor. In most cases, specifications will limit the variation in level to not more than ¼ in. in 10 ft (6 mm in 3 m), though in the case of monolithic terrazzo, for example, the variation will be limited to ⅛ in. (3 mm) in 10 ft (3 metres).

The installation of sand cushion terrazzo is a typical operation. It is normally 2½ to 3 in. (65 to 75 mm) thick, with a ½ in. (12.5 mm) terrazzo topping, a 2- to 2¼-in. (50- to 55-mm) underbed, and ¼ in. (6 mm) of sand as cushion (see Figure 11-61). To install sand cushion terrazzo, proceed as follows:

1. Cover the entire slab surface with clean, dry sand to a maximum depth of ¼ in. (6 mm) and level uniformly.
2. Install isolation membrane over sand with 3 in. (75 mm) overlap at ends and edges.
3. Install welded wire reinforcement, overlapping two squares at ends and edges and stopping the mesh 1 in. (25 mm) short of expansion joints.
4. Place concrete underbed—a mixture made from 1 part portland cement to 4½ parts sand, with just enough water to provide workability at as low a slump as possible.
5. Screed the underbed to ½ in. (12.5 mm) below the finished floor level.

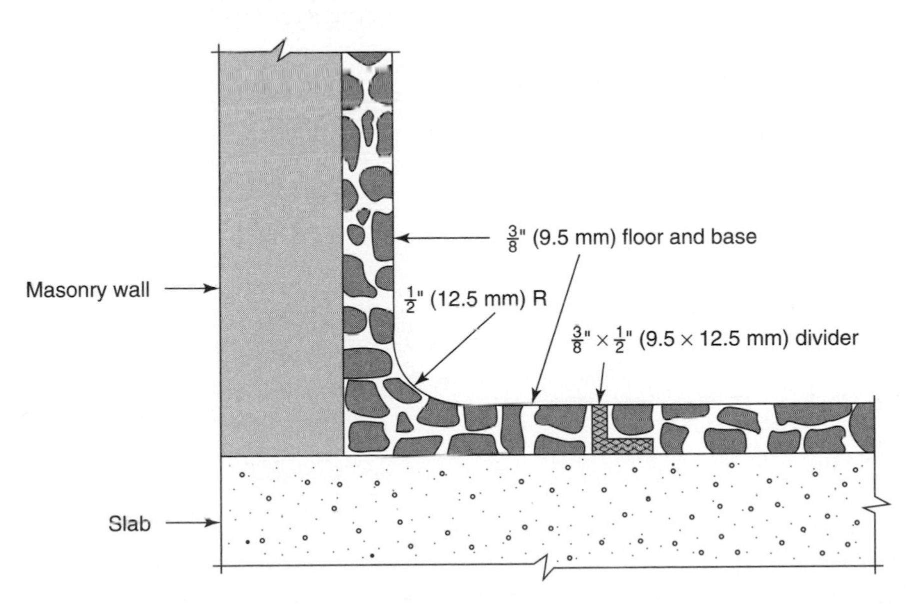


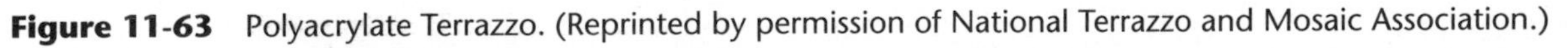
**Figure 11-63** Polyacrylate Terrazzo. (Reprinted by permission of National Terrazzo and Mosaic Association.)

6. Install control joint strips precisely over all joints in the slab and divider strips as shown in the drawings and trowel firmly along the edges to ensure positive anchorage.
7. Slush the underbed with neat cement paste, colored as specified for the topping, and broom the paste into the underbed surface.
8. Place the terrazzo mixture in the panels formed by divider strips. The topping is mixed at the rate of 1 lb of cement to 2 lb of chips (1 kg of cement to 2 kg of chips), with approximately 0.6 lb (0.6 kg) of water and color pigment as specified to produce the color required, to a uniform, workable consistency.
9. Trowel the mixture to the level of the top of divider strips.
10. Seed the troweled surface with additional chips in the same proportions as contained in the terrazzo mix, and trowel.
11. Roll the seeded surface with heavy rollers until all excess water has been extracted.
12. Trowel to a uniform surface, disclosing the lines of the divider strips.

After placement has been completed, curing material should be installed according to the manufacturer's recommendations and the topping cured until it develops sufficient strength to prevent lifting of terrazzo chips during the grinding operation.

When the terrazzo has hardened, it is wet-ground with machine grinders to a smooth, even surface. It is then washed, and a grout of cement and water, colored if necessary, is applied to fill any voids. After approximately 3 days the grout is removed, and the surface is polished by machine to a satisfactory finish.

Finally, the surface is thoroughly washed and rinsed and allowed to dry completely, and a sealer is applied according to the manufacturer's instructions to prevent dirt and stains from penetrating and marring the surface.

For complete information and installation details for all types of terrazzo, consult *Terrazzo Technical Data,* published by the National Terrazzo and Mosaic Association, Inc.

## Review Questions

**1.** Briefly describe two major types of wood subfloors.

**2.** Explain the difference between **(a)** a flat slab and a flat plate concrete floor, **(b)** a one-way joist slab and a waffle slab floor, and **(c)** structural floor slab and a slab on grade.

**3.** Outline the reasons for **(a)** isolating columns from a concrete floor around them, **(b)** providing a key between two adjacent concrete floor slab sections, **(c)** covering the subgrade with a moisture barrier before casting a concrete slab on grade, and **(d)** providing chairs for reinforcing bars on a concrete floor slab.

**4.** Explain the purpose of **(a)** metallic aggregate in a concrete floor topping, **(b)** using block flooring in a wood floor specification, **(c)** a column capital in a flat slab floor design, and **(d)** using a ribbed slab floor design.

**5.** Define each of the following terms: **(a)** *broken bond* in terrazzo flooring, **(b)** *mastic* flooring, **(c)** *sand cushion* in terrazzo flooring.

**6.** What advantage do composite floors have over ordinary concrete slab floors?

**7.** In what situation could the use of access flooring be considered as a viable alternative to other flooring systems?

# 12 Roof Systems and Industrial Roofing

The primary purpose of a roof is to protect the interior of the building from moisture and to prevent excessive heat loss. However, the roof is also an integral part of the structural frame and must be designed to sustain loads due to wind, snow, and rain, as well as to contribute to the building's exterior appearance.

The complete roof system consists of several components, including the roof frame, roof deck, roof membrane, insulation, vapor barrier, flashing and drains, construction and expansion joints, walks, parapets, and gravel stops.

A number of factors enter into the design of a roof system, such as weather, appearance, height area, and type of frame. To accommodate these diverse needs, roofing systems using various materials are continually being developed to provide economical and long-lasting protection for buildings.

## LOW SLOPE ROOF SYSTEMS

Based on design and construction objectives, the most efficient roof framing system is one that is relatively flat. Framing for low slope roofs can consist of wood, steel, or reinforced concrete, depending on the loads that must be supported and the distances between supports. The slope of the finished roof is usually in the vicinity of ¼″ in 12″ (1:50); however, that can vary. It may be less if framing spans are short or more if framing spans are long. Although low slope roofs may appear flat, some slope is necessary to facilitate roof drainage and to control *ponding*. Ponding is buildup of water on a roof during a sustained rainstorm. If the roof framing is relatively flexible, a buildup of water can occur, causing excessive roof deflections. As the weight of the water increases, the deflections become larger until the roof system fails. Although a complete collapse of the roof system may not occur, the waterproof membrane may be damaged and water leaks may result. Ponding can be prevented on low slope roofs by providing a minimum slope to the roof, using more robust framing members, and providing an appropriate number of roof drains placed in the proper locations. Low slope roofs that are enclosed by parapet walls can be drained by the use of scuppers, as illustrated in Figure 12-1. Buildings in climates that experience freezing rain must have safeguards to ensure that all roof drains remain open.

Although low slope roofs are straightforward in their design and construction, they may be susceptible to leaking due to the breakdown of the waterproof membrane, or if sufficient care is not applied when placing flashing and expansion joints.

### Wood Roof Frame

In heavy timber frame construction, the roof members for a flat roof are framed to columns as outlined in Chapter 8. A positive roof anchoring system must be employed to protect against displacement by wind. Figures 8-16 to 8-18 illustrate methods of attaching roof beams to masonry exterior bearing walls. Roof anchors are again used. Note the *anchors* on the bearing plates (Figure 8-16) to hold the roof beams in place. The framing of roof members in steel or reinforced concrete frame structures is essentially the same as that used for floors.

When greater spans between supports are required, and regular roof beams are no longer

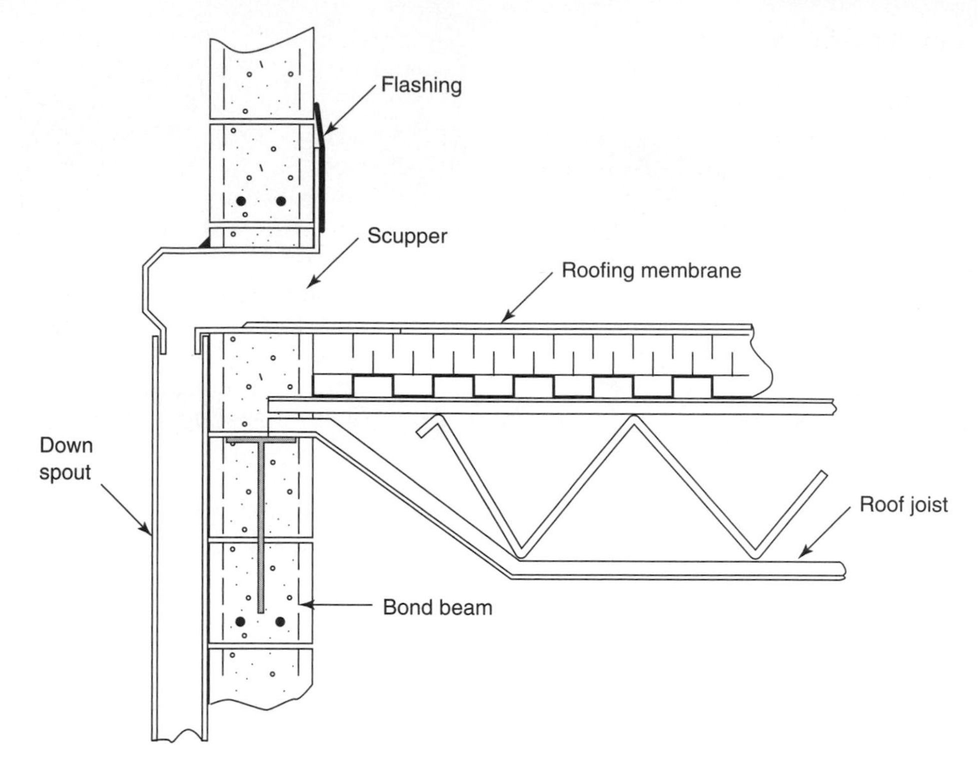


**Figure 12-1** Scupper Through Parapet Wall.

efficient as roof framing members, joists are used as framing members to support the roof decking. The spacing of the joists is usually dependent on the type of decking that is used to support the roof membrane. When regular plywood sheathing is used as decking, the joists must be spaced relatively closely, usually 12 to 16 in. (300 to 400 mm) on center. Joists supporting plank decking can be spaced up to 48 in. (1,200 mm) on center depending on the thickness of the decking being used and the loads that must be supported.

Joists can be sawn lumber sections, when relatively short spans 4 to 15 ft (3 to 5 m) are anticipated; I-shaped sections for longer spans (see Figure 12-2); and truss-shaped sections for spans reaching 24 to 30 ft (8 to 10 metres).

A third type of wood roof frame is made of prefabricated units (also known as *folded plates*) such as those illustrated in Figure 12-3. The units are 16 in. (400 mm) wide; vary in depth from 6 to 16 in. (150 to 400 mm), depending on the span; and have a maximum span of about 30 ft (10 metres). They may be handled singly or made into panels 32 or 48 in. (800 or 1,200 mm) wide. Panel ends may rest on bearing walls or be carried by supporting girders. Load-bearing walls of concrete or masonry, reinforced concrete beams, and structural steel beams must be capped with wood nailers to which the built-up panels must be secured.

**Figure 12-2** Wood I-Joist Roof Frame.

## Open-Web Joist Roof Frame

Open-web joists are light structural members composed of top and bottom chords, connected by rod or tubular steel webs. Two main types are generally used: open-web *steel* joists (see Figure 12-4) and open-web *combination* joists, in which the chords are of wood and the webs of tubular steel (see Figure 12-5).

Open-web steel joists can span to 98 ft (30 m) and depths can vary from 8 to 48 in. (200 to 1,200 mm). Joist spacing will vary depending on the type and strength of decking that is to be supported. Joists can be spaced as close as 12 in. (300 mm) on center when a light steel metal deck or plywood deck is used and up to 6 ft (2,000 mm) when relatively heavy steel cellular decking is used. To prevent lateral or vertical displacement of the joists, joist ends are usually welded to the supporting structural steel beams, or to bearing plates that are embedded in masonry bearing walls or reinforced concrete structural sections.

Combination joists span to 50 ft (15 m) with single top and bottom chords on the flat, as shown in Figure 12-5, and to 100 ft (30 m) with double top and bottom chords on edge. A wooden pad on the bearing surface of the supports simplifies the anchoring of this type of joist.

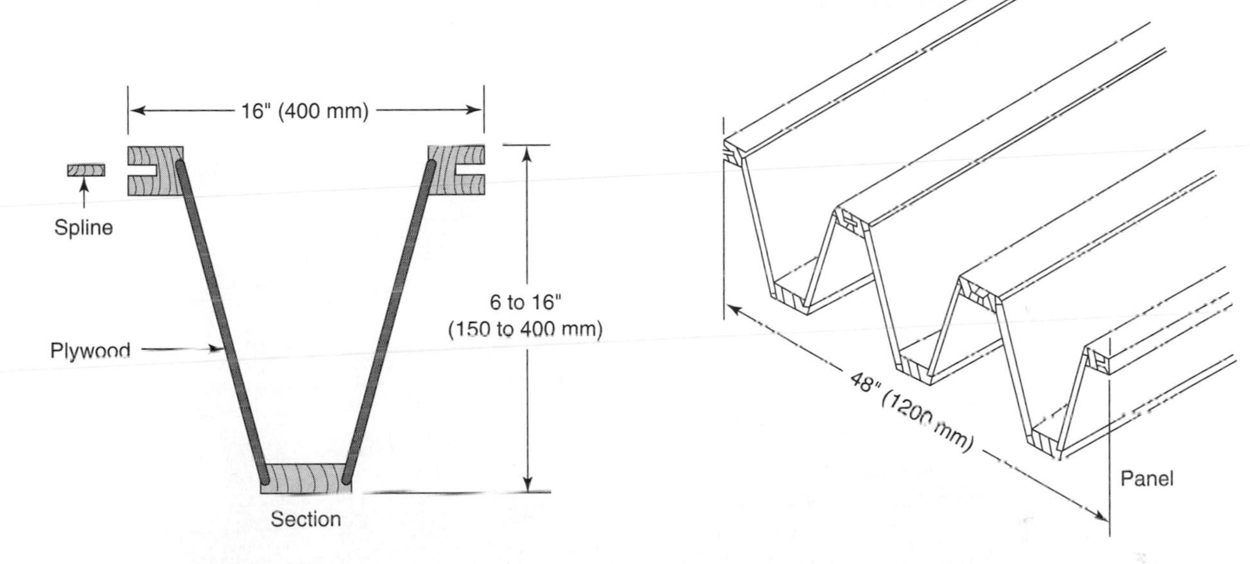


**Figure 12-3** Prefabricated Wooden Roof Panel.

**Figure 12-4** Open-Web Steel Joists.

**Figure 12-5** Open-Web Joists with Wood Chords and Round Hollow Steel Web Sections.

**Figure 12-6** Sloping Soffit Formed with Overhanging Open-Web Joists. (Courtesy of Trus Joist a Weyerhaeuser business.)

Roof overhangs may be incorporated into the system by using web joists with ends that project (cantilever) beyond the outside bearing member, as shown in Figure 12-6.

## Structural Steel Roof Frame

In a structural steel frame building, the roof frame will likely be fabricated with the same type of member as used in the rest of the building, but in a smaller size or lighter weight. In the building frame shown in Figure 12-7, the roof is framed with steel girders and light intermediate beams. Projecting eaves may be formed in conjunction with a steel frame, as indicated in Figure 12-8.

Instead of using regular rolled steel shapes, a steel roof frame may be built up with light steel framing members, such as those illustrated in Figure 12-9. The main joists are usually tack welded in place, while the cross members ride in joist hangers and are held in place with screws.

**Figure 12-7** Structural Steel Roof Frame. (Reprinted by permission of Bethlehem Steel Co.)

**Figure 12-8** Projecting Eaves with Steel Frame. (Reprinted by permission of Bethlehem Steel Co.)

## Monolithic Frame and Roof Slab Systems

Flat slabs, flat plates, ribbed slabs, and waffle slabs have been described in the discussions on floor systems in Chapter 11, and the same techniques may be used to produce roof slabs (see Figure 12-10).

Flat slabs are well suited for heavy roof loads and may be particularly useful for accommodating rooftop parking. Flat plates provide a continuous ceiling and offer complete flexibility for partition arrangement.

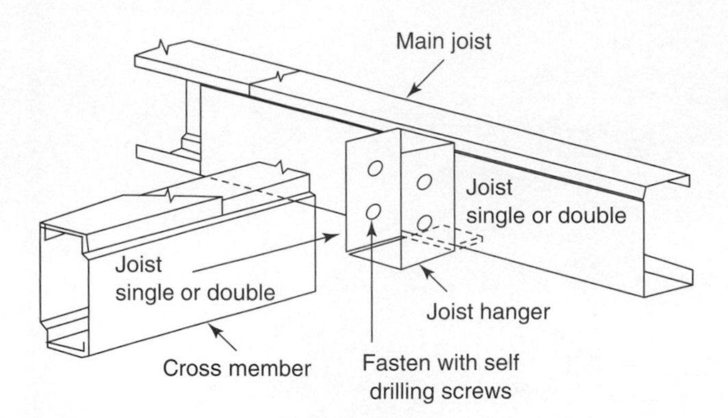


**Figure 12-9** Cold-Formed Steel Joists.

**Figure 12-10** Flat Plate Roof.

## Precast Concrete Roof Systems

1. *Prestressed concrete joists:* For projects involving relatively light loads and short spans, prestressed concrete joists may be used as a roof frame and, in combination with prestressed concrete planks or other structural decking systems, produce an economical and functional roof. The joists are available in two basic shapes: *keystone* and *tee* shape (see Figure 12-11). The keystone-shaped joist is available in depths varying from 6 to 18 in. (150 to 450 mm), with spans to 36 ft (11 m), while the tee-shaped joist comes in depths that vary from 8 to 20 in. (200 to 500 mm), with spans from 20 to 60 ft (6 to 18 m). Figure 12-12 illustrates typical roof construction with tee joists supporting prestressed concrete roof planks.

   Joists may be supported on bearing walls or on prestressed interior and exterior beams cast as part of the system (see Figure 12-13). Tables 12-1(a) and (b) give typical prestressed beam sizes for beams used in a prestressed concrete joist roof system.
2. *Single-tee roof slab:* Single tees are among the most common products of the precast concrete industry and are widely used as a roof system. They may be set edge to edge, as shown in Figure 12-14, or spaced, with the spaces between filled with cast-in-place concrete or concrete planks.

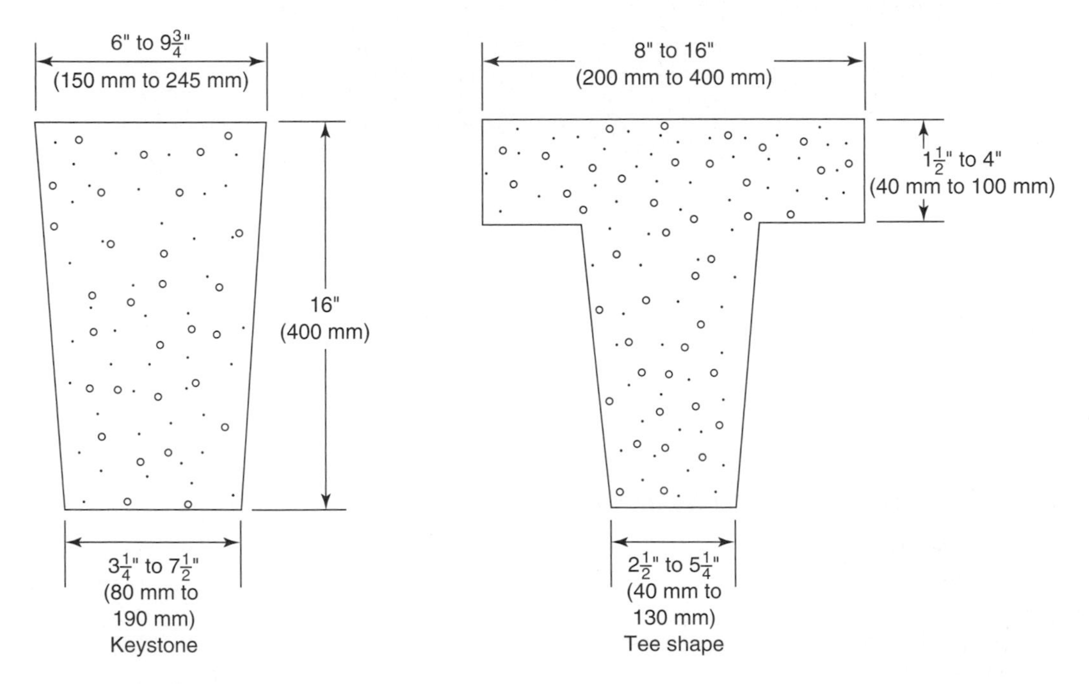


**Figure 12-11** Prestressed Precast Concrete Joists.

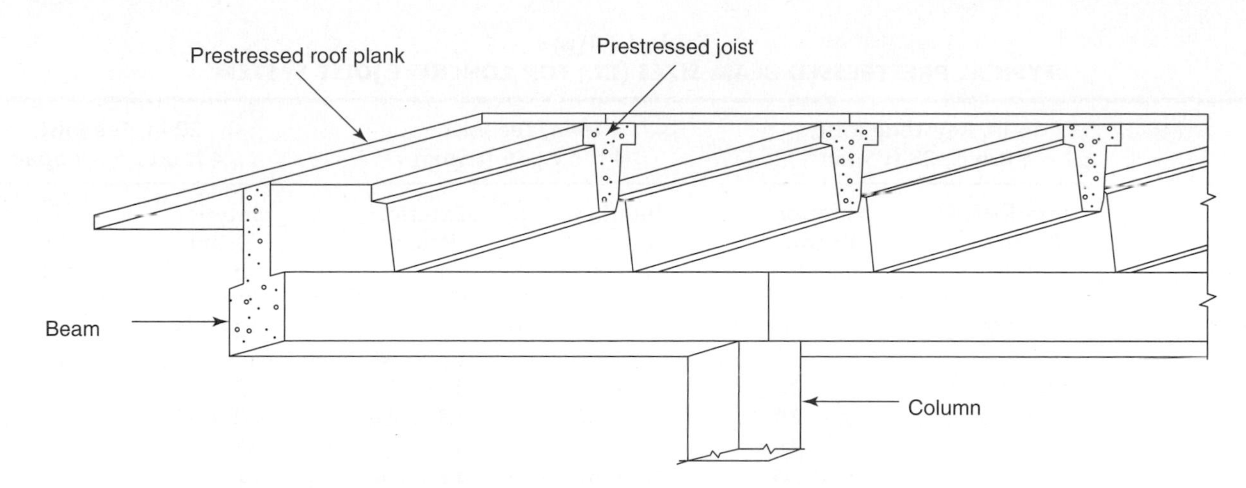


**Figure 12-12** Prestressed Precast Concrete Roof System. (Reprinted with permission of Portland Cement Association.)

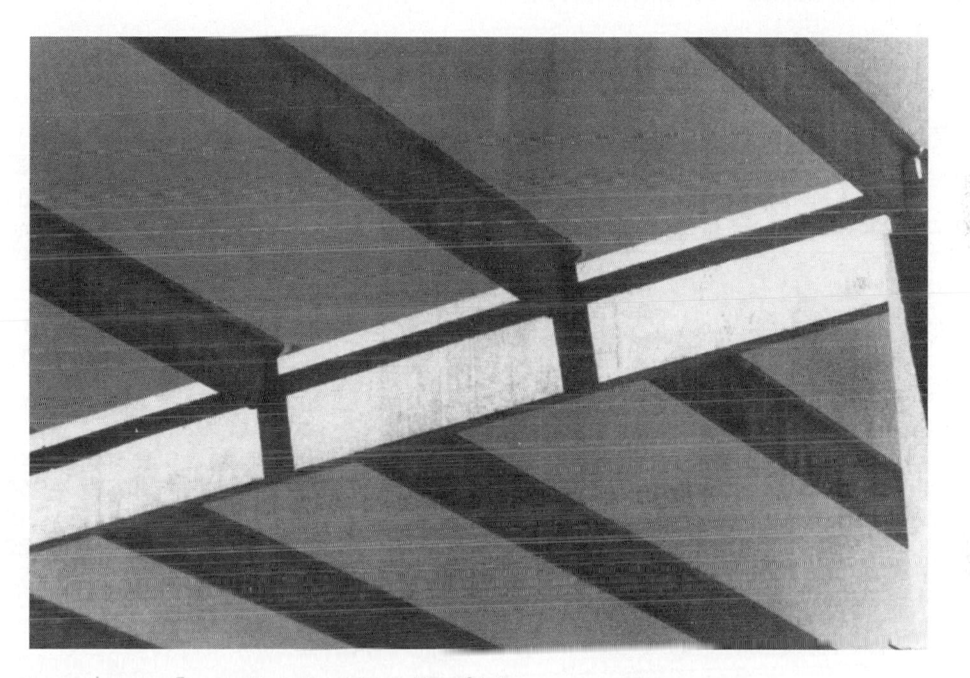

**Figure 12-13** Prestressed Beam Supporting Prestressed Roof Joists.

Flange widths range from 4 to 10 ft (1,200 mm to 3 m), although the 8-ft (2,400-mm) section is most commonly used. Depths range from 12 to 48 in. (300 to 1,200 mm) or more, with the 36-in. (900-mm) depth being a popular one. Spans are normally from 30 to 100 ft (10 to 30 m), although longer spans are possible under certain circumstances.

The ends of single-tee roof slabs are usually supported on edge beams, as illustrated in Figure 12-15, although they may be carried on bearing walls as well. Prestressed edge beams are cast in typical sizes, indicated in Tables 12-2(a) and (b).

3. *Double-tee roof slab:* Double tees are probably the most widely used prestressed concrete products in the medium-span range of buildings (see Figure 12-16). They are available in various widths and depths, depending on the manufacturer. Widths are generally 4, 5, 6, 8, 10 ft (1,200, 1,500, 1,800, 2,400, 3,000 mm), with the 4- and 8-ft (1,200- and 2,400-mm) sections probably the most popular. Depths range from 6 to 36 in. (150 to 900 mm), but in most cases, the maximum depth available for the required span will be the most economical. The 48-in. (1,200-mm) double tees are commonly used for spans up to 60 ft (18 m),

**Table 12-1(a)**
**TYPICAL PRESTRESSED BEAM SIZES (IN.) FOR CONCRETE JOIST SYSTEM[a]**

| | 16-in. Keystone Joist, 4 ft o.c., 28-ft Span | | 16-in. Tee Joist, 4 ft o.c., 36-ft Span | | 20-in. Tee Joist, 4 ft o.c., 50-ft Span | |
|---|---|---|---|---|---|---|
| Span of Beam (ft) | Interior Beam, b × h | Exterior Beam, b × h | Interior Beam, b × h | Exterior Beam, b × h | Interior Beam, b × h | Exterior Beam, b × h |
| 20 | 12 × 20[b] | 12 × 16 | 12 × 24 | 12 × 18 | 12 × 28 | 12 × 20 |
| | **12 × 18[b]** | | | **12 × 16** | | |
| 28 | 12 × 28 | 12 × 22 | 12 × 32 | 12 × 24 | 12 × 40 | 12 × 28 |
| | **12 × 26** | **12 × 20** | **12 × 30** | **12 × 22** | **12 × 38** | |
| 36 | 12 × 40 | 12 × 28 | 12 × 42 | 12 × 32 | 12 × 52 | 12 × 38 |
| | **12 × 36** | **12 × 36** | **12 × 40** | **12 × 30** | **12 × 48** | **12 × 36** |

[a]Sizes in boldface have deflected strands.
[b]Joist must be notched for this beam if tops are flush.
*Source:* Reprinted with permission of Portland Cement Association.

**Table 12-1(b)**
**TYPICAL BEAM SIZES (MM) FOR CONCRETE JOIST SYSTEM[a]**

| | 400-mm Keystone Joist, 1200-mm o.c., 8.5-m Span | | 400-mm Tee Joist, 1200-mm o.c., 11-m Span | | 500-mm Tee Joist, 1200-mm o.c., 15-m Span | |
|---|---|---|---|---|---|---|
| Span of Beam (mm) | Interior Beam, b × h | Exterior Beam, b × h | Interior Beam, b × h | Exterior Beam, b × h | Interior Beam, b × h | Exterior Beam, b × h |
| 6100 | 300 × 500[b] | 300 × 400 | 300 × 600 | 300 × 450 | 300 × 700 | 300 × 500 |
| | **300 × 450[b]** | | | **300 × 400** | | |
| 8500 | 300 × 700 | 300 × 550 | 300 × 800 | 300 × 600 | 300 × 1000 | 300 × 700 |
| | **300 × 650** | **300 × 500** | **300 × 750** | **300 × 550** | **300 × 950** | |
| 10950 | 300 × 1000 | 300 × 700 | 300 × 1050 | 300 × 800 | 300 × 1300 | 300 × 950 |
| | **300 × 900** | **300 × 900** | **300 × 1000** | **300 × 750** | **300 × 1200** | **300 × 900** |

[a]Sizes in boldface have deflected strands.
[b]Joist must be notched for this beam if tops are flush.
*Source:* Reprinted with permission of Portland Cement Association; metric conversion by author.

while the 8 ft (2,400-mm) units, which normally have a deeper web, span 70 ft (21 m) or more.

Supports for double tees consist of bearing walls and interior beams or edge beams and interior beams (see Figure 12-17). Typical prestressed beam sizes are indicated in Tables 12-3(a) and (b).

4. *Cored slabs:* The production of hollow core slabs is highly mechanized and, as a result, their cost is relatively low.

   Depths range from 4 to 12 in. (100 to 300 mm), but 6- and 8-in. (150- and 200-mm) depths are most commonly used. Widths range from 16 to 48 in. (400 to 1,200 mm), in limited increments, depending on the manufacturer. Spans may reach to 50 ft (15 m), but the general range is below 35 ft (10.5 m).

   Cored slabs are quick and easy to install and provide a finished, flush ceiling (see Figure 12-18). They may be used with almost any type of framing system, including prestressed edge and interior beams (see Figure 12-19). Typical precast beam sizes to suit various depths of slab and slab load are listed in Tables 12-4(a) and (b).

**Figure 12-14** Single-Tee Roof Slab System.

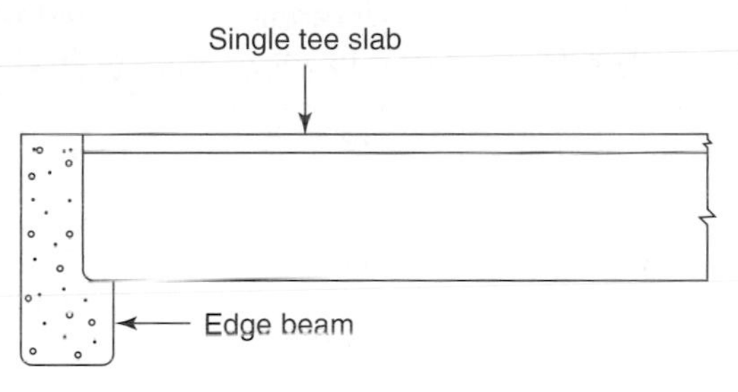


**Figure 12-15** Single-Tee Roof Slab Support.

**Table 12-2(a)**
**TYPICAL PRESTRESSED BEAM SIZES (IN.) FOR SINGLE-TEE SLAB SYSTEM**

| Span of Beam (ft) | 12-in. Single-Tee, 30-ft Span | 16-in. Single-Tee, 45-ft Span | 20-in. Single-Tee, 60-ft Span | 28-in. Single-Tee, 75-ft Span |
|---|---|---|---|---|
| 20 | 12 × 16[a] | 12 × 16[a] | 12 × 24[a] | |
| 24 | 12 × 20 | 12 × 24 | 12 × 28 | 12 × 32[a] |
| 32 | 12 × 28 | 12 × 32 | 12 × 40 | 12 × 44 |
| 40 | 12 × 36 | 12 × 40 | 12 × 48 | 12 × 56 |

[a]The stem of the tee should be notched for this beam.
*Source:* Reprinted with permission of Portland Cement Association.

**Table 12-2(b)**
**TYPICAL PRESTRESSED BEAM SIZES (MM) FOR SINGLE-TEE SLAB SYSTEM**

| Span of Beam (mm) | 300-mm Single-Tee, 9.0-m Span | 400-mm Single-Tee, 13.5-m Span | 500-mm Single-Tee, 18.0-m Span | 700-mm Single-Tee, 22.5-m Span |
|---|---|---|---|---|
| 6100 | 300 × 400[a] | 300 × 400[a] | 300 × 600[a] | |
| 7300 | 300 × 500 | 300 × 600 | 300 × 700 | 300 × 800[a] |
| 9700 | 300 × 700 | 300 × 800 | 300 × 1120 | 300 × 1100 |
| 12200 | 300 × 900 | 300 × 1000 | 300 × 1200 | 300 × 1400 |

[a]The stem of the tee should be notched for this beam.
*Source:* Reprinted with permission of Portland Cement Association; metric conversion by author.

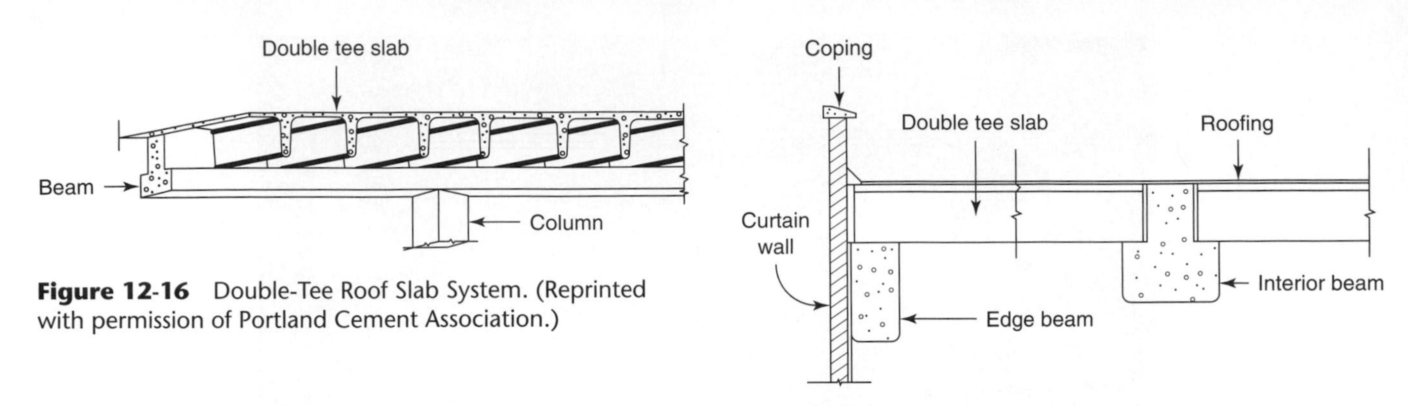


**Figure 12-16** Double-Tee Roof Slab System. (Reprinted with permission of Portland Cement Association.)

**Figure 12-17** Double-Tee Roof Slab Supports.

**Table 12-3(a)**
**TYPICAL PRESTRESSED BEAM SIZES (IN.) FOR DOUBLE-TEE SLAB SYSTEM**

| | 10-in. Double Tee, 30-ft Span | | 12-in. Double Tee, 40-ft Span | | 14-in. Double Tee, 50-ft Span | |
|---|---|---|---|---|---|---|
| Span of Beam (ft) | Interior Beam | Edge Beam | Interior Beam | Edge Beam | Interior Beam | Edge Beam |
| 20 | 12 × 20 | 12 × 16 | 12 × 22 | 12 × 18 | 12 × 26 | 12 × 20 |
| 25 | 12 × 24 | 12 × 20 | 12 × 28 | 12 × 22 | 12 × 30 | 12 × 26 |
| 30 | 12 × 28 | 12 × 24 | 12 × 32 | 12 × 28 | 12 × 36 | 12 × 32 |
| 35 | 12 × 32 | 12 × 28 | 12 × 38 | 12 × 34 | 12 × 44 | 12 × 38 |
| 40 | 12 × 38 | 12 × 34 | 12 × 44 | 12 × 38 | 12 × 50 | 12 × 42 |

*Source:* Reprinted with permission of Portland Cement Association.

**Table 12-3(b)**
**TYPICAL PRESTRESSED BEAM SIZES (MM) FOR DOUBLE-TEE SLAB SYSTEM**

| | 250-mm Double Tee, 9.0-m Span | | 300-mm Double Tee, 12.0-m Span | | 350-mm Double Tee, 15.0-m Span | |
|---|---|---|---|---|---|---|
| Span of Beam (mm) | Interior Beam | Edge Beam | Interior Beam | Edge Beam | Interior Beam | Edge Beam |
| 6 100 | 300 × 500 | 300 × 400 | 300 × 550 | 300 × 450 | 300 × 650 | 300 × 500 |
| 7 600 | 300 × 600 | 300 × 500 | 300 × 700 | 300 × 550 | 300 × 750 | 300 × 650 |
| 9 100 | 300 × 700 | 300 × 600 | 300 × 800 | 300 × 700 | 300 × 900 | 300 × 800 |
| 10 600 | 300 × 800 | 300 × 700 | 300 × 950 | 300 × 850 | 300 × 1100 | 300 × 950 |
| 12 100 | 300 × 950 | 300 × 850 | 300 × 1100 | 300 × 950 | 300 × 1250 | 300 × 1050 |

*Source:* Reprinted with permission of Portland Cement Association; metric conversion by author.

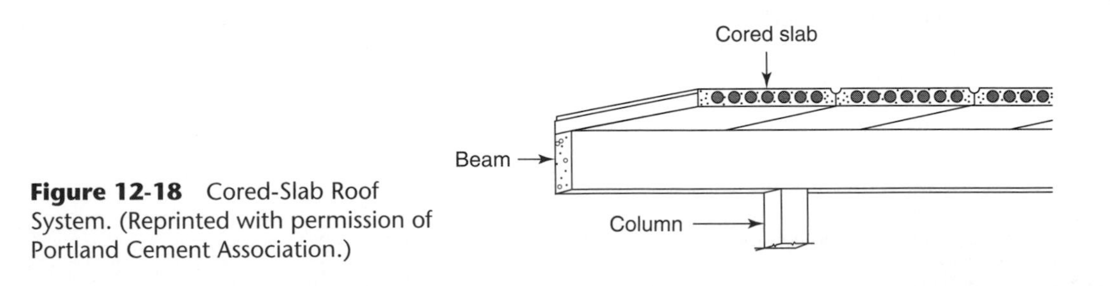


**Figure 12-18** Cored-Slab Roof System. (Reprinted with permission of Portland Cement Association.)

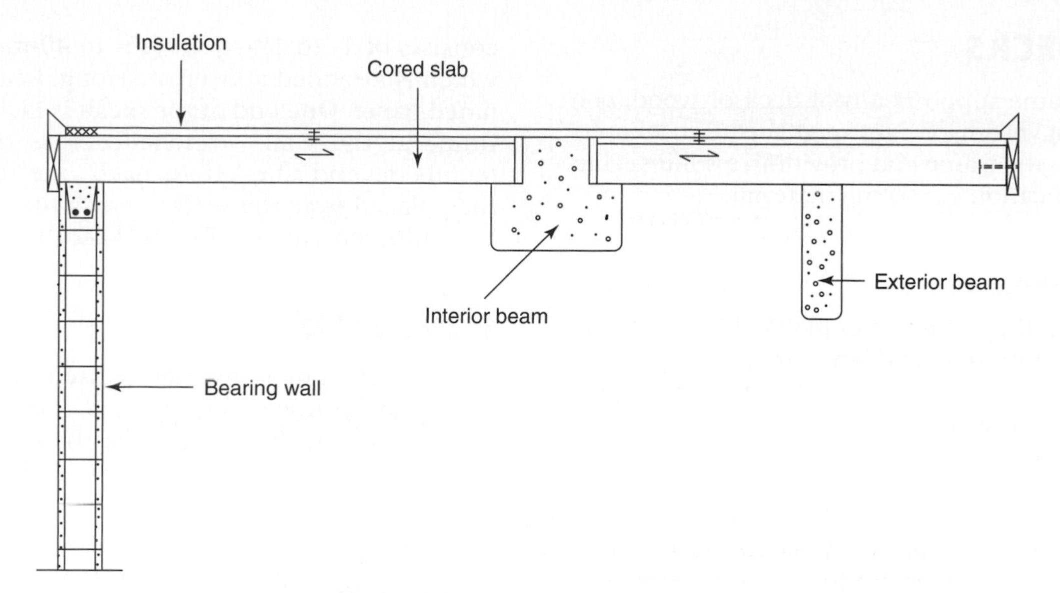

**Figure 12-19** Core-Slab Support System. (Reprinted with permission of Portland Cement Association.)

**Table 12-4(a)**
**TYPICAL PRESTRESSED BEAM SIZES (IN.) FOR CORED SLAB ROOF**

| Span of Beam (ft) | 6-in. Slab, 28-ft Span, *DL* + *LL* = 80 psf | | 8-In. Slab, 34-ft Span, *DL* + *LL* = 90 psf | | 10-in. Slab, 40-ft Span, *DL* + *LL* = 100 psf | | 12-in. Slab, 46-ft Span, *DL* + LL = 110 psf | |
|---|---|---|---|---|---|---|---|---|
| | Interior | Exterior | Interior | Exterior | Interior[a] | Exterior | Interior[a] | Exterior |
| 20 | 12 × 20 | 12 × 18 | 12 × 24 | 12 × 20 | 12 × 24 | 18 × 18 | 12 × 28 | 18 × 20 |
| 24 | 12 × 24 | 12 × 20 | 12 × 28 | 12 × 24 | 12 × 28 | 18 × 22 | 12 × 32 | 18 × 24 |
| 28 | 12 × 28 | 12 × 24 | 12 × 32 | 12 × 28 | 12 × 34 | 18 × 26 | 12 × 38 | 18 × 28 |
| 32 | 12 × 32 | 12 × 26 | 12 × 38 | 12 × 32 | 12 × 38 | 18 × 28 | 12 × 44 | 18 × 32 |

[a]Beam has heavy prestressing.
*Source:* Reprinted with permission of Portland Cement Association.

**Table 12-4(b)**
**TYPICAL PRESTRESSED BEAM SIZES (MM) FOR CORED SLAB ROOF**

| Span of Beam (mm) | 150-mm Slab, 8.5-m Span, *DL* + *LL* = 3.8 kPa | | 200-mm Slab, 8.5-m Span, *DL* + *LL* = 4.3 kPa | | 250-mm Slab, 12.2-m Span, *DL* + *LL* = 4.8 kPa | | 300-mm Slab, 14.0-m *DL* + LL = 5.3 kPa | |
|---|---|---|---|---|---|---|---|---|
| | Interior | Exterior | Interior | Exterior | Interior[a] | Exterior | Interior[a] | Exterior |
| 6100 | 300 × 500 | 300 × 450 | 300 × 600 | 300 × 500 | 300 × 600 | 450 × 450 | 300 × 700 | 450 × 500 |
| 7300 | 300 × 600 | 300 × 500 | 300 × 700 | 300 × 600 | 300 × 700 | 450 × 550 | 300 × 800 | 450 × 600 |
| 8500 | 300 × 700 | 300 × 600 | 300 × 800 | 300 × 700 | 300 × 850 | 450 × 650 | 300 × 950 | 450 × 700 |
| 9750 | 300 × 800 | 300 × 650 | 300 × 950 | 300 × 800 | 300 × 950 | 450 × 700 | 300 × 1100 | 450 × 800 |

[a]Beam has heavy prestressing.
*Source:* Reprinted with permission of Portland Cement Association; metric conversion by author.

## ROOF DECKS

The roof frame supports a roof deck of wood, concrete, steel, gypsum, or lightweight cellular concrete, anchored to the frame and providing a solid surface for the application of roofing material.

### Wood Deck

A wood deck may be made of planks laid on the flat or 2-in. (38-mm) material laminated on edge. In either case, the deck must be spiked to the supporting beams and, when they are other than wood, a wooden nailer must be provided on their bearing surface.

Solid wood decks should be kept back ½ in. (12 mm) from parapet walls or from the fascia where a flush deck is used with a cant strip to cover the gap. In the case of a flush deck, the cant strip will also act as a gravel stop.

### Concrete Roof Deck

A concrete roof deck may be placed over removable wood forms, over steel decking laid on the roof frame, or over paper-backed wire mesh stretched over roof framing members spaced 16 in. (400 mm) o.c.

The removable wood forms are used in conjunction with a steel frame and are usually supported as illustrated in Figures 11-46 and 11-48. Steel decking will be similar to that shown in Figures 11-50 and 11-51. Paper-backed wire mesh consists of 1- to 1½-sq in. (25- to 40-mm) mesh to which is attached a layer of strong, heavy, impregnated paper. One end of the mesh is fastened to the frame at the wall, stretched across the framing members, and stapled to each one. Concrete is then placed over the surface, with the mesh acting as reinforcement (see Figure 12-20).

### Steel Roof Deck

A steel deck is most frequently used in conjunction with a steel frame and is laid and covered as shown in Figure 12-21. The decking may be secured by spot welding, self-tapping screws, powder-driven pins, or nails if the frame is wood.

### Gypsum Roof Deck

Gypsum decks are made from poured-in-place gypsum or precast gypsum planks. In using poured-in-place gypsum, structural steel T-sections spaced 33 in. (825 mm) o.c. serve as supports for 32-in. (800-mm) wide, ½-in. (12-mm) thick gypsum board sheeting that is placed between the tee webs to serve as the bottom of the form (Figure 12-22). Welded wire mesh is draped over the Ts, and freshly mixed gypsum is poured and screeded to the proper level.

Gypsum planks are *plain* or *metal edged*. Both are supported on steel roof framing members, but in the case of metal-edged planks, it is not necessary for the ends to rest on a beam.

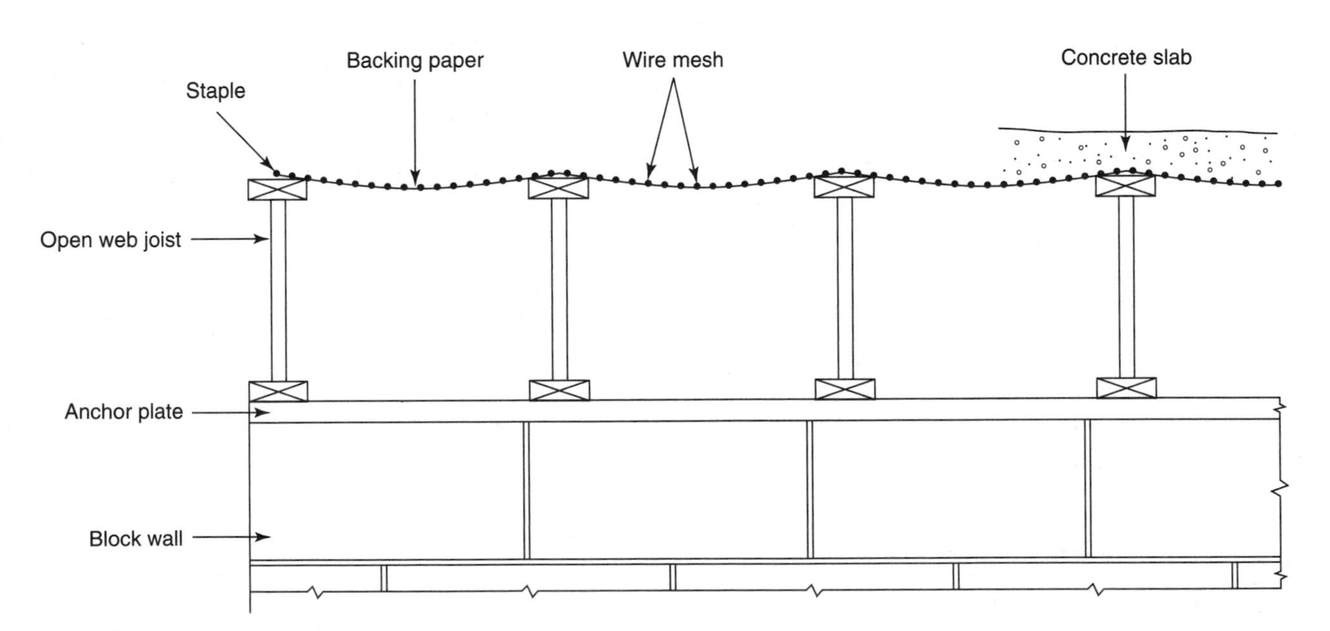


**Figure 12-20** Paper-Backed Wire Mesh over Open-Web Joists.

**Figure 12-21** Steel Deck on Open-Web Steel Joists.

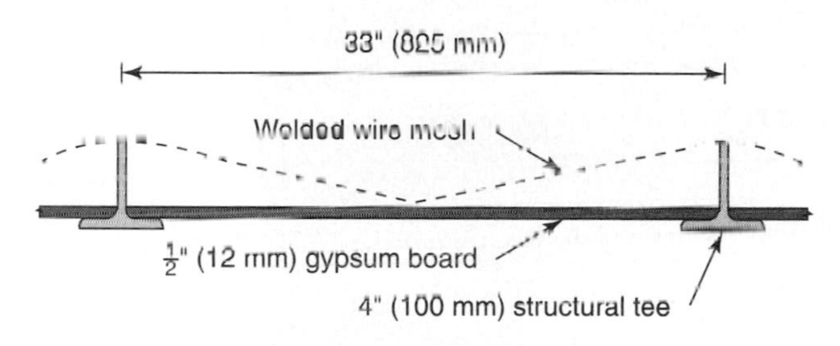


**Figure 12-22** Form for Gypsum Deck.

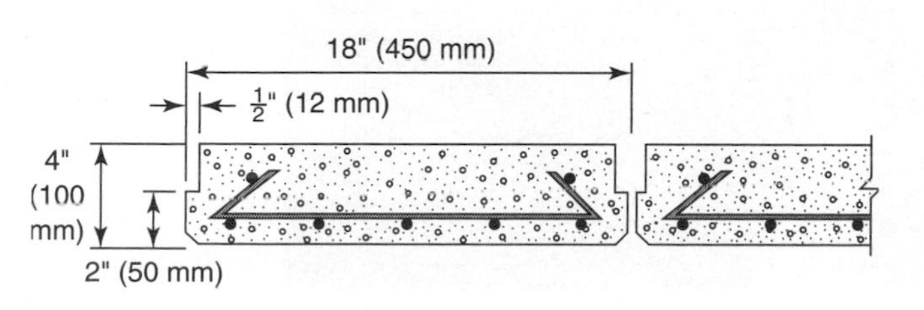


**Figure 12-23** Cellular Concrete Roof Slab.

## Cellular Concrete Roof Deck

Cellular concrete is used for roof decks as precast slabs 18 in. (450 mm) wide, 4 in. (100 mm) thick, and up to 10 ft (3 m) in length. Slabs may be used over any type of roof frame and are fastened with nails, bolts and washers, galvanized metal clips, or powder-driven steel pins. Grooves between slabs are later filled with grout (see Figure 12-23).

# TRUSS ROOF FRAMING

Another method of framing the roof of a building is by the use of *trusses*. When a building requires large open areas without partitions or columns, and the distances between walls exceed the practical limits of ordinary joists, deeper sections must be used.

Basically, trusses are deep, open-web joists designed to carry heavy loads over large spans. Trusses are normally fabricated using steel or timber sections. However, in some applications, precast and prestressed concrete sections have been used in truss-type configurations. Factors taken into account when choosing the truss material are design loads, length of clear span, ease of fabrication, and erection. However, architectural requirements may override an obvious material choice. Trusses are generally made from straight sections except in the case of bowstring wood trusses where the top chord can be constructed of a curved glued laminated section.

Simple trusses normally consist of a top and bottom chord and web members, as indicated in Figure 12-24. The web members are normally arranged in a series of triangular configurations to ensure structural stability, and the points of intersection between the web members and the chords are known as panel points. To facilitate truss analysis, the joints at the panel points are assumed to be pin-type connections. If applied loads are located at

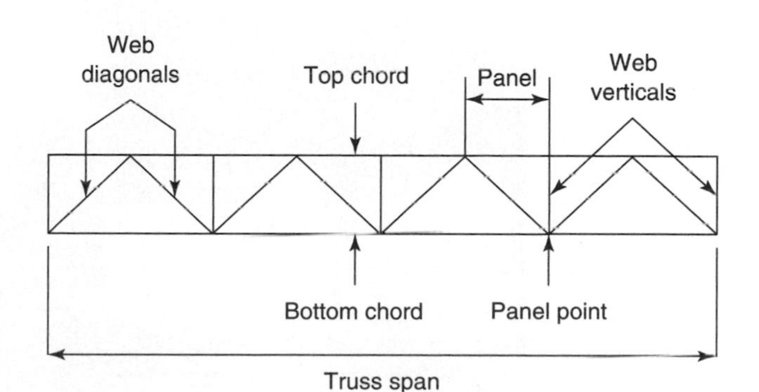


**Figure 12-24** Parts of a Truss.

**Figure 12-25** Steel Trusses Supported by Structural Steel Columns.

the panel points, the resulting forces in the chords and web members are primarily axial tension or axial compression, making the design and fabrication trusses relatively straightforward. Trusses can be supported by individual columns as illustrated in Figure 12-25, on load-bearing walls, or on roof girders (see Figure 12-26). The triangular configuration allows for a variety of truss shapes and spans as illustrated in Figure 12-27.

Sections in a steel truss may be welded, bolted, or a combination of welds and bolts. In Figure 12-28, steel trusses have been used to resist wind loads on the end of the building, and to support the roof. The roof trusses in this application are of welded construction with a posttensioned bottom chord. Using posttensioning in this manner allows for lighter truss sections and a large open area—in this case, large enough for a soccer field.

Two methods are used to connect the members of a wooden truss. In one case, members are butted together, and in the other, they are lapped. If members are butted to one another, a wood or steel *gusset plate* or *nail plate* is fastened over the joint (see Figure 12-29). A wood plate is fastened with nails and glue, a steel plate with bolts. A nail plate is a flat plate studded with spikes on one side. It is laid over the joint, and the spikes are driven or pressed into the wood. If the members overlap one another, timber connectors and bolts are used to connect them. The connectors are set in circular grooves cut in the meeting faces (Figure 12-30), and a bolt through the assembly holds the two members tightly together. Grooves are cut to a predetermined depth by a special tool that drills the bolt hole at the same time (see Figure 8-6).

Roof sheathing may be applied directly to trusses, but is more often supported by *purlins* that span from truss to truss at panel points or by joists resting on the upper chord (Figure 12-31).

Rather than using sheathing supported by purlins or joists, large trusses may be covered by stressed-skin panels, such as that shown in Figure 12-32, which will span from truss to truss and fasten directly to the upper chords.

**Figure 12-26** Timber Trusses Supported by Glued Laminated Roof Girder.

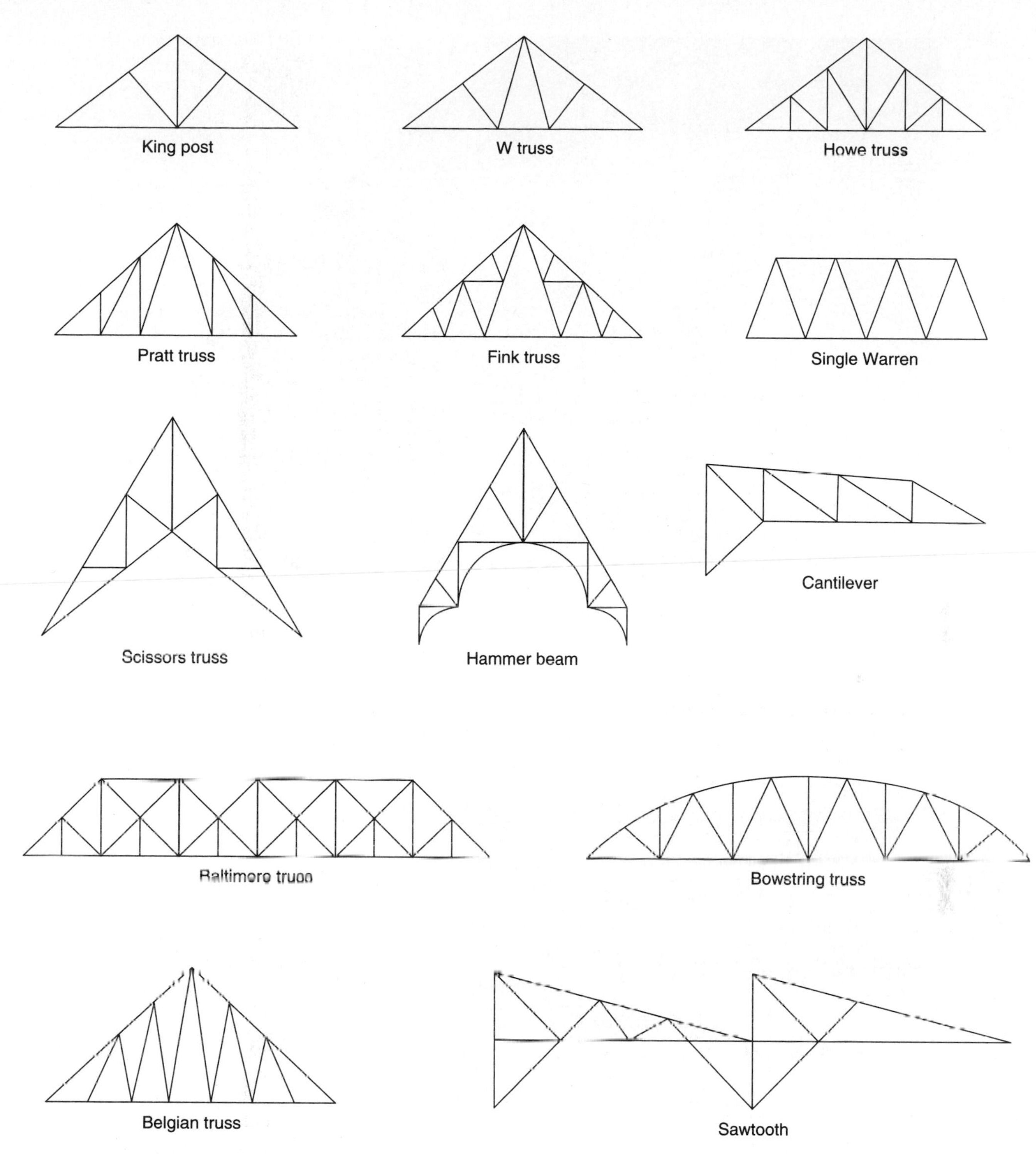


**Figure 12-27** Truss Configurations.

**Figure 12-28** Arched Posttensioned Welded Trusses. Note the Truss-Shaped Wind Columns.

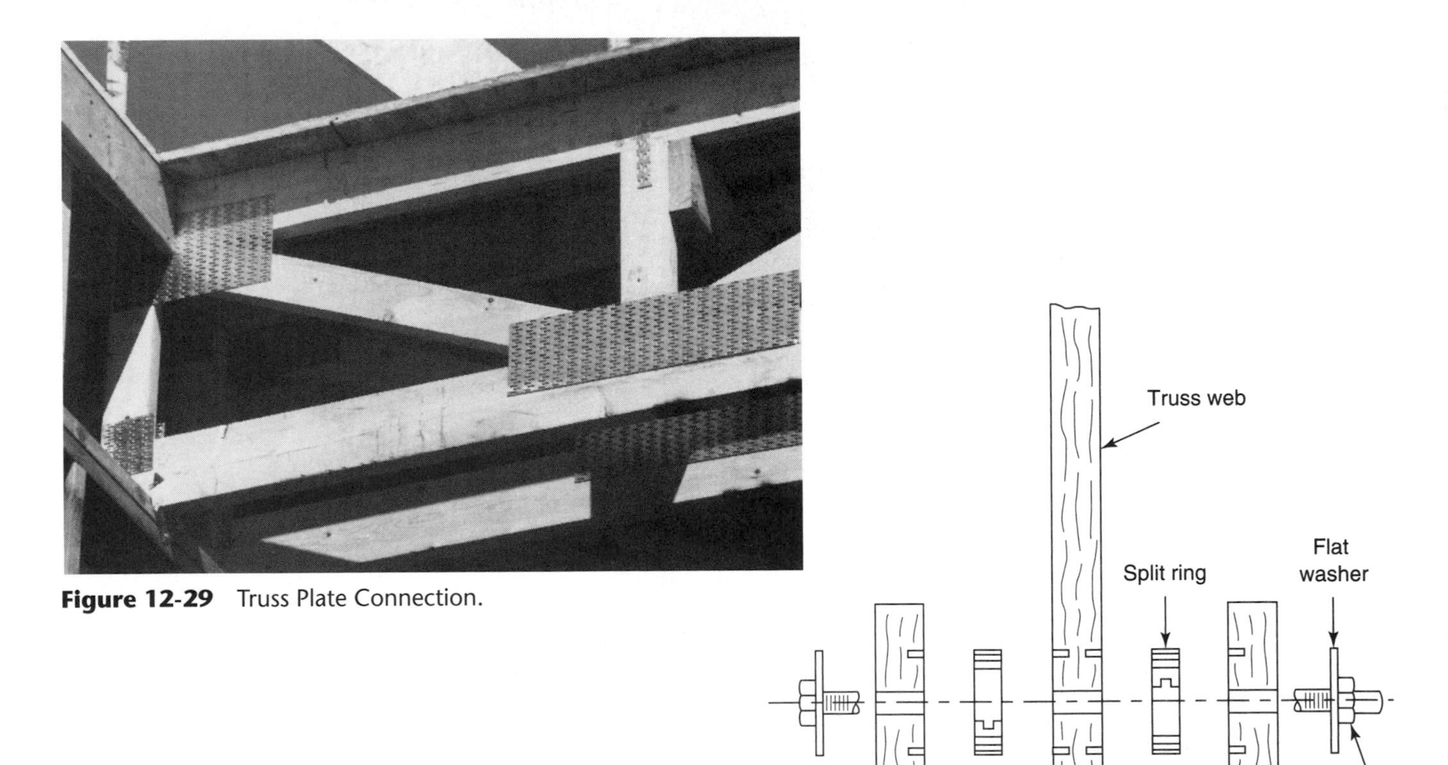


**Figure 12-29** Truss Plate Connection.

**Figure 12-30** Truss Joint Using Split Rings.

**Figure 12-31** Bowstring Trusses with Joists.

**Figure 12-32** Stressed-Skin Roof Panel. (Courtesy of Trus Joist a Weyerhaeuser business.)

## FOLDED PLATE ROOF

A folded plate roof is one in which the roof slab has been formed into a thin, self-supporting structure, usually made of either wood or concrete. Two of the most common shapes are W shape and V shape (see Figure 12-33). This style of roof is being used more and more to provide large areas of column-free space. Such roofs are capable of long spans, and their cantilever projections can be used to advantage in exterior design, as well as to counterbalance the span.

For reinforced concrete folded plates, the slope of plates is from 25° to 45° maximum, with a shell thickness of from 3 to 6 in. (75 to 150 mm). The span-to-depth ratio varies from 1:10 to 1:15.

A concrete roof of this type can be made with precast panels or may be cast in place. Figure 12-34 shows details of typical formwork for a cast-in-place concrete folded plate roof.

Span data for W- and V-shaped folded plates, based on the dimensions shown in Figure 12-33, are given in Tables 12-5 and 12-6.

A wood folded plate roof is made up of panels bolted together to form a complete roof structure. The top and bottom chords of each panel are usually laminated members, and the ribs are 2″ × 4″s (38 × 89s), 2″ × 6″s (38 × 140s), or 2″ × 8″s (38 × 184s), as the case may be, 16 in. (400 mm) o.c., with a plywood skin on both sides of the panel. Figure 12-35 shows details in the construction of a typical panel.

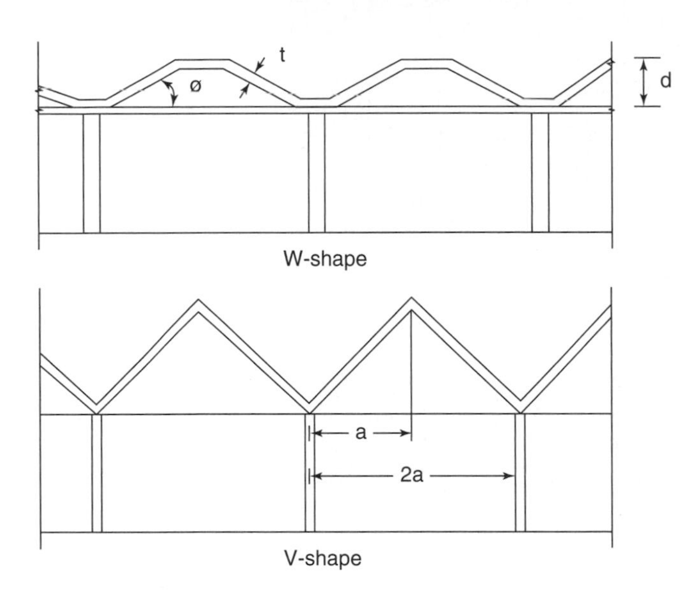


**Figure 12-33** Folded Plate Roof Shapes.

## LONG-BARREL SHELL ROOF

A long-barrel shell is a curved, thin-shell roof made from either concrete or wood, as illustrated in Figure 12-36(a). The variety of shapes range from a low-rise arc to a high-rise cylindrical arch (Figure 12-36(b)). These roofs provide long, uncluttered spans beyond the normal range of other structural roof systems, as well as unique architectural features.

Forms for concrete shells consist of a series of light bowstring trusses of the proper span and height, supported on shores or metal scaffolding and sheathed over the top with plywood. Shells are normally reinforced, but posttensioning may be introduced into the shell and transverse beams to allow a reduction in shell thickness.

Chord widths range to approximately 60 ft (18 m), while lengths vary from 40 to 160 ft (12 to 50 m). The usual span-to-depth ratio varies from 1:10 to 1:15. Shell thickness is normally 3 in. (75 mm). Long-barrel shell span data for concrete shells are given in Table 12-7, based on dimensions shown in Figure 12-36(b).

Wooden long-barrel shells are made up of a frame shaped to fit the curve (see Figure 12-37), covered inside and out with a plywood skin. Each barrel rests on *valley beams,* such as those shown in Figure 12-38.

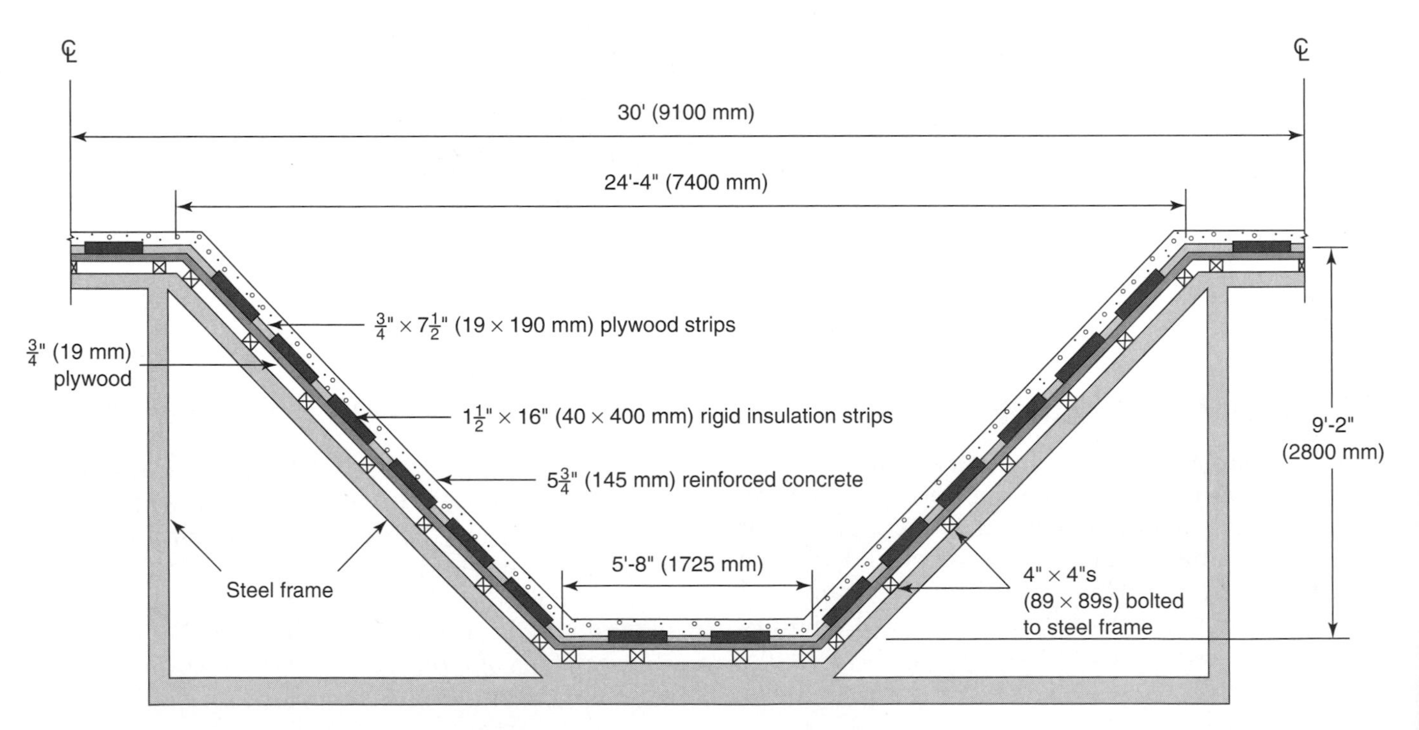


**Figure 12-34** Form for Concrete Folded Plate.

### Table 12-5
### SPAN DATA FOR W-SHAPED FOLDED PLATE

| Span | | ° (deg.) | | d (mm) | | d (ft) | | 2a[a] | | t (average) | | Reinforcement | |
|---|---|---|---|---|---|---|---|---|---|---|---|---|---|
| (m) | (ft) | Min. | Max. | Min. | Max. | Min. | Max. | (m) | (ft) | (mm) | (in.) | (kg/m² of Projected Area) | (lb/sq ft of Projected Area) |
| 12.2 | 40 | 30 | 45 | 750 | 1500 | 2.5 | 5 | 6.00 | 20 | 75 | 3 | 7.3–9.8 | 1.5–2.0 |
| 18.3 | 60 | | 45 | | 1800 | | 6 | 7.60 | 25 | 75 | 3 | 9.8–14.7 | 2.0–3.0 |
| | | 30 | | 1200 | | 4 | | | | 90 | 3.5 | | |
| 23.8 | 75 | | 45 | | 2250 | | 7.5 | 9.10 | 30 | 75 | 3 | 12.2–19.5 | 2.5–4.0 |
| | | 30 | | 1500 | | 5 | | | | 100 | 4 | | |
| 30.5 | 100 | | 45 | | 3000 | | 10 | 12.20 | 40 | 100 | 4 | 19.5–29.3 | 4.0–6.0 |
| | | 30 | | 1950 | | 6.5 | | | | 125 | 5 | | |

[a]Values shown may vary with architectural design.
*Source:* Reprinted with permission of Portland Cement Association; metric conversion by author.

### Table 12-5
### SPAN DATA FOR V-SHAPED FOLDED PLATE

| Span | | ° (deg.) | | d (mm) | | d (ft) | | 2a[a] | | t (average) | | Reinforcement | |
|---|---|---|---|---|---|---|---|---|---|---|---|---|---|
| (m) | (ft) | Min. | Max. | Min. | Max. | Min. | Max. | (m) | (ft) | (mm) | (in.) | (kg/m² of Projected Area) | (lb/sq ft of Projected Area) |
| 12.2 | 40 | 25 | 45 | 825 | 1200 | 2.75 | 4 | 6.00 | 20 | 75 | 3 | 5.9–7.8 | 1.2–1.6 |
| 18.3 | 60 | | 45 | | 1800 | | 6 | 6.00 | 20 | 100 | 4 | 9.3–13.2 | 1.9–2.7 |
| | | 25 | | 1200 | | 4 | | | | 150 | 6 | | |
| 23.8 | 75 | | 45 | | 2250 | | 7.5 | 7.50 | 25 | 100 | 4 | 12.7–18.0 | 2.6–3.7 |
| | | 25 | | 1500 | | 5 | | | | 150 | 6 | | |
| 30.5 | 100 | | 45 | | 3000 | | 10 | 9.10 | 30 | 125 | 5 | 19.5–25.4 | 4.0–5.2 |
| | | 25 | | 2025 | | 5.75 | | | | 150 | 6 | | |

[a]Values shown may vary with architectural design.
*Source:* Reprinted with permission of Portland Cement Association; metric conversion by author.

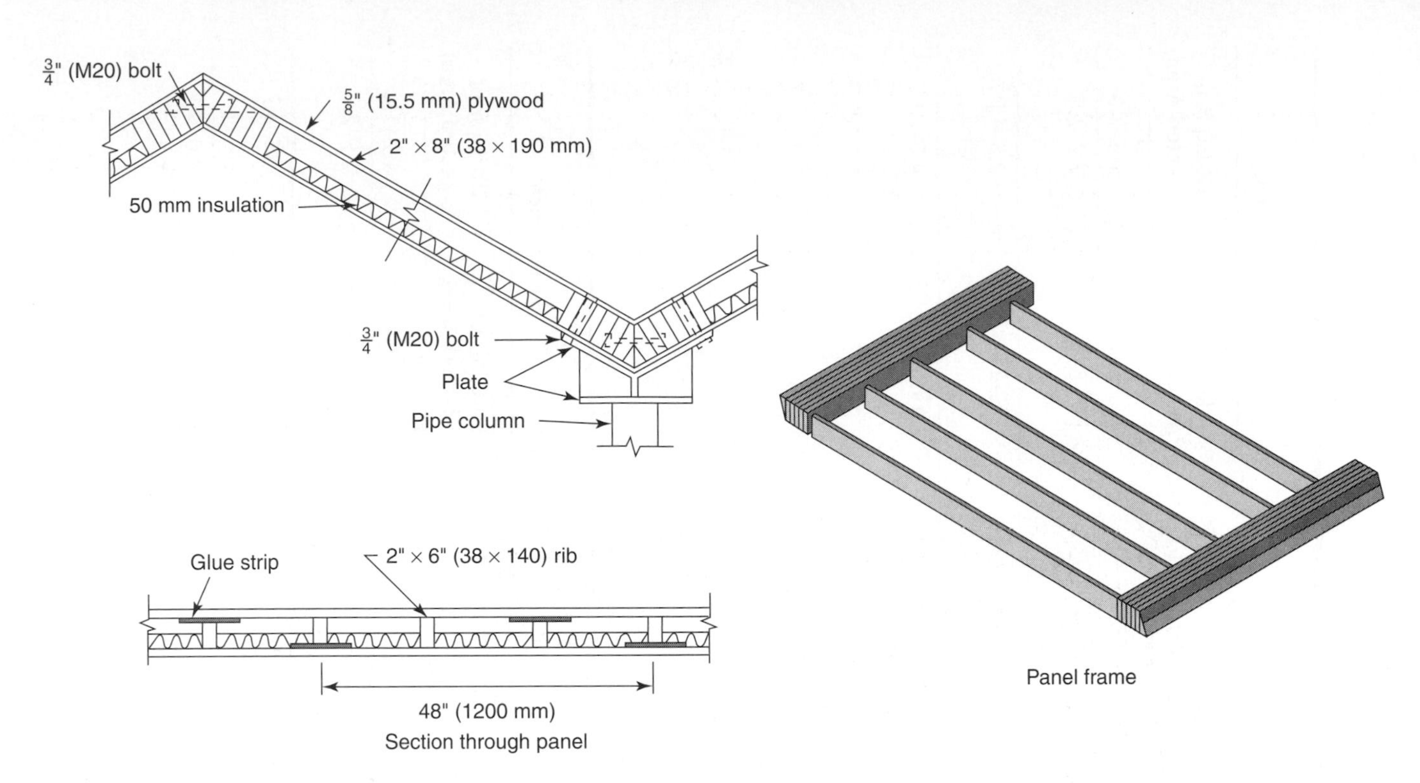


**Figure 12-35** Folded Plate Roof Panel.

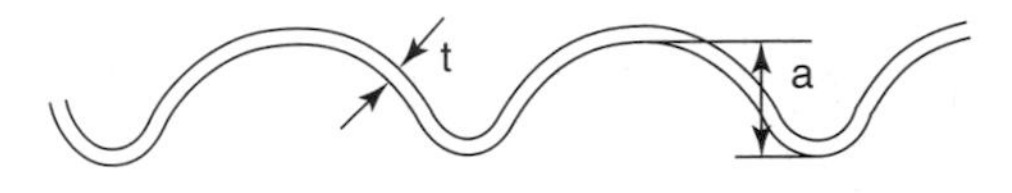


Undulating barrels

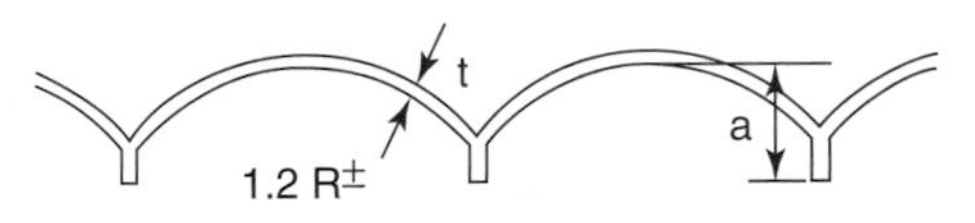


Barrels with stems

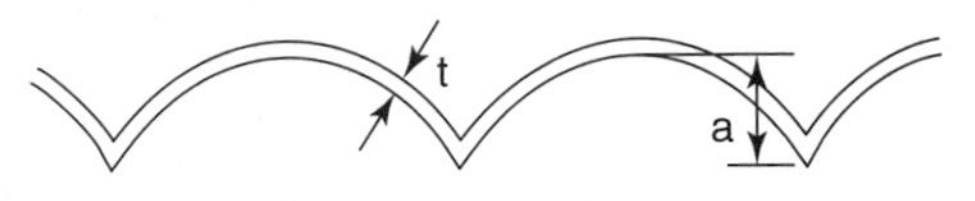


Parabolic barrels

(a)

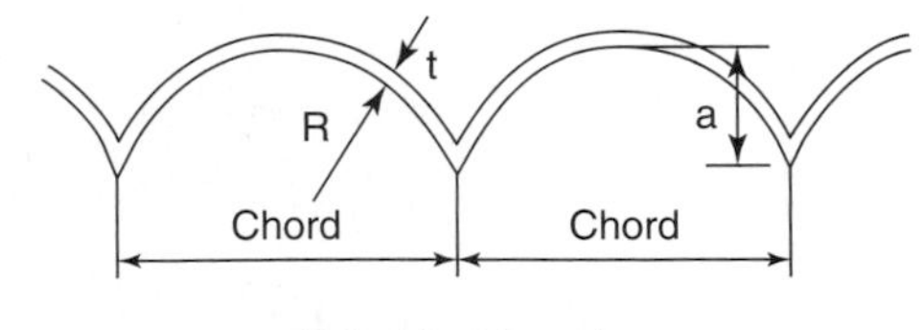


(b) Cylindrical barrels

**Figure 12-36** (a) Long-Barrel Shell Roof; (b) Multiple Long-Barrel Shells.

**Table 12-7**
**SPAN DATA FOR CONCRETE LONG-BARREL ROOF SHELLS**

| Span | | Chord Width | | $a$ | | $R$ | | $t$ | | Reinforcement | |
|---|---|---|---|---|---|---|---|---|---|---|---|
| (meters) | (ft) | (meters) | (ft) | (m) | (ft) | (m) | (ft) | (mm) | (in.) | (kg/m² of Projected Area) | (lb/sq ft of Projected Area) |
| 24.4 | 80 | 9.1 | 30 | 2.4 | 8 | 7.6 | 25 | 75 | 3 | 15.6 | 3.5 |
| 30.5 | 100 | 9.1 | 30 | 3.0 | 10 | 9.1 | 30 | 75 | 3 | 17.8 | 4 |
| 36.6 | 120 | 10.7 | 35 | 3.7 | 12 | 9.1 | 30 | 75 | 3 | 20.0 | 4.5 |
| 42.7 | 140 | 12.2 | 40 | 4.3 | 14 | 10.7 | 35 | 75 | 3 | 22.2 | 5 |
| 48.8 | 160 | 13.7 | 45 | 4.9 | 16 | 10.7 | 35 | 90 | 3.5 | 28.9 | 6.5 |

*Source:* Reprinted with permission of Portland Cement Association; metric conversion by author.

**Figure 12-37** Frame for Long-Barrel Shell. (Reprinted with permission of Council of Forest Industries of British Columbia.)

**Figure 12-38** Valley Beams under Long-Barrel Shells. (Reprinted with permission of Council of Forest Industries of British Columbia.)

## UMBRELLA-STYLE SHELL ROOFS

Umbrella-style roofs, unlike folded plates and long-barrel shells, are nearly always approximately square or circular in plan. They include such shapes as *hyperbolic paraboloid (H/P) inverted umbrella shells, H/P saddle-shaped shells,* and *dome shells.* All are thin-shell structures that, because of their shape and the manner in which they are constructed, are largely self-supporting units. Depending on the type, units require one or more supporting columns or abutments. Most are particularly useful for single-story buildings in which large spans are required for both length and width. Such shells are normally produced from either wood or concrete.

**Figure 12-39** Inverted Umbrella Roof.

### Hyperbolic Paraboloid Inverted Umbrellas

Hyperbolic paraboloid (H/P) roofs of the shape shown in Figure 12-39 use ribbed shells that spread out from a single-column support in a cantilever manner. This will produce an appealing and efficient roof framing system when several of these sections are arranged in a regular grid pattern. This type of roof is economical in material use, in that each section requires only one column for support, and stresses in the shell, although complex, are relatively small, allowing for a thin roof section to span a relatively large distance. The individual sections that comprise the roof may be square or rectangular and lend themselves to repetition, making form reuse feasible. The underside of the shell is normally finished smooth, with the ribs and edge beams extending through the top of the slab.

Normal shell thicknesses range from 3 to 3½ in. (75 to 90 mm), with spans ranging from 30 to 50 ft (9 to 15 m) between columns.

### Hyperbolic Paraboloid Saddle-Shaped Shells

A hyperbolic paraboloid (H/P) saddle-shaped shell (see Figure 12-40) is a three-dimensional slab in which strength and rigidity are attained by curving it in space, not by increasing the slab thickness. The doubly curved surface allows the transfer of loads to supports entirely by direct forces so that all parts of the shell are uniformly stressed.

Shell thickness is from 2¾ to 4 in. (70 to 100 mm) and the shell projection (see Figure 12-41) from 50 to 220 ft (15 to 65 m), with spans ranging from 50 to 160 ft (15 to 50 m). A *single-saddle* roof requires only

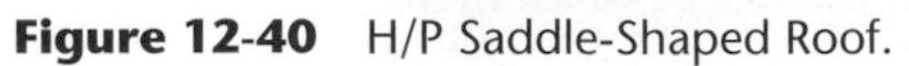

**Figure 12-40** H/P Saddle-Shaped Roof.

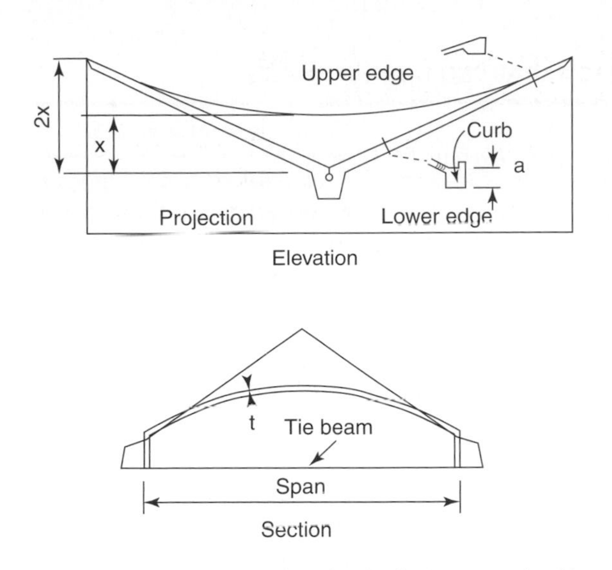


**Figure 12-41** Single Saddle H/P Shell. (Reprinted with permission of Portland Cement Association.)

two abutments (see Figure 12-41) for support. However, the shape may be varied to suit particular circumstances. In Figure 12-42, for example, there are three supports, whereas the roof in Figure 12-43 requires four.

Span data for concrete H/P saddle-shaped shell roofs are given in Table 12-8, based on the dimensions shown in Figure 12-41.

One problem involved in building H/P roofs is the forming of the roof deck. It is not in one or two planes, as is the case with more conventional roof systems, but presents a doubly curved surface, which might be likened to a warped parallelogram. This problem is overcome by using narrow board strips for the formwork sheathing as the warped plane is created by a series of straight lines.

**Figure 12-42** H/P Roof Frame with Cables.

**Figure 12-43** Four-Point Support for H/P Saddle Roof.

Table 12-8
SPAN DATA FOR CONCRETE HYPERBOLIC PARABOLOID SHELL ROOFS

| Span | | Projection | | Min. x[a] | | Average *a* | | Average *t* | | Reinforcement | |
|---|---|---|---|---|---|---|---|---|---|---|---|
| (m) | (ft) | (m) | (ft) | (m) | (ft) | (mm) | (ft) | (mm) | (in.) | (kg/m² of Projected Area) | (lb/sq ft Projected Area) |
| 15.2 | 50 | 15.2–21.0 | 50–70 | 0.9–1.5 | 3–5 | 300 | 1 | 70 | 2.75 | 9.8–14.6 | 2–3 |
| 18.3 | 60 | 18.3–25.9 | 60–85 | 1.2–1.8 | 4–6 | 300 | 1 | 70 | 2.75 | 9.8–14.6 | 2–3 |
| 22.9 | 75 | 22.9–32.0 | 75–105 | 1.8–2.7 | 6–9 | 450 | 1.5 | 75 | 3 | 14.6–19.5 | 3–4 |
| 30.5 | 100 | 30.5–42.7 | 100–140 | 2.7–4.0 | 9–13 | 600 | 2 | 80 | 3.25 | 14.6–19.5 | 3–4 |
| 38.1 | 125 | 38.1–53.3 | 125–175 | 4.0–6.1 | 13–20 | 750 | 2.5 | 90 | 3.5 | 19.5–24.4 | 4–5 |
| 45.7 | 150 | 45.7–64.0 | 150–210 | 5.2–7.6 | 17–25 | 900 | 3 | 100 | 4 | 24.4–34.2 | 5–7 |

[a]Maximum feasible limit = S/5.
*Source:* Reprinted with permission of Portland Cement Association; metric conversion by author.

**Figure 12-44** Foamed Plastic and Concrete Surface.

One solution lies in the use of strands of cable to outline the surface. These are stretched from one perimeter member to another (see Figure 12-42) and put under light tension. The surface is then produced by covering the cables with a layer of rigid material, usually foamed plastic insulation, over which is placed a thin layer of concrete (see Figure 12-44).

## Dome Shells

Dome shells may be freestanding structures or may be designed in clusters. In the latter case, adjustable segmental forms are used to create the shape, whereas in the case of a freestanding dome, it may be formed on a compacted mound of earth and lifted into final position by a series of jacks. A *tension ring* is usually required at the circumference of the shell (see Figure 12-45). Domes may be uniformly supported, as shown in Figure 12-46, or may touch the earth at as few as three points.

The thickness of the shell ranges from 3 to 4½ in. (75 to 115 mm) and spans from 50 to 500 ft (15 to 150 m) or more. The shell thickness is usually increased from 50% to 75% near the periphery of the dome. Dome span data are given in Table 12-9, based on the dimensions shown in Figure 12-47.

**Figure 12-45** Dome Shell with Tension Ring.

**Figure 12-46** Dome Shell Uniformly Supported. Note the Folded Plate Lower Roof.

**Table 12-9**
**SPAN DATA FOR DOME SHELL ROOFS**

| *D* | | *t* | | ∩ | *a* | | *R* | |
|---|---|---|---|---|---|---|---|---|
| (m) | (ft) | (mm) | (in.) | (deg.) | (m) | (in.) | (m) | (ft) |
| 30.5 | 100 | 75.0 | 3 | 30.0 | 4.1 | 13.4 | 30.5 | 100 |
| | | | | 45.0 | 6.3 | 20.7 | 21.5 | 70.7 |
| 38.1 | 125 | 75.0 | 3 | 30.0 | 5.1 | 16.8 | 38.1 | 125 |
| | | | | 45.0 | 7.9 | 25.9 | 26.9 | 88.4 |
| 45.7 | 150 | 90.0 | 3.5 | 30.0 | 6.1 | 20.1 | 45.7 | 150 |
| | | (75.0) | (3) | 45.0 | 9.4 | 31 | 32.3 | 106 |
| 53.3 | 175 | 100.0 | 4 | 30.0 | 7.2 | 23.5 | 53.3 | 175 |
| | | (90.0) | (3.5) | 45.0 | 11.0 | 36.2 | 37.7 | 123.7 |
| 61.0 | 200 | 115.0 | 4.5 | 30.0 | 8.2 | 26.8 | 61.0 | 200 |
| | | (100.0) | (4) | 45.0 | 12.6 | 41.4 | 43.1 | 141.4 |

*Source:* Reprinted with permission of Portland Cement Association; metric conversion by author.

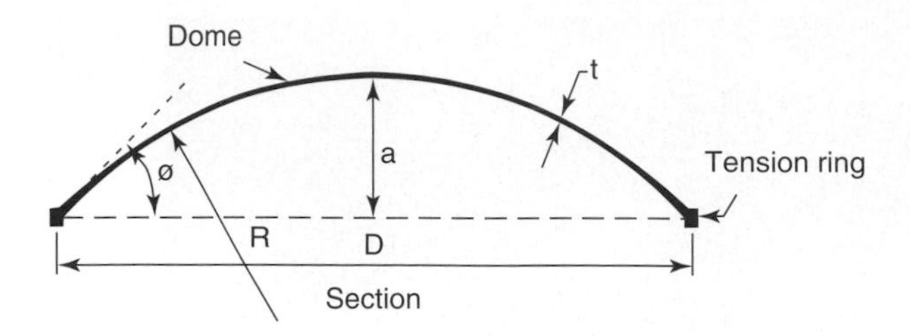


**Figure 12-47** Dome Shell Diagram. (Reprinted with permission of Portland Cement Association.)

Figure 12-48 illustrates an insulated concrete shell being raised into position by hydraulic jacks onto its supporting columns. The dome shape was produced by placing concrete over insulation and reinforcing steel supported by a compacted earth form.

## CABLE-SUPPORTED ROOFS

Large, open-area buildings such as stadiums, arenas, and airplane hangars normally require complex structural components such as trusses, space frames, or domes to support their roof membranes. This creates a large volume of space that is of minimum use but still requires heating and ventilating.

A cable-supported roof derives its structural support from the tension capacity of the steel cables. The cables are usually anchored to a *compression ring* built into the exterior walls of the building. This arrangement produces a roof support system that has minimum depth, and as the cables are flexible, the roof shape can be adapted to any building shape that may be desired. Figure 12-49 illustrates a novel shape for an arena using a cable-supported roof.

**Figure 12-49** Sports Arena with Cable-Supported Roof.

In the illustrated example, the cables were placed in two directions, producing a supporting grid. Over this grid, lightweight precast concrete panels, complete with insulation, were placed, and all voids between the panels were then filled with grout (Figure 12-50). The grouted panels were then covered with a waterproof membrane. Along with the shallow depth of the roof framing, the roof shape allows the building height to be kept to a minimum, reducing the volume of air to be heated and ventilated, and yet provides an unobstructed view of the playing surface for every seat in the building.

**Figure 12-48** Raising Dome with Hydraulic Jacks.

**Figure 12-50** Precast Panels Being Placed on Supporting Cables.

**Figure 12-51** Applying Skin to Stressed-Skin Panel. (Reprinted with permission of Council of Forest Industries of British Columbia.)

## STRESSED-SKIN PANEL ROOF

Stressed-skin panels are structures fabricated from lumber and plywood, in which the plywood skins act integrally with the lumber frame. The longitudinal framing members and the skins must be continuous in the longitudinal dimension or adequately spliced. Skins are attached to the framing members with nails and glue so as to resist the developed shear stress (see Figure 12-51).

Such panels can be produced in a variety of shapes and are often used in building a shell-type plywood roof. Plywood web beams may be introduced into such panels to provide extra stiffness in cases when only one support is to be under each section.

## ROOF MEMBRANES AND INSULATION

The essential element of any roof system is the waterproof membrane. Although its location within the roofing system may vary, its prime purpose is to provide continued protection to the underlying components of the building over a wide range of climatic conditions. To achieve this end, the membrane must be durable and physically and chemically stable, and it must resist the effects of the sun's ultraviolet rays and remain flexible. It must also be compatible with the associated insulation and the supporting decking material. To ensure a lasting and watertight installation, quality control and proper workmanship during the placement of the membrane is of utmost importance.

Many types of roofing membranes are available. The type selected for a particular building will depend on a number of factors. These factors include (1) the suitability of the membrane material (including cost) for the shape of the roof structure; (2) the necessity, location, and type of vapor barrier; (3) the

**Figure 12-52** Sheet Metal Roofing and Siding.

characteristics of the particular insulation that will be used; (4) the thermal characteristics and rigidity of the decking material; (5) the required durability of the roofing membrane; (6) the fire resistance of the roofing; and (7) the aesthetic value of the roofing material.

Roofing membranes used in commercial construction include sheet metal, asbestos-cement boards (the use of asbestos fibers in many jurisdictions is prohibited), built-up roofing consisting of asphalt and felts, single-ply flexible synthetic polymer sheets, prefabricated modified bituminous membranes, monoform roofing, liquid envelope roofing, and high strength synthetic fabrics.

## Sheet Metal Roofing

Various methods are used to make roofing sheets, but two basic types of sheet are generally available: ribbed (see Figure 12-52) and flat. Galvanized steel, plain or coated with polyester paint, and aluminum are most commonly used to make ribbed or corrugated roofing sheets. Various rib profiles are available in several different gauge thicknesses, giving the designer a good selection from which to choose.

Ribbed or corrugated sheet metal roofing sheets are normally supported by wood or steel purlins, spaced according to the gauge of the metal and the roof design loads. Table 12-10 indicates permissible spans for different gauges of typical corrugated sheet metal decking supporting various roof loads.

The two common laying orders for roofing sheets are as illustrated in Figure 12-53. When laying according to order (b), be sure to tuck the corner of the #3 sheet under #2, that of #5 under #4, and so on. Laying should start at the leeward end of the building so that side laps will have better protection from wind-driven rain. The top edges of eave sheets should extend at least 1½ in. (40 mm) beyond the back of steel purlins and 3 in. (75 mm) beyond the center line of timber purlins. At side laps (where the edge corrugation of adjacent sheets is opposite in direction), the underlapping side should finish with an upturned edge and the overlapping side with a downturned edge.

Sheets should extend at least one corrugation over the gable, and there should be 3 in. (75 mm) of overhang at the eaves. End laps between sheets should generally be 6 in. (150 mm) and side laps 1½ corrugations, but these may be increased to 9 in. (225 mm) and 2 corrugations for extreme conditions. For low-pitched roofs, it is good practice to seal the laps with a suitable caulking compound.

Special nails with a ring or screw-type shank should be used for fastening corrugated sheets to wood purlins. They should be zinc coated, with a lead underhead, and long enough to provide adequate holding power. Nails should be driven at the top of corrugations, but care must be taken not to drive them so far as to flatten the corrugations, thus preventing the next sheet from fitting properly. Sheets are fastened to steel purlins with stainless steel self-tapping screws and aluminum washers (Figure 12-54).

Flat sheets of metal (terne plate, galvanized iron, copper, lead, zinc, aluminum, Monel metal, or stainless steel) are applied over solid backing on either a flat deck or a pitched roof. Sheets are applied in strips that run up the slope of pitched roofs and are locked together by one of three types of joints (seams) in common use. These seams allow the metal to expand without buckling. Figure 12-55 illustrates these three types of seams.

## Asbestos-Cement Roofing

Although the use of asbestos fibers are prohibited due to health concerns in many countries, sheeting composed of a cement-asbestos mix provides a tough weather resistant material suitable for use as siding and roofing. The most common application is on industrial buildings.

Roofing sheets of asbestos-cement are made in several designs, four of which are shown in Figure 12-56. Corrugated board and transitile are used on sloping roofs over wood or steel purlins and are laid with a 6-in. (150-mm) end lap and one corrugation side lap.

## Table 12-10
## MAXIMUM PERMISSIBLE PURLIN SPACING FOR GALVANIZED CORRUGATED SHEET METAL[a]

| Superimposed Load | | | | | | | | | | | | | | | | | | | | | |
|---|---|---|---|---|---|---|---|---|---|---|---|---|---|---|---|---|---|---|---|---|---|
| (kN/m$^2$) | | 0.48 | | | | 0.72 | | | | 0.96 | | | | 1.2 | | | | 1.4 | | | |
| (Psf) | | 10 | | | | 15 | | | | 20 | | | | 25 | | | | 30 | | | |
| Roof Pitch | | (degrees) | | (in/ft) | | (degrees) | | (in/ft) | | (degrees) | | (in/ft) | | (degrees) | | (in/ft) | | (degrees) | | (in/ft) | |
| Gauge | Thickness | 13 | 23 | 3 | 5 | 13 | 23 | 3 | 5 | 13 | 23 | 3 | 5 | 13 | 23 | 3 | 5 | 13 | 23 | 3 | 5 |
| | (mm)* | mm | | ft | | mm | | ft | | mm | | ft | | mm | | ft | | mm | | ft | |
| 14 | 1.9 | 3429 | 4801 | 11.25 | 15.75 | 3048 | 4343 | 10 | 14.25 | 2819 | 4039 | 9.25 | 13.25 | 2667 | 4267 | 8.75 | 14 | 2515 | (3581) | 8.25 | (11.75) |
| 16 | 1.5 | 3200 | 4572 | 10.5 | 15 | 2896 | 4115 | 9.5 | 13.5 | 2667 | 3810 | 8.75 | 12.5 | 2515 | 3353 | 8.25 | 11 | 2362 | (3200) | 7.75 | (10.5) |
| 18 | 1.2 | 2972 | 4267 | 9.75 | 14 | 2667 | (3734) | 8.75 | (12.25) | 2438 | (3277) | 8 | (10.75) | 2286 | 3048 | 7.5 | 10 | 2210 | (2819) | 7.25 | (9.25) |
| 20 | 0.9 | 2819 | (3886) | 9.25 | (12.75) | (2515) | (3277) | (8.25) | (10.75) | (2210) | (2896) | (7.25) | (9.5) | (1981) | 2591 | (6.5) | 8.5 | (1829) | (2438) | (6) | (8) |
| 22 | 0.8 | 2743 | (3429) | 9 | (11.25) | (2286) | (2972) | (7.5) | (9.75) | (1981) | (2591) | (6.5) | (8.5) | (1829) | 2362 | (6) | 7.75 | (1676) | (2134) | (5.5) | (7) |
| 24 | 0.6 | 2438 | (3200) | 8 | (10.5) | (2057) | (2667) | (6.75) | (8.75) | (1753) | (2286) | (5.75) | (7.5) | (1600) | 2057 | (5.25) | 6.75 | (1448) | (1905) | (4.75) | (6.25) |
| 26 | 0.5 | (2210) | (2896) | (7.25) | (9.5) | (1829) | (2438) | (6) | (8) | (1600) | (1829) | (5.25) | (6) | (1448) | 1905 | (4.75) | 6.25 | (1295) | (1753) | (4.25) | (5.75) |

[a]Figures enclosed in parentheses are governed by stress values; the remainder are governed by deflection values.
*Approximate values.

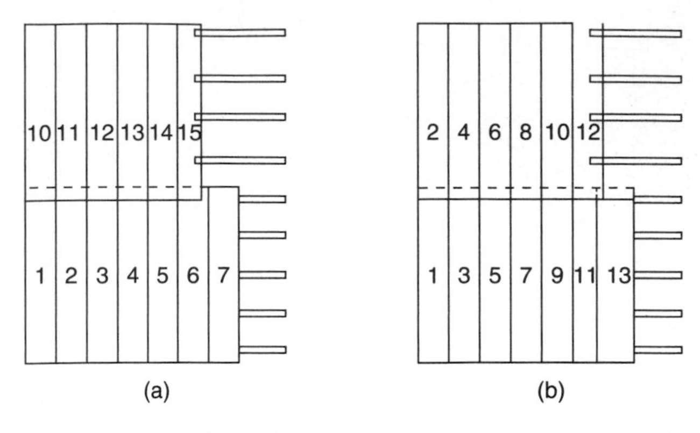


**Figure 12-53** Sheet Laying Order.

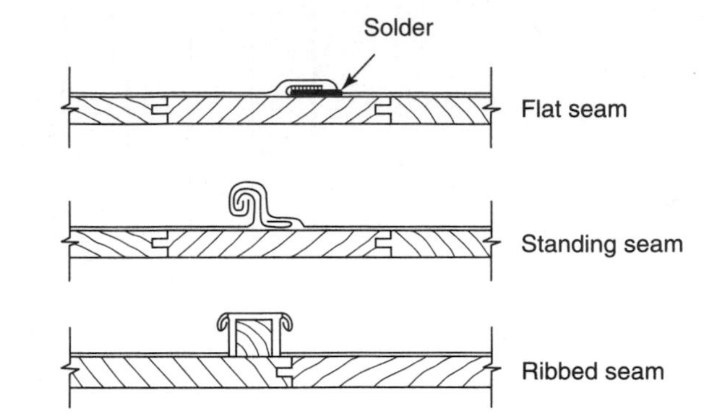


**Figure 12-55** Seams for Sheet Metal Roofing.

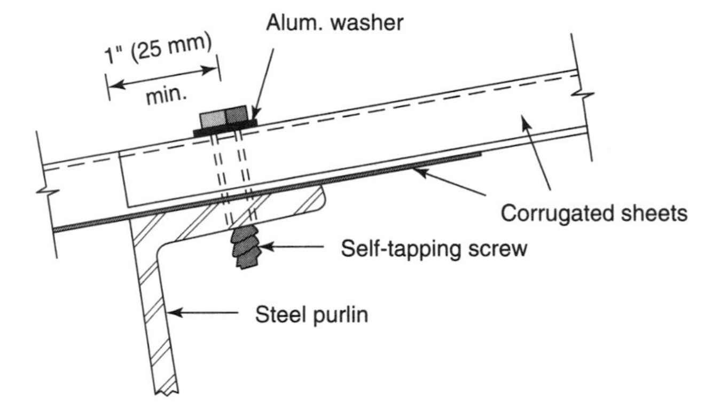


**Figure 12-54** Corrugated Roofing over Steel Purlins.

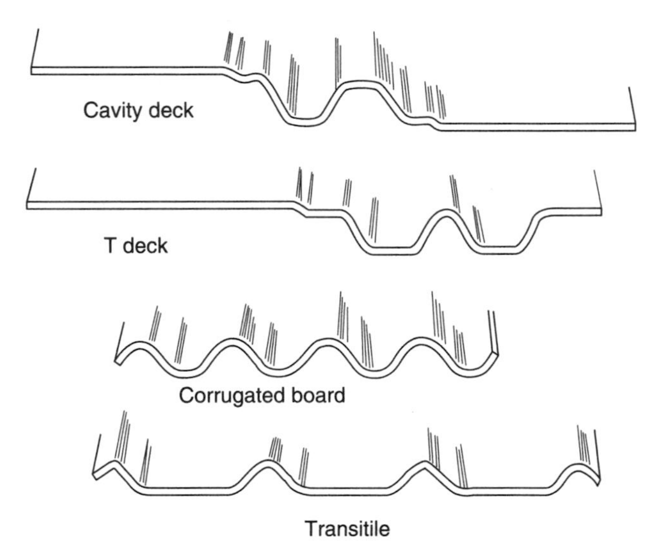


**Figure 12-56** Asbestos-Cement Roofing Sheets.

There should be expansion joints in transitile roofing, wherever they occur in the structure or every 100 ft (30 m), if possible. Figure 12-57 illustrates two types of expansion joint battens in common use, but the curved batten is preferable.

Cavity decks and T decks are attached to the frame and to one another with self-tapping screws, as in Figure 12-58.

## Built-Up Roofing

Built-up roofing membranes, until recent times, have been the choice of designers for low slope roof applications. This type of roofing membrane consists of layers of roofing felts, bonded with tar or asphalt, resulting in a flexible monolithic waterproof layer. It is laid down to conform and bond to the roof deck and it must provide a seal around all projections and openings in the roof. The membrane is turned up at vertical surfaces, such as walls and sky light curbs, as *base flashing* (see Figure 12-59). Appropriate metal flashing must be provided at the vertical surface to ensure a waterproof transition at the intersection of the horizontal and vertical surfaces. The cant strip provides a backup support to the roof membrane, as it is turned up at the vertical surface to minimize the possibility of cracking the membrane in the corner. Many variations are used for the flashing arrangement, and Figure 12-59 illustrates one example of a possible arrangement of metal flashing on the vertical surface to protect the turned up roof membrane. At roof edges with no parapet, the membrane is extended up over a cant strip to facilitate drainage, and is then covered with metal flashing (see Figure 12-60). The metal flashing may be incorporated into the fascia and serves as a finish along the roof edge. The application of all materials must be done with reasonable care to provide the necessary durability that will ensure a watertight roof and a reasonable roof life span.

Insulation and air vapor retarders are a part of most built-up roofing installations. The use of a

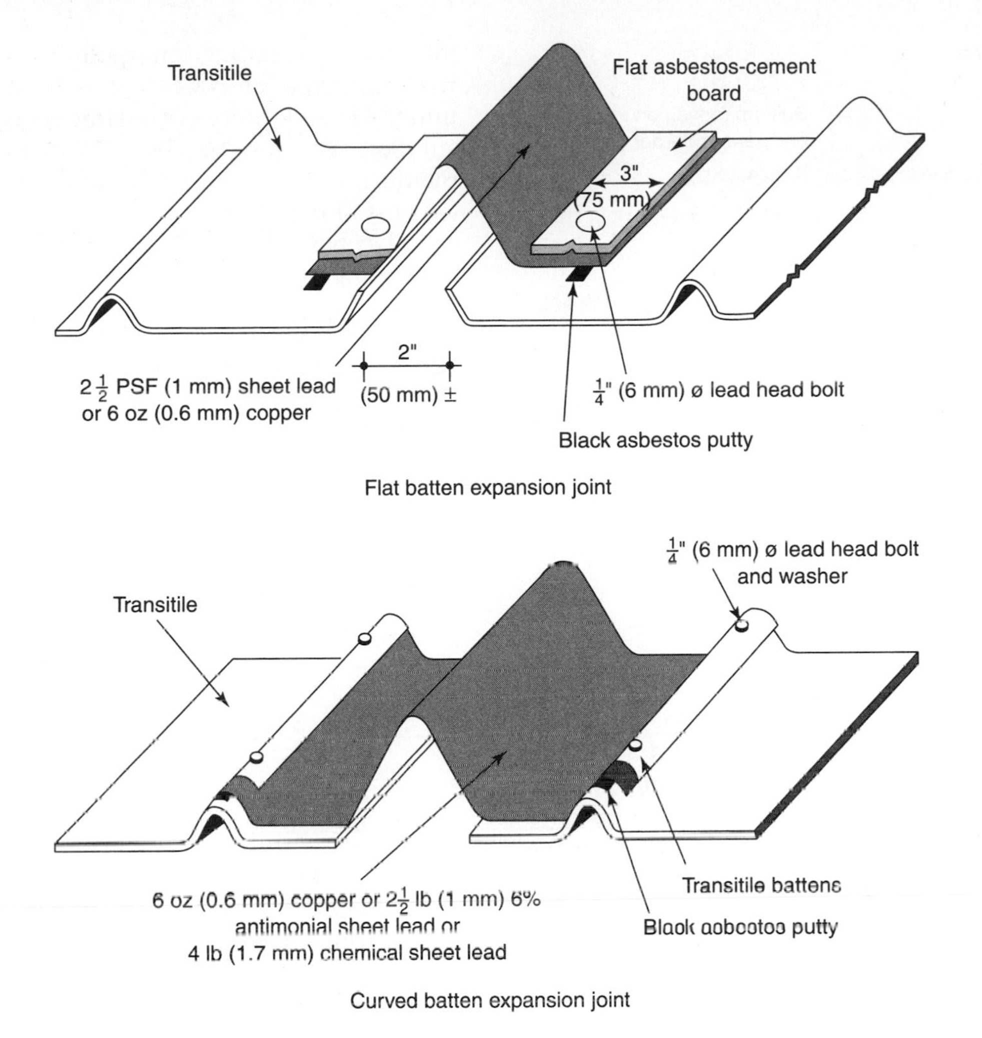


**Figure 12-57** Expansion Joints in Transitile Roofing.

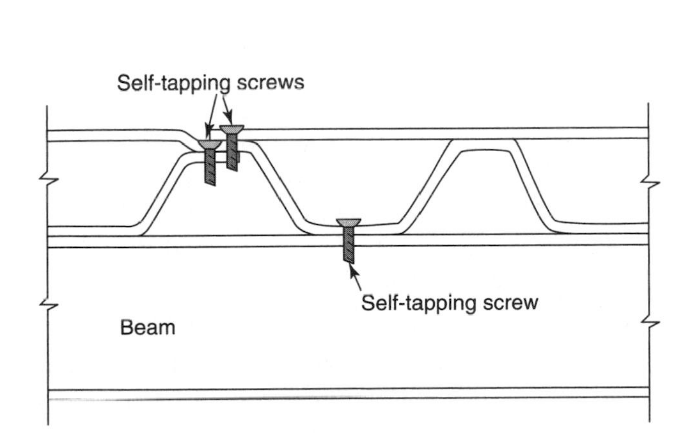


**Figure 12-58** Attaching Asbestos-Cement Decking to Roof Frame.

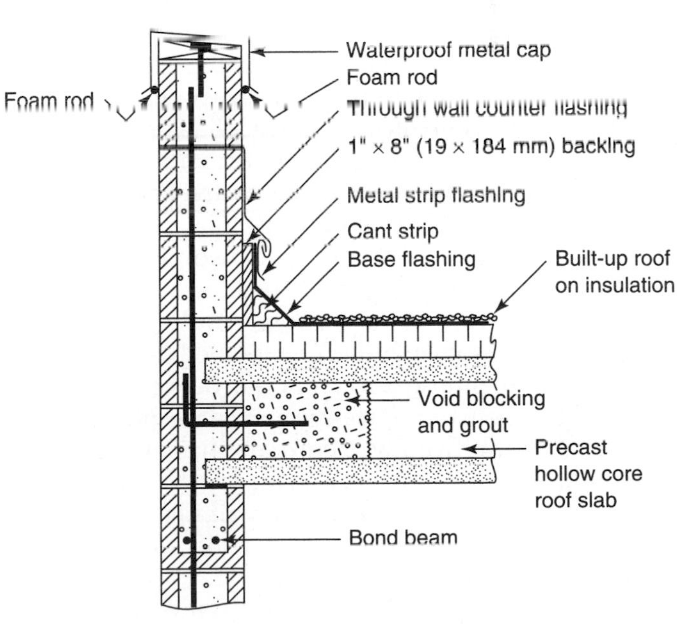

**Figure 12-59** Flashing at Parapet Wall.

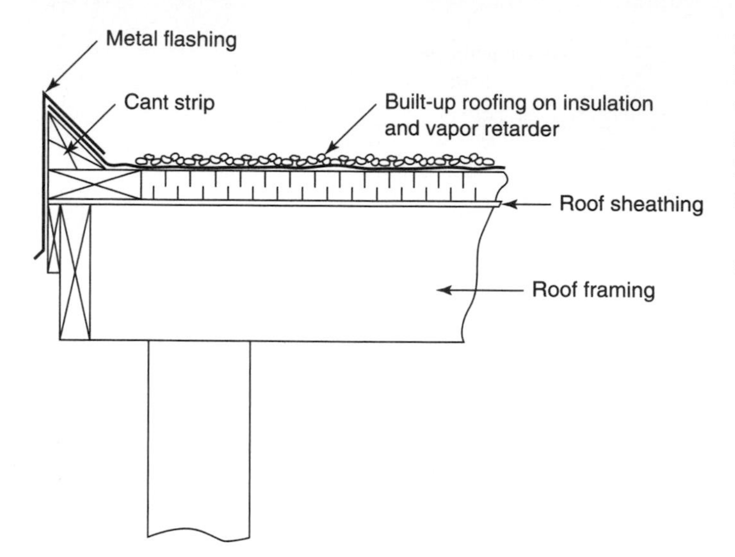


**Figure 12-60** Flashing at Roof Edge.

vapor barrier is essential wherever insulation is part of the roof structure. The proper design and installation of a vapor retarder is particularly important in northern regions. Without it, moisture vapor from within the building will penetrate the insulation, condense, destroy the insulating value, and eventually result in blistering or other damage to the roof covering.

The vapor retarder may be located at various points in the roof assembly, depending on the type of construction. It may be over a suspended ceiling, under the roof deck or, most commonly, over the roof deck just under the insulation. The insulation is sometimes completely enveloped in vapor retarder wrapping, in which case the joints between insulation boards must be adequately sealed.

Insulation is applied over the vapor retarder or over the deck by cementing it with asphalt. Roof deck insulation is of the rigid board type and may be rigid urethane, glass fiberboard, corkboard, expanded polystyrene, wood fiberboard, or cellular concrete.

Before modern single-ply and two-ply roofing membranes were in common use, two standard specifications were developed for built-up membranes: (1) a five-ply membrane for wood decks, and (2) a four-ply membrane for concrete decks. The basic difference between a five-ply and a four-ply membrane is that the five-ply membrane has sheathing paper and two plies of dry felt nailed to the deck prior to the application of the remaining layers. A typical four-ply membrane on a concrete roof is shown in Figure 12-61.

Modifications to these basic standards have been developed to meet varying conditions. The factors that are considered in the development of a membrane specification include (1) slope of the roof, (2) type of roof deck (single or assembled units), (3) nailability of roof deck, (4) presence of insulation and vapor barriers, (5) life expectancy of the membrane, (6) type of bitumen to be used, and (7) type of paper to be used.

Five basic types of membrane are recognized, depending on the types of paper and bitumen used:

1. Asphalt felts, asphalt and gravel
2. Tarred felts, pitch and gravel
3. Asbestos felts, asphalt and asphalt smooth flood coat
4. 17- or 19-in. (425- or 475-mm) selvage roofing
5. Cold process roofing

Typical specifications for the first two types of membrane using hot asphalt or pitch, based on five plies for 100 square feet of wood roof deck without insulation, would appear as follows:

1. *Incline:* ½ to 2 in./ft.
   a. *Dry sheathing:* 5 lb/100 sq ft.
   b. *Felts:* 2 layers, 15# asphalt impregnated, dry applied; 2 layers, 15# asphalt impregnated, mopped on
   c. *Asphalt:* 3 moppings, approx. 20 lb/mopping; topping, approx. 65 lb, poured; for ½- to 1-in. slope, 140° asphalt; for 1- to 2-in. slope, 170° asphalt; if over 2-in. slope, use 210° asphalt
   d. *Nails and caps:* 1 lb of 1¼-in. barbed nails; ½ lb of flat caps
   e. *Surfacing:* 400 lb of gravel *or* 300 lb of slag, ¼- to ⅝-in. size, dry, well rounded, free of dust
2. *Incline:* 0 to 1 in./ft.
   a. *Dry sheathing:* 5# dry sheathing paper
   b. *Felts:* 2 layers, 15# tarred, dry applied; 3 layers, 15# tarred, mopped on
   c. *Pitch:* standard roofing pitch, not heated over 400°F or mopped at less than 350°F; 3 moppings, approx. 25 lb/mopping; topping, 80 lb, poured
   d. *Nails and caps:* 1 lb of 1¼-in. barbed nails; ½ lb of flat caps
   e. *Surfacing:* 400 lb of gravel *or* 300 lb of slag

Based on metric units, for a 10-sq m roof area, the specifications would contain the following:

1. *Incline:* 4% to 16%
   a. *Dry sheathing:* 2.4 kg/10 m$^2$ (No. 5 sheathing paper)

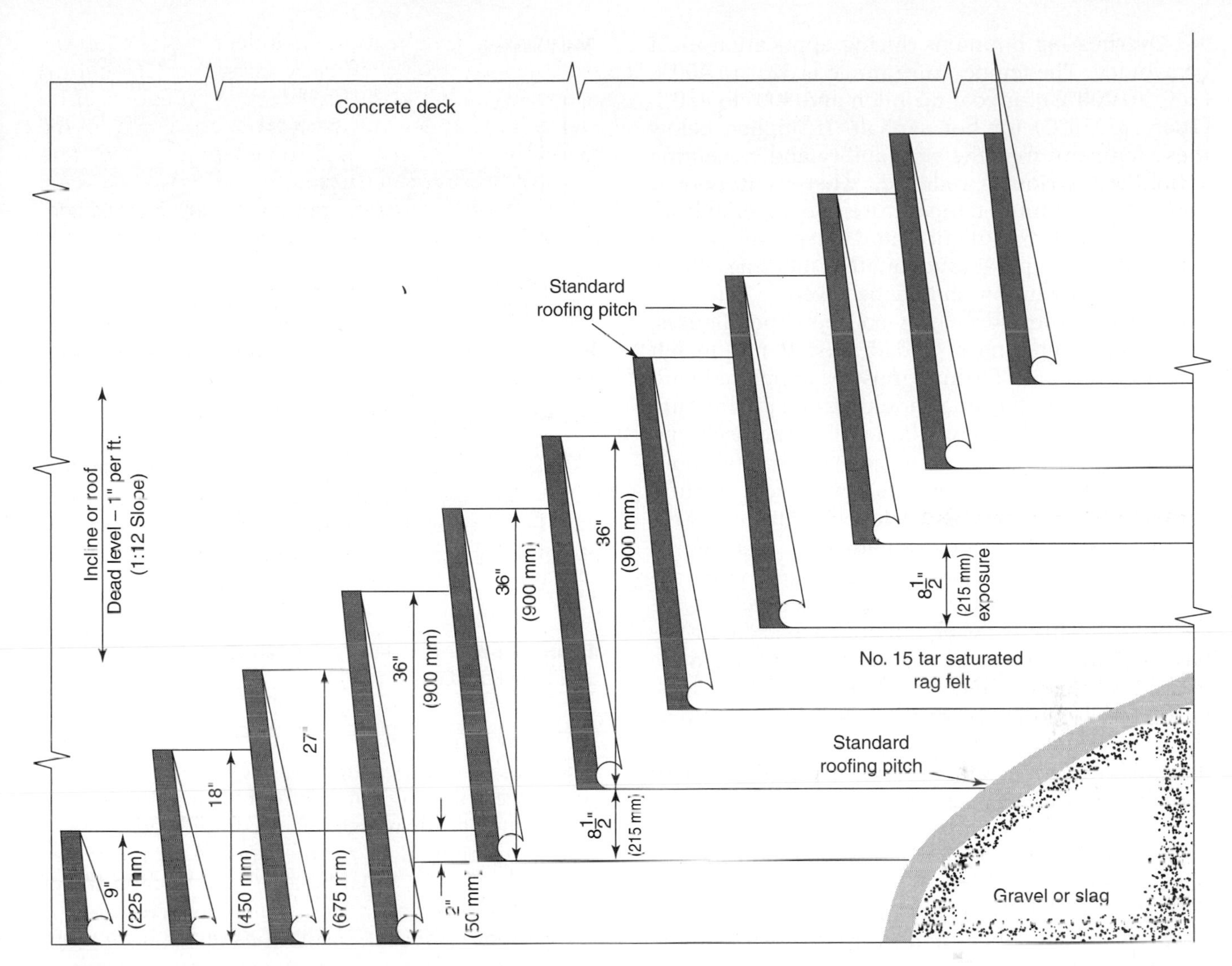


**Figure 12-61** Four-Ply Felt and Gravel Roof on Concrete Deck.

b. *Felts:* two layers, No. 15 asphalt impregnated, dry applied; two layers, No. 15 asphalt impregnated, mopped on
c. *Asphalt:* three moppings, approximately 9.75 kg/mopping; topping, approximately 32 kg poured; for 1% to 8% slope, 60°C asphalt; for 8% to 16% slope, 77°C asphalt; if over 16% slope, use 100°C asphalt
d. *Nails and caps:* 0.5 kg of 32-mm barbed nails; 0.25 kg of flat caps
e. *Surfacing:* 200 kg of gravel or 150 kg of slag, 6- to 16-mm maximum size, dry, well rounded, dust free

2. *Incline:* 0% to 8%
   a. *Dry sheathing:* No. 5 dry sheathing paper
   b. *Felts:* two layers, No. 15 asphalt impregnated, dry applied; three layers, No. 15 tarred, mopped on
   c. *Pitch:* standard roofing pitch, not heated over 200°C or mopped at less than 180°C; three moppings, approximately 12 kg/mopping; topping, 40 kg, poured
   d. *Nails and caps:* 0.5 kg of 32-mm barbed nails; 0.25 kg of flat caps
   e. *Surfacing:* 200 kg of gravel or 150 kg of slag

A double-pour top coating is desirable on low-pitched roofs. It is accomplished by embedding approximately 200 lb (90 kg) of gravel in a top pour of 50 lb (22 kg) of bitumen and later repeating the operation with 300 lb (135 kg) of gravel in 75 lb (33 kg) of bitumen on a second pour. This procedure gives added protection against standing water and melting snow or ice. The second pour may be placed after all other phases of roof construction have been completed and provides an opportunity to check for possible damage.

Overheating bitumens during application must be avoided. The temperature range is 325° to 400°F (160° to 200°C) for coal-tar pitch and 400° to 450°F (200° to 230°C) for hot asphalt. If applied below these temperatures, the workability and cementing action are seriously reduced, whereas if poured above the maximum temperature, the material loses valuable oils by distillation and forms only a thin film that cannot provide adequate cementing action.

Careful attention should be given to interruptions in the roof surface at parapets, penthouses, vents, pipes, chimneys, and drains. Bends in felt used to make base flashing must be supported with cant strips. Base flashing should extend a minimum of 8 in. (200 mm) above the level of the deck and should be protected at the top edge with metal counterflashing. Because there is danger of rupture caused by movements between horizontal and vertical surfaces, base flashing should not be fastened directly to vertical surfaces such as parapet walls. An 8-in. (200-mm) board should be placed behind the cant strip against the wall, but not fastened to it. The base flashing can then be nailed and cemented to the board and the top edge of the board and the base flashing covered by the counterflashing attached to the wall. Through-wall flashing must be provided in masonry walls extending above the roof to act in conjunction with counterflashing.

The need for expansion joints in the roof will depend on the size and design of the building. Expansion joints in the building itself must be extended to the roof. An expansion joint should be provided at each junction of the main portion of the building with a wing and on the main roof if any dimension exceeds 50 ft (15 m). It is also desirable to provide an expansion joint at each change of joist or roof direction.

## Protected-Membrane Roofs

In a conventional built-up roof the waterproof membrane, usually composed of felts and bitumen, is placed on the outer surface of the insulation. It is then covered with a layer of gravel or slag for protection. With the membrane so near the outer surface of the roof system, it must be able to resist all the stresses exerted on it by temperature changes, weather changes, and traffic-related loads. In a protected-membrane roof, the waterproof membrane and the vapor barrier are combined and placed below the insulation, giving the membrane considerably more cover than in the conventional roof (see Figure 12-62).

Using this arrangement of materials, it has been found that the membrane has a better chance of beginning its service life undamaged and being less af-

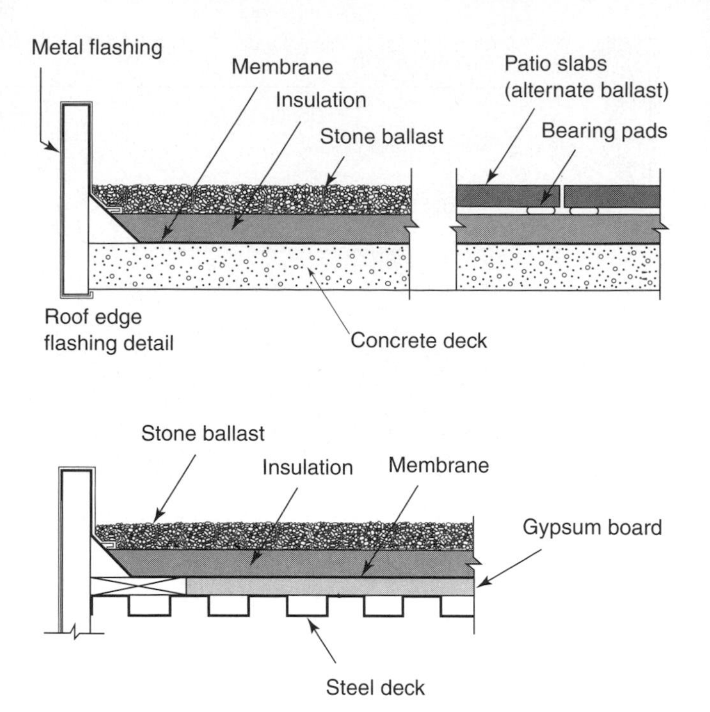


**Figure 12-62** Protected Membrane Roof on Concrete and Steel Roof Decking.

fected by temperature changes. Because large fluctuations in temperature are eliminated, control joints are normally not required in the roof.

With the advantages also come disadvantages. The insulation, now on the outside of the waterproof membrane, must be able to withstand the effects of the weather and traffic-related loads. A number of different insulations have been used: rigid-type insulations composed of polystyrene foam, PVC foam, glass foam, urethane foam, glass fiber, wood fiber, perlite fiber, and cork. Also, fill-type insulations bonded with asphalt have been considered. Although all of these served well, most required some additional treatment to minimize the absorption of moisture, the polystyrene and glass foam being the exceptions. The polystyrenes, being readily available and relatively easy to install, have become the current favorites among designers and contractors. The one drawback of polystyrene insulation is that it must be protected from the effects of ultraviolet rays. Glass foam, although very resistant to moisture absorption, is susceptible to freeze-thaw action.

The top covering of the roof system must be of sufficient weight to prevent displacement during a heavy rain or high wind. For a 2-in. (50-mm) thickness of polystyrene foam, it is recommended that a 1¾-in. (45-mm) depth of ¾- to 1¼-in. (20- to 30-mm) stone ballast with a mass of 12 lb/sq ft (60 kg/m$^2$) be used. An additional ½-in. (12-mm) depth of ballast with a mass of 5 lb/sq ft (25 kg/m$^2$) must be added for

**Figure 12-63** Patio Blocks Used as Insulation Protection.

each additional 1-in. (25-mm) thickness of insulation. In areas where wind may produce scouring, paving stones or patio slabs should be used. Where additional traffic is anticipated, wood decking or patio slabs can be used to provide additional protection for the insulation (Figure 12-63).

Inverted or protected roof systems must have good drainage to perform as intended. The covers must be applied so that they allow evaporation of the water that may seep into the insulation. Roof slopes should be maintained to encourage fast drainage, and roof drains must be placed so that ponding does not occur on the roof.

## Single-Ply Roof Membranes

The leading concern in roofing design is the ability of the waterproof membrane to resist the effects of the weather over an extended period of time while still retaining its water resistance. Until recently, the built up felt and asphalt roof membrane was the only practical means of obtaining a waterproof barrier on low slope or flat roofs. As a waterproof membrane, its success depends largely on the quality of the workmanship during placement rather than the quality of the materials used. Even if the specifications are followed explicitly, however, the life span of the roof rarely exceeds 20 years without some preventive maintenance.

The development of new weather-resistant, reinforced, modified polymer and polyvinyl chloride-based membranes has provided the roofing industry with a viable alternative to the built-up roof system. The toughness, durability, and impermeability of these new materials allow roofers to use a single membrane, permitting shorter installation times without relinquishing quality. Various thicknesses ranging from 45 to 90 mil (1.1 to 2.3 mm) are available depending on the type of application. Although initial costs may be higher for the single-ply system than the built-up membrane, the extra costs will be recovered through lower maintenance and a longer roof life.

In the roofing industry, single-ply membranes are divided into three basic categories—thermoset materials, thermoplastic materials, and modified bitumens. Thermoset membranes consist of rubber polymers and are commonly known as rubber roofing, the most common is ethylene propylene diene monomer (EPDM). Thermoset materials have proven to be resistant to the effects of sunlight and common chemicals found on roofs. Splicing of thermoset materials is done by lapping and requires the use of an additional adhesive to provide a watertight seal at the overlap. Special care must be taken when applying the adhesive to ensure that a durable and watertight bond is achieved at all splice locations.

Thermoplastic membranes are composed of plastic polymers; the most common is polyvinyl chloride (PVC). Because PVC is a semi-rigid material, additional plasticizers and stabilizers are added to enhance its flexibility and strength. In addition, this type of membrane is reinforced with glass or polyester fibers to further increase its strength and stability. Its one advantage over the thermoset membranes is that splicing is achieved by hot air welding—additional adhesives are not required. It is also very resistant to the sun's rays; however, it will become brittle when in contact with bitumen or coal tar pitch and could fail prematurely. When used as a retrofit over an existing bituminous roof, a separation sheet is recommended. PVC membranes are now available in different colors and surface textures. They are also available with liners or separation sheets bonded to the underside for applications on substrates that are non-compatible with the vinyl. Because of the ease in splicing, PVC is a popular material for complex roof shapes and as flashing material.

The modified bituminous membrane is an enhancement of the traditional built-up roofing membrane. The use of polymers improves the thermoplastic properties of the bitumen and natural or artificial fibers are added to enhance its strength. Because it is prefabricated in a factory-controlled environment, the overall thickness of the membrane is reduced to about $\frac{3}{16}$″ (4 mm) compared to $\frac{3}{8}$″ (9 mm) for a traditional built-up roof. It can be used as a single-ply application or the top membrane in a two-layer system. The membrane is attached to the supporting substrate using an asphalt adhesive or, as the roofing is unrolled, the underside of the membrane is melted down onto the supporting surface using a special propane torch. The lapped edges are sealed with additional heat or with contact cement. All joints must be sealed with a compatible asphalt adhesive to prevent delamination.

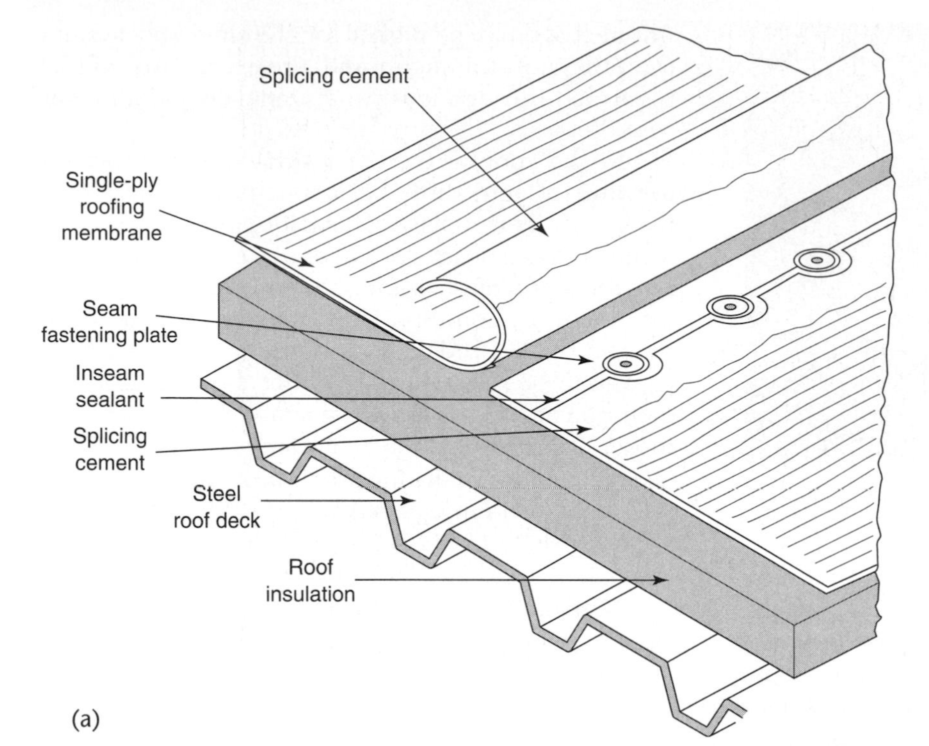

**Figure 12-64** (a) Mechanically Fastened Single-Ply Roof Membrane; (b) Self-Adhering Vinyl Membrane Used on a Curved Roof.

To prevent deterioration caused by ultra-violet rays, special protective coatings, mineral granules, or reflective metal liners are used as covers.

The single-ply membrane can be incorporated into the roof system in the same manner as the built-up membrane. The membrane can be installed using mechanical fasteners [see Figure 12-64(a)], glued directly to the insulation or roof deck material as illustrated in Figure 12-64(b), or laid loosely on top or under the insulation, provided the appropriate ballast is used. Membrane materials are available in widths to 50 ft (15 m), thus minimizing the need for splices. If seams are required, they can be sealed with adhesives or welded together with heat.

Single-ply membranes can be left exposed or covered with gravel ballast. They can be applied to flat roofs or to sloped roofs (see Figure 12-65), and if they are damaged, can be repaired with relative ease.

## Monoform Roofing

Monoform roofing is a reinforced, asphaltic membrane that is applied to a roof with a special type of spray applicator. The membrane consists of a specially formulated asphalt compound reinforced with glass fibers (see Figure 12-66).

Base felts are applied as they are to any built-up roof. A special pressure gun is used to cut the glass fiber reinforcing into designated lengths and to spray out the fibers together with liquid asphalt. The result is a one-piece reinforced membrane of great strength and flexibility that is evenly spread

**Figure 12-65** Exposed Single-Ply Roof Membrane on Curved Roof.

**Figure 12-66** Polysulfide Polymer Waterproofing Being Applied by Spray. (Reprinted with permission of ATK Thiokol Propulsion.)

over the entire surface. This method is particularly useful for contoured and steep-pitch roofs, to which a conventional built-up roof is sometimes difficult to apply.

## Liquid Envelope Roofing

Liquid envelope roofing is a vinyl-based compound, available either clear or in a range of colors. It is sprayed on in enough applications to build up a 15- to 40-mil (0.4- to 1-mm) dry film that is impervious to moisture and humidity, flexible, and elastic through a wide range of temperatures. It may be applied on wood, concrete, steel, insulated surfaces, and asphalt or tar if the surface is first coated with a suitable primer. A continuous film can be carried up the face of a parapet or over the edges of a roof to reinforce flashing.

## Fabric Roofs

Fabric roofs have been used in temporary structures for centuries. The early fabrics used were of natural materials and so had strength and durability limitations. Their use as roof material for permanent structures was never considered seriously when other more durable materials were available. Spurred on by the needs of the space program, new fabrics have been developed using new synthetic fibers of extreme toughness and durability. Because of these properties and their light weight and flexibility, these space-age fabrics are being used as an alternative to the conventional materials used for roofing. Large-area structures such as stadiums (see Figure 12-67), fieldhouses, arenas, and aquatic centers all over the world have incorporated fabric roofs with excellent results.

**Figure 12-67** Inside One of the World's Largest Air-Supported Fabric Roof Stadium. (Reprinted by permission of B.C. Place Stadium.)

The fabric is usually of a polyester, nylon, or fiberglass base coated with a polymer, vinyl, or silicone to produce an extremely strong, durable, and waterproof material. The use of additional liners produces insulation values that are comparable to conventional roofing systems. Many of the fabrics are translucent, allowing sunlight to pass through them, reducing the need for artificial lighting yet reflecting much of the sun's radiated heat. Studies have shown that initial costs of these roof systems are much the same or lower than that of conventional roof systems.

Two basic methods have been developed to support fabric roofs: *cable-suspended roofs* and *air-supported roofs*. The cable-supported roof depends on a structural frame from which the roof cables are suspended and over which the fabric is stretched. Figure 12-68 illustrates an aquatic center covered with a fabric roof. The central supporting rib is of reinforced concrete.

In the case of an air-supported structure, electric fans provide sufficient positive pressure inside the building to support the roof structure. Steel cables are usually placed over the fabric in a checkerboard fashion to provide stability for the envelope during high winds. Figure 12-69 illustrates an air-supported roof on a stadium. The total mass of the roof, including the roof panels, cables, lighting, and sound system, is 280 tons (254 tonnes). Sixteen fans rated at 100 horsepower (75 kW) each supply the necessary air pressure to keep the roof inflated.

**Figure 12-68** Aquatic Center Covered with a Fabric Roof Supported by a Reinforced Concrete Arch.

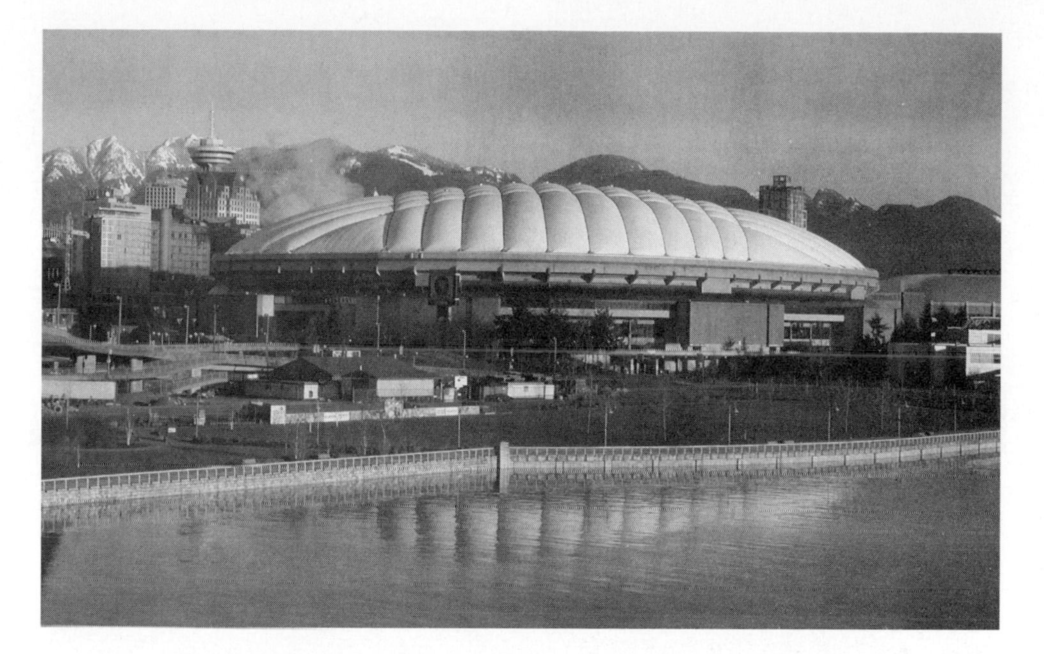

**Figure 12-69** Air-Supported Fabric Roof on a Stadium. Note the Stabilizing Cables Anchored to the Compression Ring. (Reprinted by permission of B.C. Place Stadium.)

## REVIEW QUESTIONS

**1.** Differentiate between **(a)** roof decking and roofing, **(b)** parapet and gravel stop, **(c)** prestressed concrete joist and prestressed roof slab, and **(d)** double T roof slab and cored slab.

**2.** Explain why **(a)** long spans are made possible by the use of precast roof framing members, **(b)** the cost of hollow core roof slabs is relatively low, **(c)** a solid wood roof deck should be kept back ½ in. (12 mm) from parapet walls, and **(d)** scuppers are sometimes required in parapet walls surrounding flat roofs.

**3.** In truss work, explain what is meant by **(a)** a panel point, **(b)** a split ring, **(c)** a purlin, and **(d)** a gusset.

**4.** Outline the basic features of a folded plate roof.

**5.** Describe clearly **(a)** a practical method of forming for a long-barrel shell roof, and **(b)** the form of a hyperbolic paraboloid roof.

**6.** **(a)** Describe, by means of diagrams, three types of seams used to join sheet metal roofing sheets together. **(b)** Explain why these seams are used for joining, rather than soldering or welding.

**7.** Explain what is meant by **(a)** a five-ply built-up roof, **(b)** cold process roofing, **(c)** monoform roofing, and **(d)** liquid envelope roofing.

**8.** What is the basic difference between a conventional built-up roof and a protected-membrane roof?

**9.** What are the two methods used to support fabric roofs? For what purposes are the steel cables required in each type of system?

**10.** In what three ways may a single-ply roof mem brane be applied in a roof system?

**11.** What advantages does a single-ply roof membrane have over the built-up roof system?

# 13 MASONRY CONSTRUCTION

Building with masonry (stone, brick, and the like) involves two types of construction. In one type, the exterior walls are built first and the remainder of the building is framed into them in such a way that those walls transmit the loads of the building directly to the foundations. They are called *bearing walls*. In the other type, the load-bearing framework of a building is constructed of steel, concrete, or timber, and masonry may be utilized to close the building; the walls are simply *curtain walls*.

Historically, brick and stone were the foremost masonry materials. Clay tile, architectural terra-cotta, and concrete blocks of many kinds now also enjoy wide acceptance. All are unit masonry products and require mortar to bond them together. Whether in bearing walls or in curtain walls, the basic techniques involved in building with any type of masonry are similar, though the details of construction may depend on the type of unit used or the type of wall being built.

## BUILDING WITH CONCRETE BLOCKS

### Concrete Masonry Unit Construction

The basic techniques used in building with concrete masonry units have been illustrated and described in many books and manuals and need not be repeated here. The allowable height and minimum thickness of block bearing walls are specified by building codes and depend on the type of wall and its eventual use. Masonry details can also vary from one area of the country to another, reflecting the influence of local conditions and prevailing construction practices.

Masonry blocks are available in sizes of 4-in. (100-mm) multiples. A standard block unit has a nominal height of 8 in. (200 mm), with an actual height of $7\frac{5}{8}$ in. (190 mm). This allows for a $\frac{3}{8}$-in. (10-mm) mortar joint. The nominal length is 400 mm (16 in.) and an actual length of $15\frac{5}{8}$ in. (390 mm). Nominal widths of units range from 4 in. (100 mm) to 12 in. (300 mm) in 2-in. (50-mm) increments. The actual widths are also $\frac{3}{8}$ in. (10 mm) less than the nominal widths. Pilaster blocks can be $15\frac{5}{8} \times 15\frac{5}{8}$ in. (390 × 390 mm) square, square $16\frac{3}{4} \times 16\frac{3}{4}$ in. (425 × 425 mm), or $15\frac{5}{8} \times 18\frac{5}{8}$ in. (390 × 470 mm), depending on the type. This variety allows the designer and the builder a good range of sizes with which to work. In addition to the standard sizes, many suppliers have a wide range of specialty units to deal with nonstandard situations (Figure 13-1(a)).

Load-bearing walls may be single wythe (one block thickness) or multiwythe (two or more thicknesses). Multiwythe walls may be solid or cavity type (Figure 13-1(b)). In the cavity wall, an air space is maintained between the wythes, which in many instances is filled with insulation to increase the thermal efficiency of the wall. Exterior multiwythe walls normally incorporate some other type of material such as brick or stone to act as a finishing veneer rather than to support load.

Concrete block bearing walls must be provided with horizontal or vertical supports at right angles to the face of the wall. Vertical support may be provided by cross walls, pilasters, or but-

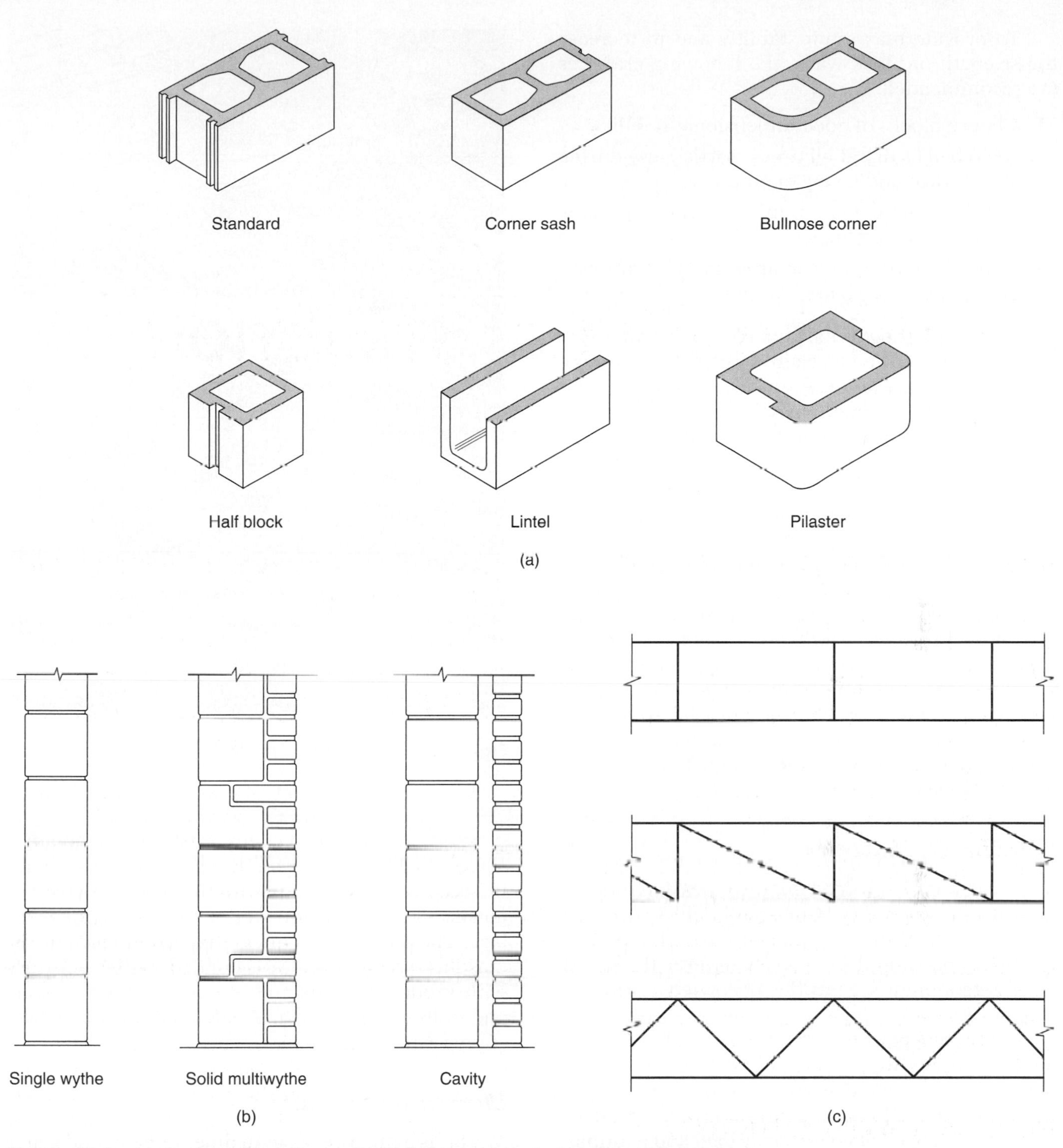


**Figure 13-1** (a) Concrete Block Types; (b) Wall Types; (c) Typical Masonry Reinforcement.

tresses; horizontal support is provided by floors or the roof. Structural codes provide guidelines for maximum distances between vertical and horizontal supports to ensure that walls do not become too slender. For example, load-bearing walls of hollow or solid block units are allowed a maximum height to thickness ratio of 20 with a minimum nominal thickness of 8 in. (200 mm). Non-load-bearing walls have the same limit; however, the minimum thickness can be reduced to 6 in. (150 mm). Cavity walls must be composed of units with a minimum nominal thickness of 4 in. (100 mm) in each wythe, and the width thickness ratio is calculated on an equivalent thickness that is two-thirds of the sum of the individual wythes making up the wall.

To provide maximum stability and to increase the strength of block walls, the following practices are recommended:

1. Choose blocks of good dimensional stability.
2. Keep blocks dry at all times, particularly during site storage and laying operations.
3. Use horizontal joint reinforcement, especially around wall openings.
4. Provide control joints at appropriate locations.
5. Use bond beams where feasible.

Continuous horizontal joint reinforcement is provided by embedding specially fabricated wire reinforcement [Figure 13-1(c)] in the horizontal mortar joints. The reinforcement may be used continuously in every or every other joint throughout the building, but should in any case be used for two consecutive joints above and below openings (unless there are control joints) and should extend at least 24 in. (600 mm) beyond the opening.

During construction, concrete block walls with no lateral support are extremely susceptible to the effects of wind. It is imperative that some temporary bracing be provided to walls (see Figure 13-2) during construction until permanent lateral support is provided by other building components such as floors and roofs. Temporary bracing must be sufficiently rigid and stable to resist the local design wind loads for the building location and should be checked by qualified design personnel.

**Figure 13-2** Temporary Bracing Supporting Masonry Wall.

## Reinforced Masonry

The concept of reinforced masonry was developed to improve the versatility of masonry units in building construction. Although plain or unreinforced masonry has been used extensively all over the world, its shortcomings are vividly portrayed during an earthquake.

Traditional masonry structures were based solely on the compressive properties of the masonry. Rules of thumb were derived, in some instances by trial and error, for wall thickness and distances between lateral supports. The lack of engineering data and a rational method of analysis relegated the use of masonry to a secondary role as a building material in present day construction. Organizations such as the International Masonry Institute, the Brick Institute of America, and the Canadian Masonry Research Institute began research programs to establish an engineered approach to masonry design and construction. Through their efforts, in conjunction with contractors and suppliers, these organizations continue to provide the latest in engineering data, design guidelines, and construction practices for masonry construction.

Reinforcment of masonry is based on guidelines similar to those used in reinforced concrete. As in reinforced concrete, the reinforcing bars provide the tensile strength which sections of plain masonry lack. The reinforcing bars are placed in voids of the masonry sections and are bonded to the masonry with grout. The outcome is that columns, beams, and walls can be sized and evaluated using standard engineering formulae.

## Mortar and Grout

Mortar is a mixture of sand, lime, cement, and water and serves as a bedding material for the masonry units. It allows the mason to make adjustments for irregularities in the masonry units, as well as variations in the supporting structural framework on which the units are being placed. Once it has set up, mortar also provides a bond between the individual masonry units, providing stability and structural integrity to the masonry section.

Two properties of mortar are paramount: its strength and its workability. The mason is primarily interested in workability, whereas the designer is concerned with strength, and some balance must be obtained to satisfy both requirements. The amount of lime used determines the workability of the mortar, whereas the amount of cement will determine the mortar's final strength. If the cement content in the mix is increased, the strength increases; however, a corresponding loss in workability results.

Mortar must not only have good workability and strength, but it must also be compatible with the masonry units that it bonds. Different masonry units have different expansion and contraction coefficients, and when used in various combinations, the mortar must be able to resist the resulting stresses that are produced. In this case, a weaker, more flexible mortar will perform better than a strong and more brittle mortar. Careful thought to the location of control joints, in this instance, will also help minimize undue stresses in the mortar.

Until recently, five mortar types were recognized by masonry codes—types M, S, N, O, and K. Type M is the strongest, but was found to be too brittle and lacked workability. Types O and K, on the other hand, were found to be too weak for the slender walls that are now being used in building envelopes. The latest masonry codes dealing with mortar specifications lean toward the use of two mortar types, S and N. These mortars have sufficient strength for the stresses that are imposed on them and yet exhibit the necessary properties such as durability, water retention, bond strength, and workability that are important to the mason and the designer.

Mortar may be mixed on site in a portable mixer or it may be obtained in a premixed fashion (see Figure 13-3) from a mortar supplier. Mortar mixed on the job site can vary in consistency because of variations in the moisture content of the sand and the accuracy of the proportions that are used in mixing. If mortar strength is a concern, random test samples may be taken and tested for strength periodically. Premixed mortar must be ordered in appropriate quantities so that it can be used before it begins to set up in the container and lose its workability. As Figure 13-3 illustrates, the premixed mortar is reworked in a mortar mixer before being used by the mason.

When reinforcing is added to the masonry, the units must be grouted (Figure 13-4). Grout consists

**Figure 13-3** Premixed Mortar and Mortar Mixer.

**Figure 13-4** Reinforcing Bars in Masonry Wall. Note the Horizontal Reinforcing Wire in the Mortar Joints.

of a high-slump concrete mix to ensure proper bonding between the units and the reinforcing steel. Masonry grout may be specified as fine or coarse depending on its maximum aggregate size. Fine grout uses sand aggregates, whereas coarse grout can have aggregates as large as 1 in. (25 mm) when the grout voids are of sufficient size.

Maximum aggregate size in the grout will depend on the size of the grout space. For grout spaces 2 in. × 3 in. (50 mm × 75 mm), for example, fine grout is recommended, whereas for grout spaces that exceed 5 or 6 in. (125 or 150 mm), coarse grout with 1-in. (25-mm) maximum aggregate size may be used. Recommended grout strength after 28 days is 2,500 psi (20 MPa), with a minimum of 2,000 psi (15 MPa).

Grout may be placed in two ways: *low-lift* grouting or *high-lift* grouting. Low-lift grouting begins when the wall or column has reached a height of approximately 4 ft (1,200 mm). With the reinforcing bars in place, grout is added to the masonry to no more than 4 ft (1,200 mm), and the mason then continues to add courses of masonry for the next lift. What is important is to ensure that the reinforcing bars extend sufficiently beyond the top of the grout to allow for a lap of 30 bar diameters with the bars in the next lift.

High-lift grouting is usually done when the wall or column section has been raised a full story. The advantages of high-lift grouting are that the steel reinforcing can be placed after the wall has been completed, and larger amounts of grout placed at one time without interrupting the mason. However, the wall must be provided with openings at the bottom of the wall to allow for the removal of mortar droppings. Once the grout spaces are cleaned and the reinforcing steel is properly positioned, the grout can be pumped in. Again, placing of grout should not exceed 4 ft (1,200 mm) without vibrating the grout and allowing for settlement and absorption of excess water. At least 15 min should be allowed before the next lift is begun to reduce the effects of hydrostatic pressure and prevent the possibility of blowouts.

## Control Joints

Control joints are vertical joints that separate walls into sections and allow freedom of movement. They should occur at intervals in long, straight walls, where abrupt changes in wall thickness take place; at openings (Figure 13-5); at intersections of main walls and cross walls; and at locations of structural columns or pilasters in main walls.

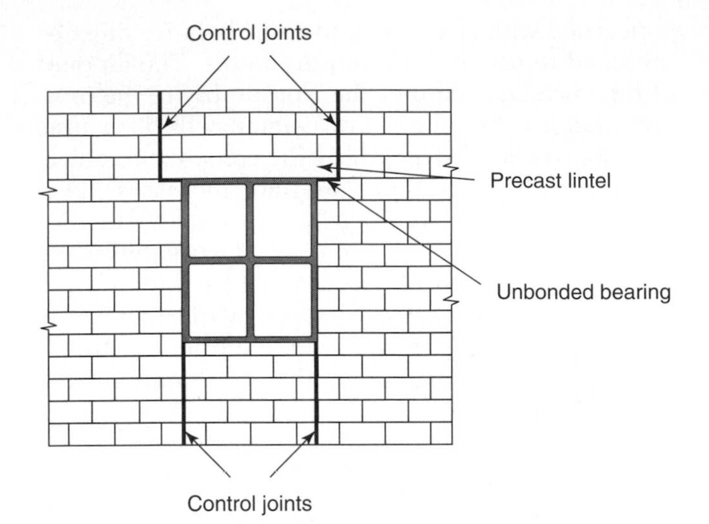


**Figure 13-5** Control Joints at Window.

**Figure 13-6** Control Joint in Block Wall. (Reprinted with permission of Portland Cement Association.)

To produce a control joint, place building paper on a coat of asphalt paint on the ends of the blocks on one side of the joint. Fill the core with concrete or mortar to provide lateral stability (Figure 13-6). Single wire ties may also be used across the joint. Rake the mortar in the control joint to a depth of ¾ in. (20 mm) and caulk with a suitable compound. Caulking may be omitted on the interior face and the joint finished with a deep groove.

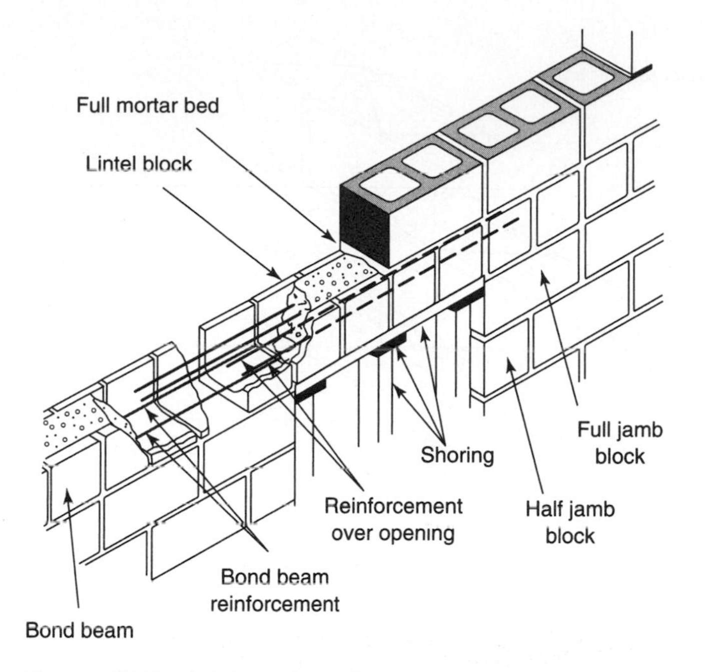


**Figure 13-7** Reinforced Bond Beam.

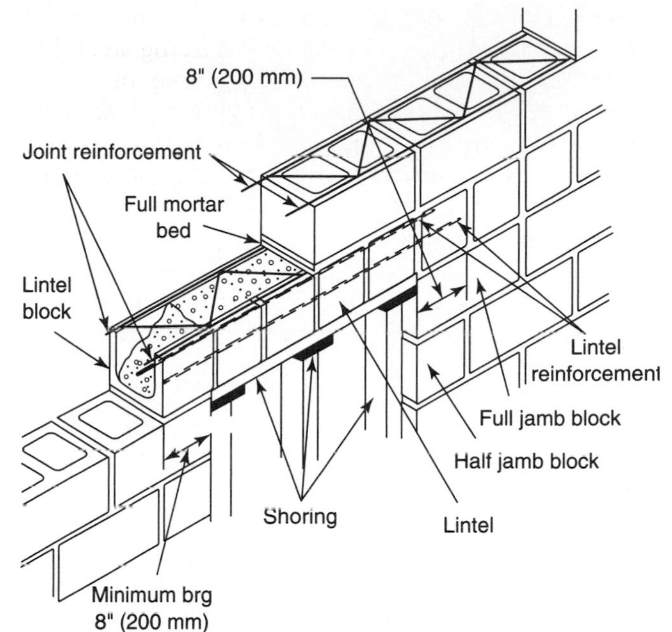


**Figure 13-9** Cast-in-Place Concrete Lintel.

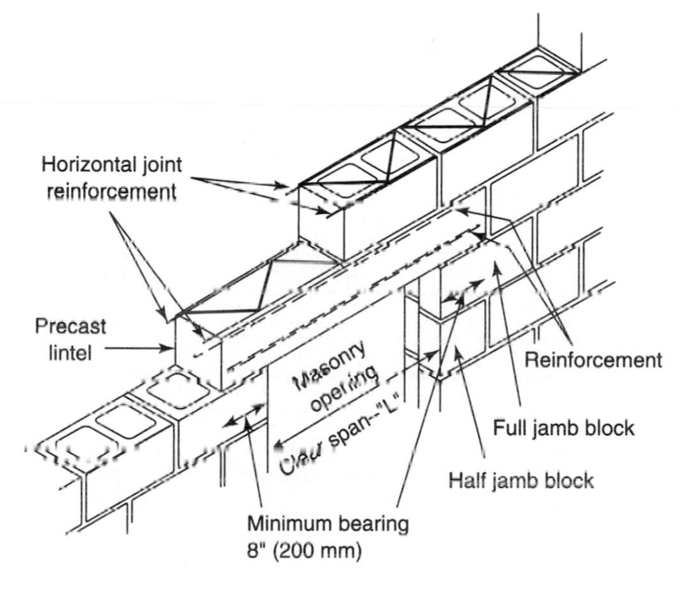


**Figure 13-8** Precast Concrete Lintel.

## Bond Beams and Lintels

The primary purpose of bond beams is to act as a continuous tie for exterior block walls where control joints are not required. They may also be used between control joints in larger buildings. Bond beams can also act as structural members, transmitting lateral loads to other structural members, and can provide bearings for beams or joists.

A bond beam is a continuous, cast-in-place, reinforced concrete beam running around the perimeter of a building or between control joints. It is formed by using standard lintel blocks, deeper bond beam blocks, or standard blocks with a large portion of the webs cut out. In Figure 13-7, the bond beam is being utilized as a lintel over an opening. Notice the shoring required in the opening.

Lintels over openings in block walls may consist of precast concrete units, as shown in Figure 13-8, or may be cast in place. In the latter case (illustrated in Figure 13-9), lintel blocks are used again and are supported by shoring. When greater strength is required, two-course lintels may be employed. In Figure 13-10, standard blocks with cutaway webs are being used to form the lintel.

The large number of shapes available makes a great variety of patterns possible for a block wall. Consult a concrete block manual for illustrations or patterns. In addition to the standard blocks, *screen* blocks are available to add to the variety or to build open walls that are to act as solar screens. Figure 13-11 illustrates three common solar screen shapes. Manufacturers' brochures will provide a complete list of such units.

Concrete blocks are also used to build backup walls for brick, tile, or stone facing. Various types of metal ties are used to bind the facing to the backup; those most commonly used are illustrated in Figure 13-12.

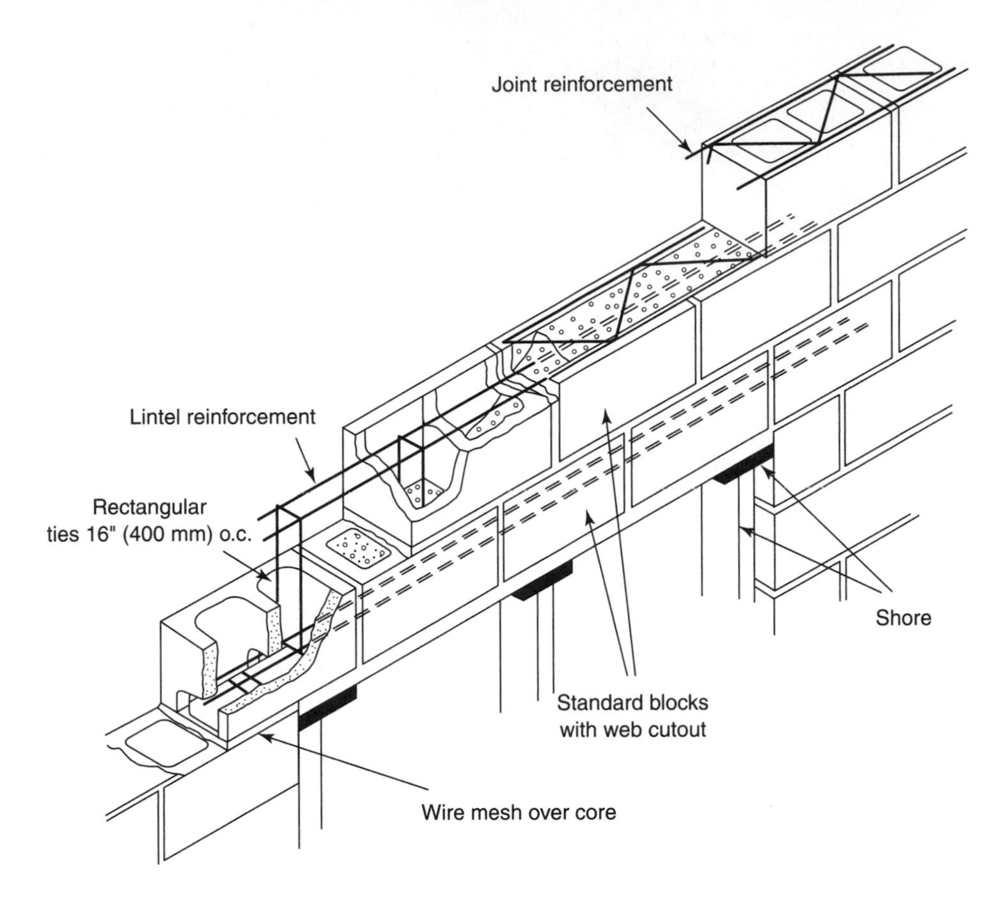


**Figure 13-10** Two-Course Lintel.

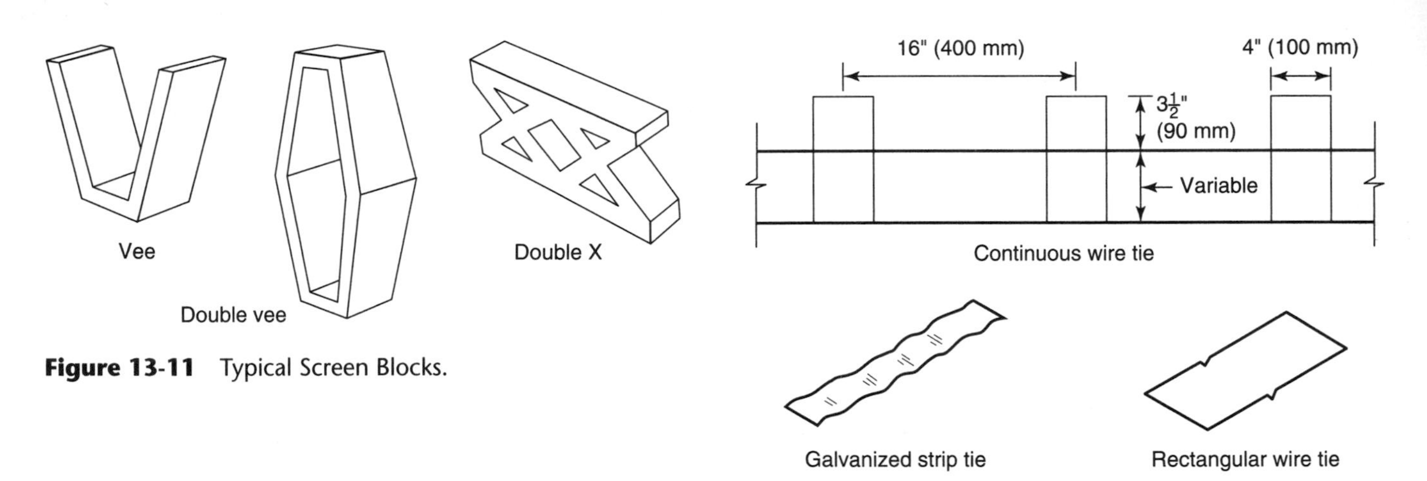


**Figure 13-11** Typical Screen Blocks.

**Figure 13-12** Typical Metal Ties.

## Customized Concrete Block Masonry

One recent development in the concrete block industry has been the production of block with a face that conforms to an architect's design for a particular building—walls of customized concrete block masonry. Figures 13-13(a) and (b) illustrate a few of the block faces available, in both standard and metric sizes. However, a wide variety of designs is possible, and a block producer should be consulted to determine what designs the local plant is capable of producing. Figures 13-14 through 13-18 illustrate some of the customized block faces that have been used in concrete block walls.

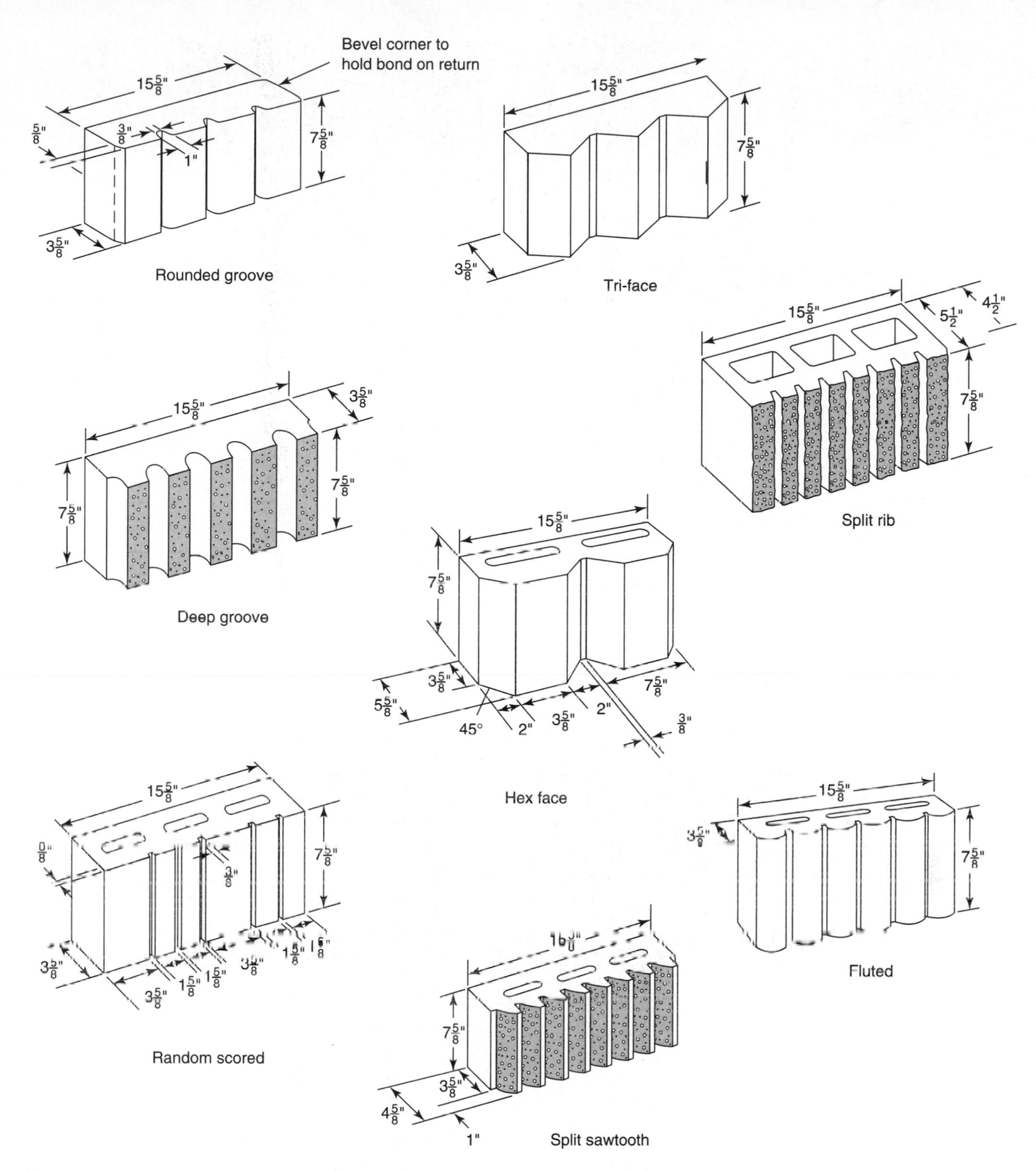


**Figure 13-13a** Customized Concrete Block Units Using Standard Dimensions.

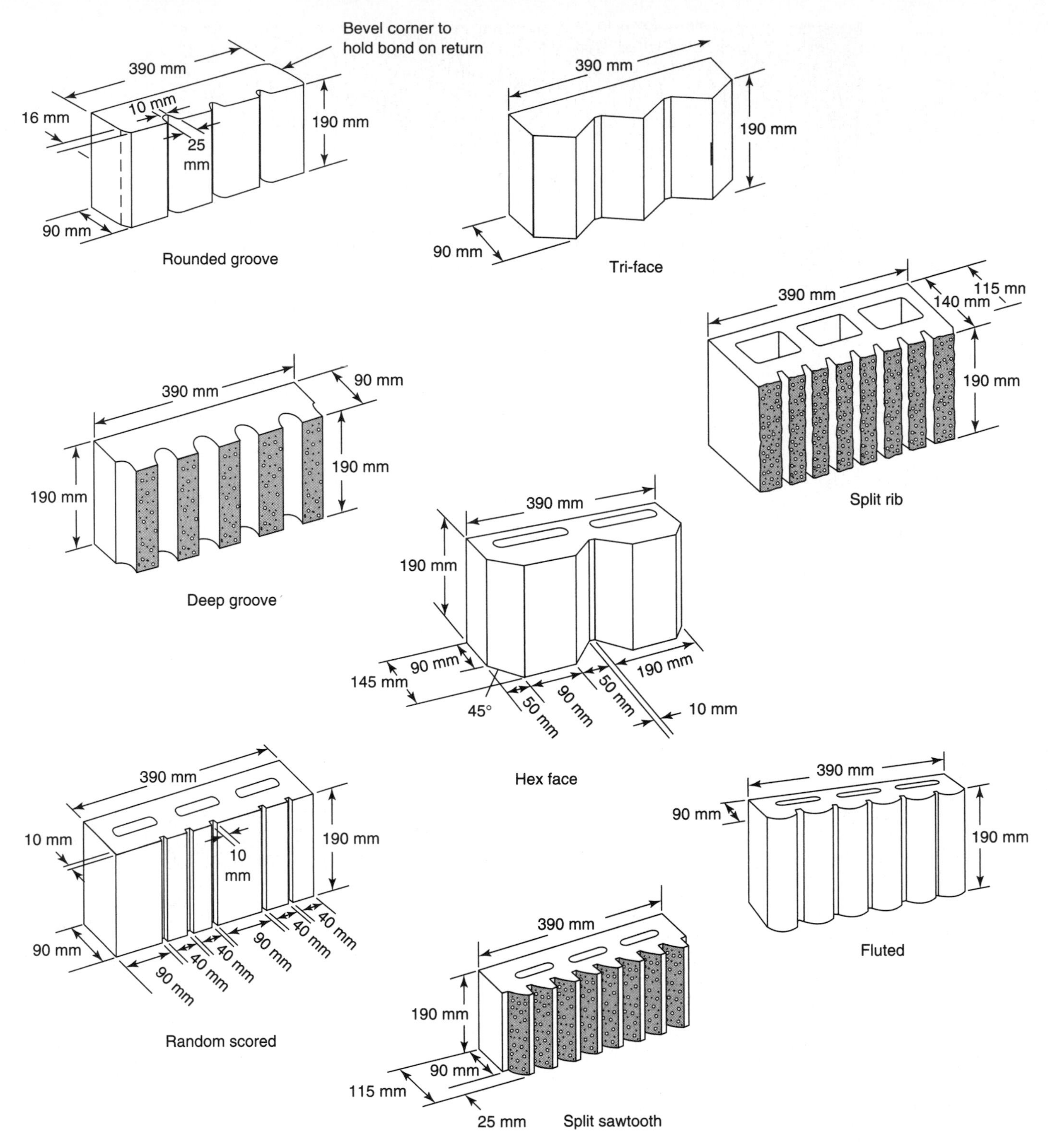


**Figure 13-13b** Customized Concrete Block Units Using Metric Dimensions.

**Figure 13-14** Screen Block Wall. (Courtesy of National Concrete Masonry Association.)

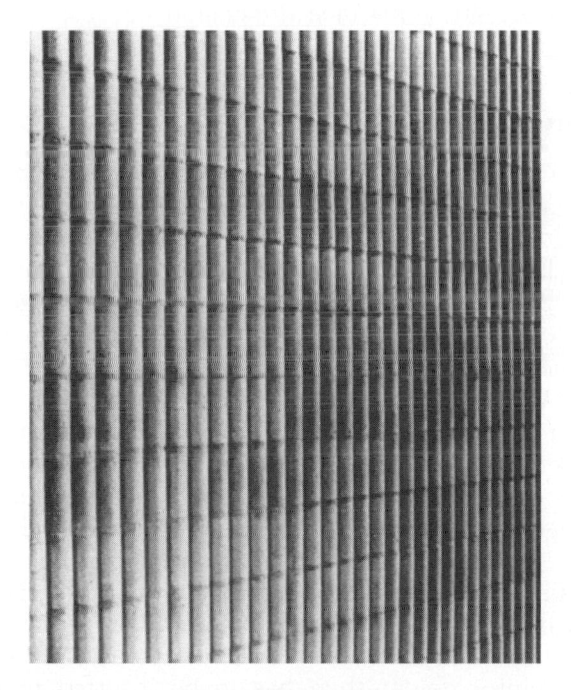

**Figure 13-15** Fluted-Face Block Wall. (Courtesy of National Concrete Masonry Association.)

**Figure 13-16** Scored-Face Block. (Courtesy of National Concrete Masonry Association.)

**Figure 13-17** Split Face with Single Score in the Center of the Block. (Courtesy of National Concrete Masonry Association.)

**Figure 13-18** Split-Rib Block. (Courtesy of National Concrete Masonry Association.)

# PREFABRICATED CONCRETE BLOCK WALL PANELS

Recent innovations in concrete masonry construction include the utilization of *new mortar systems* and the development of *prefabricated concrete block wall panels.*

## Mortar Systems

One new mortar system involves using a strong adhesive known as *organic mortar.* It is applied to masonry joints with a gun and replaces the usual portland cement mortar. The joint is approximately 1/16 in. (1.5 mm) thick and requires the use of blocks that have precise dimensions. In many cases, blocks are precision ground on their bearing surfaces to provide those close tolerances. A variation of this system is the use of an organic additive to portland cement mortar, to increase its flexural strength from 2 to 2½ times.

Another system, known as *surface bonding,* uses a mortar composed basically of portland cement reinforced by the addition of glass fibers to increase the tensile strength. Blocks are stacked without mortar in the joints, and the wall is plastered on both sides with the surface-bonding mortar.

Tongue-and-groove concrete blocks are made for use with both mortar systems to facilitate alignment and provide interlocking action between the blocks. The usual procedure calls for a full 8- by 16-in. (200- by 400-mm) face dimension because of the very thin or completely eliminated mortar joints.

## Panel Fabrication

Wall panels are fabricated by two different methods. In one, block layers (masons) position the units in the wall panels, whereas the other involves the use of a block-laying machine. The manual system can be employed either in a plant or at the job site, in all-weather enclosures on what are called *launch pads,* but the machine operates in a prefabricating plant or factory.

A block-laying machine lays blocks in either running or stack bond, inserts the joint reinforcement, makes the bed and head joints, and tools them. It thus assembles wall panels about 16 times as fast as they could be built in place, using the same personnel employed in the machine operation. The machine will handle blocks from 6 in. to 12 in. (150 to 300 mm) thick and is capable of producing panels up to 12 by 20 ft (3.5 by 6 m) in face dimensions. Vertical reinforcement then has to be inserted in the block cores and the cores grouted. When they are ready, panels are transported to the job site by truck.

Under the manual method, blocks may be laid with conventional mortar, organic mortar, or no mortar. In the last case, surface bonding is employed. Regardless of the mortar system used, panels are reinforced with

**Figure 13-19** Dovetail Masonry Anchor.

steel and grouted. Normally, panels up to 8 by 30 ft (2.4 by 9 m) in size will be laid up by this method.

At the building site, a crane with lifting devices and special equipment to prevent breakage lifts the panels into position in the structure. Walls are set and braced, and if they are load bearing, the floor panels are set directly on their top edges. Wall-to-wall and wall-to-floor connections are made through access pockets left in the panels and floor. Finally, after the connections are complete, the pockets are filled with grout.

## Tying Systems

With conventional block walls, the tying systems between walls and floors or walls and columns have been relatively simple (see Figure 13-19). The development of block panel construction has meant the need for strong connections between adjoining walls and between walls and floors. Horizontal edges (wall-to-floor edges) may be connected by spliced reinforcement, tension connections, bolted-tension connections, bolting, or welding. Vertical edges (wall-to-wall edges) may be connected by steel plates, lapped reinforcement, or formed-lapped reinforcement, among others.

## Wall-to-Floor Ties

The *spliced reinforcement system* is much the same as that used to join precast concrete panels. The bottom ends of reinforcing bars have a coupling nut attached, and when the walls are in place, a short, straight bar, threaded on one end, is attached to the upper bar by the coupling nut and to the bottom one by welding. An L-shaped bar, placed after the walls are up, is laid with one end in a pocket in the floor and the other down through the block core to tie the wall and floor together (see Figure 13-20).

The *tension connection* provides a way of tying the panels together, through the block cores, from top to bottom of the building. Panels are brought to the job site with loose vertical bars in their cores. Each bar has a threaded eyebolt attached at the top and a steel lifting plate and nut at the bottom, to act as a lifting device. After the panel has been set in place, the nut on the bottom end of the top bar is removed, and a turnbuckle is used to join the corresponding top and bottom bars together. The cores in the bottom panel are then grouted, but the cores in the upper panel are left empty until the next panel above is installed. Finally, the last panel in the tier

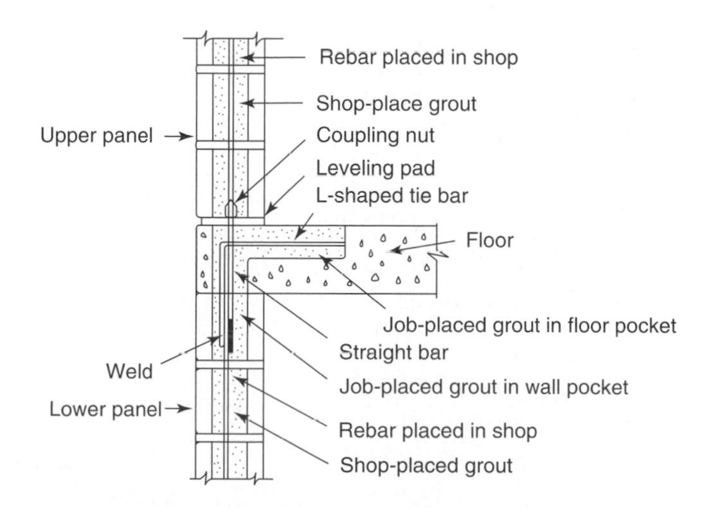


**Figure 13-20** Spliced Reinforcement Connection.

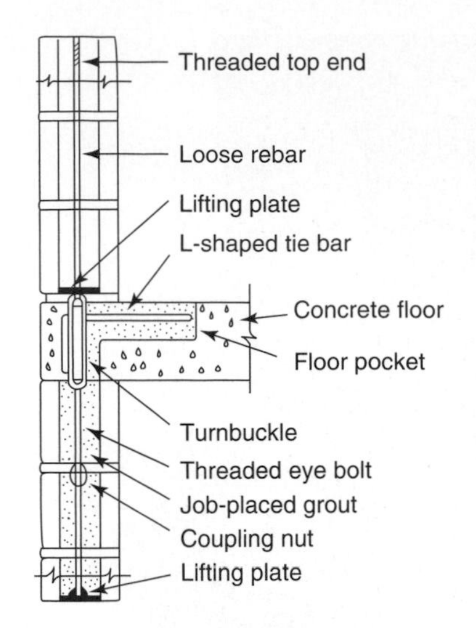


**Figure 13-21** Tension Connection.

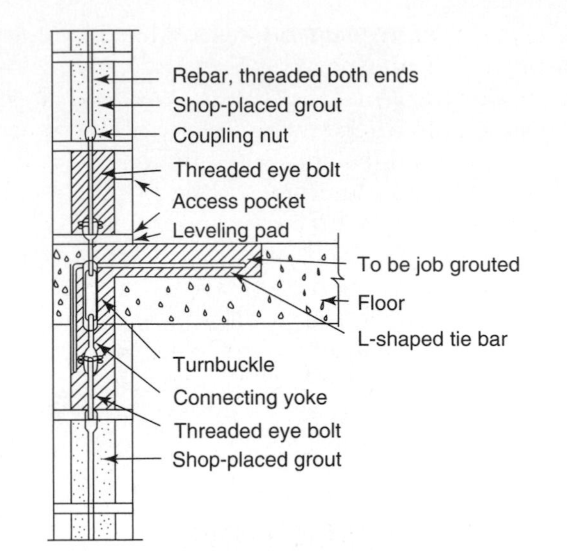


**Figure 13-22** Bolted-Tension Connection.

must have a plate at the top end of the bars to complete the tension connection (see Figure 13-21).

In the *bolted-tension connection,* vertical reinforcement bars are grouted into the panels in the plant, with threaded eyebolts attached at both top and bottom ends. When the bottom panel is in position in the wall, a connecting device consisting of a turnbuckle with a yoke at each end is attached to the end of each bar. Then, when the next panel is installed, the connection is attached to the top bars and the turnbuckle is tightened to join the two panels. Finally, grout is placed in the cores and access pockets (see Figure 13-22).

*Bolted* and *welded connections* are similar to those used with precast concrete panels. Anchor plates are set into the top and bottom of panels during fabrication, and these are welded or bolted to field-placed connections attached to the floor slabs. Access pockets are grouted after the connections are complete.

## Wall-to-Wall Ties

Several types of vertical or wall-to-wall connections are used, including a steel bar connection, a lapped reinforcement connection, and a lapped reinforcement with formed joint connection.

In the *steel bar connection,* a flat steel bar with its ends bent at right angles is preset in the top edge of one panel (see Figure 13-23). A cutout in the top end block of the adjoining panel makes installation easier. After the second panel is in place, the end core is grouted to anchor the bar.

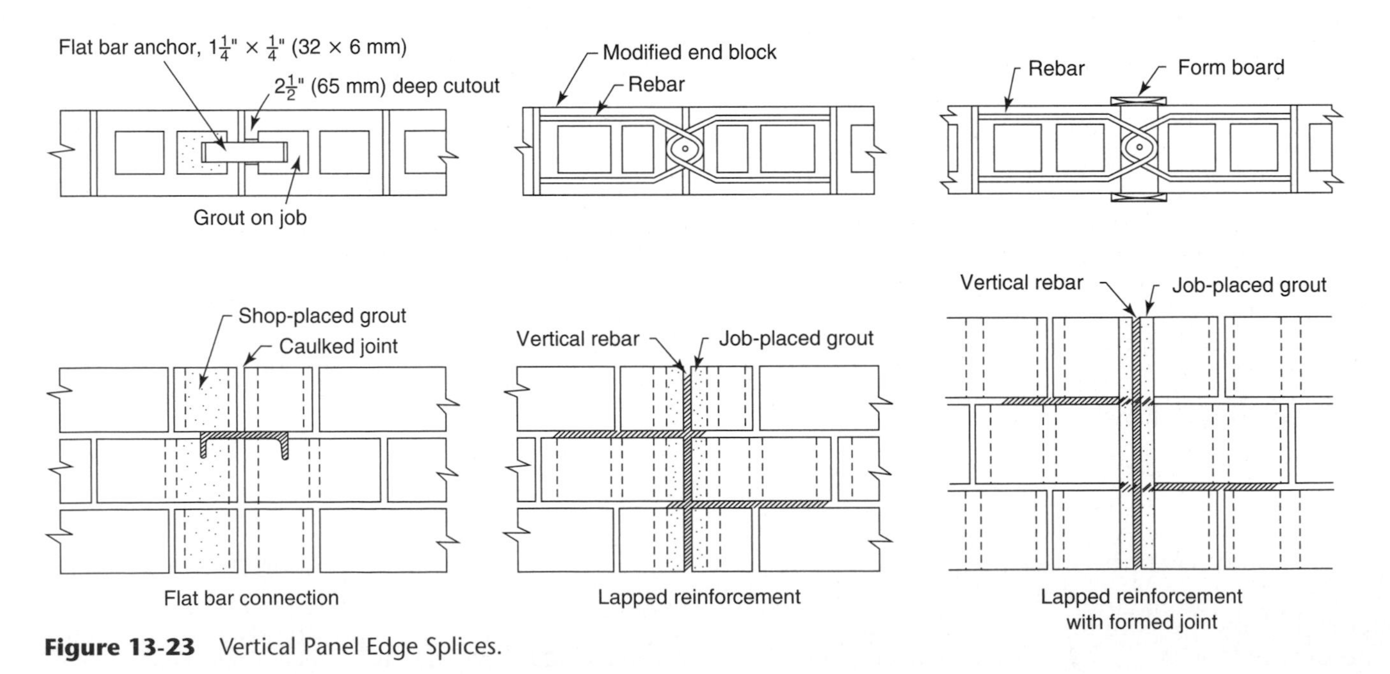


**Figure 13-23** Vertical Panel Edge Splices.

*Lapped reinforcement* consists of hairpin-shaped bars placed in alternate block courses, with their loops projecting 2 in. (50 mm) past the vertical edge of the panel. Modified end blocks are required to accommodate them, as shown in Figure 13-23. After the two adjoining panels are aligned, a vertical bar is threaded through the loops, and the core formed by the edges of the two panels is grouted.

The *lapped reinforcement with formed joint* is similar to the lapped reinforcement except that regular end blocks are used and the panels are spaced 2 to 3 in. (50 to 75 mm) apart when in position. Forms then have to be used to grout the space, after the vertical bar has been inserted.

## BRICK CONSTRUCTION

Fired clay bricks have been used as basic building units in the construction of buildings and various other structures for centuries. The strength and durability of brick made it an ideal material for the construction of load-bearing walls and as an exterior finish. In spite of the many new materials that are available today, building designers continue to incorporate brick into new structures because of these properties.

To satisfy the requirements of modern building designs, manufacturers now provide a good selection of brick sizes with varied colors and textures. Brick elements continue to serve as load-bearing structural sections (see Figure 13-24) and as interior and exterior finishes (see Figure 13-25), providing a look of permanence and elegance to even the most modern designs.

**Figure 13-24** Main Entrance to Building of Load-Bearing Brick.

**Figure 13-25** Graceful Brick Arches Enclose Second-Story Courtyard.

### Brickwork Fundamentals

Building with brick may involve the construction of solid brick walls, walls of reinforced brick, brick cavity walls, or brick veneer walls (see Figure 13-26). The terminology used in connection with various positions of brick in a wall (see Figure 13-27) applies in all cases, and almost any of the numerous *pattern bonds* available (see Figure 13-28) may be used with any one of these walls. The *structural bond* involved will depend on the type of wall. The third type of bond involved in brickwork, the *mortar bond,* is a function of the type of mortar used.

Sand mortars used in brickwork use the same components as mortars in concrete block construction: a combination of sand, cement and lime. Cements that are normally used are ordinary Portland cement or masonry cement. The use of masonry cement has the advantage that additional lime does not have to be added to the mix. When normal Portland cement is used, the cement content determines the compressive strength of the mortar and the amount of lime determines the workability of the mortar. Masonry codes such as ASTM C270 and CSA A179 provide guidelines for the proportioning

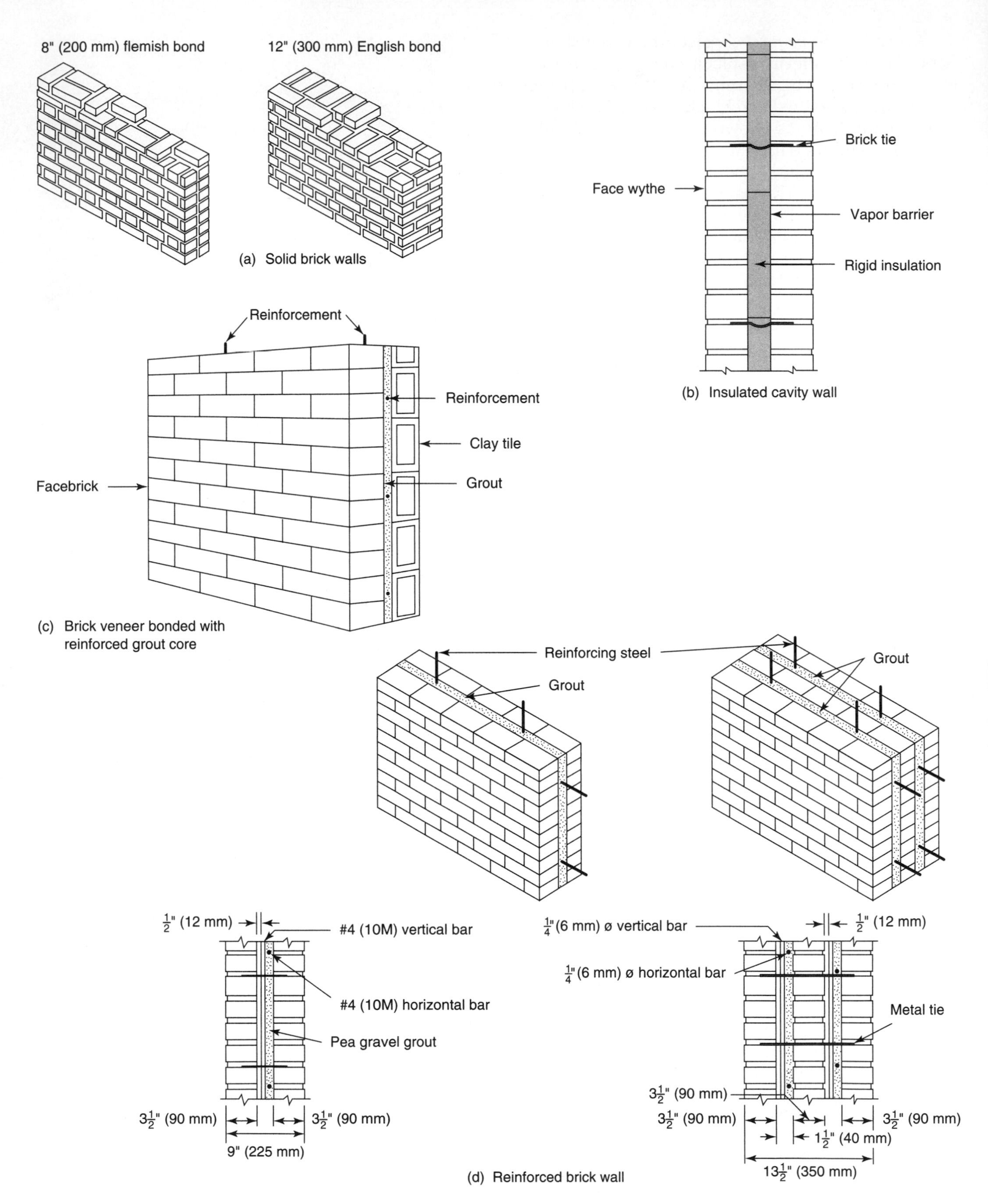
8" (200 mm) flemish bond
12" (300 mm) English bond
(a) Solid brick walls
Face wythe
Brick tie
Vapor barrier
Rigid insulation
(b) Insulated cavity wall
Reinforcement
Reinforcement
Clay tile
Grout
Facebrick
(c) Brick veneer bonded with reinforced grout core
Reinforcing steel
Grout
Grout
$\frac{1}{2}$" (12 mm)
#4 (10M) vertical bar
#4 (10M) horizontal bar
Pea gravel grout
$3\frac{1}{2}$" (90 mm)
$3\frac{1}{2}$" (90 mm)
9" (225 mm)
$\frac{1}{4}$"(6 mm) ø vertical bar
$\frac{1}{2}$" (12 mm)
$\frac{1}{4}$"(6 mm) ø horizontal bar
Metal tie
$3\frac{1}{2}$" (90 mm)
$3\frac{1}{2}$" (90 mm)
$3\frac{1}{2}$" (90 mm)
$1\frac{1}{2}$" (40 mm)
$13\frac{1}{2}$" (350 mm)
(d) Reinforced brick wall

**Figure 13-26** Typical Brick Walls.

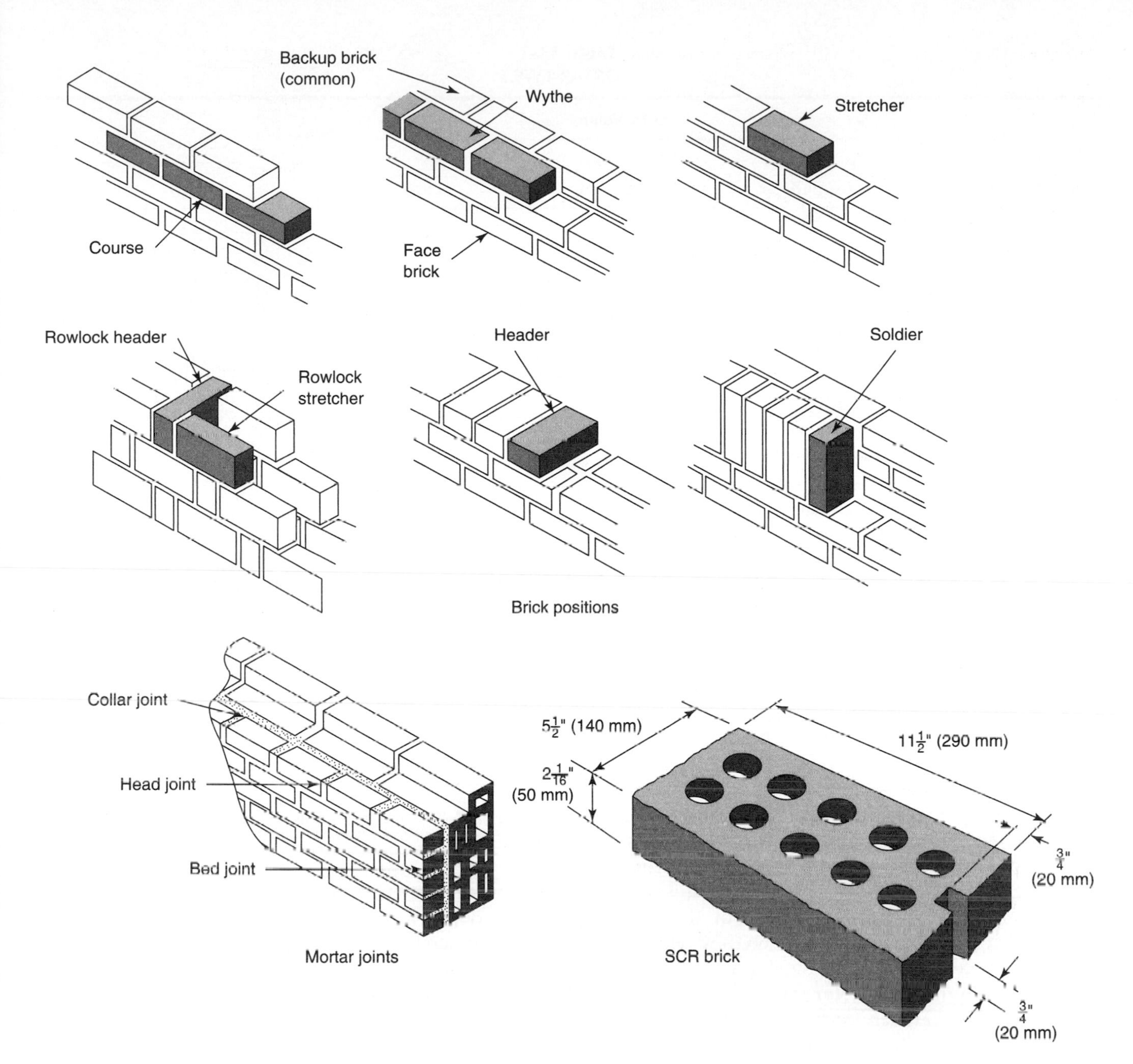

**Figure 13-27** Common Terms Used in Brickwork.

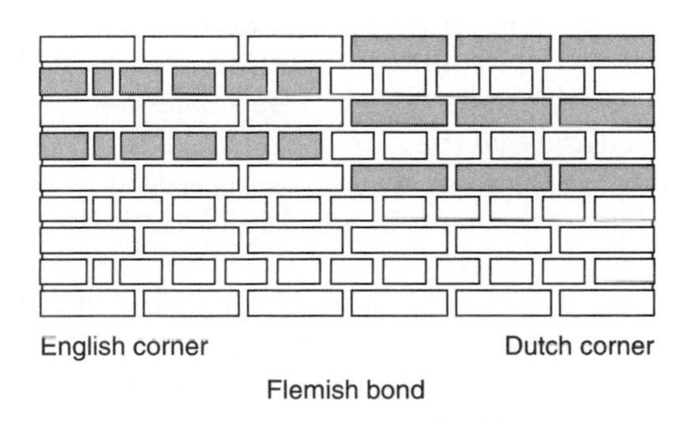


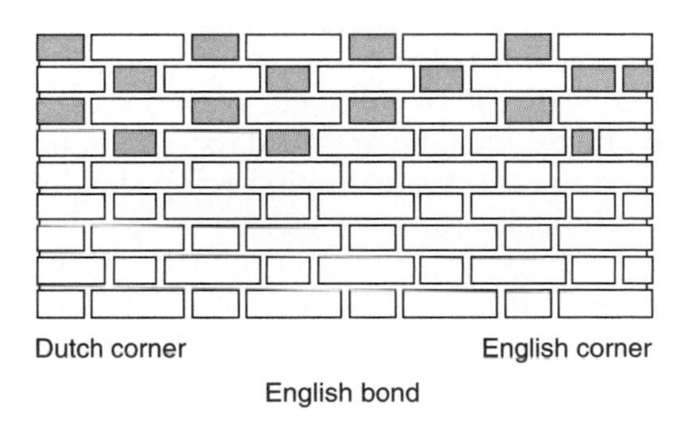


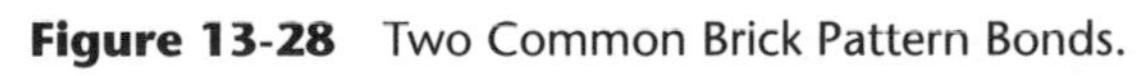

**Figure 13-28** Two Common Brick Pattern Bonds.

**Table 13-1**
MORTAR TYPES

| Type of Mortar | Parts by Volume: Portland Cement | Masonry Cement | Lime | Compressive Strength (MPa) | (psi) |
|---|---|---|---|---|---|
| S | ½ | 1 | — | 12.5 | 1800 |
| | 1 | — | ½ | | |
| N | — | 1 | — | 5 | 750 |
| | 1 | — | 1 | | |
| O | — | 1 | — | 2.5 | 350 |
| | 1 | — | 2 | | |
| K | 1 | — | 3 | 0.5 | 75 |
| | — | — | 1 | | |

of mortar components to ensure that the required strength and workability of the mortar mix is achieved. Table 13-1 illustrates typical types of mortar mixes along with their proportions of cement and lime and the resulting compressive strengths. The volume of sand that is recommended ranges between 2¼ and 3 times the sum of the volumes of cement and lime.

Types O or K may not be used where masonry is to be in direct contact with the soil, where an isolated pier is to be built, or where the wall is exposed to the elements on all sides, as in the case of a parapet wall.

When brick walls are required to have greater resistance to compressive, tensile, or shear forces than usual, they may be reinforced with steel rods, as indicated in Figure 13-26(d). Two or three wythes of brick are laid, the starting wythe to be built up not more than 16 in. (400 mm) ahead of the others. Metal ties connect each pair of wythes: Half-inch (10M) diameter reinforcing steel for a two-wythe wall or ¼ in. (6 mm) diameter for a three-wythe wall is placed in the cavities, vertically and horizontally, and the cavities are filled with pea gravel grout. The depth of any one grout placement should not exceed approximately 4 ft (1,200 mm).

Brick walls may be further strengthened by the use of bond beams (Figure 13-29). In the case of a three-wythe wall, the two outside wythes act as the form. Side forms must be provided for a bond beam in a two-wythe wall. Vertical reinforcing should extend through the beam.

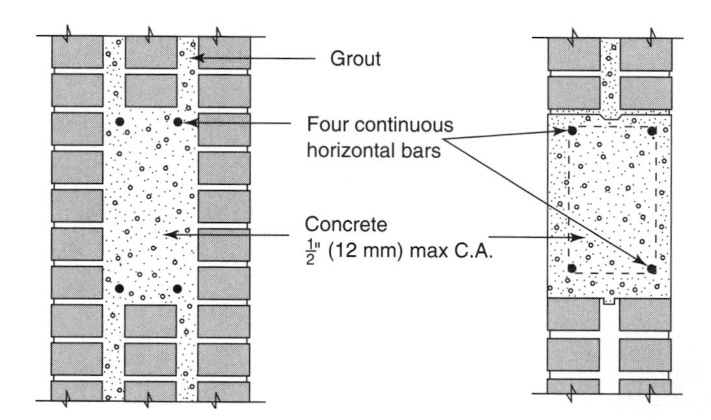


**Figure 13-29** Reinforced Bond Beams.

A brick cavity wall consists of two wythes of brick separated by a continuous air space not less than 2 in. (50 mm) wide. Metal ties are used to connect the two wythes. They should be spaced 36 in. (900 mm) o.c. horizontally, with rows of ties spaced 16 in. (400 mm) vertically. The spacing should be staggered in alternate rows. The facing wythe is always a nominal 4 in. (100 mm) thick, while the interior wythe may be 4, 6, or 8 in. (100, 150, or 200 mm) thick, depending on the building height, loads, and the distance between lateral supports.

If insulation is required in the wall, it is usually installed in the cavity, and 2-in. (50-mm) rigid insulation is normally used. Regardless of whether insulation is required, a vapor barrier in the form of an asphalt coating is usually applied to the cavity face of the inner wythe.

Weep holes must be provided at the bottom of the outer wythe to allow the cavity to drain. During construction, ensure that the weep holes are properly installed and that they are not blocked by falling mortar. Plastic inserts, oiled rods, sash cord, or flexible plastic foam strips can be used to form the drain holes. The plastic inserts are preformed and are left in place.

Brick veneer construction consists of applying a 4-in. (100-mm) wythe of brick as facing material over a backup wall. The backup wall may be of frame construction or a masonry wall of common brick, structural clay tile, concrete block, cellular concrete blocks, or cast-in-place concrete. In any case, the load-bearing properties of brick are not utilized.

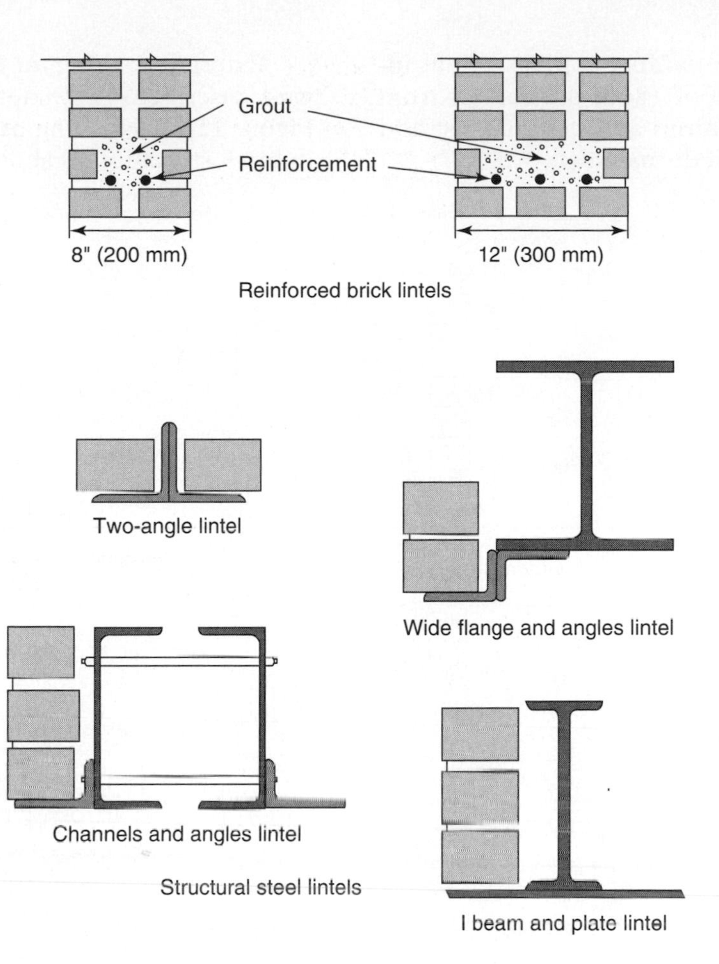


**Figure 13-30** Brick Lintels.

Structural stability of the facing wythe is obtained by anchoring the veneer to the backup with metal ties or by grouting it to the wall with a reinforced grout core, usually 1 in. (25 mm) thick.

## Lintels

The support of the masonry over openings in brick walls is an important consideration. This support will probably be provided by a lintel, which may be a reinforced concrete unit, a reinforced brick structure, or a built-up structural steel member (see Figure 13-30).

Reinforced brick lintels are similar to a bond beam. A support must be made on which to build the lintel, and the brick is laid in running bond with full bed and head joints. Horizontal reinforcing steel may be laid in the mortar joints as the units are laid, or the lintel may be built as illustrated in Figure 13-30. In the latter case, the grout is placed to the level of only one brick course at a time.

A number of built-up structural steel shapes may be used as lintels, depending on the thickness of walls, the length of spans, and the loads involved.

## Flashing

One item of vital importance in the construction of brick walls is the installation of flashing. The purpose of flashing is to exclude moisture or to direct any moisture that may penetrate the wall back to the exterior. To perform satisfactorily, flashing must be a permanent material and must be properly installed. Most flashing is made of sheet metal (copper, lead, aluminum, or galvanized iron), fabric saturated with asphalt, or pliable synthetic material. Copper and bituminous material are sometimes combined to form flashing material.

Two types of flashing are in common use: *external* and *internal*. External flashing prevents the penetration of water at points where walls intersect flat surfaces such as roofs. Internal flashing is built into and usually concealed in the wall to control the spread of moisture and to direct it to the outside. This is sometimes called *through* flashing.

The points at which flashing should be installed are (1) above grade in exterior walls; (2) under and behind window sills; (3) over lintels; (4) over spandrel beams; (5) at projections or recesses from the face of a wall; (6) under parapet copings; (7) at intersections of wall and roof; (8) at projections from the roof, such as ventilators and penthouses; and (9) around chimneys and dormers.

External flashing for roof and wall intersections was described in Chapter 12 and illustrated in Figure 12-60. Flashing at various vulnerable points in brick structures is shown in Figure 13-31.

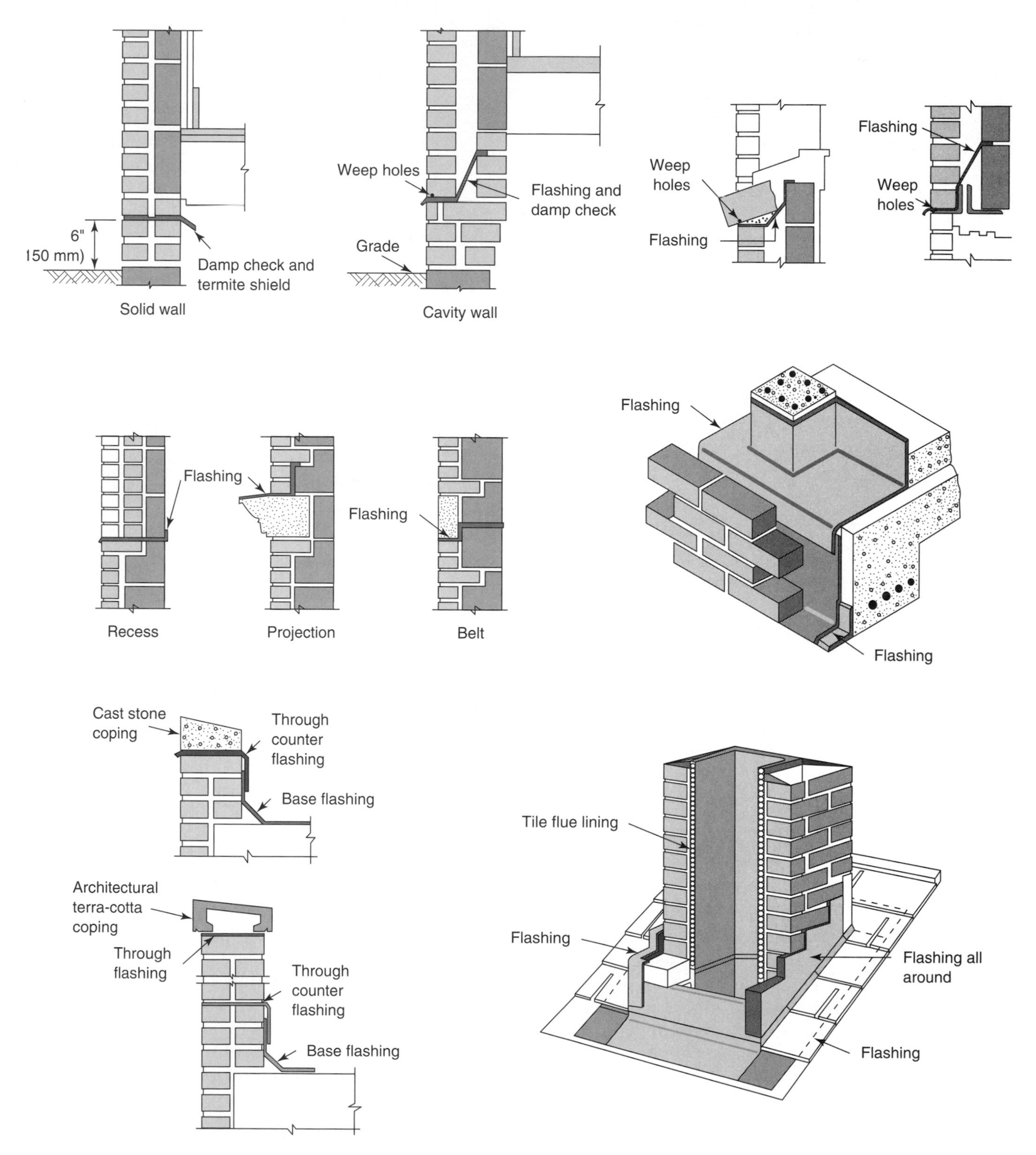


**Figure 13-31** Flashings in Brick Construction.

## Expansion Joints

Expansion joints are needed in brick construction under certain conditions. Their purpose is to provide separations in a structure so as to relieve the stresses set up by changes in temperature and moisture conditions or caused by settlement. They should generally be located at offsets, if the wall running into the offset is 50 ft (15 m) or more in length, and at junctions in walls. Figure 13-32 illustrates typical expansion joints in combination brick and tile walls. Notice the molded copper water stop used in the vertical joint on the outer face. Caulking is later applied to seal the joint. A premolded expansion joint material is used to separate the sections of backup wall, and the vertical joint on the face is covered with a metal plate, which is fastened to one section only.

## Load-Bearing Brick Walls

Until recent years, the design of brick buildings was based on the empirical requirements of building codes with regard to the minimum allowable wall thickness and maximum height. These regulations often placed economic restrictions on load-bearing brick walls for buildings over four or five stories in height.

New developments are now taking place in which the design of load-bearing brick walls is based on a realistic structural analysis of the proposed building. As a result, relatively thin, load-bearing brick walls are being extended to as much as 18 stories (see Figure 13-33). Designs are including the interior and the exterior walls for load-bearing purposes; consequently, the thickness of all walls can be reduced. In addition, brick construction

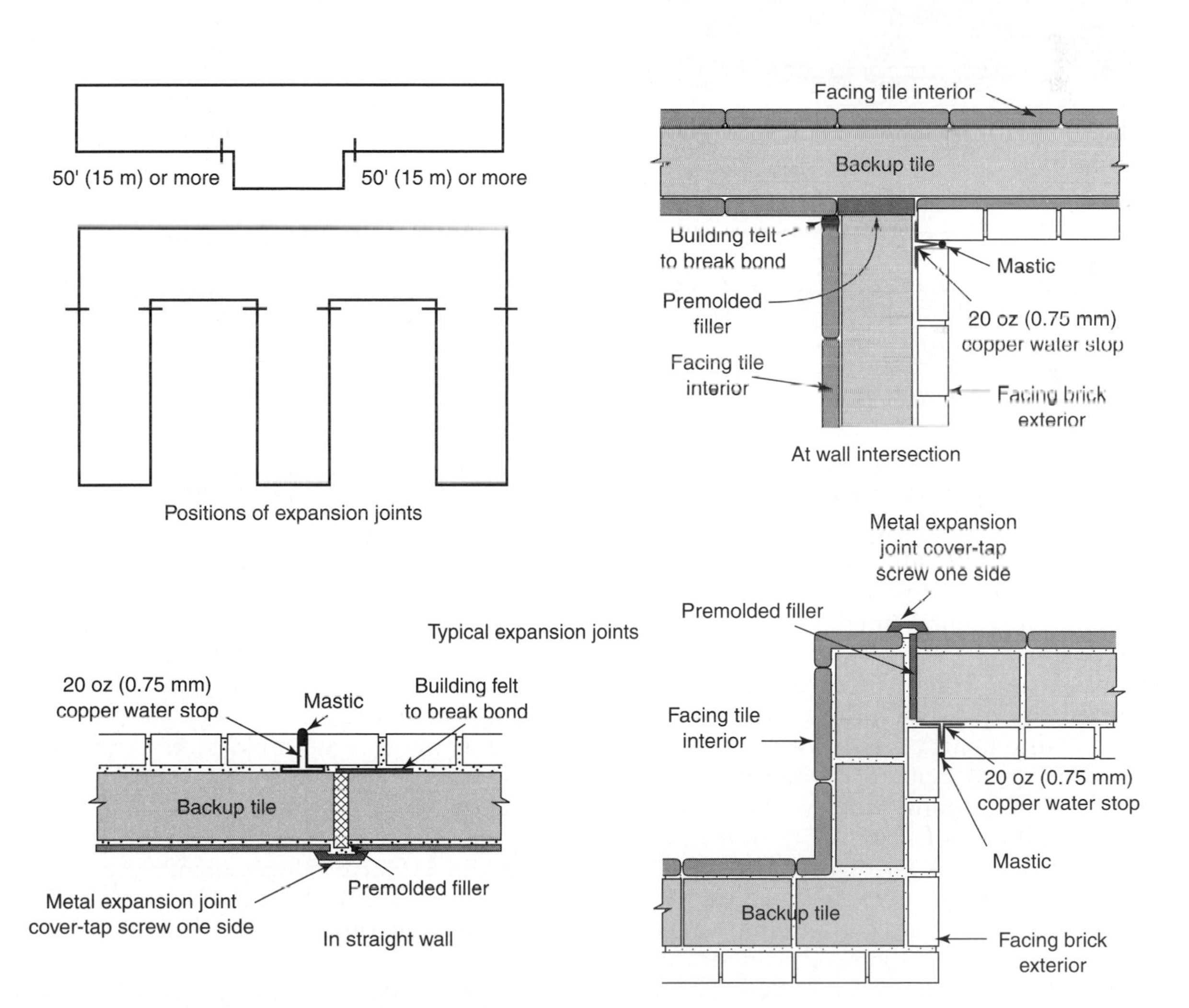

**Figure 13-32** Expansion Joints in Brick Construction.

**Figure 13-33** Typical Load-Bearing Brick Wall Structure. (Reprinted by permission of International Masonry Institute.)

allows a considerable degree of flexibility in design, and designers take advantage of this flexibility to distribute the floor loads over as many bearing members as possible. For example, in the buildings shown in Figure 13-34, the load-bearing masonry walls radiate from a central utility core like the spokes of a wheel.

Some examples of the structural details of these relatively thin load-bearing walls and the variety of floor systems possible with them are shown in Figures 13-35(a) and (b), 13-36(a) and (b), and 13-37. The building represented in Figure 13-35(i) is eighteen stories tall, with 15¼-in. (385-mm)-thick bearing walls. Part of this thickness is attributable to thermal insulation rather than to structural requirements. In Figure 13-35(ii), the floor loads of the sixteen-story building are carried on the 6-in. (150-mm) inner wythe of the exterior cavity wall and the 6-in. (150-mm) interior brick bearing partitions. In Figure 13-35(iii), the fourteen-story building has exterior bearing walls 7¼ in. (185 mm) in thickness. The building represented by Figure 13-35(iv) is seventeen stories in height, with 11-in. (275-mm) reinforced brick masonry walls.

**Figure 13-34** Brick Building with Interior Bearing Walls. (Reprinted by permission of International Masonry Institute.)

Figure 13-36 illustrates the exterior and interior bearing walls of an eight-story apartment building in which the apartment floors are supported by open-web steel joists, while the corridor floors are simply 5-in. (125-mm) reinforced concrete slabs, spanning from one load-bearing corridor wall to the other. Figure 13-37 illustrates the use of precast hollow core concrete floor slabs with thin masonry bearing walls.

## Prefabricated Brick Panels

The same techniques used in the prefabrication of concrete block wall panels are applied to the production of prefabricated brick panels. Some are veneer-type panels, just one brick thick, whereas others constitute load-bearing walls of combined brick and block, bonded by grout-filled collar joints and wire reinforcement (see Figure 13-38).

The development of high-bond mortars is a key factor in the increasing use of prefabricated brick panels, because the new mortars will produce 300 to 400 psi (2 to 3 MPa) of flexural strength, compared with approximately 150 psi (1 MPa) for conventional mortars. However, both kinds are used in manufacturing such units.

As is the case with block, brick panels may be made in masonry panel plants or at the job site, under all-weather enclosures on launch pads. In either case, most panels are made with mortar containing an additive that produces higher compressive and tensile strength. Such mortar is also more resistant to water penetration, thus improving the weather ability of the panels.

After about 7 days' curing, panels are ready for installation. They are lifted into place by crane

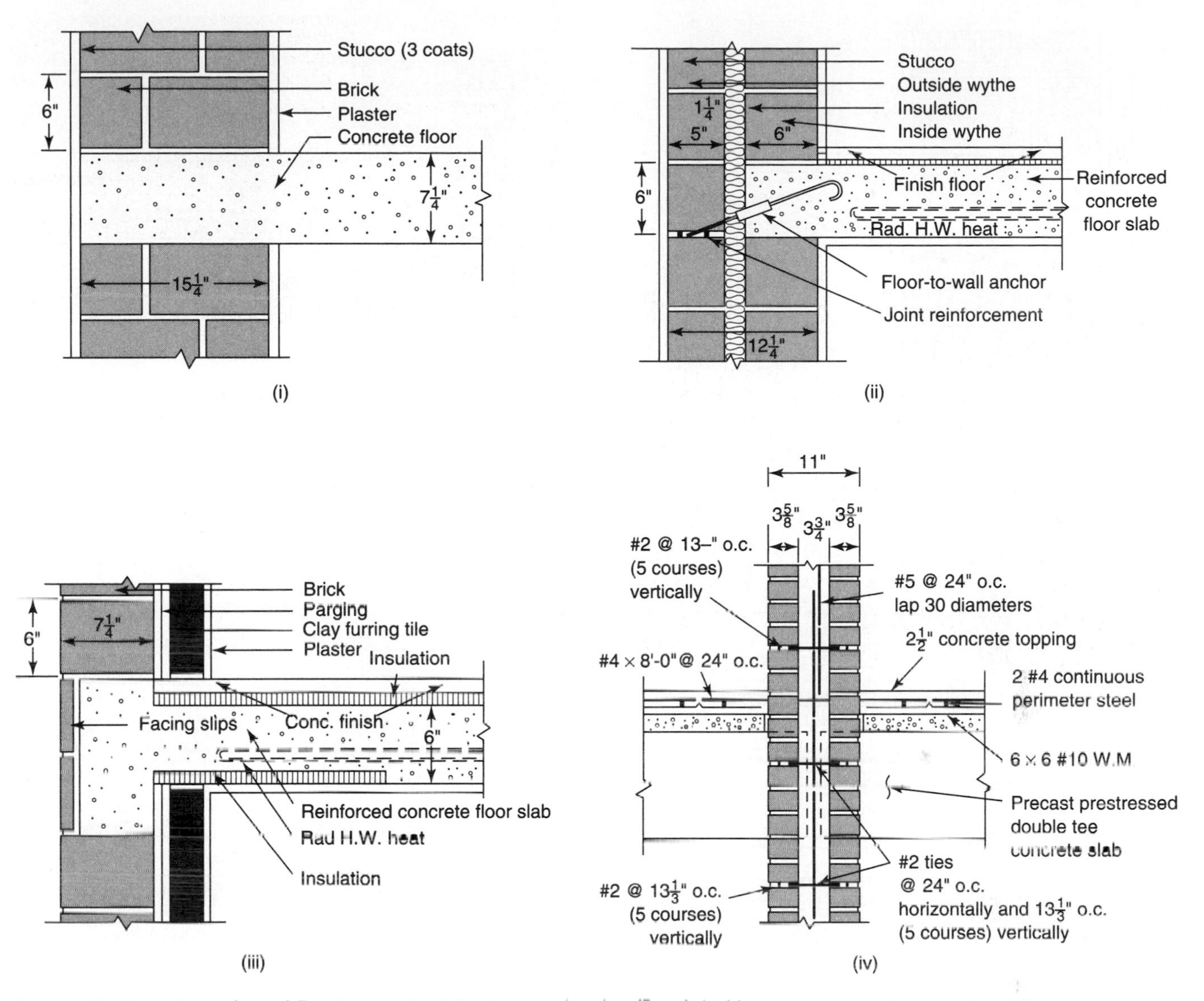


**Figure 13-35a** Typical Load-Bearing Brick Walls, Standard Units. (Reprinted by permission of International Masonry Institute.)

(Figure 13-39) and anchored, usually by welding prelocated weld plates into the structural frame (see Figure 13-40).

## BUILDING WITH STONE

Although stone has been a major structural material in the past, modern builders use it almost always as a veneer or curtain wall material. As a veneer, stone is laid up to a masonry backup wall in *ashlar* or *rubble* patterns. Ashlar patterns feature cut stone laid with well-defined course lines, whereas rubble patterns use uncut or semicut stone laid with few or no course lines. Stones for ashlar work are produced in thicknesses of from 2 to 8 in. (50 to 200 mm), in heights of from 1 in. to 48 in. (25 to 1,200 mm), and in lengths of from 1 in. to 8 ft (25 to 2,400 mm).

Methods of bonding stone veneer to backup walls vary somewhat, depending on the size of stones used. For small stones, ties similar to those used for brick and tile facing work are commonly used. For larger stones, and particularly when the backup wall is nonbearing, a variety of anchor types is available (Figure 13-41).

Two methods are used to support stone veneer on nonbearing walls. One method involves the use of shelf angles at the spandrel beams, as shown in Figure 13-42(a). The stone is held in place on the shelf by a ½-in. (12.5-mm) rod welded to the horizontal leg of the angle or by dowels protruding

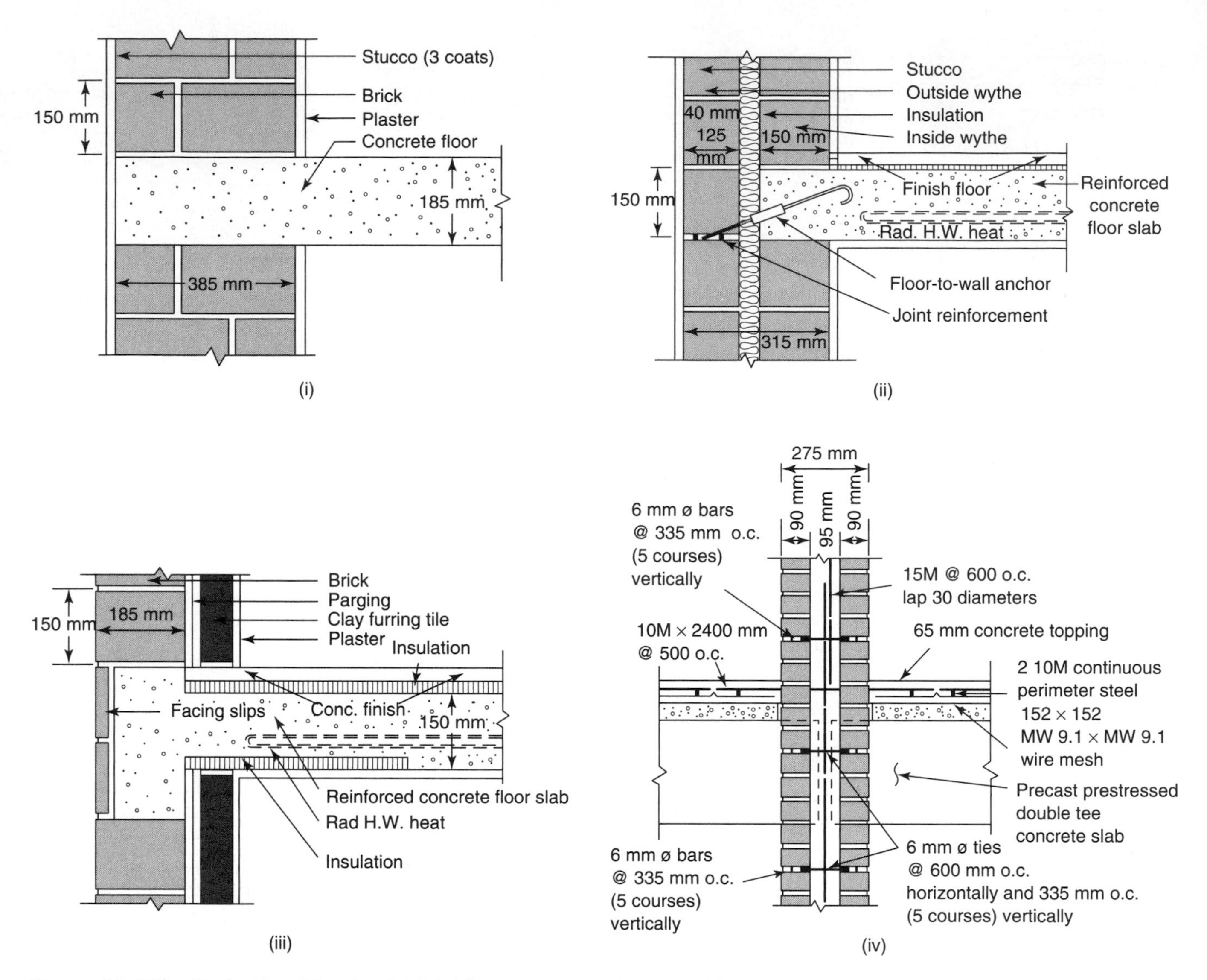


**Figure 13-35b** Typical Load-Bearing Brick Walls, Metric Units. (Reprinted by permission of International Masonry Institute.)

through the horizontal leg. In the first instance, the rod lies in a groove cut in the bottom edge of the stone slab, and in the second the dowel fits into a hole in the edge of the stone.

In the second method, bond stones rest on the spandrel beam and provide support for the rest of the veneer, as shown in Figure 13-42(c). Anchors are not required in the bond stones themselves, but are used between bond stones and regular veneer slabs and between veneer and the backup wall. If the stones are not more than 30 in. (750 mm) in height, two anchors per stone should be supplied in both the top and bottom beds. When stones are under 24 in. (600 mm) in width, they require only one anchor per stone.

An alternative method of applying stone veneer, shown in Figure 13-42(d), is particularly applicable if insulation is used in the wall.

Stone is also used to face soffits and columns or to trim brick and tile walls. Trim stones include sills, lintels, quoins, belts, and copings. Stone soffits are hung from the frame by strap hangers that fit into slots in the edges of the soffit slabs (Figure 13-43). Several of the methods used to face columns with stone are illustrated in Figure 13-44.

Belts and copings are secured by strap anchors or dowels, and sills, lintels, and quoins are built into and supported by the masonry surrounding them. Figure 13-44 shows one method of anchoring coping stones to a parapet wall.

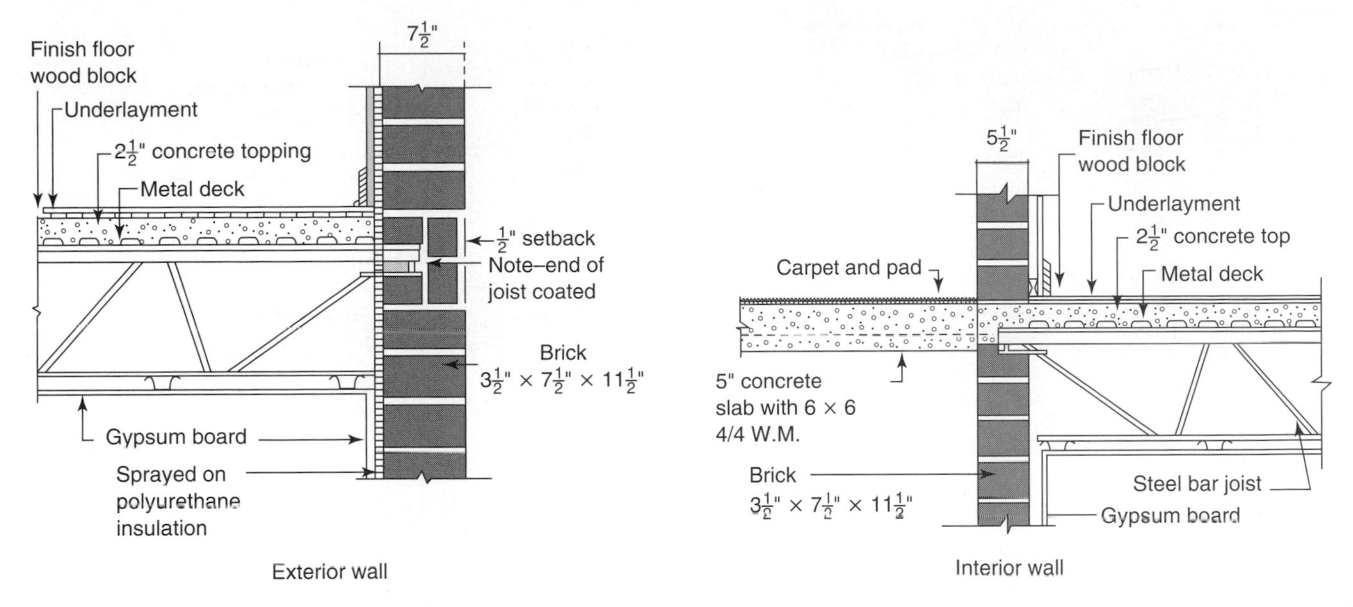


**Figure 13-36a** Open-Web Steel Joist Floor Frame with Load-Bearing Brick Walls, Standard Units. (Reprinted by permission of International Masonry Institute.)

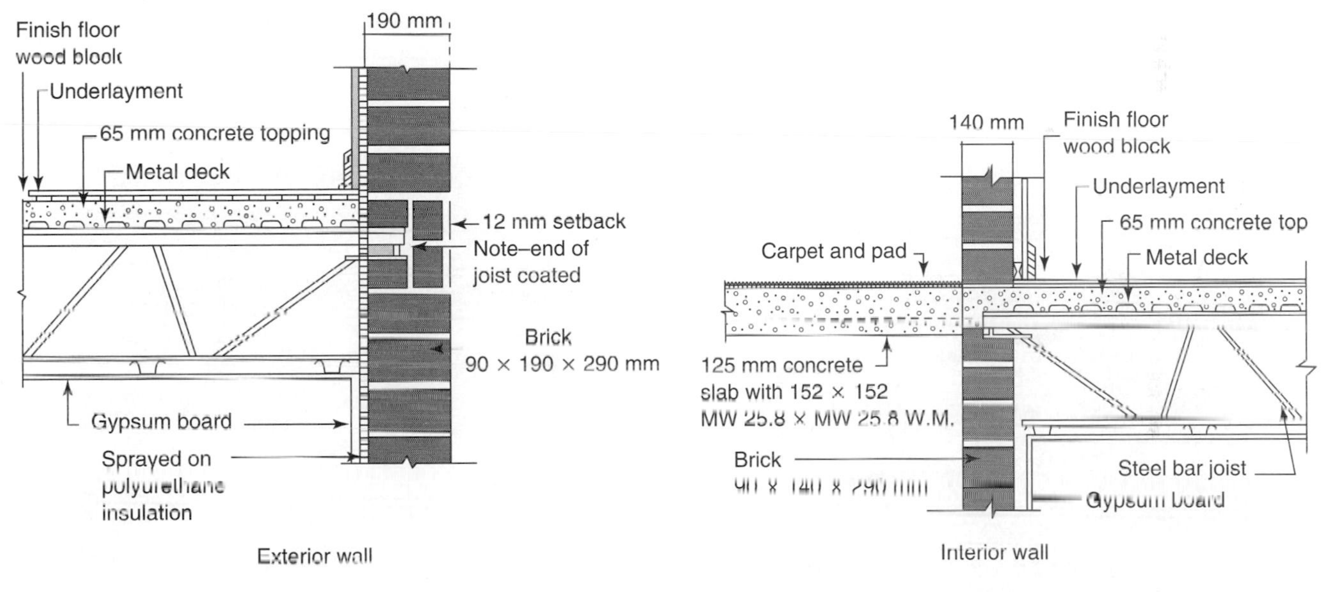


**Figure 13-36b** Open-Web Steel Joist Floor Frame with Load-Bearing Brick Walls, Metric Units. (Reprinted by permission of International Masonry Institute.)

Stone ashlar is anchored to solid backup walls with noncorroding corrugated wall ties and with bond stones where the backup material will allow it. If stone course lines do not coincide with those of the backup material, toggle bolts are used to secure the wall ties to the block or tile (Figure 13-45).

Mortar used for setting stone or for pointing joints should be made with a nonstaining cement. Pointing consists of raking out joints to a depth of ¾ in. (20 mm) and filling them with pointing mortar, packed into place and rubbed smooth to the indicated level. Joints should be wet before the pointing mortar is applied. Pointing mortar is omitted if joints are to be caulked.

Wire brushes and acids should not be used to clean stones. As soon as possible after the mortar has set, they should be cleaned with water, soap powder, and fiber brushes. Approved machine-cleaning processes may also be used.

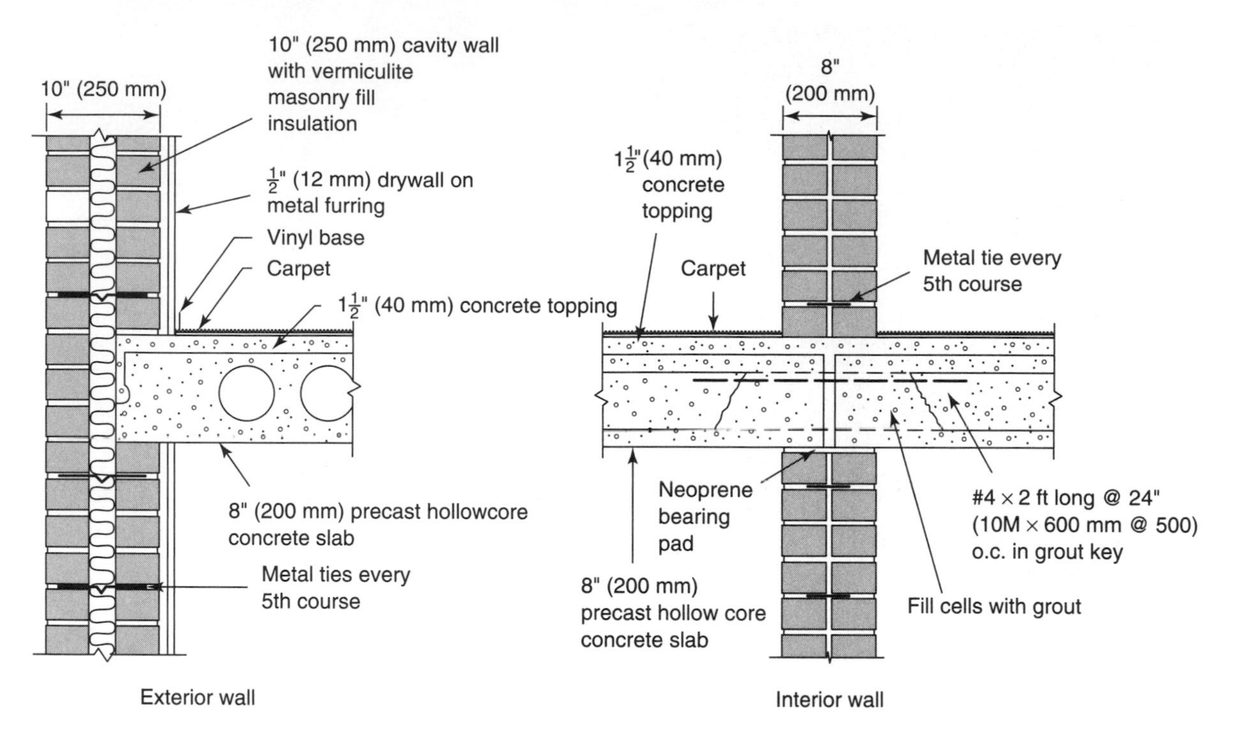


**Figure 13-37** Brick Load-Bearing Walls with Precast Concrete Floor Slabs. (Reprinted by permission of International Masonry Institute.)

**Figure 13-38** Precast Brick Facing Panels Being Set in Place.

**Figure 13-39** Curved Precast Brick Section Being Hoisted by Crane.

**Figure 13-40** Weld Plate for Anchoring Brick Panels to a Structural Frame. (Reprinted by permission of International Masonry Institute.)

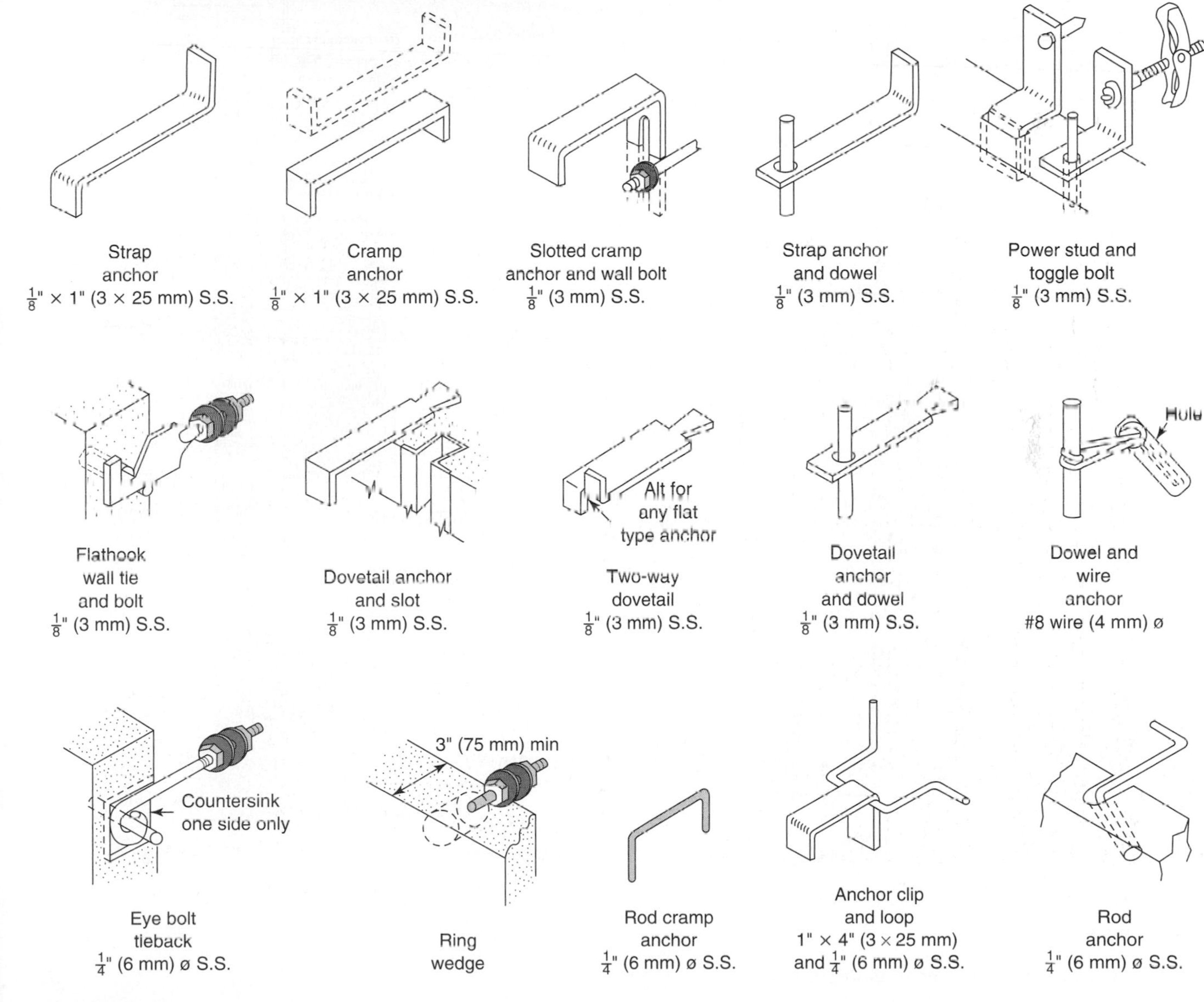


**Figure 13-41** Stone Anchors.

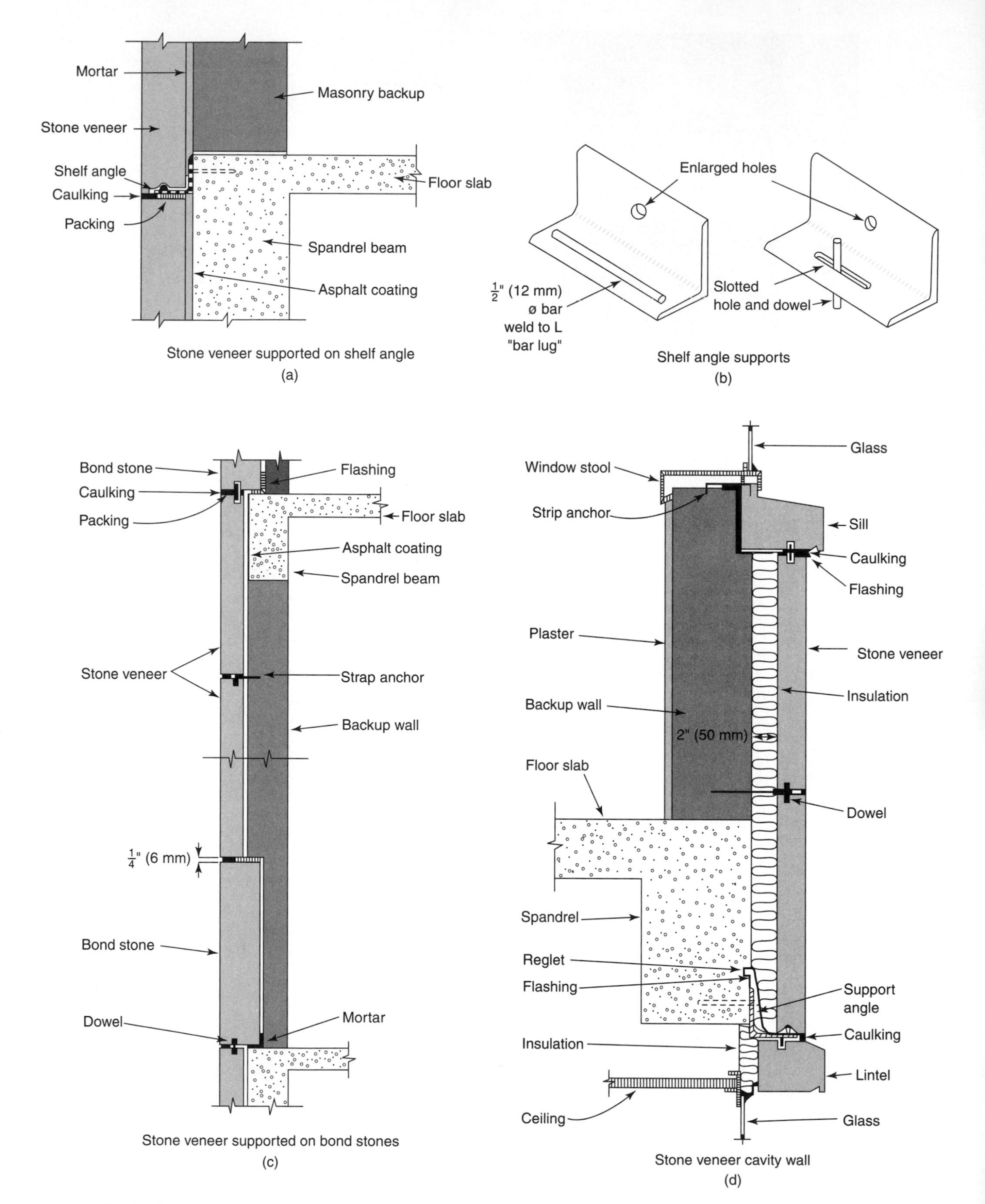

**Figure 13-42** Stone Veneer Support Systems.

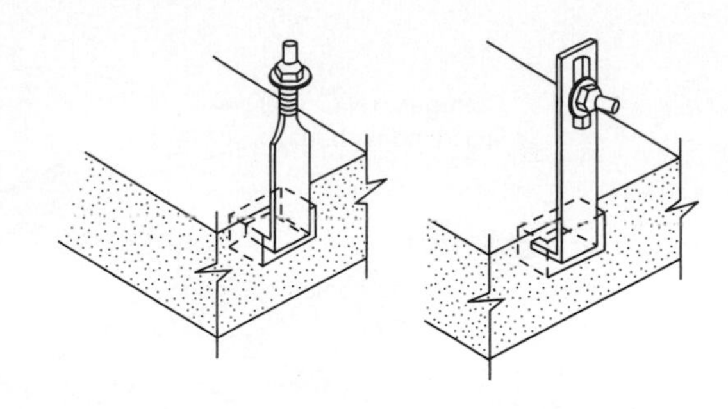

**Figure 13-43** Stone Soffit Strap Anchors.

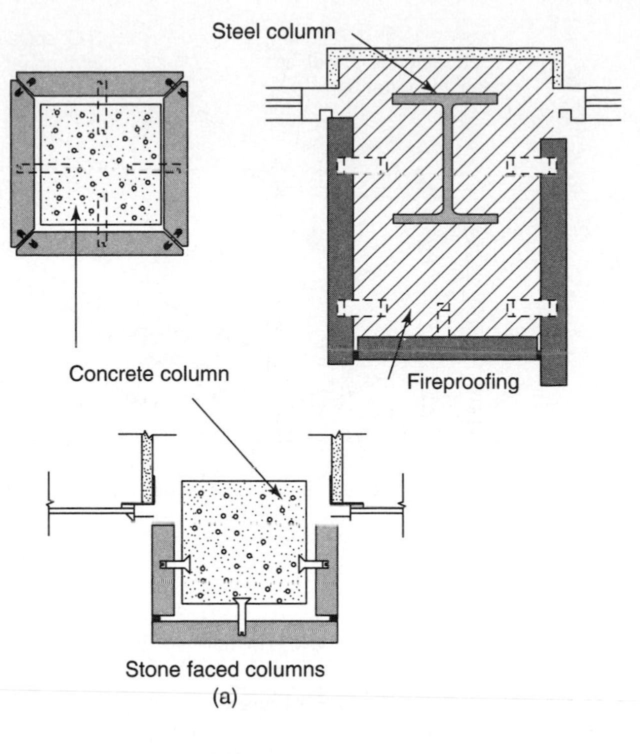

## COLD WEATHER MASONRY WORK

The precautions that must be taken when placing concrete in cold weather also apply to masonry work. Unless protection is provided, the mortar bond may be very poor, resulting in inadequate strength and high permeability. Adequate protection should be provided for at least 48 hours.

Warm mortar is the first requirement. Sand may be kept warm by piling it around a large pipe heated by a gas jet, and it should be turned over periodically to maintain an even distribution of heat throughout the pile (see Figure 13-46). Warm water should be used also, but the temperature of the water will depend on the sand temperature.

But warm mortar is not enough; masonry units should also be warm to prevent the sudden cooling of warm mortar as it comes in contact with cold masonry. It is often difficult to store quantities of bricks or blocks in heated enclosures, and one alternative is to heat a smaller quantity of units just before they are to be used. This may be done by covering a small pile of bricks or blocks with a heavy tarpaulin (see Figure 13-47). With warm mortar and warm masonry, work can go on within a shelter and full bond strength can be developed (see Figure 13-48). For best results, heat should be maintained for as long as possible.

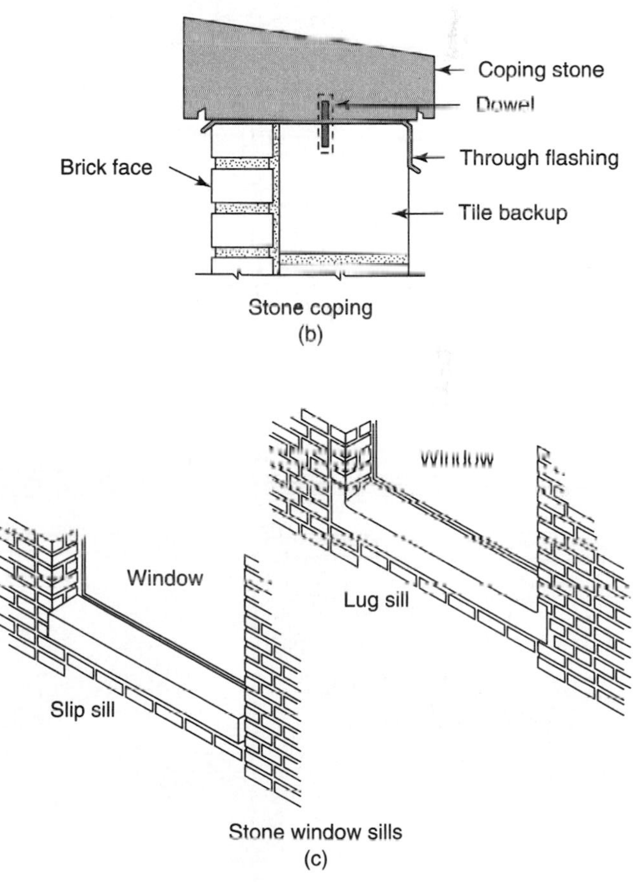


**Figure 13-44** Stone Trim.

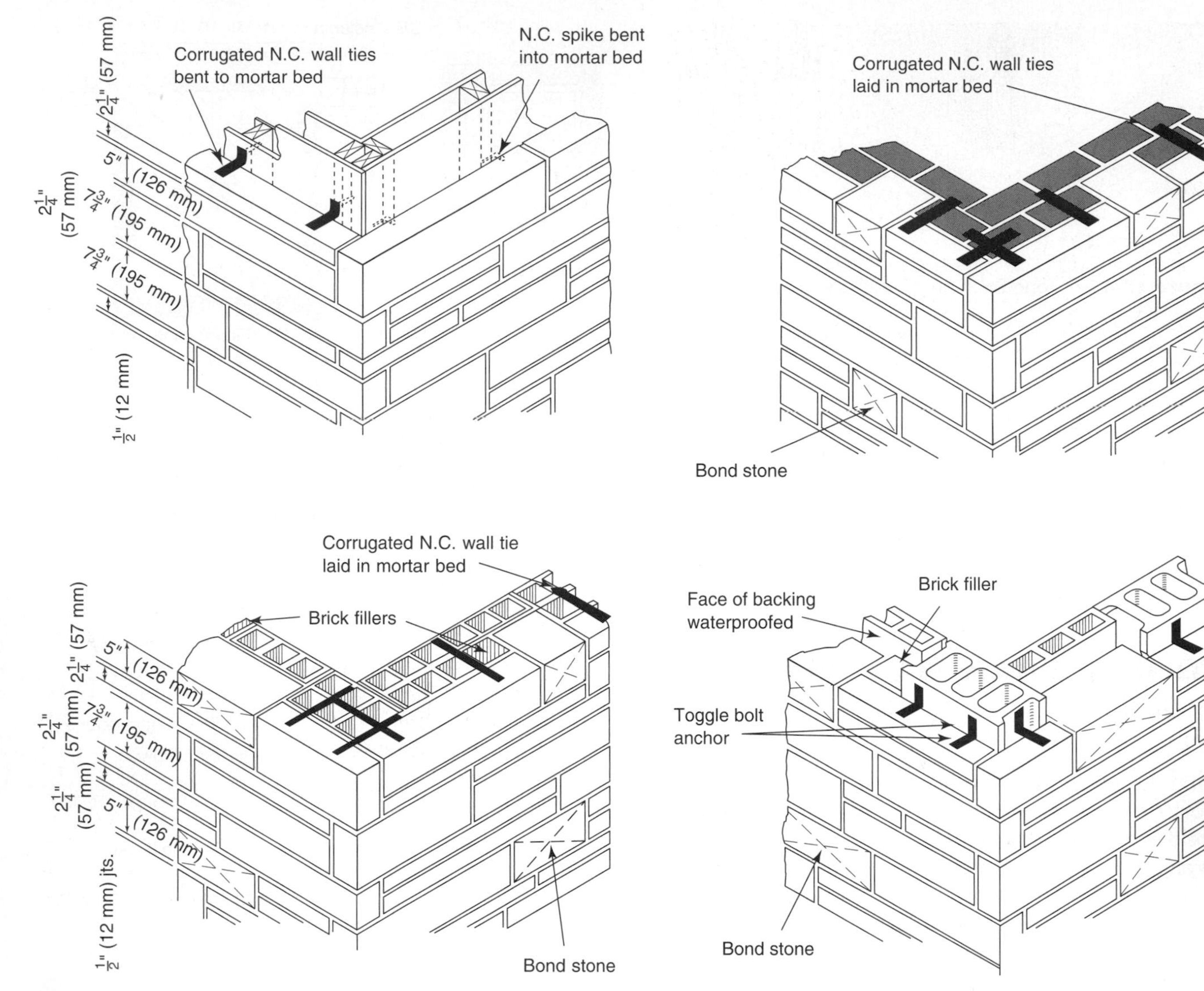


**Figure 13-45** Anchoring Stone to Various Wall Backings.

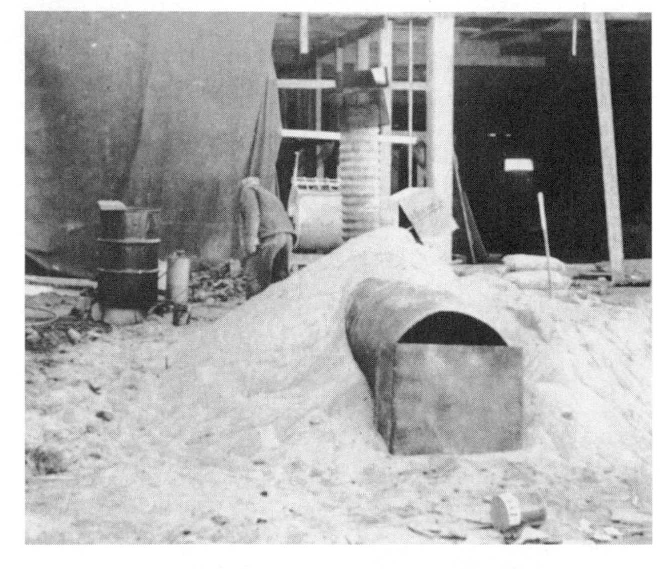

**Figure 13-46** Keeping Mortar Sand Warm. (Reprinted with permission of Portland Cement Association.)

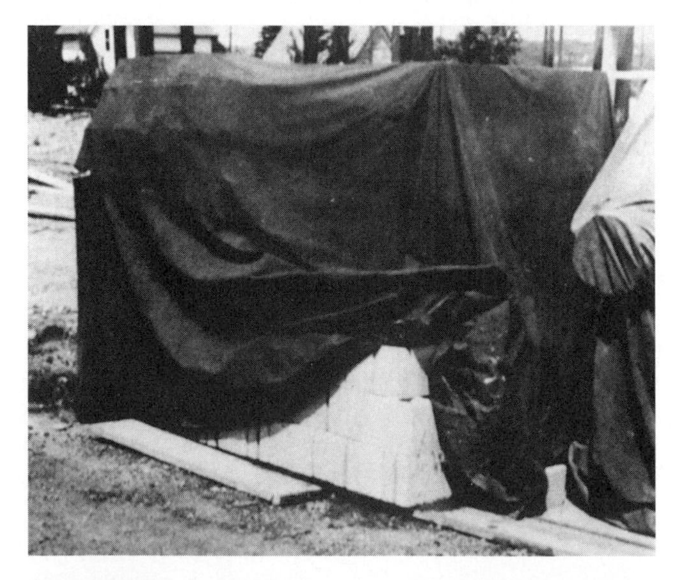

**Figure 13-47** Heating Blocks Just Prior to Use. (Reprinted with permission of Portland Cement Association.)

**Figure 13-48** Laying Block Behind a Protecting Tarpaulin. (Reprinted with permission of Portland Cement Association.)

## Review Questions

**1.** What two properties of mortar must be considered when specifying mortar for a particular application?

**2.** Explain the difference between low-lift grouting and high-lift grouting in masonry. Where might one method be more advantageous than the other?

**3.** Explain briefly **(a)** the purpose of a control joint in a block wall, **(b)** the reason for caulking a control joint on the exterior face of the wall, **(c)** the purpose of horizontal joint reinforcement in a block wall, and **(d)** the purpose of a bond beam.

**4.** By means of neat diagrams, indicate three methods of providing a lintel over an opening in a block wall.

**5.** Outline concisely what you understand by *customized concrete block masonry.*

**6.** Explain clearly why tying systems are so important when prefabricated block panels are being used in a building.

**7.** Explain the differences among the terms *pattern bond, structural bond,* and *mortar bond,* as used in connection with brickwork.

**8.** Explain what is meant by **(a)** a brick wythe, **(b)** a weep hole in a brick wall, **(c)** a collar joint, **(d)** common brick, and **(e)** brick veneer.

**9.** Describe why and where you would use expansion joints in brick construction.

**10.** Write a brief summary of load-bearing brick wall construction, indicating **(a)** why the system has generated interest, **(b)** the inherent advantages to using such a system, **(c)** the type of building for which it is best suited, and **(d)** how the cost should compare with that of a similar building using curtain wall construction.

**11.** Explain briefly **(a)** the difference between ashlar and rubble patterns in stone veneer, **(b)** the purpose of a bond stone, **(c)** the reasons for using a stone belt course in a masonry wall, and **(d)** how a stone soffit may be anchored when using stone for interior finish.

# 14 Curtain Wall Construction

The development of steel and concrete structural frames for multistory buildings has led to the adoption of non-load-bearing exterior walls known as *curtain walls*. Their primary functions are to protect the interior of the building from the elements and to enhance its appearance (Figure 14-1).

The original need for curtain walls arose when masonry walls required some method to improve their resistance to moisture penetration, especially during periods of high wind and rain. One method that was developed was the *cavity wall*. In the design of a cavity wall, the inside wythe is normally the load-bearing portion of the wall, whereas the exterior wythe serves as the curtain wall. Because the outer wythe is not required to support vertical loads, its thickness is governed by the amount of lateral support provided by the load-bearing wythe. The main function of the exterior wythe is to act as

**Figure 14-1** Typical Curtain Wall.

a buffer, sufficiently reducing the velocity of the rain to minimize its penetration through the exterior wythe and prevent it from reaching the inner wythe. Any moisture that does penetrate the exterior wythe will run harmlessly down the void between the two wythes and be drained out the bottom. This concept works well if freezing does not occur in the wall cavity.

To ensure proper performance of the cavity wall, its design and construction must be carefully considered and executed. The minimum width of the cavity must be 2 in. (50 mm) and the cavity must be kept free of mortar droppings to prevent moisture migration to the inner wythe. Weep holes must be provided at the bottom of the external wythe and kept free of mortar droppings during construction to ensure proper drainage of the cavity.

The modern curtain wall must be capable of more than the ability to resist moisture penetration. To meet present-day energy-efficiency requirements, insulation values and control of air leakage are concerns that must be addressed when selecting materials and establishing details for the design of the curtain wall. This also must be accomplished inexpensively and without compromising the aesthetic value of the building.

The methods and materials used in curtain wall design and construction vary depending on the type of building and the amount of money available. To meet these criteria, a wide variety of colors, textures, designs, and materials is now available to the designer to enhance the appearance of a building.

From a construction point of view, curtain walls provide the following advantages over traditional construction:

1. The reduction in the exterior wall thickness results in more floor space.
2. Because the curtain wall must support only its dead weight, it needs to be designed for wind loads only, which results in a lighter wall section. In high-rise structures, the savings in total weight can be substantial, and savings can be achieved through smaller foundations and supporting beams (see Figure 14-2).
3. The application of the curtain wall can be coordinated with the construction of the structural frame, which results in shorter construction times.
4. Because curtain walls usually come in prefabricated panels, they can be well insulated and sealed, which prevents excessive air and moisture leakage.

**Figure 14-2** Curtain Wall on Modern High-Rise.

## MASONRY CURTAIN WALLS

Masonry curtain walls were traditionally composed of individual masonry units such as brick, clay tile, and concrete block. To improve the versatility of the masonry, the use of individual units has been augmented by engineered sections such as panelized masonry, stone panels, and precast concrete panels. In general, masonry curtain walls may be separated into two basic types: load-bearing and non-load-bearing. Load-bearing curtain walls (see Figure 14-3) are required to support gravity loads as well as resist lateral loads due to wind. Non-load-bearing curtain walls resist lateral loads only and depend on the structural frame of the building for lateral support.

### Unit Masonry

There are two methods of incorporating unit masonry into a building as a curtain wall. One is to build the units into the wall openings formed by columns and beams so that the frame may provide

**Figure 14-3** Load-Bearing Lightweight Concrete Block Curtain Wall.

**Figure 14-4** Non-Load-Bearing Lightweight Concrete Block Curtain Wall.

both vertical and lateral support (see Figure 14-4). The other is to erect the exterior walls independently of any vertical support. The walls carry their own dead load to the foundation (Figure 14-5), but the

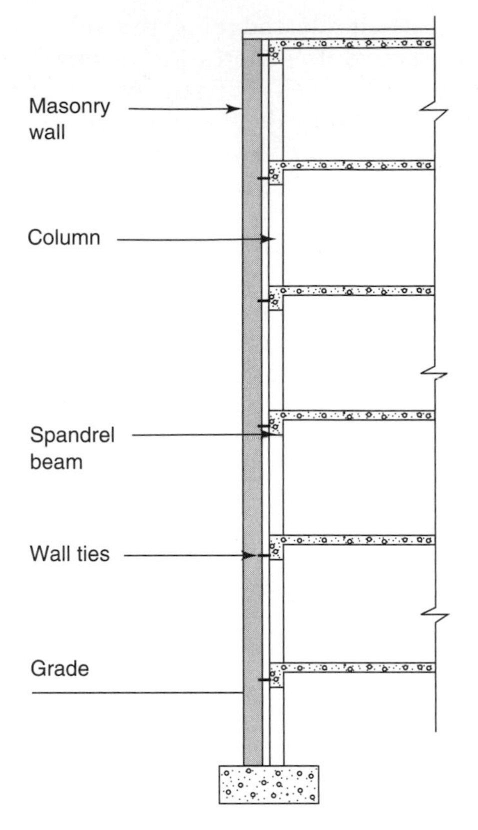


**Figure 14-5** Curtain Wall Carried on Foundation.

structural frame provides the wall with lateral support and carries all other vertical loads. This system allows considerable flexibility between walls and frame. Lateral support is provided by flexible anchors called *looped rods* (Figure 14-6).

Curtain walls supported entirely by the building frame may be laid and bonded to the spandrel beams with mortar or may be keyed to the beams as shown in Figure 14-7. Notice that lateral support is provided here by flat metal ties cast into the column and laid in the mortar joints of the masonry. Also note that the concrete structural frame has been left exposed, as part of the overall design. The faces of a steel structural frame may likewise be left exposed. Figure 14-8 shows a two-wythe brick curtain wall erected in a steel frame so as to leave the column flanges exposed. In Figure 14-9, a brick cavity wall is used as a curtain wall, while leaving the steel frame exposed. Notice the channels used to receive the brick and the cover plates used to hold the channels. Figure 14-10 illustrates the same type of construction at a beam.

The structural frame is usually concealed by a masonry curtain wall. In Figure 14-11, a two-wythe brick curtain wall conceals the exterior of the steel frame. The inner flange is sheathed in furring tile. In this type of construction, the face wythe is carried over the spandrel beams on lintel angles riveted or

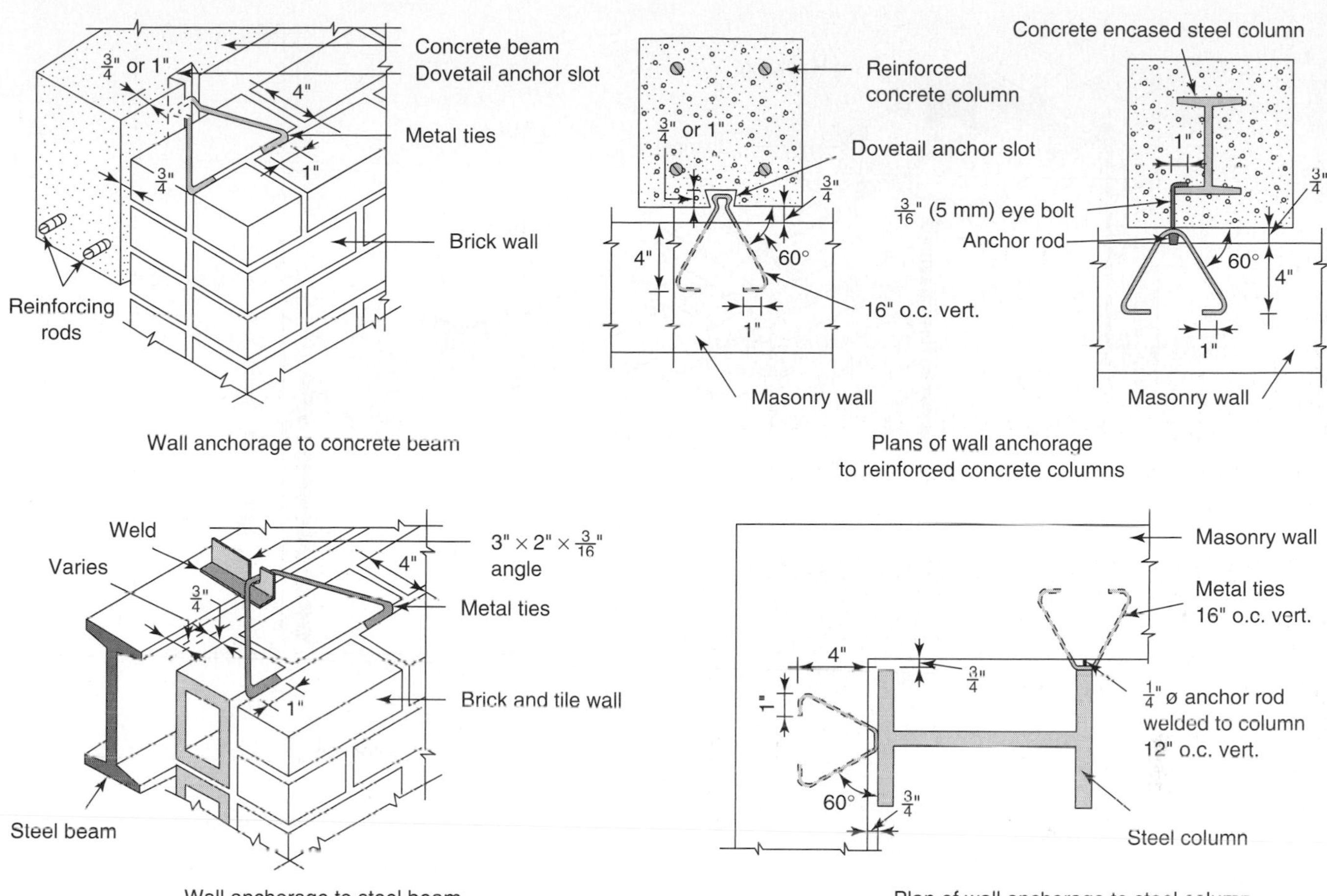

Wall anchorage to steel beam

Plan of wall anchorage to steel column

**Figure 14-6** Lateral Supports to Beams and Columns.

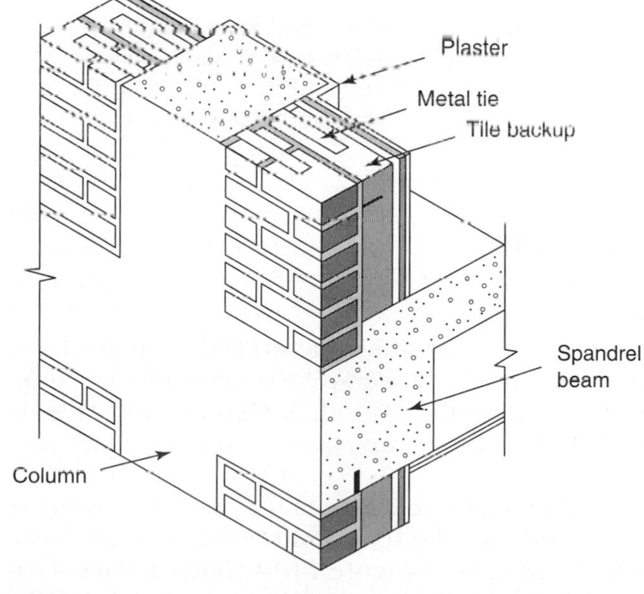


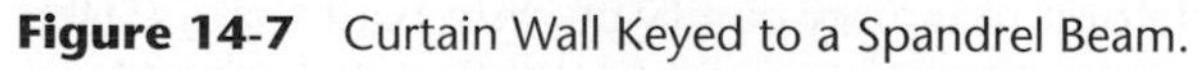
**Figure 14-7** Curtain Wall Keyed to a Spandrel Beam.

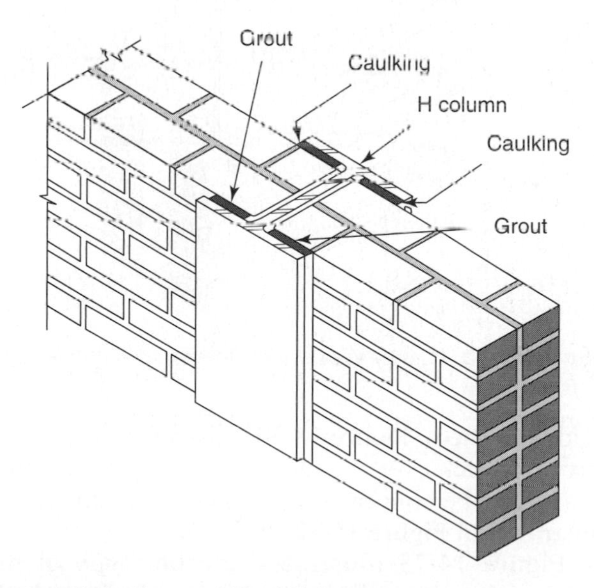

**Figure 14-8** Exposed Steel Column.

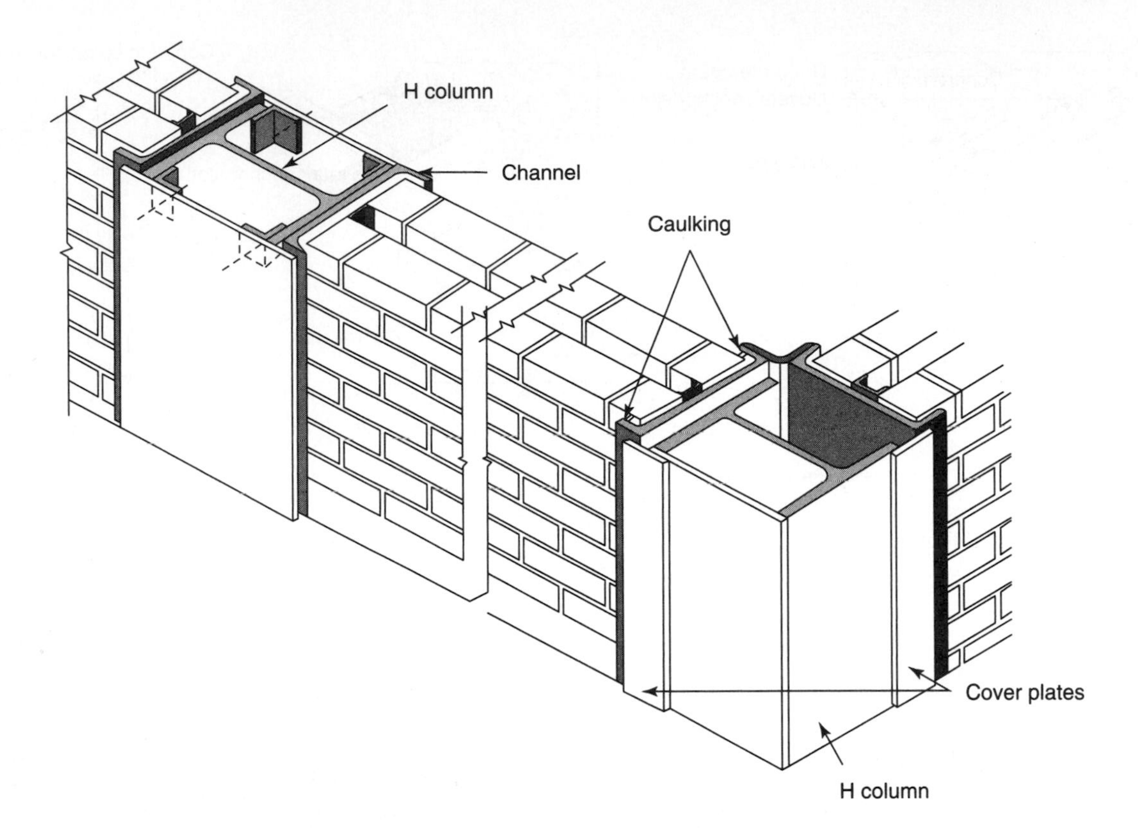


**Figure 14-9** Cavity Curtain Wall with Exposed Frame.

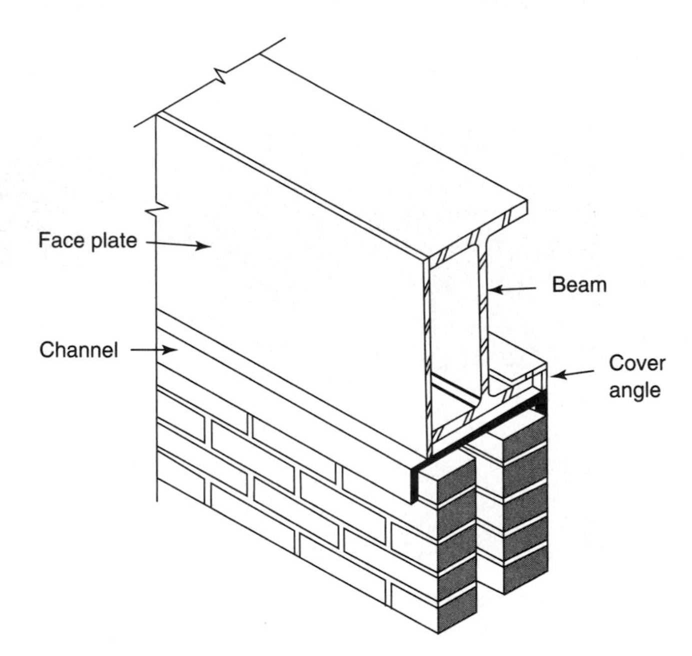


**Figure 14-10** Cavity Wall Under Beam.

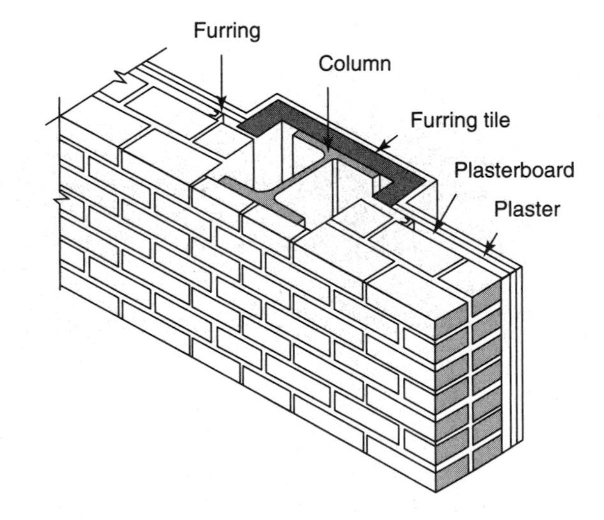


**Figure 14-11** Steel Frame Concealed.

welded to them. Figure 13-30 illustrates two ways of carrying brick over the face of a beam. A structural frame may be concealed by a cavity curtain wall, as illustrated in Figure 14-12.

Figure 14-13 illustrates another type of masonry curtain wall, in which a structural tile backup wall is faced with architectural terracotta. In this case, the structural frame is steel and fireproofed with concrete. Concrete blocks (standard, lightweight, or cellular concrete) are also widely used as backup material for a masonry curtain wall.

Lightweight concrete blocks are often used as structural backing for curtain walls that are faced with brick veneer. Figure 14-14 shows a typical example of a masonry cavity curtain wall that is used in buildings where good insulation values are re-

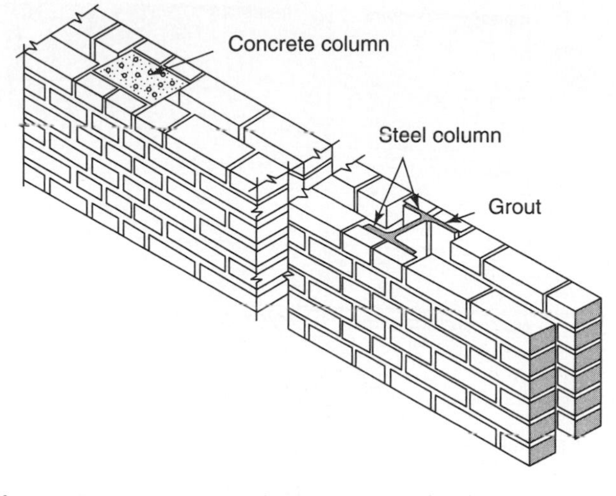


**Figure 14-12** Structural Frame Concealed by Cavity Wall.

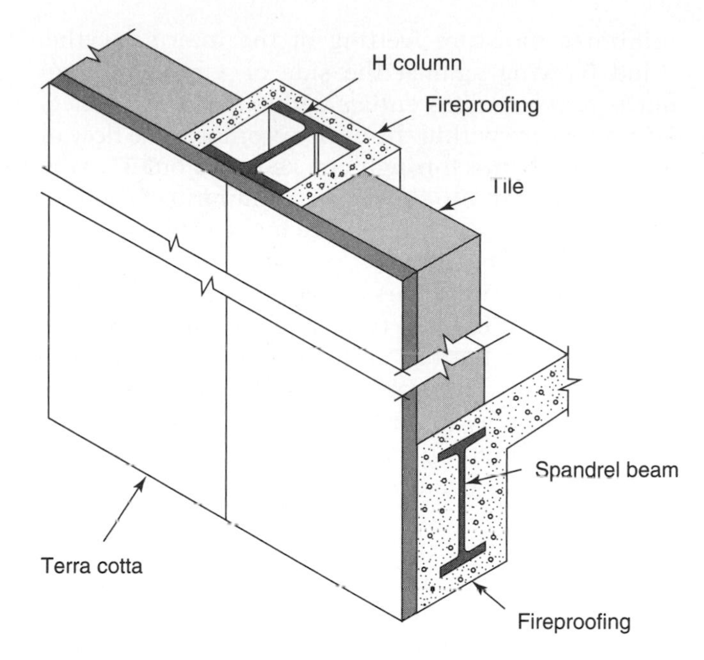


**Figure 14-13** Terra-Cotta-Faced Curtain Wall.

**Figure 14-14** The Construction of a Brick-Faced Curtain Wall. The Brick Anchor is Fastened through the Insulation into the Structural Frame Which Will Provide the Lateral Stability for the Brick Veneer. In this Application the Brick Veneer is Being Placed Directly on the Reinforced Concrete Grade Beam.

quired and a durable exterior is desired. In this application, the masonry blocks absorb the lateral wind loads imposed on the brick veneer through the brick ties providing the necessary structural stability, while the brick veneer serves as the rain screen and provides a durable exterior finish. The insulation between the two wythes provides an unbroken insulation blanket on the exterior of the building ensuring consistent insulation values while an air barrier on the warm side of the insulation prevents air leakage through the building envelope.

The rain screen principle is an enhancement of the normal cavity wall and an attempt to further minimize moisture wetting of the interior wythe. Wind blowing against the side of a building produces pressure differentials within the cavity. A negative pressure within the cavity promotes the flow of air and moisture through cracks in the outer wythe of the wall and into the cavity. To minimize such infiltration of water, the pressure in the cavity must be equalized to the exterior pressure on the wall as quickly as possible. This is usually accomplished by providing vent holes at the top of the brick veneer as well as at the bottom and dividing the cavity into compartments of various sizes. Compartments within the cavity prevent lateral air movement due to pressure differentials in the wall cavity and ensure faster pressure equalization of the cavity with the wall exterior.

To ensure that the brick veneer is securely fastened to the masonry blocks, the ties used to attach the brick veneer to the masonry blocks become a vital structural component of the wall construction. To transfer the loads from the brick to the block without adverse effects to the brick veneer, the ties must be structurally adequate and placed in the wall according to the manufacturer's specifications.

In high-rise construction, allowances must be provided for shrinkage and movements in the structural frame. Curtain walls supported by reinforced concrete frames must have adequate control joints (see Figure 14-15) to ensure that vertical loads from the building frame are not transferred to the curtain wall as the building shortens over time. Curtain walls that have the brick veneer supported on ledger angles (Figure 14-16) are especially susceptible to failure in buckling if subjected to vertical loads.

**Figure 14-15** Vertical and Horizontal Control Joints Between Precast Brick Curtain Walls.

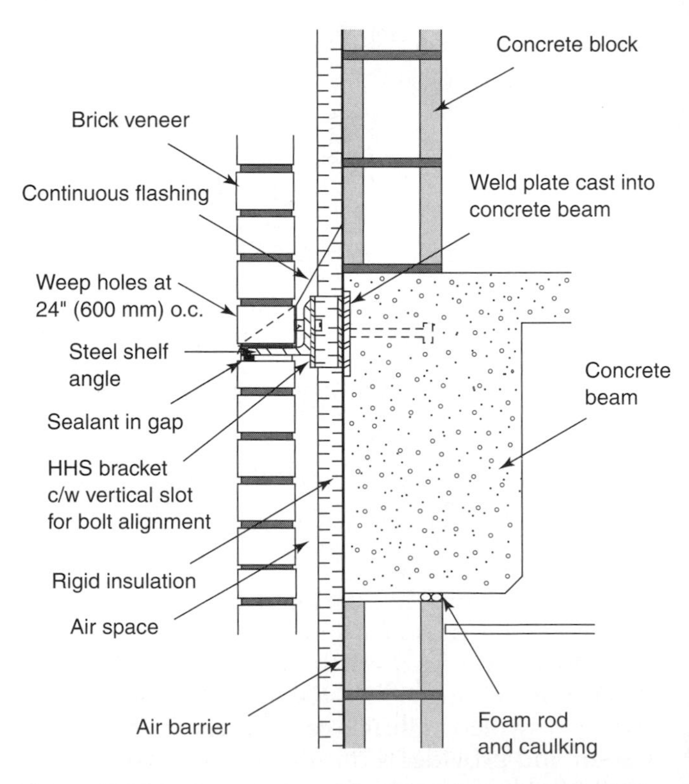


**Figure 14-16** Support Detail for Brick Veneer Curtain Wall.

## STONE PANELS

Several types of stone panels are available to be used as curtain walls. One is a 3- or 4-in. (75- or 100-mm) thickness of stone dimensioned to fit the particular job. One method of attaching this type of panel is illustrated in Figure 14-17. In this case, the panels are held in position at the bottom by rods ½ in. (12.5 mm) in diameter and approximately 8 in. (200 mm) long, welded to the lintel angle at joints in the stone. Two alternative methods of support are shown in Figure 14-18.

Another commonly used type of stone panel is a *sandwich panel,* consisting of a stone face, a layer of rigid insulation, and a sheet metal or hardboard back, all cemented together. This type of panel is relatively light and is often supported by a metal subframe or *grid,* which is in turn connected to the

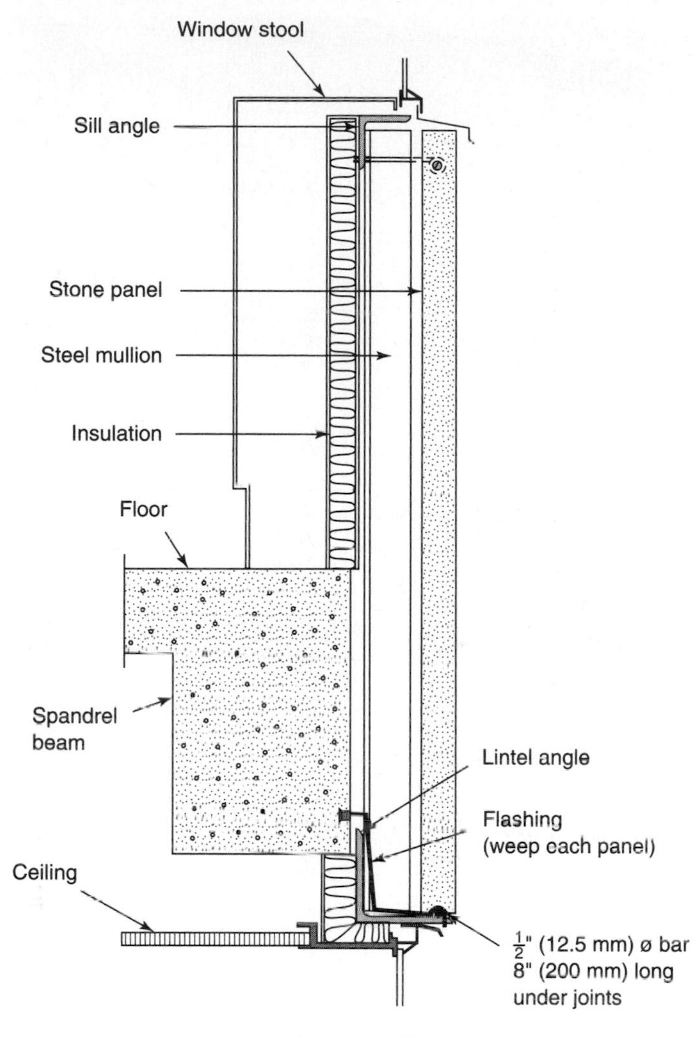


**Figure 14-17** Stone Panel Curtain Wall.

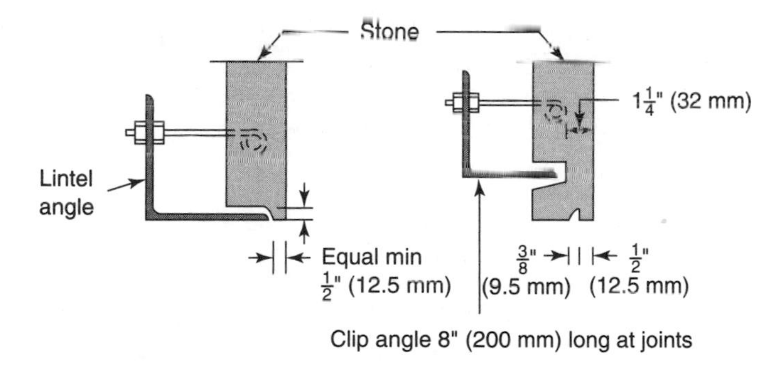


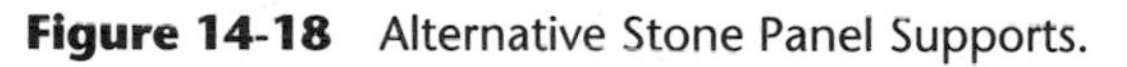

**Figure 14-18** Alternative Stone Panel Supports.

structural frame (Figure 14-19). A great variety of metal grids is available for use with stone or other types of curtain wall panels. Figure 14-19 shows a typical aluminum grid being used with stone sandwich panels.

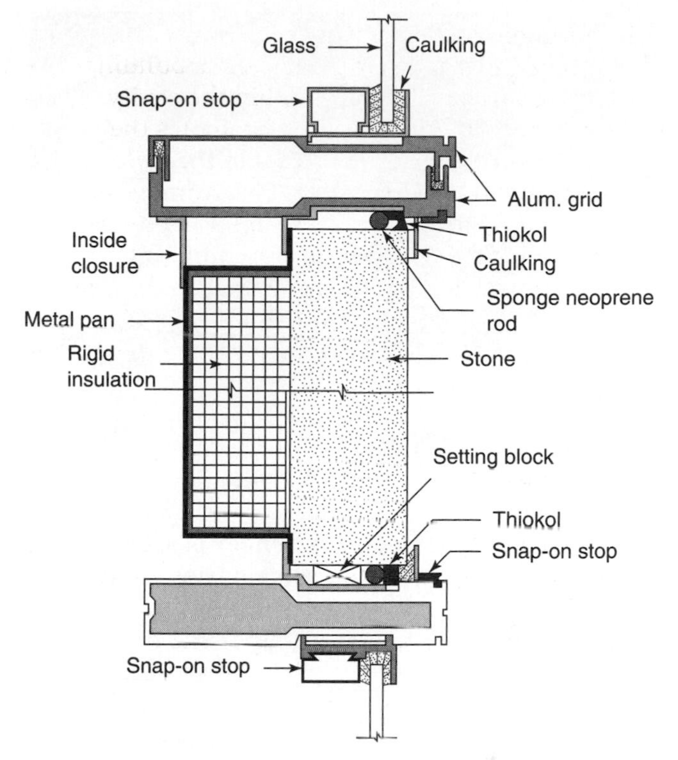


**Figure 14-19** Details of Stone Sandwich Panel Subframe.

## PRECAST CONCRETE CURTAIN WALLS

Two types of precast concrete units are used in curtain wall construction. One is a panel that is cast, cured, and finished in a precasting plant (see Figure 14-20) and brought to the job site for erection in much the same way that stone or masonry panels would be used. Such precast panels are custom made to the size and with the surface texture specified for a particular building. See Chapter 9 for details of architectural precast concrete wall panels.

Figure 14-21 shows typical precast concrete panels being erected over a structural steel frame, and Figure 14-22 indicates how those panels are anchored to the frame. Many other anchoring systems are in use, some of which are illustrated in Figure 9-64. Another method is shown in Figure 14-23, in which steel dowels project into recesses in the top and bottom edges of panels. Joints should be sealed as described in Chapter 9, using a waterproof, elastic caulking at the exterior face.

The other type of precast concrete unit is produced by the tilt-up method, also described in Chapter 9. The structural frame for such panels may be steel or concrete, cast in place, or precast. In the case of a structural steel frame, it will be erected before the panels are set up. A precast concrete frame may be erected before

**Figure 14-20** Factory-Made Precast Concrete Wall Panels. (Reprinted with permission of Portland Cement Association.)

**Figure 14-21** Erecting Precast Concrete Panels. (Reprinted with permission of Portland Cement Association.)

**Figure 14-22** Panels Anchored to Frame. (Reprinted with permission of Portland Cement Association.)

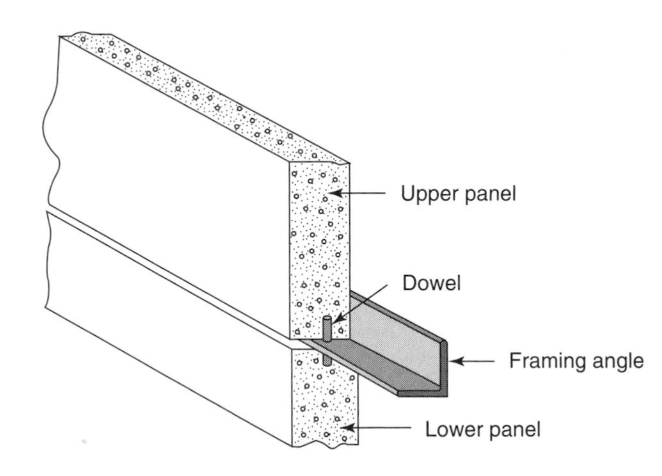


**Figure 14-23** Dowel Anchor for Precast Concrete Panel.

or at the same time as the panels, whereas in other cases, a cast-in-place concrete frame may be placed after the panels are erected (see Figure 14-24).

Panels may be designed to produce a flush tilt-up wall, usually the case with a steel frame (see Figure 14-25), or columns may be exposed (see Figure 14-26).

Forms for the panels are built on the job site, often on the floor slab; the necessary reinforcing is installed, and the concrete is placed (see Figure 14-27). Finishing produces the specified surface, and the concrete is properly cured. When the panels have gained

**Figure 14-24** Panels Erected before Frame. (Reprinted with permission of Portland Cement Association.)

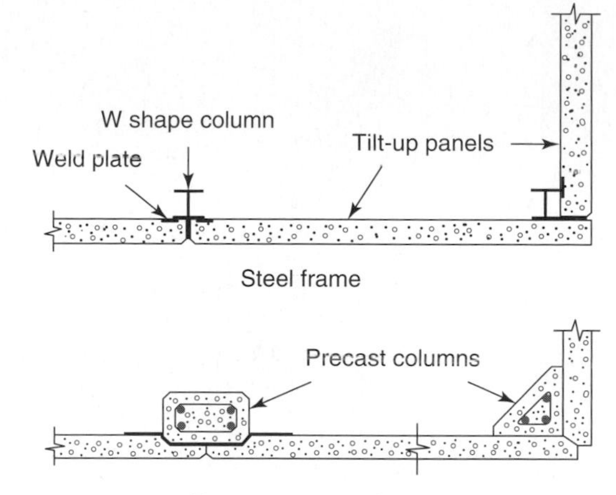


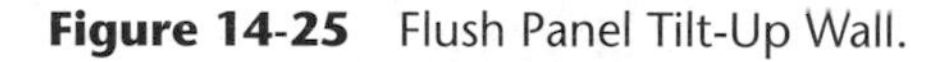

**Figure 14-25** Flush Panel Tilt-Up Wall.

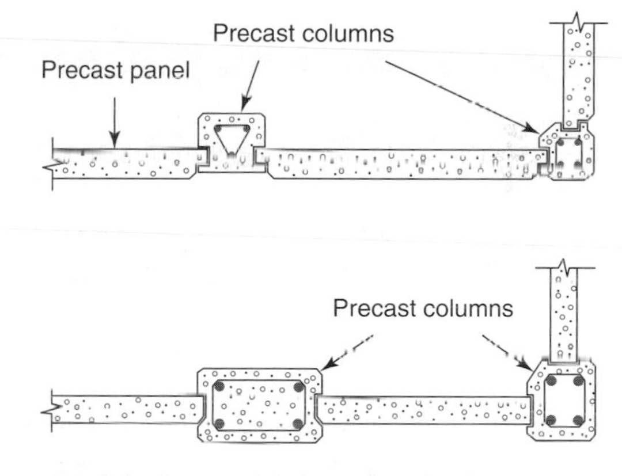

**Figure 14-26** Exposed Columns with Tilt-Up Panels.

sufficient strength, they are lifted or tilted into position [see Figure 14-28(a)]. They are held in position at the bottom by dowels projecting from the foundation or by welding plates that have been cast into the bottom edge of the panel and the top of the foundation [see Figure 14-28(b)]. When panels are to be set before the columns, they will be cast with reinforcing bars projecting from their edges to be cast into the columns (see Figure 14-24). After plumbing and bracing the panels, simple wood forms are built around the openings and concrete is placed. Columns may be made the width of the opening or wide enough that the edges of the panels will be cast into them.

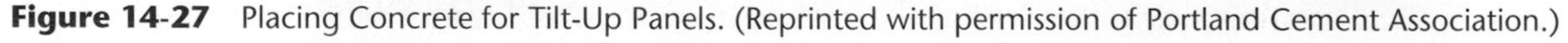

**Figure 14-27** Placing Concrete for Tilt-Up Panels. (Reprinted with permission of Portland Cement Association.)

(a)

(b)

**Figure 14-28** (a) Raising Tilt-Up Panel into Place. (Reprinted with permission of Portland Cement Association.) (b) Tilt-Up Panels Held in Place by Temporary Braces.

Sandwich wall panels are specified for many buildings where curtain wall construction is being used. Sandwich panels consist of a core of rigid, low-density insulation, covered on both sides with a thin layer of reinforced, high-strength concrete. The panel may be cast with solid outer edges so that the core is completely enclosed, or if adequate shear connectors are provided, the insulation may be extended to the panel edges (see Figure 14-29).

Surface treatment of tilt-up panels is an important part of a whole operation. Basically, six types of surfaces may be produced: smooth, colored, textured, patterned, ribbed, or exposed aggregate.

A smooth surface may require only screeding and floating, or if a more uniform appearance is needed, the surface may be troweled (see Figure 14-30). If the surface is to be painted, a bond breaker must be used that will not stick to the surface of the concrete and prevent paint from adhering.

Colored surfaces are achieved by using white cement or color pigments. A white surface will usually be produced by adding a thin layer of concrete made with white cement and white sand to the face of the normally cast panel. Color pigments are troweled into a normal concrete surface at specified rates in pounds per square foot to produce the color required.

A textured (roughened) surface may be produced by brooming, sandblasting, using a cork float, or lining the form with rubber matting or some type of foamed plastic.

A patterned surface may be made by forming the pattern with strips or blocks of foamed plastic cemented to the casting surface or by using wood strips in the same manner. More complicated patterns may be made by producing the shapes desired in a bed of damp, hard-packed sand laid over the casting surface.

Ribbed surfaces are formed by placing ribbed, corrugated, or striated liners on the casting surface as a bottom form. In some cases, wedge-shaped wood strips may be used if there is some way to fasten them down. Such wood strips should be well oiled to prevent them from swelling and coated with bond breaker for easy stripping.

Exposed aggregate surfaces are produced in several ways. One method is to cast the panel with the face side up and seed the aggregate evenly over the fresh concrete surface by shovel or by hand. It is then embedded in the surface by hand float or straightedge until all aggregates are entirely embedded and mortar surrounds and slightly covers all of the particles. When the concrete has almost attained its final set, the surface is washed with water and a stiff brush to remove the cement paste and expose the aggregate.

When larger, uniformly sized aggregate is to be used, a common method is to cover the casting surface with a uniform layer of sand to the depth of ap-

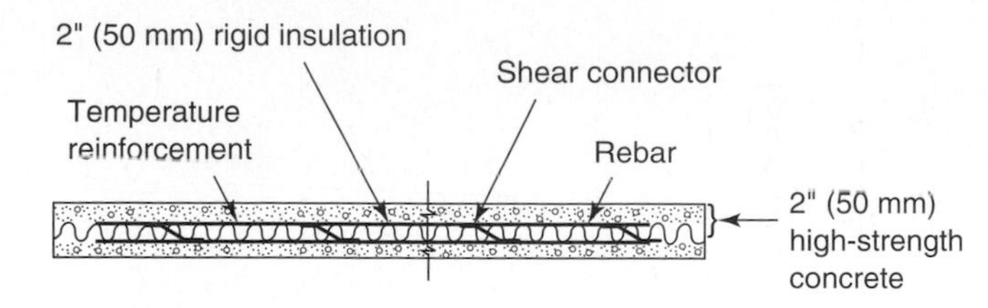


**Figure 14-29** Sandwich Panel.

**Figure 14-30** Finishing Tilt-Up Panel. (Reprinted with permission of Portland Cement Association.)

proximately one-third the diameter of the aggregate. Pieces of aggregate are then pushed into the sand as close together as possible and dampened down with a water spray to settle the sand around each piece. The reinforcement is then set into the form on chairs and the concrete placed.

Still another method is used with flat, irregular shapes and sizes of stone. They are laid face down in the form, and the spaces between them are filled with dry sand to a depth of approximately one-third to one-half the thickness of the stone. The stone is moistened and the remainder of the spaces filled with mortar, after which the regular reinforcing and placing of concrete occurs. Excess sand may be removed after the panels are raised by brushing or air blasting.

## LIGHTWEIGHT CURTAIN WALLS

The use of various materials such as reinforced plastic, steel coated with porcelain enamel, stainless steel, aluminum, copper, and bronze as exterior finishes has led to a wide variety of lightweight, single-skin, insulated sandwich panels of various designs (Figure 14-31).

Basic construction (see Figure 14-32) for this type of panel consists of an interior liner, an insulating layer, and an exterior finish. The amount of insulation and the type of interior finish can vary, as can the outer finish. Two basic approaches are available in the construction of curtain walls: (1) the curtain wall panel can be preassembled and supplied to the site ready for installation, or (2) the sandwich can be made up on the site by installing the components individually, creating a layered wall system.

An example of the latter is shown in Figure 14-33. The construction of the curtain wall for this building consists of an interior liner, fiberglass insulation

**Figure 14-31** Molded Reinforced Plastic Curtain Wall Panels.

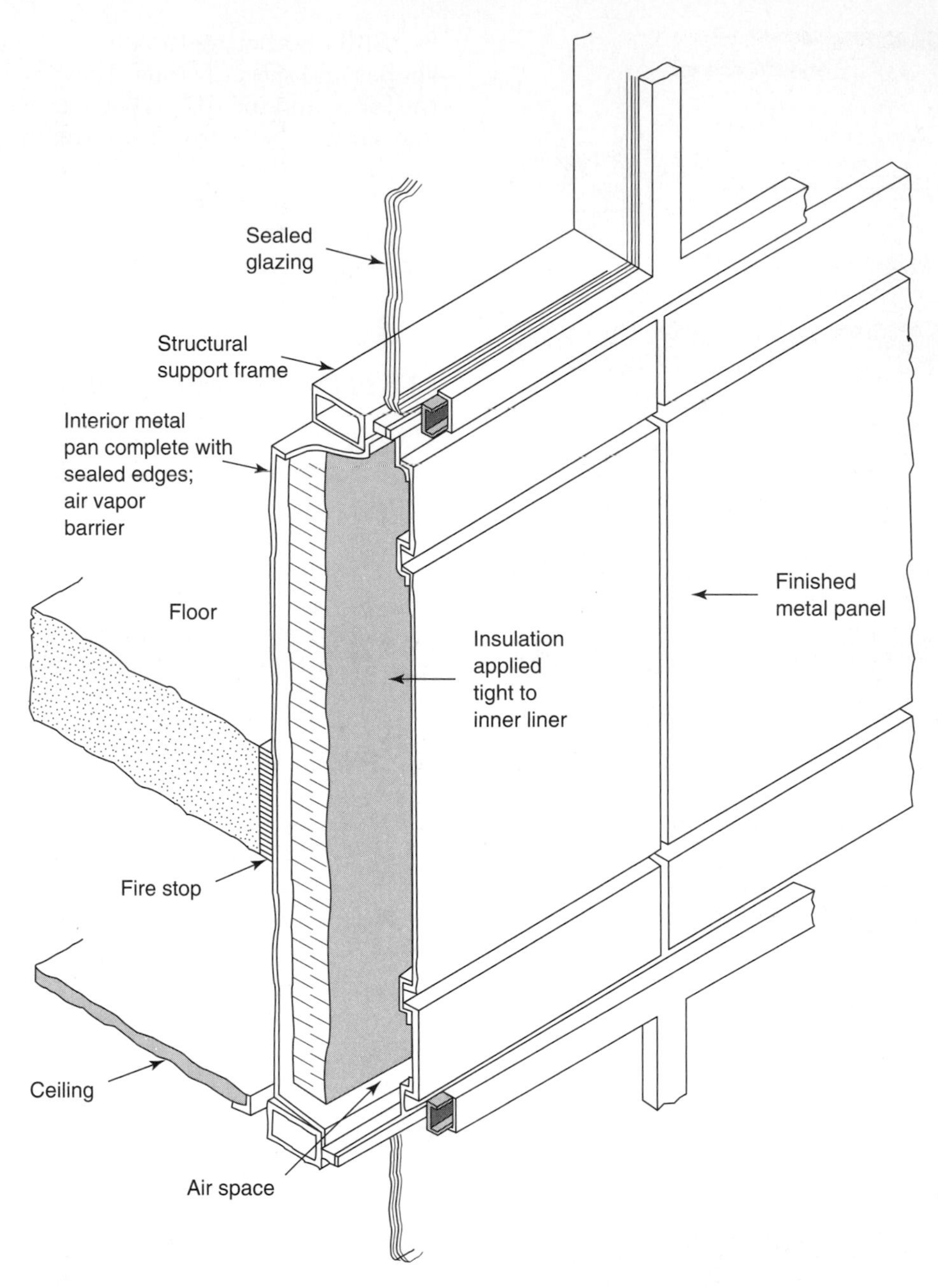


**Figure 14-32** Lightweight Sandwich Curtain Wall.

with attached vapor barrier on the inside, covered with a prefinished metal siding attached to the structural steel frame. Another approach is shown in Figure 14-34. In this application, sheet metal studs are used as the basic frame for the curtain wall. Exterior-rated drywall on the studs will serve as the exterior sheathing, to which insulation will be attached and covered by a brick veneer. Additional insulation may also be provided between the studs before the vapor barrier and the interior wall finish are installed.

Prefabricated panels can be attached directly to the building frame or to subframes (Figure 14-35) that are attached to the exterior of the building frame (Figure 14-36). Panels that are designed to act as stressed-skinned panels (such as that shown in Figure 14-37) require no subframe and are attached directly to the structural frame of the building by various types of clips, some of which are illustrated in Figure 14-38.

Figures 14-39 and 14-40 illustrate typical applications of curtain walls. In Figure 14-39, two sizes of panels are mounted together to an aluminum subframe. Notice the swing stage scaffold used in erecting the panels. In Figure 14-40, panels and wide mullion strips are mounted on the backup wall without the use of a subframe, and a complete scaffolding system is being used during the erection of the panels.

**Figure 14-33** Field-Assembled Curtain Wall with Prefinished Metal Siding as the Outer Skin.

**Figure 14-34** Sheet Metal Studs Provide Support for Curtain Wall Sheathing.

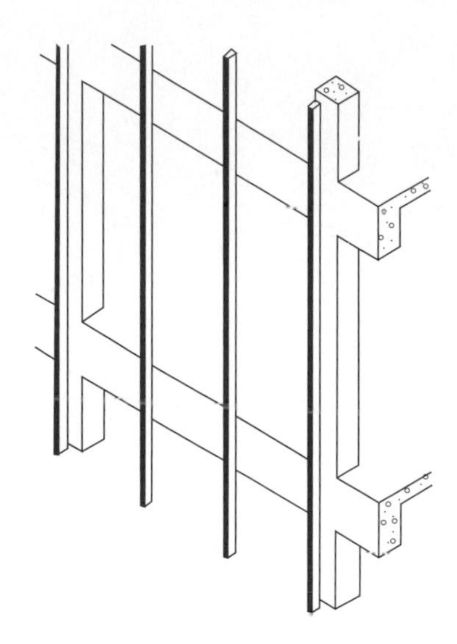

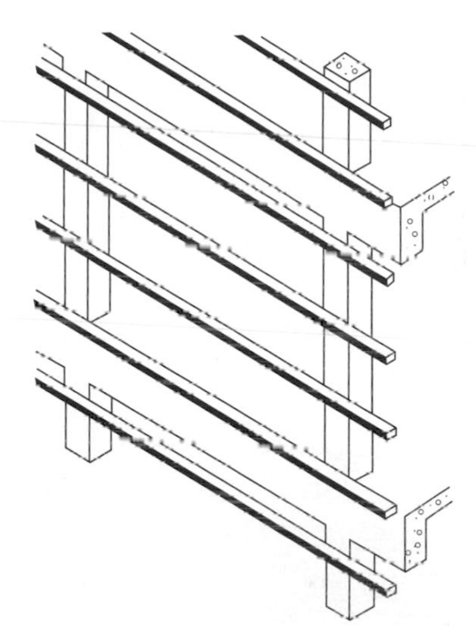

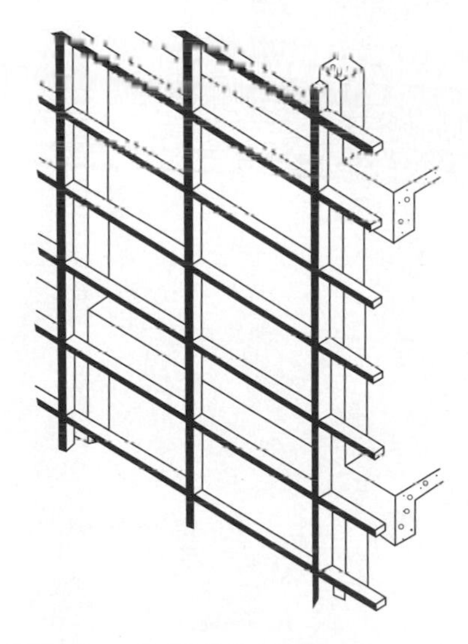

**Figure 14-35** Curtain Wall Subframe Systems.

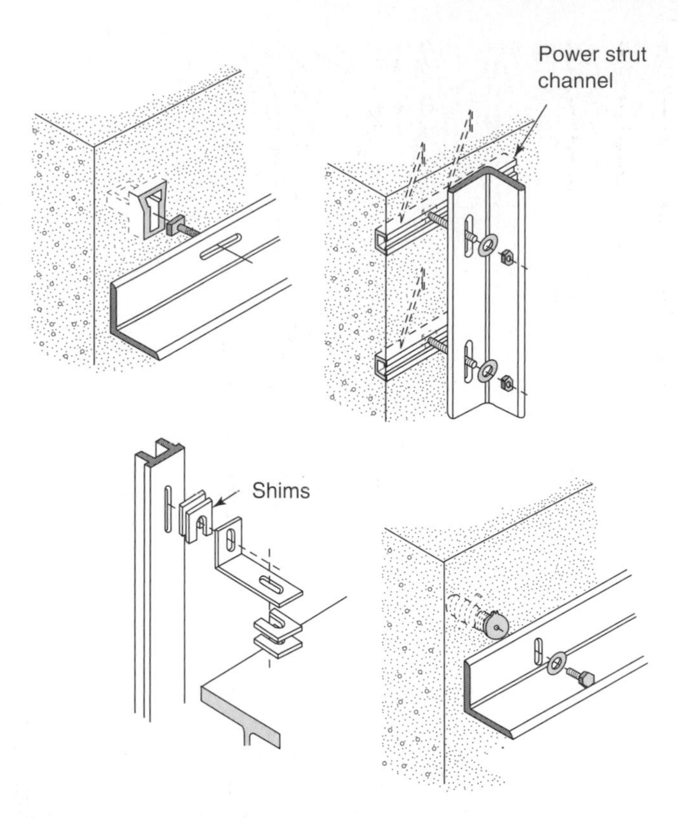

Figure 14-36 Methods of Attaching a Subframe to a Concrete or Steel Structural Frame.

## GLASS CURTAIN WALLS

Many types of subframes may be used to mount glass panels in curtain walls. The type used will depend on whether the wall is single or double glazed, on whether the units are fixed or movable, on the thickness of the glass, and on the size of each unit. Figure 14-41 illustrates the use of an extruded aluminum shape as a framing member to support a double-glazed glass curtain wall section.

All curtain walls require some type of structural framing to absorb the lateral forces due to

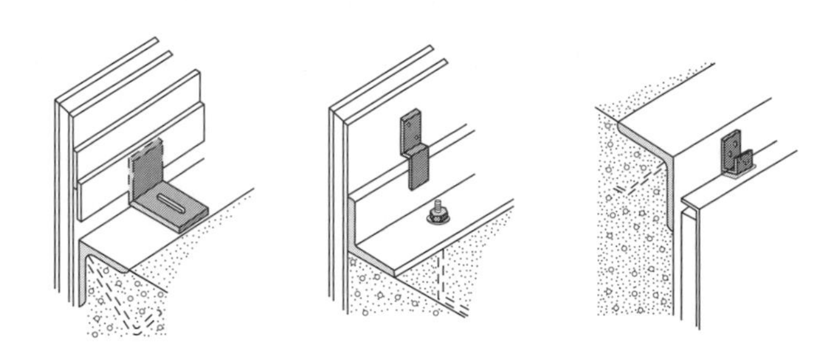

Figure 14-38 Curtain Wall Panels Clipped to a Structural Frame.

Figure 14-37 Steel Sandwich Panels Being Installed.

Figure 14-39 Curtain Wall with a Subframe.

**Figure 14-40** Curtain Wall Clipped to a Concrete Block Backup Wall.

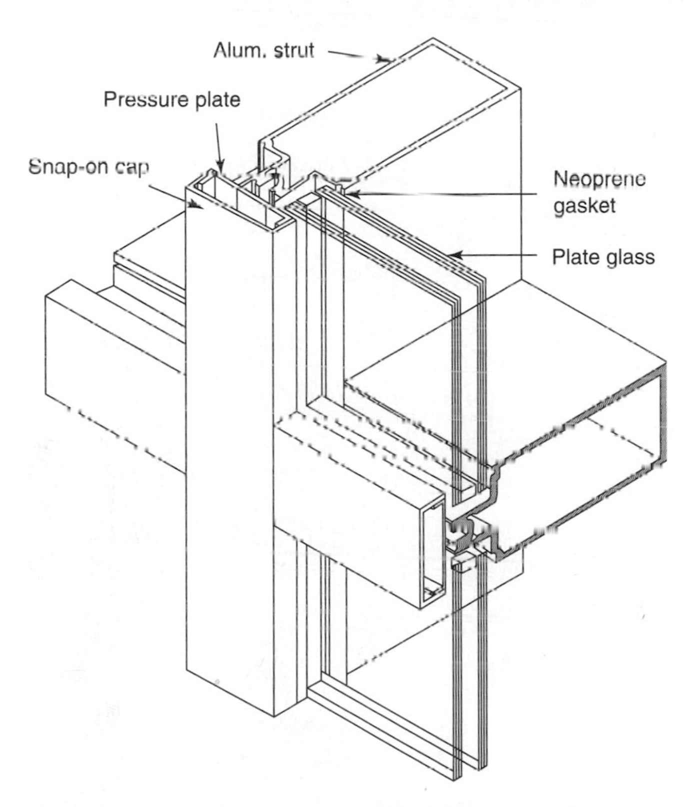


**Figure 14-41** Glass Curtain Wall Subframe Details.

wind. The more slender the curtain wall cross section, the more critical is the need for this additional framing. When glass panels are used, the flexural properties of the glass limit the distance between lateral and vertical supports. The structural framing must be sufficiently rigid to ensure that the lateral deflection caused by the force of the wind does not exceed the flexural limits of the glass panel. Should excessive deflections be encountered, the glass panels may develop breaks within the structural sealant between the glass panel and the panel frame. Air leaks may develop, reducing the efficiency of the curtain wall, and in many cases, moisture penetration can result. In extreme cases, whole glass panels can literally pop out of the supporting frame.

To ensure proper support for the glass panels, various types of extruded aluminum sections have been developed as the basic components for the construction of the frames. These frames can be prefabricated in the shop and trucked to the construction site in relatively large sections (see Figure 14-42), or the extruded sections can be precut and assembled on site, depending on the complexity of the framing (see Figure 14-43).

Once the main building frame has been constructed, the curtain wall framing sections are lifted into place and attached to the building frame (see Figure 14-44). The larger the opening to be enclosed by the glass panels, the more substantial are the requirements of the curtain wall frame. In many cases, structural steel sections are used to provide additional lateral stability to the aluminum framing (see Figure 14-45).

Figure 14-46 illustrates a section of reflective glass paneling being installed on a concrete building frame. In this example, the portion of panel covering the floor beam is a translucent section that incorporates an insulated panel covered with reflecting glass. Once completed, the building exterior resembles a giant mirror (see Figure 14-47) that is transparent from the inside but reflective from the outside during daylight hours.

**Figure 14-42** Prefabricated Aluminum Frames for Glass Curtain Wall.

**Figure 14-43** Precut Extruded Aluminum Sections for Glass Curtain Wall Framing Assembled on Site.

## Review Questions

**1.** What are the two primary functions of curtain walls?

**2.** What was the original need for a cavity wall?

**3.** Explain the meaning of each of the following terms: **(a)** curtain wall, **(b)** panelized masonry, **(c)** sandwich panel, and **(d)** spandrel beam.

**4.** Outline four advantages of using curtain walls in large buildings.

**5.** Explain the purpose of each of the following: **(a)** a backup wall, **(b)** *looped rods* in curtain wall-to-frame connections, **(c)** a *cavity wall* in curtain wall construction, and **(d)** a *setting block* used in stone panel erection.

**6.** By means of neat diagrams illustrate one practical method of providing **(a)** a textured surface, **(b)** a

**Figure 14-44** Glass Curtain Wall Framing Attached to Structural Frame.

**Figure 14-45** Hollow Structural Steel Sections Used as Lateral Support Framing for Glass Curtain Wall.

**Figure 14-46** Reflective Glass Panel Curtain Wall Being Attached to Reinforced Concrete Building Frame.

**Figure 14-47** Reflective Glass Curtain Wall.

ribbed surface, and **(c)** a patterned surface to a tilt-up concrete panel.

**7.** Outline two basic reasons for using a subframe to carry curtain walls over a structural frame.

**8. (a)** Illustrate by a horizontal section how you think the steel sandwich panels shown in Figure 14-39 are constructed. **(b)** Outline, with the aid of diagrams, a practical method for attaching those panels to the building.

# 15 Building Insulation

In building construction, the use of insulation is an essential component in the design and construction of the building envelope and is important for the protection of structural members within the building frame. The need for appropriate fire protection of structural members within buildings has been vividly illustrated by recent world events. As wall systems have become more sophisticated and building codes have become more stringent, a thorough understanding of insulation application and the evaluation of its performance within the building envelope have become crucial to the long-term performance of all buildings.

Three requirements in building design are addressed by the application of insulation or insulating materials within the building envelope assembly: (1) maintaining temperature control within the building without using excess energy; (2) control of flame spread during a fire (fire ratings of building components and assemblies); and (3) the control of sound transmission. Because of their unique properties, insulating materials (when correctly selected) will resolve any combination of the three.

## THERMAL INSULATION

Thermal insulation is generally considered to be any material that has a relatively low coefficient of thermal conductivity; that is, it impedes the transfer of heat from one face to the other. Good thermal insulation is illustrated in Figure 15-1, where a slab of cellular glass successfully retards the passage of heat through its thickness.

In buildings, thermal insulation is used to enclose the entire building shell to minimize heat loss (see Figure 15-2) when the outside temperature is low and to keep the interior of the building cool when the outside temperature is high. The amount of insulation applied to a structure is normally based on the number of days when supplemental energy must be used to heat or cool the building and the payback period required to recover the cost of the insulation with the money that is saved in heating and cooling costs.

### Insulation against Temperature Changes

Another use for thermal insulation is to protect exposed structural sections from large temperature changes. Large fluctuations in temperature cause expansion and contraction of the exposed sections, which can cause unnecessary stresses in the building frame. Figure 15-3 illustrates an exposed concrete column being insulated against temperature changes. Cellular glass, rigid glass fiberboard, and extruded polystyrene foam panels are the usual materials used for this type of application. Panels are normally secured to structural sections using a combination of mastic and pins (Figure 15-4).

### Insulation of Heat Conveyors

Heat ducts and water lines are also covered with insulation to prevent heat loss. Rigid glass fiberboard is normally used to insulate heating ducts, whereas

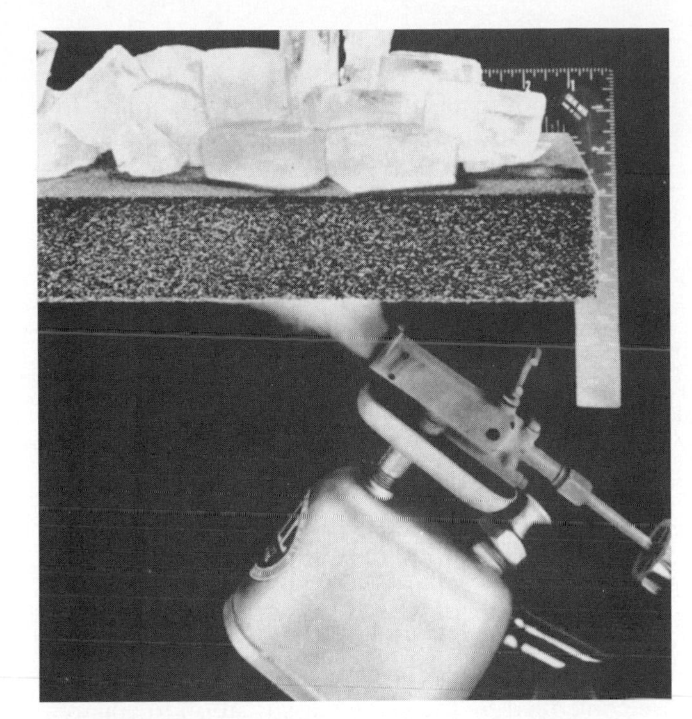

**Figure 15-1** Good Thermal Insulator. (Reprinted by permission of Pittsburgh Corning Corp.)

**Figure 15-2** Expanded Polystyrene Insulation Is Used to Enclose Building Frame.

**Figure 15-3** Insulating a Column against Temperature Changes. (Reprinted by permission of Pittsburgh Corning Corp.)

**Figure 15-4** Rigid Glass-Fiber Batts Attached with Pins to Concrete Wall.

rigid insulation of cellular glass, glass fiber, and foamed plastics, wrapped with metallic foil, is used around pipelines (Figure 15-5). In some applications, heat ducts are lined with rigid insulation rather than placing the insulation on the outside (Figure 15-6).

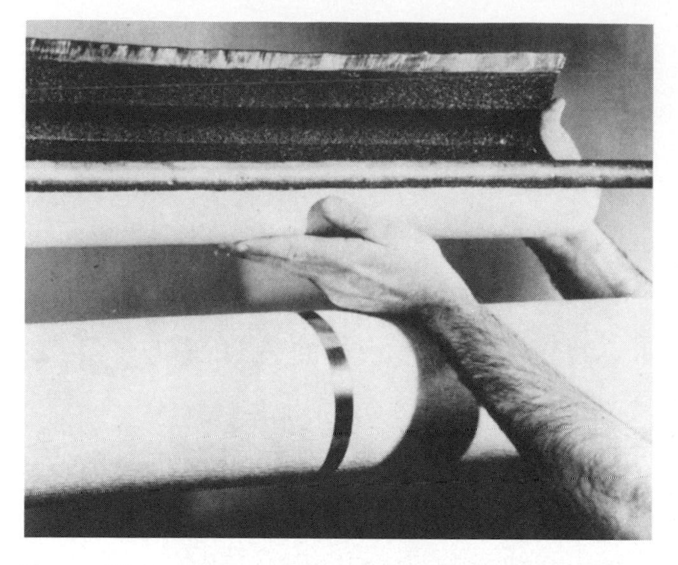

**Figure 15-5** Insulating a Pipe with Cellular Glass. (Reprinted by permission of Pittsburgh Corning Corp.)

**Figure 15-6** Air-Conditioning Ducts Lined with Glass Fiber Insulation.

## FIREPROOFING

Two material designations are recognized by fire codes when dealing with fire-resistance ratings of building assemblies: noncombustible and combustible. Noncombustible materials are mineral based and include concrete, asbestos cement, gypsum plaster, glass, and natural rock such as slate, sandstone, and marble. Combustible materials are those that contain or are formulated from animal or vegetable matter. Wood, fiberboard, paper, felt, plastics, asphalt, and pitch are typical examples. Construction materials that are a combination of combustible and noncombustible materials are usually considered combustible unless the quantity of combustible matter is quite small.

The amount of fireproofing that is applied to a structural section depends on the fire-resistance rating required and the type of material that is to be protected. Materials in general use for the protection of a structural steel frame, for example, include regular and lightweight concrete, cellular concrete, gypsum wallboard, plaster, and sprayed-on mineral wool (Figure 15-7).

### Fire Ratings

All modern building codes provide guidelines for minimum fire-resistance ratings of building components. The initial criterion for establishing building component fire ratings is the anticipated occupancy or combinations of occupancies for the building. During the initial design stages of a building, a *major occupancy classification* is assigned to the building in question and from this point on, all design decisions are controlled by the given classification to ensure a safe and fire resistant structure. Typical occupancy classifications include assembly occupancies (arena type and theatre type occupancies, for example), care and detention type occupancies, residential, business and personal services, mercantile, and industrial occupancies. Industrial occupancies, because of their variability, range from low hazard to high hazard conditions.

Once the building size, location, and the occupancies are established, fire ratings for all structural components must be determined. These can vary from one to three hours, or more, depending on the required fire separations. Every portion of the building assembly is evaluated according to its location and use in the building. Roof assemblies, floor assemblies, stairwells, partitions, exposed beams and columns, load-bearing walls, and exterior walls must all be evaluated on their ability to resist the effects of fire for a predetermined length of time. Research laboratories such as Underwriters Laboratories Incorporated and government institutions conduct fire tests on all building materials and various assemblies and provide design data such as flame-spread ratings, smoke developed classifications, and fire ratings.

### Methods of Fireproofing

Fireproofing steel beams and columns with concrete is accomplished by encasing the members in cast-in-place concrete. This method involves building a form around the structural steel section or using a prefabricated form. When concrete floor slabs are used with the steel frame, fireproofing is usually cast as part of the slab.

**Figure 15-7** Structural Steel Beams and Decking Fire Protected with Sprayed-On Insulation.

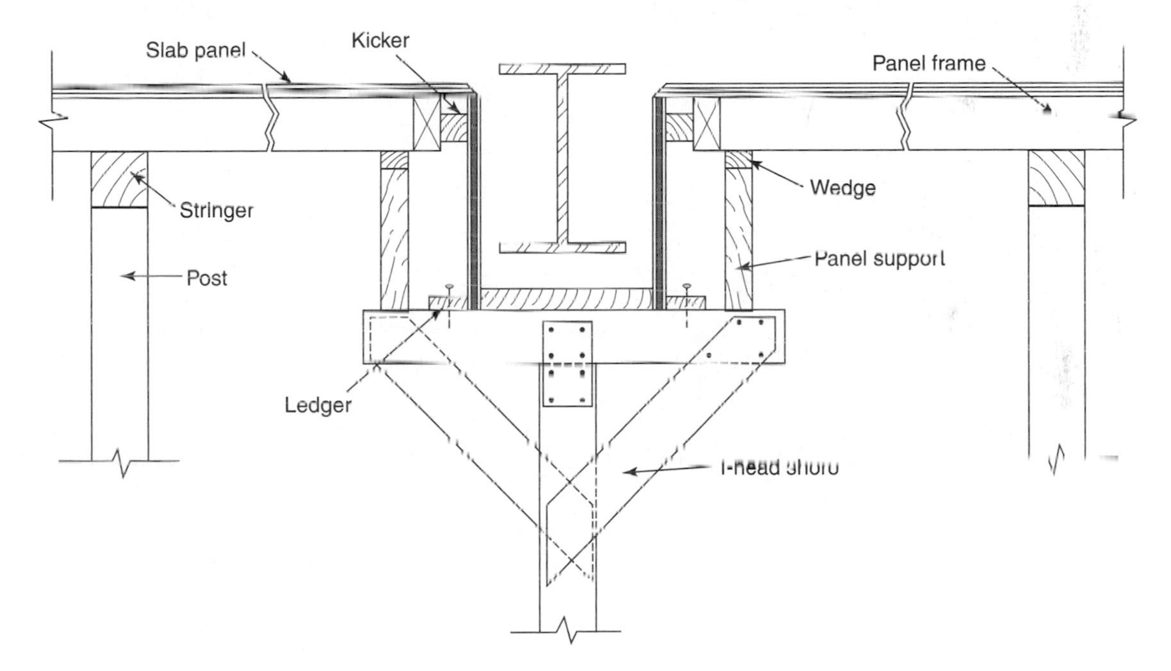


**Figure 15-8** Fireproofing Form Supported from Below.

Many systems of forming are available, but three common ones are illustrated in Figures 15-8, 15-9, and 15-10. In Figure 15-8, the forms are supported from below by posts. Notice that the fireproofing form is built in such a way that the sides may be removed without interfering with slab form supports or beam bottom supports. This is done by using kickers near the top edges of the sides and ledger strips along their bottom edges.

The form shown in Figure 15-9 is supported by the steel frame, which is accomplished by hangers placed over the beams. The formwork is suspended from the legs of these hangers and, in this particular case, is held in place by standard wedges. Fixed spreaders eliminate the need for soffit spacers.

The formwork shown in Figure 15-10 is also hung from the steel frame, but heavy-duty steel strapping is used in this case to support the form and

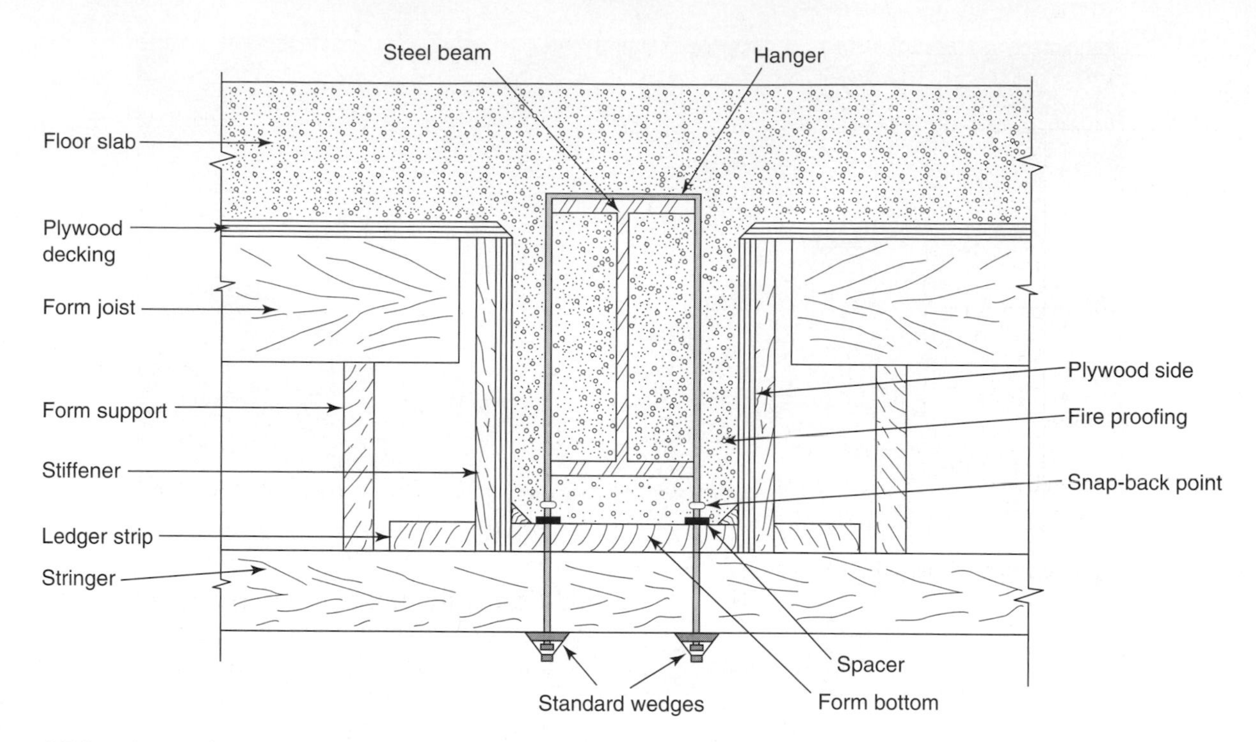


**Figure 15-9** Fireproofing Form Supported by Hangers.

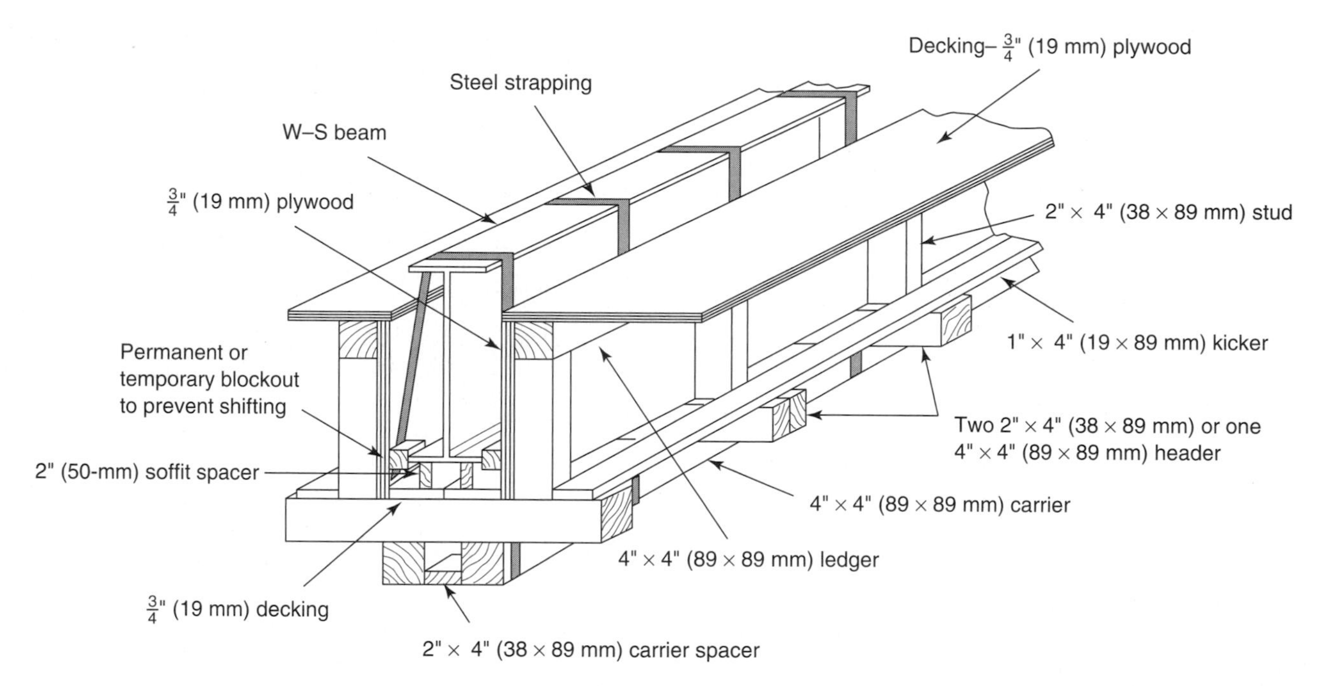


**Figure 15-10** Fireproofing Form Supported by Steel Strapping.

concrete load. Note that soffit spacers and blockouts are used here to hold the form in its proper position.

Columns are fireproofed with concrete by first building or fitting a form around them and then filling it with concrete. It is sometimes advantageous to build the form in two sections, upper and lower. Place and fill the bottom section first and then the top section.

Cellular concrete makes an excellent fireproofing material because of its high insulation value and its light weight. It is used in 2-, 3-, or 4 in. (50-, 75-, or 100-mm) slabs, which may be fabricated to suit any particular situation. Figure 15-11 illustrates a typical application of cellular concrete fireproofing to a steel girder, and Figure 15-12 shows the same material used to fireproof a column.

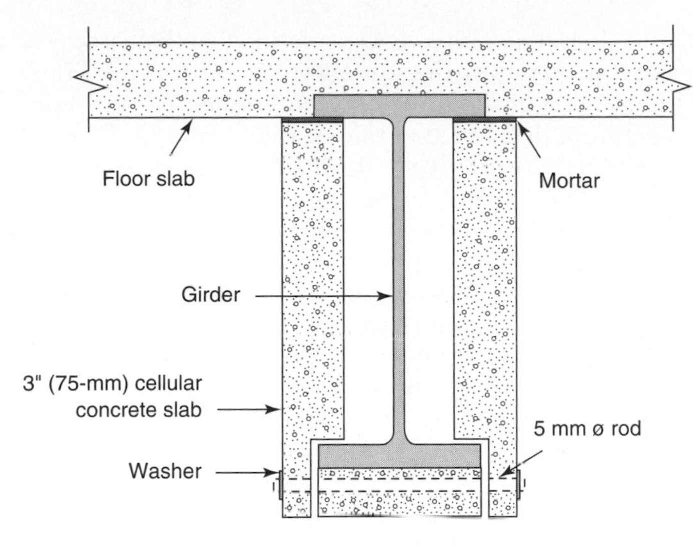


**Figure 15-11** Fireproofing a Girder with Cellular Concrete.

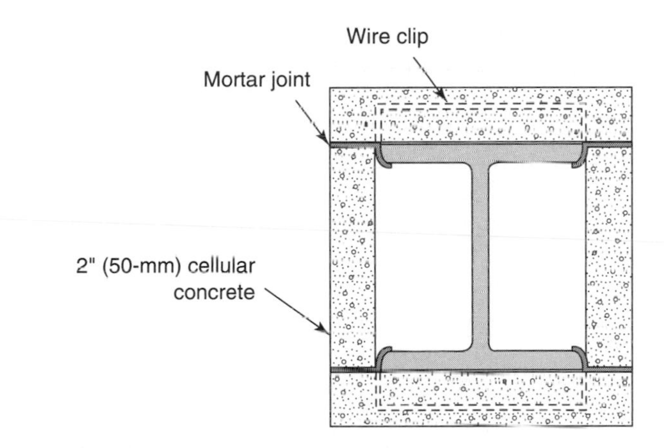


**Figure 15-12** Column Fireproofed with Cellular Concrete.

Gypsum wallboard is one of the most common materials used as a covering over framing members to provide fire protection. A ⅝-in. (16-mm) thick, fire-rated wallboard can provide up to a 1-hr fire rating when properly applied. Figure 15-13 illustrates a steel column covered with two layers of ⅝-in. (16 mm) fire-rated gypsum wallboard. An assembly of this kind can provide a 2-hr fire rating if the steel column is of sufficient thickness. In addition to being a good fire retarder, gypsum wallboard is economical, light in weight, easy to install, and provides a smooth surface that is easily finished. It is used as a covering on partitions, walls, and ceilings and can be used with wood framing, metal stud framing, concrete, or masonry.

A large amount of fireproofing is done with vermiculite or perlite plaster. Not only do these plasters resist heat transmission, but they are very light and can be rapidly applied. The steel members are sheathed in wire mesh or metal lath, stiffened by the

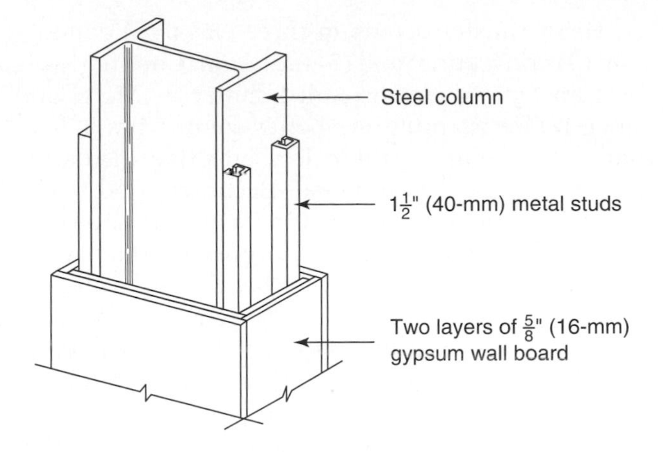


**Figure 15-13** Steel Column Protected with Gypsum Wallboard.

addition of light rods if necessary, and the plaster is sprayed over the wire to the required depth.

Asbestos-free, sprayed-on insulations such as mineral wool and cellulose fiber are also used as fire retardants. Applied with special spraying equipment, they are often used where irregular surfaces require protection, such as structural steel members and open-web joists. Being asbestos free, these insulations can be left exposed, as in the case of ceilings, where they provide a pleasing textured finish.

## BUILDING INSULATION

Because heat always seeks lower temperature levels, whenever a difference in temperature exists between two surfaces of a material, heat will flow to the colder side. This means that, when the outside temperature is low, heat from a building will flow to the outside. Conversely, during hot weather, heat tends to flow inward and warm the inside of a building.

The thermal efficiency of a building is based on the insulation values of the building systems that make up the protective blanket or outer skin of the building. Present-day designers must give careful consideration to the amount of insulation used in floors, roofs, and outer walls and the impact of its location on the various components within the assemblies. The same attention to detail must be given to the application of insulation as is given to the structural analysis of the building frame or in the assignment of correct fire ratings to building components. The heating and ventilation costs make up a large percentage of the operating budget for all modern buildings, and any savings that may be achieved benefit not only the owner but also the tenant. On a wider scope, because most buildings are heated and cooled using fossil fuels, the negative impact on the environment lessens if less fuel is used.

Heat transfer occurs in three ways: (1) conduction; (2) convection; and (3) radiation. Conduction is the transfer of heat through a material. Metal and concrete, for example, are better conductors of heat than wood, making them less effective insulating materials compared to wood. Because wood is less dense than concrete or steel (and it also has more air voids), it has better insulating properties than either steel or concrete. On the other hand, insulating materials such as expanded polystyrene or fiberglass batts have superior insulation values when compared to wood. Good insulating materials have a high percentage of air voids and physical properties that restrict the movement of air through them. Convection is the transfer of heat through the movement of air. Cold air is denser than warm air, and so air rises when heated. This movement of air makes the room feel cooler than it really is. Air movement such as this will occur within the cavity of a wall system that is not properly sealed and results in heat loss that can also contribute to moisture problems within the wall cavity.

Radiation is the transfer of heat energy by absorption. Darker materials absorb more energy than light materials. A shiny surface will reflect more energy than it absorbs. This concept can be used to contain heat energy within a contained space or it can be used to reflect unwanted heat energy. Special coatings on window glass used in buildings, for example, block out heat energy from the sun and reduce the amount of cooling required within a building. These coatings can also prevent the loss of heat through the window on cold days. An insulation that is covered with a reflective surface is capable of controlling all three forms of heat loss.

*Thermal conductivity* is defined as that property that determines the quantity of heat that will flow through a unit of area of a material based on a temperature difference of one degree. Units for thermal conductivity are Btu·in./hr·ft$^2$·°F (in metric units this becomes watts per metre per degree Celsius, or W/m·°C). The *thermal conductance, C,* of material is defined as the amount of heat in watts transmitted through a thickness of material or a combination of materials over a unit area producing a temperature difference of one degree Fahrenheit and is calculated by dividing the thermal conductivity by the material thickness. The resulting units become Btu/hr·ft$^2$·°F (in metric units this becomes watts per metre squared per degree Celsius, or W/m$^2$·°C). The less heat a material allows to pass, the better the insulating quality of the material.

The resistance to heat flow through a material, or *R* factor, is the inverse of the conductance (1/*C*). Each material has its own particular *R* value, and the greater the *R* factor, the better the insulating qualities of the material. Typical *R* values are listed in Table 15-1 for some of the more common materials used in construction. To differentiate the value for the resistance to heat flow of a material (*R*), used with standard units, the symbol *RSI* is often used for metric units. The resulting units for thermal resistance *R* become hr·ft$^2$·°F/Btu. In metric, the units for *RSI* are metres squared degrees Celsius per watt, or m$^2$·°C/W.

The heat transmission coefficient, or *U* factor, is defined as the amount of heat, in Btu, transmitted through 1 square foot of a building section (roof, wall, or floor system) in one hour for each degree Fahrenheit of temperature difference between air on the cold side and air on the warm side. Expressed

**Table 15-1**
**TYPICAL MATERIAL THERMAL RESISTANCE FACTORS***

| Material | *RSI* Factor | *R* Factor |
|---|---|---|
| 1 in. (25 mm) rigid glass fiber | 0.770 | 4.40 |
| 1 in. 25 mm polystyrene | 0.850 | 4.86 |
| 1 in. (25 mm) polyurethane | 1.030 | 5.88 |
| ½ in. (12.5 mm) gypsum wallboard | 0.078 | 0.45 |
| 3½ in. (90 mm) fiberglass batts | 2.100 | 12.00 |
| 8 in. (200 mm) concrete block (lightweight aggregate) | 0.352 | 2.00 |
| 4 in. (100 mm) clay brick | 0.074 | 0.42 |
| ¾ in. (20 mm) stucco | 0.028 | 0.15 |
| Outside air surface at 15-mph (24-km/hr) wind | 0.030 | 0.17 |
| Inside air, vertical surface, heat flow horizontal | 0.120 | 0.68 |
| Vertical air space, ½ in. (13 mm) minimum, heat flow horizontal | 0.171 | 0.97 |
| Double glazing, ½ in. (13 mm) air space | 0.360 | 2.04 |

*Values obtained from standard reference documents such as the American Society of Heating Refrigeration and Air-conditioning Engineers (ASHRAE) *Handbook of Fundamentals* and from data of tests done by the National Research Council of Canada.

in metric units, the heat transmission coefficient, or $U$ factor, is the amount of heat, in watts, transmitted through 1 square metre of a building section for each degree Celsius of temperature difference between air on the cold side and air on the warm side. The overall $U$ factor of a building section is calculated by summing the individual $R$ factors of the various components in the building section and taking the inverse of the total. In equation form,

$$U_{\text{factor}} = \frac{1}{R_{\text{total}}}$$

---

***Example:***

A building wall section is composed of the following materials:

- 4-in. face brick
- 2-in. rigid glass fiber insulation
- 8 in. hollow concrete block
- ½-in. drywall

Calculate the overall heat transmission factor, $U$, for the given wall.

***Solution:***

First, determine the total resistance using the $R$ values given in Table 15-1.

| | | |
|---|---|---|
| Outside air | = | 0.17 |
| 4-in. face brick | = | 0.12 |
| 2 in. rigid insulation | = | 8.80 |
| 8-in. concrete block | = | 2.00 |
| ½-in drywall | = | 0.45 |
| Inside air | = | 0.68 |
| $R_{\text{total}}$ | = | 12.52 |

The $U$ value for the wall is the inverse of the $R$ value or

$$U = \frac{1}{R_{\text{total}}} = \frac{1}{12.52} = 0.080$$

---

***Example:***

A building wall section is composed of the following materials:

- 100-mm face brick
- 50-mm air space
- 200-mm hollow concrete block
- 12.5-mm drywall

Calculate the overall heat transmission factor, $U$, for the given wall.

***Solution:***

First, determine the total resistance using the $RSI$ values given in Table 15-1.

| | | |
|---|---|---|
| Outside air | = | 0.030 |
| 100-mm face brick | = | 0.074 |
| 50-mm air space | = | 0.171 |
| 200-mm concrete block | = | 0.352 |
| 12.5-mm drywall | = | 0.078 |
| Inside air | = | 0.120 |
| $RSI_{\text{total}}$ | = | 0.825 |

The $U$ value for the wall is the inverse of the $RSI$ value or

$$U = \frac{1}{RSI_{\text{total}}} = \frac{1}{0.825} = 1.212$$

---

If the 50-mm air space is filled with 50 mm of polystyrene insulation, the $RSI$ factor is increased from 0.171 to 1.70, giving the wall a total $RSI$ factor of 2.354 and a resulting $U$ factor of 0.425, making the wall almost three times more resistant to heat loss.

A well-constructed and well insulated wall will ensure that the dew point temperature occurs at a point in the wall section where the resulting moisture can be eliminated. The best scenario, of course, is to have the wall constructed in such a fashion so that the dew point is never reached. However, this may not be practical. The use of an appropriate arrangement of *insulation, air vapor diffusion retarder,* and *air barrier,* along with the use of mechanical equipment to control humidity levels and air pressures, is the current approach to obtaining optimum thermal efficiency within the building. Through hard experience, it has become evident that the *wide brush approach* to building envelope design is no longer adequate. Experts in the field of building science, such as the Building Science Corporation (Westford, MA), through on-site testing and evaluation, have developed strategies for workable building envelopes based on various climate conditions, materials, and construction techniques. They, as well as others in the field of building science, continue to provide workable details for designers and professional support to the construction industry in the evaluation of existing structures.

An important requirement of a well-constructed building wall is to minimize warm moist air from entering the wall, cooling and condensing, and potentially causing damage to the wall components. The temperature at which condensation occurs is known as the *dew point.* The dew point can be defined as the temperature at which air, having a certain relative humidity, will condense when cooled. Good building envelope design dictates that any moisture entering a wall cavity must be allowed to escape by drying or by drainage. The ideal situation, of course, is to keep the wall cavity perfectly dry. However, in reality, this is usually difficult to achieve. Figure 15-14 illustrates the use of insulation in two different locations within the wall cross section as possible options. Note that no additional air vapor diffusion retarder or air barrier has been used.

Wall 1 has the insulation on the warm side of the wall, whereas Wall 2 has insulation on the exterior face of the masonry block. Note the range of temperature gradients between summer and winter conditions which the wall materials must endure. The brick veneer serves as a rain screen and the air space between the brick and block serves as the drainage plane. In each case, the amount of insulation ensures that the dew point will occur on the cold side of the insulation. In the case of Wall 1, the potential for condensation is at the interface between the insulation and the concrete block wall. Assuming a batt-type insulation is used, moisture passing through the insulation could condense on the cold side of the insulation. If an air vapor diffusion retarder is placed on the warm side of the insulation, as is the case for walls in cold climates, the flow of moisture from the warm side to the cold side will be minimized. Any moisture caught within the wall must pass through the block to escape. If an air barrier is used on the exterior face of the concrete block, it must also be permeable to facilitate the drying process.

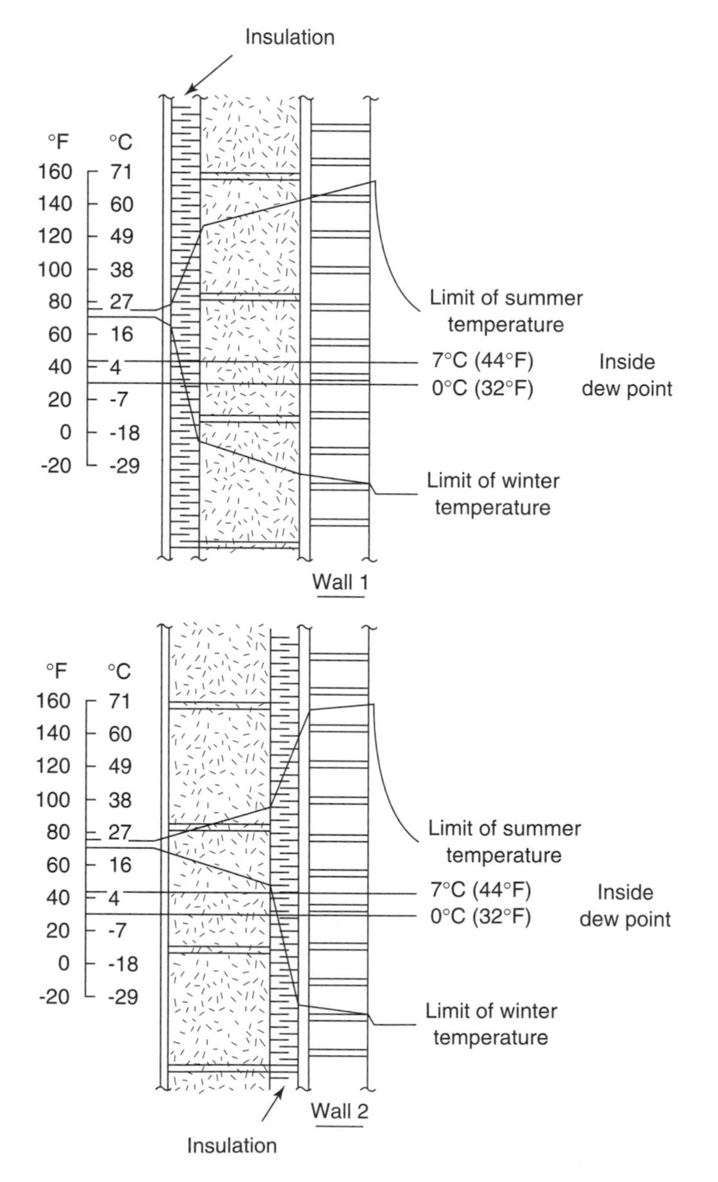

**Figure 15-14** Temperature Gradients in Insulated Walls.

In the case of Wall 2, the dew point occurs on the exterior face of the insulation. Drying is more easily attained in that the air vapor does not have to pass through any other material to get to the wall cavity where removal of any moisture by drying or drainage can occur. In this instance, the block wall (if sealed) would act as the air vapor retarder, restricting the flow of water vapor. Now suppose that an air barrier were to be applied to the exterior face of the insulation to control airflow due to infiltration. Although it would resolve the airflow problem, it would be the surface on which condensation could occur and bring about potential problems within the wall.

The foregoing is just one example of the wide variety of climatic-imposed conditions that can occur within wall assemblies and demonstrates the need for a comprehensive approach to their design. If the effects of mechanical air conditioning are imposed onto the climatic conditions, further evaluation of the wall assembly must be made to insure that moisture buildup does not occur in an inappropriate location within the wall cross section. Buildings that are colder on the inside than on the outside, such as refrigeration plants, ice arenas, and air-conditioned buildings in very hot climates, experience conditions that are the reverse of buildings in cold climates. It quickly becomes evident that the choice of materials, construction details, and a high level of workmanship all play a part in achieving a workable and durable building envelope.

## Insulation Application

1. *Perimeter insulation.* Buildings constructed on surface foundations are subject to perimeter heat loss and footing movements. Perimeter insulation helps control this heat loss and protects the footings from frost action. Figure 15-15 illustrates two methods using rigid insulation to prevent frost penetration under footings of shallow foundations. Polystyrene foam is preferred for use as perimeter insulation because of its

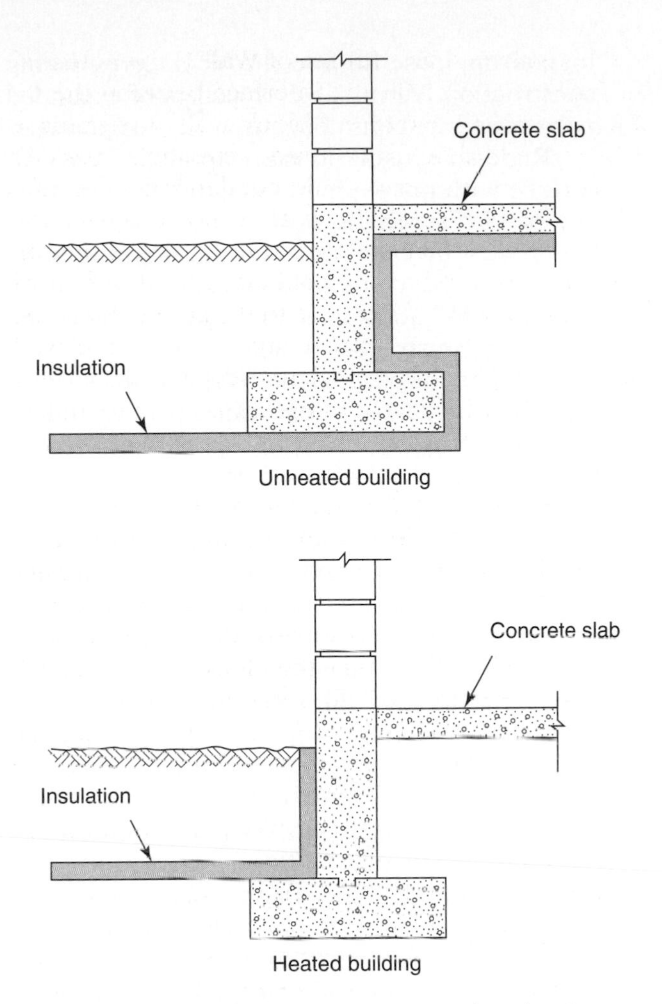


**Figure 15-15** Shallow Footings Protected against Frost Action.

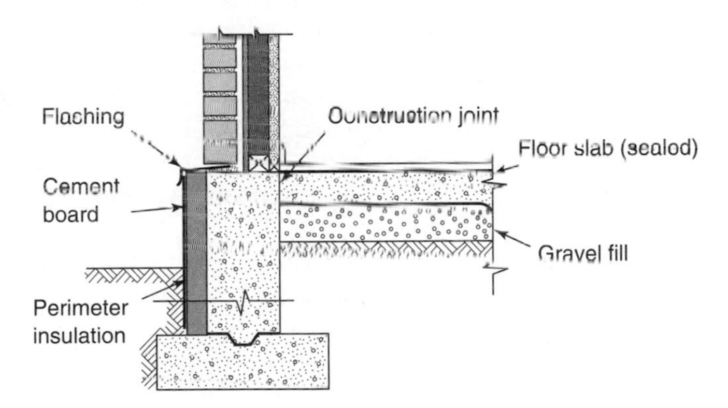


**Figure 15-16** Perimeter Insulation Outside.

high thermal resistance, low water absorption, and high compressive strength.

Insulation around the perimeter of a building may be placed on the inside or the outside face of the foundation. When applied to the outside (Figure 15-16), the portion above grade must be protected from physical damage with a covering of treated wood, stucco on metal lath,

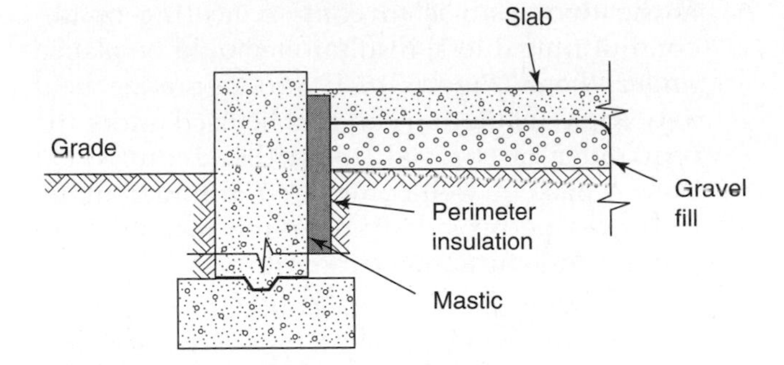


**Figure 15-17** Perimeter Insulation with Slab on Grade.

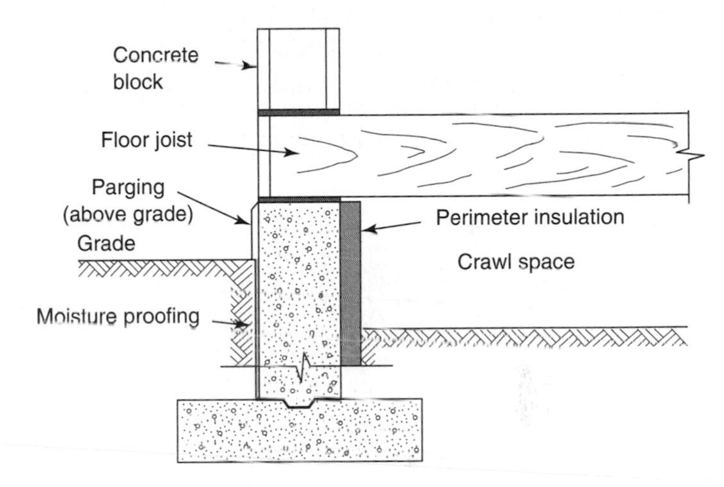


**Figure 15-18** Perimeter Insulation in Crawl Space.

or asbestos-cement board. If polystyrene foam is used, it must be protected from the sun's rays to prevent deterioration. Many codes recommend that the insulation be carried a minimum of 2 ft (600 mm) below finished grade, but better results can be obtained by carrying the insulation down to the top of the footings. When slab on-grade construction is involved, the perimeter insulation should be installed prior to backfilling of the foundation trench. The insulation should be carried down to the frostline or down to the top of the footing (Figure 15-17). Notice that the insulation reaches the top of the floor slab. If the area under the floor slab is to be used as a crawl space, several steps are necessary to install perimeter insulation. If the area under the floor slab is to be used as a crawl space, the insulation must cover the entire surface of the crawl space walls to at least the frost line (see Figure 15-18), preferably down to footing level. Ventilators, either natural or mechanical, are usually installed to provide air circulation in the crawl space. However, the type of venting and the amount of venting will depend on building conditions such as heating requirements and insulation of the floor assembly.

If the floor slab is to contain heating or air-conditioning ducts, insulation should be placed under them (Figure 15-19) to minimize heat loss. Rigid insulation can also be used under the remainder of the slab to further reduce heat loss. A base of well-compacted granular material must be provided under the insulation to prevent moisture migration into the slab.

2. *Foundation wall insulation.* Foundation walls are usually insulated with some type of rigid insulation, including insulating boards and rigid slabs or blocks. The insulating boards include wood and cane fiberboards, straw board, and laminated fiberboard. Rigid slab insulations include cellular glass, foamed plastic rigid insulation (see Figure 15-20), and cellular concrete. Additional insulation may be introduced into concrete block walls by pouring loose fill into the block cores during construction. Vermiculite or pellets of expanded polystyrene are commonly used for this purpose.

Rigid slab insulations are applied to the wall surface with hot asphalt, portland cement mortar, special synthetic adhesives, or wire clips. Walls must be clean and smooth before insulation is applied to ensure a good bond. It may be necessary to backplaster the wall to achieve the necessary finish.

If asphalt is used as the adhesive, the surface must first be coated with asphalt primer and allowed to dry. Adhesive is then applied to the insulation by dipping the back face and two adjacent edges of each block in hot asphalt. Blocks are applied to the primed surface while the asphalt is still molten. Cellular glass and cellular concrete may be applied by the hot asphalt method.

When synthetic adhesive is specified, the back face and two adjacent edges of each slab are given a trowel coating of adhesive approximately ⅛ in. (3 mm) thick. Sufficient pressure must be applied to the blocks to ensure tight joints and good contact with the wall. Both cellular glass and foamed plastic may be applied with synthetic adhesives.

Rigid slab insulation may be applied to block or brick walls by either of the methods just described or may be held in place by wire clips, one end of which has been embedded in a mortar joint (Figure 15-21). The edge joints should still be sealed with hot asphalt or some other adhesive.

Modified Portland cement mortars are used for adhering foamed plastic insulation to masonry walls where the insulation is part of an exterior

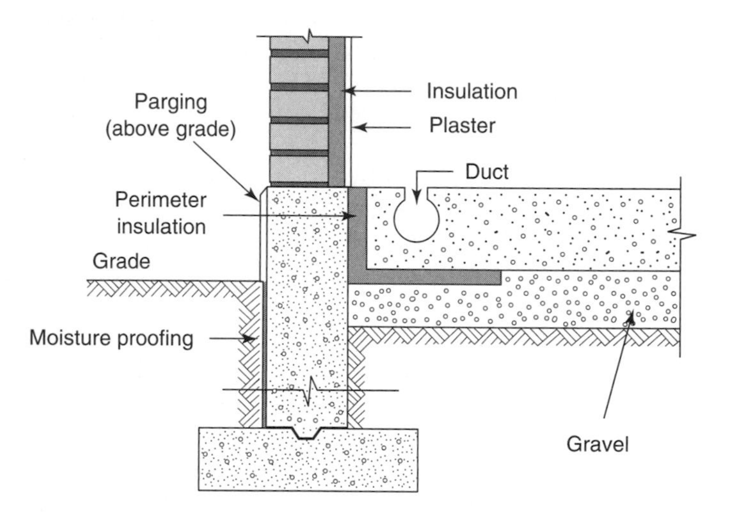


**Figure 15-19** Perimeter Insulation with Slab Ducts.

**Figure 15-20** Foamed Plastic Rigid Insulation on Foundation Wall.

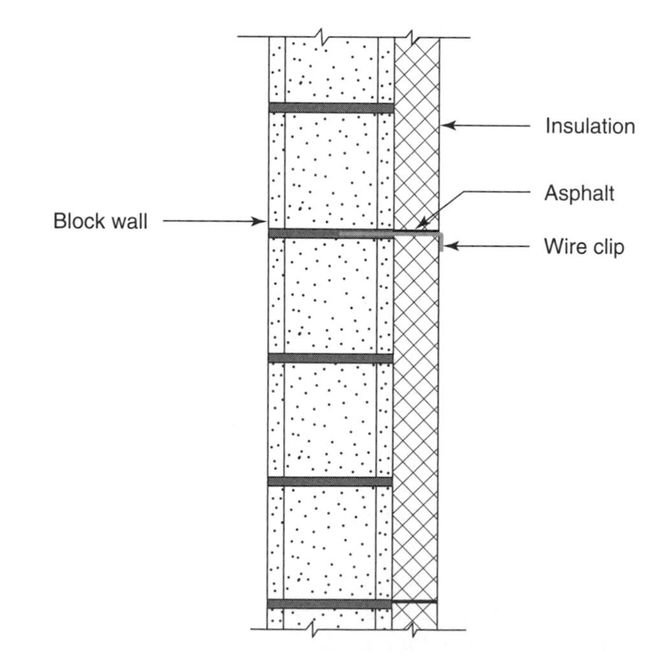


**Figure 15-21** Insulation Held by Wire Clips.

**Figure 15-22** Foundation Wall Strapped for Insulation.

finish, such as stucco. Normal Portland cement mortars are used when cellular concrete is the insulating medium. The application of the mortar is by hand trowel or it can also be sprayed on. Normal time for such adhesives to set is 24 hours.

Rigid boards may be applied to foundation walls in several ways. One method is to strap the wall, as shown in Figure 15-22, and attach the board insulation to the straps with nails. If the foundation wall is cast in place, insulation may be attached inside the inner form face. This may be done with nails or, if the boards are thick, with wires through the form sheathing and insulation. The concrete adheres to the face of the insulation, and the insulation remains permanently attached to the wall when the forms are removed. The hot asphalt or the synthetic adhesive method may also be used to secure board insulation to foundation walls.

3. *Curtain wall insulation.* Masonry curtain walls may be insulated by any of the methods applicable to foundation walls. Brick or block curtain walls are often built as cavity walls, in which case insulation may be installed within the cavity. It may be loose fill, such as vermiculite, perlite, cellulose fiber, or gypsum fiber, or it may be rigid slab insulation thick enough to fill the cavity. Care should be taken that mortar does not protrude from joints so as to restrict the flow of insulation being poured or blown into place. When using slabs, insulation must be installed as construction proceeds, because the metal ties between the inner and outer wythes of the wall should occur at joints in the insulation.

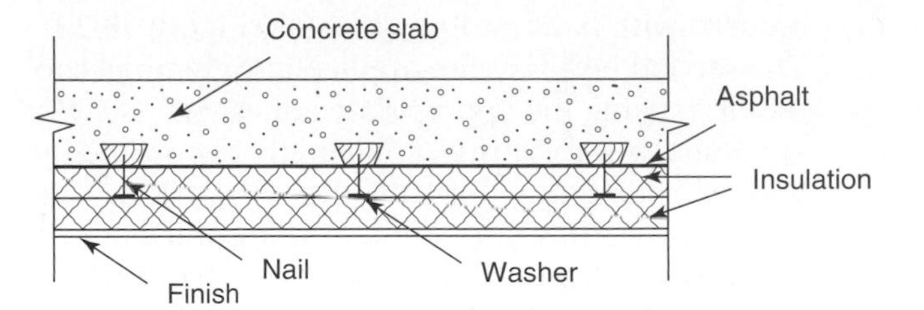


**Figure 15-23** Insulation on Solid Ceiling.

Another insulating technique for masonry walls involves the use of cellular concrete blocks. These are used to build a backup wall, which is faced with a veneer of brick, tile, terracotta, or stone.

Curtain walls having a stud frame may be insulated with batts placed between the studs, with rigid insulation applied over plywood or other sheathing, or by *foamed-in-place* insulation. Insulation is foamed in place by placing a specified amount of plastic resin and activating agent in each space to be insulated. The two materials react to form a thermosetting, insulating foam that will fill each space to a predetermined depth. The operation is repeated until each space is completely full.

Prefabricated panels for curtain wall construction are often manufactured with the insulation built in as the core of a sandwich. Such walls usually require no further insulation. If single-skin panels are used, provision must be made for applying the insulation to the back of the panels or to a rigid sheath erected behind them.

Insulation for glass curtain walls is provided by the use of sealed units. In the case of glass blocks or tiles, the two halves of each unit are fused together, enclosing a dry, low-pressure air space. Sheets or plates are hermetically sealed around the edges, in pairs, leaving a space between them.

4. *Ceiling insulation.* Insulation may be applied directly to the underside of floors, or a suspended ceiling system may be employed. In either case, many kinds of insulation may be used, including insulating boards, rigid slabs, or sprayed-on insulation consisting of vermiculite plaster or cellulose.

When insulation is to be applied under concrete floors, nailing strips should be cast into the bottom of the slab on appropriate centers for the type of insulation being used (Figure 15-23). The surface must be primed, and the first layer of insulation is applied at right angles to the nailing strips, using hot asphalt or an approved adhesive on the contact face and edges. This layer is further secured with nails and washers (see Figure 15-23). The second layer is then applied at right angles to the first, using the appropriate adhesive.

Suspended ceilings are usually supported by hanger rods from the roof, floor frame, or the slab above. They support a system of T-bars or metal pans, which in turn support the rigid insulation (Figure 15-24) or the material that will act as a base for sprayed-on insulation. The space above the insulation must be thoroughly ventilated by free or forced circulation of air.

T-bars of steel or aluminum are suspended on 18- to 24-in. (450- or 600-mm) centers by rod hangers placed close enough together to prevent any deflection of the completed ceiling. The first layer of rigid insulation is laid between the T-bars after they have been coated with asphalt priming paint. The second layer is applied on top of the first, using a flood coat of hot asphalt or a coating of approved adhesive on the contact surface and edges of each slab. The insulation should be laid in parallel courses, staggering joints between courses and between layers. Finally, the upper surface of the insulation should receive a flood coat of hot asphalt or a ⅛-in. (3-mm) coating of asphalt emulsion. The hanger rods should also be insulated to a distance of approximately 18 in. (450 mm) from the surface of the insulation (see Figure 15-24) with at least half the thickness of insulation used on the ceiling.

When sprayed-on insulation is specified, the T-bar or metal pan support system may be used to carry some type of rigid board or gypsum lath (see Figure 15-25). Another alternative is to suspend a system of runner channels and furring channels and to wire metal lath to the underside of this framework (see Figure 15-26). Sprayed-on insulation may be applied to either of these two bases. Figure 15-27 illustrates the application of sprayed-on insulation on a concrete ceiling.

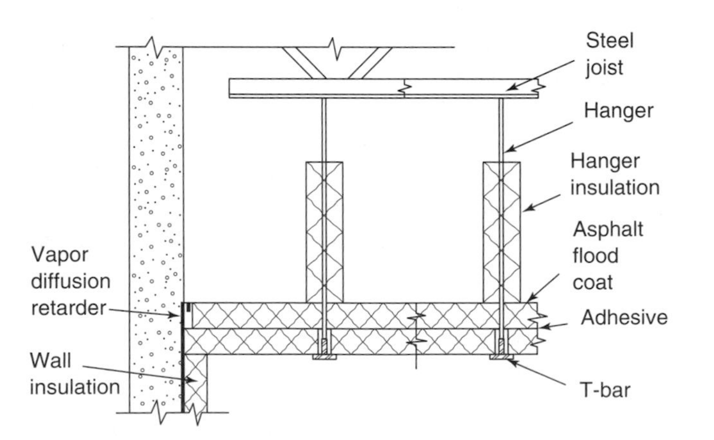

**Figure 15-24** Suspended T-Bar Ceiling.

Framed ceilings may be insulated with batt insulation between the framing members, and any rigid ceiling with accessible space above it may be insulated with loose fill poured or blown into place.

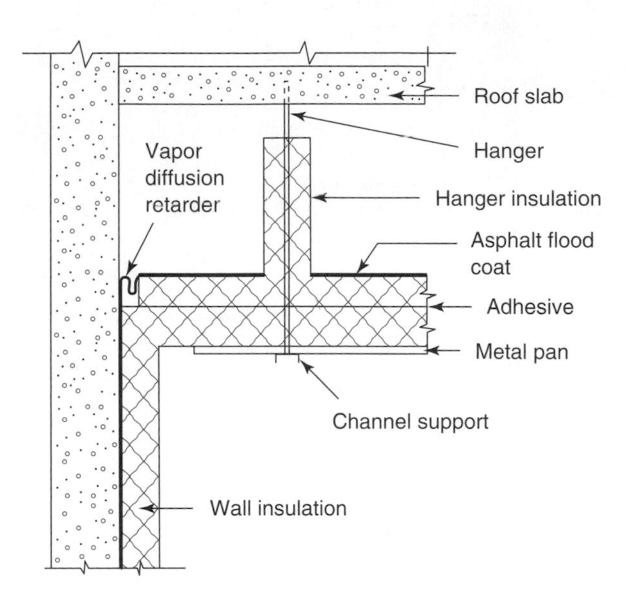


**Figure 15-25** Suspended Metal Pan Ceiling.

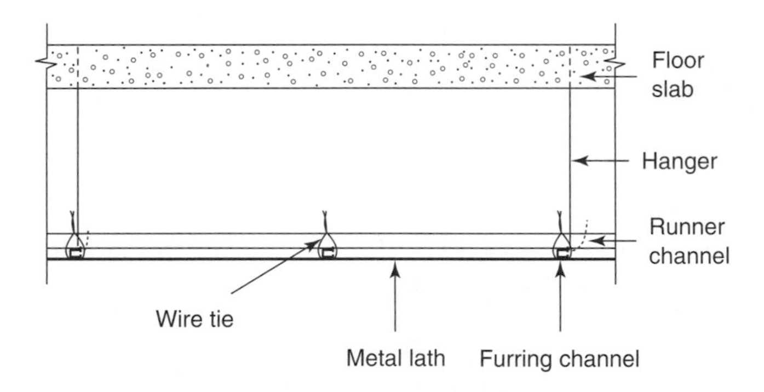


**Figure 15-26** Suspended Ceiling Frame.

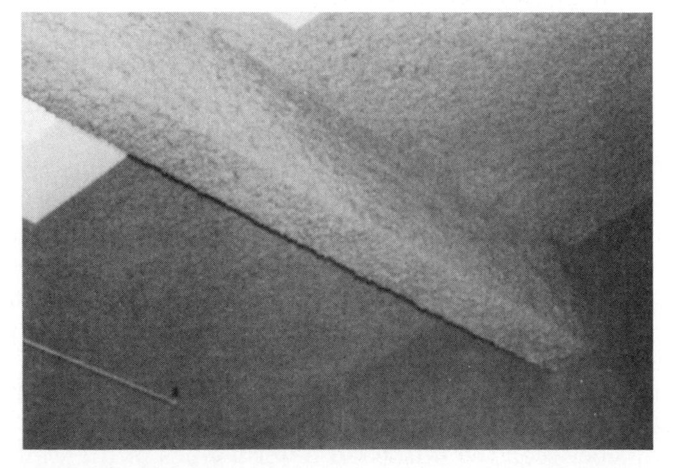

**Figure 15-27** Sprayed-On Insulation on Concrete Ceiling.

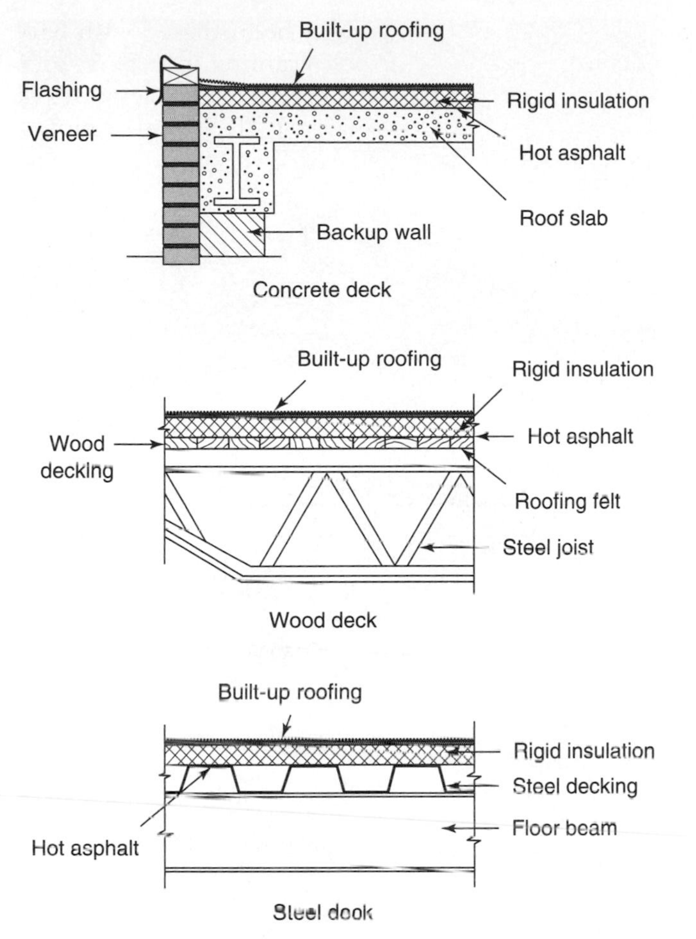


**Figure 15-28** Insulation Applications to Roof Deck.

5. *Roof insulation.* Roof decks may be insulated with any type of rigid insulation, including cellular glass, rigid foamed plastic, straw board, fiberboards (usually double or triple laminated), and cellular concrete. With concrete, gypsum, precast slab, or steel decks, no undercoating is normally required, but sheathing paper or roofing felt is nailed over a wood deck (see Figure 15-28).

   To obtain a proper bond between insulation and decking, all surfaces must be dry before and during the application of the insulation. It is applied to the prepared surface by embedding it in hot asphalt, used at the rate of approximately 25 lb/sq ft (12 kg/10 $m^2$). On steel decks, the long dimension of the insulation boards should be parallel to the ridges of the deck, with the edges of the boards resting on the ridges.

### Use of Reflective Insulation

Reflective insulation may be produced from any metallic substance that has the ability to reflect infrared rays. Aluminum foil has proved to be the most practical and economical material for this purpose. It is produced in several forms, including foil-backed gypsum lath, rigid insulation with foil back, foil laminated to kraft paper, corrugated paper with foil on both sides, and plain sheet foil in rolls.

To be effective, reflective insulation must have an air space of at least ¾ in. (20 mm) in front of it. Figure 15-29 illustrates a few typical applications of foil insulation in which the air space is provided.

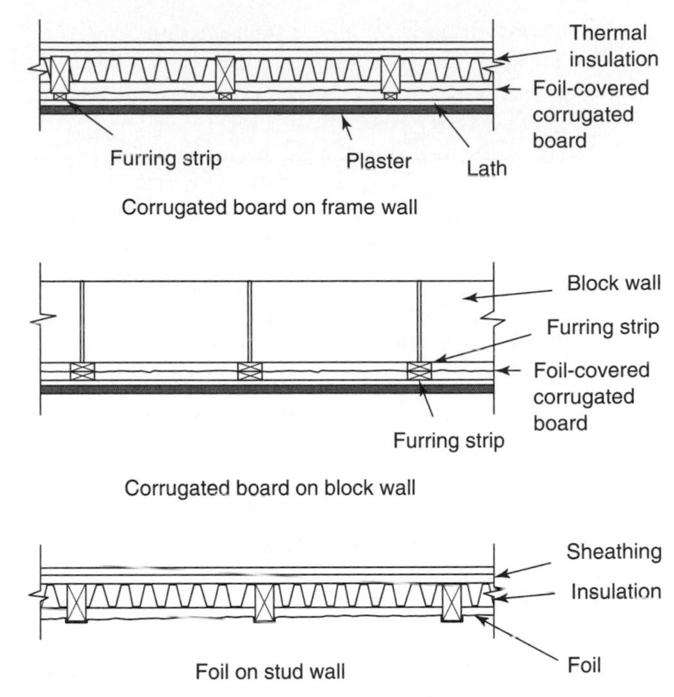


**Figure 15-29** Typical Applications of Reflective Insulation.

## SOUND INSULATION AND CONTROL

There are two main problem areas concerning sound and its control in old and new buildings. One is the improvement of hearing conditions and the reduction of unwanted noise in any given room, and the other is the control of sound transmission from one room to another through walls, floors, and ceilings.

Sound waves travel through the air in the form of small pressure changes alternately above and below the normal atmospheric pressure (see Figure 15-30). The average variation in pressure above and below the normal is called *sound pressure,* which is related to the loudness of a sound. A sound wave is one complete cycle of pressure variation, as illustrated in Figure 15-30.

The number of times this cycle occurs in 1 sec is the *frequency* of the wave, and the unit of measurement of frequency is called a *hertz,* abbreviated *Hz,* which represents one cycle per second.

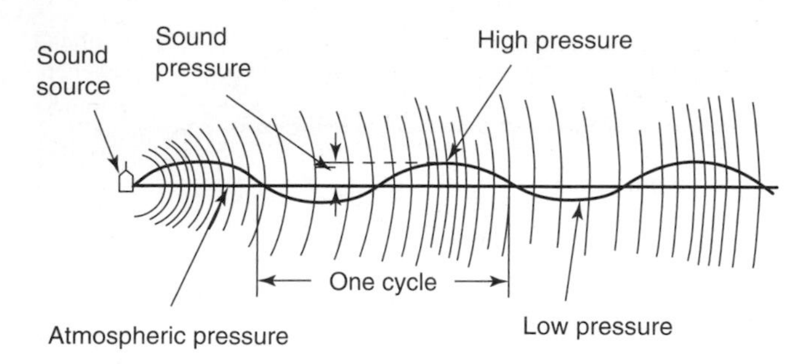


**Figure 15-30** Diagram of a Sound Wave.

The improvement of hearing conditions and the reduction of unwanted noise are accomplished by the proper design of inner walls and ceilings and control of the amount of sound reflected from walls and ceilings. Reflected sound causes reverberations or echoes, which often distort the original sound.

The control of reverberations is accomplished by the use of products that have a much greater ability to absorb sound waves than most building materials. The amount of sound absorbed is measured in *sabins,* 1 sabin being equal to the sound absorption of 1 sq ft of perfectly absorptive surface (in practice, no such surface exists). The fraction of sound energy absorbed, at a specific frequency, during each reflection is called the *sound absorption coefficient* of that surface. Some surfaces are primarily reflective (e.g., glass, concrete, masonry), absorbing perhaps 5% or less of the sound energy. Such materials have sound absorption coefficients of 0.05 or less. On the other hand, some acoustical materials can absorb 90% or more of the sound energy and so have a coefficient of 0.90 or better.

Because most sounds contain a range of frequencies, it is necessary to use an average of the absorption coefficients when considering sound absorption. A means of comparing reductions for noise that is mainly in the middle frequencies has been developed. It is called the *noise reduction coefficient* (NRC) and represents the average amount of sound energy absorbed over a range of frequencies between 250 and 2,000 Hz.

The exact amount of sound absorption required for various rooms is difficult to state because of individual differences in noise interpretation, but three general guidelines may be used:

1. In larger rooms with low ceilings and widely spaced, medium-density noise sources such as general offices, the *minimum average absorption coefficient* after acoustical treatment of all surfaces should be 0.20.
2. In smaller rooms containing closely spaced or high-level noise sources such as business machines, the *maximum average absorption* coefficient after treatment should be 0.50.

**Figure 15-31** Acoustic Panels Suspended from Roof Framing in a Large, Open Area.

3. To create a definite, noticeable improvement in noise reduction, the average absorption coefficient before treatment, or the *total room absorption,* should be increased at least three times. However, the application of this rule is subject to the minimum and maximum values of the first two rules.

The ceiling shown in Figure 15-31 is one example of how part of a room may be designed to affect acoustics.

## Sound-Absorbing Materials

Sound-absorbing substances are called *acoustic materials* and can generally be classified in three groups: *acoustic tiles, assembled acoustic units,* and *sprayed-on acoustic material.*

Acoustic tile is intended primarily for ceilings but may be used on upper wall surfaces as well. When used over a solid surface, such as plaster, plywood, or hardboard, the surface must be smooth, even, and clean. Tiles may be secured with nails or

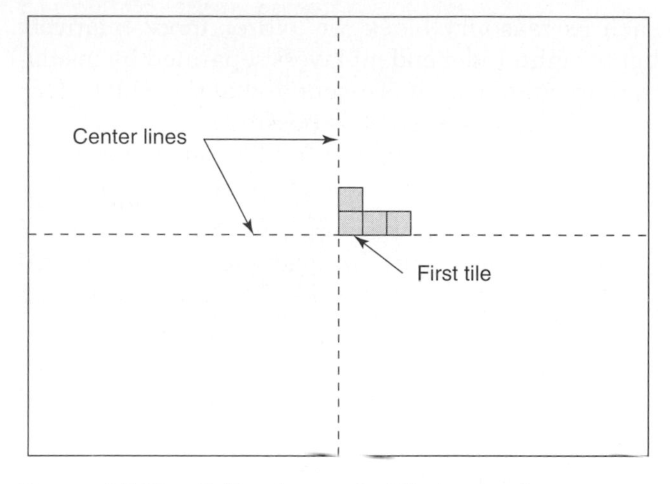


**Figure 15-32** Ceiling Layout for Tile Application.

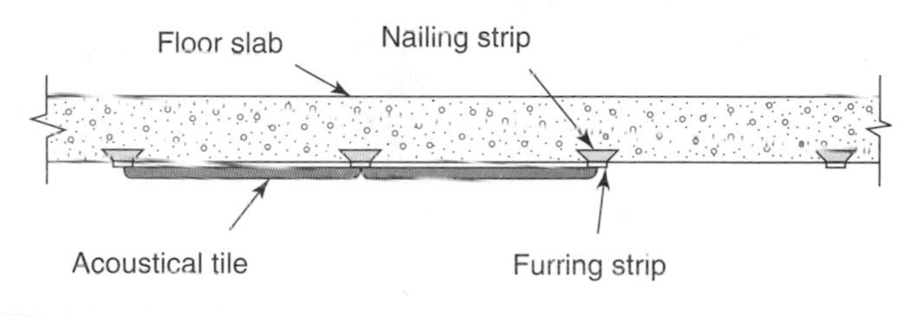


**Figure 15-33** Acoustical Ceiling Tile on Furring.

adhesive, depending on the type of backing used. In any case, the two center lines of the ceiling should be drawn first, and tiling should begin at the intersection of these lines (see Figure 15-32). If adhesive is to be used, some should first be rubbed on the corners of each tile to act as a prime. A walnut-sized dab of adhesive is then applied to each corner. The tile should be placed close to its final location and forced to slide into place to spread the adhesive more completely and ensure a good bond with the backing.

Tiles may be nailed to furring strips attached to the undersides of solid decks or framed ceilings (Figure 15-33). As usual, tiling should begin along a center line snapped across the furring strips.

Acoustic tile may also be installed in a suspended ceiling framework. Some tiles are made with square edges that rest on T-bars, whereas others fit over H-bars (Figure 15-34).

Assembled acoustic units consist of sound-absorbing material such as mineral wool and fiberglass insulation fastened to hardboard, asbestos board, or sheet metal facing that has been perforated to allow sound waves to penetrate. These units may be fastened to the wall face on furring strips, suspended in front of the wall or from the ceiling, or incorporated into the wall paneling.

Sprayed-on acoustic materials are widely used for sound control because of the relative ease of application on surfaces of almost any shape (see Figure 15-27). Vermiculite, perlite, and cellulose fiber mixed with adhesive are used for this purpose, and

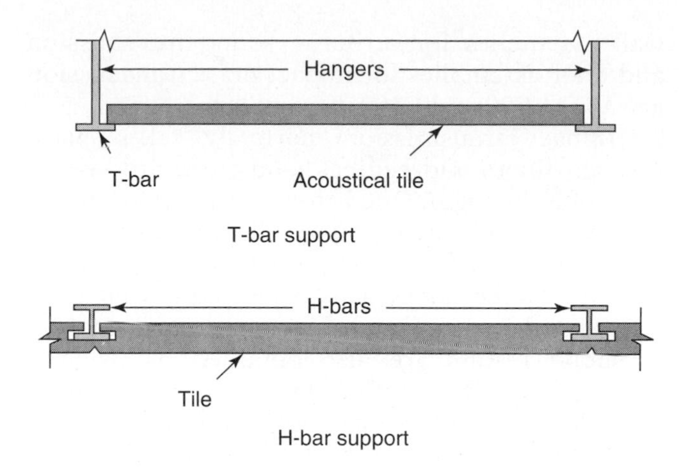


**Figure 15-34** Acoustical Tile in a Suspended Ceiling.

new, sprayed-on types of insulation that do not have an asbestos-fiber base. This is a result of safety requirements that place restrictions on the use of such materials as asbestos-fiber-containing toxic dusts.

Vermiculite plastic is a ready-mixed product requiring only the addition of water. It can be applied over any firm, clean surface such as plaster base coats, masonry, and galvanized metal. It is applied in two coats if no more than ½ in. (12.5 mm) is required or in three coats if a greater thickness is specified. The first coat should be approximately ⅜ in. (9.5 mm) thick, straightened with a darby, and allowed to dry. The second coat is then sprayed on to the desired thickness and texture. The temperature should not be less than 55°F (13°C) for a week prior to the application of the plastic, during its application, and for long enough after its application to allow it to set thoroughly. Adequate ventilation must be provided.

Asbestos-fiber insulation was a prepared material containing an adhesive mixed with the fiber. It was sprayed on any solid surface or on metal lath backing in several thin coats to a total depth of from ½ to ¾ in. (12.5 to 20 mm). The surface to be coated was first primed with an adhesive (except in the case of metal lath).

The ceiling shown in Figure 15-27 has been sprayed with acoustic insulation. Additional sound absorption within a room can be provided by the use of carpet and drapes.

## Sound Transmission Control

In nearly all types of construction, sound transmission from room to room takes place as a result of diaphragmatic vibration through walls, floors, or ceilings. The vibrations may be initiated by impact, such as a footstep, or by the action of sound waves striking the surface. One vibration is known as *impact noise transmission* and the other as *airborne sound transmission*. Standard tests used to evaluate wall assemblies for airborne sound transmission and floor assemblies for impact noise transmission are ASTM E90 and E492, respectively.

Impact transmission normally takes place through floors and ceilings, and there are several ways of reducing it. One is to cover the floor with a resilient material such as cork tile or a heavy carpet to absorb the impact. Another is to use a suspended ceiling between floors. This is particularly effective if resilient hangers are used.

Good results are also obtained by using a *floating* floor. The floating floor may be a separate 2-in. (50-mm) concrete slab, isolated from the structural slab by a 1-in. (25-mm) foamed plastic or paper-covered fiberglass quilt. It may be a wood floor nailed to sleepers that are supported on flat steel springs (Figure 15-35), or it may be a wood floor laid on sheets of resilient material such as fiberboard or foamed plastic (see Figure 15-36).

The sound insulating efficiency of a wall or floor assembly is designated as its transmission loss and is measured in decibels (dB). Based on this loss (and using standard methods of calculation such as those outlined in ASTM E413), each assembly can be assigned a sound transmission class (STC) rating. Typical STC rating values for floors and partition wall assemblies range from 35 to 70. An STC rating value of 50 is considered to be a minimum acceptable value for most building applications.

An acceptable STC rating in a wall can be achieved by using a single layer of dense material such as masonry block, or two or more relatively lightweight, independent layers separated by insulation. In either case it is essential that the wall be free of cracks and as airtight as possible.

Several types of wall assemblies can achieve STC ratings of 50 or better. Among them are the following:

1. A single-wythe masonry wall having a mass of 80 psf (390 kg/m$^2$), for example, 8 in. (200 mm) of brick or 7 in. (175 mm) of concrete, including plaster if any.
2. A masonry cavity wall consisting of two wythes having a mass of 20 psf (98 kg/m$^2$) each, held together with metal ties, and enclosing a 2-in. (50-mm) cavity.
3. A composite wall, consisting of a basic masonry wall having a mass of at least 22 psf (107 kg/m$^2$). For example, 4 in. (100 mm) of hollow clay tile or 3 in. (75 mm) of solid gypsum tile, with a sheath made of ½-in. (12.5-mm) gypsum lath mounted with resilient clips and plastered with ¾-in. (20-mm) sanded gypsum plaster (see Figure 15-37).
4. A stud wall consisting of 2 × 4-in. (38 × 89-mm) studs with ½-in. (12.5-mm) gypsum lath mounted on resilient clips on each side and plastered with ½-in. (12.5-mm) gypsum plaster. Mineral wool or fiberglass batts are placed between studs (see Figure 15-37).
5. A staggered stud wall composed of 2 × 4 in. (38 × 64 mm) studs set at 16-in. (400-mm) centers on a common 2 × 6 in.-(38 × 140-mm) plate; ½-in. (12.5-mm) gypsum lath nailed on both sides and plastered with ½-in. (12.5-mm) sanded gypsum plaster. Paper-backed fiberglass batts are hung between studs on one side only.

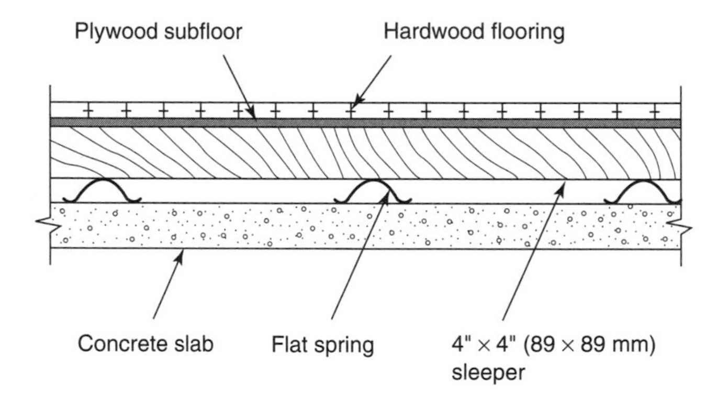


**Figure 15-35** Wood Floor on Springs.

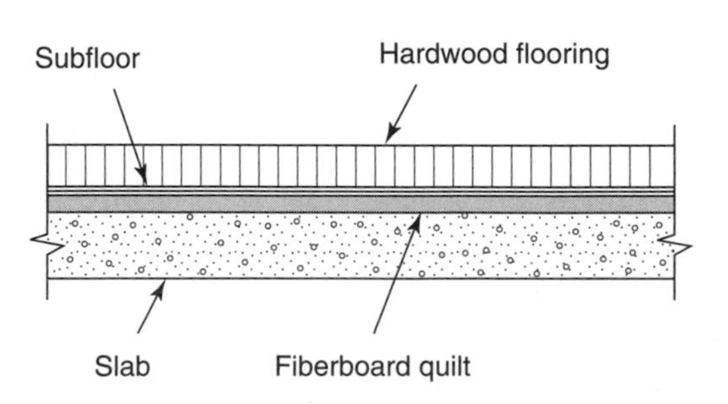


**Figure 15-36** Wood Floor on an Insulation Cushion.

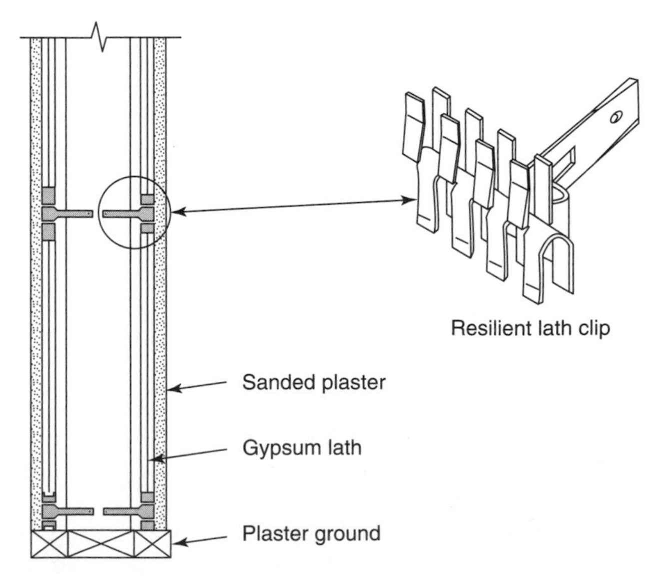


**Figure 15-37** Sound Insulation with Lath Clips.

Wall constructions that will produce STC ratings of from 45 to 49 include the following:

1. A single-wythe masonry wall having a mass of at least 36 psf (176 kg/m²), for example, 4-in. (100 mm) solid gypsum tile or a 4-in. (100-mm) brick wall including plaster, if any.
2. A composite wall similar to that discussed in the foregoing step Number 3 but with furring strips to support the gypsum laths.
3. A staggered stud wall faced with drywall, consisting of two sets of 2 × 3-in. (38 × 64-mm) studs at 16-in. (400-mm) centers on a common 2 × 4-in. (38 × 89-mm) plate. Two layers of ⅝-in. (16-mm) gypsum wallboard are applied to each face, the first nailed and the second cemented. The joints must be staggered and both surfaces must be sealed. A mineral or glass wool blanket is placed inside the wall cavity (Figure 15-38).

A single-wythe masonry wall having a mass of at least 22 psf (107 kg/m²) including plaster, if any, should have an STC rating of 40 to 44.

Walls with ratings of from 35 to 40 include the following;

1. A 2 × 3 or 2 × 4 in. (38 × 64 or 38 × 89 mm) stud wall with ⅜-in. (9.5-mm) gypsum lath and ½-in. (12.5-mm) sanded gypsum plaster on both faces.
2. A 2 × 3 or 2 × 4 in. (38 × 64 or 38 × 89 mm) stud wall with two layers of ⅜-in. (9.5-mm) drywall on each side, the first nailed, the second cemented, with staggered joints.

Floors that will have an STC rating of 50 or better include the following:

1. A 4-in. (100-mm) solid concrete slab or its equivalent, having a mass of at least 50 psf (240 kg/m²), plastered directly on the underside and covered with wood flooring on wood furring strips. Its impact rating, however, will not be more than 30.
2. A 4-in. (100-mm) concrete slab, as discussed in number 1, covered with a 1-in. (25-mm) quilt of foamed plastic or paper-covered fiberglass supporting a 2-in. (50-mm) concrete topping. The impact rating, again, will not be more than 30.

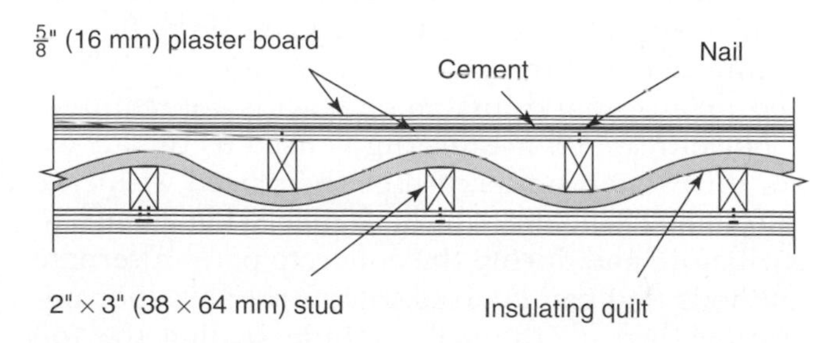


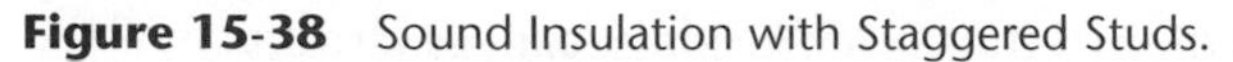
**Figure 15-38** Sound Insulation with Staggered Studs.

3. An open web steel joist framework covered by a paper-backed fiberglass or foamed plastic quilt supporting a 2-in. (50-mm) concrete topping. The ceiling consists of ½-in. (12.5-mm) gypsum lath mounted on resilient clips and plastered with ½-in. (12.5-mm) sanded gypsum plaster. As with the foregoing, this floor will also have an impact rating of not more than 30.

## MOISTURE CONTROL

The prevention and control of moisture penetration in building assemblies remains a constant challenge to building designers and builders. Moisture may migrate through building envelope assemblies in either a liquid or a vapor form. As a liquid, moisture penetration is usually from the exterior to the interior. However, in the case of water vapor, it can move in either direction depending on the relative conditions existing between the interior and exterior of the building. Differences in vapor pressure, temperature, and air pressure all influence the direction of water vapor flow. Because both forms of moisture can exist at the same time, methods that control moisture in a liquid form do not always serve as a solution for the control of moisture in a vapor form. Building envelope experts now conclude that some moisture will always be present within building assemblies in one form or another and building assemblies must be designed and constructed in a fashion that will allow for drainage and drying. Excess moisture within a wall assembly contributes to building material deterioration, the loss of insulation values, and the growth of mold.

Water vapor moves through building assemblies in two ways: (1) by vapor diffusion, and (2) by air movement. Water vapor diffusion can be defined as the passage of water in vapor form through a material. The amount of water vapor that passes through a material is related to the permeability of the material. The more permeable a material, the greater the amount of vapor that will pass through it—and vice versa. The best approach to the control of vapor diffusion through wall assemblies is the use of an *air vapor diffusion retarder.* These are materials that have low permeability and minimize the diffusion process. Materials commonly used as vapor diffusion retarders include polyethylene film, waxed kraft paper, foils of aluminum or other metals, and latex paint.

Air vapor diffusion retarders are normally applied on the warm side of the insulation and, to be most effective, they must be one unbroken surface over the area that they cover. Air vapor diffusion retarders produced in narrow widths such as waxed paper and aluminum foil should be applied vertically over wall studs with edges lapped on the studs and sealed with caulking. To facilitate the splicing of the air vapor diffusion retarder at the ceiling and floor, the ends should extend at least 6 in. (150 mm) under the ceiling finish and floor sheathing. Polyethylene film is usually available in sheets wide enough to reach from floor to ceiling, and it is the material of choice. Latex paint serves a dual purpose; it is decorative and also serves as an air vapor diffusion retarder.

Although the diffusion of water vapor through a wall assembly is an important consideration in building envelope design, the control of air movement through the assembly is equally important. Air movement through openings in a wall assembly is due to an air pressure difference across the wall section. Pressure differences can result from wind blowing against the side of the building (*infiltration*) or air conditioning equipment producing a positive pressure inside the building and forcing air through the building envelope to the outside (*exfiltration*).

In many applications, the air vapor diffusion barrier doubles as the air barrier to control exfiltration. To be effective as an air barrier, the vapor retarder must have sufficient strength to withstand air pressure differences due to mechanical systems and pressure due to wind. All edges of the air vapor diffusion retarder must be sealed for it to be totally effective. That includes all openings around electrical outlets, around mechanical piping and ducting, and windows and doors. All edges and laps must be completely sealed with caulking or adhesive to ensure that no air enters into the wall assembly.

To control air infiltration, air barriers commonly known as building wraps are applied under the exterior finish or siding as shown in Figure 15-39. Building wraps are composed of permeable materials such as asphalt impregnated paper and, more recently, perforated polyolefin membranes. These special membranes are air barriers only and are not intended as water barriers. Proper protection against ongoing contact with liquid water is a requirement when these membranes are used in a wall assembly. Manufacturers and local codes provide the necessary specifications and guidelines to ensure long-term performance of these new materials.

The prevention of infiltration of liquid water into a roof system should employ waterproof barriers composed of rubber, vinyl, or other impermeable membranes combined with a positive roof slope as the main line of defense. However, moisture in the form of water vapor may be trapped under the waterproof membrane, condense, and cause damage in the form of water stains on the underside of the roof. In extreme situations the amount of water that condenses can make the building interior uninhabitable for its intended use. Appropriate levels of insulation, the addition of an *air vapor diffusion retarder* and the regulation of air

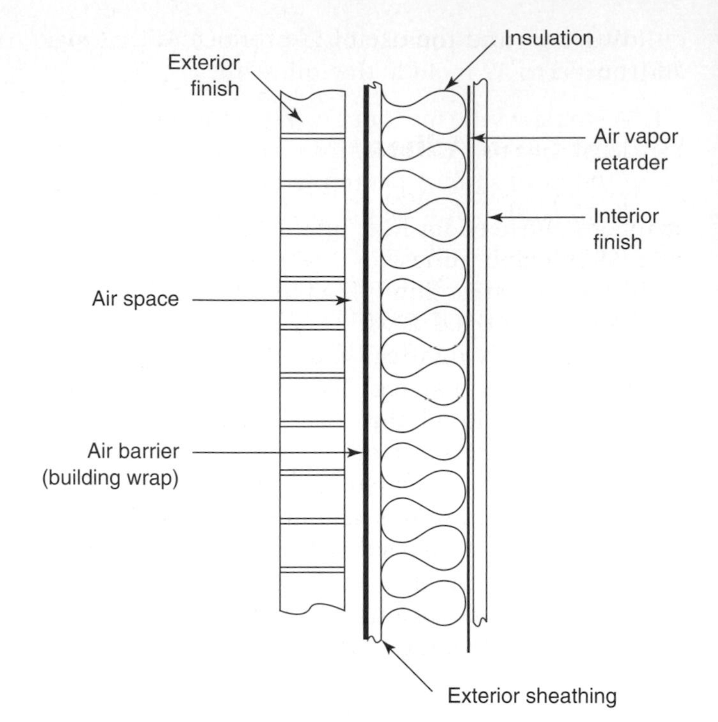


**Figure 15-39** Insulated Wall Section.

humidity levels can be combined to provide the required protection against such problems. Roofs are no less susceptible to condensation problems than are other parts of the building envelope, and corrective measures similar to walls are used to alleviate the problem. Waterproofing of walls below grade is usually done by applying some type of impermeable membrane to the exterior face of the wall using an appropriate adhesive.

Masonry and concrete walls below grade are dampproofed by the application with a coating of cutback asphalt or with a cement-based polymer-modified coating. Masonry walls should be smoothed with ¼-in. (6-mm) thick parging before the dampproofing is applied. Above ground level, masonry walls may be given a coating of transparent sealer to provide additional protection against water penetration.

Concrete slabs on ground may or may not need an air vapor diffusion retarder depending on their end use. Concrete floors sealed on the top surface with an impermeable floor finish require an air vapor diffusion retarder under the floor slab to control vapor migration and prevent moisture from being trapped under the floor finish. If the floor is unfinished, and the migration of moisture is limited, an air vapor diffusion retarder is not required. If polyethylene film sheeting is used as the air vapor diffusion retarder, it must be placed with care to minimize the amount of damage that it will incur before and during the concrete pour. Alternate methods that can be used to control air vapor migration through the slab include sealing the top surface of the concrete slab, proper drainage of the building site, and the use of granular material under the slab.

## Sealants and Adhesives

As building codes and standards become more stringent, the need for high quality, environmentally safe sealers and adhesives has never been greater within the construction industry. Chemical companies have developed a wide variety of compounds that are virtually unaffected by sunlight, moisture, and temperature changes. The various formulations exist in either liquid that can be poured or as a caulking for gun application. They come in one-part, two-part, or three-part formulations in liquid or powder form. Many sealants can be applied directly to the underlying substrates without a primer. To aid building material specifiers, designers, and tradespeople, the various manufacturers provide comprehensive technical support for sealant specification and application.

Typical applications include saw cuts in concrete floors (control joints), precast concrete panel joints, exterior insulation and finish systems, curtain wall joints, mullion joints, and glazing. The effective sealing around mechanical piping and ducting passing through fire rated walls, floors, roofs, and building envelope membranes is achieved by using special formulations that will retard flame spread during a fire. In modern building design and in the retrofitting of existing buildings, the application of these sealants has evolved as a separate trade.

Sealants composed of silicone formulations are used in structural applications such as in curtain wall construction as well as in building joints that are subject to movement. Silicone formulations double as sealants and adhesives and come in two-part and single-part formats. Silicone sealants can tolerate joint movement of ±50% and have excellent adhesion to nonporous materials such as glass, metals and plastic laminates.

Two-part epoxies provide a tough, abrasion resistant joint filler ideal for control joints in concrete floors. Special non-shrink formulations that can be poured easily are used as grout material under steel bearing plates resting on concrete foundations. Three-component polysulfide formulations are used for underwater conditions and in areas exposed to chemicals. Polyurethane sealants come in one-part and two-part formulations in a variety of grades that can be applied to almost any material. Typical applications are expansion joints in concrete floors and decks, industrial floors, and airport runways.

In general, all sealants and adhesives require some preparation of the substrate materials to ensure proper bonding. Joints must be free of deleterious materials such as concrete formwork release agents, surface dirt or rust, old surface treatments, and any other substances that may inhibit proper adhesion. Elements to consider when specifying a particular sealant are as follows:

- The type of substrate
- The amount of movement at the joint
- Exposure conditions and thermal ranges
- Joint loading conditions (structural or nonstructural application)
- Required curing, tooling, and tack-free times
- Color of the finished joint

Properties of the sealant, such as structural strength, modulus of elasticity, adhesion, cohesion, durability, elongation, and stress reversals can be tempered to ensure compatibility with specific materials, exposure conditions, and temperature ranges.

## Review Questions

**1.** What three concerns in building design can be dealt with by the use of insulation?

**2.** What property in a material determines its ability to insulate?

**3.** What two kinds of materials are recognized by fire codes for use in building assemblies?

**4.** Explain clearly why **(a)** cellular concrete is a good fireproofing material, **(b)** hot air ducts must be insulated, and **(c)** vermiculite is a good insulator.

**5.** Define the term *dew point* and explain its importance when considering the location and quantity of insulation in exterior walls.

**6.** Outline three important steps that must be taken when applying perimeter insulation to the exterior of a foundation wall and explain why the steps are important.

**7.** Outline the steps that must be taken to ensure the effectiveness of reflective insulation and explain why they are important.

**8.** Differentiate between **(a)** *airborne* transmission and *impact* transmission of sound, and **(b)** a *moisture* barrier and a *vapor* barrier.

**9.** Outline three methods of reducing impact transmission of sound through floors.

**10.** Determine the *U* factor of a wall having the following components: 4-in. brick, 2-in. of polyurethane insulation, 8-in. concrete block, metal studs with 3½ in. fiberglass batts, and ½-in. drywall.

**11.** Determine the *U* factor of a wall having the following components: stucco, 25-mm polystyrene insulation, 200-mm concrete block, 25-mm polystyrene insulation, and 12-mm drywall.

# 16 FINISHING

During the construction of the structural frame, a number of other elements in the building must be considered. Among these are the heating and ventilating, sewer and water, fire protection systems, and electrical requirements. In high-rise and retail structures, elevators and escalators must also be incorporated into the structural frame. Many good references are available on the various systems currently in use, and it is beyond the scope of this text to deal with the many details.

To ensure that these various elements are dealt with at the appropriate time, some method of job planning and control must be available. Job planning and scheduling can be accomplished in a number of ways, depending on the size and nature of the project. On small projects, the manager may plan the activities from day to day using only a simple bar chart and experience. On a large construction site, a whole group of people may be responsible for the planning and control of the job. Sophisticated methods such as the critical path method of scheduling used in conjunction with a computer allow for close monitoring of every aspect of the construction process.

As work on the structural frame progresses, the contractors who are responsible for the supply and installation of nonstructural items begin to arrive on the job site. Their arrival must be coordinated by the project manager to ensure that their part of the work can be done with a minimum amount of delay and interference with other personnel on the site. Office space, tool storage, and material storage areas must be available and ready for each subcontractor's arrival.

Finishing can be considered in two parts: exterior finishing and interior finishing. Exterior finishing can include cladding, exterior doors and windows, painting, exterior trim such as cornices, landscaping, sidewalks, and parking areas. Interior finishing includes numerous items such as partitions, wall finishes, floor finishes, stairs, ceilings, trim around doors and windows, cabinets, and fixtures.

Once the building frame has reached the point when the floor and roof are in place or, in the case of multistory buildings, several floor levels have been completed, work on the exterior curtain walls (Figure 16-1) and the interior partitions can begin. At this time the mechanical and electrical trades begin installing their portion of the work to ensure that the heating, air-conditioning, and lighting systems are in place before the wall panels and roof systems are in place and access to piping and ducting becomes limited.

Exterior finishing can begin as soon as the structural frame has advanced to the point when cladding, exterior glazing, and painting can be applied without conflict with the construction of the frame. The safety of the various trades personnel must be one concern of the project coordinator when scheduling the activities of the finishing subcontractors. Work on exterior finishing depends to some extent on weather conditions, and allowances must be provided in the scheduling for delays due to inclement weather conditions.

The amount of interior finishing done in a building will depend on the needs of the owner. In buildings that lease space, much of the interior may be left unfinished until a tenant specifies the amount

**Figure 16-1** Curtain Wall Framing Ready for Insulation and Interior Finish.

and type of finishing required. However, most buildings require some basic interior finishing, particularly in the areas open to the public. Lobbies must be finished to attract potential tenants; public washroom areas must be completed; stairwells, elevators, fire doors, and ceilings must be installed; and floors must be finished. When buildings offer other services, such as conference rooms, recreational areas, and retail outlets, these areas must be fully completed and equipped before the public can be admitted.

The selection of finishes is done by the architect during the design stage of the building. The choice of the exterior finish must be considered carefully, from both practical and aesthetic points of view. When considering the performance of the material, selection is usually based on guidelines such as availability, cost, durability, insulation value, maintenance, and cost of installation. When considering aesthetic value, color, texture, and compatibility with the surrounding environment must be considered.

Interior finishes must also complement the building interior and meet the requirements of the client. Materials selected must be durable and easily cleaned. Because most buildings are a mini-environment in themselves with their own air-circulation systems, interior finishes must be chemically inert so that undesirable fumes are not recirculated by the air-circulation equipment. Many excellent products made from wood, metal, natural aggregates, and plastic are now available for use as interior finishes. The use of water-based adhesives, stains, and paints ensures that all exposed surfaces are free of toxic residues.

## EXTERIOR FINISHES AND FACINGS

The exterior finish is the most visible part of a building and contributes to its overall appearance. Choosing an exterior facing for a building is based on such elements as the building budget, climatic conditions, and the aesthetic that the architect wishes to achieve. In general, exterior facings must be resistant to solar radiation, rain penetration, the effects of air pollution, and the effects of extreme temperature cycles. Colors must resist fading and the surface should be washable. The choice of materials and finishes for the building exterior has never been better, and it is beyond the scope of this text to deal with them all. However, an overview of the more common types is warranted.

### Siding

*Siding* is a term used for the outer skin of a wall assembly (Figure 16-2). Its prime purpose is to shed water as well as provide a pleasing appearance. Originally, all siding was cut from wood, with cedar being the preferred choice, as cedar contains a natural preservative. More widespread softwood species such as fir, pine, or spruce are also used. When covered with an exterior quality stain or paint, they serve very well as an exterior finish.

Wood siding normally comes in widths varying from 4 in. to 12 in. (100 mm to 300 mm). The profile can vary from a plain surface to one that provides some additional relief to the overall appearance (Figure 16-3). Wood siding is normally used in

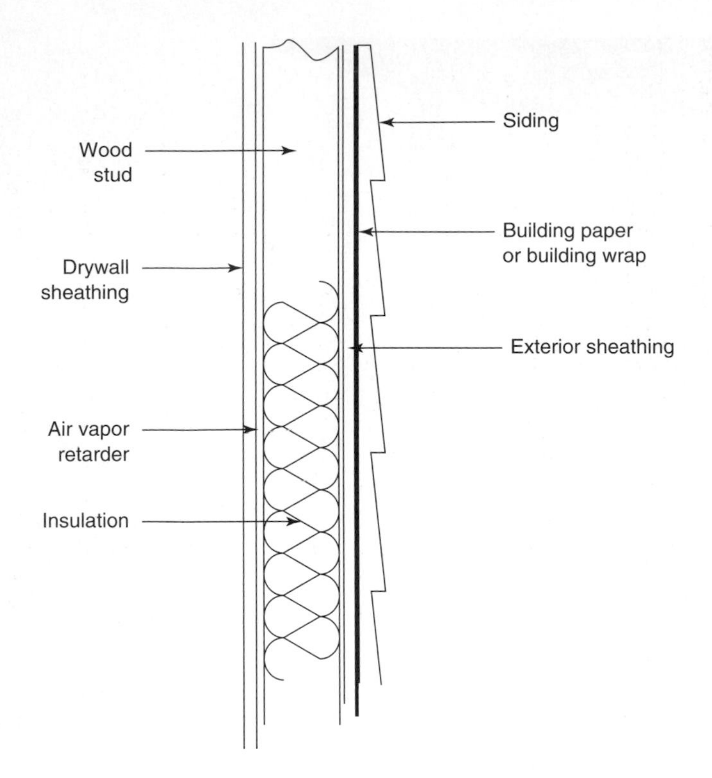


**Figure 16-2** Wall Cross Section with Siding.

**Figure 16-4** Stained Wood Siding Provides a Warm Exterior Finish.

conjunction with a wood frame. It can be applied over sheathing such as plywood or it can be nailed to furring strips. The siding can be applied horizontally, vertically, or diagonally (Figure 16-4). Wood siding may be finished with latex or alkyd paints or stains. Stains may be solid or semitransparent to allow the natural grain of the wood to show.

In today's market, siding products are manufactured from materials such as vinyl, steel, aluminium, and fiber cement. Products from any of these materials, when applied according to manufacturers' specifications, produce a pleasing and long-lasting exterior covering. Sheet metal and vinyl sidings come in various profiles and colors. Ease of installation and minimum maintenance make these materials a popular choice for many applications (Figure 16-5).

Due to environmental concerns and the shortage of large trees, manufacturers have developed new materials, known as engineered wood products,

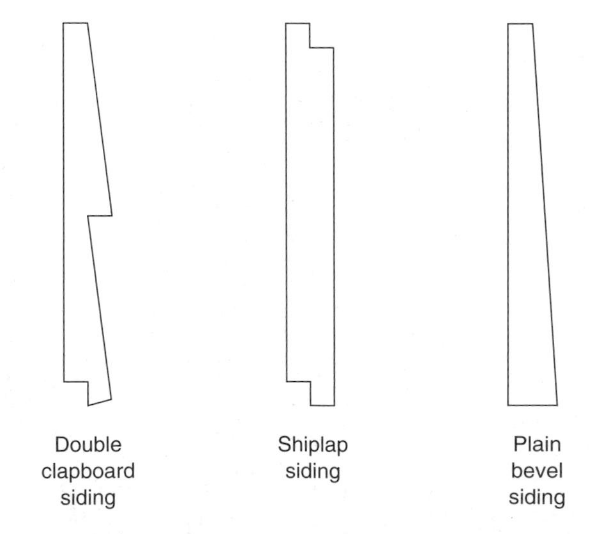


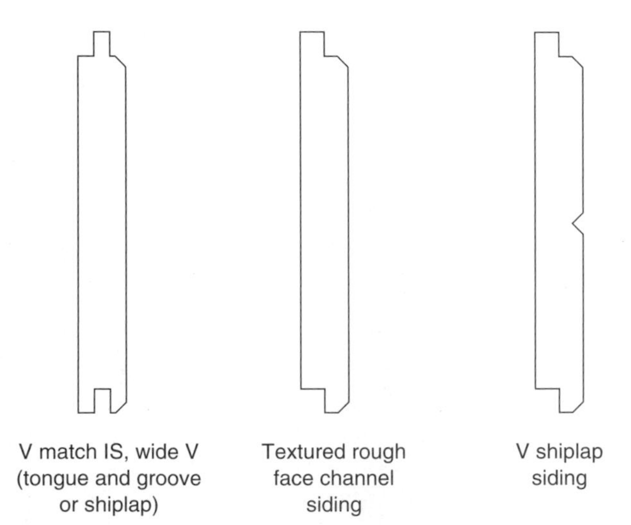


**Figure 16-3** Some Examples of Wood Siding Profiles.

**Figure 16-5** Prefinished Metal Siding is Available in Different Colors and is Easily Applied.

which make more efficient use of the available wood. These products use wood fibers combined with resins and preservatives to produce a homogeneous material that is more stable than natural wood. These products cut easily, do not split, take paint very well, and are resistant to decay. Siding made from these materials has the look of real wood but is more durable and requires less maintenance.

Siding manufactured from fiber cement is a relatively new application for a material that has been used as sheathing for damp applications. It is a very stable material and is naturally resistant to decay and insects. It is also very durable and it takes paint well.

To minimize moisture problems within the wall cavity, a permeable building paper or building wrap is applied over the exterior sheathing before the siding is applied. The building paper or wrap serves as an air barrier and as a drainage plane for any moisture that may penetrate the siding. In some applications, furring strips are used under the siding to provide an additional air space between the exterior sheathing and the siding.

## Stucco

Stucco provides a durable, weather resistant, and cost-effective exterior finish that is comparable to a brick finish (Figure 16-6). Colors can be added to the finish coat to ensure a uniform non-fading finish. It can be textured and enhanced with bits of colored glass or stone. Traditional stucco consists of a lime, sand, and Portland cement mix that is applied in two or three layers depending on the supporting substrate. Three layers are used when the stucco is applied to wood sheathing and two layers are normally used when the stucco is applied to a masonry surface.

When applying Portland cement stucco on wood sheathing, the supporting sheathing is covered with building wrap or paper, and a furred wire mesh is nailed over it (Figure 16-7). The wire mesh provides the necessary anchorage for the base (first) coat of stucco. The minimum thickness of this first coat is

**Figure 16-6** Clean Lines of a Stucco Finish Complement the Brick.

**Figure 16-7** Exterior Wall Covered with Building Paper and Stucco Wire.

½ in. (12.5 mm). The second coat, also known as a brown coat, is ¼ in. (6 mm) in thickness, and the finish coat is ⅛ in. (3 mm) in thickness. The brown coat may also be used as the bedding for exposed aggregate finishes. Colored finish coats are normally applied in two applications to ensure complete coverage and the desired texture. When stucco is applied to masonry or concrete surfaces, the rough surface of the supporting material provides an ideal surface and no additional anchorage is required. For this type of application, the stucco is applied in two layers. As no additional reinforcing mesh is required, the total thickness of the stucco is reduced to ⅝ in. (16 mm)—the base coat is just ⅜ in. (9 mm) thick.

To minimize cracking in the stucco surface, it is recommended that control joints be provided every 20 ft (6 m). Materials used for control joints must be weatherproof and corrosion resistant. Control joints must also be provided where joints occur in the supporting material, such as in masonry block walls.

The development of synthetic stucco, consisting of latex-modified cementitious materials, has added a new dimension to traditional stucco finishes (Figure 16-8). Used as a base coat, they can be applied directly to masonry or concrete substrates. They are tougher, more crack resistant, and easier to mix and trowel than the traditional stucco mix.

These new materials have led to a new type of wall system known as the exterior insulation and finish system (Figure 16-9). In this type of wall system, insulation board is bonded to the exterior wall sheathing using the base coat as the adhesive. To

**Figure 16-8** Synthetic Stucco Application.

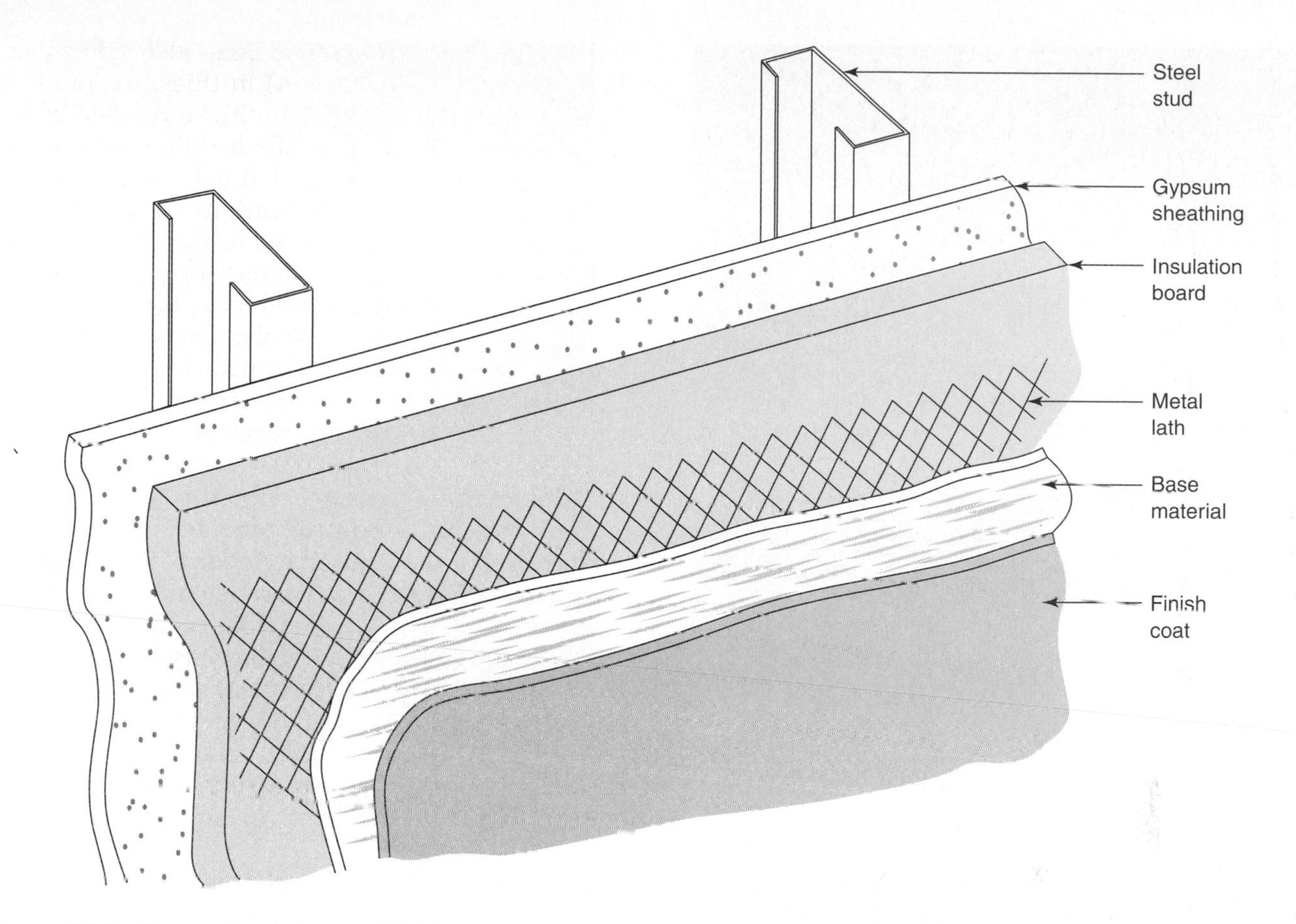


**Figure 16-9** Exterior Insulation and Finish System Detail.

provide extra strength and durability, metal lath is then bonded to the insulation using the same base material. The exterior face is then covered with an acrylic resin finish coat. A wide variety of colors and textures are available.

## Stone and Brick

Natural and polished stone exterior finishes have been used in buildings since the beginning of time. Although stone finishes are relatively costly to fabricate and install, their durability and aesthetic qualities cannot be matched. Stone can be applied to the exterior as a veneer in a variety of patterns and textures (Figure 16-10). Polished marble (Figure 16-11) is a popular choice for many large stately buildings.

Brick exterior finishes are chosen for their durability and appearance (Figure 16-12). Although the initial costs may be higher than other exterior finishes, over the life of the building they prove to be cost effective. To ensure good performance from a brick exterior, attention to detail such as the placement of flashing, control joints, and the use of appropriate ties are of prime importance. Good workmanship is paramount.

**Figure 16-10** Ashlar Stone Exterior Facing.

**Figure 16-11** Polished Marble Facing.

**Figure 16-12** Brick Is a Traditional Material for Use as an Exterior Finish.

## Precast Concrete

Precast concrete panels are a favorite choice of many designers, not only for their durability, but also for their variability. They can be manufactured to close tolerances in almost any shape and size (see Figure 16-13). Surface finishes can vary from a polished surface to one with exposed aggregates. They are erected easily and, with good detailing, they need very little maintenance. Because concrete can be cast into any shape, precast concrete sections of lightweight concrete are popular for building trims and cornices (Figure 16-14).

**Figure 16-13** Lightweight Precast Concrete Panels Can Be Cast in Almost Any Shape.

## Masonry Blocks

Many buildings have load-bearing walls constructed of concrete block units. Because of the durability of masonry construction, the block units can be left exposed to the elements. Paints developed for masonry provide an inexpensive solution (Figure 16-15) to the exterior appearance of a building constructed in this manner.

## Entrances and Exterior Doors

Entrances must be designed to allow for unobstructed movement of people and yet somehow minimize the loss of heat in the winter months. In the

**Figure 16-14** Lightweight Trim Gives a Traditional Appearance to a Building Facade.

summer, entrances must also serve as a barrier between the cool interior of the building and the outside heat. The concept of an air-lock system using two sets of doors (Figure 16-16) is used to control drafts during cold weather and provides a transition between the outside and the inside of a building. Revolving doors, and doors equipped with automatic closers, are effective in reducing air loss from a building (Figure 16-17) without impeding pedestrian traffic.

Exterior doors and entranceways must also be durable to ensure the security of the building. Doors are usually mounted in steel (Figure 16-18) or aluminum door frames, ensuring maximum security. Various locking systems are available, many complete with alarms to deter forced entry.

**Figure 16-16** Double Sets of Doors Provide Good Access Yet Control Air Movement.

**Figure 16-15** Painted Concrete Blocks as Exterior Finish.

**Figure 16-17** Revolving Doors Are Efficient in Controlling Air Loss from an Air Conditioned Building.

**Figure 16-18** Steel Door Frames Being Set into Building Frame.

It is common practice for the manufacturer to fit the doors into their frames so that they are brought to the job site ready for installation. The doors must be removed from their frames and stored until building has progressed to the finishing stage. The frames can be placed in their openings as required.

When wood doors have to be hung on the job, it is important that hinges be properly installed to ensure easy operation. Figure 16-19 indicates hinge clearance details for various thicknesses of door and size of hinge.

Steel swinging doors are hung and operated in similar fashion to wood doors. Hinge recesses and lock holes are formed into the door to match hinge and strike recesses in the frame.

The hardware for glass swinging doors is installed in the shop, and the door is delivered to the job ready to be set in place. Figure 16-20 shows the typical hardware on a plate glass door hung in an aluminum frame.

Various items of hardware may be specified for any particular door. Figure 16-21 shows a number of them and indicates their proper location.

Installation of glass panels flanking outside doors in entranceways will be similar to that of other glass curtain walls. Figure 16-22 illustrates the installation of a glass panel in both wood and aluminum frames.

## WINDOWS

Buildings without windows would be unthinkable. In addition to their aesthetic qualities, windows serve a number of important needs in buildings. Specifically, these are (1) to provide natural light, thereby reducing the need for artificial lighting (Figures 16-23 and 16-24); (2) to provide the occupants with a view to the outside, meeting the need of humans to keep in touch with the natural environment; (3) to provide fresh air; and (4) to take advantage of the radiant heat of the sun during winter months. Many building designs incorporate atriums and solariums enclosed with glass or acrylic panels to produce a green space within the building enclosure (Figure 16-25).

Window styles for commercial buildings include conventional wood sashes (see Figure 16-26), similarly styled windows with aluminum sashes (Figure 16-27), and window walls with sealed units (Figure 16-28). The style that is used depends primarily on the anticipated use of the building and the design effect that the architect wishes to create. Figures 16-29 and 16-30 illustrate typical details of wood and aluminum sashes produced by window manufacturers. Most commercial window applications use

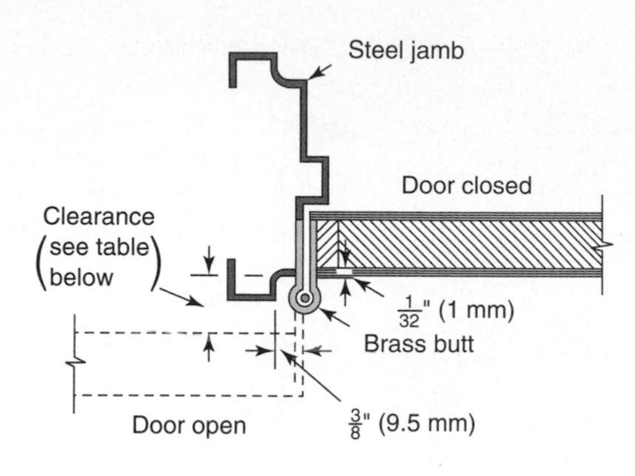


Clearance of stock size butts

| Thickness of door (mm) | (ins) | Size of butt (mm) | (ins) | Maximum clearance (mm) | (ins) |
|---|---|---|---|---|---|
| 35 mm | 1 3/8 | 76 × 76<br>89 × 89<br>102 × 102 | 3 × 3<br>3 1/2 × 3 1/2<br>4 × 4 | 19<br>22<br>41 | 3/4<br>7/8<br>1 5/8 |
| 45 mm | 1 3/4 | 102 × 102<br>114 × 114<br>127 × 127 | 4 × 4<br>4 1/2 × 4 1/2<br>5 × 5 | 25<br>35<br>51 | 1<br>1 3/8<br>2 |
| 50 mm | 2 | 114 × 114<br>127 × 127<br>152 × 152 | 4 1/2 × 4 1/2<br>5 × 5<br>6 × 6 | 32<br>44<br>70 | 1 1/4<br>1 3/4<br>2 3/4 |
| 57 mm | 2 1/4 | 127 × 127<br>152 × 152<br>152 × 178<br>152 × 203 | 5 × 5<br>6 × 6<br>6 × 7<br>6 × 8 | 32<br>57<br>83<br>108 | 1 1/4<br>2 1/4<br>3 1/4<br>4 1/4 |

**Figure 16-19** Hinge Clearance Details.

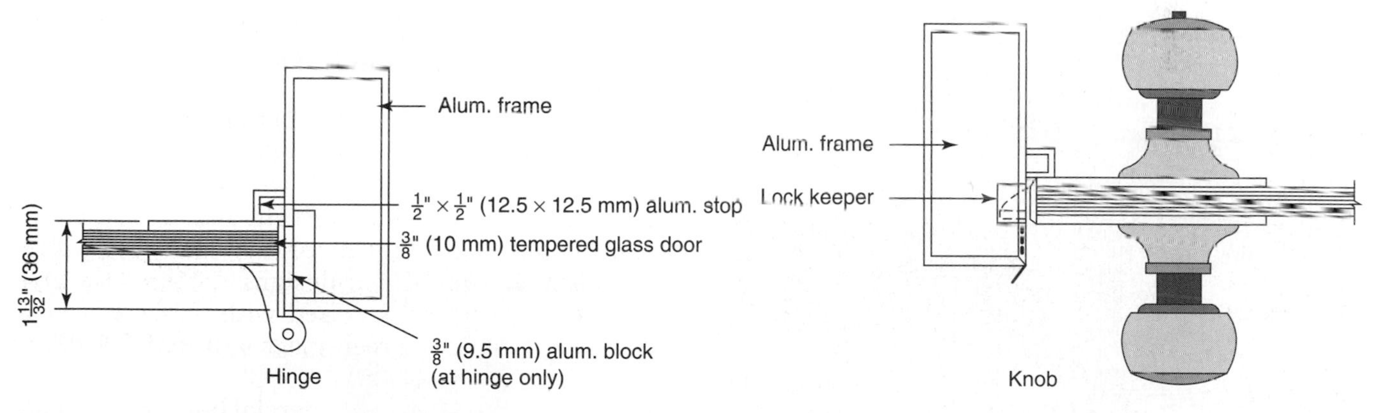


**Figure 16-20** Hardware on a Glass Door.

extruded aluminum frames that have anodized or powder coated finishes. Window manufacturers that use wood frames apply aluminum or vinyl cladding to the exterior face of the frame to improve weather resistance and eliminate the need for paint. The interior face is primed or left unfinished to allow for painting or staining.

To ensure the best possible performance of the window, three areas of concern must be examined when designing and installing windows: (1) The gain and loss of heat through the window; (2) air movement between the window frame and the building frame; and (3) the possibility of moisture penetration around the window opening.

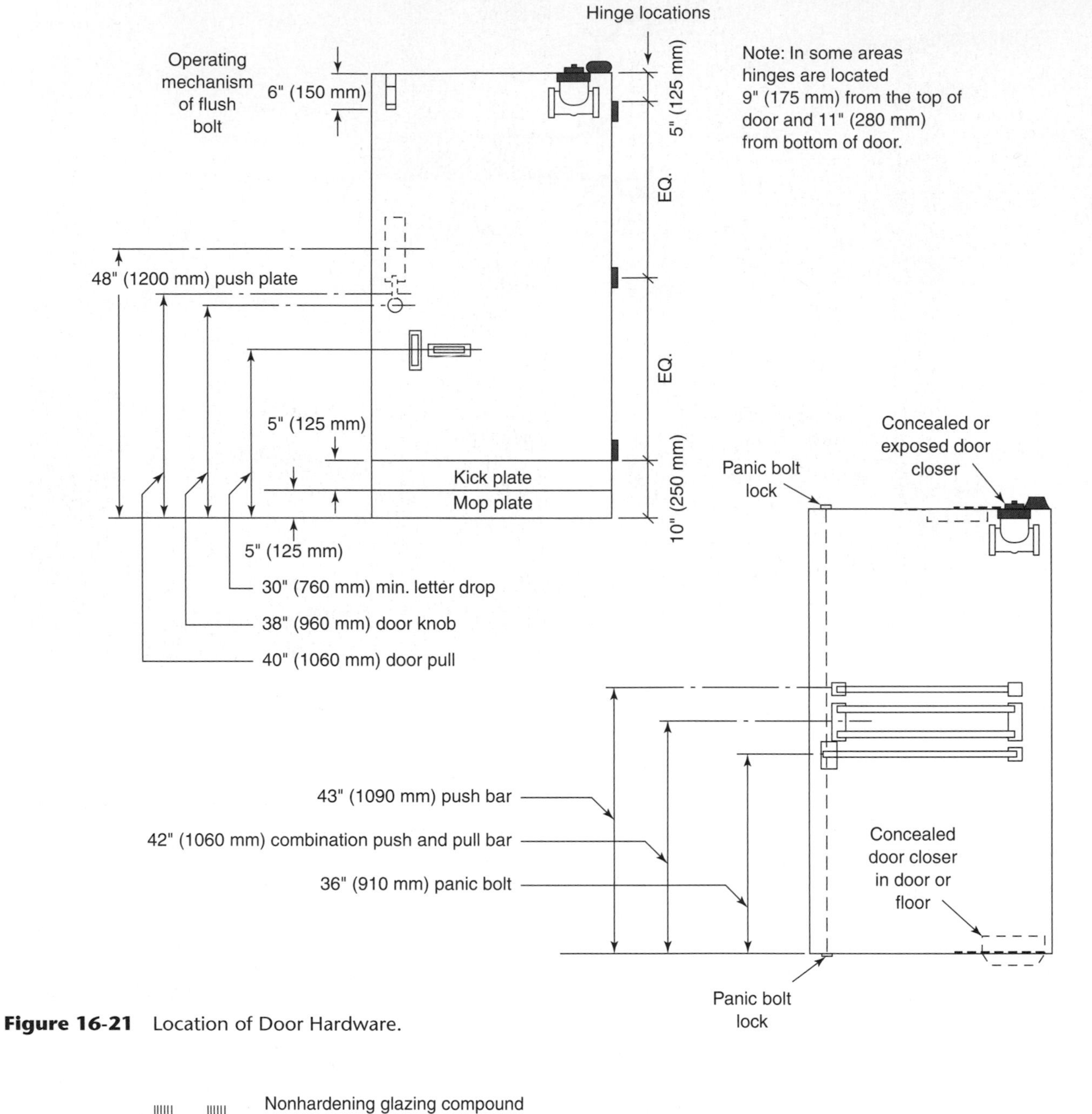


**Figure 16-21** Location of Door Hardware.

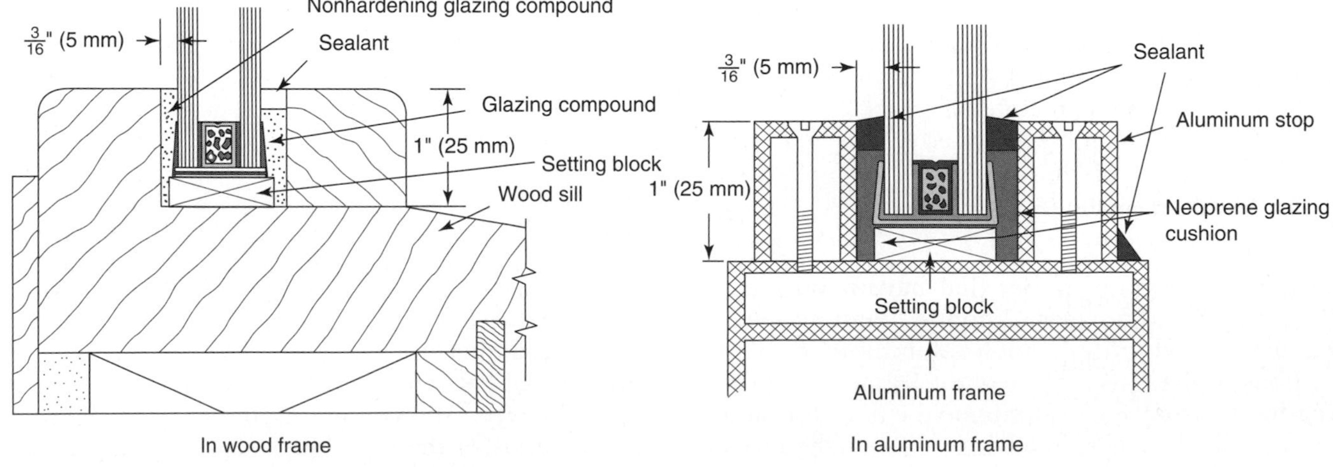

**Figure 16-22** Installation of Sealed Glass Unit.

**Figure 16-23** Glass Blocks Provide Good Light Transmission Without Loss of Privacy.

**Figure 16-24** Skylights Consisting of Glass or Acrylic Panels Are an Excellent Source of Natural Light.

To minimize the heat loss or heat gain through a window, a number of options have been developed. The most effective approach is to use multiple layers of glass, separated by an air space, bonded together at the perimeter to produce a sealed unit. Double and triple layers of glass assembled in this manner are the most common. Sealing the glass layers around the perimeter provides a dead air space between the layers and adds to the insulation value and prevents condensation within the air space. The sealed unit is then mounted in a frame of extruded aluminum, extruded plastic, steel, or wood. Each framing material has its benefits and drawbacks and the selection of a framing material is usually based on thermal efficiency, maintenance, and aesthetic requirements. Durability, long-term maintenance, and compatibility of the window frame with the structural frame of the building must also be considered. The use of thermally broken frames adds to the overall thermal efficiency of the window.

To further improve the thermal efficiency of windows, reflective and tinted coatings can be applied to the glass panes. In addition to the coatings, the air space between the glass panes can be filled with an inert gas such as argon. The most recent development is to use argon gas in conjunction with glass panes that are sprayed with spectrally selective coatings which reduce heat loss in the winter and reduce heat gain in the summer. These coatings are known as low emittance or low-E coatings. Low-E coatings are transparent to visible light and short-wave infrared light waves and reflect long-wave infrared light waves. As these coatings are microscopically thin, they do not significantly reduce the transmission of visible light. The amount of heat gain and heat loss can be varied depending on the climate, allowing for a customized approach to window selection depending on the needs of the building and its location.

Because every window opening represents a discontinuity in the building envelope, the placement of the window unit is crucial to the overall performance of the building. Proper sealing of air barriers and vapor retarders around the window opening ensures elimination of air movement through the space between the window unit and the building frame. Warm air moving through the gap between the window frame and the building wall and condensing within the wall cavity can produce premature deterioration of building components. Water penetration due to rain adds to the problem. Proper flashing must be installed to ensure that any moisture that does enter the wall cavity is allowed to escape either by drying or drainage.

**Figure 16-25** Atrium in Building Provides Excellent Light for the Entire Height of a Building.

**Figure 16-26** Typical Wood Sash Apartment Window.

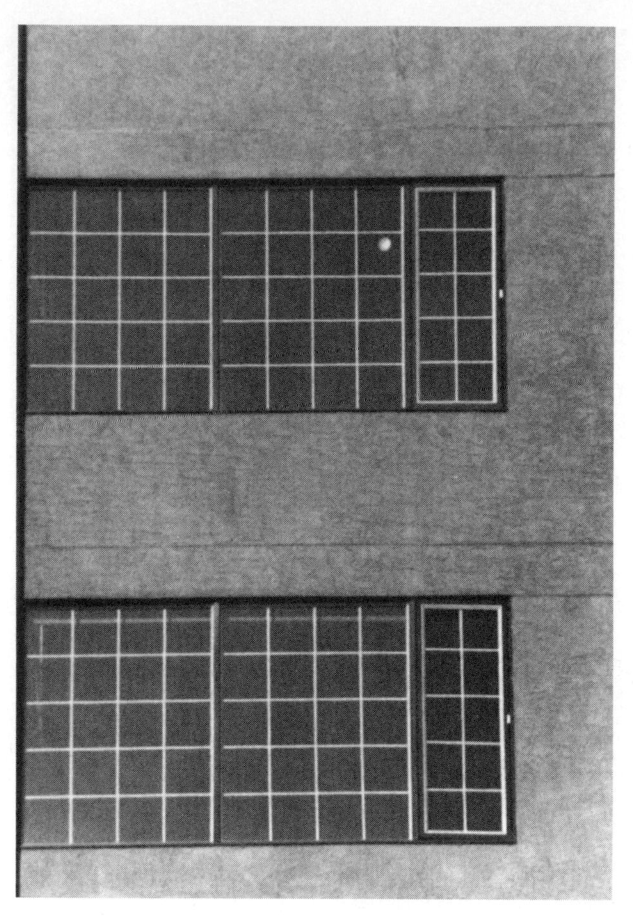

**Figure 16-27** Metal Frame Sash Windows.

From a construction point of view, allowances must be made for the installation of the window unit in the structural frame. The size of the rough opening will determine the size of the window unit. The rough opening must be large enough to ensure sufficient clearance for the installation of the window unit. Window manufacturers normally provide data on their products that indicate the size of rough opening required. In many applications, window units are custom made to meet the requirements of the building.

To ensure the best possible performance of a window after installation, rough openings must be within manufacturers' tolerances. If rough openings are too large, heat loss and moisture penetration may become problems. In fixed window installations, building frame expansion may compound resulting in the loosening of the entire window unit. In some instances, whole windows have fallen out of the building frame. Windows installed in rough openings that are too small may experience excessive compressive forces causing failure of the glass lite. Windows equipped with openers will jamb should a shift in the building frame occur.

When large areas must be covered with sloped glazing (Figure 16-31) or domed areas are enclosed with glazing (Figure 16-32), structural framing

**Figure 16-28** Aluminum Frame Glass Curtain Wall.

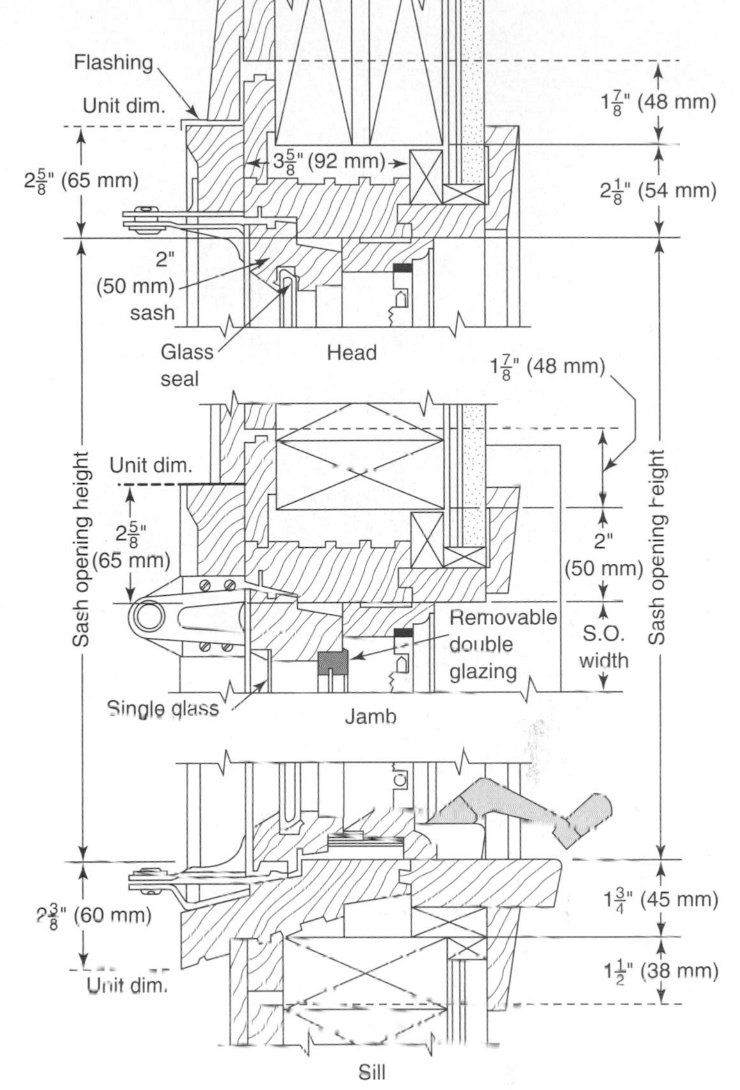


**Figure 16-29** Typical Details on a Wood Sash Window (Reprinted with permission of Andersen Corporation)

must be incorporated into the *lite* supports, and the whole assembly must be designed in much the same way as a roof framing system. Special details are required for the support of the glass or acrylic lites (Figure 16-33) to ensure that no moisture leakage or condensation occurs and that the whole system is as energy efficient as possible.

## INTERIOR FINISHING

The main purpose of interior finishing is to provide the building occupants with a pleasing environment in which to work and live. When choosing materials for interior finishing, designers consider many factors, including color, durability, soundproofing, washability, chemical stability, fire resistance, rate of flame spread, cost, and availability. All of the foregoing must be considered to some degree, depending on the anticipated use of the building and budget considerations. Figure 16-34 shows the use of drywall and vinyl floor coverings in a hallway to facilitate cleaning and maintenance.

With the many different products now available, the choices for the designer are limitless; however, the traditional materials and their by-products still form the basis for most finishes. Stone, clay, cement, wood, glass, and now plastic are being combined to provide beautiful, easily maintained, and long-lasting finishes. It is beyond the scope of this text to cover all the many different finishes presently available; however, some basic finishing materials and their applications are dealt with in the following pages.

## PLASTERING

Plaster is well known and widely used for finishing walls and ceilings in many types of buildings. Either portland cement or gypsum plaster may be used, depending on the composition of the wall and the conditions to which the surface will be subjected.

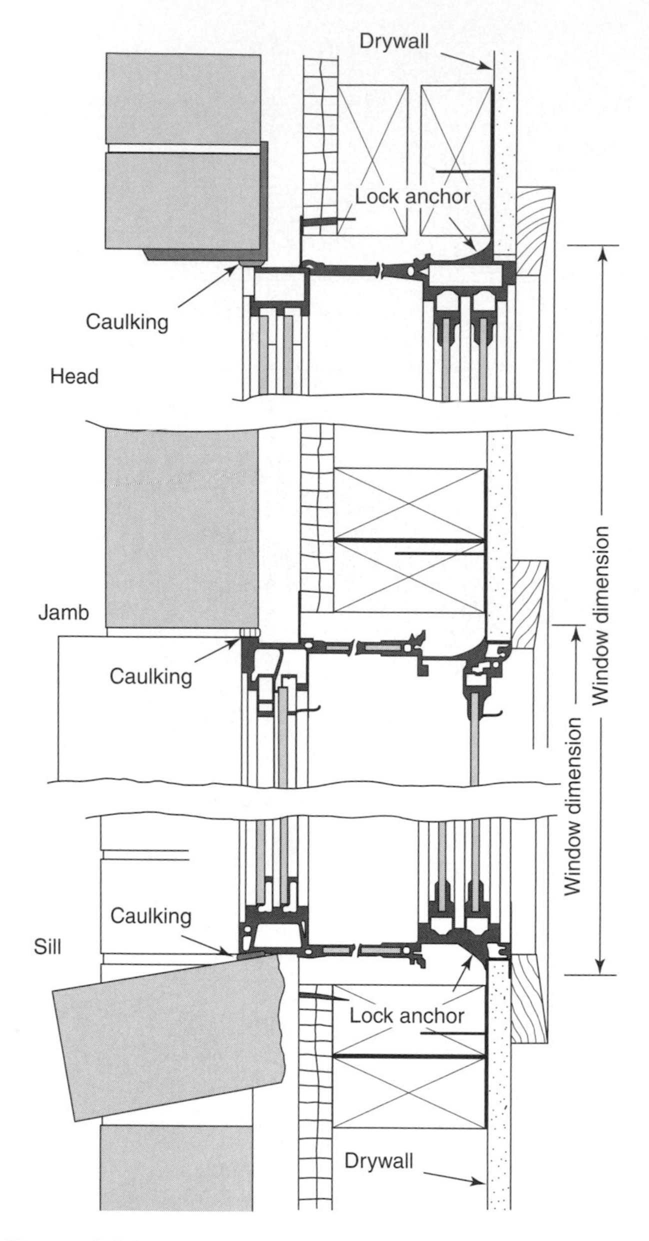


**Figure 16-30** Typical Details on an Aluminum Sash Window. (Reprinted with permission of Humphrey Products Inc.)

**Figure 16-31** Large Areas of Sloped Glazing Require Structural Framing for Support.

Portland cement plaster may be applied to concrete, masonry, and metal lath bases and is used where walls, ceilings, and partitions are subject to rough use or extreme moisture conditions. Gypsum plaster bonds well to gypsum lath, metal lath, fiberboard lath, and gypsum or clay tile. Plaster is usually applied in three coats: the first, or *scratch,* coat; the second, or *brown,* coat; and the finish coat.

## Portland Cement Plaster

Portland cement base plaster is made by mixing 1 part of normal portland or masonry cement with 3 to 4 parts of plaster sand. Sand grading should be as follows:

- Passing #4 (5-mm) sieve: 100%
- Passing #8 (2.5-mm) sieve: 80% to 98%
- Passing #16 (1.25-mm) sieve: 60% to 90%
- Passing #30 (630-μm) sieve: 35% to 70%
- Passing #50 (315-μm) sieve: 10% to 30%
- Passing #100 (160-μm) sieve: not more than 10%

Larger proportions of aggregate to cement may be used when the aggregate is well graded, with a good proportion of coarse particles. When normal portland cement is used, a plasticity agent may be used to increase the workability of the mortar.

The scratch coat should be approximately ⅜ in. (10 mm) thick. It should be *dashed* on concrete walls unless the surface is sufficiently rough to ensure an adequate bond for a troweled coat. The scratch coat must be troweled on clean, dry masonry walls. Before plastering begins, walls should be evenly dampened to control suction. Hair or fiber at the rate of approximately 1 lb/bag (0.5 kg/bag) of cement should be used in a scratch coat applied to metal lath. The scratch coat must be crosshatched before it hardens to provide a mechanical bond for the brown coat. It should be kept damp for at least 2 days immediately following its application and should then be allowed to dry thoroughly.

The surface of the scratch coat should be dampened evenly before beginning the application of the brown coat, which is approximately ⅜ in. (10 mm) thick, brought to a true, even surface, and then either roughened with a wood float or crosshatched lightly. The brown coat should be damp cured for at least 2 days and then allowed to dry.

**Figure 16-32** Glazed Domes Require Structural Frames for Support.

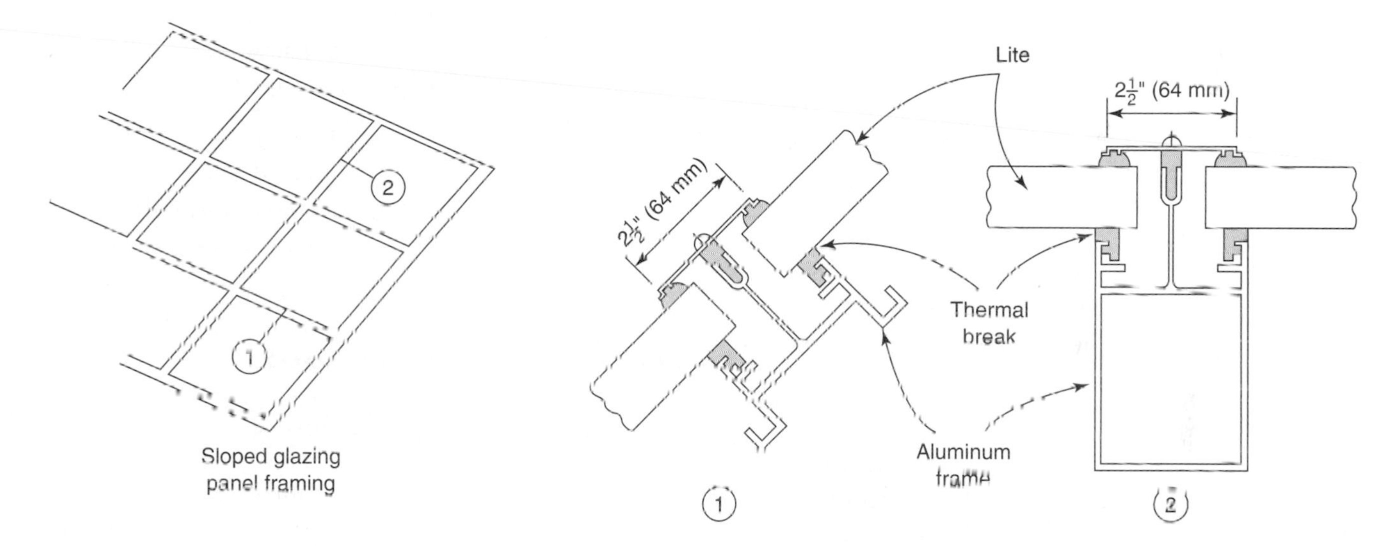


**Figure 16-33** Thermally Broken Glazing for Use in Sloped Glazing Applications.

Finish coats may require a slightly finer aggregate, but excessive fineness should be avoided. Color may be added in the form of mineral pigments, at the rate of not more than 6% of the cement, by weight. The usual finish on interior portland cement plaster is a sand float or low-relief texture finish, but plaster may be troweled smooth, if specified. Moist curing should take place for at least 7 days. During plastering operations, the temperature should be maintained at or above 50°F (10°C).

## Gypsum Plaster

Several types of gypsum plaster are manufactured, each for a specific use. They include hardwall plaster for scratch and brown coats; cement bond plaster for base plaster (to be applied to a concrete surface); and finish plasters for the finish coat over a gypsum plaster base.

Either sand or vermiculite may be used as aggregate with gypsum hardwall plaster for scratch and brown coats. Use 2 parts of dry sand by mass to 1 part of plaster for the scratch coat over gypsum lath, wood lath, or metal lath. For the scratch coat over gypsum tile, brick, or clay tile, use 3 parts of sand by mass to 1 part of plaster. The same general specifications apply for sand used in gypsum plaster as for that used in cement plaster.

The scratch coat is approximately ⅜ in. (10 mm) thick and is crosshatched to receive the brown coat.

**Figure 16-34** Walls Covered with Drywall and Vinyl Floors Provide Smooth, Durable Surfaces for Institutional Buildings.

Curing is carried out in the same manner as for cement plaster.

For the brown coat, use 3 parts of dry sand by mass to 1 part of plaster. This coat is also approximately ⅜ in. (10 mm) thick, floated, and broomed to a straight, even surface, ready for the finish coat.

If vermiculite aggregate is specified, use 1 cu ft (0.03 $m^3$) of aggregate to 1 sack of hardwall plaster over gypsum lath. If the plaster is to be applied over masonry, 1½ cu ft (0.04 $m^3$) of aggregate may be used per sack of hardwall. For the brown coat, use approximately 1½ cu ft (0.04 $m^3$) of vermiculite per sack of plaster.

Cement bond plaster requires only the addition of water. It is applied in two coats to a total thickness of approximately ⅜ in. (10 mm) on ceilings and ⅝ in. (16 mm) on walls. Before it has set, the surface must be broomed to receive the finish coat.

Several types of gypsum plaster are available for the finish coat; one is the widely used gypsum finish plaster. It is mixed with lime putty to make the putty coat. Lime putty is produced either by slaking quick lime or by soaking hydrated lime for about 12 hr. The two ingredients are mixed in the following ratios: 1 part finish plaster to 2 parts dry hydrated lime by mass or 1 part plaster to 3 parts lime putty by volume. Water is added to the mixture to attain the desired degree of plasticity.

This putty finish is applied in two coats over a fairly dry brown coat base. A thin first coat should be ground into the base thoroughly, and a second coat should then be used to fill in the imperfections. The surface is troweled to a smooth finish, sprinkling water on with a brush to provide workability.

A sand float finish is produced by using screened (hairs or fibers have been removed) hardwall plaster and sand in the proportion of 1 part plaster to not more than 2 parts of sand by mass. The maximum size of sand particles will determine the coarseness of the finish. This plaster is applied in two coats to a total thickness of not more than ⅛ in. (3 mm) over a set but not fully cured hardwall base coat. After leveling, the surface is floated with a carpet or cork float to produce the desired texture.

A prepared finish plaster is available that requires only the addition of water. It is mixed at the rate of approximately 2 parts plaster to 1 part water by volume. The plaster should be allowed to soak for at least 30 min before it is used. Application and finishing are essentially the same as for a lime putty coat. This plaster does not produce as white a surface as lime putty, but because it contains no lime, it may be decorated as soon as it is dry.

Plaster of paris and Keene's cement are two other types of gypsum finish plaster with special applications. Plaster of paris is used where a rapid set is required, and Keene's cement is used where sanitary conditions or high humidity necessitate a hard, impervious surface.

### Vermiculite Finish Plaster

A finish plaster in which vermiculite is used as the aggregate is made by mixing vermiculite and screened hardwall at the rate of 100 lb of gysum to 1 cu ft of vermiculite finish aggregate (1,600 kg of gypsum to 1 $m^3$ of vermiculite finish aggregate). The plaster is first mixed with water and the aggregate is added to the putty. More water may be added to make the plaster workable. This finish may be applied over any gypsum-base plaster to a depth of ⅛ in. (3 mm). The procedure for applying, leveling, and troweling is the same as that used for other finish plasters.

## WALL TILE

Several kinds of wall tile are produced in various sizes for interior finishing, including ceramic, glass, steel, copper, and plastic tile. Accessories such as base, cap, triangles, and corners are available for most of them. Figures 16-35(a) and (b) illustrate the various shapes available in plastic and ceramic wall tile in standard and metric units.

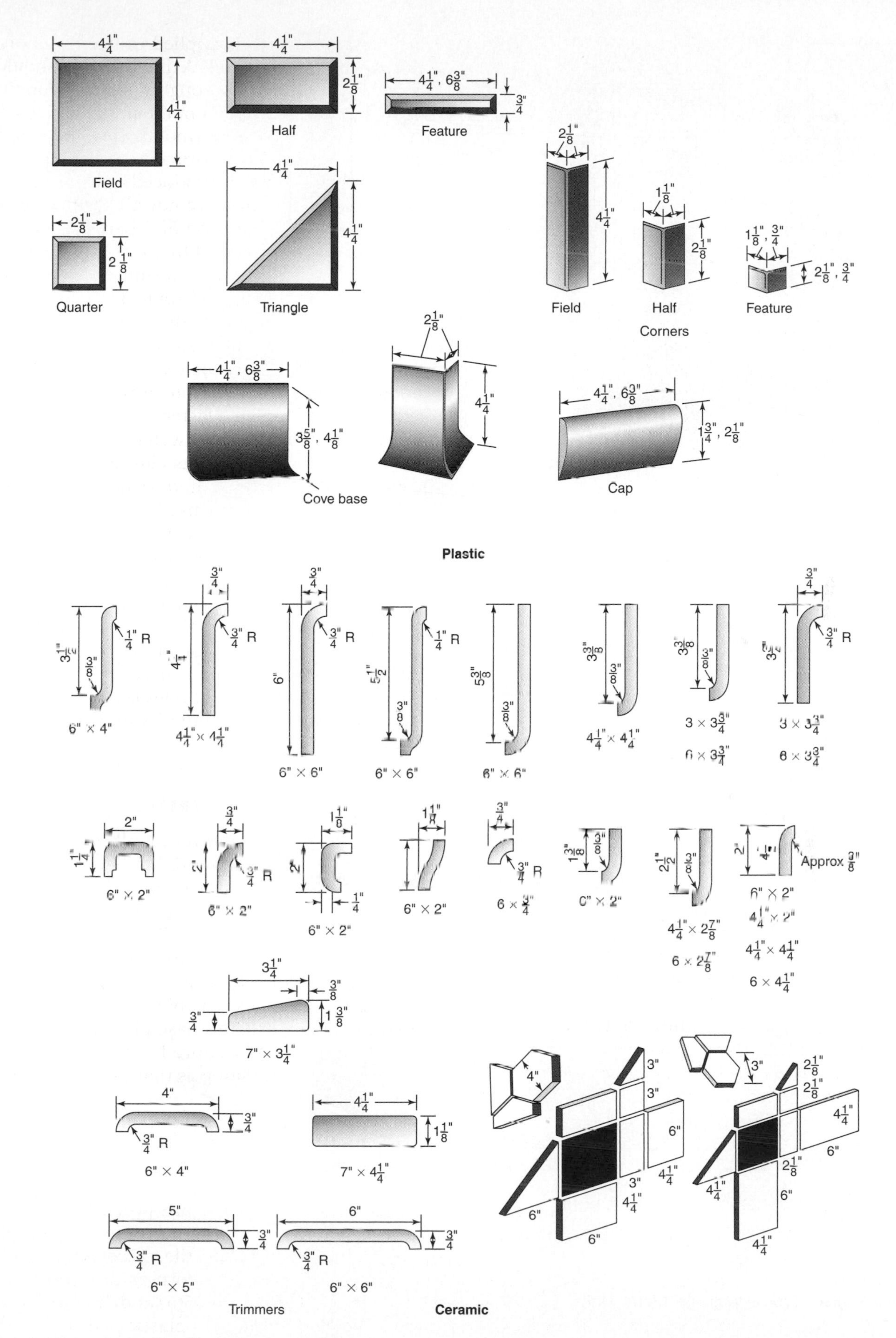


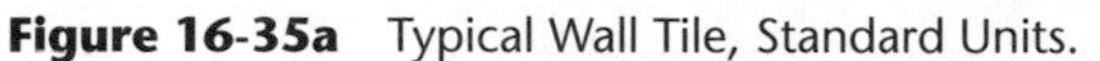

**Figure 16-35a** Typical Wall Tile, Standard Units.

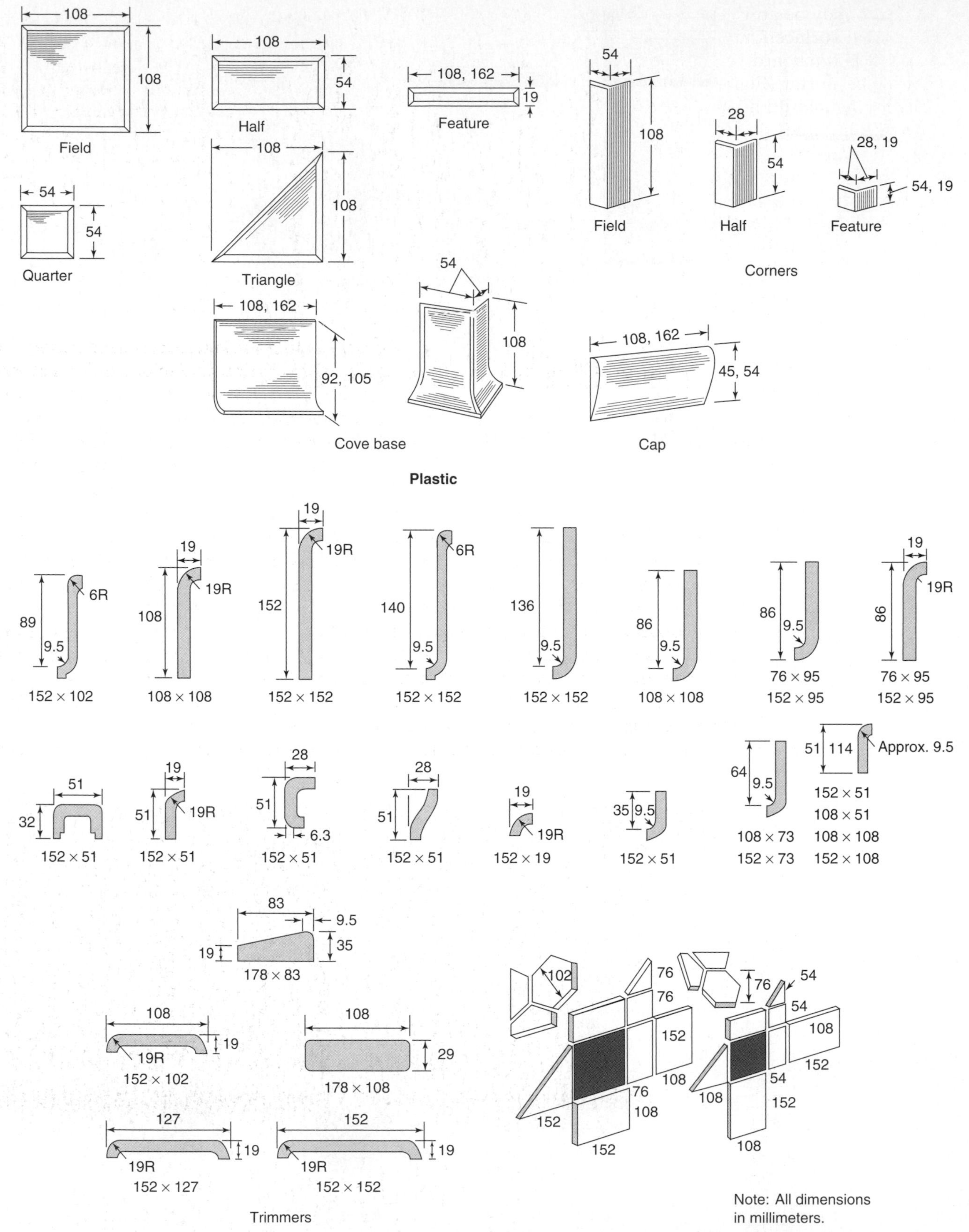


**Figure 16-35b** Typical Wall Tile, Metric Units.

Wall tile may be applied with adhesives to most hard, smooth surfaces such as concrete, hardboard, plywood, gypsum board, or plaster. Ceramic tile may also be set in mortar (Figure 16-36).

Tile laying should begin at an established level and on vertical lines on the wall. If tile is to cover the entire wall from floor to ceiling, a level line should be drawn at a convenient height from the floor, measured to accommodate the base and a specified number of tiles. If the tiled surface is to be a *dado,* draw the top line of the tile and a plumb line down the center of the wall. Tile laying should begin at the intersection of these lines. All materials should be at room temperature, 60° to 70°F (16° to 21°C). Apply the adhesive to the wall with a notched trowel using a wavy motion. The coating should be of sufficient thickness so that when tile is pressed against it, the ridges of cement will flatten and contact at least 60% of the back surface of the tile. Do not force the edges of tiles tightly against one another; grouting compound, made for that purpose, is applied to the joints by tube or by narrow spatula after the tiles are in place. Joints are then wiped out to the desired depth.

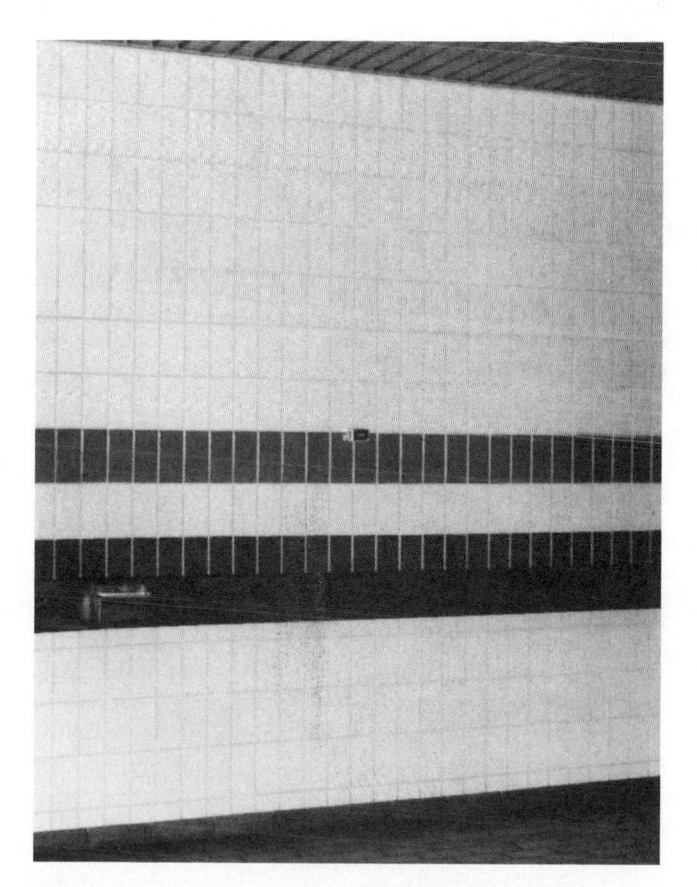

**Figure 16-36** Ceramic Tile Wall Finish.

## HARDBOARD

Hardboard with various face designs, such as embossed leather or wood grain, striated, grooved, plastic coated, and plain or cut into squares like tile, is used for interior finish.

Panels should be conditioned before being applied to the walls by allowing each to stand separately on its long edge for at least 24 hr. Do not try to install hardboard panels in abnormally damp areas.

Hardboard may be attached to the wall with nails, batten strips, or adhesives. If nails are used, drill a shallow hole at nail locations with a drill slightly smaller than the shank of the nail. Nails should be spaced approximately 8 in. (200 mm) apart at intermediate supports and 4 in. (100 mm) apart around the edges. A solid backing must be provided when the board is to be applied with adhesive. Spread adhesive over the entire back surface with a saw-toothed applicator and brace the panels in place until the adhesive sets.

## GYPSUM BOARD

Gypsum board has largely replaced plaster as an interior finishing material for walls and ceilings, mainly because of the ease of installation. It is commonly known as drywall construction. New developments in production have led to high-density gypsum boards that are moisture resistant and provide excellent fire ratings. Sheets that are 4-ft (1,200 mm) wide are available in 16-ft (4,800-mm) lengths; the large size coupled with a smooth, paintable surface (Figure 16-37), allows for time-effective installation over regular stud framing, concrete, or masonry construction.

It may be applied in a single layer of ½-in. or ⅝-in. (12.5- or 16-mm) board or a double thickness of ⅜-in. (9.5-mm) board. The single layer is secured by drywall nails or screws or by gluing to studs or furring strips. When a double thickness is used, the inner layer is applied vertically and nailed, and the outer layer is applied horizontally and cemented with a gypsum cement to the inner layer. The outer layer is held in place with double-headed nails that are removed once the cement has set. Nail holes are filled with gypsum joint filler at the same time as the joints are taped and filled. The joints are sized after sanding, and the surface is ready to be painted.

Gypsum board is produced with printed wood grain patterns and with a plastic-fabric-coated surface. The former is applied with nails whose heads have the same color as the pattern, and the fabric-surfaced board is usually held by aluminum battens, producing a paneled effect. The battens are placed at

**Figure 16-37** Gypsum Board Is an Easily Applied Material That Provides a Smooth Finish.

studs, held secure by screws to the stud, and are capped with a plastic batten strip (Figure 16-38).

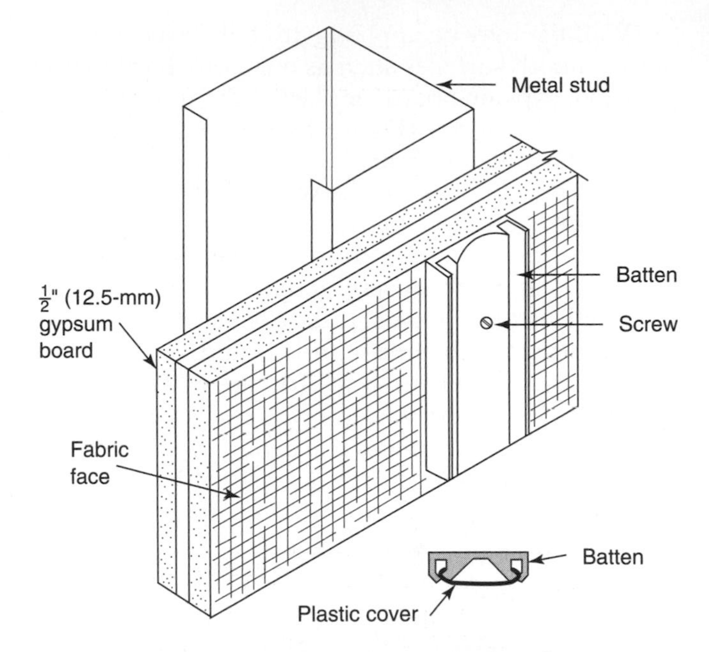


**Figure 16-38** Gypsum Board over Metal Studs.

## PLASTIC LAMINATES

Plastic laminates in thicknesses ranging from 1/16 to 1½ in. (1.5 to 38 mm) are used in various ways for interior finishing. The thin sheets, 1/16, 1/10, 1/8 in. (1.5, 2.5, 3 mm), are used alone or laminated to plywood backing to line interior walls. Thin sheets are applied directly by the use of contact adhesives. Edges may be butted together, but aluminum edge moldings are generally used (Figure 16-39). These moldings may be used with other types of wall paneling and plastic laminates.

The application of a thin, plastic laminate sheet to plywood backing is usually done in a factory or shop. The plastic laminate may be applied to a flat sheet of plywood or may be postformed over a curved surface. Faced plywood sheets or strips are used as paneling, base, or ceiling cover. Figure 16-40 illustrates two methods used to attach panels to walls for a flush joint fit. Paneled effects may be produced in several ways, some of which are illustrated in Figure 16-40.

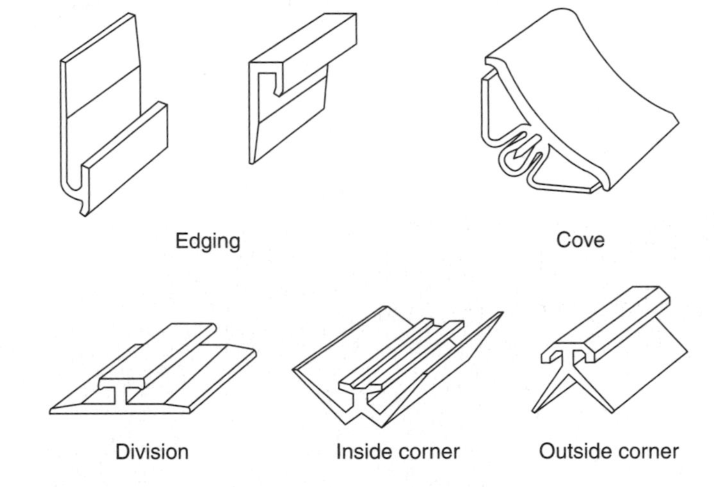


**Figure 16-39** Aluminum Panel Moldings.

Thick plastic laminates are used for partitions, sliding doors, baseboards, window sills, table tops, and the like, and are manufactured to specifications.

## MASONRY FINISHES

Interior walls may be faced with standard 100-mm-wide brick or with brick veneer units. Standard brick may be either common or face brick, depending on the appearance that the designer wishes to create. For example, the texture on a face brick can add warmth to an interior wall (Figure 16-41) and provide a pleasing contrast with the floor finish. Interior partition walls or load-bearing walls faced with brick provide good

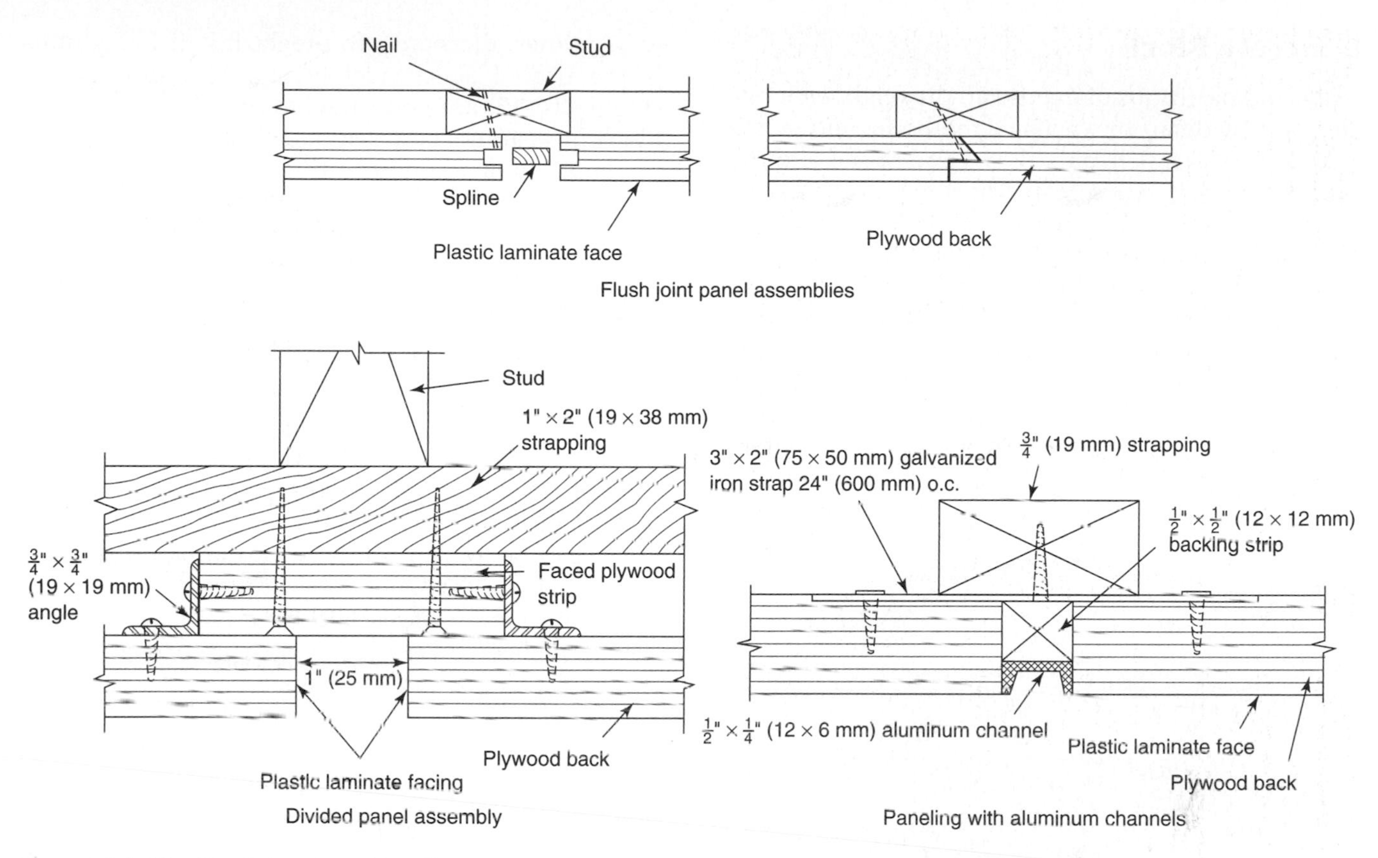


**Figure 16-40** Plastic Laminate Paneling.

**Figure 16-41** Brick and Tile Provide a Durable Interior Finish to Walls and Floors.

**Figure 16-42** Interior Brick Curtain Wall.

sound insulation and add to the fire rating of the wall system, in addition to providing a durable wall finish.

The procedure for applying face brick to interiors is similar to that used for exterior facing brick except that stack bond is more commonly employed. The brick may be bonded to the backup wall with metal ties or by a mortar coat between backup and face brick. Particular attention should be paid to the mortar joints. Sand should be fine enough to pass a #16 (1.25-mm) sieve. Where non-staining mortar is specified, ammonium or calcium stearate is added at the rate of 3% of the mass of cement used.

Interior curtain walls or load-bearing walls faced with brick provide good sound insulation and a high degree of fire protection, in addition to providing a durable wall finish (Figure 16-42).

## Concrete Block

Walls and partitions of exposed lightweight concrete blocks provide an inexpensive method of providing a durable wall finish, as well as good sound and fire ratings. Several types of finishes are available depending on the requirements of the building. Storage and manufacturing areas of buildings can have unfinished blocks, whereas areas that need some color are painted with latex paint (Figure 16-43) and can be made to look very attractive.

In areas that require a more durable finish, preglazed units can be used (Figure 16-44). Prefinished units such as these provide a wide range of colors and patterns and a smooth durable surface that is resistant to chemical action and is easily cleaned.

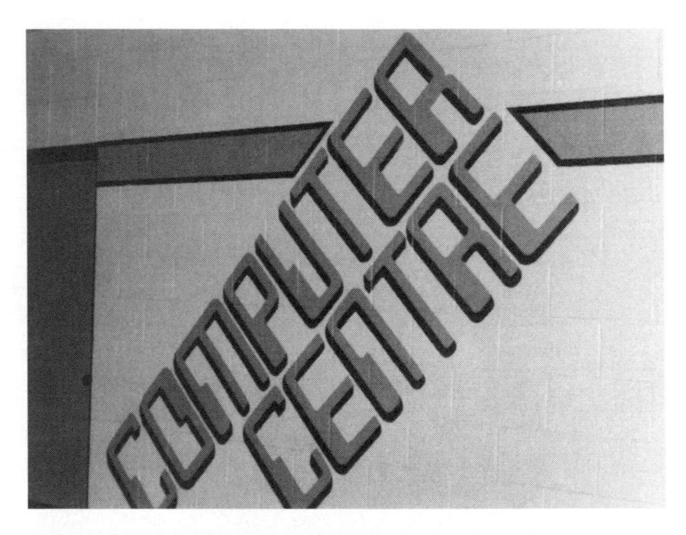


**Figure 16-43** Masonry Block Walls Can Be Painted to Suit Any Taste.

**Figure 16-44** Wall of Block Units with Preglazed Finish.

Another alternative to preglazing of block units is the use of sheet steel facing. The steel panels (Figure 16-45) are preformed to fit over the concrete block face. Joints between blocks must be raked out to allow the facing unit to fit around the edges. Adhesive is applied to the back of the panel, and it is then pressed into place over the face of the block.

## Facing Tile

Facing tiles are available in thicknesses of from 2 in. to 8 in. (50 mm to 200 mm) and may be laid with either horizontal or vertical cells. Tile laid with horizontal cells should have divided bed and head joints. Enough mortar should be used so that the excess will ooze from the joints as the unit is pressed into place. Excess mortar is then struck off and the joints are tooled to a flush or concave surface.

When facing tile is laid with vertical cells, the mortar bed should not carry through the wall on the connecting webs and end shells except at corners and at ends. The cells should not be filled with mortar but should instead be left open to allow drainage of any moisture penetrating from the surface. Weep holes should be provided at the bottom of the wall to allow moisture to escape.

Mortar for tile is similar to that used for face brick. If pointing mortar is specified, regular mortar

**Figure 16-45** Steel Facing Units for Concrete Block.

is raked back, and pointing mortar is applied with a tuck pointer's trowel. Pointing mortar should consist of 1 part portland cement and ⅛ part of hydrated lime to 2 parts of fine sand (#50, 315 $\mu$m) mesh or finer. Ammonium or calcium stearate is added at the rate of 2% of the mass of cement used.

Brick and tile surfaces should be cleaned with burlap as work progresses and later scrubbed with a stiff brush and water when the walls are completed. Mortar stains on unglazed brick or tile may be removed by using a wash consisting of 1 part hydrochloric acid to 9 parts of water by volume. Do not use this solution on glazed brick or tile.

### Ceramic Mosaic

Ceramic mosaics are small, flat tile units that have been assembled into sheets in the plant and held by cement on a polyethylene coated backing material or by a paper sheet over the tile faces. Back-mounted units are applied to the wall with tile cement, and face-mounted ones are usually applied by pressing the sheets of tile into a wall coating of mortar. When the mortar has set, the paper mounting is scrubbed from the face of the tile.

Joints are grouted with either fine mortar or prepared grouting compound once the tiles are in place.

### Stone Facing

Stone used for interiors is normally applied as a veneer over walls and around columns (Figure 16-46). Cut stones in thicknesses of from 1 in. to 4 in. (25 to 100 mm) and of various face dimensions are employed. The procedure for laying stone in interior work is essentially the same as that used for stone exteriors. Small stone units are laid in mortar beds and anchored to the wall with corrugated metal ties (see Figure 13-45). Larger slabs may be secured by any of the stone anchors illustrated in Figure 13-41. Because of the stones' mass, two wedges are placed under the bottom edge of each stone in a course and are allowed to remain until the mortar bed hardens. The wedges are then removed and the holes are filled with mortar (see Figure 16-47).

**Figure 16-46** Marble Finish on Walls and Floor of Lobby in Large Office Building.

## CEILINGS

The type of ceiling to be used will depend on a number of factors, including the type of structural floor in the building, the location of the mechanical services, the intended use of the building, and whether acoustical treatment is required.

Some designers will specify that, for a particular use, the underside of prestressed concrete floor slabs such as single or double Ts, ribbed slabs, or waffle slabs may be painted and left exposed, as illustrated in Figure 16-48.

When the building's mechanical services are located below the floor slabs (see Figure 16-49), they will have to be hidden by suspending the ceiling below them. Heavy wires or small rods attached to the underside of the floor slab support a system of T-bars. Careful attention must be paid to this operation in order to have all the bars hanging perfectly level. The T-bars, in turn, support some type of ceiling

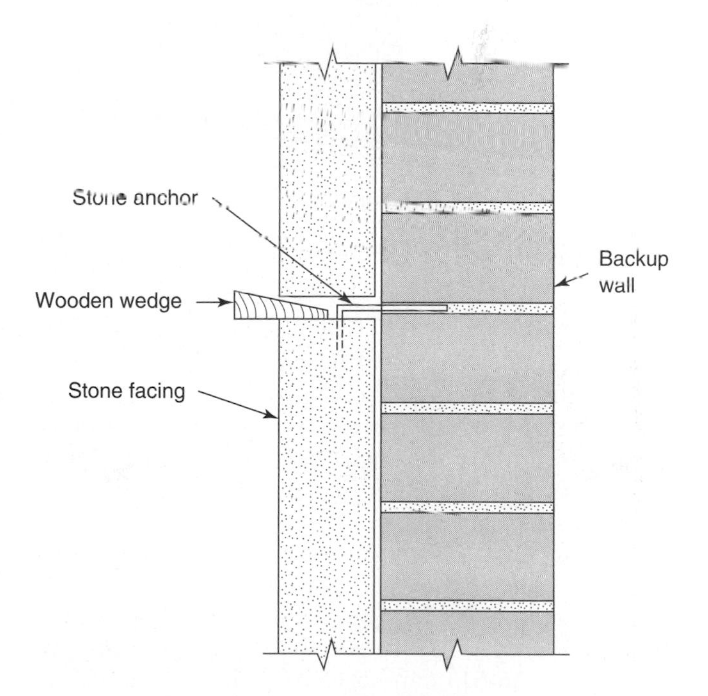


**Figure 16-47** Stone Panel Wedged in Place.

**Figure 16-48** Underside of Waffle Slab as Finished Ceiling.

**Figure 16-49** Suspended Ceiling System.

panels. In Figure 16-49, for example, workers are installing steel-faced ceiling panels.

Ceiling materials also contribute to the overall effect that a designer wishes to create within a room. Figure 16-50 illustrates the special effect produced by mirrored ceiling tiles, giving the room an extra sense of space.

In some applications, the space above the ceiling becomes part of the heating and ventilation system and is used for the distribution of air throughout the building space. Prepainted ribbed sheet metal panels (Figure 16-51) can serve as grilles to distribute the air uniformly and provide an interesting, durable, and noncombustible ceiling finish.

The acoustical treatment of the ceiling may involve plastering with acoustical plaster or spraying the structural frame members with sound-absorbing insulation (Figure 16-52). If suspended ceilings are required, fire-resistant acoustical panels composed of glass fiberboard, ceramic, or mineral wool fiberboard with various textures and embossed designs are used (Figure 16-53).

## STAIRS

Stairs of one kind or another are a necessary requirement in any building of more than one floor. Plans sometimes call for prefabricated stairs, and tempo-

**Figure 16-50** Mirrored Ceilings Add a Sense of Space to a Room.

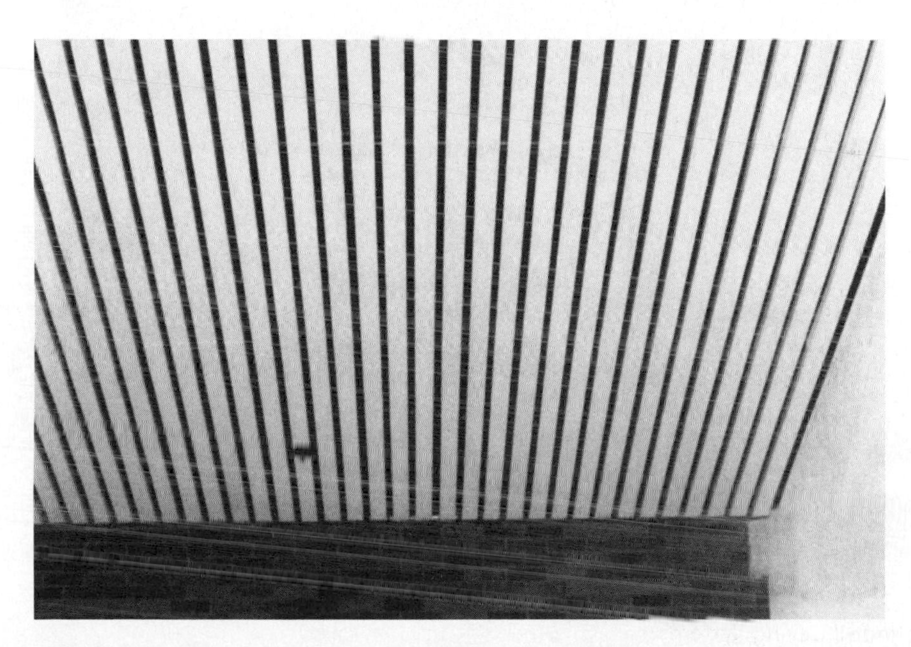

**Figure 16-51** Ribbed Metal Ceiling Panels Serve as Grilles for Air Distribution.

rary stairs are then required until the permanent ones have been installed. These will usually be made of wood. In other cases, stairs will be roughed in as construction progresses from floor to floor and are later finished as part of the interior finishing schedule.

Stairs may be of wood, concrete, stone, or steel, but the same general principles apply to their planning and design:

1. The stairwell opening must be long enough to allow ample headroom under the opening header. The usual minimum is 80 in. (2,050 mm).
2. The stair must be at a slope that makes climbing as easy as possible. The angle of the line of flight should be from 25° to 35°.
3. The height of risers must be kept within reasonable limits. In many public buildings the maximum rise is 6 in. (150 mm), and in most cases risers should not be allowed to exceed 7 in. (180 mm).
4. The tread must be wide enough to provide safe footing (11 in. [280 mm]) is the minimum width that should be considered).

**Figure 16-52** Sprayed Insulation Provides Acoustical Treatment to Exposed Concrete Ceiling.

**Figure 16-53** Acoustical Tile in Suspended Ceiling.

5. All risers must be of uniform height.
6. Long, straight flights of stairs without a break should be avoided. Most codes provide a maximum allowable length of stair without a landing.
7. Stairs must be 44 in. (1,100 mm) wide when serving buildings more than three stories above grade or more than one story below grade and 36 in. (900 mm) wide when serving buildings with three stories or less above grade or one story below grade.
8. Nonslip treads should always be provided.
9. Handrails must be provided at the proper height, 32 in. to 38 in. (800 to 965 mm).
10. Stairs with treads of varying width from one end to the other (*winders*) should be avoided whenever possible.

## Wood Stairs

Wood stairs may be made with an open or cutout stringer, a semihoused, a housed, or a built-up stringer (Figure 16-54).

The exact riser height is determined by measuring the vertical distance from one finished floor to another and dividing that distance by the desired riser height. The result, to the next whole number above or below that figure, will provide two alternative numbers of risers that may be used. Those two numbers divided into the total rise provide a choice in the exact height of riser to be used. The width of treads is determined by the total run available for the stairs (keeping in mind that there is normally one less tread than riser in the complete flight) and by the requirement for headroom (if any).

With all types of wood stairs except those with housed stringers, the height of the bottom riser must be reduced by an amount equal to that of the thickness of tread being used (see Figure 16-54). If stairs are resting on a subfloor, the thickness of the finished floor must be added to the height of the bottom riser.

Stringers for housed stairs may be laid out from either the back edge or the front, and edges must be straight and parallel. The top and bottom ends of housed and semihoused stringers may be extended to provide the easement required (see Figure 16-54).

## Steel Stairs

Steel stairs may be made in several ways. One simple method is to use two structural channels, back to

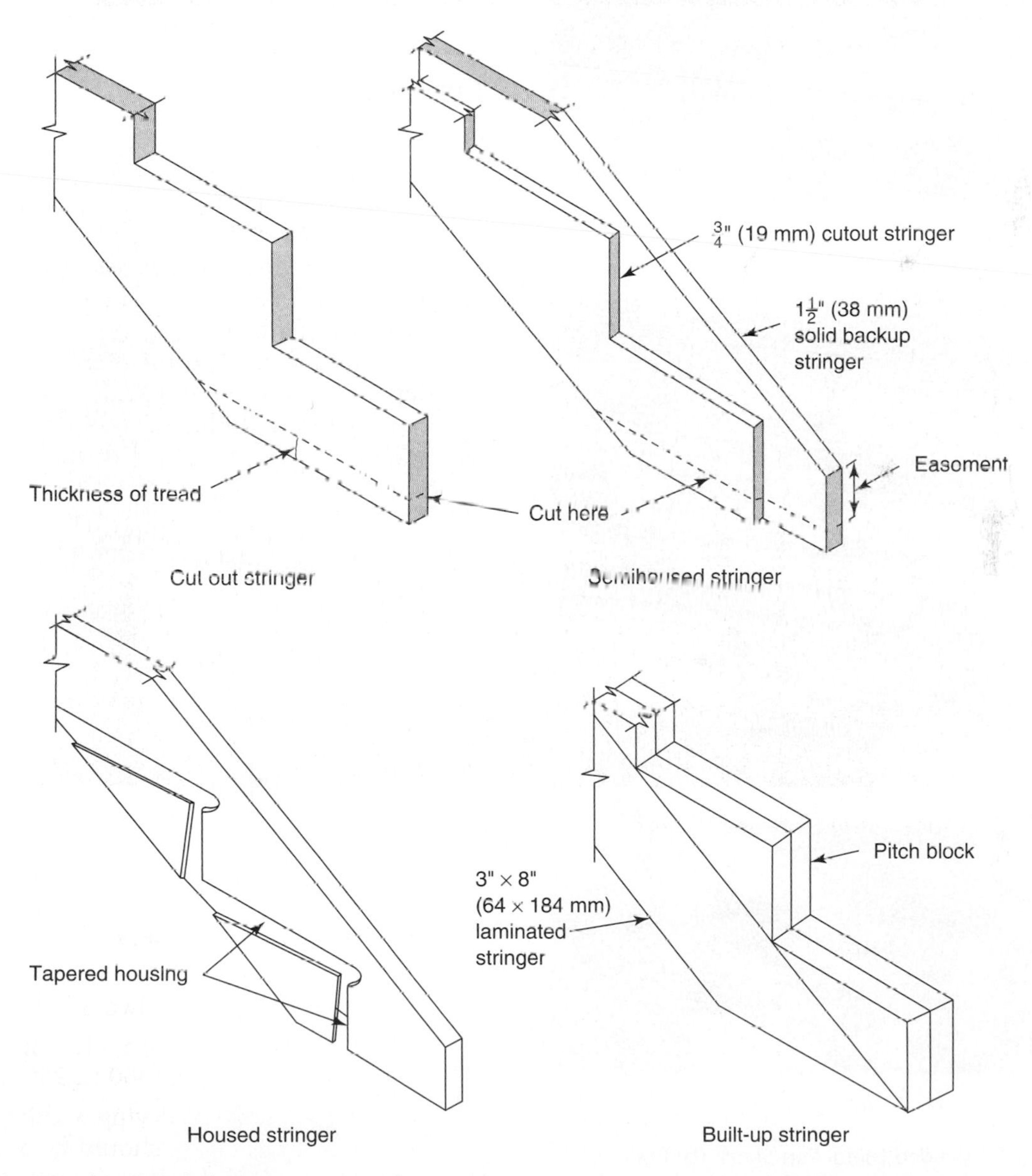

**Figure 16-54** Wooden Stair Stringers.

back, and weld or bolt steel treads between them, as illustrated in Figure 16-55(a).

Another method of building a steel stair is illustrated in Figure 16-55(b). Two or more structural steel members (depending on the width of the stair) are used as stringers. They may be channels, I beams, or heavy plates. Tread brackets are welded to the top edge or to the inside face of the stringers, and steel treads or *subtreads* are then welded to the brackets. In the stair shown in Figure 16-55, the subtreads are hollow steel pans that will later be filled with concrete, terrazzo, or magnesite.

A somewhat similar type of stair is made by bolting pressed steel risers and treads to steel stringers (Figure 16-56). The tread mold may receive a stone slab or may be a pan to be filled with poured-in-place tread material.

A spiral steel stair is made by welding nonslip treads or subtreads in rising succession around a central axis such as a steel pipe. This is one type of stair in which the tread is wider at one end than the other, and every effort should be made to build them as wide as possible in the normal line of travel.

Spiral stairs are gaining increasing attention for buildings such as apartment blocks because they take up less space than conventional stairs and, when properly designed and finished, they can be quite attractive (see Figure 16-57).

(a)

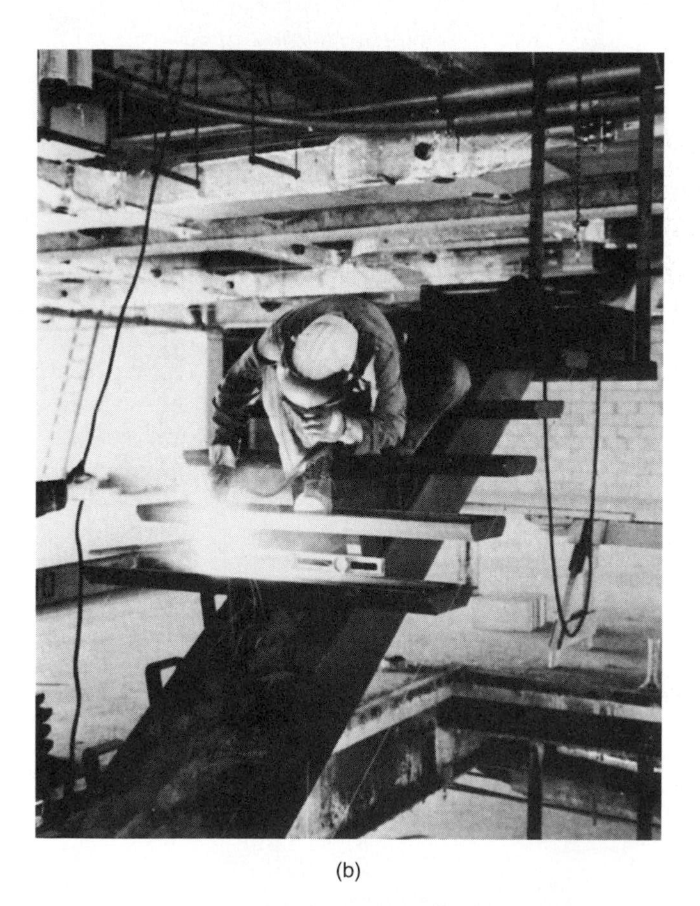

(b)

**Figure 16-55** (a) Welded Metal Pan Stairs. (b) Built-Up Steel Stair.

## Concrete Stairs

Reinforced concrete stairs may be precast or cast in place. Precast units must be tied into the structure, which may be done by casting the top slab around the lower end, as in Figure 16-58. Anchor plates cast into the end of the stair may be welded to a matching plate in the lower floor. The top end of the stair usually rests on a beam ledge.

The method used to form a cast-in-place stair will depend on the type of stair to be constructed. A *closed* string or flight (between enclosing walls) will require forms that are different from those

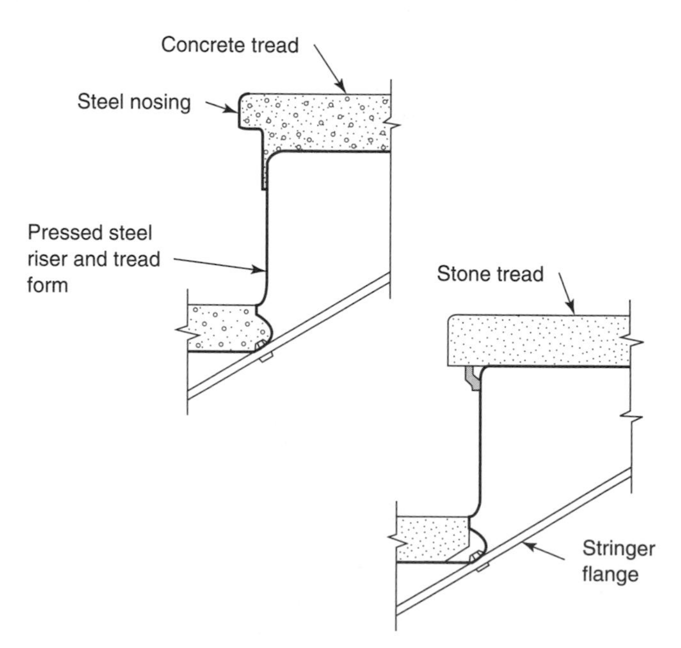


**Figure 16-56** Stone or Concrete Treads on a Steel Stair.

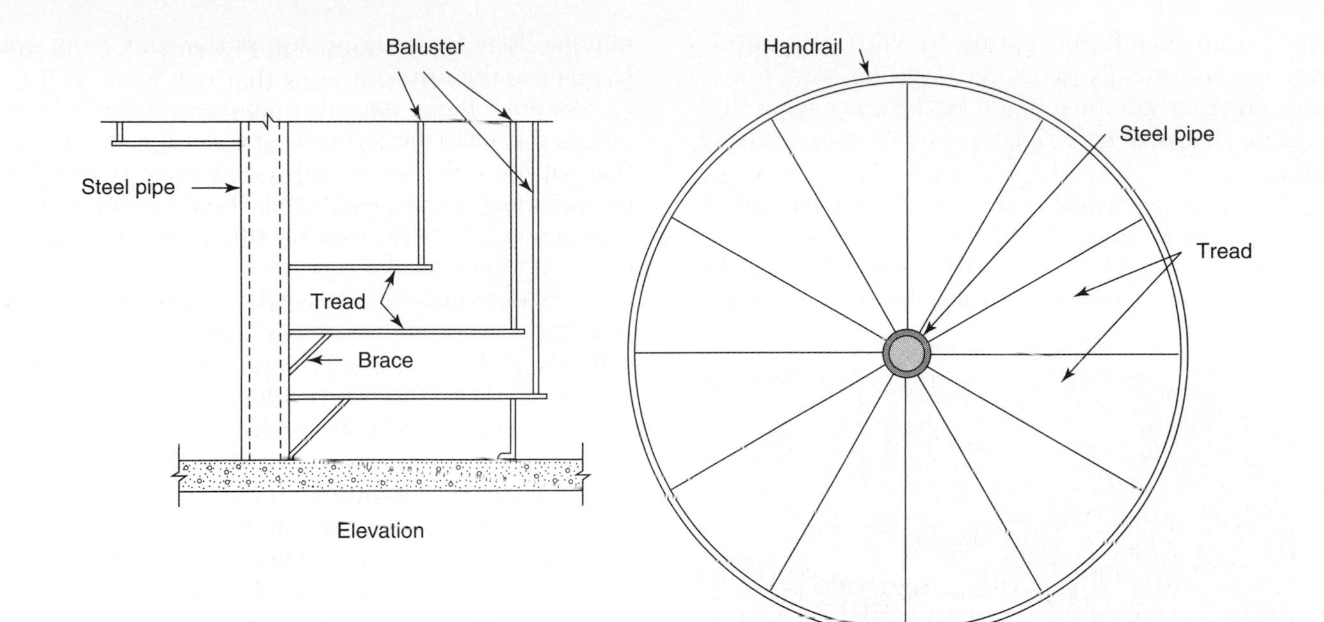

**Figure 16-57** Spiral Stair.

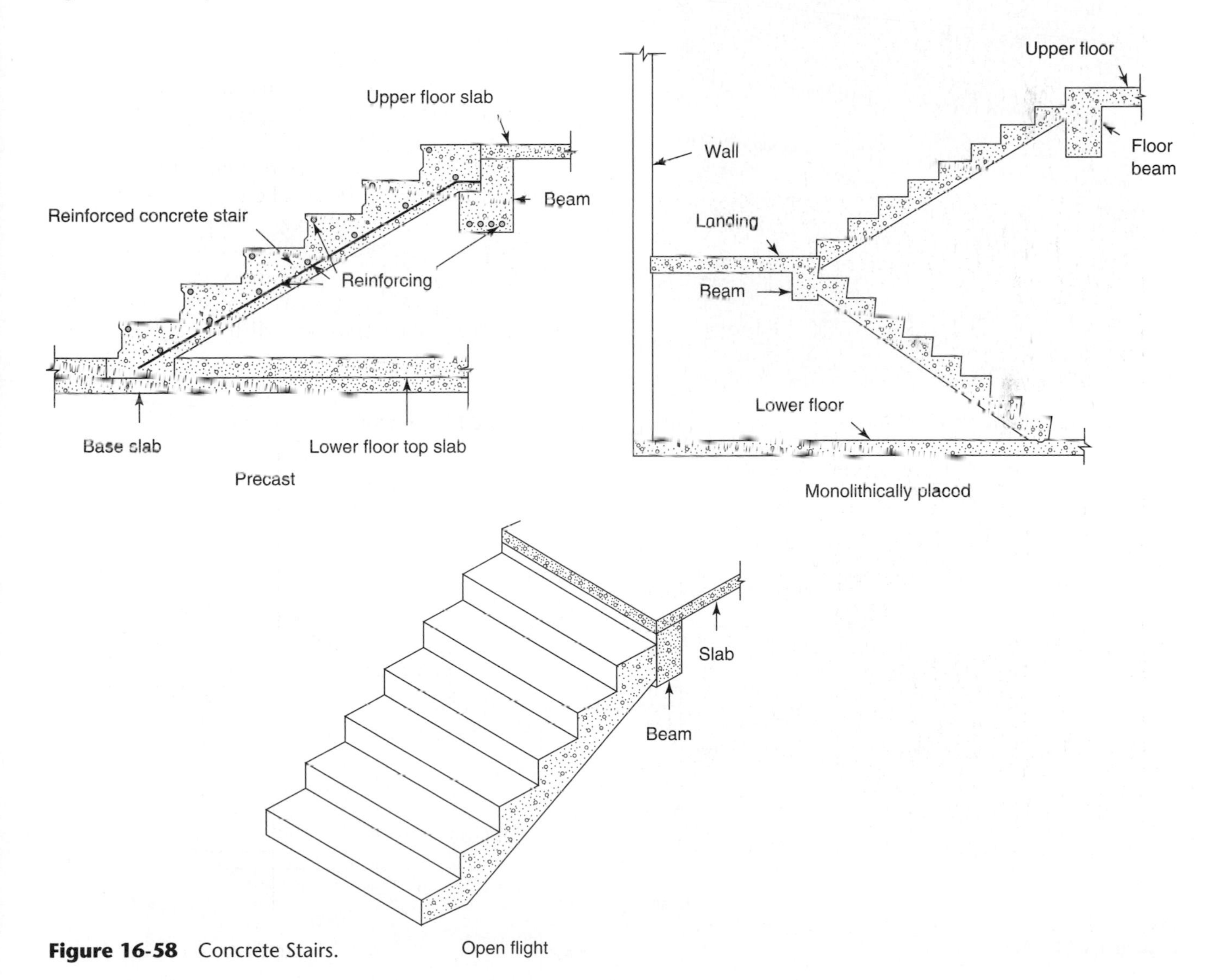


**Figure 16-58** Concrete Stairs.

used for an *open* flight (Figure 16-59). If the stair is cast monolithically with the landing and upper floor, the method must again be different from that used to form a stair cast between two existing floors.

A stair to be cast between existing walls requires a soffit form (Figure 16-60) and riser forms held in place by inverted stringers. The stringers must be supported independently of the soffit form, and this may be done by wedging across the stair, from one stringer to the other.

An open flight stair also requires the soffit form, but in this case the stringers rest on the framework that supports the soffit form (see Figure 16-61). The bottom edge of the riser forms should be beveled so that the surface of the treads may be troweled back to their junction with the next riser. A wide stair may require extra support at the center of the risers to prevent them from bowing under the pressure of the concrete. A 2″ × 6″ (38 × 140 mm) cut out as a stringer and inverted may be placed against the face of the risers and braced at the bottom to provide that support.

Figure 16-62 illustrates one method of forming a stair being cast monolithically with floor or landing.

**Figure 16-59** Cantilevered Concrete Stair.

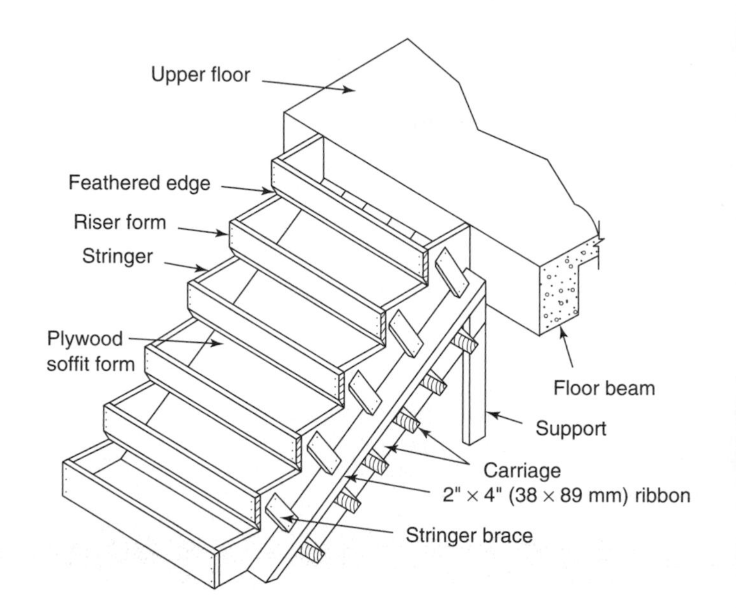


**Figure 16-61** Form for an Open Stair.

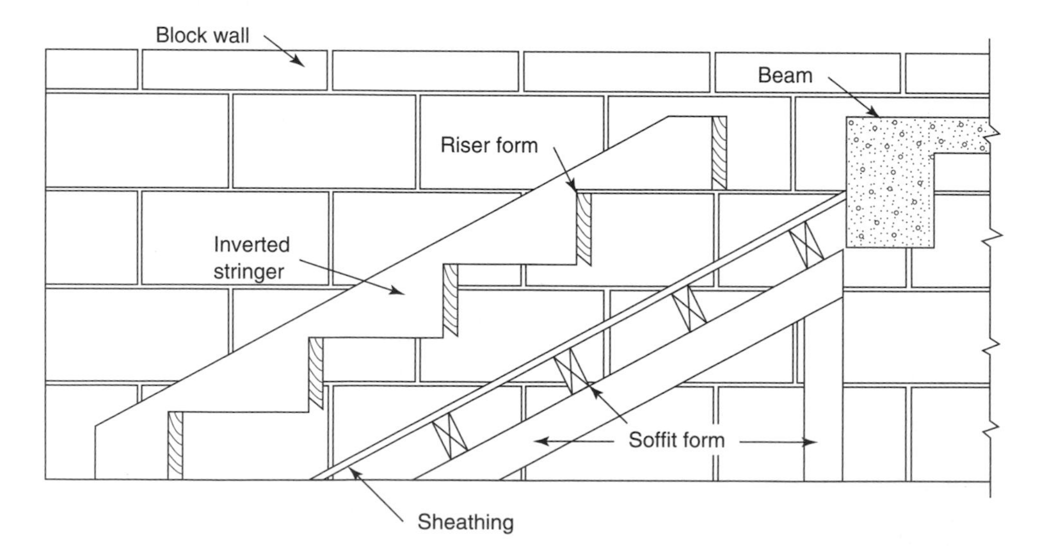

**Figure 16-60** Form for a Closed Stair.

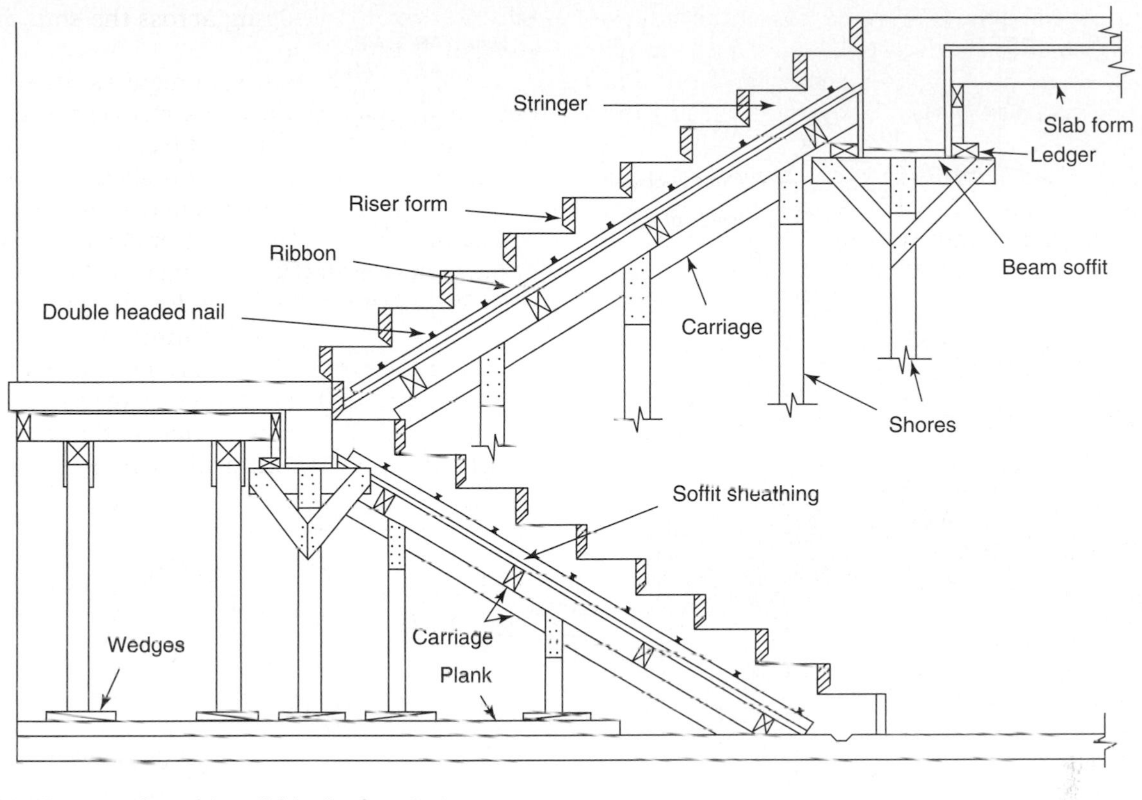


**Figure 16-62** Form for a Monolithically Cast Stair.

Circular cast-in-place concrete stairs present special forming problems. In the first place, the stringers must be flexible enough to be bent. Second, although the total rise of both inside and outside stringers is the same, the total run of each varies because of the difference in radius of the circles circumscribing the inside and outside edges of the stair. Finally, the soffit must be made in sections of relatively thin material if it is to fit the warped surface of the underside of the stair. To lay out and build the forms for a circular concrete stair, proceed as follows:

1. Draw circles representing the inside and outside circumferences of the stair to scale (Figure 16-63).
2. Measure the total rise and also determine the number and height of risers required. This will be the same for both inside and outside stringers.
3. Determine the number of treads and calculate the width of each on the outside circumference. This is the tread run for the outside stringer.
4. Step off these tread widths around the outside circle and join each point to the center of the circles. The distance between points at which the inside circle is cut represents the tread width at the inside of the stair and is the tread run for the inside stringer.
5. Calculate the total run for both stringers and lay both out against a common total rise (Figure 16-63).
6. Make the vertical distance from the junction of the rise and the run to the bottom edge of the stringer (distance $x$, Figure 16-63) the same for both stringers.
7. Build inside and outside circular walls to carry the stair form. Make the necessary allowance for the thickness of the stringer material, as shown in Figure 16-63.
8. Bend the stringers around the face of the walls and fasten them in place.
9. Make a template for one section of soffit. Lay out a full-size plan view of one tread and draw in the center line. On either side of this center line, lay out half the length of distance $a$ at one end and half the length of distance $b$ at the other, at right angles to the center line. Join the ends of these lines. The figure *LMNO* represents the approximate shape of the soffit section.
10. Cut this template out of thin material and shape its ends until the piece fits snugly against the form walls under the stringers. Cut as many soffit sections of this shape as there are treads in the stair.

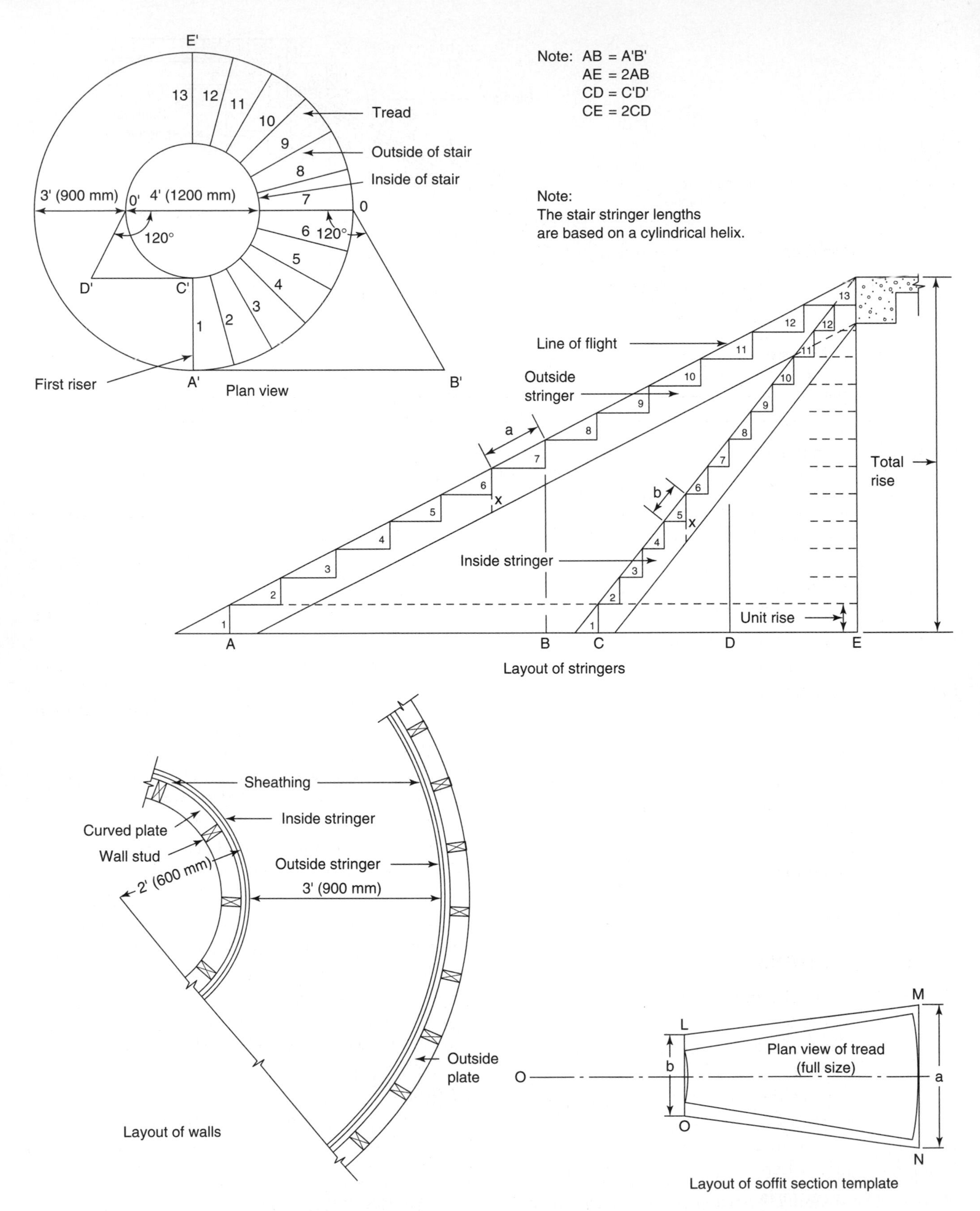


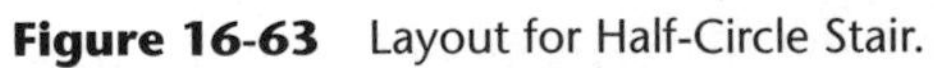

**Figure 16-63** Layout for Half-Circle Stair.

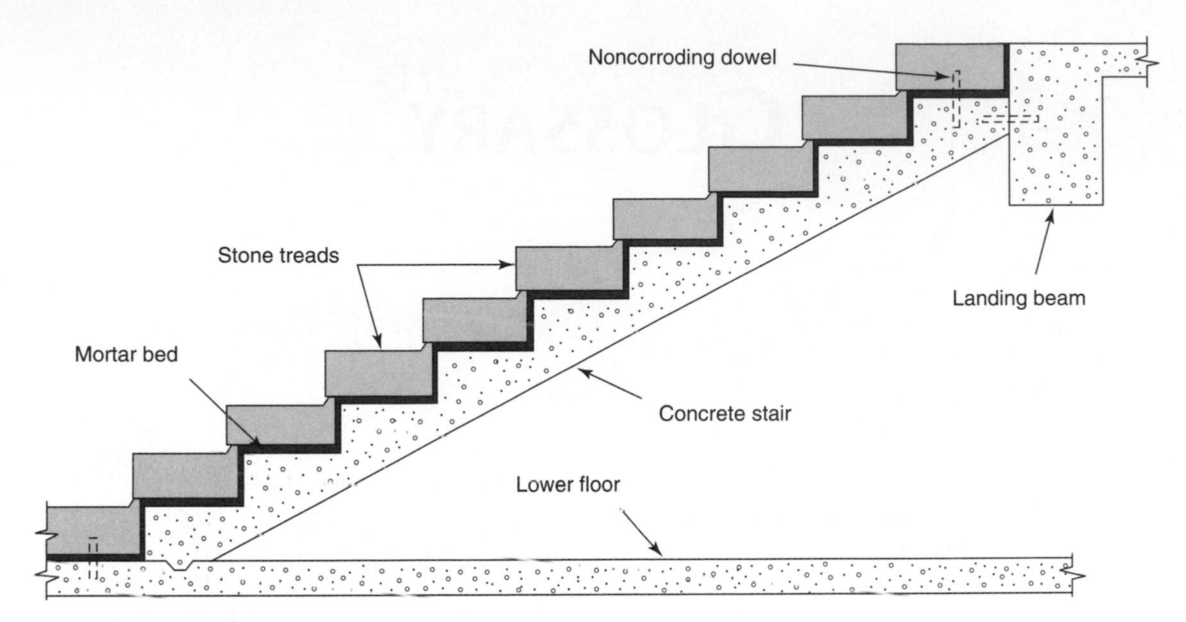


**Figure 16-64** Stone Stair.

11. Nail the soffit sections to the bottom edge of the stringers, nailing a ribbon under them for support.
12. Nail riser forms with tapered bottom edges to the stringer risers, and nail blocking in front of each to provide further support.
13. Place the necessary shoring under the center of the soffit to keep the soffit sections in line.
14. Place the specified reinforcement in the stair form.

## Stone Stairs

Stone stairs are usually a combination of either stone and steel or stone and reinforced concrete. In the stone and steel combination, steel stringers and subtreads are capped with stone treads (see Figure 16-56). In the second case, the stair is formed by concrete, and stone slabs provide the finished surface (see Figure 16-64).

## Review Questions

**1.** List five areas of a building that are considered as exterior finishing.

**2.** List five areas of a building that are considered as interior finishing.

**3.** At what time in the construction schedule may interior and exterior finishing begin?

**4.** What three features of exterior finishing may a designer consider when specifying exterior finishes?

**5.** What four criteria may a designer consider when selecting materials for interior finishes?

**6.** What three properties of materials are usually considered when selecting exterior finishes?

**7.** Why are two sets of doors used in the main entrance of many buildings?

**8.** What are four reasons for using windows in a building?

**9.** What two materials are most often used in manufacturing window frames?

**10.** Explain: **(a)** under what conditions portland cement plaster is preferred to gypsum plaster, **(b)** why masonry walls should be dampened before applying a plaster scratch coat, **(c)** what purpose hair or fiber serves in a scratch coat, **(d)** what is meant by cross-hatching a scratch coat, and **(e)** why more sand or lightweight aggregate may be used in mixing the brown coat than the scratch coat.

**11.** What two aspects of drywall make it appropriate for interior finishing?

**12.** What two types of finishes can be applied to masonry blocks and where might each finish be used?

**13.** What two methods or materials can be used to provide acoustical treatment to a ceiling?

**14.** Define the following terms used in stair design and construction: **(a)** line of flight, **(b)** headroom, **(c)** tread run, **(d)** total rise, **(e)** baluster, **(f)** newel post, **(g)** winder, **(h)** easement, and **(i)** soffit.

**15.** Outline the reason why **(a)** the vertical edges of wood doors should be dressed to a slight bevel, **(b)** riser forms for cast-in-place concrete stairs should have their bottom edges tapered to the back face, **(c)** hardboard panels should be conditioned prior to installation, and **(d)** setting blocks are used in the installation of plate glass windows.

# GLOSSARY

**Acceleration Due to Gravity** The increase in velocity per second of an object in free fall. On earth the approximate acceleration of an object in free fall is 32.2 ft/sec$^2$ (9.81 m/sec$^2$).

**ACI** American Concrete Institute.

**Acoustic Tile** A product used as a finish for ceilings that has the capability to absorb sound.

**Asphalt** A petroleum product that is used as a binder for roofing membranes and in the making of pavements.

**ASTM** American Society for Testing and Materials.

**Backsight** In surveying, it is the reading taken on the last control point in a series, which has been established before sighting on the next point.

**Batten** A cover strip that provides protection against the penetration of water at a splice point in exterior sidings.

**Beam** A structural section that spans between two supports while supporting loads applied normal to its longitudinal axis.

**Bearing Capacity** The property of soil that is used to establish the size of footings. It is based on various tests, depending on the type of soil being tested.

**Bearing Pile** A slender foundation section set vertically in the ground that develops its load capacity by its end resting on a stable layer of soil.

**Belt** A line of stone trim encircling the exterior of a building.

**Bentonite** A special clay material used in soil stabilization and in waterproofing of foundation walls.

**Blasting Cap** A device used in detonating dynamite.

**Bond Beam** A section of masonry wall that has been reinforced with steel bars and grout to provide additional stability to the wall.

**Bulkhead** A temporary barrier placed inside a form at some point where the daily concrete pour is to be terminated.

**Bulking** A resulting increase in volume of a material that has been disturbed.

**Burden** Unusable material that requires removing, usually in a mining operation.

**Caisson** A containment vessel used to provide protection for the construction crew during the placement of foundations in unstable soil conditions or in areas having a high water table.

**Chamfer Strip** A strip placed in the corner of formwork to round off the corner and facilitate the removal of the form without damaging the concrete.

**Clip Angle** A small section of angle used in connecting structural steel sections.

**Column** A structural member used primarily to resist axial compressive loads.

**Conductivity** The ability of a material to transmit electricity or heat.

**Construction Joint** A break in a concrete pour, usually at a bulkhead location.

**Control Joint** A break provided in a concrete component to allow for the change in length due to temperature changes and shrinkage.

**Coping** Overhanging trim at the top of a wall.

**CSA** Canadian Standards Association.

**Dado** In architecture a term used for the lower portion of an interior wall covered with a special finish such as tile.

**Darby** A wide, flat blade on a long handle used to level off the surface of a poured concrete slab before the finishing begins.

**Decibel** A unit used to describe the intensity of sound.

**Decking** That portion of a structural frame that spans between supporting beams or walls and serves as a floor or roof.

**Density** The amount of material contained within a unit volume, excluding all voids in the material. In metric units it is expressed as mass per unit volume.

**Dewatering System** An arrangement of pipes and pumps used in the lowering of the water table in the vicinity of an excavation.

**Dowel Bars** Reinforcing bars used to connect two concrete sections.

**Drift Pin** A pin used to align structural sections in preparation for bolting.

**Dry Concrete** A concrete mix that has little or no water added to the cement and aggregate mixture.

**Elevation** The height of a point with respect to a datum point of known elevation.

**End Plates** Plates welded to the ends of structural steel sections for connection purposes.

**Fascia** The vertical finish on a rafter overhang or roof edge.

**Flashing** Light gauge metal used to cover areas of discontinuity between different building elements.

**Float** A tool used in the leveling of poured concrete prior to final finishing.

**Footings** That portion of a structural frame used in the transfer of building loads to the supporting soil.

**Force** A vector quantity representing the interaction between two objects or masses. A vector quantity has both magnitude and direction.

**Foresight** In surveying, it is the sighting taken on a point in a traverse to establish its elevation with respect to a known elevation.

**Form Buck** A frame placed within concrete wall forms to allow for openings within the concrete wall.

**Formwork** A temporary structure used as a mold for plastic concrete.

**Friction Pile** A foundation section that develops its load capacity from the frictional forces that are developed between the surrounding soil and the surface of the section.

**Furring Channels** Light steel or plastic sections used in the support of suspended ceiling panels.

**Girder** A main structural section that supports secondary roof or floor beams.

**Grade** A term used to describe the slope of a roadway, pipeline, or drainage ditch. Usually expressed as a percent.

**Gram** A basic unit of measure in the metric system representing a quantity of matter.

**Grout** A nonshrink mixture of cement and fine aggregate used in filling fractures in hardened concrete, in rock formations, and under beam-and column-bearing plates.

**Gusset Plate** A plate welded to structural steel sections, normally used for connecting cross-bracing.

**Hairpin Anchor** Anchors embedded in the concrete foundations for attaching temporary bracing during the erection of the steel frame.

**Hardpan** A densely compacted mixture of clay and granular material.

**Isolation Joint** A joint that separates two structural components and allows for movement between them.

**Kicker** A continuous wood spacer between the sheathing at the top of beam formwork and the edge of the slab formwork framing.

**Kilo-** A prefix used in the metric system to designate a factor of 1,000 units. For example, 1 kilogram equals 1,000 grams.

**Laser** Light rays that are concentrated into a controlled beam of light.

**Lintel** A small beam spanning over a door or window opening.

**Lite** A sheet or pane of glass enclosed by a frame.

**Litre** A basic unit for volume in the metric system; approximately 1 American liquid quart.

**Mass** A term depicting a quantity of matter. In the metric system, a common unit of mass is the kilogram.

**Mass Density** The amount of mass occupying a unit volume. In the metric system, the units for mass density are usually given in kg/m$^3$.

**Mastic** A sticky resin used in caulking compounds.

**Mega-** A prefix used in the metric system to represent a factor of one million. For example, one meganewton is equal to one million newtons.

**Metre** A basic unit of length in the metric system (approximately 39.37 in).

**Metric Ton** Also known as tonne. Equals 1,000 kg.

**Milli-** A prefix used in the metric system to designate 1/1,000. For example, 1 millimeter equals 1/1,000 of a meter.

**Monolithic** Concrete sections that are poured together without any breaks.

**Monument** A permanent marker used in designating a point of known elevation.

**Mudsill** A bearing pad used under a shore to distribute the shoring load over a larger surface.

**Newton** A term representing a unit of force in the metric SI system. One newton equals one kilogram meter per second per second (kg m/sec$^2$).

**Parapet** That portion of a perimeter wall that extends above the roof of a building.

**Pascal** A unit of pressure (force per unit area) in the metric system. It represents one newton per square meter (N/m$^2$).

**PCA** Portland Cement Association.

**Pier** A short column used in transferring building loads to the foundation pads.

**Pile** A term used to describe a structural section placed in the ground for the purpose of supporting a superimposed load.

**Plywood** A sheathing product made of several wood veneers with their grain normal to one another, producing a panel that has uniform properties in both directions.

**Portland Cement** A finely ground mixture of lime, silica, alumina, and iron components that, when mixed with water, will harden into a stonelike mass.

**Purlin** A secondary structural section that supports a roof deck.

**Quick Condition** A loss of bearing capacity that occurs in a saturated, fine-grained soil when it is disturbed.

**Quoin** Stone trim placed to produce a raised or exposed edge, providing relief to the surface of the wall.

**Rebound** The property of a material to increase in volume when an existing load has been removed.

**Resistivity** The nonconductance of a material.

**Rustication Strips** Strips placed in the formwork to produce an attractive design in the face of the finished concrete.

**Saturated, Surface-Dry Aggregates** Aggregates that have all pores completely filled with water, yet are dry on the surface.

**Screeds** Temporary supports used to maintain the correct elevation of a poured concrete slab during the leveling off period of the finishing process.

**Scupper** A channel cut into the parapet of a building at the roof elevation to provide a means for roof drainage.

**Shear Connector** Any connector used to resist forces developed between the common surfaces of two sections working as a single unit.

**Shore** A vertical structural member that is used for supporting formwork at a predetermined elevation.

**Sill** The horizontal ledge at the bottom of a window or a flat section serving to distribute concentrated loads over a larger area.

**SI System of Units** System International d'Unites; the latest version of the metric system.

**Slab on Grade** A concrete slab that is poured on prepared soil and depends on the base material for its strength.

**Slurry** A watery mixture of some cementous material used for stabilizing soils or for sealing fractures in walls.

**Soffit** The underside of a roof overhang.

**Soldier Piles** Structural sections driven into the ground at regular intervals to which sheeting is attached for the purpose of restraining embankments, usually around an excavation.

**Spandrel Beam** Beams that are located around the perimeter of a building, usually supporting roof or floor loads and wall loads.

**Split Spoon Sampler** A special sampling mechanism that is used to extract relatively undisturbed soil samples from the bottom of a test hole for the purpose of lab analysis.

**Stirrups** Reinforcing bars placed vertically at some angle to the main tension steel in a reinforced concrete beam to resist shear stresses.

**Stressed-Skin Panels** Structural sections having ribbed frames covered with some relatively thin material that contributes to the load-carrying capacity of the section.

**Structural Lightweight Concrete** A concrete that has a 28-day compressive strength in excess of 15 MPa (2,175 psi) and an air-dry density of no more than 1,850 kg/m$^3$ (115 lb/cu ft).

**Structural Slab** A concrete slab that must be able to support its own weight plus any other superimposed load between two points of support.

**Subfloor** Sheathing that is used to provide a sound surface for finished flooring materials.

**T-Bars** Light, metal, T-shaped sections used in supporting acoustic panels in suspended ceilings.

**Three-Hinged Arch** An arch that is broken into two parts and joined together at the peak with a pin-type connection, as well as having pin-type connections at the base.

**Till** A mixture of clay, sand, and boulders.

**Turnbuckle** A steel bar with threaded portions at both ends fitted with eyebolts, so that when turned, will shorten the distance between the two eyebolts. Used in tightening bracing cables in structural steel frames.

**Underpinning** The process of providing additional support to existing foundations.

**Vapor Diffusion Retarder** Any membrane that retards the migration of water vapor from one surface to another.

**Vernier** A secondary scale that allows for the reading of fractional divisions on a fixed scale. In the case of a scale marked off in degrees, the vernier allows for a reading in some fraction of the angle.

**Waterstop** A rubber or vinyl insert placed at a bulkhead to bridge the gap in the concrete at the point where two concrete pours meet.

**Well-Point** A perforated pipe, protected by a screen, that is driven into the ground for the purpose of collecting water.

**Yoke** A metal or wooden section placed around a column form to provide support to the form sheathing.

# INDEX